Agrobacterium tumefaciens

From Plant Pathology to Biotechnology

Edited by

Eugene Nester
University of Washington, Seattle

Milton P. Gordon
University of Washington, Seattle

Allen Kerr
University of Adelaide, Adelaide, Australia

APS PRESS
The American Phytopathological Society
St. Paul, Minnesota

This book has been formatted from computer files submitted
to APS PRESS by the editors of the volume. No editing or proofreading has been done by the Press.

Cover photographs: photomicrograph by Martha C. Hawes, University of Arizona, Tucson; cotton plant courtesy of U.S. Department of Agriculture

Library of Congress Control Number: 2004110645
International Standard Book Number: 0-89054-322-4

Printed in the United States of America on acid-free paper

The American Phytopathological Society
3340 Pilot Knob Road
St. Paul, Minnesota 55121-2097, U.S.A.

We dedicate this volume to Jeff Schell, friend, colleague, and sometimes competitor. Jeff was a pioneer in studies on *Agrobacterium* and his many seminal contributions through basic and applied research form a cornerstone of this anthology. His scientific legacy will continue not only through his own writings but also in the publications of the numerous individuals whose careers he influenced.

Obituary

Jeff Schell (1935–2003)

What can we learn from a soil bacterium that induces tumours in plants? This was the question that Jeff Schell set out to tackle when, in 1967, he became an associate professor of genetics at what was then called the Rijksuniversiteit Gent. Schell died on 17 April 2003. In the intervening quarter-century, the lessons that he and others extracted from work with the bacterium concerned, *Agrobacterium tumefaciens*, were to provide the basis for the plant biotechnology industry.

In the mid-1960s, it had just become generally accepted that some viruses cause tumours, as was acknowledged with the award of a 1966 Nobel prize to Peyton Rous. But bacteria capable of inducing tumours in animals were unheard of. And although there had long been an awareness of plant tumours called 'crown galls', they generated only minor research interest. Schell, however, wanted to explore at the molecular level what, 20 years previously, Armin Braun had dubbed "tumour-inducing property": the ability of *Agrobacterium* to induce crown galls.

Schell had been trained in Gent as a bacterial taxonomist, and had then had the good fortune to do postdoctoral work with Bill Hayes, the father of bacterial genetics, at Hammersmith Hospital in London. Later, with Lou Siminovitch in Toronto, he discovered the fascinating world of bacteriophage — viruses that infect bacteria. Along with this grounding in bacterial and phage genetics, he had an especial talent for formulating hypotheses and designing experiments to test them. All of this was the perfect background for solving the crown gall puzzle.

Jeff and I had both been active in the same student movements. When he became a professor, his lab was next to mine, and we decided to join forces. A good collection of oncogenic and non-oncogenic strains of *Agrobacterium* was available to us at the taxonomy laboratory where he had done his PhD, and we set to work comparing the strains. We found that tumour-inducing strains of *Agrobacterium* harboured a large element, which we called the Ti-plasmid. And we proposed that some DNA of this Ti-plasmid might become integrated into the genome of its host and induce the crown gall. This hypothesis met with scepticism from most plant physiologists as a seemingly wild and untestable idea — of course, molecular biology was then only just coming into existence.

SCHELL FAMILY

Pioneer of plant genetic engineering

It was not long, however, before the appropriate techniques were developed. With the advent of gene cloning and DNA hybridization by Southern blots, proof came fast that a copy of a segment on the Ti-plasmid was preferentially inserted in expressed DNA sequences of the plant. But why were genes in a bacterium active only in a plant cell?

The answer stemmed from work by Georges Morel and Jacques Tempé, who had shown that the Ti-plasmid causes a plant to produce certain unique amino-acid-derived molecules. From this, Schell developed the concept of 'genetic colonization', according to which the whole *raison d'être* of the Ti-plasmid is making plant cells secrete these molecules, which constitute a source of carbon and nitrogen that only *Agrobacterium* can use. Crown galls, then, constitute a unique ecological niche for the microbe.

The implication of these findings was clear: if DNA was being transferred by *Agrobacterium*, it should be possible to replace the tumour-inducing genes by others that conveyed new traits to plants. The race to turn *Agrobacterium* into a reliable gene vector was on, with fierce competition from Gene Nester at the University of Washington in Seattle, and Mary-Dell Chilton at Washington University and Monsanto in St Louis. Schell was a keen sportsman, with a particular interest in sailing, and he relished competition (especially when he felt he could win). Nonetheless, this was a scientific race conducted on amicable terms, with information being exchanged and synchronized publication of many of the notable papers.

In 1978, Schell had become director of the Max Planck Institute for Plant Breeding Research in Cologne, Germany. This he turned into an international leader in plant research, nurturing generations of students and young scientists from around the world. He continued co-directing the team in Gent, however, and our work culminated in 1983 with the announcement, simultaneously with Chilton and Monsanto, of the *Agrobacterium*-mediated creation of the first genetically modified plants that expressed a new trait.

With this development, plant molecular biology had a key tool with which to flourish: from then on, transgenes and gene regulatory sequences could be routinely transferred into plants. Financiers and agrochemical companies became keen to invest in the new discipline of plant biotechnology through start-up companies focusing on crop improvement. Schell was instrumental in helping many of these industrial initiatives to integrate the findings of fundamental research into their activities. Throughout his career he was concerned about injustice and suffering; he was particularly alarmed by environmental problems, and was convinced that this new technology would be instrumental in achieving more sustainable agricultural practices. He believed that, for all their good intentions, those who wish to ban genetic modification in crops, or who argue that developing countries should reject food-aid with a genetically modified component, are profoundly misguided.

Even though Schell was suffering from a neurodegenerative disease, right up to his death he continued to put the case for the need to apply plant biotechnology to help farmers in developing countries. He was also particularly active in promoting German–Israeli cooperation in the plant sciences, through the Otto Warburg Minerva Center in Jerusalem, which runs joint programmes with the Weizmann Institute and the Hebrew University.

Jeff Schell was not only a thoughtful and innovative scientist, but also an outstanding speaker and a charismatic mentor. And in one sense, the best may be yet to come, for the enduring influence of his work is likely to be seen in many of the advances in agriculture in the decades ahead.

Marc Van Montagu

Marc Van Montagu is at the Instituut Planten Biotechnologie voor Ontwikkelingslanden (IPBO), Universiteit Gent, KL Ledeganckstraat 35, B-9000 Gent, Belgium.
e-mail: marc.vanmontagu@rug.ac.be

CONTENTS

Quorum Sensing

DNA Sequencing

Biological Control of Crown Gall

PREFACE

The journey from plant pathology to biotechnology began in 1904, one hundred years ago, when Erwin F. Smith started detailed work on crown gall, a disease that affects a wide range of plants. The disease is so called because affected plants have large swellings or galls, usually at the crown of the plant, at or just below soil level. Smith showed that crown gall was caused by a soil-inhabiting bacterium that he called *Bacterium tumefaciens,* later renamed *Agrobacterium tumefaciens.* At that time, crown gall was entirely a plant pathological problem but many years later, plant physiologists became interested in the disease, largely due to the work of Armin C. Braun. He made the remarkable discovery that crown gall was really a plant cancer by showing that the causal bacteria could be eliminated from galled tissue but the tissue continued to grow and divide out of control. Also, sterile galls could be grown in culture in the absence of the plant hormones, auxin and cytokinin, which all normal plant tissues require. Braun coined the phrase "tumor inducing principle," which implied that the bacteria transmitted something to the plant. This work was done in the 1940s. Gradually, evidence built up that the tumor inducing principle had to be nucleic acid, either RNA or DNA. A new phase of understanding began when it was observed that the ability to induce tumors could be transmitted from a pathogenic strain of *A. tumefaciens* to a non-pathogenic strain. Final proof of the nature of the tumor inducing principle had to wait for the development of techniques in molecular biology.

In 1977, a small piece of DNA, called transferred DNA, or T-DNA, located on a large plasmid in *A. tumefaciens* was shown to be the elusive tumor-inducing principle. But how did it work and even more intriguing, how did it get from bacterium to plant cell? Details of the mechanism are still under investigation. By the mid-1980s, we understood how T-DNA caused plant tumors and we had a fair idea of how the T-DNA was transferred from the bacterium to the plant cell. Could the system be manipulated? Could foreign DNA be transferred by the same system? The answer was a resounding "YES." Vectors were developed to simplify the transfer process and make it more efficient. Useful genes were located, cloned and transferred to plants. This was the start of the genetic engineering of crops, now popularly (or unpopularly) known as GM crops. Last year, 167.2 million acres (67.7 million hectares) of genetically engineered crops were cultivated in 18 countries, but the cultivation of GM crops is still very controversial. Thus far, insect and herbicide resistance are the principle traits incorporated into plants, although genes coding for many other useful characters, such as nutrition and flavor, and cold and drought-resistance are waiting in the wings. What started out as an investigation of a plant disease has been developed into a revolutionary tool for agriculture.

In this book, we trace the change from plant pathology to biotechnology. We chose what we considered to be the seminal scientific papers in the biology and application of *Agrobacterium.* This was not an easy task and others would undoubtedly have made a different selection. It is certainly true that many excellent papers are not included because of space limitations.

Once we agreed on the papers, we asked several people, who, as far as possible, had been directly involved in the original work, to write a commentary on one or more paper(s). The commentators were asked to give a background to the papers, explain the problems faced and the techniques used. The papers chosen are reprinted here along with the commentaries. We believe they provide an interesting insight into the way fundamental research progresses, which is not readily gleaned from the inherently formal styles required for scientific papers. We hope you find the journey interesting.

However, the journey is not over. As the commentary by Paul Hooykaas indicates, it looks as though T-DNA will insert into any cell, be it plant, fungal or even mammalian. Is there a possibility of using *Agrobacterium* in gene therapy? Will *Agrobacterium* prove to be as useful a tool in fungal genetics as it has been in plant genetics? Its potential is mind-boggling.

As is inevitable with the progress of science, important minor themes developed out of the studies on crown gall. Three are represented here, each by two papers and one commentary. First, we have the intriguing phenomenon known as "quorum sensing" in bacteria, a term coined by Stephen Winans to describe population density dependence. Although first discovered in another organism, *Vibrio fischeri,* it was only when quorum sensing was discovered in *Agrobacterium* that the widespread nature of the phenomenon was realized. Second, the complete DNA sequencing of one strain of *A. tumefaciens* by two independent groups was a major achievement, which we thought deserved to be included. The third minor theme in this book is the biological control of crown gall. This control is probably the best known and most efficient biological control system for any plant disease. The control organism, strain K84, had to be genetically engineered to prevent loss of effectiveness and the resulting organism, strain K1026, was the first genetically engineered organism to be released for commercial use in 1988.

The crown gall saga is a classic case of curiosity-driven research performed by small groups of widely dispersed scientists that ultimately led to a major revolution in agriculture. We suggest that *Agrobacterium* still contains mysteries that remain to be explored and exploited by future investigators.

The assembling of this anthology was greatly aided by the dedicated efforts of numerous individuals who contributed enormously to all stages of the project. Doug Vollgraff, Donna Stewart and Gail Rawson were responsible for typing emails to commentators, locating addresses, requesting permission from journals and commentators and maintaining order to the inflow and outflow of countless email messages from editors, commentators and journal representatives. Gail Rawson deserves special thanks for her yeoman efforts in getting the last version of the manuscript into its final camera-ready form. We also extend our gratitude to Karen Scholthof for carefully reading two drafts of the commentaries, clarifying unclear statements and catching innumerable trivial mistakes in the format. Karen also identified two anonymous reviewers who provided helpful critiques. Martha Hawes provided the photomicrograph reproduced on the front and back covers.

The project would not have been initiated without an invitation from Randy Ploetz and Karen Scholthof representing APS Press. It could not have been completed without the willing participation of all of the commentators who took the time to recall and then write down their own personal significant experience with *Agrobacterium tumefaciens.* We hope that the finished product will make everyone's efforts seem worthwhile.

Eugene Nester
Milton P. Gordon
Allen Kerr

June 2004

SOURCES

Bevan, M. W., Flavell, R. B., and Chilton, M.-D. 1983. A chimaeric antibiotic resistance gene as a selectable marker for plant cell transformation. Reprinted by permission from *Nature* 304: 184–187. Copyright 1983 Macmillan Publishers Ltd.

Braun, A. C., and Mandle, R. J. 1948. Studies on the inactivation of the tumor-inducing principle in crown gall. Reprinted by permission from *Growth* 12:255–269. Copyright 1948 Growth Publishing Co., Inc.

Bundock, P., den Dulk-Ras, A., Beijersbergen, A., and Hooykaas, P. J. J. 1995. Trans-kingdom T-DNA transfer from *Agrobacterium tumefaciens* to *Saccharomyces cerevisiae.* Reprinted by permission from *EMBO Journal* 14:3206–3214.

Chilton, M.-D., Drummond, M. H., Merlo, D. J., Sciaky, D., Montoya, A. L., Gordon, M. P., and Nester, E. W. 1977. Stable incorporation of plasmid DNA into higher plant cells: The molecular basis of crown gall tumorigenesis. Reprinted from *Cell* 11:263–271. Copyright 1977, with permission from Elsevier.

de Framond, A. J., Barton, K. A., and Chilton, M.-D. 1983. Mini-Ti: A new vector strategy for plant genetic engineering. Reprinted by permission from *Bio/Technology* 1:262–269.

Drummond, M. H., Gordon, M. P., Nester, E. W., and Chilton, M.-D. 1977. Foreign DNA of bacterial plasmid origin is transcribed in crown gall tumors. Reprinted by permission from *Nature* 269:535–536. Copyright 1977 Macmillan Publishers Ltd.

Feldmann, K. A., and Marks, M. D. 1987. *Agrobacterium*-mediated transformation of germinating seeds of *Arabidopsis thaliana:* A non-tissue culture approach. Reprinted by permission from *Molecular and General Genetics* 208:1–9. Copyright 1987 Springer-Verlag.

Fraley, R. T., Rogers, S. G., Horsch, R. B., Sanders, P. R., Flick, J. S., Adams, S. P., Bittner, M. L., Brand, L. A., Fink, C. L., Fry, J. S., Galluppi, G. R., Goldberg, S. B., Hoffmann, N. L., and Woo, S. C. 1983. Expression of bacterial genes in plant cells. Reprinted by permission from *Proceedings of the National Academy of Sciences of the United States of America* 80:4803–4807.

Garfinkel, D. J., Simpson, R. B., Ream, L. W., White, F. F., Gordon, M. P., and Nester, E. W. 1981. Genetic analysis of crown gall: Fine structure map of the T-DNA by site-directed mutagenesis. Reprinted from *Cell* 27:143–153. Copyright 1981, with permission from Elsevier.

Goodner, B., Hinkle, G., Gattung, S., Miller, N., Blanchard, M., Qurollo, B., Goldman, B. S., Cao, Y., Askenazi, M., Halling, C., Mullin, L., Houmiel, K., Gordon, J., Vaudin, M., Iartchouk, O., Epp, A., Liu, F., Wollam, C., Allinger, M., Doughty, D., Scott, C., Lappas, C., Markelz, B., Flanagan, C., Crowell, C., Gurson, J., Lomo, C., Sear, C., Strub, G., Cielo, C., and Slater, S. 2001. Genome sequence of the plant pathogen and biotechnology agent *Agrobacterium tumefaciens* C58. Reprinted by permission from *Science* 294:2323–2328. Copyright 2001 American Association for the Advancement of Science.

Grimsley, N., Hohn, T., Davies, J. W., and Hohn, B. 1987. *Agrobacterium*-mediated delivery of infectious maize streak virus into maize plants. Reprinted by permission from *Nature* 325:177–179. Copyright 1987 Macmillan Publishers Ltd.

Hamilton, R. H., and Fall, M. Z. 1971. The loss of tumor-initiating ability in *Agrobacterium tumefaciens* by incubation at high temperature. Reprinted by permission from *Experientia* 27:229–230.

Hernalsteens, J.-P., Van Vliet, F., De Beuckeleer, M., Depicker, A., Engler, G., Lemmers, M., Holsters, M., Van Montagu, M., and Schell, J. 1980. The *Agrobacterium tumefaciens* Ti plasmid as a host vector system for introducing foreign DNA in plant cells. Reprinted by permission from *Nature* 287:654–656. Copyright 1980 Macmillan Publishers Ltd.

Herrera-Estrella, A., Van Montagu, M., and Wang, K. 1990. A bacterial peptide acting as a plant nuclear targeting signal: The amino-terminal portion of *Agrobacterium* VirD2 protein directs a β-galactosidase fusion protein into tobacco nuclei. Reprinted by permission from *Proceedings of the National Academy of Sciences of the United States of America* 87:9534–9537.

Herrera-Estrella, L., Depicker, A., Van Montagu, M., and Schell, J. 1983. Expression of chimaeric genes transferred into plant cells using a Ti-plasmid-derived vector. Reprinted by permission from *Nature* 303:209–213. Copyright 1983 Macmillan Publishers Ltd.

Hiei, Y., Ohta, S., Komari, T., and Kumashiro, T. 1994. Efficient transformation of rice (*Oryza sativa* L.) mediated by *Agrobacterium* and sequence analysis of the boundaries of the T-DNA. Reprinted from *The Plant Journal* 6:271–282, by permission from Blackwell Publishing Ltd.

Hoekema, A., Hirsch, P. R., Hooykaas, P. J. J., and Schilperoort, R. A. 1983. A binary plant vector strategy based on separation of *vir*- and T-region of the *Agrobacterium tumefaciens* Ti-plasmid. Reprinted by permission from *Nature* 303:179–180. Copyright 1983 Macmillan Publishers Ltd.

Holsters, M., Silva, B., Van Vliet, F., Genetello, C., De Block, M., Dhaese, P., Depicker, A., Inzé, D., Engler, G., Villarroel, R., Van Montagu, M., and Schell, J. 1980. The functional organization of the nopaline *A. tumefaciens* plasmid pTiC58. Reprinted from *Plasmid* 3:212–230. Copyright 1980, with permission from Elsevier.

Htay, K., and Kerr, A. 1974. Biological control of crown gall: Seed and root inoculation. Reprinted from *Journal of Applied Bacteriology* 37:525–530, by permission from Blackwell Publishing Ltd.

Kerr, A. 1969. Transfer of virulence between isolates of *Agrobacterium.* Reprinted by permission from *Nature* 223:1175–1176. Copyright 1969 Macmillan Publishers Ltd.

Otten, L., De Greve, H., Hernalsteens, J.-P., Van Montagu, M., Schieder, O., Straub, J., and Schell, J. 1981. Mendelian transmission of genes introduced into plants by the Ti plasmids of *Agrobacterium tumefaciens.* Reprinted by permission from *Molecular and General Genetics* 183:209–213. Copyright 1981 Springer-Verlag.

Perlak, F. J., Fuchs, R. L., Dean, D. A., McPherson, S. L., and Fischhoff, D. A. 1991. Modification of the coding sequence enhances plant expression of insect control protein genes. Reprinted by permission from *Proceedings of the National Academy of Sciences of the United States of America* 88:3324–3328.

Petit, A., Delhaye, S., Tempé, J., and Morel, G. 1970. Recherches sur les guanidines des tissus de crown gall: Mise en évidence d'une relation biochimique spécifique entre les souches d'*Agrobacterium tumefaciens* et les tumeurs qu'elles induisent. Reprinted from *Physiologie Végétale* 8:205–213.

Piper, K. R., Beck von Bodman, S., and Farrand, S. K. 1993. Conjugation factor of *Agrobacterium tumefaciens* regulates Ti plasmid transfer by autoinduction. Reprinted by permission from *Nature* 362:448–450. Copyright 1993 Macmillan Publishers Ltd.

Roberts, W. P., Tate, M. E., and Kerr, A. 1977. Agrocin 84 is a 6-*N*-phosphoramidate of an adenine nucleotide analogue. Reprinted by permission from *Nature* 265:379–380. Copyright 1977 Macmillan Publishers Ltd.

Smith, E. F., and Townsend, C. O. 1907. A plant-tumor of bacterial origin. Reprinted from *Science,* New Series, 25:671–673.

Stachel, S. E., Messens, E., Van Montagu, M., and Zambryski, P. 1985. Identification of the signal molecules produced by wounded plant cells that activate T-DNA transfer in *Agrobacterium tumefaciens.* Reprinted by permission from *Nature* 318:624–629. Copyright 1985 Macmillan Publishers Ltd.

Stachel, S. E., Timmerman, B., and Zambryski, P. 1986. Generation of single-stranded T-DNA molecules during the initial stages of T-DNA transfer from *Agrobacterium tumefaciens* to plant cells. Reprinted by permission from *Nature* 322:706–712. Copyright 1986 Macmillan Publishers Ltd.

Thomashow, M. F., Nutter, R., Montoya, A. L., Gordon, M. P., and Nester, E. W. 1980. Integration and organization of Ti plasmid sequences in crown gall tumors. Reprinted from *Cell* 19:729–739. Copyright 1980, with permission from Elsevier.

Vaeck, M., Reynaerts, A., Höfte, H., Jansens, S., De Beuckeleer, M., Dean, C., Zabeau, M., Van Montagu, M., and Leemans, J. 1987. Transgenic plants protected from insect attack. Reprinted by permission from *Nature* 328:33–37. Copyright 1987 Macmillan Publishers Ltd.

Van Larebeke, N., Genetello, C., Schell, J., Schilperoort, R. A., Hermans, A. K., Hernalsteens, J.-P., and Van Montagu, M. 1975. Acquisition of tumour-inducing ability by non-oncogenic agrobacteria as a result of plasmid transfer. Reprinted by permission from *Nature* 255:742–743. Copyright 1975 Macmillan Publishers Ltd.

Van Montagu, M. 2003. Jeff Schell (1935–2003). Reprinted by permission from *Nature* 423:934. Copyright 2003 Macmillan Publishers Ltd. Photograph of Jeff Schell reprinted by permission of Betsy Schell.

Watson, B., Currier, T. C., Gordon, M. P., Chilton, M.-D., and Nester, E. W. 1975. Plasmid required for virulence of *Agrobacterium tumefaciens.* Reprinted by permission from *Journal of Bacteriology* 123:255–264.

White, P. R., and Braun, A. C. 1941. Crown gall production by bacteria-free tumor tissues. Reprinted by permission from *Science* 94:239–241. Copyright 1941 American Association for the Advancement of Science.

Wood, D. W., Setubal, J. C., Kaul, R., Monks, D. E., Kitajima, J. P., Okura, V. K., Zhou, Y., Chen, L., Wood, G. E., Almeida, N. F., Jr., Woo, L., Chen, Y., Paulsen, I. T., Eisen, J. A., Karp, P. D., Bovee, D., Sr., Chapman, P., Clendenning, J., Deatherage, G., Gillet, W., Grant, C., Kutyavin, T., Levy, R., Li, M.-J., McClelland, E., Palmieri, A., Raymond, C., Rouse, G., Saenphimmachak, C., Wu, Z., Romero, P., Gordon, D., Zhang, S., Yoo, H., Tao, Y., Biddle, P., Jung, M., Krespan, W., Perry, M., Gordon-Kamm, B., Liao, L., Kim, S., Hendrick, C., Zhao, Z.-Y., Dolan, M., Chumley, F., Tingey, S. V., Tomb, J.-F., Gordon, M. P., Olson, M. V., and Nester, E. W. 2001. The genome of the natural genetic engineer *Agrobacterium tumefaciens* C58. Reprinted by permission from *Science* 294: 2317–2323. Copyright 2001 American Association for the Advancement of Science.

Yadav, N. S., Vanderleyden, J., Bennett, D. R., Barnes, W. M., and Chilton, M.-D. 1982. Short direct repeats flank the T-DNA on a nopaline Ti plasmid. Reprinted by permission from *Proceedings of the National Academy of Sciences of the United States of America* 79:6322–6326.

Yanofsky, M. F., Porter, S. G., Young, C., Albright, L. M., Gordon, M. P., and Nester, E. W. 1986. The *virD* operon of *Agrobacterium tumefaciens* encodes a site-specific endonuclease. Reprinted from *Cell* 47:471–477. Copyright 1986, with permission from Elsevier.

Zambryski, P., Joos, H., Genetello, C., Leemans, J., Van Montagu, M., and Schell, J. 1983. Ti plasmid vector for the introduction of DNA into plant cells without alteration of their normal regeneration capacity. Reprinted by permission from *EMBO Journal* 2:2143–2150.

Zhang, L., Murphy, P. J., Kerr, A., and Tate, M. E. 1993. *Agrobacterium* conjugation and gene regulation by *N*-acyl-L-homoserine lactones. Reprinted by permission from *Nature* 362:446–447. Copyright 1993 Macmillan Publishers Ltd.

A CENTURY OF PIONEERING WORK BEGINS

Paul D. Peterson
Pee Dee Research & Education Center, Clemson University
Florence, South Carolina, United States
E-mail: ppeters@clemson.edu

Commentary on the paper:

Smith, E. F., and Townsend, C. O. 1907. A plant-tumor of bacterial origin. Science, New Series, 25:671–673.

In 1902 the United States Department of Agriculture reported that crown gall was "one of the most serious enemies to fruit growers . . . from California to the Alleghenies" (Galloway 1902). This announcement may have been news to some readers, but judging from the amount and frequency of references to the disease on fruit trees and many other economic plants in the late nineteenth and early twentieth century both in America and Europe, crown gall hardly went unnoticed. Various theories had been advanced as to its cause, particularly frost damage, mechanical injury, and fungi. Some researchers suspected that bacteria might be the culprit, but their studies had been largely disregarded due to questionable methodology and to the limited acceptance of bacteria as plant pathogens (Smith and Townsend 1911). Circumstances changed, however, when Erwin Frink Smith, the foremost authority on phytobacteriology, with the assistance of fellow USDA Plant Pathologist Charles Orrin Townsend, began a series of critical investigations in 1904. Three years later Smith and Townsend published the preliminary results from these studies providing well-established evidence for the bacterial origin of crown gall.

Much had changed in the career of E. F. Smith during the twelve years between 1892 when he first examined tumor-like growths on the roots of peach trees and 1904 when he decided to study a similar gall of Paris daisy (*Chrysanthemum frutescens*). It was this research that would lead directly to his classic paper establishing the etiology of crown gall. Smith, in 1892, had been a fairly unknown American plant pathologist trying desperately to find his scientific footing in a recently created division of the USDA devoted to solving plant disease problems. The late nineteenth century was an exciting time for advances in plant pathology at the USDA. Departmental researchers had rapid success solving fungal life cycles, and efforts to control several major plant diseases with fungicides were fundamentally altering agricultural practices. Smith had joined the USDA in 1886 with a B.S. in biology from the University of Michigan and spent the first year or so as an assistant in what would later be regarded as seminal work to understand and control grape diseases caused by fungi (Campbell et al. 1999).

But while Erwin F. Smith shared in the exciting advances in agricultural science at the USDA, he also suffered his share of frustrating disappointments during his early career. In 1887, he received a new assignment to study two mysterious peach diseases, yellows and rosette. Thus began one of the most arduous periods of his long career. The discovery of the causal agents and controls for these two maladies resisted even Smith's most thorough observations and meticulous research over the next six years.

Smith's fortune began to change, however, with a reawakened interest in bacteriology. His connection to bacteriology actually extended back to the mid-1880s when he briefly worked in the field of human hygiene and sanitation in Michigan. During this time, Smith took a passionate interest in reviewing the exciting advances in bacteriology coming from the European laboratories of Louis Pasteur, Robert Koch, and Joseph Lister. There also was important work being done in the United States during the period, particularly in the areas of animal and plant bacteriology, and Smith had followed it closely after joining the USDA. He especially was impressed with the landmark study of hog cholera published in 1889 by Theobald Smith and Veranus A. Moore of the USDA's Bureau of Animal Industry. This paper appears to have inspired his immutable commitment to bacteriological technique. Smith also paid close attention to the groundbreaking research on fire blight of pome fruits by the Americans Thomas J. Burrill and Joseph C. Arthur in the 1880s that had led to the first discovery of plant pathogenic bacteria (Campbell et al. 1999, Griffith et al. 2003).

By focusing on bacteriology, Smith discovered his forté. He devoted considerable time to mastering, improving, and communicating information on bacteriological methodology and technique, before applying his increasing expertise to plant diseases. In 1893, he began a series of seminal studies that would lead to the demonstration of the bacterial origins of a number of infamous diseases such as cucumber wilt and southern bacterial wilt (Campbell et al. 1999). Eight years later he had given Alfred Fischer, a distinguished professor of botany at the University of Leipzig, an intellectual thrashing during the well-known Fischer-Smith controversy between 1898–1901. Fischer, along with other prominent European scientists, refused to accept either the existing proof that myriad plant diseases were caused by bacteria and thus, that plant bacteriology was a credible field of study, or that American scientists had attained the scientific maturity to be leading the movement. In a series of articles in the *Centralblatt für Bakteriologie,* Smith championed the scientific evidence based on experimental proof using the tools of bacteriology. His characteristic attention to detail and brilliant summations of the state of knowledge led to a victory that was complete and conpicuous (Campbell 1981). He had come a long way from those frustrating days with peach yellows.

It was as the acknowledged leader of the field of phytobacteriology that Smith undertook research on the mysterious disease known as crown gall. He had first examined crown gall on the roots of peach trees in 1892, but more pressing research on peach yellows and rosette prevented anything beyond cursory observations. During the 1890s James W. Toumey, a botanist and entomologist at the University of Arizona Agricultural Experiment Station had investigated crown gall of almond and speculated that it might be caused by a myxomycete similar to *Plasmodiophora brassicae,* known to be the causal agent of club root of cabbage. Toumey and other researchers had demonstrated that crown gall was an infectious disease by obtaining symptomatic tumors on healthy plants either through close contact with diseased tissue or by grafting. Around the turn of the twentieth century, other scientists in the United States and Europe continued to study the disease, but its cause remained baffling (Braun 1982).

It was Smith, his colleague C. O. Townsend (an expert on sugar-beet diseases), and their laboratory assistants, Alice Haskins and Nellie A. Brown, who finally discovered the cause of crown gall. Smith's interest in the disease was aroused when a shipment of tumor-bearing Paris daisy plants arrived at the USDA Bureau of Plant Industry in 1904 from a commercial grower in New Jersey. Unlike his brief observations in 1892, this time Smith was prepared for sustained investigations into the nature of the crown gall disease. Characteristic of his scientific practice, he began with an intensive review of the literature available on the subject. He paid particular attention, for example, to the research that had been done in France and Italy on gall-forming diseases, especially olive knot caused by *Bacillus oleaetuberculosis* (*Pseudomonas savastanoi*), a disease with symptoms similar in appearance to crown gall. After familiarizing himself with other work, he moved on to a series of thorough investigations in his USDA laboratory (Braun 1982, Campbell et al. 1999).

Smith and his colleagues found their work painstakingly slow over the next two years. It did not take long for them to eliminate insects and fungi from the list of possible causal agents, but initial attempts to isolate suspect bacteria detected in the interior of undecayed galls went nowhere. Only after it was discovered that bacteria were not "uniformly distributed" in gall tissues and that isolations worked "from that part of the stem where the tumor joins the healthy tissues," did their investigations produce results. With the ability "to isolate the organism and to demonstrate it unmistakenly in stained sections," Smith and his coworkers proceeded to produce "subcultures from poured plate colonies" and to inoculate healthy Paris daisy plants with needle punctures, pricking some plants without inoculation as checks. From subsequent galls produced, the bacterium was "reisolated in pure culture," its identity confirmed, and the experiments repeated.

In April, 1907, Smith and Townsend announced their success in *Science* with "A Plant-Tumor of Bacterial Origin." They proposed that "the organism causing these tumors" be named *Bacterium tumefaciens,* while they carefully described the causal agent as "a schizomycete causing rapid multiplication of the young tissues of *Chrysanthemum frutescens, Prunus persica,* etc." The authors were anxious to present their discovery of a new bacterial disease, yet these were only the early experiments on the host range of crown gall that Smith and Townsend soon broadened to include a wide variety of cultivated plants (Smith and Townsend 1911).

As a matter of note, Smith later credited Fridiano Cavara of Italy with isolating the crown gall bacterium from grape galls a decade before and successfully reinoculating the organism into healthy plants reproducing the disease symptoms (Smith and Townsend 1911). Cavara's limited studies, however, were less than convincing and seem generally to have been ignored. It appears almost certain that Smith was unaware of this work before publishing his paper in 1907 (Braun 1982, Campbell et al. 1999). This does not detract from the importance of the Smith and Townsend publication. Its merits firmly rest on careful methodology and technique lacking in Cavara's incomplete studies.

Along with the paper in *Science,* Smith and Townsend also published a translated version of the information on their crown gall studies in the prestigious *Centralblatt für Bakteriologie* (20: 89–91, 1907). As it turned out, these were only Smith's first forays into what became a fervid focus on plant galls and the histology of tumors over the rest of his career. Before long he was expanding his idea, hinted at in the 1907 paper, of comparisons between the growth and structure of plant tumors and human cancer. Yet, even with his long and celebrated career in plant bacteriology that followed, it is the classic 1907 paper on the etiology of crown gall that perhaps is best remembered. For with it, a century of pioneering work on *Agrobacterium* had begun.

References

Braun, A. C. 1982. A history of the crown gall problem. Pages 155–210 in: Molecular Biology of Plant Tumors. G. Kahl and J. S. Schell, eds. Academic Press, New York, U.S.A.

Campbell, C. L., ed. and trans. 1981. The Fischer–Smith Controversy: Are There Bacterial Diseases of Plants? Phytopathological Classics 13. American Phytopathological Society, St. Paul, MN, U.S.A.

Campbell, C. L., Peterson, P. D., and Griffith, C. S. 1999. The Formative Years of Plant Pathology in the United States. American Phytopathological Society, St. Paul, MN, U.S.A.

Galloway, B. T. 1902. Report of the chief of the Bureau of Plant Industry. Pages 47–108 in: Annual Reports of the Department of Agriculture. GPO, Washington, D.C., U.S.A.

Griffith, C. S., Sutton, T. B., and Peterson, P. D., eds. 2003. Fire Blight: The Foundation of Phytobacteriology. American Phytopathological Society, St. Paul, MN, U.S.A.

Smith, E. F., and Townsend, C. O. 1911. Crown Gall of Plants; Its Cause and Remedy. USDA Bureau of Plant Industry Bulletin No. 213. GPO, Washington, D.C., U.S.A.

PAUL D. PETERSON: I graduated with a B.A. degree in history, an M.A. in public history/archival management, and a Ph.D. in plant pathology from North Carolina State University. I began my plant pathology studies under the direction of the late Dr. C. Lee Campbell, whose infectious passion for plant pathology and epidemiology cultivated my unique pursuit of merging a career in history and science. I am currently a postdoctoral fellow at Clemson University where I am working with Dr. Bruce Martin on a new disease of turfgrass referred to as Rapid Blight. Also, I am a research associate in the Department of Plant Pathology at North Carolina State University where I direct several history projects. I have written numerous journal articles and three books on the history of plant pathology. I also serve as the historian for the American Phytopathological Society.

SPECIAL ARTICLES

A PLANT-TUMOR OF BACTERIAL ORIGIN

THE number of vegetable galls known positively, *i. e.*, by exact experiment, to be due to bacteria, is not very great. The discovery of a new one of undoubted bacterial origin is, therefore, of considerable interest to plant pathologists, and may be of some interest to animal pathologists, especially to those interested in determining the origin of cancerous growths.

For two years the writers have been studying a tumor or gall which occurs naturally on the cultivated marguerite, or Paris daisy. It has been difficult to isolate the organism and to demonstrate it unmistakably in stained sections. Recently the bacteria (seen in small numbers in the unstained tissues on the start) have been plated out successfully. With subcultures from poured plate colonies, thus obtained, the galls have been reproduced abundantly and repeatedly during the last few months, the inoculations having been made by needle-pricks. From galls thus produced the organism has been reisolated in pure culture and the disease reproduced, using subcultures from some of the colonies thus obtained and puncturing with the needle as before. More than 300 galls have been produced by puncture inoculations. Under the most favorable conditions (young tissues) the swellings begin to be visible in as short a time as four or five days, and are well developed in a month, but continue to grow for several months, and become an inch or two in diameter.

In some of our experiments one hundred per cent. of the inoculations have given positive results (40 punctures out of 40 in one series; 62 punctures out of 62 in another), while the check plants have remained free from tumors, and also, in nearly every case, the check punctures on the same plant. In the two series just mentioned there were 110 check punctures on the same plants, all of which healed normally and remained free from galls. Old tissues are not very susceptible. The tumors grow rapidly only in young fleshy organs. The organism attacks both roots and shoots. It frequently induces abnormal growths on the wounded parts of young cuttings. Its power to produce hyperplasia is not confined to the marguerite. Well-developed small tumors have been produced in a few weeks on the stems of tobacco, tomato and potato plants and on the roots of sugar beets. More interesting economically is the fact that galls closely resembling the young stages of crown-gall have been produced on the roots of peach trees by needle-pricks, introducing this organism. In eighteen days these growths have reached the size of small peas, the checks remaining unaffected. It is too early, perhaps, to say positively that the cause of the wide-spread and destructive crown-gall of the peach has been determined by these inoculations, but it looks that way. Of course, the most that can be affirmed absolutely at this writing is that we have found an organism which when inoculated into the peach produces with great regularity galls which in early stages of their growth can not be distinguished from the crown-gall. The matured daisy galls also look astonishingly like the peach gall. Numerous experiments which ought to settle the matter definitely in course of the next three months are now under way. In the best series of experiments on peach roots (that inoculated from a standard nutrient agar culture five days old) 14 groups of needle-punctures (5 in each group) were made on nine trees, 13 tumors resulting. The fourteenth group was on a weak tree which did not leaf out, and might therefore be left out of the count. In that case we have 100 per cent. of infections. On the roots of nine young trees from the same lot, held as checks, 75 punctures were made, using a sterile needle, but no galls resulted. In another series of 9 peach trees inoculated at the same time as the preceding and examined on the twenty-third day, 75 per cent. of the punctures had yielded galls (9 tumors on 7 plants). These roots were inoculated by needle-pricks from a culture believed to be rather too old (glycerin agar streak 6 days), but the plants were set out again, and it is not unlikely that galls will finally develop on the roots of the other two plants. The plants, inoculated and uninoculated, were set, immediately after making the needle-punctures, in good greenhouse soil, in new ten-inch pots, and have been subject to the same conditions as to light, heat and water.

That crown-gall of the peach is due to a myxomycete the writers have never been willing to admit, because the inoculation experiments described by Professor Toumey do not clearly establish such fact. He saw often in the tissues of the galls what he interpreted to be the protoplasm of a slime mold mixed in with the protoplasm of the host plant, and he obtained sparingly what he supposed to be the fruiting bodies of this organism on the cut surface of the galls. He made, however, only two series of inoculations with the spores of his *Dendrophagus globosus,* four trees in the first case and six trees in the second, one developing the disease in the first instance and two in the second. Why did not the other seven trees contract the disease when the spores were thrust into the wounded tissue? He did not fully exclude the possibility that the three infections were due to some other cause accidentally introduced on his needle point. The *Dendrophagus* sporangia furnishing spores for the inoculations grew not on culture media but on the cut surface of a gall (an infectious substance). What if a few bacteria had been carried up from the surface of the gall, contaminating the surface or interior of the sporangia? Then the needle might occasionally have introduced two organisms into the wounds instead of one, as believed, and the unsuspected one might have been the cause of the disease. This supposition is not excluded by any of Professor Toumey's experiments.

The fact remains well established, however, by experiments of various persons: Thaxter, Halsted, Selby, Toumey, Smith, Von Schrenk and Hedgcock, etc., that when minced galls are buried in the earth near the roots of sound trees, the latter develop galls. The disease is therefore a communicable one, but the cause, in spite of much study by many persons, is still in dispute.

For the organism causing these tumors the name *Bacterium tumefaciens* is proposed with the following brief characterization: *B. tumefaciens* n. sp., a schizomycete causing rapid multiplication of the young tissues of *Chrysanthemum frutescens, Prunus persica,* etc., the result being the production of tumors or galls. The organism is motile, especially in young cultures; it is non-gas-forming and aerobic (twelve days) with all of the sugars and alcohols tried (dextrose, saccharose, lactose, maltose, mannit and glycerin). It is white on standard nutrient agar and potato and in peptonized bouillon. In tubes of bouillon it grows best at the top, producing a stringy ragged rim easily separable on shaking. It does not cloud bouillon heavily. The surface colonies on agar, 25° C., are small, round, smooth and rather dense. In agar streak cultures the organism is inclined to pile up along the track of the needle rather than to spread widely. It is inclined to be viscid on agar, after three days. It gradually blues litmus milk, throwing down the casein by means of a lab ferment, or at least not by the production of any acid, finally the litmus is reduced. It does not liquefy standard nutrient gelatin (fifteen days) and does not grow in the thermostat at blood heat (agar, bouillon). In young agar streak cultures it is a medium-sized, short rod, with rounded ends, often in pairs with a plain constriction, the elements usually being 1μ or less in diameter and two to three times as long as broad. The one to three flagella are polar. It is not yellow on any medium, or green fluorescent, nor does it brown the agar. It is rather short-lived on agar. It does not grow in Cohn's solution and does not infect olive shoots. It occurs principally at the bottom of the tumor rather than uniformly distributed in its tissues. It is best isolated from that part of the stem where the tumor joins the healthy tissues. There are slight indications of metastasis. Non-pathogenic yellow organisms are frequently obtained on plates made from older portions of the galls.

ERWIN F. SMITH,
C. O. TOWNSEND
BUREAU OF PLANT INDUSTRY,
U. S. DEPARTMENT OF AGRICULTURE,
April 4, 1907

ARMIN C. BRAUN AND THE DISCOVERY OF *AGROBACTERIUM*-MEDIATED TRANSFORMATION OF PLANT CELLS

Andrew N. Binns
Department of Biology and the Plant Science Institute
University of Pennsylvania, Philadelphia, Pennsylvania, United States
E-mail: abinns@sas.upenn.edu

Commentary on papers:

White, P. R., and Braun, A. C. 1941. Crown gall production by bacteria-free tumor tissues. Science 94:239–241.

Braun, A. C., and Mandle, R. J. 1948. Studies on the inactivation of the tumor-inducing principle in crown gall. Growth 12:255–269.

As a result of his landmark studies beginning in the 1940s and continuing into the early 1980s, Armin C. Braun is widely viewed as the father (well—maybe grandfather, at this point in time!) of the modern studies concerning *Agrobacterium*-mediated transformation of plant cells. His work, particularly in the 1940s and 1950s challenged strongly held beliefs. Through careful experimentation, Braun redirected an entire field of study, one that led ultimately to advances in areas ranging from plant genetic engineering to studies of infectious diseases in both plants and animals. How did this happen? What were these landmark discoveries? Some of them are represented in the papers presented below. Others are not included in this volume. I would argue that three fundamental insights developed as a result of Braun's work: 1) a "tumor inducing principle" (TIP), elaborated by virulent strains of *Agrobacterium* at a wound site on the plant, causes a 'transformation' of the plant cells to a tumorous nature and this phenotype persists in the absence of the inciting bacteria; 2) tumorous growth of the plant cells results from activities that are separable from the actual cause of the transformation; and 3) transformation by *A. tumefaciens* does not necessarily result in the irreversible loss of totipotency of the transformed plant cell. My objective in the next few paragraphs is, in an admittedly personal style, to discuss Braun's career and research findings. His findings framed the biological questions that future cellular, molecular and genetic approaches have tackled, yielding so many intriguing results, many of which are included in this volume.

How did Braun end up working on *A. tumefaciens* and, subsequently, discover that the persistent growth of crown gall tumors do not require the continued presence of the bacteria? According to his last paper—a reflection on his scientific career (Braun 1986)—Braun gained interest in the crown gall tumor problem as a Ph.D. student when he was studying Plant Pathology at the University of Wisconsin. Although the laboratory of A. J. Riker at Wisconsin was actively studying crown gall tumorigenesis, Braun's thesis work was on the wildfire and leafspot diseases of tobacco. His interest in *Agrobacterium*-mediated tumorigenesis was actually sparked by a visit to the laboratory of C. Stapp in Germany where he met a student studying crown gall on apple trees. The student had an incredibly difficult time recovering virulent *Agrobacterium* strains from actively growing tumors, an observation that surprised and impressed Braun. Indeed, it led him to postulate that the dogma of the time—virulent strains at the infection site were producing substances that stimulated tumor growth—might be incorrect. Could it be that the bacterium was not required for the persistent growth of the tumor?

Braun applied for a postdoctoral fellowship to tackle this question at the Department of Plant Pathology at the Rockefeller Institute (then in New Jersey) headed by L. O. Kunkel. During his initial studies he confirmed earlier reports of "secondary" tumors of sunflower, that is, tumors that arise at a considerable distance from the inoculation site. Braun showed that while these secondary tumors were devoid of bacteria they could be grafted onto a healthy host plant and form what appeared to be typical crown gall tumors. This was inconsistent with earlier studies indicating that bacterially produced plant growth substances—such as indole-3-acetic acid—were responsible for tumorous growth. As fate would have it, P. White also worked in the same department and had recently developed methods for *in vitro* root organ culture. With Kunkel's consent, Braun sought out White to initiate a collaboration on the crown gall issue: would the secondary tumors grow in aseptic culture (as opposed to the non-tumorous stem tissues that, as White had earlier shown, could not) and if so, would they retain the tumorous phenotype, that is the capacity to grow as tumors when grafted onto healthy host plants?

The results of this collaboration came in the form of three papers, the first of which, the 1941 *Science* paper, is reprinted in this volume. (Also see White and Braun, 1942; 1943.) These studies clearly demonstrated that secondary crown gall tumors were bacteria free, but could grow *in vitro* on medium that did not support growth of normal stem tissues and could form transplantable tumors. Ultimately, these findings revolutionized the study of crown gall—somehow the bacterium must actually cause a persistent transformation of the plant cells into tumor cells.

Beyond the fact that the plant tumors can persist in the absence of *Agrobacterium,* the studies on secondary tumors led Braun to focus on another issue—the physiological basis of the tumorous phenotype. In what I have always considered an amazing insight, Braun and Laskaris (1942) took advantage of bacterial mutants to distinguish between tumor initiation and tumorous growth. These studies showed that a mutant strain (A66) of *A. tumefaciens* could initiate extremely slow growing (attenuated) tumors on decapitated tomato plants, but that supplementation of these tumors with an auxin such as indole-3-acetic acid would stimulate tumor proliferation to a level observed in wild type strain A6 induced tumors. Treatment of uninfected but wounded plants with auxin would not cause such growth. Thus, the initiation step and the tumor progression (growth) steps were distinct. Later studies (Braun 1958) demonstrated that the physiological basis of crown gall tumor growth was the development of the capacity for the transformed cells to grow continuously in the absence of exogenously supplied auxin and cytokinin. We now know that the initiation step is, in fact, the transfer of the T-DNA into the plant cell genome whereas the tumor growth results from the expression of various plant hormone biosynthetic genes on that T-DNA. Characterization of the molecular basis of the A66 mutant phenotype was achieved some forty years later (Binns et al. 1982) and shown to be the result of an insertion element (IS66) that had jumped into the one of the auxin biosynthesis genes of the T-DNA.

The second paper reprinted in this volume (Braun and Mandle 1948) focuses our attention on the initiation phase of the crown gall transformation. This was the culmination of several years of work—interrupted by Braun's service in the Army during World War II—that utilized heat treatments to disrupt the transformation process. These studies took advantage of Kunkel's observation that *Vinca rosea* could withstand elevated temperatures (e.g., 46°C) for a week or more without serious injury. In contrast, *Agrobacterium* was killed under such conditions. Furthermore, Riker found that tumor formation was disrupted at 32°C, a temperature that is not lethal to either the host or the bacterium. Through a series of temperature shift experiments Braun deduced that during a period of time 24–96 hours after inoculation, *Agrobacterium* must elaborate a "material active in initiating tumor development" (Braun 1943), which he later described as "the active principle that is responsible for the conversion of normal cells to neoplastic cells." (Braun 1947). Interestingly, Braun suggested that a possible candidate for this principle was "a chemical fraction of the bacterial cell that is capable of initiating, as in the case of the transforming substance (desoxyribonucleic [sic] acid) of the pneumococci, a specific alteration in the host cell" (Braun, 1947). As we see in the Braun and Mandle paper, this principle was named the "Tumor Inducing Principle" (TIP). Temperature shift experiments were used in that study to show that the TIP can be elaborated by the bacterium and transform host plants inoculated and initially held at 32°C for 24 hrs, if the inoculated plants were then moved to 25°C for as short a time period as 10 hr. We now know that the molecular bases for the 32°C effect on *Agrobacterium*-mediated tumor formation is that *vir* gene expression from the Ti plasmid is significantly decreased and that the *Vir*B complex necessary for T-DNA transfer into the plant cell is apparently unstable at this temperature.

The nature of the TIP was the subject of numerous studies from 1948 on, by numerous investigators. As noted above and discussed in detail elsewhere in this volume, the Ti plasmid is crucial and the T-DNA can now be viewed as the TIP. I've always found it intriguing that, despite his earlier proposal, Braun was amongst the last to acknowledge that bacterial DNA transferred into the host was the TIP. Significant reasons for this were that 1) other studies of his (see below) indicated that certain types of crown gall tumor cells remained totipotent and 2) numerous reports of dubious quality, particularly during the 1960s, suggesting bacterial DNA was the TIP. This attitude was maintained in the face of all the work related to the opines and the presence or absence of particular Ti plasmids. However, in the fall of 1977 I was fortunate enough to sit in on a meeting of Braun with Mary-Dell Chilton. She brought Braun up to date on the intricacies of solution hybridization analysis of nucleic acids and then presented her data that aseptic crown gall tumor cells did, indeed, have a small piece of the Ti plasmid in their genome. In the face of his extremely strong held beliefs to the contrary, Braun said "Mary-Dell—you have convinced me that Ti plasmid DNA is in the tumor cell and this is probably the TIP." This was no small admission and speaks not only to Braun's ultimate openness as a scientist but also to the strength of Chilton's data and the clarity of her presentation. It also confirms something that Braun would repeat any number of times in the years I knew him—"Andy, the great part of science is that the truth will emerge, eventually."

Finally, no discussion of Braun's career, particularly from my personal perspective, is complete without acknowledging his work on "recovery" from the tumorous state. In a 1951 paper, Braun demonstrated that the spontaneously appearing shoots from tobacco teratomas induced by *A. tumefaciens* strain T37 would not root but could be grafted onto healthy host plants and grow into normal-appearing, fertile shoots. These set seed and the progeny were completely normal, indicating a recovery from the tumorous state. However, because the original teratomas may have contained a mixture of tumorous and non-tumorous cells the possibility existed that the observed "recovery" was simply a selection of the non-transformed cells during the regeneration process. To overcome this criticism, Braun took advantage of the then recently developed methods for cloning single cells, and derived cloned lines of T37 induced teratomas. Grafting studies (Braun 1959) of these lines yielded the same results as in the 1951 paper: normal appearing, fertile shoots could form spontaneously and, again, produce progeny that were normal in all respects. Thus, transformation did not necessarily result in the loss of totipotency so characteristic of plant cells. In later work, Braun and colleagues demonstrated that the original grafted shoots retained the capacity to synthesize nopaline and to reinitiate the tumorous phenotype when returned to culture (Braun and Wood 1976, Wood et al., 1978) whereas the 'recovered' progeny did not have these capacities. These studies were particularly important to the then fledgling concept of plant genetic engineering: insertion of the T-DNA and expression of genes on it, were compatible with the differentiation of plant cells into normal appearing, fertile shoots. But why did the T-DNA not appear in the progeny? Was there some mechanism whereby plants could identify and eliminate foreign DNA? Or were gene products from the T37 T-DNA sufficient to disrupt gametogenesis or fertilization of T-DNA containing gametes? A collaborative effort between Mary-Dell Chilton, Ken Barton, and Tony Matzke at Washington University, and myself (then a new faculty member at Penn), sought to answer these questions by engineering the T-DNA of T37 so as to knock out the so-called "rooty locus" (the cytokinin biosynthetic gene) and insert other genes at that site. The resultant engineered strains could still transform tobacco cells and appropriate hormone manipulations of these cells allowed them to regenerate shoots that could be rooted via standard procedures. More importantly, the resultant normal-appearing plants set fertile seed that transmitted the engineered T-DNA to their progeny (Barton et al., 1983). Thus, Braun's original observations on the suppression of and recovery from the tumorous state of T37-induced tumors led directly to the experiments demonstrating that the *Agrobacterium* system could be used for the genetic engineering of plants.

There is no doubt that Braun looked back upon his career with both satisfaction and surprise. The last sentence of his final paper summarizes this best: "It is most gratifying, indeed, for one who has spent a lifetime studying plant tumors to see the concepts developed with the use of those relatively uncomplicated systems find application, and not only contribute to an understanding of cancer as a fundamental biological problem, but provide as well a model for the genetic engineering studies now being so actively and enthusiastically pursued in laboratories around the world." (Braun 1986).

References

Barton, K. A., Binns, A. N., Matzke, A. J. M., and Chilton, M.-D. 1983. Regeneration of intact tobacco plants containing full length engineered T-DNA and transmission of T-DNA to R1 progeny. Cell 32:1033–1043.

Binns, A. N., Sciaky, D., and Wood, H. N. 1982. Variation in hormone autonomy and regeneration potential of cells transformed by strain A66 of *Agrobacterium tumefaciens.* Cell 31:605–612.

Braun, A. C. 1943. Studies on tumor inception in the crown-gall disease. Am. J. Bot. 30:674–677.

Braun, A. C. 1947. Thermal studies on tumor inception in the crown gall disease. Am. J. Bot. 30:674–677.

Braun, A. C. 1951. Recovery of crown-gall tumor cells. Cancer Res. 11:839–844.

Braun, A. C. 1958. A physiological basis for autonomous growth of the crown-gall tumor cell. Proc. Natl. Acad. Sci. U.S.A. 44:344–349.

Braun, A. C. 1959. A demonstration of the recovery of the crown-gall tumor cell with the use of complex tumors of single cell origin. Proc. Natl. Acad. Sci. U.S.A. 45:932–938.

Braun, A. C., 1986. The crown gall tumor problem: Contributions to experimental oncology and genetic engineering. Pages 13–39 in: Handbook of Plant Cell Culture: Techniques and Applications. D. A. Evans, W. R. Sharp, and P. V. Ammirato, eds. MacMillan Publishing Co., New York, U.S.A.

Braun, A. C., and Laskaris, T. 1942. Tumor formation by attenuated crowngall bacteria in the presence of growth promoting substances. Proc. Natl. Acad. Sci. U.S.A. 28:468–477.

Braun, A. C., and White, P. R. 1943. Bacteriological sterility of tissues derived from secondary crown-gall tumors. Phytopathology 33:85–100.

Braun, A. C., and Wood, H. N. 1976. Suppression of the neoplastic state with the acquisition of specialized functions in cells, tissues and organs of crown gall teratomas of tobacco. Proc. Natl. Acad. Sci. U.S.A. 73:496–500.

White, P. R., and Braun, A. C. 1942. A cancerous neoplasm of plants: Autonomous bacteria-free crown gall tissue. Cancer Res. 2:597–617.

Wood, H. N., Binns, A. N., and Braun, A. C. 1978. Differential expression of oncogenicity and nopaline synthesis in intact leaves derived from crown gall teratomas of tobacco. Differentiation 11:175–180.

ANDREW N. BINNS: In 1971, I graduated B.A. (Magna cum laude) in Biology from Lawrence University, Appleton, Wisconsin, and in 1979 obtained a Ph.D. from Princeton University. After a postdoctoral period at the Rockefeller University, New York (1977–1980), I was appointed to the faculty of the University of Pennsylvania.

As an undergraduate, I first examined crown gall tumors in a laboratory exercise of an introductory botany course taught by Dr. Nicholas C. Maravolo. Later, in Maravolo's plant growth and development course, I read and was intrigued by Armin Braun's paper on recovery from the crown gall tumorous state via regeneration (his 1959 paper). I went on to do research in Maravolo's lab on polarity and regeneration in tissues from the bryophyte, *Marchantia polymorpha.* In graduate school, I carried out my Ph.D. studies in the laboratory of Dr. Fred Meins, Jr., demonstrating that the conversion of cultured tobacco cells from a hormone requiring to hormone independent growth state was the result of epigenetic changes in those cells. Fred, who did his Ph.D. work in Braun's lab, was really the person who helped me the most in the development of my career as a scientist and a teacher. My studies were of interest to Armin Braun at the Rockefeller University, who was then carefully examining the possible epigenetic basis of tumors. Because we had communicated several times regarding these studies, I asked him if he would write letters of reference for my postdoctoral applications. In the return mail came a letter saying that he would be happy to do so, but would I be interested in carrying out postdoctoral studies in his lab? Needless to say, I accepted the offer! My time there was spent learning as much as I could about *Agrobacterium* and crown gall through both experiments and long, cherished conversations with both Dr. Braun and his colleague, Dr. Henry Wood. Finally, though I never formally worked in her laboratory, Dr. Mary-Dell Chilton has been a friend, collaborator and informal mentor whose support and advice has always been treasured.

CROWN GALL PRODUCTION BY BACTERIA-FREE TUMOR TISSUES

CROWN gall has in the past been produced only by inoculation of a host plant with *Phytomonas tumefaciens* (Smith and Town.) Bergey *et al.* either as a pure culture or in the form of a preparation of tissues infected therewith. Although crown-gall tissues have not always yielded cultures of the organism, it has been presumed that the bacteria were present or at least had been present at some stage in the development of the tumor. The production of tumors without direct stimulation from the bacteria, although envisioned as a possibility by certain workers, has not previously been demonstrated, nor has that possibility been given serious credence by other investigators.

One of us has recently confirmed the work of Smith *et al.*,[1] showing that true secondary tumors developed in sunflower plants at a distance of several internodes from the site of the primary tumor, even when inoculations were made into fully elongated internodes at a distance from the apical bud.[2] Many of the secondary tumors were found to be sterile by the usual bacteriological criteria. The bacteria-free secondary tumors appeared, therefore, to be admirably suited for an experimental study of the question as to whether host cells under the influence of the bacteria acquire the capacity for autonomous growth.

Recently developed techniques for the *in vitro* cultivation of excised plant tissues[3] were applied to the study of secondary tumors. Tissue fragments from the interior of a series of such tumors were removed aseptically and placed in 125 ml Erlenmeyer flasks, each containing 50 ml of White's standard glycine-thiamin nutrient[4] stiffened with 0.6 per cent. of thoroughly leached agar. Out of 107 original isolations from tumors, 2 were contaminated with bacteria and 4 with molds. Out of 50 control cultures taken from healthy plants, 4 were contaminated with bacteria and 3 with molds. The larger tissue masses available in the tumors and the consequent greater ease of manipulation were doubtless responsible for the lower incidence of contaminations in tumor cultures.

Cultures capable of continued rapid growth were not established from any of the small petiolar tumors nor from the petioles of healthy plants. On the other hand, out of 37 isolations from large secondary tumors from the stems at nodes, 19 grew successfully. Six isolations were carried through 3 or more passages, being divided at each passage. Two strains, one isolated on February 10, 1941, the other on March 25, 1941, were retained for detailed study. The first of these isolations has, at the time of writing (August, 1941), been maintained through 13 successive passages, has been divided into 482 pieces, each several times as large as the original, and has undergone a theoretical increase[5, 3] in volume of approximately 450,000 times. Throughout this manipulation, it has been constantly in contact, both at intact and at freshly cut surfaces, with a nutrient shown by ourselves and by others to be capable of supporting a profuse growth of *Phytomonas tumefaciens;* yet, out of the 482 cultures not a single one has developed any bacterial growth. Attempts to isolate bacteria from these cultures by grinding and plating on nutrient agar or in broth have consistently failed. When cultures were ground and the paste injected into sunflower or tomato plants, no galls were produced such as regularly appeared when a paste prepared in a similar manner from young primary tumors was injected. The results seem to furnish almost unquestionable evidence that these tissues, which multiply rapidly, nevertheless do so without continued stimulation from crown-gall bacteria. Their capacity for autonomous growth appears established.

Tumor tissues grown *in vitro* are colorless or, especially just after being divided, slightly tinged with brown from oxidative products, scattered remnants of dead cells, etc. The surfaces are rough, covered with irregular unorganized protuberances, and may put forth unorganized outgrowths either along the agar or up into the air. They are firm but easily cut and of uniform texture. Histologically they are mostly parenchymatous with scattered scalariform elements. These elements form a structure closely resembling that of crown-gall tissue stimulated by *Phytomonas tumefaciens* when inoculated into parenchymatous tissues of the host.

Tissues from cambial and procambial regions of healthy sunflowers have likewise been isolated in culture but present a very different picture. In the 9 weeks that they have been maintained, they have been carried through 5 passages, but their growth rate has been so slow that the volume increase during this period has been of the order of 30 times as compared to about 400 times for the gall tissues over a corresponding period. Cultures of normal tissue maintain their outlines, whereas gall cultures grow as irregular or subglobular masses. Normal tissue cultures regularly contain chlorophyll, are woody in texture and frequently produce roots, a phenomenon not observed to date in cultures of gall tissues.

At the end of 5 successive passages *in vitro,* 10 tumor cultures were grafted back into young healthy sunflower plants. Of these, 5 implants had, at the end of 7 weeks, grown into typical crown-gall tumors having diameters up to 1 cm, one plant was accidentally broken, and 4 implants failed to grow. Similar results have been obtained with a second series of grafts using cultures from the 6th passage and a third series using cultures from the 10th passage. Attempts to isolate bacteria from one of the tumors by grinding and plating in nutrient dextrose agar failed. The results seem to indicate that the tumor-inducing capacity has been retained by these bacteria-free tissues through at least 10 successive passages *in*

[1] E. F. Smith, N. A. Brown and L. McCulloch, *U. S. Dept. Agr., Bur. Pl. Ind., Bull.* 255, 1912.
[2] A. C. Broun, *Phytopath.*, 31: 135–149, 1941.
[3] P. R. White, *Am. Jour. Bot.*, 26: 59–64, 1939.
[4] P. R. White, *Plant Physiol.*, 14: 527–538, 1939.
[5] P. R. White, *Plant Physiol.*, 9: 585–600, 1934.

vitro. Preliminary examination has shown the induced tumors to have a histological structure considerably more uniform than that of most crown galls, with extensive hyperplasia but relatively little disorganization. This last feature is quite marked and may possibly be due to the uniform distribution of the mechanism of hyperplasia in contrast to the scattered and localized centers of stimulus characteristic of bacterial galls. Fusion with the host tissue was excellent, although growth was mostly if not entirely a function of the transplant, as is the case with transplanted animal neoplasms. It is clear from these results that the affected tissues have undergone a drastic change which is indicated, first, by their capacity to produce galls, a quality not found in normal tissue, and, second, by their behavior *in vitro,* where their growth habits differ markedly from those of normal tissues under identical conditions. That this change was originally brought about by some stimulus from the crown gall organism seems clear. That its maintenance is not dependent on the continued presence of the bacteria is equally clear. Further details will be published elsewhere.

Philip R. White
Armin C. Braun

The Rockefeller Institute for Medical Research,
Princeton, N. J.

Growth, 1948, **12**, 255-269.

STUDIES ON THE INACTIVATION OF THE TUMOR-INDUCING PRINCIPLE IN CROWN GALL

ARMIN C. BRAUN[1] AND R. J. MANDLE[2]

(Received for publication July 16, 1948)

Normal plant cells may be completely and permanently modified by the brief and localized presence of the crown-gall bacterium in susceptible plant tissues. Following the action of the bacteria, the proliferation of altered plant cells becomes an automatic process which is independent of the growth restraining influences of the host and the continued presence of the inciting bacteria. Aside from its biological activity, little is known concerning the nature of the principle responsible for the transformation of normal plant cells to tumor cells. Recently, however, a method has been described (2) for studying the process of tumor inception and development in those plant species that will not tolerate high temperatures for periods sufficiently long to kill the inciting bacteria of the crown-gall disease. This method is based on a finding which demonstrated that, although a temperature of 32° C. completely and abruptly halts the process of the transformation of normal cells to tumor cells, that temperature does not exert an inhibitory influence on the factors responsible for the multiplication of the tumor cells after the cellular transformation has been fully accomplished. The crown-gall bacteria were found to multiply as well in the host at 32° C. as at 25° C. By the use of this procedure it was possible to define precisely the period in which the cellular transformation is accomplished in *Vinca rosea* L., the Madagascar periwinkle. The results obtained in these experiments confirmed earlier work (1) in showing that there is a lag of somewhat more than one day between inoculation of plants with bacteria and the consummation of the cellular transformation. The results indicated further that the bacteria were capable for the most part of bringing about the alteration of normal cells to tumor cells only during the four-day period immediately following the wounding of the

[1]*Department of Animal and Plant Pathology, The Rockefeller Institute for Medical Research, Princeton, N. J., and* [2]*The University of Pennsylvania, Philadelphia, Pa.*

host. This was true despite the fact that many virulent bacteria were in intimate contact with the host cells after the fourth day.

By use of this technique it appeared possible to define more adequately the conditions which are critical to the neoplastic alteration and to characterize certain properties of the transforming principle. It is with these phases of the crown-gall problem that the present paper is concerned.

Methods and Materials

In the experiments reported here *Kalanchoe daigremontiana* Hamet and Perrier was used as the test object throughout. This host was selected because it characteristically produces small plantlets at the edges of its leaves. When mature these plantlets drop from the leaves and if environmental conditions are favorable they develop into plants typical of the species. The propagation of unlimited numbers of genetically uniform plants was therefore possible. The plants were grown in 4-inch pots and used when about 3 inches tall. Despite the fact that genetically similar plants were used in the experiments, some variation in the response of the plants to bacterial infection was noted. Different portions of the same stem sometimes responded differently to the inciting agent as measured by the size of the resulting tumors. Plants treated under similar conditions also showed at times some variation in their response to infection. As a result each experiment was performed on at least three different occasions during a period of more than a year and representative data from these are presented in the tables that follow.

The virulent B2 strain of *Agrobacterium tumefaciens* (Smith and Town.) Conn was used as the test organism. A moderately heavy tap water suspension of a 24-hour culture was inoculated into the plants by means of the needle puncture method.

Thermostatically controlled glass chambers in which the temperature was maintained at 32-33.5° C. were used throughout the work. The chambers were placed in a greenhouse and protected from the direct rays of the sun by shading. The relative humidity of the chambers was about 90 per cent. When not in the chambers, the plants were held on a greenhouse bench at 25°-26° C.

All results unless otherwise noted were recorded 14-16 days after the plants had been placed at 32° C. for the final incubation at that temperature. The criterion used here to determine whether normal plant cells had been transformed to tumor cells by the bacteria was based on the fact that a perceptible mass of tumor tissue resulted from the growth of those cells after a suitable period.

Experimental Results

Exploratory experiments demonstrated that tumors were produced on inoculated *Kalanchoe* plants at a temperature of 25° C. but not at 32° C. In its response to heat, therefore, this host behaved as did the periwinkle (2). Since the period during which normal cells are transformed to tumor cells had not yet been defined for *Kalanchoe*, it was necessary at the outset to determine the minimum period necessary for the bacteria to accomplish the cellular alteration at 25° C. as well as the maximum time following wounding in which tumors could be initiated in this host.

As in the earlier work, *Kalanchoe* plants were inoculated at exactly 24-hour intervals for a period of 5 days. In certain experiments inoculations were made at intervals of 30, 32, 34, and 36 hours. Following inoculation the plants were permitted to remain on a greenhouse bench at a temperature of about 25° C. Six days after the first inoculations were made all plants were placed in a chamber at 32° C. for a period of 16 days. This was done to halt the process of cellular transformation and to permit development of tumors from cells that had been altered during the incubation period at 25° C. The results of one typical experiment which are shown in Table 1 and Figure 1 demonstrate that the cellular alteration was not accom-

TABLE 1

Representative experiment showing the effect of time on tumor inception and development. Inoculated plants were held at 25° C. for periods indicated below. The plants were then placed and held at 32° C. to stop the process of cellular transformation and permit the development of altered cells into tumors.

Incubation period at 25° C.	Relative size of resulting tumors
24 hours	—
30 hours	—, occ. ±
32 hours	—, ±
34 hours	—, ±
36 hours	±
48 hours	+, ++, occ. +++
72 hours	++, +++
96 hours	++, +++

occ. = occasionally.

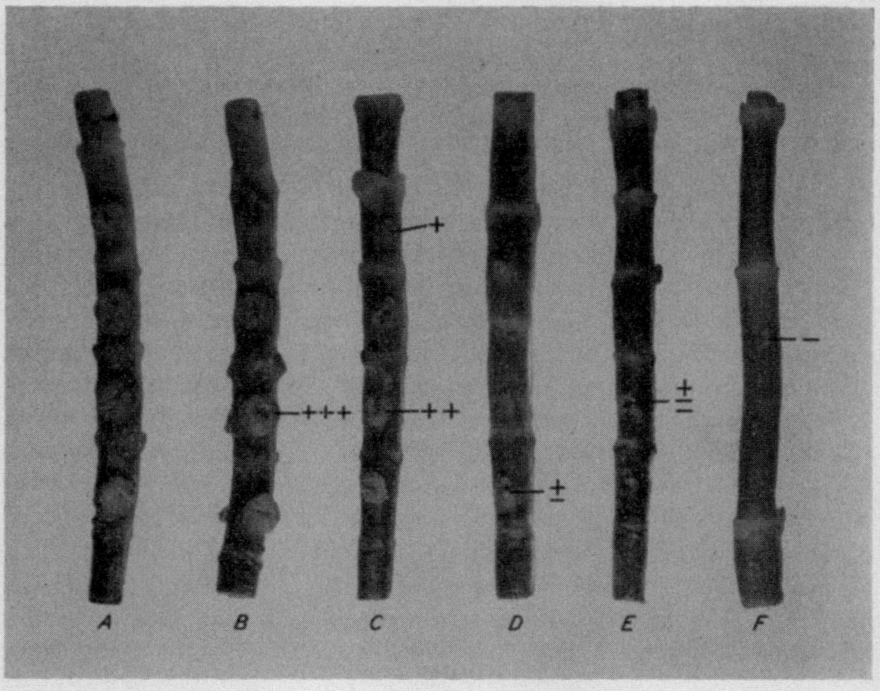

FIGURE 1
Kalanchoe stems inoculated with B2 strain of *Agrobacterium* tumefaciens. Following inoculation the bacteria were permitted to act on the host at 25° C. for (A) 96 hours, (B) 72 hours, (C) 48 hours, (D) 34 hours, (E) 30 hours, (F) 24 hours before the plants were placed at 32° C. to stop the process of cellular transformation and permit development of the tumors. Tumor size was graded as indicated in the Figure. (Photographs by J. A. Carlile.)

plished in 24 hours but was consummated during a 36-hour or longer incubation period at 25° C. Further experiments showed that the minimum period necessary for the bacteria to change normal cells to tumor cells was 32±2 hours at 25° C.

In order to determine the maximum time in which tumors can be initiated in *Kalanchoe* after wounding, the host plants were inoculated with the crown-gall bacteria and placed immediately into a chamber at 32° C. At intervals of exactly 24 hours, plants were removed from the chamber and placed on a greenhouse bench for a period of 25 days. The results which are shown in Table 2 demonstrate that not only the percentage of tumors initiated but also

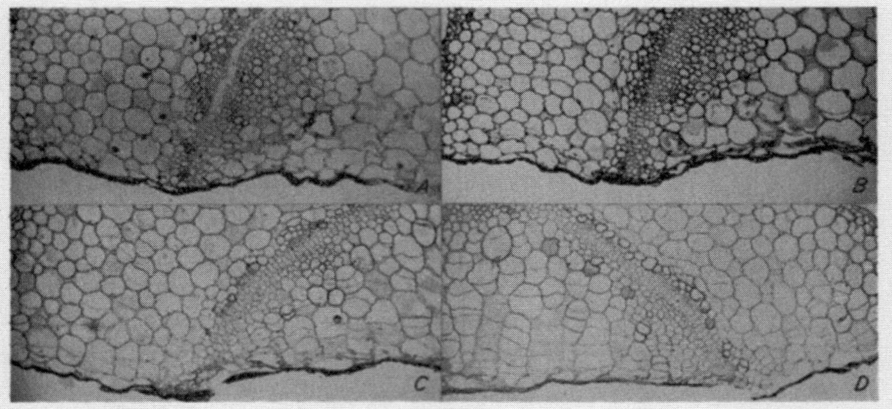

FIGURE 2
Four stages in the process of wound healing as it occurred in *Kalanchoe* at 25° C. Material was fixed (A) 24 hours, (B) 48 hours, (C) 72 hours, (D) 96 hours after wounding. Note that wound healing as determined by active cell division begins between the second and third day after wounding. (Photographs by J. A. Carlile.)

TABLE 2
One representative experiment showing the effect of incubation time at 32° C. on tumor inception and development. Plants were inoculated and placed immediately at 32° C. for periods indicated below. Following incubation at 32° C. the plants were placed and held at 25° C. The results were recorded 25 days after inoculation.

Incubation period at 32° C.	Relative size of resulting tumors	Percentage
24 hours	++, +++	100
48 hours	++, +++	100
72 hours	+, ++	67
	±	33
96 hours	±	82
	—	18
120 hours	±	15
	—	85

the size of the resulting tumors becomes progressively smaller between the second and fifth days after wounding. Plants held at 32° C. for 24 to 48 hours produced large tumors in all instances. On the other hand, about 85 per cent of the punctures were negative and the remainder gave rise to small tumors in those plants that had been held at 32° C. for a 5-day period. Thus we find an inception period which corresponds closely to that of the periwinkle (1, 2).

There is a lag of 32±2 hours between inoculation and the beginning of transformation of normal cells to tumor cells. The host, for the most part, does not respond 5 days or more after wounding.

These results suggested that wound healing might serve as a limiting factor in the ability of the bacteria to convert normal cells to tumor cells. A histological study was therefore undertaken to determine the rate of wound healing in *Kalanchoe* at 25° C. and at 32° C. Needle punctures were made through internodes at 24-hour intervals for a period of 5 days. Six days after the first punctures were made the desired stem segments were placed in formalin-acetic acid-alcohol fixative, dehydrated and cleared, embedded in paraffin, sectioned and stained with dilute carbol fuchsin and orange G. Representative sections of the stages of wound healing as they occurred at 25° C. are shown in Figure 2. It is clear from these photographs that gross wound healing, as measured by the number of cell divisions following the puncture, becomes active between the second and third days after wounding. Thereafter the amount of healing becomes progressively greater until the process is complete. Wound healing in *Kalanchoe* progressed at about the same rate at 25° C. and at 32° C. The results reported in Table 1 demonstrate that normal cells are changed to tumor cells in 32±2 hours when the plants are held at 25° C. following inoculation. Tumors of the size designated as ++ and +++ were initiated in 48 and 72 hours. This suggests that the process of cellular transformation is most active between the second and third days after inoculation. Thus in *Kalanchoe* it appears that it is just before or during the earliest stages of active wound healing that the cellular alteration is most readily accomplished.

Since the minimum period needed to accomplish the cellular transformation was found to be approximately 32±2 hours at 25° C., an attempt was made to determine whether this period could be reduced appreciably by permitting the bacteria an initial incubation in the plants at 32° C., a temperature at which both the host and the organism develop well but at which tumors are not initiated. For these experiments an incubation period of 24 hours at 32° C. following inoculation was used. Preliminary studies had shown that this period approached the optimum for the experiments reported below. After the initial incubation at 32° C. the plants were placed at 25° C. for

TABLE 3
Plants were inoculated and placed immediately at 32° C. for 24 hours. Following incubation at 32° C. the plants were placed at 25° C. for the periods indicated below. After the desired incubation at 25° C. the plants were again placed and held at 32° C. to stop the cellular transformation process and permit the development of altered cells into tumors

Incubation period at 25° C.	Relative size of resulting tumors
6 hours	—
8 hours	—
10 hours	—, ±
12 hours	±
14 hours	+
16 hours	+
20 hours	++

the periods indicated in Table 3, after which time they were held at 32° C. to stop the process of cellular transformation and to permit development of the tumors. The results reported in Table 3 demonstrate that under the conditions of the experiment normal cells were not converted to tumor cells during a single exposure of 8 hours or less at 25° C. but were transformed in 10 hours or more. As the incubation period at 25° increased from 10 hours to 20 hours the relative size of the resulting tumors also increased. Thus, an exposure of 10 hours at 25° C. following an initial incubation of 24 hours at 32° C. yielded approximately the same results as did a total incubation time of 32±2 hours at 25° C. These findings suggest, therefore, that the bacteria require an incubation period of roughly 20 to 24 hours for environmental adjustment plus an additional 10-hour period at 25° C. to produce the active principle in amounts necessary to transform normal to neoplastic cells. This does not mean, of course, that small amounts of the transforming principle may not be produced during the period of environmental adjustment.

It has been previously suggested (3) that the transforming principle either *1*) is very heat labile and is destroyed at 32° C. under the conditions that exist in the plants tested, or 2) is rendered inactive by an inhibitor that is activated at or above 30° C., or possibly *3*) is not produced by the bacteria at the higher temperature. Experiments were next conducted, therefore, in an attempt to learn which, if any, of the three possibilities suggested above was the determining factor involved in the inability of the bacteria to accomplish the cellular transformation at 32° C. The problem was approached by means of an experiment in which the plants were alternately subjected to temperatures of 25° C. and 32° C. following an initial incubation of inoculated plants for 24 hours at 32° C. A single 10-hour exposure at 25° C. was shown to be the minimum period in which the cellular transformation was accomplished under the conditions of these experiments. The alternate exposure times at 25° C. and 32° C. used here are listed in Table 4. In all instances a total of 30 hours at

TABLE 4

Plants were inoculated and placed immediately at 32° C. for 24 hours. Following incubation at 32° C. all plants were treated alternately at 25° C. and 32° C. for periods indicated below. Intermittent exposures totalling 30 hours at 25° C. were given in each instance. Following treatment all plants were held at 32° C.

Type of treatment given: Seven and one-half 4-hour periods at 25° C. alternated with *seven*	Relative size of resulting tumors	Calculated No. of units of transforming principle available
1-hour periods at 32° C.	++	23
2-hour periods at 32° C.	+	16
4-hour periods at 32° C.	—	2
Five 6-hour periods at 25° C. alternated with *four*		
1-hour periods at 32° C.	+++	26
2-hour periods at 32° C.	+, ++	22
4-hour periods at 32° C.	±, +	14
6-hour periods at 32° C.	—, ±	6

25° C. or three times the minimum 10-hour period necessary to convert normal to tumor cells at that temperature was given.

The results presented in Table 4 show that the transforming principle is cumulative. This is evidenced by the fact that, whereas a single exposure of 4 or 6 hours at 25° C. is inadequate in itself for the bacteria to bring about the cellular alteration, a series of exposures of these durations permits such an alteration. The data also demonstrate that a direct relationship exists between the size of the tumors produced and the total duration of exposure times at 32° C. alternated with exposures at 25° C. for an accumulated period of 30 hours at the lower temperature. For example, large tumors were produced when the bacteria were permitted to act for periods of 4 or 6 hours, totalling in each instance 30 hours at 25° C. when these periods were alternated with one-hour exposures at 32° C. As the intervals that the plants were held at 32° C. were increased in relation to the total time that they were held at 25° C., the size of the resulting tumors decreased. When the total exposure at 32° C. approached the exposure at 25° C. small tumors or no tumors at all were produced. The results suggest, therefore, that the activity of the transforming principle was destroyed at a temperature of 32° C. at roughly the same rate that this principle was produced by the bacteria during a comparable period at 25° C.

This finding suggests that the residual activity of the transforming principle can be computed on a semi-quantitative basis. For convenience in these calculations, 10 units of transforming principle are arbitrarily considered necessary to alter irreversibly normal *Kalanchoe* cells to tumor cells. A single 10-hour exposure period has been established as being the minimum time necessary to accomplish the cellular alteration under the conditions of the experiment. If it is assumed that the transforming principle is being produced at a constant rate by the bacteria after they have established themselves in the host, then it follows that one unit of transforming principle is produced every hour during the incubation at 25° C. The inactivation of this principle at 32° C. was found to be roughly equal to its production at 25° C. Thus for every unit produced at 25° C. in one hour, approximately one unit was destroyed at 32° C. in the same period. In the series of experiments shown in Table 4 in which 4-hour intervals at 25° C. were alternated with 1-hour intervals at 32° C., a total of 30 units was produced and 7 were destroyed leaving 23 units available for the cellular transformation. These 23 units are equal to somewhat more than twice the minimum of 10 units necessary to produce the irreversible cell change. The size of the tumors produced experimentally corresponded to a ++ tumor. Other results expressed in units are shown in Table 4. These data show that roughly 25 units or more of the transforming principle appear to be necessary to initiate tumors of the size designated as +++. The availability of 20 to 24 units resulted in ++ tumors, while 14 to 19 units gave a + reading. About 10 units are the minimum necessary to accomplish the irreversible cell change under the conditions of these experiments. These figures, of course, represent merely a rough approximation and are not to be taken as absolute values. It should be pointed out here that once the concentration of transforming principle necessary for the conversion of normal cells to tumor cells is reached or passed, the reaction becomes irreversible and the resulting tumor cells develop into a neoplasm independently of the bacteria. In these experiments only the amount of cellular alteration that had been accomplished during a 30-hour period at 25° C. was considered.

If the results shown in Table 4 are compared with those in Table 3, it is found that a single exposure of 10 to 12 hours at 25° C. is sufficient to give a ± reading, while 14-hour and 20-hour periods permit the bacteria to initiate tumors of the size designated as + and ++, respectively. Thus, in this single exposure experiment, in which the number of hours of incubation at 25° C. is expressed directly in units, the results obtained correspond very well with the results found in the experiments in which the transforming principle was intermittently produced at 25° C. and either partially or wholly inactivated at 32° C.

The unit hypothesis was checked experimentally in another way. Instead of permitting the bacteria to act on the host for alternate periods at 25° C. and 32° C. for a total of 30 hours at the lower temperature, a series of shorter total exposures at 25° C. were used. In these experiments 4-hour periods at 25° C. were alternated with 1 and 2-hour periods at 32° C. The total exposures at 25° C. in each experiment are indicated in Table 5. The results of these experiments demonstrated again that the relative size of the tumors initiated corresponds within the limits of biological experimentation with the calculated number of units of transforming principle produced.

Discussion

An observation reported earlier demonstrated (2) that normal plant cells may be converted into tumor cells by the crown-gall bacteria at a temperature of 25° C. but not at 32° C. Once the cellular transformation has been fully consummated at 25° C., however, the tumor cells develop into a neoplastic growth at both temperatures. The bacteria and the host develop well at 32° C. By utilizing this finding, it was possible to define certain conditions which are critical to the neoplastic alteration.

The experiments reported here demonstrated that normal cells of the host plant *Kalanchoe daigremontiana* were converted into tumor cells in 32±2 hours at 25° C. The plants were for the most part no

TABLE 5

Plants were inoculated and placed immediately at 32° C. for 24 hours. Following incubation at 32° C. the plants were held alternately at 25° C. and 32° C. for periods indicated below. Following treatment all plants were held at 32° C.

No. of 4-hour periods at 25° C.	No. of alternate 1-hour periods at 32° C.	Total time at 25° C. in hours	Total time at 32° C. in hours	No. of available units	Results	No. of 4-hour periods at 25° C.	No. of 2-hour periods at at 32° C.	Total time at 26° C. in hours	Total time at 32° C. in hours	No. of available units	Results
2	1	8	1	7	—, occ. ±	2	1	8	2	6	—
3	2	12	2	10	—, ±	3	2	12	4	8	—, occ. ±
4	3	16	3	13	±, +	4	3	16	6	10	—, ±
5	4	20	4	16	±, +	5	4	20	8	12	±
6	5	24	5	19	+, ++	6	5	24	10	14	±, +
7	6	28	6	22	+, ++	7	6	28	12	16	±, +
7.5	7	30	7	23	++	7.5	7	30	14	16	±, +

occ. = occasionally.

longer susceptible 5 days after wounding. When, however, the plants were inoculated with bacteria and held at 32° C. for 24 hours and then placed for as short a period as 10 hours at 25° C., the cellular transformation was accomplished. Periods of 8 hours or less at 25° C. following a 24-hour incubation period at 32° C. were inadequate to bring about the cellular alteration. In each instance described above a total of 30 to 34 hours is necessary to accomplish the alteration. These results suggest, therefore, that the bacteria require an incubation period of roughly 20 to 24 hours either at 25° C. or at 32° C. for environmental adjustment plus an additional 10-hour period at 25° C. to produce the transforming principle in the amounts necessary to bring about the cellular alteration. It has also been established from these results that a definite threshold concentration of the transforming principle is necessary to convert normal plant cells to tumor cells. Once this threshold is reached or passed the reaction becomes irreversible and the tumor cells develop into a neoplastic growth.

Three possibilities have previously been presented in an attempt to account for the inability of the crown-gall bacteria to alter normal plant cells to tumor cells at a temperature of 32° C. It was suggested (3) that either the transforming principle (*1*) is not produced by the bacteria at 32° C., or (*2*) is rendered inactive by an inhibitor that is activated at that temperature, or (*3*) is very heat labile and is destroyed at 32° C. under the conditions that exist in the plants tested.

It has not been possible to determine from the data presented here whether the transforming principle is produced by the bacteria at 32° C. Whatever the effect of this temperature on the ability of the organisms to produce this principle, the results nevertheless suggest that the transforming principle itself is inactivated at 32° C. This is shown by the fact that large tumors were produced when the bacteria were permitted to act for periods of 4 or 6 hours totalling in each instance 30 hours at 25° C. when these periods were alternated with one-hour exposures at 32° C. As the intervals that the plants were held at 32° C. were increased in relation to the time that they were held at 25° C. the size of the resulting tumors decreased. When the interval at 25° C. approached or equaled the interval at 32° C. small tumors or no tumors at all resulted. This suggests that the transforming principle was completely or almost completely inactivated under the conditions of this experiment. If no inactivation of this principle had occurred at 32° C. the resulting tumors would all have been equal in size because a total of 30 hours at 25° C. was given in each instance. Other data reported here confirm this conclusion.

The question, however, as to whether the inactivation of the transforming principle at 32° C. was the result of the activation of an inhibitor at that temperature or whether the transforming principle itself is heat labile is impossible to answer from the data presented.

In postulating an inhibitor mechanism several assumptions have to be made to account for the observed results. It would have to be assumed, for example, that the inhibitor itself is free to act at 32° C. but that it is bound at 25° C. and as a result is rendered ineffective against the transforming principle at that temperature. This reaction would either have to be reversible or, if the substrate to which the inhibitor is bound is heat labile and is destroyed at 32° C., that material would have to be promptly replaced at 25° C. That this postulated inhibitor is not cumulative to any appreciable extent at 32° C. is evidenced from the fact that when plants were placed at 32° C. for 24 hours immediately after inoculation and then held at 25° C. the cellular alteration was accomplished in as short a period as 10 hours at the lower temperature. Thus, the incubation period at 32° C. for 24 hours plus 10 hours at 25° C. yielded the same result as did total incubations of 32 to 34 hours at 25° C. This demonstrates, therefore, that there was no appreciable accumulation of an inhibitor following a 24-hour incubation period at 32° C. While such an inhibitor system cannot at present be ruled out it appears that all the results obtained here can be explained more directly on the basis that inactivation at 32° C. is due to a thermal destruction of the transforming principle itself. Further work on this phase of the problem is now being carried out.

It has been indicated that in these experiments the tumor-inducing principle was inactivated at 32° C. at roughly the same rate that it was produced by the bacteria at 25° C. By shifting the balance in favor of production, on one hand, or inactivation on the other, it was possible to predict within the limits of biological experimentation the eventual outcome by merely adding the number of hours that the plants were held at 32° C. and subtracting this figure from the number of hours that the plants were alternately kept at 25° C. It could be determined from the value received not only whether tumors would be initiated, but, if they were produced, the relative size of the resulting tumors could be predicted with some accuracy. The results suggest, furthermore, that the relative size of the resulting tumors, in the early stages of their development, appears to be dependent upon the amount of tumor-inducing principle available at the cellular alteration.

Summary

This investigation is based on the finding that normal *Kalanchoe* cells may be transformed to tumor cells by the crown-gall bacteria at a temperature of 25° C. but not at 32° C. Once the cellular transformation has been accomplished the tumor cells develop into a neoplastic growth at both temperatures.

The results obtained demonstrate that the minimum period necessary for the bacteria to alter normal cells to neoplastic cells is 32 ± 2 hours at 25° C. The plants were for the most part no longer susceptible 5 days after wounding. The period necessary for the bacteria to accomplish the cellular alteration could be reduced from 32 ± 2 hours to 10 hours at 25° C. when the plants were permitted an incubation period of 24 hours at 32° C. following inoculation.

A comparison of histological findings and plant inoculation studies indicated that it is before or during the early stages of active wound healing that normal cells are most readily converted into tumor cells.

The studies reveal, furthermore, that inactivation of the tumor-inducing principle itself occurs at 32° C. The possible nature of this inactivation is discussed.

The principle responsible for the alteration of normal cells to tumor cells was inactivated at 32° C. in these experiments at about the same rate that it was produced by the bacteria during a comparable period of time at 25° C. As a result of this finding a semi-quantitative method was developed which made it possible to predict within the limits of biological experimentation not only whether tumors would be initiated under a given set of experimental conditions, but, if they were produced, the relative size of the resulting tumors could be estimated in advance with some accuracy.

The relative size of the tumors appears to be dependent upon the amount of transforming principle available at the time of the cellular alteration.

REFERENCES

1. BRAUN, ARMIN C. 1943. Studies on tumor inception in the crown-gall disease. *Am. J. Bot.*, **30**, 674-677.

2. ———. 1947. Thermal studies on the factors responsible for tumor initiation in crown gall. *Am. J. Bot.*, **34**, 234-240.

3. ———. 1947. Recent advances in the physiology of tumor formation in the crown-gall disease of plants. Growth Symposium, *Growth*, **11**, 325-337.

TREASURE THE UNEXPECTED

Allen Kerr
Emeritus Professor of Plant Pathology
University of Adelaide, Adelaide, Australia
E-mail: kerra@senet.com.au

Commentary on the paper:

Kerr, A. 1969. Transfer of virulence between isolates of *Agrobacterium.* Nature 223:1175–1176

I started to investigate crown gall in 1967, having spent the previous three years in Ceylon (now Sri Lanka) where I studied diseases of tea. As no tea is grown in South Australia, I had to find another project. After consulting with other plant pathologists and extension workers, I decided to work on crown gall, which was a serious economic problem in stone fruit orchards and nurseries. I knew it was also a very interesting scientific problem because I had been lecturing to undergraduates on crown gall for several years. The hypothetical tumor inducing principle of Armin C. Braun (see preceding papers) held great fascination.

My work started as an ecological study of *Agrobacterium tumefaciens.* Where did it live? Was it ubiquitous or was it confined to plant nurseries? Did it live on plant roots or was it widely distributed through soil? How abundant was it in these various locations? If we could find answers these questions, we would perhaps be in a position to control the disease by concentrating on the problem area. I was confident that the questions could be answered because a selective medium for *A. tumefaciens* had recently been described by Schroth et al. (1965). This would enable the distribution and abundance of agrobacteria to be measured, even in soil, which commonly contains 10^8 organisms per gram. The first task was to establish that the selective medium worked. Soil samples from various sources were tested along with samples of crown gall tissue. Bacteria grew on the selective medium but, much to my surprise, none could cause crown gall when reinoculated into susceptible plants. This continued for several months until I realised that the medium must be at fault, not me.

I changed to a semi-selective medium and pathogenic agrobacteria grew in profusion. When inoculated into tomato seedlings, fast-growing galls developed. Then something very unusual happened and I knew I was on to something interesting. When I tried to re-isolate bacteria from the tomato galls induced by the pathogenic isolates (to fulfill Koch's postulates), both on semi-selective and highly selective media, pathogens were detected on both media. For six months, I had been trying unsuccessfully to isolate pathogens on the highly selective medium of Schroth et al. and now, when I didn't expect it, there they were. The solution depended on two separate phenomena. First, we (my students, Philip Keane and Peter New, and I) established that there were two distinct biovars of *Agrobacterium,* which could be distinguished by nutritional characteristics, by serology and by protein patterns following gel electrophoresis. Biovar 2 (now recognized as a separate species) was the predominant pathogen in Australia (Keane et al. 1970) and, as shown later, in many other countries. It will not grow on Schroth's medium; only biovar 1 strains will grow there. The second phenomenon was even more interesting. The reason we could now isolate biovar 1 pathogens was because the ability to cause crown gall was being transferred from pathogenic biovar 2 strains to previously non-pathogenic biovar 1 strains, which had been splashed from soil on to the gall surfaces.

As an undergraduate at Edinburgh University, Scotland, I had taken a one-term course in genetics given by Charlotte Auerbach but I needed to learn a lot more genetics before I could understand the transfer phenomenon I had discovered. Fortunately, a marvellous textbook by William Hayes (Hayes 1968) had recently been published. This became my bible and I soon realised that pathogenicity transfer in *Agrobacterium* was almost certainly the result of plasmid transfer. (I had not even heard of a plasmid before reading Hayes' textbook.) Unfortunately, in 1969, my skills and those of my colleagues were not sufficient to distinguish large plasmid DNA from chromosomal DNA, although we did try. It was another five years before this was achieved. However, the ability to transfer pathogenicity from one strain of *Agrobacterium* to another was of crucial importance in the demonstration by Van Larebeke et al. (1975) and by Watson et al. (1975), that pathogenicity in *Agrobacterium* is dependent on the presence of a large plasmid, the tumor-inducing or Ti plasmid. The hypothetical tumor-inducing principle of A. C. Braun had been found.

References

Hayes, William. 1968. The Genetics of Bacteria and Their Viruses. 2nd ed. Blackwell Scientific Publications, Oxford and Edinburgh, U.K.

Keane, P. J., Kerr, A., and New, P. B. 1970. Crown gall of stone fruit II. Identification and nomenclature of *Agrobacterium* isolates. Aust. J. Biol. Sci. 23:585–595.

Schroth, M. N., Thompson, J. P., and Hildebrand, D. C. 1965. Isolation of *Agrobacterium tumefaciens–A. radiobacter* group from soil. Phytopathology 55:645–647.

Van Larebeke, N., Genetello, C., Schell, J., Schilperoort, R. A., Hermans, A. K., Hernalsteens, J.-P., and Van Montagu, M. 1975. Acquisition of tumour-inducing ability by non-pathogenic agrobacteria as a result of plasmid transfer. Nature 255:742–743.

Watson, B., Currier, T. C., Gordon, M. P., Chilton, M.-D., and Nester, E. W. 1975. Plasmid required for virulence of *Agrobacterium tumefaciens.* J. Bacteriol. 123:255–264.

ALLEN KERR: In 1947, I graduated B.Sc. from the University of Edinburgh with second class honors in Botany. My first job was as assistant mycologist at the North of Scotland College of Agriculture. With my superior Dr. Betty Gray, I had to advise farmers about crop diseases in the north of Scotland, an area that stretched from just north of Dundee to the Shetland Islands and from Peterhead in the east to the Outer Hebrides. I was also encouraged to research soft rot of potatoes caused by *Erwinia carotovora,* which had become an important problem. In 1951, I was appointed assistant lecturer in the University of Adelaide. As a staff member, I was allowed to study for a Ph.D. that was awarded in 1956 for studies on lawn and cereal diseases. From 1963–1966, I worked at the Tea Research Institute, Ceylon, studying diseases of tea.

I didn't really have any mentors except for my contemporaries, both students and colleagues. The person who has most influenced my work is Max Tate; combining his chemical skills with my biological skills, we have published some interesting papers. Others who have influenced me greatly were Jacques Tempé and Stephen Farrand.

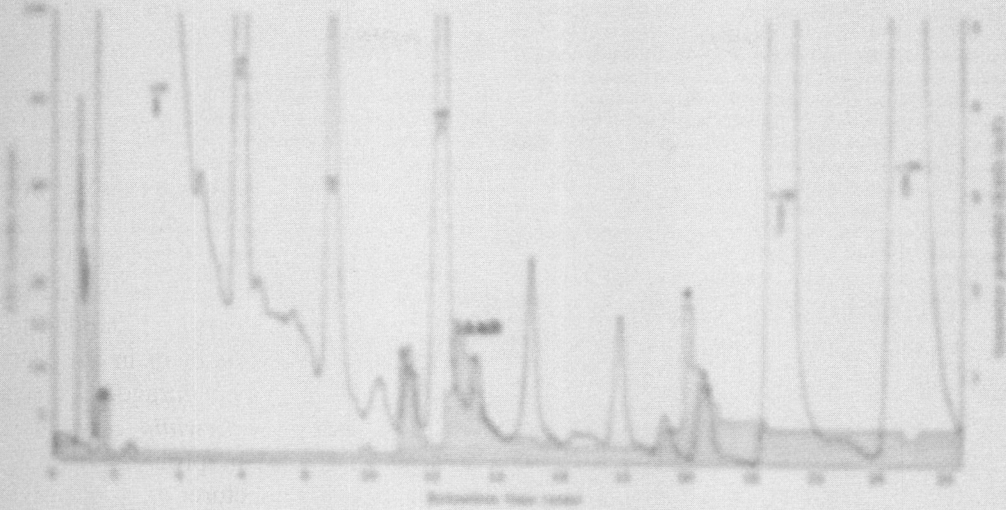

Transfer of Virulence between Isolates of *Agrobacterium*

In South Australia, at least three distinct forms of *Agrobacterium* can be isolated from crown galls on stone fruit trees[1]. These forms will be referred to as *Agrobacterium* sp., *A. tumefaciens* and *A. radiobacter*. *Agrobacterium* sp. and *A. tumefaciens* induce the formation of galls when inoculated into peach or tomato, but can be distinguished by various biochemical tests including 3-ketolactose production[2], by serology (our unpublished data) and by protein patterns following gel electrophoresis (our unpublished data). *A. radiobacter* is non-pathogenic but otherwise cannot be distinguished by any of these tests from *A. tumefaciens*.

I now report transfer of virulence from *Agrobacterium* sp. to *A. radiobacter* which is converted to *A. tumefaciens*. Virulence transfer was first observed when tomato plants inoculated with *Agrobacterium* sp. became contaminated with water-splashed soil containing *A. radiobacter*. Six weeks after inoculation, approximately 50 per cent of the *A. radiobacter* contaminants were pathogenic and indistinguishable from *A. tumefaciens*. Transfer of virulence has now been achieved between a single-cell isolate of *Agrobacterium* (sp.) (donor) and an isolate of *A. radiobacter* (recipient) labelled with an antibiotic resistance marker to distinguish it from the natural soil population of *A. radiobacter*.

Agrobacterium sp. was isolated from a gall on a young peach tree. Cultures of single cells were obtained[3] and one of these was used as the donor. It is 3-ketolactose negative, cannot utilize nitrate as a source of nitrogen and will not grow on Schroth's medium[4]. *A. radiobacter* was isolated from partially sterilized potting soil and a heavy suspension was plated on yeast-mannitol agar containing 25 p.p.m. of chloramphenicol. An isolate resistant to chloramphenicol was obtained and used as recipient. It is 3-ketolactose positive, can utilize nitrate as a source of nitrogen, grows readily on Schroth's medium and is serologically distinct from the donor.

Stems of 4 week old tomato plants (Cultivar SA early dwarf red) growing in partially sterilized potting soil were inoculated with a 48 h culture of the donor isolate grown on yeast-mannitol agar. Inoculated plants were placed in a galvanized iron tray and sub-irrigated to prevent soil splash during watering. In the second and third week after inoculation 1 ml. of the recipient culture was suspended in sterile distilled water (2×10^3 cells/ml.) was pipetted onto the surface of the developing galls. Three inoculated plants were contaminated with the recipient and three were left uncontaminated as controls. Six weeks after inoculation the galls were surface-sterilized in 0·5 per cent sodium hypochlorite for 10 min, macerated in 100 ml. of sterile distilled water, diluted 1 : 10 and 0·1 ml. was plated on Schroth's medium which inhibits growth of the donor but allows growth of the recipient. No colonies developed on the plates from control plants. From the plates from treated plants twenty individual colonies were selected at random, transferred to yeast-mannitol agar slants and tested for pathogenicity. Six of the isolates were pathogenic, inducing the formation of galls when inoculated into tomato plants. All six were keto-3-lactose positive, could utilize nitrate and grow on Schroth's medium, were serologically indistinguishable from the parent recipient and were resistant to chloramphenicol. Single-cell cultures of these isolates were also pathogenic.

This seems to be the first unequivocal evidence for transfer of virulence from a plant pathogenic to a saprophytic species of bacterium, though DNA transformation of unmarked isolates of *Agrobacterium* and *Rhizobium* has been reported[5,6].

In the field, galls are always contaminated with soil containing the ubiquitous saprophyte *A. radiobacter*. It seems likely that virulence transfer is a common field phenomenon. As yet, there is no indication of the mechanism of transfer.

A. KERR

Waite Agricultural Research Institute,
Glen Osmond, South Australia 5064.

Received July 7, 1969.

[1] Kerr, A., *Austral. J. Biol. Sci.*, **22**, 111 (1969).
[2] Bernaerts, M. J., and De Ley, J., *Nature*, **197**, 406 (1963).
[3] DeVay, J. E., and Schnathorst, W. C., *Nature*, **199**, 775 (1963).
[4] Schroth, M. N., Thompson, J. P., and Hildebrand, D. C., *Phytopathology*, **55**, 645 (1965).
[5] Klein, D. T., and Klein, R. M., *J. Bact.*, **66**, 220 (1953).
[6] Kern, H., *Arch. Mikrobiol.*, **51**, 140 (1965).

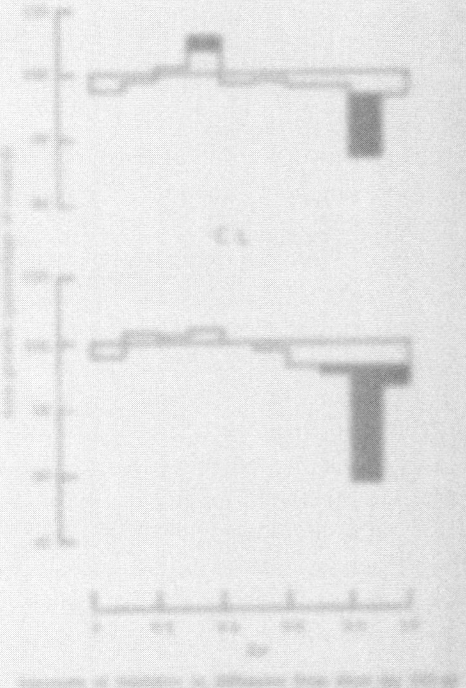

GEORGE MICHEL MOREL (1916–1973): CHEMIST, BOTANIST, PLANT TISSUE CULTURE PIONEER

Jacques Tempé
Oustal/Viel, Route de Marmande
47180 Meilhan-sur-Garonne, France
E-mail: JTEMPE@aol.com

Commentary on the paper:

Petit, A., Delhaye, S., Tempé, J., and Morel, G. 1970. Recherches sur les guanidines des tissus de crown gall: Mise en évidence d'une relation biochimique spécifique entre les souches d'*Agrobacterium tumefaciens* et les tumeurs qu'elles induisent. Physiol. Vég. 8:205–213.

In 1963, I joined Georges Morel's laboratory on a temporary basis, at the INRA research center of Versailles, to complete some biochemical work on mutagenic substances I had previously synthesized. The project was the continuation of work performed during my student years.

During his studies in chemical engineering, Morel, after attending a lecture on plant tissue culture by Roger Gautheret, was so fascinated by this emerging technique that he decided to engage in this field. He convinced Gautheret to take him as a Ph.D. student. The son of an horticulturist, Morel was a keen and knowledgeable botanist. He probably considered plant tissue culture the ultimate form of his father's trade which the elder Morel could not continue due to his premature death. After Morel obtained his Ph.D. (1948) he went to the United States with a Rockefeller Fellowship, to work with W. C. Robbins, the root culture specialist. He also worked at Harvard University with R. Wetmore and K. Thimann during two stays, in 1949–1950 and 1952.

Back in France, Morel started his own laboratory at Versailles, where he rapidly established a renowned plant tissue culture collection. He was soon engaged on several fronts of research that had, as a common denominator, *in vitro* plant tissue culture. He had already acquired worldwide recognition for his work in this field and especially on meristem culture, which enabled him to cure several plant species (e.g. dahlia, potato) propagated as clones of endogenous viruses. Applying this technique to orchids, Morel discovered he could propagate some of the most valuable varieties by meristem culture. He is thus directly responsible for the popularizing of many orchid cultivars that, are now mass-produced *in vitro,* at low cost. His main interest, however, was crown gall. While he was at Harvard, Morel had several opportunities to meet with Armin C. Braun who had been engaged in studies on crown gall for many years. In collaboration with Philip White, Braun had been able to confirm Erwin Smith's hypothesis that crown gall is a plant cancer. He showed this by producing aseptic tissue cultures of sunflower crown gall tumors, which were shown to develop into tumors when subsequently grafted onto a plant of the same species.

As a chemist, Morel was skilled in analytical techniques and had set up a unit for paper chromatography. Together with Arlette Goldman-Ménagé, a doctoral student, he applied this technique to the analysis of crown gall tissue cultures in which they had found an unknown guanidine, which gave a strong reaction with the Sakaguchi reagent. A few months earlier, C. Lioret had identified lysopine in similar material. Misled by the guanidine specialist in the Collège de France, who failed to identify octopine on their paper chromatograms, Goldman and Morel undertook the purification of the unknown substance. From several kilograms of cultures they obtained, more than a year later, a few hundred milligrams of a pure compound which turned out to have the chemical composition of octopine, the reductive condensation product of arginine and pyruvic acid. In some molluscs (e.g. octopus, squid, scallop) devoid of lactic dehydrogenase, octopine serves as a sink for pyruvic acid generated by muscle activity. The role of octopine in these animals was studied by a group in the Collège de France.

Opinions on Opines

I had already been in the lab for some time and had established a good relationship with Morel and several members of his group. In order to fully confirm the identity of the substance present in crown gall tissues, we decided to synthesize some octopine. During this work, which took place mostly after hours and during weekends, the relationship with Morel became closer and I decided to apply for a permanent position in his laboratory.

To account for the presence of new compounds (later known as opines) in crown gall tumors Morel proposed that in cancer cells the metabolism of organic acids is upset–resulting in the production of imino acids derived from pyruvic acid and basic amino acids. While I was visiting Armin Braun in his laboratory and discussing various hypotheses, we came up with the idea that, if Morel was correct, the level of octopine in crown gall tissues could be a reflection of the level of malignancy of tumor cells and, therefore, of the virulence of *Agrobacterium tumefaciens* strains. Braun had previously characterized a tobacco line which he considered to be attenuated. This teratoma line constantly produced shoots from which plants could be regenerated. Therefore, this line seemed to be the ideal material for checking the current hypothesis. This idea could easily be tested since, in order to have a more reliable analytical technique and to be able to quantitate levels of octopine in crown gall tissues, I guanidine which he named nopaline, from nopal, another name for the prickly pear. In crown gall tumors of cacti, such as *Opuntia,* with crassulacean acid metabolism, the perturbation of organic acid metabolism would lead to the synthesis of nopaline, or so he thought. To test this idea, Morel had gathered a collection of cacti from which he was trying to isolate crown gall tissue cultures. However, analysis of the material provided by Braun showed the presence of huge amounts of nopaline. This material had been isolated from a tobacco tumor inoculated with *A. tumefaciens* strain T37. This finding led us to conclude that the strain of *Agrobacterium* and not the plant metabolic type, was responsible for the type of guanidine accumulated in tumors. Since plant cells acquired, upon interaction with the bacterium, a character that was permanently maintained in bacteria-free tissue cultures, we proposed that, upon interaction with the bacterium, plant cells acquired new genetic information. To test this idea, a collection of strains was inoculated individually to *Datura stramonium* plants. Analysis of the tumors produced showed that while most *A. tumefaciens* strains initiated octopine tumors, a few strains were responsible for the synthesis of nopaline. Two strains elicited tumors which contained neither octopine nor nopaline. This was clear evidence that the previous hypothesis was wrong and that the type of guanidine present in tumors was determined by the strain that had been inoculated to produce the tumor. Since the property was maintained in aseptic tissue cultures, this was indeed suggestive of transmission of genetic information from the bacterium to the plant cells upon tumor induction.

At that point, a paper published by B. Lejeune and M. F. Jubier (1968), students of Lioret who were trying to elucidate the biosynthesis of lysopine, attracted our attention. After incorporating lysopine in the culture medium of several bacterial species they showed that *A. tumefaciens* B6, which resulted in tumors that synthesized octopine and lypopine, degraded lysopine. With Annik Petit, we decided to repeat the same type of experiment with octopine and nopaline containing culture media inoculated with various *Agrobacterium* strains. The result was clear cut: strains which elicited octopine- or nopaline-containing tumors showed the same specificity with respect to degradation of these compounds. Further degradation studies with radioactive octopine showed that most of the label was recovered as arginine or the products of arginine metabolism. This suggested that octopine was degraded by a reverse pathway of its synthesis in tumors. I presented these results at a meeting of the French Society for Plant Physiology in 1969. In my presentation I suggested one could predict which new metabolite (octopine or nopaline) would be synthesized in a tumor, by knowing the metabolic type (opine type) of the bacterium that elicited the tumor. This indicated that DNA transfer from bacterium to plant cell occurred during the process of tumor induction. I further suggested that the same enzyme could be at work in opposite directions in tumors and *Agrobacterium* strains. Indeed, at that time, I had discarded as improbable the idea that, in a tumor, the bacteria might be feeding on the crown gall-specific metabolites produced by tumor cells. The hypothesis that crown gall tumors could be the result of DNA transfer from *Agrobacterium* to plant cells had been formulated previously by others, including Braun, and had received support from hybridization experiments. However, serious doubts were later cast on the validity of the hybridization data.

Our work on the specificity of opine degradation by *Agrobacterium* strains was confirmed (Lippincott et al. 1973). However, the specificity of opines as biochemical markers of crown gall cells was questioned (Johnson et al. 1974). Lioret himself, citing work by Seitz and Hochster, had admitted that lysopine was probably a normal plant product (Lioret 1966). So, early in 1974, the situation on the opine front was rather grim; Morel had died at the end of 1973 and I was the only one left to defend the specificity of opine synthesis in crown gall tumors. Because I had a precise analytical tool with which I was sure identification of octopine or nopaline was error-proof and because I knew the chemistry of these compounds, I was sure our critics were wrong. I was fortunate to convince M.-D. Chilton, during a meeting in England in 1974, to re-evaluate the work of Johnson et al. (1974). A few months later she confirmed by letter that their analytical work was most certainly wrong (a color-blind analyst had wrongly identified octopine on paper electrophoretograms) but the erroneous results were never formally retracted. The discovery of the Ti plasmid in the Schell–Van Montagu laboratory came as a break-through since it could be shown that octopine degradation was a Ti plasmid borne property. Our results were then quickly confirmed. Yet as late as 1978, Lippincott et al. (1978) maintained they were able to detect octopine in normal plant tissues, with elevated levels in embryonic tissues. (An analogy with animal cancers must have inspired these workers.) However, our supposition that the same enzyme was responsible for opine synthesis in tumors and degradation by bacteria was no longer tenable since mutants could be found that were affected in one of these properties only. We had to come back to the formerly discarded hypothesis that opines in crown gall tumors serve to feed tumorigenic *Agrobacterium* strains which was then formally expressed as the opine concept or the genetic colonization concept formulated by Schell et al. (1979).

References

Johnson, R., Guderian, R. H., Eden, F., Chilton, M.-D., Gordon, M. P., and Nester, E. W. 1974. Detection and quantitation of octopine in normal plant tissues and in crown gall tumors. Proc. Natl. Acad. Sci. U.S.A. 71: 536–539.

Lioret, C. 1966. Recherches sur la signification physiologique de la lysopine dans les tissus de crown gall de scorsonère cultivés *in vitro.* Physiol. Vég. 4: 89–103.

Lippincott, J. A., Beiderbeck, R., and Lippincott, B. B. 1973. Utilization of octopine and nopaline by *Agrobacterium.* J. Bacteriol. 116:78–383.

Lippincott, J. A., Chang, C. C., Creaser-Pence, V. R., Birnberg, P. R., Rao, S. S., Margot, J. B., Whatley, M. H., and Lippincott, B. B. 1978. Genetic determinants governing enhancement of tumor initiation by avirulent agrobacteria, *Agrobacterium*-host adherence and octopine synthesis. Proc. 4th Int. Conf. Plant Path. Bact. Angiers. 189–197.

Schell, J., Van Montagu, M., De Beuckeleer, M., De Block, M., De Picker, A., De Wilde, M., Engler, G., Genetello, J. P., Holsters, M., Seurinck, J., Van Vliet, F., and Villaroel, R. 1979. Interactions and DNA transfer between *Agrobacterium tumefaciens,* the Ti plasmid and the plant host. Proc. R. Soc. London B 204:251–266.

JACQUES TEMPÉ: After high school I trained for the competitive entrance exam to the Institut National Agronomique (INA), "Grande École" in agricultural sciences and biology. The training program (mathematics, physics, chemistry and biology) had been very interesting; however, I found the teaching at INA rather uninspiring. Except for a few courses, it was obsolete, and I was very disappointed to have to absorb outdated science. I quickly learned how to skip courses and to prepare overnight for end of term exams. While I was still at INA, I enrolled at the organic chemistry laboratory of École Polytechnique. It was here that I was approached by Henri Heslot, a professor of genetics at INA, who asked me to prepare mutagenic compounds for his studies on yeast and higher plant genetics. He later suggested I should continue this work in Georges Morel's laboratory at the Department of Agriculture Research Centre of Versailles, which I joined in 1963. At a time when money was becoming easily available for biology, Morel was happy to have me help him develop a modern lab.

In the early seventies, crown gall studies in the lab were on three fronts: trying to elucidate the reason for the synthesis of unusual amino acid derivatives (opines) in crown gall tumors (A. Goldmann-Ménagé), pursuing the study of the correlation between opine synthesis in tumors and degradation by *Agrobacterium tumefaciens* (A. Petit) and, seemingly the most promising, studying the putative role of bacteriophages in oncogenicity of *Agrobacterium* strains (J. Tourneur). The sudden death of Morel, in 1973, left Annik without a thesis advisor and she asked me to replace him. Our collaboration was the continuation of the work reported in the following paper; it resulted in a description of the nutritional relationship between *A. tumefaciens* and its host plant for which we coined the name "opine concept."

I obtained my degree from INA in 1959, degrees in chemistry in the following years and a doctorate in science in 1982. Besides Henri Heslot and George Morel who had a prominent influence on the development of my career, I was supported by François Jacob and Pierre Schaeffer.

In a book dedicated to Jeff Schell, it is appropriate to mention that he helped me when, due to my uncompromising character, times were difficult. But the memory of him I most treasure is that of our many skiing vacations in the Alps.

Physiol. Vég., 1970, 8 (2), 205-213.

Recherches sur les guanidines des tissus de Crown Gall. Mise en évidence d'une relation biochimique spécifique entre les souches d'*Agrobacterium tumefaciens* et les tumeurs qu'elles induisent (1)

Annik PETIT, Sonia DELHAYE, Jacques TEMPÉ et Georges MOREL

Station Centrale de Physiologie Végétale,
Centre National de Recherches Agronomiques, route de Saint-Cyr, 78 - Versailles

(Manuscrit reçu le 30 décembre 1969)

RÉSUMÉ

L'étude détaillée des guanidines des tumeurs de crown gall obtenues par inoculation de 43 souches d'*Agrobacterium tumefaciens* (Conn) montre que ces guanidines sont spécifiques de la souche inductrice.

La dégradation par les bactéries de ces substances est également spécifique et sa première étape est probablement catalysée par une déshydrogénase du même type que celles qui sont responsables de la synthèse des guanidines dans les tumeurs.

SUMMARY

The systematic study of the guanidines of crown gall tumors obtained from 43 different strains of Agrobacterium tumefaciens *(Conn) shows that these are specific of the bacterial strains.*

The metabolism of the guanidines in the bacteria is equally specific and its first step is probably catalysed by deshydrogenases of the same type as the ones which are responsible for the synthesis of the guanidines in the tumors.

INTRODUCTION

Les travaux de BRAUN sur la transformation tumorale induite par *Agrobacterium tumefaciens* tels qu'ils sont résumés par lui-même et STONIER (1958) suggèrent l'existence d'un principe inducteur de tumeur, doué de continuité génétique, élaboré par la bactérie au contact d'une blessure de l'hôte et transmis à quelques cellules de celui-ci au cours de l'induction tumorale.

(1) Communication présentée au Colloque de la Société française de Physiologie végétale : Transformations tumorales chez les végétaux. Paris, 29 novembre 1969.

Si la transformation tumorale est assurée par un mécanisme de ce type, on doit pouvoir mettre en évidence dans les cellules cancéreuses la présence permanente de matériel génétique d'origine bactérienne. De nombreux auteurs ont cherché à identifier le principe inducteur de tumeur en essayant d'obtenir une préparation acellulaire oncogène à partir de bactéries virulentes ; les résultats de ces travaux ont été analysés par BRAUN et STONIER (1958). Dans aucun cas apparemment les hyperplasies provoquées n'étaient de nature tumorale (MANIGAULT, 1968).

Récemment cependant, KOVOOR (1967, 1969) a pu obtenir un tissu de scorsonère tumoral avec du DNA d'*Agrobacterium tumefaciens,* AARON - DA CUNHA (1969) a transmis le caractère tumoral du crown-gall de tabac par greffe de tissus irradiés aux rayons X et LEFF et BEARDSLEY (1969) ont produit des tumeurs sur des boutures de tournesol avec le DNA d'un phage tempéré d'*Agrobacterium tumefaciens* B_6 que PARSONS et BEARDSLEY (1968) avaient précédemment isolé à l'état végétatif dans des tissus de crown-gall de tournesol.

Dans les deux premiers cas, il s'agit bien de tumeurs ; celles-ci sont en effet cultivables *in vitro* en l'absence de régulateurs de croissance et greffables. Dans le troisième cas, nous ne savons pas si le caractère tumoral a été établi rigoureusement.

Cependant, il n'a pas été démontré que les tissus tumoraux obtenus par ces auteurs sont effectivement des tumeurs de crown-gall. On sait en effet que ces tissus ont un métabolisme azoté particulier qui les différencie de *tous* les autres types de tumeurs : ils accumulent des acides aminés nouveaux — la lysopine (LIORET, 1956 ; BIEMANN et *al.,* 1960) ; l'octopine, l'acide octopinique et la nopaline (MÉNAGÉ et MOREL, 1964, 1965*a*, 1965*b* ; GOLDMANN, THOMAS et MOREL, 1969). L'accumulation d'octopine et de lysopine est caractéristique de *toutes* les tumeurs, ainsi que de leurs tissus en culture, provoquées par *A. tumefaciens* B_6 sur les plantes hôtes étudiées par les auteurs cités ci-dessus.

Si donc ce caractère est absent, l'induction tumorale obtenue avec le DNA bactérien ou celui du phage, ou par greffe de tissus tumoraux irradiés, n'est pas identique à celle provoquée par la bactérie et ne peut pas constituer une preuve de l'existence du principe inducteur de tumeur.

On peut aborder le problème d'une façon moins directe que celle qui consiste à tenter de mettre en évidence *a priori* le principe inducteur de tumeur. En effet, si une entité génétique provenant de la bactérie est responsable de la transformation tumorale, on doit retrouver dans les cellules des tissus tumoraux des produits dont la synthèse est sous le contrôle de cette entité.

Les résultats de nos travaux montrent que cette hypothèse de travail n'est pas sans fondement. Ils indiquent en effet qu'au cours de la transformation tumorale un système enzymatique propre à la bactérie est probablement transmis à la cellule végétale, ce qui suppose la présence d'une fraction du génôme bactérien, douée de continuité génétique, dans les cellules transformées, ainsi que son expression dans ces cellules.

RÉSULTATS

Après avoir montré (GOLDMANN et *al.,* 1968) que les souches B_6 et T_{37} d'*Agrobacterium tumefaciens* induisent respectivement l'accumulation d'octopine et de nopaline dans les tumeurs qu'elles provoquent sur de nombreux hôtes, nous avons entrepris une étude systématique des guanidines des tumeurs obtenues par inoculation de diverses souches.

Nous avons utilisé pour ce travail principalement *Datura stramonium* et *Datura tatula*. Les plantes étaient cultivées en superserre, au Phytotron du CNRS, à Gif-sur-Yvette et inoculées 6 à 8 semaines après le semis, à un stade de développement comportant de 6 à 8 feuilles. Après avoir décapité la tige au-dessus de la quatrième feuille on dépose sur la plaie quelques gouttes de suspension de bactéries en croissance exponentielle. L'apparition des tumeurs est rapide et celles-ci atteignent déjà une taille importante après 3 à 4 semaines.

Lorsqu'elles sont récoltées, 6 à 8 semaines après l'inoculation, des tumeurs ont un poids de 40 à 50 grammes. Les acides aminés libres de ces tumeurs sont séparés des autres constituants cellulaires par fixation sur résine cationique et élués à l'ammoniaque, ils sont ensuite chromatographiés sur un analyseur d'acides aminés équipé pour le dosage des guanidines (TEMPÉ et *al.*, 1967). Les résultats de ces travaux sont groupés dans le tableau I. Sur 43 sortes de tumeurs examinées, 20 contenaient de l'octopine, 19 de la nopaline ; les tumeurs inoculées avec les souches A_{66} et Rubi contenaient ces deux substances, celles obtenues avec Eu_6 et R_{10} n'ont pas d'autres guanidines que l'arginine.

Bien que les résultats obtenus ne soient pas rigoureusement constants, la quantité des guanidines des tumeurs variant dans d'assez larges proportions, notamment en fonction du temps, il est important de remarquer que la spécificité observée est nécessairement le reflet d'une particularité de la souche bactérienne inductrice, qui s'exprime dans les tumeurs par l'apparition des guanidines. L'observation de ce phénomène ne donne cependant aucun renseignement sur la nature de l'information, bien qu'elle soit certainement d'origine bactérienne. L'accumulation de ces substances pourrait simplement n'être qu'une conséquence lointaine d'un caractère spécifique du principe inducteur de tumeur, liée au désordre métabolique consécutif à la transformation tumorale. Le fait que SEITZ et HOCHSTER (1964) aient montré la présence de lysopine dans des tissus sains semble venir à l'appui de ce genre d'hypothèse. Cependant, nous n'avons jamais pu confirmer le résultat de ces auteurs.

L'existence dans les cellules tumorales de déshydrogénases pour la lysopine l'octopine et la nopaline n'implique donc pas nécessairement l'acquisition d'une information génétique nouvelle codant pour ces enzymes et provenant de la bactérie.

Nous avons tenté de retrouver dans les bactéries l'équivalent biochimique de cette spécificité. Pour cela, nous avons étudié la dégradation des guanidines des tumeurs par diverses souches d'*Agrobacterium tumefaciens*. Dans une première expérience, nous avons cultivé les souches B_6 et T_{37} sur le milieu S décrit par MONOD (1958), contenant un pour mille de glucose et additionné de quantités variables d'octopine. Ce milieu contient une source d'azote minéral, ce qui permet d'éviter une utilisation non spécifique de l'azote organique ajouté au milieu. Les bactéries sont ensemencées dans des erlenmeyers de un litre, contenant 300 millilitres de milieu avec 2 millilitres d'une pré-culture sur le même milieu en croissance exponentielle. La culture est faite sur une table d'agitation à la température de 28 °C. Après 24 heures de croissance, les bactéries sont centrifugées et on dose les guanidines du surnageant avec le réactif au diacétyl-α-naphtol (MÉNAGÉ et *al.*, 1966). On constate la disparition de l'octopine dans le milieu ayant supporté la croissance de la souche B_6, et ce pour des concentrations allant jusqu'à 250 mg par

TABLEAU I

souche (*a*)	guanidine des tumeurs		souche	guanidine des tumeurs	
	octopine	nopaline		octopine	nopaline
A6	+		S4 (4)		+
A66 (1) (*b*)	+	+	S31 (1)		+
Apple 185 (2)		+	S56 (4)	+	
B2	+		S341 (1)	+	
B6	+		S353 (1)		+
B58 (3)	+		S378 (1)	+	
B78 (4)	+		SH (3)		+
B79 (4)	+		SP (3)	+	
IIBNV6 (1) (*c*)	+		T37 (1)		+
IIBV7 (1)		+	W1 (3)	+	
Delft (4)		+	42 IV (1)		+
EU6 (2)	0	0	71 (4)		+
H5 (3)		+	176 (3)	+	
H100 (3)		+	806 (3)		+
9P (5)		+	181 (3)		+
10 (5)	0	0	Incense cedar (2)		+
31 (5)	+		S379 (1)	+	
32 (5)	+		ATV (3)		+
61 (5)	+		BP (3)	+	
62 (5)	+		Chr 10 (4)		+
63 (5)	+		179 (3)		+
Rubi (6)	+	+			

(*a*) Le chiffre entre parenthèses indique le nom de la personne ou la collection auprès de laquelle nous nous sommes procuré la souche : (1) Pr. MANIGAULT, Institut Pasteur, Paris ; (2) Pr. BRAUN, Rockefeller University, New York ; (3) Pr. BEARDSLEY, Manhattan College, New York ; (4) Mr FAIVRE AMYOT, C.N.R.A., Versailles ; (5) Dr. RAFAILA ; (6) ATCC, Rockville, Maryland.

(*b*) Souche atténuée produisant des tumeurs volumineuses lorsqu'on applique de l'auxine au voisinage de la plaie d'inoculation.

(*c*) Souche décrite comme avirulente, dérivée par BRAUN de II BV 7 qui provoque des tumeurs lorsqu'elle est inoculée dans les mêmes conditions que la précédente.

litre. On retrouve par contre toute l'octopine dans le surnageant de la culture inoculée avec T_{37} bien que la croissance ait été tout à fait comparable à celle de B_6.

La même expérience, refaite avec 6 souches, sur le même milieu contenant 50 mg/l d'octopine a donné, au bout de 40 heures de culture, les résultats indiqués dans le tableau II.

L'examen des résultats indique que la spécificité déjà démontrée pour B_6 et T_{37} s'étend aux autres souches étudiées.

Nous avons ensuite répété cette expérience en introduisant de la nopaline dans le milieu de culture, sous forme d'extrait d'acides aminés libres de tératomes de tabac (BRAUN, 1948) qui en contiennent une quantité abondante. Les résultats, figurés dans le tableau III, sont aussi nets que dans l'expérience précédente : les deux souches à nopaline et A_{66} utilisent complètement le substrat alors que les souches à octopine le laissent pratiquement intact.

TABLEAU II

Souche	D.O. du surnageant de culture après coloration	Guanidines des tumeurs induites avec la souche
T_{37}	0,28	nopaline
SH	0,29	nopaline
B_6	0,04	octopine
R_{31}	0,00	octopine
A_6	0,00	octopine
A_{66}	0,00	nopaline + octopine
Témoin non inoculé	0,30	–

TABLEAU III

Souche	D.O. du surnageant de culture après coloration	Guanidines des tumeurs induites par la souche
T_{37}	0,01	nopaline
SH	0,02	nopaline
B_6	0,62	octopine
R_{31}	0,56	octopine
A_6	0,52	octopine
A_{66}	0,02	nopaline + octopine
Témoin non inoculé	0,62	–

Il convient de remarquer que la souche A_{66}, qui est un mutant spontané à virulence atténuée de A_6 (HENDRICKSON et *al.*, 1934) et qui induit l'accumulation des deux guanidines dans les tumeurs de *Datura stramonium,* utilise les deux substrats.

Nous avons préparé, à partir d'arginine tritiée, par une méthode dérivée de celle d'IZUMYA et *al.* (1957), de l'octopine radioactive et nous avons étudié sa dégradation par la souche B_6 dans des conditions où toute l'octopine du milieu est utilisée.

La culture a été faite dans un erlenmeyer de 1 litre contenant 300 ml du milieu suivant :

KH_2PO_4	10^{-1} M	$FeSO_4$	$2 . 10^{-5}$ M
$(NH_4)_2SO_4$	$1,5 . 10^{-2}$ M	$MnCl_2$	10^{-5} M
$MgSO_4$	$8 . 10^{-3}$ M	octopine ^3H	$2 . 10^{-4}$ M, 20 μCi
$CaCl_2$	$7 . 10^{-5}$ M	glucose	$1,1 . 10^{-2}$ M

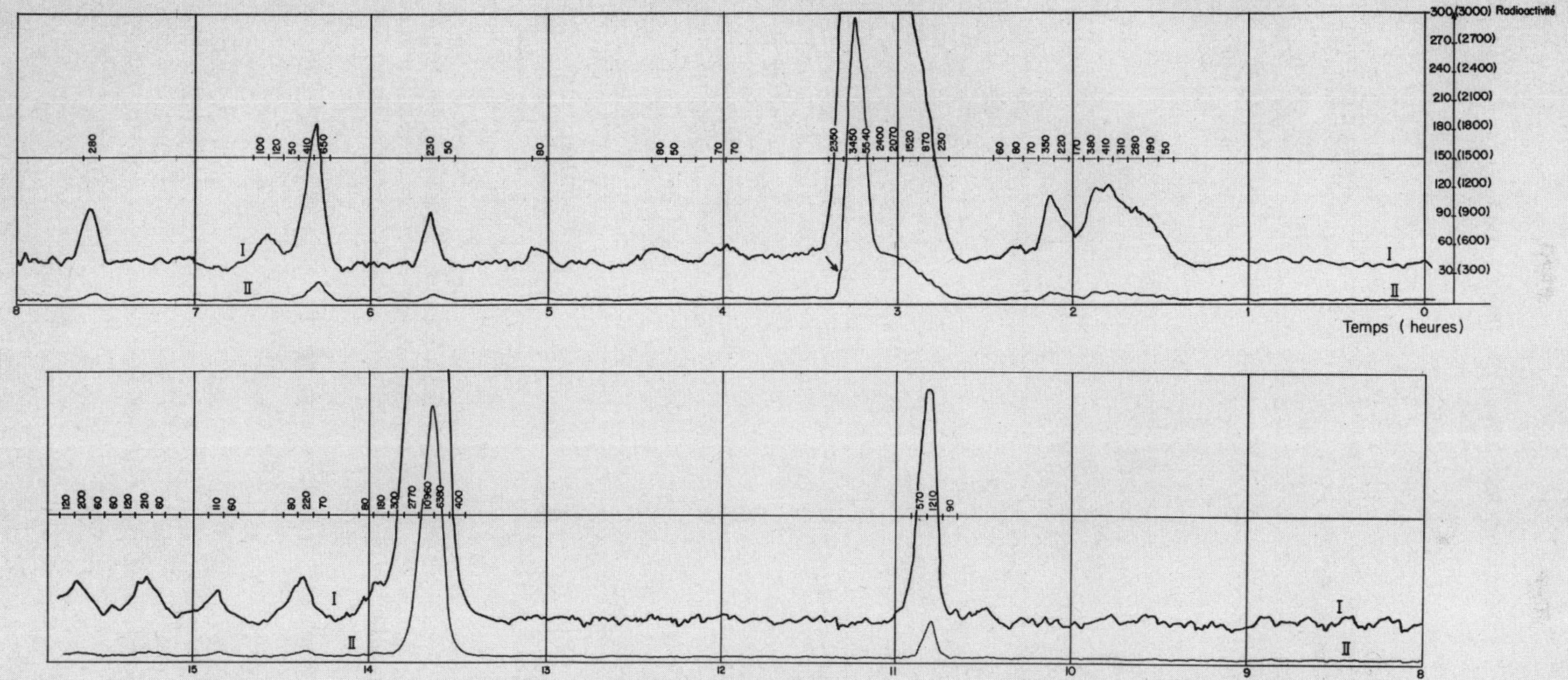

FIG. 1. — Répartition de la radioactivité des acides aminés protéiques d'*Agrobacterium tumefaciens* B_6 cultivé sur octopine tritiée. En ordonnées, impulsions par minute, enregistrement à deux sensibilités différentes ; en abscisses, temps en heures. On a porté sur un axe parallèle à celui des temps le nombre de désintégraton enregistrées en 5 minutes, dépassant significativement le bruit de fond (80 impulsions en 5 mn) après soustraction de celui-ci.

Le pH de la solution minérale est ajusté à 7,1 avec de la potasse 6 N. Le glucose et l'octopine sont stérilisés à part dans 30 ml d'eau distillée. L'inoculation a été faite avec 1 ml d'une culture en phase exponentielle. Après 17 h 30 de culture, les bactéries ont été centrifugées, lavées avec le même milieu et lysées par du lauryl sulfate de Na.

Nous avons retrouvé le marquage uniquement dans les protéines cellulaires, à l'exclusion du surnageant et des acides aminés libres. Après traitement de la fraction protéique par l'acide chlorhydrique 6N bouillant, pendant 7 heures et évaporation de l'acide, nous avons chromatographié l'hydrolysat sur un analyseur d'acides aminés (Technicon) en interposant, entre la sortie de la colonne échangeuse d'ions et le système analytique une cellule de comptage de 2 ml à flot continu, remplie de cristaux d'anthracène et placée dans le puits d'un compteur à scintillation (Packard Tricarb 3365). Ce compteur étant équipé d'un dispositif d'enregistrement (Potentiomètre Speedmax MECI) et d'une imprimante (IBM) totalisant le nombre de désintégrations perçues toutes les cinq minutes. L'enregistrement de la radioactivité de l'éluat est reproduit dans la figure (1). Sur un total de 57 600 désintégrations dépassant significativement le bruit de fond, 28 400 correspondaient à l'acide glutamique (49 %) et 21 000 à l'arginine (37 %). A côté de ces deux pics principaux on note la présence de deux pics, l'un élué au voisinage de l'isoleucine (1700 cpm, 3,1 % du marquage), l'autre avant la lysine (1000 cpm, 1,8 % du marquage) ; le reste de la radioactivité (9 %) était élué en pics mal définis au début du chromatogramme et, après la séparation, au cours du lavage de la colonne par la soude, 0,2 N. On sait (Greenberg, 1961, p. 191 ; Meister, 1965, p. 707) que l'arginine est convertie dans les cellules des microorganismes, aussi bien que dans celles des organismes supérieurs, en acide glutamique, par une suite de réactions enzymatiques réversibles, faisant intervenir l'ornithine et l'aldéhyde semi-glutamique ; il est donc normal de penser que l'octopine est d'abord scindée en arginine et acide pyruvique ou lactique. Cette hypothèse est confirmée par le fait que l'activité spécifique de l'arginine est environ cinq fois plus forte que celle de l'acide glutamique. On s'attendrait également à trouver une fraction du marquage dans la proline qui dérive également de l'aldéhyde semi-glutamique ; en fait, à cause de la quantité importante d'acide glutamique dans l'hydrolysat, celle-ci est mal séparée et apparaît sur le chromatogramme comme un épaulement qui permet d'estimer que la quantité de proline est au moins cinquante fois plus faible que celle d'acide glutamique. L'enregistrement de la radioactivité présente également un épaulement, signalé par une flèche sur la figure 1, qui pourrait correspondre à la proline marquée. La très grande proportion de la radioactivité retrouvée dans l'arginine et l'acide glutamique (86 %) après un marquage long, indique que le mécanisme de scission de l'octopine est unique. Ces résultats suggèrent fortement que cette scission est catalysée par une octopine déshydrogénase.

DISCUSSION

Dans une étude du catabolisme de la lysopine, entreprise dans le but d'étudier la répartition du marquage de la lysopine synthétisée *in vivo* par les tissus de crown-gall, Lejeune et Jubier (1967) ont montré que cet acide aminé peut être utilisé comme source d'azote par divers microorganismes appartenant à des genres différents, et notamment *Agrobacterium tumefaciens* B_6. Ils ont montré qu'une fraction, bien que peu importante, de la lysopine, est transformée par les bactéries en lysine et ont suggéré que le deuxième fragment de la réaction de dégradation est le lactate ou le pyruvate.

Les travaux de LEJEUNE et JUBIER, ainsi que les nôtres, indiquent la présence, dans les bactéries, de déshydrogénases spécifiques dont les substrats sont précisément les acides aminés anormaux des tumeurs induites par ces microorganismes.

L'acquisition, à la suite de la transformation tumorale, par les tissus, d'un caractère biochimique propre à la bactérie inductrice est un phénomène qui présente une analogie frappante avec les phénomènes d'échange de matériel génétique bien connus chez les bactéries.

Cette analogie est en effet très marquée : en même temps que nous introduisons dans la cellule végétale le caractère tumoral, nous y introduisons un caractère biochimique qui ne dépend que de la souche bactérienne utilisée pour provoquer la transformation et dont nous avons retrouvé l'expression dans la bactérie. A l'appui de cette hypothèse on peut ajouter qu'il y a un certain degré de liaison entre le marqueur biochimique et la virulence de la bactérie, ainsi A_{66}, souche à nopaline et octopine, est un mutant atténué de A_6, souche à octopine, isolé par HENDRICKSON et *al.* (1934) qui ne produit de tumeurs que si l'on applique de l'auxine au voisinage de la plaie d'inoculation ; les souches à nopaline T_{37} et II B V7 sont décrites comme peu virulentes par BRAUN et STONIER (1958) et MANIGAULT (1968). Par contre, II BNV 6, qui produit des tumeurs quand on l'inocule dans les mêmes conditions que A_{66}, induit la formation d'octopine.

Nos résultats sont donc très fortement en faveur de l'hypothèse de BRAUN. L'isolement des octopine et nopaline déshydrogénases des tissus tumoraux et des bactéries et leur comparaison au niveau moléculaire permettraient, si l'on peut établir leur identité, de confirmer l'existence d'un principe inducteur de tumeur élaboré par la bactérie, transmis à la cellule végétale et doué de continuité génétique.

BIBLIOGRAPHIE

AARON, DA CUNHA, 1969. — Sur la libération, par les rayons X, d'un principe tumorigène contenu dans les tissus de crown-gall de tabac. *C. R. Acad. Sci.*, Paris, **268,** 318-321.

BIEMANN K., LIORET C., ASSELINEAU K., LEDERER E. et POLONSKI J., 1960. — Sur la structure chimique de la lysopine, nouvel acide aminé isolé du tissu de crown-gall. *Bull. Soc. Chim. biol.,* **42,** 979-991.

BRAUN A.C., 1948. — Studies on the origin and development of plant teratomas incited by the crown-gall bacterium. *Amer. J. Bot.,* **35,** 511-519.

BRAUN A.C. et STONIER T., 1958. — Morphology and physiology of plant tumors. *Protoplasmatologia. Handbuch der Protoplasmaforschung,* **10,** Vienne, Springer Verlag 1-53.

GOLDMANN A., TEMPÉ J. et MOREL G., 1968. — Quelques particularités de diverses souches d'*Agrobacterium tumefaciens. C. R. Soc. Biol.,* Paris, **162,** 630-631.

GOLDMANN A., THOMAS D.W. et MOREL G., 1969. — Sur la structure de la nopaline, métabolite anormal de certaines tumeurs de crown-gall. *C. R. Acad. Sci.,* Paris, **268,** 852-854.

GREENBERG D.M., 1961. — Biosynthesis of Amino Acids and Related Compounds. In *Metabolic Pathways,* D.M. GREENBERG, ed., Academic Press, New York et Londres, p. 191.

HENDRICKSON A.A., BALDWIN I.L. & RIKER A.J., 1934. — Studies on certain physiological characters of *Phytomonas tumefaciens, Phytomonas rhizogenes and Bacillus radiobacter.* Part. II, *J. Bacteriol.*, **28**, 597-618.

IZUMYA N., WADE R., WINITZ M., OTEY M.C., BIRNBAUM S.M., KOEGER R.J. & GREENSTEIN J.P., 1957. — Studies on diastereoisomeric α-amino acids and corresponding α-hydroxy acids. VIII. Configurations of the isomeric octopines. *J. Amer. chem. Soc.*, **79**, 652-658.

KOVOOR A., 1967. — Sur la transformation de tissus tumoraux de Scorsonère provoquée *in vitro* par l'acide désoxyribonucléique d'*Agrobacterium tumefaciens. C. R. Acad. Sci.*, Paris, **265**, 1613-1626.

KOVOOR A., 1969. — Cité par GAUTHERET R.J., *in* Les phénomènes tumoraux chez les végétaux. *Ann. Inst. Pasteur,* Lille, **XX**, 19-21.

LEFF J. et BEARDSLEY R.E., 1969. — The induction of crown-gall tumors by viral DNA. *XIth International Botanical Congress, Abstracts,* p. 125.

LEJEUNE B. et JUBIER M.F., 1967. — Etude de la dégradation de la lysopine par *Agrobacterium tumefaciens. C. R. Acad. Sci.,* Paris, **264**, 1803-1805.

LIORET C., 1956. — Sur la mise en évidence d'un acide aminé non identifié particulier aux tissus de crown-gall. *Bull. Soc. fr. Physiol. vég.*, **2**, 76.

MANIGAULT P., 1968. — *Transformations tumorales.* Masson et Cie éd., Paris.

MEISTER A., 1965. — Intermediary Metabolism of Amino Acids. I. Proline. In : *Biochemistry of the Amino-Acids,* Academic Press, New York et Londres, p. 707-712.

MÉNAGÉ A. et MOREL G., 1964. — Sur la présence d'octopine dans les tissus de crown-gall cultivés *in vitro. C. R. Acad. Sci.*, Paris, **259**, 4795-4796.

MÉNAGÉ A. et MOREL G., 1965 *a.* — Sur la présence d'un acide aminé nouveau dans le tissu de crown-gall. *C. R. Soc. Biol.*, Paris, **159**, 561.

MÉNAGÉ A. et MOREL G., 1965 *b.* — Sur la présence d'un acide aminé nouveau dans les tissus de crown-gall. *C. R. Acad. Sci.*, Paris, **261**, 2001-2002.

MÉNAGÉ A., TEMPÉ J. et MOREL G., 1966. — Analyse des acides aminés par chromatographie. II. Détection automatique des guanidines. *Physiol. Vég.*, **4**, 317-322.

MONOD J., 1958. — *Recherches sur la croissance des cultures bactériennes.* Hermann éd., Paris, p. 32.

PARSONS L.C. & BEARDSLEY R.E., 1968. — Bacteriophage activity in homogenates of crown-gall tissue. *J. Virol.*, **2**, p. 651.

SEITZ E.W. & HOCHSTER R.M., 1964. — Lysopine in normal and in crown-gall tumor tissue of tomato and tobacco. *Can. J. Bot.*, **42**, 999-1004.

TEMPÉ J., MÉNAGÉ A. et MOREL G., 1967. — Adaptation de l'autoanalyzer à une double coloration permettant la révélation simultanée des acides aminés et des guanidines. In : *Automation in analytical chemistry. Technicon Symposia* 1966, Vol. **II.** Mediad Incorporated ed., White Plains, N.Y., U.S.A., 421-425.

LOSS OF TUMOR-INDUCING ABILITY

Robert H. Hamilton
State College, Pennsylvania, United States
E-mail: rhh3@psu.edu

Commentary on the paper:

Hamilton, R. H., and Fall, M. Z. 1971. The loss of tumor-initiating ability in *Agrobacterium tumefaciens* by incubation at high temperature. Experientia 27:229–230.

This paper was a result of my attempt to prove the general idea that plant tumors induced by *Agrobacterium tumefaciens* must involve transfer of the so-called tumor inducing principle (Braun and Mandle 1948) from bacteria to plant cells. It was presumed that this factor must be self-replicating in the plant cells, since it was known that bacteria-free tumor tissue cultures could be maintained. After discussing these ideas with Dr. Christine Pootjes who had just come to Penn State, we decided to look again at bacteriophages. My first Ph.D. student, Robert P. Zimmerer, worked with Dr. Pootjes on this subject and we were forced to conclude, as had others, that there was no relationship between phages and tumor induction (Zimmerer et al. 1966). This was the beginning of a long collaboration with Dr. Pootjes, which was the most important outcome of this project, since I am not a microbiologist.

Plant pathologists knew that *Agrobacterium* strains were not always stable in laboratory cultures. After prolonged culture, the bacteria sometimes failed to induce tumors. In the mid-1960s, I attended a meeting of the American Society of Plant Physiology at the University of Illinois and heard a paper by James and Barbara Lippincott on their bean leaf tumor assay (Lippincott and Haberlein 1966). High temperature reduced the number of tumors on bean leaves. I remember discussing this with them and I wondered if this might have an effect on the bacteria.

As chance would have it, the late professor Robert Dickey of Cornell University had come to Penn State on sabbatical leave. He had done fieldwork on crown gall disease and had many field isolates of *A. tumefaciens* that he agreed to give us for heat stability tests. I decided to screen these strains for tumor-inducing ability after incubation for various times at 36°C.

I asked my research assistant to do a weekly bean leaf assay on these strains. Only a few of the strains were temperature sensitive and became avirulent. Two of the strains (C58 and Ach5) were especially sensitive and were used in the more detailed study reported in this short paper.

We realised that there might be some extra-chromosomal DNA (plasmid) which was eliminated by heat treatment of the heat-sensitive strains. Dr. Pootjes and I spent much time looking for this extra DNA and we found it in cesium chloride gradients of both virulent and avirulent strains of C58 heated for 38 hours at 36°C. After Schell and co-workers found heating C58 eliminated the tumor-inducing plasmid, we were at a loss to explain our results.

Dr. Kerr had shown that with the use of appropriate strains, virulence could be transferred between strains (Kerr 1969). He was kind enough to send us his virulent strain TT133 and we found we could convert our avirulent C58 back to virulent. This work was done by my graduate student, Mike Chopin, for his MS thesis (Hamilton and Chopin 1975).

I still wanted to explain our plasmid results. When Kathy Falkenstein came to my lab as a Ph.D. student, I proposed that she characterize the plasmids in our virulent strain of C58. She found that this strain contained two large plasmids running on 0.6% agarose gels as a doublet. All avirulent strains had only the slower band. Electron microscope measurements indicated the presence of a 126 megadalton (Ti) plasmid plus a 153 megadalton plasmid in the virulent strains; avirulent strains had only the larger plasmid (Falkenstein 1979).

Since most of this took place more than twenty years ago, my memory may be selective or even faulty but this is as I recall it.

References

Falkenstein, K. F. 1979. Studies on crown gall tumor induction by *Agrobacterium tumefaciens*. Ph.D. thesis, Pennsylvania State University, U.S.A.

Hamilton, R. H., and Chopin, M. N. 1975. Transfer of the tumor inducing factor in *Agrobacterium tumefaciens.* Biochem. Biophys. Res. Commun. 63:349–354.

Kerr, A. 1969. Transfer of virulence between isolates of *Agrobacterium.* Nature 223:1175–1176.

Lippincott, J. A., and Haberlein, G. T. 1966. The induction of leaf tumors by *Agrobacterium tumefaciens.* Amer. J. Bot. 52:396–403.

Zimmerer, R. P., Hamilton, R. H., and Pootjes, C. 1966. Characterization of a phage from *Agrobacterium tumefaciens.* J. Bacteriol. 92:746–750.

ROBERT H. HAMILTON: I grew up in southern Illinois where my mother was a teacher and my father a school administrator. I attended one-room schools and graduated from McLeansboro High School. I attended the University of Illinois and graduated in 1950. In 1953, I completed my M.S. in Agronomy at Rutgers University, New Jersey with Dr. Richard Aldrich. My thesis was on the on the use of chlorinated benzoic acid derivatives as herbicides. I then spent two years in the U.S. Army at Camp Detrick working on herbicides under Dr. James Brown and Dr. Robert Weintraub. In 1955, I took a job with the USDA at Michigan State University working on weed control in sugar beet. I was attached to the Botany and Plant Pathology Department where I was permitted time to work on my Ph.D. My thesis work was with Professor Robert Bandurski who had just come to Michigan State. It was an exciting time and he greatly influenced my entire career. I graduated in 1960. Following one year with the USDA at North Carolina State University working with Dr. Donald Moreland, I took a position at Pennsylvania State University in the Department of Botany and Plant Pathology (now Biology). I continued there until my retirement in 1991. I spent a sabbatical leave in 1970–71 at the Swiss Federal Station for Horticulture in Wadenswil, Switzerland.

Separatum EXPERIENTIA *27*, 229 (1971)
Birkhäuser Verlag, Basel (Schweiz)

The Loss of Tumor-Initiating Ability in *Agrobacterium tumefaciens* by Incubation at High Temperature

Avirulent strains of *Agrobacterium tumefaciens* have frequently appeared in laboratory stock cultures, but the nature of this change is not understood. Temporary loss of virulence has also been observed following heating[1,2] or prolonged subculture in a glycine-rich medium[3–5]. A number of UV-induced auxotrophic mutants of *A. tumefaciens* have been isolated[6], which are fully virulent only if supplemented with the required nutrient.

We wish to report the conversion of cells of a highly virulent strain of *A. tumefaciens* (C-58) into stable nutritionally sufficient avirulent cells. To our knowledge, this is the first report of the conversion of virulent cells of *A. tumefaciens* into stable avirulent cells under controlled conditions. The derived avirulent cells consistently fail to initiate tumors on tomato stems, sunflower stems, and pinto bean leaves. These cultures have been maintained from 2 to 4 years without regaining virulence.

Our initial experiments involved the cultivation and weekly transfer of 8 virulent, single colony isolates from the C-58 or Ach strains on nutrient agar slants for 4 weeks at 36 °C or at room temperature. After 4 weeks, cultures (Group I) incubated at 36 °C were unable to induce tumors on pinto bean leaves and sunflower stems.

Avirulent cultures have been isolated from strain C-58 following 36 °C incubation for much shorter time intervals (Group II). An overnight broth culture of a virulent, single colony isolate was streaked on nutrient agar and incubated at 36 °C. After various time intervals (Table I) 8 colonies were removed and each inoculated into nutrient broth and grown overnight at 25 °C on a shaker (Parent or Pt isolate). The cultures were adjusted to a selected concentration (turbidometric readings) by dilution with nutrient broth and applied to carborundum-injured pinto bean leaves[7]. Following inoculation, the plants were kept in a dark high humidity chamber at 25 °C for 72 h. Tumor counts or qualitative estimates of the number of tumors per leaf were made after 7 days.

A decrease in tumor induction was noted in many of the Pt isolates incubated 48 h or longer at 36 °C (Table I). That the reduction in virulence observed for these cultures was due to the number of avirulent cells present in the colony, was verified by plating a number of these parent isolates on nutrient agar, and selecting 8–12 colonies. These progeny (Pg) subcultures (shaken in nutrient broth overnight at 25 °C) were tested on bean leaves. Avirulent subcultures (Pg) were obtained from one of the Pt isolates which had been grown at 36 °C for 48 h (Table II). Progeny subcultures indicated that 72 h or 84 h Pt isolates were a mixture of weakly virulent and avirulent cells, while progeny from 120 h Pt isolates were avirulent.

Since loss of virulence could be due to nutritional deficiencies, growth was tested on 3 minimal media which support the growth of the parent C-58 strain. Liquid shake cultures were used and growth was observed after 24 and 48 h. Group I avirulents (4-week heated) grew only on USCHINSKY's medium[8] (glycerol, ammonium lactate, asparagine and salts), and thus appear to be auxotrophs. Group II avirulents (48 h) grew on both MANASSE's[9] (glutamic acid and salts) and the NITSCH and NITSCH (sucrose, nitrate, and salts) medium[10]. Neither Group II virulents nor avirulents grew on the richer USCHINSKY's medium, thus toxicity may be involved in this case. The growth of Group II isolates was also determined in the bean host. Leaves were inoculated in the usual manner and bacteria washed off 10–15 min after inoculation. 3 leaves were harvested at intervals, rinsed in sterile water, and ground with sand in nutrient broth. The homogenate was diluted and streaked on nutrient agar for plate counts. The results indicated that both virulent and avirulent Group II

Table I. Virulence on bean leaves of selected isolates (Pt) of *Agrobacterium tumefaciens* (strain C-58) following incubation at 36 °C for 36–120 h on nutrient agar[a]

Incubation at 36 °C (h)	Relative virulence[b] of 8 selected isolates (Pt cultures)				
	++++	+++	++	+	0
0	5	3			
36	4	4			
48		1	5	2	
60		2	4	2	
72			5	3	
84				5	3
120					8

[a] Each of the 8 colonies selected for bioassay at each time was grown in nutrient broth shake culture overnight (Pt culture) and 4×10^7 cells applied on each of 6 bean leaves. [b] ++++, 200+ tumors; +++, 100–200 tumors; ++, 50–100 tumors; +, less than 50 tumors per leaf; 0, none, but at 4×10^8 cells per leaf an occasional tumor was observed.

Table II. Virulence on bean leaves of subcultures (Pg) derived from parent isolates (Pt) of *Agrobacterium tumefaciens* (C-58) which had been grown at 36 °C for 48–120 h[a]

Parent isolates (Pt)		Number of subcultures (Pg) with relative virulence[b]				
Incubation at 36 °C (h)	Virulence[b]	++++	+++	++	+	0
48	+++		7	1		
48	+++		1	10	1	
48	++			7	2	7
72	+				4	4
84	+				5	4
120	0					9

[a] Progeny subcultures (Pg) were selected from plates streaked with cells from the 36 °C grown parent isolates (Pt). The subcultures (Pg) were grown in nutrient broth shake overnight at 25 °C and bioassayed by inoculating 6–12 leaves with $0,4 - 4 \times 10^8$ cells per leaf depending on the expected virulence. [b] Rating same as for Table I.

[1] J. A. LIPPINCOTT and B. B. LIPPINCOTT, Science *147*, 1578 (1965).
[2] T. STONIER, R. BEARDSLEY, L. PARSONS and J. MCSHARRY, J. Bact. *91*, 266 (1966).
[3] B. J. LONGLEY, T. O. BERGE, J. M. VANLANEN and I. L. BALDWIN, J. Bact. *33*, 29 (1937).
[4] J. M. VANLANEN, I. L. BALDWIN and A. J. RIKER, J. Bact. *41*, 95 (1941).
[5] W. C. SCHATHHORST, J. DEVAY and T. KOSUGE, Phytopathology *54*, 1436 (1964).
[6] B. B. LIPPINCOTT and J. A. LIPPINCOTT, J. Bact. *92*, 937 (1966).
[7] J. A. LIPPINCOTT and G. T. HEBERLEIN, Am. J. Bot. *52*, 856 (1965).
[8] C. STAPP, *Plant Pathogenic Bacteria* (Oxford University Press 1961).
[9] R. MANASSE and W. A. CORPE, Can. J. Microbiol. *13*, 1951 (1967).
[10] J. P. NITSCH and C. NITSCH, Am. J. Bot. *43*, 839 (1956).

isolates grow in an apparent log phase at the about same rate for 28 h, but subsequently the avirulent leveled off at a higher concentration per leaf.

The avirulent Group II cells give the keto sugar test[11] characteristic of many isolates of *Agrobacterium* and have a typical *A. tumefaciens* colony morphology at room temperature. Preliminary observations by electron microscopy indicated no gross differences between room temperature grown Group II avirulent or virulent cells.

Although the growth rates were about the same at 25 °C, the avirulent cultures grew 4–8 times as fast as virulent cultures at 37 °C in nutrient broth shake culture. When grown on nutrient agar plates at 37 °C for 24 h, the larger avirulent colonies are easily distinguished from the pinpoint virulent colonies. This growth characteristic has proven to be a much more sensitive and precise method for the isolation of avirulents. If colonies of strain C-58 are grown at 37 °C for 24 h, the cells washed from the plates, and replated at 37 °C, 2–4% of the cells produce the large colony type avirulents. Preliminary data indicate that many of the colonies grown at 37 °C contain a few avirulent cells. Thus the conversion appears to be a high frequency event. All attempts to restore virulence by incubation of the avirulent cultures with lysates from the virulent cells, isolated DNA from virulent cells, Lv-1 phage DNA[12], or the intact Lv-1 phage have so far failed.

Although our results could be explained by the occurrence of a high frequency mutation, perhaps a more likely possibility is the loss of a virulence factor. In other strains this factor may become integrated more tightly with the bacterial DNA. It may be significant that strain C-58 of *A. tumefaciens* is also one of the most virulent strains we have tested. In any event these strains should be extremely useful in the investigation of virulence and tumor induction[13].

Zusammenfassung. Durch Inkubation bei hohen Temperaturen (37 °C) verliert das *Agrobacterium tumefaciens* C-58 allmählich die Fähigkeit zur Gallen- oder Tumorbildung in Pflanzen.

R. H. HAMILTON and M. Z. FALL

Department of Biology,
The Pennsylvania State University,
University Park (Pennsylvania 16802, USA),
14 July 1970.

[11] M. J. BERNAERTS and J. DELEY, J. gen. Microbiol. *22*, 129 (1960).

[12] R. P. ZIMMERER, R. H. HAMILTON and C. POOTJES, J. Bact. *93*, 746 (1966).

[13] We wish to thank Prof. R. DICKEY, Department of Plant Pathology, Cornell University, for his suggestions and interest, as well as for the strains C-58, Ach, and CG. We also thank Dr. CHRISTINE POOTJES for Phage samples. Contribution No. 48 from the Dept. of Biology and 3777 from The Pennsylvania Agricultural Experiment Station.

DISCOVERY OF THE Ti PLASMID

Marc Van Montagu
Department of Molecular Genetics, Ghent University, Gent, Belgium
E-mail: mammon@gengenp.rug.ac.be

Commentary on the paper:

Van Larebeke, N., Genetello, C., Schell, J., Schilperoort, R. A., Hermans, A. K., Hernalsteens, J.-P., and Van Montagu, M. 1975. Acquisition of tumour-inducing ability by non-oncogenic agrobacteria as a result of plasmid transfer. Nature 255:742–743.

Around 1968, my friends and colleagues, Walter Fiers, Jeff Schell and I felt that it was time to study the molecular biology of eukaryotic organisms, as the new concepts and tools of prokaryotic molecular genetics had become well established. Jeff Schell, who graduated in bacterial taxonomy was interested in *Agrobacterium tumefaciens* because of the early work of Armin C. Braun on plant tumor induction. In the taxonomy lab, there was a collection of more than 100 strains of agrobacteria and as nobody was currently studying the molecular basis of tumor induction, Jeff and I decided to join forces and take up the challenge and make crown gall research our long term project.

From the start, Jeff and I had the working hypothesis that it was an extra-chromosomal element that was conferring tumor-inducing ability. Indeed a thorough comparison by numerical taxonomy of non-oncogenic strains (usually named *A. radiobacter*) and oncogenic strains (usually named *A. tumefaciens*), indicated a very close relationship. Since we were both experienced in working with bacteriophage λ, we favored research on prophages of agrobacteria. To test oncogenicity, we needed plants and a knowledge of the plant sciences. Jeff hesitated to introduce to the lab too many unknowns at the same time. He proposed to contact Rob Schilperoort at the University of Leiden, who was already an authority on crown gall research.

In the early seventies, most leading research in bacterial molecular genetics involved work with extra-chromosomal elements of *E. coli.* The new knowledge and concepts came from studies on the mechanisms of lysogeny of phages like λ that insert at specific attachment sites, or insert randomly like Mu or establish as a plasmid like P1. Mechanisms of the plasmid transfer, particularly the antibiotic resistance-conferring R-plasmids, was also a common topic, along with the most intriguing and just discovered insertion elements (IS). As often in life sciences, progress came from the availability of new tools, in this case, new methods of DNA purification by ultracentrifugation and DNA visualisation by electron microscopy.

We sent a MD/Ph.D. graduate student, Nick Van Larebeke and a technician, Ivo Zaenen to Rob Schilperoort's lab to learn tumor induction of plants and most of all to use the excellent facilities for analysing DNA by ultracentrifugation.

I myself had learned electron microscopy of DNA at an EMBO course in Geneva (Edouard Kellenberger's lab). Already in those days, long before the start of the Heidelberg laboratories, the European Molecular Biology Organisation (EMBO) was instrumental in bringing the best of U.S. and European molecular biologists together in summer courses, workshops, seminar tours and post-doc exchange fellowships.

I had full-time access to an EM to study *E. coli* infections with conditional lethal mutants of RNA phages, as well as DNA spreading (Kleinschmidt technique), and the DNA-heteroduplex technique developed by Ross Inman for studying deletions, insertions, and DNA exchanges between the lamboid coli phages. This led us to study the DNA peaks discovered by Ivo Zaenen upon ultracentrifugation of DNA from agrobacteria. We demonstrated that these peaks were the super-coiled form of very large plasmids. Indeed in those days, 200 Kbp plasmids were considered unusually large and very difficult to isolate. The correlation between the presence of such a plasmid in different tumor-inducing agrobacteria, and its absence in non-oncogenic strains was sufficiently intriguing that Sydney Brenner, then Editor in Chief of the *Journal of Molecular Biology,* allowed the publication of the "tour de force" of identifying these plasmids and establishing a correlation between them and oncogenicity (Zaenen et al. 1974). However, these findings were far from proof that the plasmids carried genes for oncogenicity. Later on it was demonstrated that both oncogenic and non-oncogenic strains contained even larger plasmids. But these observations comforted us in our working hypothesis and guided our future research to find the "tumor-inducing principle" and how it functioned.

A hint of how to proceed came from the work of Allen Kerr (1969). He had established that the tumor-inducing capacity could be transferred between agrobacteria. Could this transfer of oncogenicity be linked to transfer of the plasmid we had located in oncogenic strains? In those days, there was no PCR or Southern blotting. But we had identified other traits that were correlated with the presence of the plasmid such as catabolic properties, agrocin 84 sensitivity and phage sensitivity/exclusion. Using a plasmid-free non-oncogenic recipient and the "Kerr cross" conjugation system, we demonstrated that acquisition of oncogenicity and associated traits was the result of plasmid transfer. The tumor-inducing (Ti) plasmid had been found.

References

Kerr, A. 1969. Transfer of virulence between isolates of *Agrobacterium.* Nature 223:1175–1176.

Zaenen, I., Van Larebeke, N., Teuchy, H., Van Montagu, M., and Schell, J. 1975. Supercoiled circular DNA in crown-gall inducing *Agrobacterium* strains. J. Mol. Biol. 86:109–127.

MARC VAN MONTAGU: In 1955, I graduated B.Sc. (licentiaat) in Chemistry, majoring in Organic Chemistry/Biochemistry from the University of Ghent. In 1965 I graduated from the same university with a Ph.D. in Organic Chemistry/ Biochemistry.

My only real mentors were my future colleagues in Gent, Walter Fiers and Jeff Schell. As a high school student, I had excellent courses in organic chemistry and through my English class, I became acquainted with "Life Sciences and Biochemistry" as a result of reading *Arrowsmith* by Sinclair Lewis. I became fascinated by the chemistry of living organisms. In those days in Gent, biochemistry was not accepted as a topic for major study, but I was the first chemistry student allowed to major in organic chemistry with an enzyme (β-amylase in barley) as a subject for research.

In 1954, my final year of undergraduate studies, I met both Walter Fiers, an agricultural engineer involved in a Biochemistry Master's degree, and Jeff Schell, an undergraduate student in Zoology. Apart from being interested in the politics of how to construct a better and more humane society, we all three were fascinated by this emerging science of "molecular biology" and particularly "molecular genetics". Together we discovered the fascinating world of bacteriophages and extrachromosomal elements. Walter Fiers graduated from Caltech with Robert Sinsheimer, where he was the first to demonstrate the existence of a circular genome (phage ΦX174). This discovery made the front page of the *New York Times* and the *Wall Street Journal* and made us realise the competitive nature of research.

(Reprinted from Nature, Vol. 255, No. 5511, pp. 742–743, June 26, 1975)

Acquisition of tumour-inducing ability by non-oncogenic agrobacteria as a result of plasmid transfer

IT has been suggested that the tumour-inducing capacity of *Agrobacterium* is determined by plasmid genes[1,2]. The acquisition of virulence by non-pathogenic agrobacteria from pathogenic agrobacteria has been demonstrated by Kerr[3,4]. We investigated whether this phenomenon could be the result of transfer of a plasmid. Besides providing further evidence for the hypothesis that plasmid genes carry the genetic information for the tumour-inducing ability of *Agrobacterium*, the system offers the possibility of studying the genetics and functions of the plasmid.

The donor and recipient strains of *Agrobacterium* must be well-characterised. Donor strain *A. rhizogenes* 223, used in a first set of experiments, has the typical biochemical properties of cluster 2 (ref. 5). It cannot produce 3-ketolactose, does not grow anaerobically on nitrate containing medium, grows slowly and is resistant to all of the seven different types of virulent phages of our typing set (see Table 1). The cured non-tumour-inducing receptor strain C58-C1 was derived from a single colony of *A. tumefaciens* C58 by growth at 37 °C, which resulted in both the loss of its plasmid and the ability to induce tumours[2]. From C58-C1 a double mutant was isolated which showed resistance to 100 μg ml^{-1} rifampicin and to 300 μg ml^{-1} streptomycin. This C58-C1 *rif str* was repeatedly tested for plasmid and tumour-inducing ability and was always found to be negative for both. *A. tumefaciens* C58-C1 *rif str* belongs to the TT9 group of cluster 1 (ref. 5), can produce 3-ketolactose, can grow anaerobically and shows, among our laboratory strains, a unique and typical sensitivity pattern to the virulent phages of our typing set (see Table 1). We crossed the oncogenic *A. rhizogenes* strain 223 with the non-oncogenic *A. tumefaciens* strain C58-C1 *rif str* on *Kalanchoë daigremontiana* as host plant. The experimental procedures were as described by Kerr[4]. Eight weeks after inoculation 15 out of 30 reisolated single colonies of the antibiotic-resistant recipient bacteria seemed to be oncogenic on 1-week-old sunflower seedlings. A pathogenic colony was then subjected to further extensive purification. The resulting bacteria showed the antibiotic resistance, positive 3-ketolactose test, anaerobic growth and the phage sensitivity pattern characteristic of the avirulent C58-C1 *ref str* strain used as recipient (Table 1). They were repeatedly found to induce crown gall tumours on pea seedlings, sunflower seedlings and *Kalanchoë daigremontiana*. Using ultracentrifugation[1,2] and subsequent electron microscopy we demonstrated, in the purified newly oncogenic recipient bacterium, a large plasmid with a mean length of 58.0 μm±1.3 μm s.d., which lies in the range 61.1 μm±1.4 μm s.d. that we found for a sample of plasmid molecules isolated from the donor strain 223. *A. tumefaciens* C58-C1 *rif str* also acquired other characteristics that are known to be coded for by the plasmid, namely sensitivity to agrocins 84 and 396 (G. Engler, J.P.H., M. Holsters, M.V.M., R.A.S., and J.S., unpublished), the ability to degrade an arginine derivative nopaline (ref. 6 and G. M. Bomhoff, R.A.S., J.P.H., H. Kester, and J.S., unpublished) and the ability to exclude the virulent *Agrobacterium* phage AP1 (J.P.H., M.V.M., and J.S., unpublished). Strain C58-C1 *rif str* (TI223) differs from the parental oncogenic strain C58 in that it cannot be cured of its plasmid by growth at 37 °C, even after 75 generations, whereas a culture of the parental strain C58 is cured completely after 25 generations of growth at 37 °C. This property apparently depends on the nature of the plasmid present in the bacterium.

To confirm these results we performed a better controlled experiment on otherwise sterile pea seedlings in sterile test tubes containing agar medium. This cross involved the crown gall-inducing *Agrobacterium tumefaciens* strain Kerr 14 and the naturally occurring non-oncogenic *Agrobacterium radiobacter* S1005, made resistant against 100 μg ml^{-1} rifampicin and 300 μg ml^{-1} streptomycin. The pea seedlings were inoculated with S1005 *rif str* 12 d after inoculation with *A. tumefaciens* Kerr 14. As controls we used pea seedlings that were only inoculated with either Kerr 14 or S1005 *rif str*. Using sensitivity to agrocin 84 as a screening test we were able to pick up a single colony that was still resistant to 100 μg ml^{-1} rifampicin and to 300 μg ml^{-1} streptomycin, had the typical sensitivity pattern of *A. radiobacter* S1005 in a test with our typing set of seven different virulent *Agrobacterium* bacteriophages (Table 1), still produced the typical agrocin S1005, effective against *A. tumefaciens* B6-S3, but had become oncogenic as shown by tests on the pea seedlings[7]. By ultracentrifugation and subsequent electron microscopy this oncogenic *A. radiobacter* S1005 (TI Kerr 14) was shown to harbour a large plasmid. The strain had also become sensitive to agrocins 84 and 396, showed an efficient exclusion of phage AP1 and could degrade nopaline. On control pea seedlings inoculated with the recipient strain S1005 *rif str* no tumours were found. From uninoculated pea seedlings no bacteria could be isolated after 6 weeks. From tumours induced by *A. tumefaciens* Kerr 14 but not contaminated with S1005 *rif str* no bacteria could be isolated that were resistant to rifampicin and streptomycin.

Table 1 Properties of *Agrobacterium* strains used

Strain	Resistance to 100 μg ml^{-1} rifampicin	Resistance to 300 μg ml^{-1} streptomycin	Ability to produce 3-ketolactose[12]	Ability to grow anaerobically on nitrate-containing medium[5]	Sensitivity to virulent phages of typing set: GS1	GS2	GS3	GS5	GS6	GS18	AP1
223	—	—	—	—	—	—	—	—	—	—	—
C58-C1	+	+	+	+	—	+	—	+	+	—	+
C58-C1 (TI 223)	+	+	+	+	—	+	—	+	+	—	+
Kerr 14	—	—	+	+	+	—	+	—	+	+	+
S1005 *rif str*	+	+	+	+	+	+	—	—	+	—	+
S1005 *rif str* (TI Kerr 14)	+	+	+	+	+	+	—	—	+	—	+

Strain	Ability to degrade nopaline	Exclusion of phage AP1	Sensitivity to agrocins 84 and 396	Oncogenicity	Presence of a large plasmid
233	+		+	+	+
C58-C1	—	—	—	—	—
C58-C1 (TI 223)	+	+	+	+	+
Kerr 14	+	+	+	+	+
S1005 *rif str*	—	—	—	—	—
S1005 *rif str* (TI Kerr 14)	+	+	+	+	+

These results show that the transfer of virulence as observed by Kerr, rests on a transfer of a large plasmid from the oncogenic strain to the avirulent recipient. By the acquisition of such a plasmid from a taxonomically distant strain, a strain that had been made non-oncogenic by curing of its plasmid as well as a naturally occurring non-oncogenic *A. radiobacter* strain could be rendered oncogenic. Crown gall tumour-inducing ability, ability to degrade nopaline, the exclusion of virulent phage AP1 and the sensitivity to agrocins 84 and 396 are all phenotypic properties coded for by one large plasmid. As tumour-inducing plasmids from two distantly related oncogenic agrobacteria conferred these same properties to their new host, some of these properties may be related to the process of tumour induction. It has been suggested[8] that the receptor sites for agrocin 84 could be part of a specific molecular configuration on the bacterial cell surface necessary for interaction with a proposed specific receptor site on the plant cell wall[9-11]. The correlation between oncogenicity and agrocin 84 sensitivity could however, be simply a result of genetic linkage. The ability to degrade arginine derivatives could be part of a biochemical pathway which, when operating in a plant cell, leads to the production of these amino acids, which in itself could be important in tumour development, or, alternatively or additionally, could sustain further growth of the infecting pathogenic bacteria.

N. Van Larebeke
Ch. Genetello
J. Schell

Laboratorium voor Genetica,
Rijksuniversiteit Gent, B9000 Gent, Belgium

R. A. Schilperoort
A. K. Hermans

Biochemisch Laboratorium,
Rijksuniversiteit Leiden,
Leiden, Netherlands

J. P. Hernalsteens
M. Van Montagu

Laboratorium voor Genetische Virologie,
Vrije Universiteit Brussel,
1640 St.-Genesius Rode, Belgium

Received April 2; accepted April 18, 1975.

1 Zaenen, I., Van Larebeke, N., Teuchy, H., van Montagu, M., and Schell, J. *J. molec. Biol.*, **86**, 109-127 (1974).
2 Van Larebeke, N., *et al.*, *Nature*, **522**, 169-170 (1974).
3 Kerr, A., *Nature*, **223**, 1175-1176 (1969).
4 Kerr, A., *Physiol. Pl. Pathol.*, **1**, 241-246 (1971).
5 Kersters, K., de Ley, J., Sneath, P. H., and Sackin, M., *J. gen. Microbiol.*, **78**, 227-239 (1973).
6 Bomhoff, G. H., thesis, Univ. Leiden (1974).
7 Manigault, P., Kurkdjian, A., *C.r. hebd. Séanc., Acad. Sci. Paris., Sér D.*, **264**, 2304-2306 (1967).
8 Roberts, W. P., and Kerr, A., *Physiol. Pl. Pathol.*, **4**, 81-91 (1974).
9 Lippincott, B. B., and Lippincott, J. A., *J. Bact.*, **97**, 620-628 (1969).
10 Schilperoort, R. A., thesis, Univ. Leiden (1969).
11 Beiderbeck, R., *Z. für Naturforschung*, **28c**, 198 (1973).
12 Bernaerts, M. J., and de Ley, J., *Nature*, **197**, 406-407 (1963).

Printed in Great Britain by Henry Ling Ltd., at the Dorset Press, Dorchester, Dorset

A PLASMID WAS PRESENT AFTER ALL

Bruce Watson[1] and Eugene W. Nester[2]

[1]Columbia Winery, Woodinville, Washington, United States
Email: bruce.watson@cwine.com

[2]Department of Microbiology, University of Washington
Seattle, Washington, United States
Email: gnester@u.washington.edu

Commentary on the paper:

Watson, B., Currier, T. C., Gordon, M. P., Chilton, M.-D., and Nester, E. W. 1975. Plasmid required for virulence of *Agrobacterium tumefaciens*. J. Bacteriol. 123:255–264.

When I (Bruce Watson) started working with *Agrobacterium tumefaciens* strain C58 as a graduate student in Milt Gordon's lab, we hypothesized that the ability of *Agrobacterium* to form crown gall tumors might involve a plasmid. In large part, this suspicion rested on the surprising observation made by Bob Hamilton and his assistant, Linda Fall, whose paper is included in this anthology (Hamilton and Fall 1971). They observed that heating *A. tumefaciens* C58 to 37°C resulted in the loss of virulence in a high proportion of the cells. I repeated their experiment and readily confirmed their observation. Since heating cells to an elevated temperature was a recognized method for curing bacteria of their plasmids, we thought it quite possible that strain C58 was being cured of a plasmid necessary for tumor formation. In hindsight, it was fortunate that Hamilton studied strain C58 because we later found that heating other virulent strains did not lead to a ready loss of virulence. The possibility of a plasmid being involved was also consistent with a previous observation made by Kerr, namely that virulence could be transferred from a virulent to an avirulent strain, another seminal paper included in this volume (Kerr 1969). However, repeated attempts to find a plasmid in C58 failed. Of course all my attempts used the conventional techniques of plasmid isolation which worked for the more usual size plasmids (approximately 20 Kbp).

In the course of these studies, Gene Nester, in the Department of Microbiology, who was a member of our crown gall group in Seattle, was on an eight-month sabbatical leave at the University of Leiden in the Netherlands, working in the laboratory of Rob Schilperoort, one of the modern pioneers in crown gall research. In the course of his stay, Rob told him of a preprint from the laboratory of Jeff Schell and Marc Van Montagu in Gent, Belgium, on which Rob was an author. They reported that a large plasmid was present in virulent strains of *Agrobacterium,* but was absent in avirulent strains (Zaenen et al. 1974). This paper provided not only the technical information that I needed to demonstrate the presence of a plasmid, but it convinced me that I was on the right track in looking for plasmids. When Gene told Rob about the Hamilton and Fall paper, Rob was skeptical initially, in large part because there were so many unconfirmed reports in the crown gall literature. However, Gene convinced Rob that the report was true and that I would be willing to send him the putative "cured" avirulent strain from Seattle if, in turn, he would allow Gene to send the preprint to Seattle. Rob agreed. Both the European and Seattle laboratories then did the obvious experiments; namely, to compare the plasmid content of the wild type and the heated avirulent derivative. Both groups reached the same conclusion, namely that a large plasmid (approximately 200 Kbp) was missing from the avirulent strain. Using old-fashion renaturation kinetics and Cot curve analyses, since Southern hybridization techniques were not yet invented, I demonstrated that the plasmid was not integrated into the bacterial chromosome. Transfer of the plasmid from the virulent to the avirulent strain using the "Kerr cross" technique, conferred virulence on the formerly avirulent strain. These data, together with the data from the Gent and Leiden groups (Van Larebeke et al. 1975) provided definitive evidence that a large plasmid was necessary for tumor formation. This conclusion set the stage for the experiments which defined the role of the large plasmid, termed the tumor-inducing, or Ti plasmid, in tumor formation.

In hindsight, my initial efforts to identify a plasmid in the wild type C58 strain were foiled because I was not looking for such a large molecule. Apparently, the large plasmid got tangled up with chromosomal DNA during the isolation procedures, since I took great pains not to subject the DNA to harsh conditions which might break the circular DNA molecules. However, controlled shearing of the chromosomal DNA was exactly what was required to free the large plasmid. Initially we used a vortexing technique to accomplish the shearing. Then, Tom Currier, a graduate student in Gene Nester's group, devised a purification procedure based on shearing the DNA by gently forcing it through a syringe needle followed by cesium chloride–ethidium bromide centrifugation (Currier and Nester 1976). This technique led to a more reproducible procedure to isolate the Ti plasmid.

Throughout all of my studies on plasmids, I received a great deal of help from Stanley Falkow, a professor of microbiology and world authority on plasmids of medical importance. He and his colleagues were especially helpful with electron microscope techniques. Their enthusiastic help, I think was in part a result of their interest in following our studies on such a large plasmid, certainly the largest one of which they were aware. I recall with mixed emotions the tedious tracing of innumerable images of very large plasmids on paper using a photographic enlarger, and then laying string on them to measure their length. Seems that technology has advanced during the last 30 years!

References

Currier, T. and Nester, E. W. 1976. Isolation of covalently closed circular DNA of high molecular weight from bacteria. Anal. Biochem. 66:431–441.

Hamilton, R. H. and Fall, M. Z. 1971. The loss of tumor-initiating ability in *Agrobacterium tumefaciens* by incubation at high temperature. Experentia 27: 229–230.

Kerr, A. 1969. Transfer of virulence between isolates of *Agrobacterium.* Nature 233:1175–1176.

Van Larebeke, N., Genetello, C., Schell, J., Schilperoort, R. A., Hermans, A. K., Hernalsteens, J.P. and Van Montagu, M. 1975. Acquisition of tumor-inducing ability by non-oncogenic agrobacteria as a result of plasmid transfer. Nature 255:742–743.

Zaenen, I., Van Larebeke, N., Teuchy, H., Van Montagu, M., and Schell, J. 1975. Supercoiled circular DNA in crown-gall inducing *Agrobacterium* strains. J. Mol. Biol. 86:109–127.

BRUCE WATSON: I was raised in Eugene, Oregon, and acquired an interest in chemistry while still in grade school. This was encouraged by my parents and also by the University of Oregon Department of Chemistry, Junior Scientist Program, which made it possible for me to buy a lot of chemicals that I shouldn't have had. I survived many organic syntheses, rare earth element separations, etc. to enter undergraduate school in the University of Oregon Chemistry Department and was fortunate enough to work in the laboratory of enzymologist Sidney Bernhard. I finished my B.S. in chemistry in 1970. After attending one year of graduate school at the University of California, Berkeley, I moved back to the Pacific Northwest and into the laboratory of Milton Gordon in the University of Washington Department of Biochemistry, where I completed my Doctoral degree in 1975.

After graduate school I moved away from crown gall and worked as a postdoc in the University of Washington Botany department on a variety of organisms. In the autumn of 1986 I worked harvest at Columbia Winery and realized that I had found what I really wanted to do. I am still with Columbia Winery today, although the winery has become part of a much larger company. I am in charge of Quality Control for the NW Operations of Canandaigua Wine Company, and thus work with Covey Run and Ste. Chapelle wineries as well. Blending wines for the brand Alice White takes me to Australia on a regular basis.

EUGENE W. NESTER: I graduated from Cornell University with a B.S. degree and a major in Bacteriology. After spending two years in the U.S. Army, mostly at Fort Detrick in Maryland, I entered the graduate program in Microbiology at Western Reserve University, College of Medicine. There I worked with John Spizizen on the biochemistry of the interaction between bacteriophage T2 and *E. coli.* While I was in his laboratory, Spizizen discovered DNA mediated transformation in *Bacillus subtilis,* a significant accomplishment, which made possible a genetic approach to many interesting physiological processes. Following my Ph.D. degree, granted in 1959, I became a postdoctoral trainee with Joshua Lederberg who had just assumed the Chair of Genetics in the School of Medicine at Stanford University. Because I came from the Spizizen lab, Josh suggested that I develop a DNA-mediated transformation system in *E. coli.* In the course of these unsuccessful studies, I became interested in the mechanism by which DNA is taken up by *B. subtilis* in DNA-mediated transformation. I also identified and mapped several genes of aromatic and histidine biosynthesis in *B. subtilis.* This led to a biochemical-genetic analysis of aromatic acid synthesis in this organism.

Upon coming to the University of Washington as a Assistant Professor in the Department of Microbiology, I continued to study aromatic acid synthesis and competence in *B. subtilis* for almost 10 years before deciding that studying the transformation of plants by *Agrobacterium* was more exciting and challenging than investigating the transformation of *B. subtilis.* I successfully justified this change in direction to funding agencies by pointing out that my studies still involved the genetic transformation of cells and that the formation of plant tumors involved aromatic acid (auxin) synthesis. I have never regretted my decision to switch my field of research.

One of the most enjoyable aspects derived from my studies on crown gall have been the number of truly outstanding graduate students and post doctoral trainees who have passed through the lab. Many of their contributions are included in this volume.

People who helped guide my career include John Spizizen, my Ph.D. mentor, and Charley Yanofsky, who taught me the importance of genetics in analyzing biological systems, and the importance of maintaining a focus in research; Josh Lederberg, who taught me to think "outside the box." In all of the studies on *Agrobacterium* carried out in Seattle, Milt Gordon has played an indispensable role. His deep knowledge and enthusiasm for science in general—and plant biology and chemistry—in particular contributed enormously to the success of our joint research program.

JOURNAL OF BACTERIOLOGY, July 1975, p. 255–264
Vol. 123, No. 1

Plasmid Required for Virulence of *Agrobacterium tumefaciens*

BRUCE WATSON, THOMAS C. CURRIER, MILTON P. GORDON, MARY-DELL CHILTON, AND EUGENE W. NESTER*

Departments of Biochemistry and Microbiology, University of Washington, Seattle, Washington 98195*

Received for publication 24 January 1975

The irreversible loss of crown gall-inducing ability of *Agrobacterium tumefaciens* strain C-58 during growth at 37 C is shown to be due to loss of a large plasmid (1.2×10^8 daltons). The gene responsible for this high rate of plasmid loss at elevated temperatures seems to be located on the plasmid. In addition, another spontaneous avirulent variant, *A. tumefaciens* strain IIBNV6, is shown to lack the virulence plasmid which its virulent sibling strain, IIBV7, possesses. Deoxyribonucleic acid reassociation measurements prove that the plasmid is eliminated, not integrated into the chromosome, in both of the avirulent derivatives. Transfer of virulence from donor strain C-58 to avirulent recipient strain A136 results from the transfer of a plasmid, which appears identical to the donor plasmid by deoxyribonucleic acid reassociation measurements. The transfer of virulence in another cross, K27 × A136, was also shown to result from the transfer of a large plasmid. These findings establish unequivocally that the large plasmid determines virulence. Two additional genetic determinants have been located on the virulence plasmid of *A. tumefaciens* strain C-58: the ability to utilize nopaline and sensitivity to a bacteriocin produced by strain 84. The latter trait can be exploited for selection of avirulent plasmid-free derivatives of strain C-58. The trait of nopaline utilization appears to be on the virulence plasmid also in strains IIBV7 and K27.

Crown gall, a malignancy affecting dicotyledonous plants, is incited by inoculation of a wound site with *Agrobacterium tumefaciens*. The mechanism by which living bacteria transform plant cells in the healing wound is unknown. Several lines of evidence suggest that a bacterial plasmid may play an important role in crown gall tumor induction. Zaenen et al. (25) reported that 11 virulent strains of *A. tumefaciens* contain a large plasmid, whereas 8 avirulent strains lack such plasmids. Hamilton and Fall (4) demonstrated that virulent *A. tumefaciens* strain C-58 could be converted efficiently to stable avirulent derivatives by passage at elevated temperature (37 C). As the findings of Van Larebeke et al. (23) suggest, this loss of virulence could be explained by loss of a plasmid, a phenomenon observed in other bacterial systems upon growth at high temperature (1, 5, 14, 17, 22). Kerr (8, 9) and R. Hamilton (personal communication) have recently demonstrated the transfer of virulence to avirulent *Agrobacterium* strains by an unidentified mechanism of genetic exchange. This mechanism might reasonably involve plasmid transfer. Further, Kerr and his collaborators (10, 18) have found a high correlation in Australian strains between the virulence trait and sensitivity to a bacteriocin produced by strain K84, whereas Lippincott et al. (12) observed a high correlation between virulence. These observations might imply that these loci are located on the virulence plasmid.

Techniques for identifying plasmid deoxyribonucleic acid (DNA) in *A. tumefaciens* (25) made it possible for us to demonstrate directly that plasmid loss and transfer were responsible for the loss (4) and transfer of virulence (8, 9) observed by earlier investigators. This paper presents evidence which demonstrates unequivocally that the large plasmid studied by Zaenen et al. (25) does indeed play a key role in determining virulence of strain C-58.

(This work was presented in part at the Second Annual John Innes Symposium, Norwich, England, 16 to 19 July 1974.)

MATERIALS AND METHODS

Culture media. Stock cultures were maintained at 4 C on 1.5% agar slants containing 0.8% nutrient broth (Difco). Cultures were routinely streaked from the stock cultures onto nutrient agar plates, and single colonies were used to inoculate liquid media. A 0.8% nutrient broth solution was routinely used as a complex liquid culture medium.

The synthetic medium most commonly used was AB medium (3). AB medium was sometimes supplemented with 10 ml of yeast extract-peptone broth (3) per liter when used for isotope incorporation. Biotype 2 organisms (7) were routinely grown in the mannitol glutamate medium of Roberts and Kerr (18). On occasion, cells were grown in a glutamate-minimal medium which consisted of the following (per liter): $CaCl_2$, 0.05 g; $MgSO_4 \cdot 7H_2O$, 0.2 g; NaCl, 0.2 g; NH_4NO_3, 2.7 g; $Fe(NO_3)_3$, 5.0 mg; $MnCl_2$, 1.0 mg; $ZnCl_2$, 0.5 mg; sodium citrate, 2.0 g; sodium glutamate, 2.0 g; $CoCl_2$, 0.5 mg; NaH_2PO_4, 0.37 g; and K_2HPO_4, 1.1 g. The carbon source, 0.25% glucose, was sterilized separately. This medium is modified from one described by Stonier (20).

Bacterial strains. *A. tumefaciens* strain C-58 was received from R. H. Hamilton, Pennsylvania State University; strains IIBV7 and IIBNV6, as well as B6, from A. Braun, Rockefeller University; strains T37 and 15955 from J. Lippincott, Northwestern University; and biotype 2 strains 27 (K27) and 84 (K84) from A. Kerr, Waite Agricultural Research Institute, Australia. Strains NT1, NT2, and BR1 were derived from strain C-58 in the course of these experiments.

Strain A136 is a derivative of strain NT1 which is resistant to 5 μg of rifampin and 50 μg of nalidixic acid per ml. The two independent mutations were selected in two steps, first on rifampin (5 μg/ml) containing AB minimal medium and then on AB medium supplemented with nalidixic acid (50 μg/ml).

Bacteriocin K84 and bacteriocin T37 sensitivity assay. A modification of the procedure used by Stonier (21) and Kerr and co-workers (10, 18) was used. An agar plate (AB or mannitol-glutamate medium) was inoculated with a loopful of the bacteriocin-producing strain, T37 or K84, so as to form a lawn (about 1 cm in diameter) in the center of the plate. Plates were incubated 4 days at room temperature to allow growth of the bacteriocin-producing strain and diffusion of the bacteriocin. The strain to be tested was either cultivated in liquid AB medium or on nutrient agar, AB, or mannitol-glutamate plates. Several colonies of the strain to be tested were suspended in 2 ml of 1/15 M phosphate buffer, pH 7, and a loopful was streaked near the lawn of the bacteriocin-producing strain. After 2 days of incubation at 29 C, sensitive strains revealed a zone of inhibition extending several centimeters from the center of the plate. Within this zone, resistant clones arose as single colonies. Resistant strains grew uniformly along the streak, exhibiting no inhibition.

Assay for virulence. *Kalanchoe daigremontiana* was obtained from R. A. Schilperoort, University of Leiden. Virulence assays were performed on kalanchoe leaves, which were wounded in four to six places with a sterile toothpick and inoculated either with a bacterial colony or a drop of yeast extract-peptone broth culture. Virulence was scored 10 to 20 days after inoculation. Virulence assays on sunflower plants (*Helianthus annuus* cv. Mammoth, W. Atlee Burpee Co., Riverside, Calif.) were performed using stem sites wounded and inoculated with toothpicks. *Nicotiana tabacum* var. *Xanthi* n.c. were inoculated in stem wounds with toothpicks or by syringe.

Genetic transfer experiments. These experiments were performed by the procedures described by Kerr (9). Cherry tomato plants (*Lycopersicon esculentum* Mpl. var. *cerasiforme* Hort) were used as host. The details of the cross are described below.

Radioactive labeling of bacterial cultures. Overnight cultures grown in AB medium were inoculated at a concentration of 5×10^7 cells/ml into AB medium containing 1 to 30 μCi of [^{3}H]thymidine per ml. Cultures were incubated at 29 C with aeration and harvested at a titer of about 5×10^8 cells/ml.

Preparation of bacterial lysates. Several methods of preparing lysates were used in the course of these experiments, the basic procedure being the neutral lysis technique of Zaenen et al. (25). On occasion, sodium dodecyl sulfate was substituted for Sarkosyl NL97 in the procedure. As an alternate method of lysis, the bacteria were treated with ampicillin prior to treatment with Sarkosyl and Pronase (S. K. Farrand, manuscript in preparation). All of these methods yielded plasmid-containing lysates which behaved identically. Prior to plasmid isolation on cesium chloride-ethidium bromide gradients or alkaline sucrose gradients, all lysates were sheared by agitation with a Vortex mixer for 1 min at top speed in a 16-mm (outer diameter) test tube. Very small volumes of lysate were sheared by pipetting in a 2-ml serological pipette 20 to 25 times.

Analytical cesium chloride-ethidium bromide density gradient centrifugation. The lysate in 4.8 ml of TES buffer [0.05 M tris(hydroxymethyl)aminomethane, 0.05 M NaCl, 5 mM ethylenediaminetetraacetate, pH 8.0] (27) was added to 4.90 g of CsCl, and to the solution was added 0.125 ml of ethidium bromide solution (10 mg/ml in TES buffer). The solution was centrifuged to equilibrium in polyallomer or cellulose nitrate tubes (50 to 70 h, Spinco type 65 rotor, 40,000 rpm, 14 C). Approximately 50 fractions were collected, and radioactivity was assayed by spotting aliquots on Whatman 3MM filters. Dry filters in 10 ml of toluene-liquifluor scintillant were assayed in a Beckman CPM-100 liquid scintillation counter.

Alkaline sucrose gradients. Sheared lysates (0.1 ml) were layered onto 5 ml of a 5 to 20% alkaline sucrose gradient (0.5 M NaCl, 0.3 M NaOH, 0.02 M ethylenediaminetetraacetate) in a cellulose nitrate tube (2 by ½ inch, ca. 5.1 by 1.3 cm). The gradients were centrifuged for 20 min at 35,000 rpm in a Spinco SW50.1 or SW39 rotor at 16 C in a Beckman model L centrifuge. Approximately 50 fractions were collected from the bottom of the tube directly onto Whatman 3MM filter disks which were assayed as described above. In some cases, the fractions were collected directly into vials, to which 1 ml of water and 10 ml of Triton cocktail (3) were added.

Preparative plasmid isolation. The details of this technique will be described in a separate publication (T. C. Currier and E. W. Nester, manuscript in preparation). Briefly, the procedure consists of shearing and alkaline denaturation of the lysate followed by neutralization, treatment with phenol, precipitation, and centrifugation of the DNA on cesium chloride-ethidium bromide gradients. Gradients containing labeled plasmid were fractionated from the bottom, and radioactivity was assayed by one of the methods described above. Gradients containing unlabeled plasmid was illuminated with 360-nm-wavelength light to visualize the fluorescent plasmid-ethidium bromide complex. This band was removed from the tube by piercing with a syringe (18-gauge needle) 1 to 2 mm below the band and withdrawing 1 ml of solution.

Ethidium bromide was removed by four extractions with isopropyl alcohol saturated with 6 M NaCl and 0.6 M trisodium citrate. Plasmid DNA was dialyzed versus 0.15 M NaCl-0.015 M trisodium citrate-1 mM tris(hydroxymethyl)aminomethane-1 mM ethylenediaminetetraacetate, pH 8. After treatments with ribonuclease and Pronase B, the DNA was extracted with phenol and subjected to gel filtration (Sephadex G25 coarse, 1 by 23 cm column).

Electron microscopy. Grids of plasmid DNA from cesium chloride-ethidium bromide gradients were prepared by the Kleinschmidt procedure (11), using preparations which had been stored at 4 C for about a week to allow single-strand nicking. A JEOL electron microscope was used to visualize the DNA. Photographs were taken, and projections were traced on paper. Silk suture was laid on the tracings to allow determination of the contour length.

DNA isolation. DNA was isolated from bacterial lysates by a modification of the Marmur procedure (13).

DNA renaturation kinetics. All DNA samples were sheared with the French pressure cell at 12,000 lb/in^2 to give a double-strand molecular weight of approximately 1×10^6. The separation of single- from double-stranded DNA was performed on hydroxylapatite as previously described (3).

Isolation of [^{3}H]octopine [N^2-(D-1-carboxyethyl)-L-arginine]. White Burley tobacco tumor line A6Sch, a gift of R. A. Schilperoort, University of Leiden, was labeled with 250 μCi of [^{3}H]arginine and cultivated for 4 weeks as described (6). Tumor tissue (48 g) was harvested and ground in an ice-cold mortar with 48 ml of 95% ethanol, and the homogenate was dialyzed versus 500 ml of 50% ethanol overnight at 5 C. The dialysate was applied to an 8-ml column of Bio-Rad AG50W-X8, 50 to 100 mesh, H^+ form, packed in 50% ethanol. All ninhydrin-positive material was retained. After washes with 200 ml of 50% ethanol and 20 ml of distilled water, the column was eluted with 2 M NH_4OH (100 ml), and the eluate, containing amino acids including octopine, was lyophilized and redissolved in 2 ml of water. Half of this material was subjected to preparative electrophoresis on a full sheet of Whatman 3MM paper (in a buffer consisting of pyridine-acetic acid-water, 100:8:2,5000, pH 6.1; 1 h; 2,000 V). The octopine region, identified by streaking phenanthrene-quinone reagent (23) along one edge of the paper, was near the origin on the cathode side, well removed from arginine, near the cathode end of the sheet. Labeled octopine was eluted, lyophilized, and redissolved in 1 ml of water. More than 90% of the radioactivity in this material appeared to be in octopine, as judged by electrophoretic mobility at pH 6.1 and 2, as well as paper chromatography in butanol-acetic acid-water (12:3:5) and in methyl ethyl ketone–2-butanol–*t*-butanol–diethylamine–water (40:20:20:1:25). No radioactive arginine was detectable in this octopine fraction. The yield of labeled octopine, determined by the colorimetric guanidine assay previously described (6), was 0.0167 μmol per g of fresh tumor tissue. The specific activity, determined by diluting an aliquot into 1 ml of water and adding to 10 ml of Triton cocktail (3), was 1.2×10^7 counts per min/μmol.

Isolation of [^{3}H]nopaline [N^2-(1,3-dicarboxypropyl)-L-arginine]. A Xanthi tobacco tumor line, which was determined previously to be a high nopaline producer (6), was labeled as described above. After 3 weeks, tumor tissue (13 g) was ground with 13 ml of 95% ethanol, and the homogenate was centrifuged (10,000 × *g* 10 min). The clear supernate was subjected to gel filtration (Sephadex G25 coarse) in 50% ethanol to remove high-molecular-weight material, and the amino acid fraction was isolated by ion-exchange batch adsorption and elution as described above. A sample (40%) was subjected to preparative electrophoresis as described above, the nopaline region (guanidine-positive zone on the anode side) was cut out and eluted, the eluate was lyophilized, and the residue was redissolved in 1 ml of water. By the four criteria described for octopine, approximately 90% of the radioactivity in this fraction was found identical in behavior to both synthetic nopaline (a gift of W. Zdybak and W. S. Chilton, University of Washington) and natural nopaline (purified from the same tumor line by preparative electrophoresis). The yield of labeled nopaline was 1.14 μmol per g of fresh tumor tissue. The specific activity; determined as above, was 9.5×10^5 counts per min/μmol.

Chemicals. [*methyl*-^{3}H]thymidine (50 to 60 Ci/mol), L-arginine (10.4 Ci/mM), and liquifluor were purchased from New England Nuclear Corp. Calbiochem was the source of nalidixic acid and Pronase B. The latter enzyme was made up at a concentration of 5 mg/ml in 0.05 M tris(hydroxymethyl)aminomethane-0.02 M ethylenediaminetetraacetate (pH 8.0) and incubated at 37 C for 90 min prior to use. Lysozyme was a 2×-crystallized product of Worthington Biochemical Corp. Sarkosyl NL97 was obtained from Geigy Chemical Co., and rifampin from Mann Laboratories. Ampicillin was a product of Bristol Laboratories. Octopine and ethidium bromide were purchased from Sigma Chemical Co. Fresh solutions of the latter were prepared immediately prior to use. Hydroxylapatite was a product of Bio-Rad. Cesium chloride (radiotracer grade) was purchased from Harshaw Chemical Co. All other chemicals were commercial products of the highest purity available and were not purified further.

RESULTS

Isolation of avirulent derivatives of strain C-58. Avirulent derivatives of *A. tumefaciens* strain C-58 were isolated by the technique of

Hamilton and Fall (4). This strain was grown confluently on nutrient agar for 15 days at 37 C, with transfers every 3 days. After this time, a loopful of the culture was streaked onto nutrient agar for single-colony isolation, and the plate was incubated at 29 C. Six colonies chosen at random were recloned two additional times at 29 C on nutrient agar. The parent C-58 strain is 3-ketolactose positive (2) and induces tumors on kalanchoe. All six heat-treated isolates are 3-ketolactose positive and do not induce tumors on kalanchoe. Two of these avirulent isolates, NT1 and NT2, were chosen for further study.

Covalently closed circular DNA in strain C-58 and its absence in avirulent derivatives. Lysates of strains C-58, NT1, and NT2 were analyzed for the presence of covalently closed circular DNA species by centrifugation in cesium chloride-ethidium bromide gradients. Confirming the findings of Van Larebeke et al. (23), we detected a small, higher-density peak of covalently closed circular DNA in the case of C-58, but not NT1 (Fig. 1) or NT2 (data not shown). The predominant species of DNA in the dense peak from strain C-58 is a large circular DNA molecule of contour length 57.9 μm, determined from nine electron photomicrograph tracings. This corresponds to a molecular weight of 120×10^6 (11). A few small circular DNA molecules comprising only a few percent of the total population were also observed. The contour length of one was found to be 8.55 μm, corresponding to a molecular weight of 17.7×10^6. Plasmids are known occasionally to be comprised of smaller subunits which under certain conditions segregate from the larger parent plasmid (15). It is thus possible that these smaller plasmids are cast-off components of the large plasmid.

Centrifugation of lysates on alkaline sucrose gradients again clearly demonstrates the presence of a rapidly sedimenting radioactive peak attributable to plasmid DNA in the lysate of strain C-58 (Fig. 2). None is observed in lysates of strains NT1 (Fig. 2) or NT2 (data not shown). All of these findings taken together demonstrate that the loss of virulence in strain C-58 is associated with the loss of the covalently closed circular form of the plasmid.

Absence of plasmid DNA sequences in strains NT1 and NT2. The absence of covalently closed circular DNA in strains NT1 and NT2 could be due either to elimination of the plasmid from the cell or to integration of the plasmid into the chromosomal DNA. Examples of both situations have been described in *Escherichia coli* (5). To distinguish between these possibilities, the total DNA of NT1 and NT2 was examined for DNA sequences homologous to plasmid DNA. C-58 plasmid DNA alone renatures with ideal second-order kinetics (Fig. 3) and has $C_0t_{1/2}$ = 0.10, close to the expected value for its genome size of 1.2×10^8 daltons. The addition of 820 μg of total C-58 DNA per ml decreases the observed $P_0t_{1/2}$ (3) value of 2.9 μg of labeled plasmid DNA per ml by a factor of 10. The unlabeled DNA has therefore increased the real concentration of plasmid DNA by 9×2.9 = 26.1 μg/ml, per 820 μg of total C-58 DNA added per ml. Plasmid DNA therefore constitutes 3.2% of the total C-58 DNA, similar to the value observed in another *A. tumefaciens* strain (M. D. Chilton, S. K. Farrand, F. C. Eden, T. C. Currier, A. J. Bendich, M. P. Gordon, and E. W. Nester, *In* R. Markham, D. Davies, D. Hopwood, and R. Horne (ed.), *Modification of Information Content of Plant Cells*, in press).

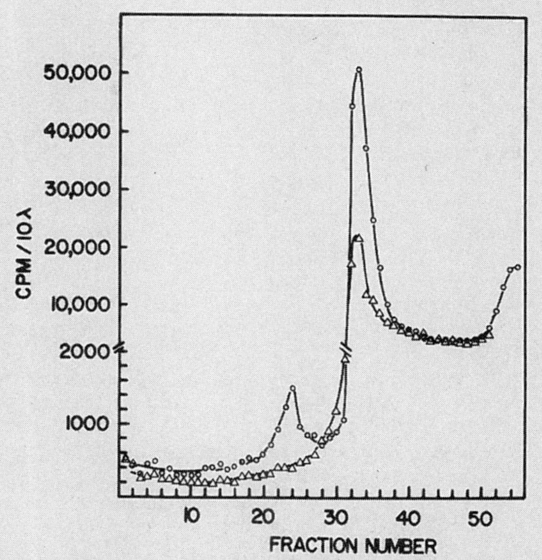

FIG. 1. *Cesium chloride-ethidium bromide gradients of DNA of virulent and avirulent strains. Five-milliliter cultures of C-58 and NT1 were grown in glutamate-minimal medium supplemented with 50 µg of arginine per ml, 0.01% Casamino Acids, and 0.2 µCi of [³H]thymidine. The cells were lysed by the penicillin procedure, and the lysate was adjusted to density in cesium chloride-ethidium bromide as described. Gradients were established, in nitrocellulose tubes, by centrifugation for 70 h at 35,000 rpm in a Beckman type 65 rotor at 15 C. The tubes were punctured, and 50 fractions were collected. Aliquots (10 µl) were spotted on Whatman 3 MM filter paper, and the samples were counted in 10 ml of toluene-liquifluor scintillant.* (O) *Strain C-58;* (Δ) *strain NT1.*

Unlabeled total DNA from strains NT1 and NT2 does not affect the rate of renaturation of labeled plasmid DNA (Fig. 3). Thus, these bacterial strains do not contain any detectable

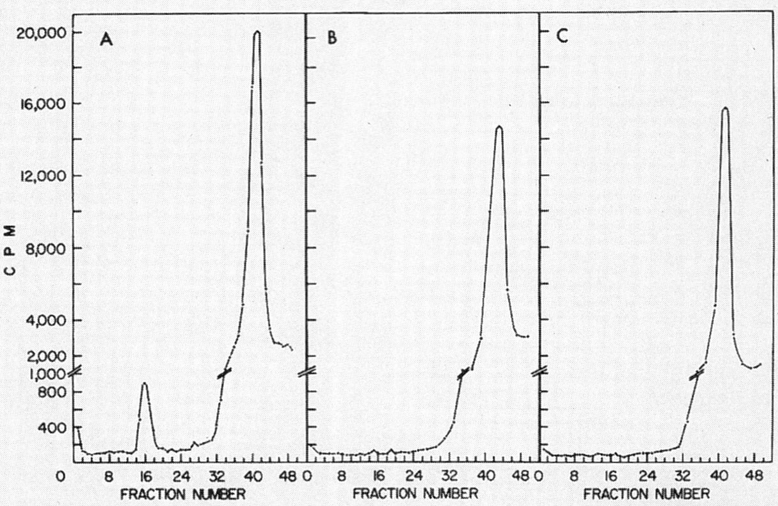

FIG. 2. *Alkaline sucrose gradient profiles of DNA from virulent and avirulent strains. One-milliliter cultures of strains C-58, BR1, and NT1 were lysed with Sarkosyl, and 0.1 ml was layered on a 5 to 20% gradient of alkaline sucrose. The gradients were centrifuged in an SW39 rotor at 35,000 rpm for 20 min at 14 C. The bottom of the tube was punctured, and fractions were collected on Whatman 3MM filter paper prior to counting in a toluene-liquifluor scintillant. (A) Strain C-58; (B) strain NT1; (C) strain BR1.*

plasmid DNA sequences. We conclude that avirulent strains NT1 and NT2 have eliminated the plasmid. These data further demonstrate that chromosomal DNA of C-58 has no detectable nucleotide sequence homology with C-58 plasmid DNA.

Restoration of virulence by plasmid transfer. Kerr (8, 9) has shown that the virulence trait can be transferred from virulent donor bacteria to avirulent recipients in a tomato crown gall tumor. The following experiment demonstrates that this transfer of virulence involves the transfer of the large plasmid from donor to recipient bacteria.

Strain A136, a drug-resistant derivatives (see above) of strain NT1, was used as a recipient, whereas strain C-58 was employed as donor. A 4-week-old tomato plant was wounded and inoculated with the donor and 6 days later was reinoculated without wounding at the same site with the recipient bacteria. After 5 weeks, tumor tissue was excised, surface sterilized with 0.5% sodium hypochlorite, ground with sterile water in a sterile mortar, and plated on drug-containing medium selective for the recipient phenotype. Control platings of donor bacteria from tomato tumors lacking recipients gave no colonies on the counterselective medium.

Single colonies of the recipient phenotype were picked from counterselective plates and recloned twice on nutrient agar. Clones were then retested for recipient phenotype (drug resistance, 3-ketolactose production) and finally were tested for virulence on kalanchoe. Of 80 clones of recipient phenotype tested for virulence, 3 were virulent and had received a plasmid from the donor strain C-58.

Two converted strains from this cross were investigated further (Table 1). Both strains contained plasmid, as detected in labeled lysates by sedimentation in alkaline sucrose gradients and/or by centrifugation in cesium chloride-ethidium bromide density gradients. The presence of circular DNA was verified in strain A174 by electron microscopy. To confirm that the plasmid in the recipient cells actually origi-

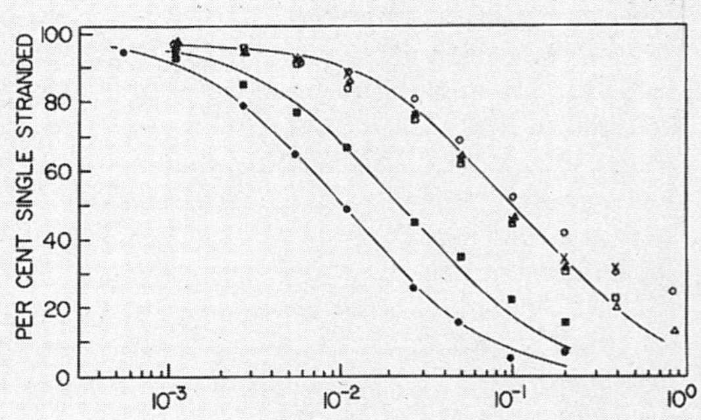

FIG. 3. *Renaturation studies of plasmid DNA from A. tumefaciens strain C-58. DNA was prepared, denatured, and renatured as described. ³H-labeled C-58 plasmid DNA (2.9 μg/ml) was renatured alone (○) or in the presence of 820 μg of C-58 bacterial DNA per ml (●), 700 μg of NT1 bacterial DNA per ml (×), 850 μg of NT2 bacterial DNA per ml (□), 530 μg of BR1 bacterial DNA per ml, (△), and approximately 20 μg of plasmid DNA from strain A174 per ml (■).*

TABLE 1. *Characteristics of parental and converted strains*

Strain	Resistance to 5 μg of rifampin and 50 μg of nalidixic acid per ml	Growth on Schroth agar	3-Ketolactose production	Virulence	Presence of plasmid	Sensitivity to bacteriocin K84
C-58	−	+	+	+[a, b]	+[c, d]	+
K27	±	−	−	+	+[d]	+
A136	+	+	+	−[a, b]	−[c]	−
A174[e]	+	NT	+	+[a]	+[c, d, g]	+
A175[e]	+	NT	+	+[a]	+[c]	+
A177[h]	+	+	+	+[a, b]	+[d]	+
A178[h]	+	+	+	+[a]	+[c, d, g]	+
A181[h]	+	+	+	+[a, b]	+[c, d, g]	+
A197[h]	+	+	+	+[a, b]	+[c, d, g]	+

[a] Tested on kalanchoe.
[b] Tested on sunflower.
[c] Assayed by alkaline sucrose gradients.
[d] Assayed by cesium chloride-ethidium bromide gradient.
[e] Progeny of the cross between C-58 and A136.
[f] NT, Not tested.
[g] Visualized by electron microscopy.
[h] Progeny of the cross between K27 and A136.

nated in the donor bacteria, DNA homology with the donor plasmid was established by hybridization studies. Unlabeled plasmid DNA from converted strain A174 was added to labeled donor plasmid DNA from strain C-58, and the kinetics of renaturation were compared to those of labeled plasmid DNA alone (Fig. 3). The plasmid DNA from the converted strain accelerated the renaturation of the donor plasmid, and the renaturation data fit theoretical curves for second-order reactions. These data indicate that in this case the conversion to virulence is mediated by transfer of the entire plasmid. However, neither electron microscopy nor DNA renaturation kinetic analysis would detect loss of less than 5% of the plasmid genome.

One other cross was performed, using strain K27 as the donor and strain A136 as the recipient. All details of the mating procedure were the same as described above except that the primary selective medium was Schroth agar (21). Clones were tested for their drug resistance, ability to produce 3-ketolactose, growth on Schroth agar (21), and finally their virulence. Nineteen of the 92 colonies which originally grew on Schroth agar proved to be converted by these selective criteria. Lysates were prepared from four such clones. All four were shown to have large plasmids by either alkaline sucrose gradients, centrifugation in gradients of cesium chloride-ethidium bromide, or electron microscopy (Table 1).

Bacteriocin 84 sensitivity as a plasmid-borne trait. A number of strains of *A. tumefaciens* are known to produce diffusible bacteriocins which kill or prevent the growth of other *A. tumefaciens* strains (10, 18, 21). Kerr and his collaborators (10, 18) found that strain K84 produces a bacteriocin to which a number of virulent strains of *A. tumefaciens* are sensitive. Further, when mutants were selected which were esistant to the bacteriocin of strain 84, they had lost their tumor-inducing ability. We reasoned that sensitivity to bacteriocin 84 might therefore be a trait carried on the virulence plasmid.

Accordingly, strains C-58, NT1, and NT2 were tested for sensitivity to bacteriocin 84. Consitent with our supposition, strain C-58 was sensitive to bacteriocin 84, whereas NT1 and NT2 were resistant. A bacteriocin 84-resistant derivative of C-58, BR1, was found to be avirulent on kalanchoe and 3-ketolactose positive. As suspected, it does not contain plasmid demonstrable by alkaline sucrose gradient centrifugation (Fig. 2), nor are plasmid DNA sequences detectable in this strain by renaturation kinetic analysis (Fig. 3).

The above results strongly suggest that genetic information determining sensitivity to bacteriocin 84 is carried on the C-58 virulence plasmid. To confirm genetic linkage between virulence and bacteriocin 84 sensitivity, independent clones of C-58 were selected for loss of each trait and then tested for loss of the other. We observed concomitant loss of the two traits in 20 out of 20 cases, 10 selected by growth at 37 C and 10 selected by resistance to bacteriocin of strain 84. Three of each set of ten avirulent isolates, chosen at random, were shown to be plasmid free by alkaline sucrose gradient analysis of labeled lysates (data not shown). These results establish convincingly that bacteriocin 84 sensitivity is a trait carried on the C-58 virulence plasmid. In further support of this conclusion, we found that transfer of virulence from C-58 to A136 was accompanied by transfer of bacteriocin 84 sensitivity (Table 1). This was also true of crosses involving *A. tumefaciens* K27 as donor as reported by Roberts and Kerr (18). This suggests that bacteriocin 84 sensitivity is also borne by the K27 virulence plasmid.

Utilization of nopaline as a plasmid-borne trait. Lippincott and his collaborators (12) have reported a high correlation within the genus *Agrobacterium* between virulence and the ability to metabolize the unusual amino acids octopine or nopaline or both. This correlation suggests that genes conferring these abilities might be situated on virulence plasmids. Consistent with this view, R. A. Schilperoort, J. J. M. Dons, and H. Ras (*In* R. Markham, D. Davies, D. Hopwood, and R. Horne (ed.), *Modification of Information Content of Plant Cells*, in press) have reported that an avirulent derivative of strain C-58 which lacks the covalently closed circular form of the plasmid has also lost the ability to grow on nopaline as a sole source of nitrogen. We have examined the parents and progeny described in the foregoing genetic crosses for the ability to utilize labeled octopine and nopaline (Table 2). Confirming the results of Schilperoort et al. (in press), we find that the avirulent strain NT1 cannot metabolize nopaline, whereas its virulent parent, C-58, does so readily. Neither strain utilizes octopine significantly. The very slow utilization of octopine by C-58 has been observed in most nopaline-utilizing strains thus far examined (unpublished data). Converted strains which have received the C-58 virulence plasmid also acquire the ability to utilize nopaline (Table 2). These data indicate that a determinant for nopaline utilization is borne by the C-58 virulence plasmid.

TABLE 2. *Octopine and nopaline utilization traits of virulent, avirulent and converted strains*[a]

Strain	[^{3}H]octopine incorporated (counts/min)	[^{3}H]nopaline incorporated (counts/min)
C-58	276	1,960
NT1	54	68
K27	2,261	1,747
A177	100	2,218
A174	130	1,816
A178	67	1,690
IIBV7	52	1,821
IIBNV6	61	46

[a] Overnight cultures in mannitol-glutamate medium were diluted fivefold into fresh medium and 5 μl of labeled octopine or nopaline (10,000 counts/min) was added per ml of culture. After 4 to 6 h of growth with aeration at 30 C, a 0.2-ml sample was pipetted into 2 ml of ice-cold trichloroacetic acid, and insoluble material was collected on Whatman GF/C filters. Dried filters were assayed for radioactivity as described.

Strain K27, which utilizes both octopine and nopaline, transfers only the nopaline utilization trait to progeny which receive its virulence plasmid (Table 1). This finding suggests that a transferable K27 virulence plasmid determines nopaline but not octopine utilization.

Frequency of plasmidless C-58 in a population. Assuming that bacteriocin 84 resistance can be equated to loss of the virulence plasmid, one can use this trait to measure the frequency of plasmidless bacteria in a C-58 colony. Individual colonies of strain C-58 grown on nutrient agar at 29 C were suspended and plated on bacteriocin 84 plates and on nutrient agar. The frequency of bacteriocin 84-resistant (plasmidless) bacteria in both 3- and 8-day-old colonies was 1.1×10^{-4} to 1.2×10^{-4}. This estimate is an upper limit because any slight growth of sensitive bacteria on the bacteriocin-containing plates could produce additional plasmidless bacteria.

Relative growth rate of C-58 and NT1. In an effort to assess whether growth at elevated temperatures confers a selective advantage on plasmidless organisms, we investigated the relative growth rates of strain C-58 and its plasmidless derivatives at various temperatures. At 29 and 32 C, NT1, NT2, and BR1 all produce colonies of similar size on nutrient agar. However, at 37 C all three avirulent strains form distinctly larger colonies than does virulent strain C-58. Growth of all four strains is slower at 37 than at 29 or 32 C.

We were unable to detect a difference in generation time between C-58 and NT1 growing in aerated nutrient broth cultures. However, a small difference would not be evident in the few generations observable by this method.

Location of gene responsible for heat sensitivity in C-58. The most reasonable explanation for the phenomenon of plasmid curing in strain C-58 is that the plasmid either does not replicate in complete synchrony with the chromosomal DNA or is not partitioned efficiently to daughter cells during cell division at elevated temperatures. Since it is possible to transfer the plasmid from a wild-type strain into NT1, it is possible to determine whether the gene responsible for curing is plasmid or chromosome borne in C-58. To this end, two virulent converted clones from the cross K27 × A136 were streaked onto nutrient agar, and five colonies from each clone were subjected to the 37 C curing regimen described above for strain C-58. Controls were passaged similarly on plates at 29 C. At the end of 16 days, a single colony from each of the 10 original colonies was isolated and recloned at 29 C. For virulence assay, the cells from each colony were inoculated into 10-day-old sunflower plants. All 10 colonies had retained their virulence. As a control, an identical experiment was done using five virulent converted clones derived from the cross C-58 × A136. In this case, colonies from each of the five clones had lost virulence. Control clones passaged at 29 C retained virulence.

These results indicate that the introduction of the plasmid of strain K27 into a background of strain C-58 results in a virulent strain which is not heat curable. The genetic trait responsible for loss of plasmid upon growth at elevated temperature therefore appears to be situated on the C-58 plasmid, and absent from the K27 plasmid.

Plasmid-borne virulence in another strain. Variants IIBV7 and IIBNV6, isolated by Armin Braun in the course of transferring strain IIB, appear to be an isogenic pair of strains differing only by the virulence trait. Strain IIBV7 is virulent on kalanchoe, tobacco, and sunflower, whereas strain IIBNV6 is avirulent on all three plants. Both strains are ketolactose positive, sensitive to a bacteriocin produced by *A. tumefaciens* (18), and resistant to bacteriocin from strain K84. To confirm that these strains are isogenic, the extent of DNA homology between the two was measured. A trace of labeled DNA from strain IIBV7 was allowed to renature alone and in the presence of a high concentration of homologous DNA or IIBNV6 DNA. Both unlabeled DNA preparations accelerate the renaturation of labeled probe DNA, and the kinetics are identical for at least 95% of the reaction. The duplexes formed in this experiment were subjected to thermal denaturation, and the melting profiles (Fig. 4) of homo- and heteroduplexes were identical. Thus IIBV7 and IIBNV6 are indeed closely related strains, although differences in a few percent of the DNA sequences could not have been detected by these techniques.

To determine whether IIBV7 and IIBNV6 differed by the presence of a plasmid, lysates of labeled cultures were sedimented through alkaline sucrose gradients. Strain IIBV7 exhibited a fast-sedimenting peak of radioactivity typical of covalently closed circular DNA, whereas strain IIBNV6 did not (data not shown). The IIBV7 lysate also exhibited a dense band in a cesium chloride-ethidium bromide gradient, and the presence of circular DNA molecules in this peak was confirmed by electron microscopy. The latter findings confirm the data of Van Larebeke et al (23).

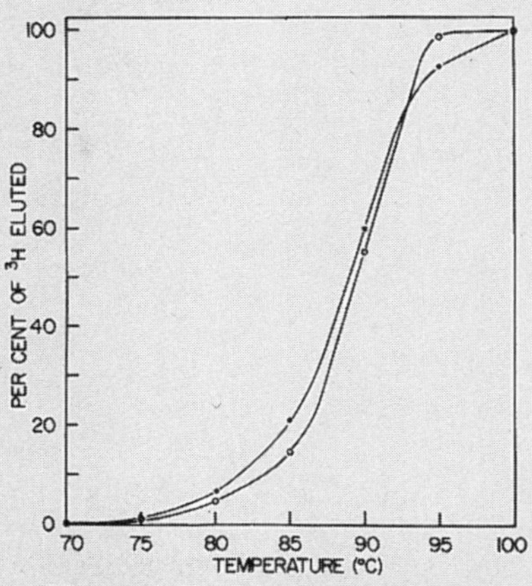

FIG. 4. *Thermal dissociation profiles of homo- and heterduplexes formed with ^{3}H-labeled NA of A. tumefaciens strain IIBV7. Duplexes were formed between 11.5 μg of ^{3}H-labeled IIBNV6 DNA per ml by incubation (0.15 M phosphate buffer, pH 6.8, 70 C) to C_0t of 1.24 × 10^1. Duplexes (a total of 300 counts/min) diluted into 0.15 M phosphate buffer were applied to hydroxylapatite columns at 60 C and washed with this buffer at 60, 65, and 70 C to elute single-stranded DNA (less than 10% of applied radioactivity). Melting profile of duplexes was determined by washing columns with 0.15 M phosphate buffer (4 ml) at 5 C temperature increments up to 100 C. Finally, columns were washed with 2 ml of 0.3 M phosphate buffer at 100 C. Samples were assayed for radioactivity in Triton cocktail as previously described. Symbols:* ○, *IIBV7 duplexes;* ●, *IIBNV6 duplexes.*

By DNA renaturation kinetic analysis, we found that the DNA sequences of the IIBV7 virulence plasmid are missing in the DNA of avirulent strain IIBNV6. A trace of labeled IIBV7 plasmid DNA was allowed to renature alone or in the presence of high concentrations of total DNA from IIBV7 or IIBNV6 bacteria (Fig. 5). Strain IIBV7 DNA increased the rate of renaturation of labeled plasmid, whereas DNA of strain IIBNV6 had no effect. We conclude that the virulence plasmid has been eliminated in the latter strain, resulting in loss of tumor-inducing ability.

Nopaline utilization as a plasmid marker in IIBV7. Confirming the results of Lippincott

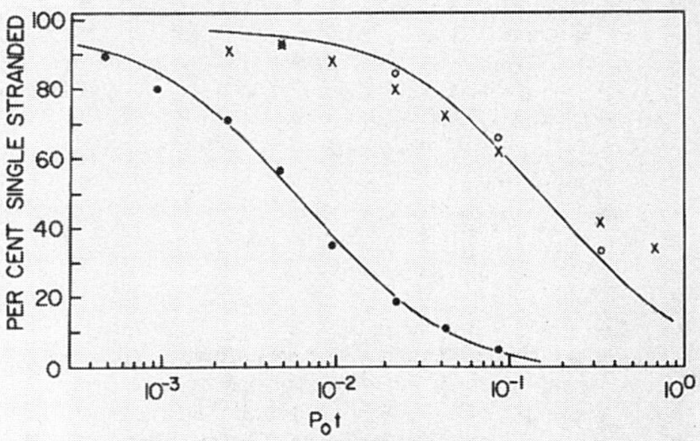

FIG. 5. *Renaturation studies of plasmid DNA from A. tumefaciens strain IIBV7. DNA was prepared, denatured, and renatured as described. ³H-labeled IIBV7 plasmid DNA (1.5 μg/ml) was renatured alone (○) or in the presence of 710 μg of IIBV7 bacterial DNA per ml (●) or 920 μg of IIBNV6 bacterial DNA per ml (×).*

et al. (12) and Petit et al. (16), we find that strain IIBV7 utilizes nopaline, whereas its avirulent plasmidless derivative IIBNV6 does not (Table 2). This observation suggests that nopaline utilization is a trait on the virulence plasmid of strain IIBV7. Neither strain IIBV7 nor IIBNV6 utilizes octopine significantly.

Growth rate of IIBV7 and IIBNV6 at elevated temperature. Strains IIBV7 and IIBNV6 both give rise to colonies on nutrient agar after 2 days of incubation at 29 C. However, at 37 C IIBV7 yields colonies in 2 to 3 days, whereas IIBNV6 produces colonies only after a week or more of incubation. Thus, proficient growth at 37 C may be a plasmid-associated trait in this pair of strains, quite in contrast to strain C-58 and its avirulent derivatives. Thus, there appears to be no correlation between the presence of a large plasmid and the temperature optimum for cell growth.

Attempts to eliminate plasmids from other strains by growth at 37 C. Strains B6, 15955, and T37 all grow at 37 C but retain virulence after 15 days of growth at this temperature. Another strain examined, K27, does not grow at 37 C on nutrient agar.

DISCUSSION

The results presented in this paper provide an understanding at the molecular level of several intriguing observations on *A. tumefaciens* published in recent years. The production of avirulent derivatives of strain C-58 upon growth at 37 C (4) and also upon selection for resistance to bacteriocin 84 (10, 18) are due to the loss of a large plasmid. Further, the transfer of the virulence trait (8, 9) to avirulent strains results from transfer of this virulence plasmid. These results demonstrate unequivocally that the plasmid is essential to the expression of virulence.

The strong correlation between the virulence of a strain and its ability to catabolize octopine or nopaline or both (12) may have its basis in the localization of such genes on the virulence plasmid. Experimental evidence certainly is consistent with this view in the case of nopaline. In two nopaline-utilizing strains examined in the present study, C-58 and IIBV7, this trait appears to be plasmid borne. The data which support this contention are strong, since the conclusion rests on the observation that the loss of a plasmid results in the loss of the character. In the case of C-58, this conclusion is further supported by the observation that the introduction of the virulence plasmid into the avirulent derivative NT1 results in the regaining of the ability to utilize nopaline. The observation that strain K27 can transfer its ability to utilize nopaline, but not octopine, to nonutilizing strains suggests that the gene(s) for nopaline but not octopine utilization is present on the transferred virulence plasmid in this strain. However, this interpretation must be viewed with some caution, since it is not clear whether chromosomal genes as well as plasmid genes are transferred. Further, in that cross, it is not certain whether the entire virulence plasmid is transferred. For these reasons, the identification of gene loci on the virulence plasmid is far more conclusive when a correlation exists between the loss of the plasmid and the loss of a marker.

It is not known how many genes are involved in the utilization of octopine and nopaline. It is possible that specific permeases as well as the specific degradative enzymes are required for the breakdown of these unusual amino acids. If more than one gene is involved, then it is possible that one is located on the virulence plasmid, and the other on the chromosome. Only if both are present in the same strain can the unusual amino acids be degraded.

The search for gene loci located on the virulence plasmid has not been very successful. We have screened strains C-58 and NT1, as well as IIBV7 and IIBNV6, for variations in their resistance to 19 different antibiotics and 12 inhibitory metals and there were no differences. In addition, all of these strains were identical in the ability to grow on 24 different carbohydrate as a sole carbon source (R. Levin, unpublished observations). So far, only the genes concerned with virulence, sensitivity to the bacteriocin produced by strain K84, and nopaline utilization can be assigned to the same virulence plasmid. It seems clear that it will be necessary to identify and characterize many more loci before the plasmid becomes amenable to a genetic and biochemical analysis. Such studies should also provide insight into the role of the plasmid in tumor formation.

ACKNOWLEDGMENTS

The outstanding technical assistance of Alice Montoya is gratefully acknowledged. We thank Stanley Falkow for extensive assistance with electron microscopy as well as numerous illuminating discussions.

B. W. was supported by National Science Foundation fellowship 807681 and by Public Health Service grant GM-0052 from the National Institute of General Medical Sciences. T. C. was a predoctoral trainee supported by Public Health Service grant 5T01GM00511 from the National Institute of General Medical Sciences. These studies were also supported by Public Health Service grant CA-13015 from the National Cancer Institute.

LITERATURE CITED

1. Asheshov, E. H. 1966. Loss of antibiotic resistance in *Staphylococcus aureus* resulting from growth at high temperature. J. Gen. Microbiol. **42**:402–410.
2. Bernaerts, M. J., and J. Deley. 1963. A biochemical test for crown gall bacteria. Nature (London) **197**:406–407.
3. Chilton, M. D., T. C. Currier, S. K. Farrand, A. J. Bendich, M. P. Gordon, and E. W. Nester. 1974. *Agrobacterium tumefaciens* DNA and PS8 bacteriophage DNA not detected in crown gall tumors. Proc. Natl. Acad. Sci. U.S.A. **71**:3672–3676.
4. Hamilton, R. H., and M. Z. Fall. 1971. The loss of tumor-initiating ability in *Agrobacterium tumefaciens* by incubation at high temperature. Experentia **27**:229–230.
5. Jacob, F., S. Brenner, and F. Cuzin. 1963. On the regulation of DNA replication in bacteria. Cold Spring Harbor Symp. Quant. Biol. **28**:329–348.
6. Johnson, R., R. H. Guderian, F. Eden, M. D. Chilton, M. P. Gordon, and E. W. Nester. 1974. Detection and quantitation of octopine in normal plant tissue and in crown gall tumors. Proc. Natl. Acad. Sci. U.S.A. **71**:536–539.
7. Keane, P. J., A. Kerr, and P. B. New. 1970. Crown gall of stone fruit. II. Identification and nomenclature of *Agrobacterium* isolates. Aust. J. Biol. Sci. **23**:585–595.
8. Kerr. A. 1969. Transfer of virulence between strains of *Agrobacterium*. Nature (London) **223**:1175–1176.
9. Kerr. A. 1971. Acquisition of virulence by non-pathogenic isolates of *Agrobacterium radiobacter*. Physiol. Plant Pathol. **1**:241–246.
10. Kerr, A., and K. Htay. 1974. Biological control of crown gall through bacteriocin production. Physiol. Plant Pathol. **4**:37–48.
11. Kleinschmidt, A. K., D. Lang, D. Jaeherts, and R. K. Zahn. 1963. Preparation and length measurements of the total deoxyribonucleic acid content of T2 bacteriophages. Biochim. Biophys. Acta **61**:857–867.
12. Lippincott, J. A., R. Beiderbeck, and B. B. Lippincott, 1973. Utilization of octopine and nopaline by *Agrobacterium*. J. Bacteriol. **116**:378–383.
13. Marmur, J. 1961. A procedure for isolation of deoxyribonucleic acid from microorganisms. J. Mol. Biol. **12**:468–487.
14. May, J. W., R. H. Houghton, and C. J. Perrett. 1964. The effect of growth at elevated temperatures on some heritable properties of *Staphylococcus aureus*. J. Gen. Microbiol. **37**:157–160.
15. Nisioka, T., M. Michiko, and R. Clowes. 1969. Composite circular forms of R-factor deoxyribonucleic acid molecules. J. Bacteriol. **97**:376–385.
16. Petit, A., S. Delhaye, J. Tempé, and G. Morel. 1970. Recherches sur les quanidines des tissues de crown gall. Mise en evidence d'une relation biochimique specifique entre les souches d'*Agro bacterium tumefaciens* et les tumeurs qu'elles induisent. Physiol. Veg. **8**:205–213.
17. Richmond, M. H. 1972. Plasmids and extrachromosomal genetics in *Staphylococcus aureus*, p. 159–186, *In* J. O. Cohen (ed.), The staphylococci. Wiley-Interscience, New York.
18. Roberts, W. P., and A. Kerr. 1974. Crown gall induction: serological reactions, isozyme patterns and sensitivity to mitomycin C and to bacteriocin, of pathogenic and non-pathogenic strains of *Agrobacterium radiobacter*. Physiol. Plant Pathol. **4**:81–91.
19. Schroth, M. N., J. P. Thompson, and D. C. Hildebrand. 1965. Isolation of *Agrobacterium tumefaciens-A. radiobacter* group from soil. Phytophatology **55**:645–647.
20. Stonier, T. 1956. Labeling crown gall bacteria with ^{32}P for radioautography. J. Bacteriol. **72**:259–268.
21. Stonier, T. 1960. *Agrobacterium tumefaciens* Conn. II. Production of an antibiotic substance. J. Bacteriol. **79**:889–898.
22. Terawaki, Y., T. Hisao, and T. Alchiba. 1967. Thermosensitive replication of a kanamycin resistance factor. J. Bacteriol. **94**:687–690.
23. Van Larebeke, N. G. Engler, M. Holsters, S. Van den Elsacker, I. Zaenen, R. A. Schilperoort, and J. Schell. 1974. Large plasmid in *Agrobacterium tumefaciens* essential for crown gall-inducing ability. Nature (London) **252**:169–170.
24. Yamada, S., and H. A. Itano. 1966. Phenanthrenequinone as an analytical reagent for arginine and other mono substituted quanidines. Biochim. Biophys. Acta **130**:538–540.
25. Zaenen, I., N. Van Larebeke, H. Teuchy, M. Van Montogu, and J. Schell. 1974. Supercoiled circular DNA in crown gall inducing *Agrobacterium* strains. J. Mol. Biol. **86**:109–127.

MY LIFE WITH PHYTOHORMONES

Milton P. Gordon
Department of Biochemistry, University of Washington
Seattle, Washington, United States
E-mail: sunnieg@u.washington.edu

Commentary on the paper:

Chilton, M.-D., Drummond, M. H., Merlo, D. J., Sciaky, D., Montoya, A. L., Gordon, M. P., and Nester, E. W. 1977. Stable incorporation of plasmid DNA into higher plant cells: The molecular basis of crown gall tumorigenesis. Cell 11:263–271.

My career in science began with the synthesis of phytohormones when I was a teenager at school and ended with the molecular explanation of abnormal growth due to over synthesis of these hormones. My work with crown gall started in 1971 when Eugene Nester, recognizing my interests and my facilities to work with plants, suggested we collaborate on studies of a "plant cancer" caused by *Agrobacterium tumefaciens.* Gene had read extensively on crown gall disease and believed that the field was ripe for a molecular approach. Intrigued by cancer studies since serving at Sloan Kettering, I agreed to combine this work with the plant nucleic acid research underway in my laboratory. From the literature, Gene and I learned that a large piece of bacterial DNA responsible for the crown gall tumor was supposed to have been incorporated into transformed plant tissue. There were also reports that bacterial genes, including an RNA polymerase, were expressed in the tumor. One group found an agrobacterial phage in the gall tissue. To make the situation more confusing, there was a widely quoted report that a "normal tissue" was available, which grew in the absence of phytohormones and produced an abnormal amino acid, octopine, which was characteristic of some crown gall tissues. At this point, we teamed up with a post-doctoral worker, Mary-Dell Chilton who was experienced in renaturation kinetics of DNA.

Our working hypothesis was that crown gall tumors contained bacterial DNA. At first, we relied on the laborious procedure of Cot curve analysis to detect foreign DNA in crown gall tissue. This procedure measured the kinetics of renaturation of a radioactive single-stranded DNA probe in the presence of a large excess of complementary strands of DNA isolated from the tissue under investigation. These experiments were laborious and time-consuming. We had to grow large amounts of aseptic gall tissue and prepare highly purified DNA, which was placed in small capillary tubes. I recall loading hundreds of capillary tubes with the viscous DNA by dropping them down a one-meter tube on to the desktop. The initial experiments to detect bacterial DNA in tumor tissue were negative (Chilton et al. 1974). We felt discouraged that we could not find the reported large pieces of foreign DNA in crown gall tissues as indicated in the literature. We then toyed with the idea that there was a progressive hierarchy of loss of phytohormone control, starting with "normal" tissue which required auxins and cytokinins, to crown gall which grew as hormone independent chaotically disorganised tumors. We considered the possibility that crown gall was due to an epigenetic phenomenon in which the exposure of plant tissue to the bacterium increased the expression of the endogenous cytokinin and auxin genes in the plant tissue.

At this time, Jeff Schell, Marc Van Montagu and Ivo Zaenen (Zaenen et al. 1974) discovered a large plasmid in tumorigenic strains of *Agrobacterium.* Based on this finding, our laboratory explored strain C58, which lost tumorigenicity upon heating. We found that this loss of tumorigenicity was associated with the loss of a large plasmid (Watson et al. 1975). The plasmid and tumorigenicity could be reacquired by a "Kerr" cross.

We then looked for the presence of large pieces of plasmid DNA in the crown gall tumors, but again, results were negative. Almost in desperation, we increased the sensitivity of our experiments, by cutting the plasmid into small fragments with the restriction enzyme *Sma*I which we purified in our lab. We then used the highly-labelled individual pieces of plasmid as probes to detect these fragments in tumor tissue. We were very excited when we observed that a specific small fraction of the plasmid increased the kinetics of renaturation. After several successful repeats of the experiment, we felt certain that this fragment of the plasmid was present in the crown gall tissue.

In order to insure that the tumor lines we were studying were lines derived from a single transformation event and contained no untransformed cells, I visited George Melcher's laboratory in Tübingen, Germany to clone the crown gall tissue. We are indebted to his laboratory for their hospitality and for demonstrating their cloning techniques from single cells. Maria Dolores Sacristan and L. Schilde-Renschler helped us to clone the tumors so that we had a number of lines derived from single cells. We also found that the so-called "normal" tissue that synthesized octopine contained T-DNA and was probably mislabeled sometime during its years in laboratory culture.

To those scientists who understood the technique, the Cot analysis provided unequivocal evidence for the presence of Ti-plasmid DNA in tumor tissue. However for many classical botanists, the evidence was not convincing. A "little wiggle" in the Cot plot had such profound implications that a more definite proof was demanded. It was soon forthcoming.

I was frequently asked when presenting our kinetic data to more classically inclined scientists: "Do you really believe you have foreign DNA in your plant just because the line wiggles a bit?" In 1980, Southern blotting was used in our laboratory (Thomashow et al. 1980) and in Gent, Belgium by Jeff Schell and Marc Van Montagu (Genetello et al. 1979); this technique provided direct visual evidence for the presence of bacterial plasmid DNA in crown gall tissue. These experiments were entirely convincing to those interested.

My life in research has made a complete cycle and I was back at my high school days. I worked with and made wonderful friends along the way.

The social impacts that resulted from these experiments, while still being debated, are having a profound effect on world agriculture.

References

Chilton, M.-D., Currier, T. C., Farrand, S. K., Bendich, A. J., Gordon, M. P., and Nester, E. W. 1979. *Agrobacterium tumefaciens* DNA and PS8 bacteriophage DNA not detected in crown gall tumors. Proc. Natl. Acad. Sci. U.S.A. 71:3672–3676.

Genetello, C., Hernalsteens, J.-P., Holsters, M., Seurinck, J., Silva, B., Van Vliet, F., and Villarroel, R. 1979. Interactions and DNA transfer between *Agrobacterium tumefaciens,* the Ti-plasmid and the plant host. Proc. R. Soc. Ser. B 204:251–266.

Thomashow, M. F., Nutter, R., Montoya, M. P., Gordon, M. P., and Nester, E. W. 1980. Integration and organization of Ti plasmid sequences in crown gall tumors. Cell 19:729–739.

Watson, B., Currier, T. C., Gordon, M. P., Chilton, M.-D., and Nester, E. W. 1975. Plasmid required for virulence of *Agrobacterium tumefaciens.* J. Bacteriol. 123:255–264.

Zaenen, I., Van Larebeke, N., Teuchy, H., Van Montagu, M., and Schell, J. 1974. Supercoiled DNA in crown gall inducing *Agrobacterium* strains. J. Mol. Biol. 86:109–127.

MILTON P. GORDON: As a teenager, I bought chemicals with money I earned and synthesized several phytohormones, including naphthalene acetic acid, derivatives of substituted phenoxy and naphthalene acetic and indole acetic acids, using the Fischer indole synthesis.

In 1947, I started my undergraduate studies at the University of Minnesota. After three years, I graduated summa cum laude with majors in mathematics and chemistry. In 1950, Herbert E. Carter, chair of Biochemistry at the University of Illinois, Urbana, offered me a fellowship that paid for my graduate work. I graduated in 1953 with a Ph.D. in Biochemistry and minors in organic chemistry and physiology.

Following a four-year job working on purine analogs with George Brow at the Sloan Kettering Institute, I joined Wendell Stanley at the University of California, Berkeley, in 1957 where I studied *Tobacco mosaic virus* (TMV) and worked independently. This experience introduced me to research with plants.

In 1959, I joined the Department of Biochemistry at the University of Washington, Seattle. My projects included the action of chemotherapeutic agents on the synthesis of TMV and other plant viruses. I designed a state-of-the-art growth-room with controlled temperature, lighting and purified air, allowing me to set up ideal conditions for plant propagation. The facility, one of only a few in the world, attracted the attention of visiting scientists and brought me into contact with Gene Nester and crown gall.

Cell, Vol. 11, 263-271, June 1977, Copyright © 1977 by MIT

Stable Incorporation of Plasmid DNA into Higher Plant Cells: the Molecular Basis of Crown Gall Tumorigenesis

Mary-Dell Chilton,* Martin H. Drummond,* Donald J. Merlo,* Daniela Sciaky,* Alice L. Montoya,* Milton P. Gordon† and Eugene W. Nester*
*Department of Microbiology and Immunology and
†Department of Biochemistry,
University of Washington,
Seattle, Washington 98195

Summary

Evidence is presented that crown gall tumors are caused by the incorporation of part of a virulence plasmid carried by the inciting bacterium, Agrobacterium tumefaciens. The rate of reassociation of labeled plasmid DNA was slightly accelerated in the presence of tobacco crown gall tumor DNA, but not normal tobacco DNA. Treatment of tumor DNA with DNAase abolished the acceleration. To determine whether all plasmid sequences are represented in tumor DNA, the labeled plasmid DNA was separated into specific fragments after digestion with restriction endonuclease Sma I. Renaturation rates for DNA from bands 1, 2, 7, 8, 9, 12 and 14 were not affected by tumor DNA. DNA from band 3 showed a slight rate increase in the presence of tumor DNA, indicating 21-27 copies of 14-18% of the DNA sequences in this (doublet) band. The band 3 doublet was separated by electrophoresis into bands 3a and 3b. Tumor DNA had little effect on the rate of reassociation of labeled band 3a DNA. Band 3b DNA renatured rapidly in the presence of tumor DNA, and its rate increase indicated that approximately 18 copies of 40% of band 3b DNA sequences are present per diploid tumor cell. This amounts to 3.7×10^6 daltons of foreign genetic information and represents a contribution of 0.0011% to the DNA content of the tumor cell. The relationship between this plant tumor and virally induced animal tumor systems is discussed.

Introduction

Crown gall tumors are induced in a wide variety of dicotyledonous plants by inoculation of wound sites with virulent Agrobacterium tumefaciens. Zaenen et al. (1974) first noted that virulent strains of A. tumefaciens harbor large plasmids, and there is now good evidence that the virulence trait is plasmid-borne. Virulence is lost when the bacteria are cured of plasmid (Chilton et al., 1974a; Van Larebeke et al., 1974; Watson et al., 1975) and is acquired by avirulent strains when a virulence plasmid is introduced by conjugation in plants (Watson et al., 1975; Van Larebeke et al., 1975; Bomhoff et al., 1976) or in vitro (Gordon et al., 1976; Schell et al., 1976; Chilton et al., 1976; Bomhoff et al., 1976). DNA homology studies have demonstrated that in some instances, the large virulence plasmids of different A. tumefaciens strains are only distantly related to one another (Currier and Nester, 1976a). Fingerprinting of plasmid restriction endonuclease digests confirms this conclusion (Sciaky, 1977). If all virulence genes are closely related, they must therefore constitute only a minor fraction of the plasmid genome. Only a few genetic markers have thus far been found to reside on virulence plasmids.

Several kinds of evidence suggest that A. tumefaciens, in transforming normal plant cells, confers new heritable traits on the tumor cells by DNA transfer. First, tumor tissue stably exhibits altered growth requirements in the absence of the inciting bacterium. The tumor cells grow in vitro in the absence of viable bacteria on agar lacking auxin and cytokinin hormones, a medium on which normal plant tissue is unable to grow. Hormone independence persists indefinitely in cultured tumor tissue. Acquisition of new genetic information from the bacterium could explain this heritable alteration. Second, crown gall tumor cells produce unusual amino acid derivatives. The best studied of these are octopine, N2-(D-1-carboxyethyl)-L-arginine (Menagé and Morel, 1964) and nopaline, N2-(1,3-dicarboxypropyl)-L-arginine (Goldman, Tempé and Morel, 1968; Goldman, Thomas and Morel, 1969), derivatives or arginine which are not found in normal plant tissue (Gordon et al., 1976; Kemp, 1976; Holderbach and Beiderbeck, 1976). Which, if either, of these compounds the tumor will produce is determined by the bacterial strain which incites the tumor (Petit et al., 1970; Bomhoff, 1974; Schilperoort and Bomhoff, 1975; Gordon et al., 1976; Schell et al., 1976; Montoya et al., 1977) and is specified somehow by genetic information on the virulence plasmid (Bomhoff et al., 1976; Gordon et al., 1976; Schell et al., 1976; Nester et al., 1977; Montoya et al., 1977). Since most lines of axenic cultured tumor tissue persistently produce either octopine or nopaline, acquisition of new genetic information from the bacterium seems a plausible explanation for these new biochemical traits. The plasmid location of both the virulence gene(s) and the octopine/nopaline production determinant suggests plasmid genes as the most probable candidates for the putative transferred DNA.

The question of the presence of plasmid DNA sequences in crown gall tumor DNA has been investigated in three laboratories. Tobacco tumors were found not to contain whole plasmid genomes under conditions allowing detection of one copy per diploid tumor cell (Chilton et al., 1974a; Dons, 1975; Gordon et al., 1976). A recent report (Matthysse and Stump, 1976) of detection of 0.1-0.2%

plasmid DNA in a Vinca rosea crown gall tumor was not repeatable in this laboratory (Sciaky, 1977; S. K. Farrand, D. J. Merlo and M.-D. Chilton, unpublished data). Our studies rule out the presence of as much as 0.006% plasmid DNA in DNA isolated from this Vinca rosea tumor line.

All previous published investigations of the possibility of plasmid DNA in crown gall tumors have used the entire plasmid genome as the labeled probe. None of these experiments rules out the possibility that a specific small fraction of the plasmid might be present in tumor DNA. We now report DNA hybridization studies of higher sensitivity which demonstrate the presence of multiple copies of a small part of the plasmid genome in a tobacco crown gall tumor line.

Results

Hybridization of Tumor DNA with Total Plasmid DNA as Probe

Plasmid DNA from A. tumefaciens strain A277 was used as probe in hybridization studies of DNA from clone E9 of a tobacco tumor incited by strain B6-806. Figure 1 presents a Pot plot (percentage of single-stranded probe DNA against probe DNA concentration times time) (Chilton et al., 1974b) of the kinetics of reassociation of a low concentration

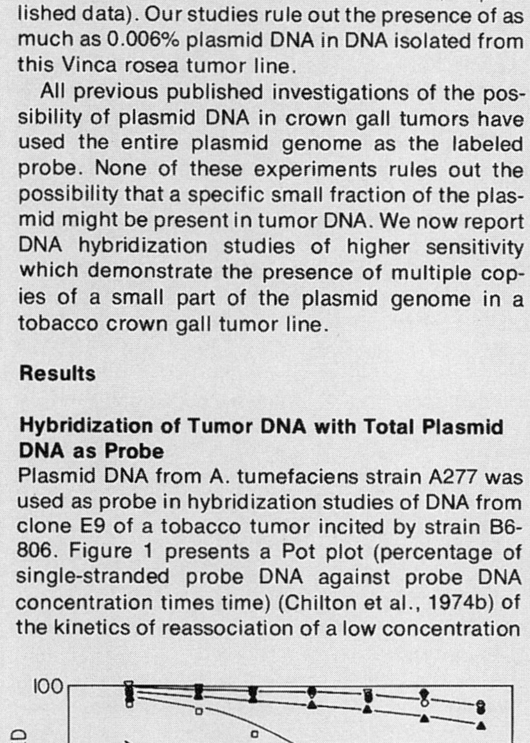

Figure 1. Renaturation Kinetics of A. tumefaciens A277 Plasmid DNA in the Presence of Tumor DNA and Control DNAs

^{32}P-labeled plasmid DNA (33×10^6 cpm/μg, 9×10^{-4} μg/ml) was allowed to reassociate in the presence of the following sheared DNA samples: (○) 2.2 mg/ml salmon DNA; (●) 1.9 mg/ml tobacco normal callus DNA; (▲) 2.3 mg/ml crown gall tumor E9 DNA; (▼) DNAase-treated crown gall tumor E9 DNA (originally 2.3 mg/ml); (□) 1.9 mg/ml salmon DNA + 0.038 μg/ml A277 plasmid DNA (0.8 copy model); (△) 2.3 mg/ml salmon DNA + 0.46 μg/ml A277 plasmid DNA (10 copy model). Ideal second-order reaction curves are drawn for the two reconstruction reactions only. All DNA concentrations were determined as described previously (Chilton et al., 1974b). Pot is in units of moles of nucleotide·liter^{-1}·sec.

of labeled plasmid DNA in the presence of approximately 2 mg/ml tumor DNA and control (salmon or normal tobacco callus) DNA. A small but reproducible acceleration of probe reassociation was seen when tumor DNA was present. Three separate experiments with different preparations of tumor DNA and different isolates of plasmid gave similar results. Two controls demonstrated the specificity of the acceleration of plasmid reassociation by tumor DNA. First, treatment of tumor DNA with DNAase abolished its ability to effect acceleration (Figure 1). Second, the rate of reassociation of a heterologous control probe, ^3H-labeled Serratia marcescens DNA (labeled by nick translation with ^3H-thymidine triphosphate), was the same in the presence of tumor, salmon and normal callus DNA (data not shown). The small rate change observed with tumor DNA and plasmid probe was thus not due to impurities in tumor DNA which accelerate DNA reassociation nonspecifically (Merlo and Kemp, 1976).

Reconstruction mixtures of salmon DNA plus unlabeled whole plasmid were prepared to simulate the kinetics expected if tumor DNA contained 0.8 or 10 copies of the entire plasmid genome per 6×10^{12} daltons of tumor DNA [the diploid cell content for tobacco (Siegel et al., 1973)]. These results (Figure 1) show that there is not as much as 0.8 copy of the entire plasmid per diploid tumor cell. They suggest either that plasmid DNA is present in the tumor at a much lower level (<1 copy of the whole plasmid per 10 diploid tumor cells), or that only part of the plasmid genome is in the tumor DNA. It is difficult to distinguish between these possibilities by any method of analysis of the kinetics (Sharp, Pettersson and Sambrook, 1974). We have therefore used a different experimental approach, which has been used by Sharp et al. (1974) for the analysis of adenovirus 2 DNA fragments incorporated into transformed rat cell DNA.

Isolation of Specific Fragments of Labeled Plasmid DNA for Use as Probe

If all the plasmid DNA sequences are present in tumor DNA, then every restriction endonuclease digest fragment of the plasmid when used as probe in tumor DNA-driven reactions should reassociate at an accelerated rate. On the other hand, if only part of the plasmid genome is in the tumor, the reassociation rate of some fragments should be accelerated in the presence of tumor DNA, while other fragments should show no rate increase.

The virulence plasmid of A. tumefaciens strain A277 was digested to completion with restriction endonuclease Sma I, and the digestion products were separated. The molecular weights of resolvable bands, together with estimated multiplicity, are listed in Table 1. The total molecular weight of the fragments (120 × 10^6 daltons) is in close agreement with the molecular weight of the A277 plasmid, as calculated from contour length measurements (120 × 10^6 ± 2 × 10^6 daltons). Thus the 19 bands of the Sma I digest account for essentially all of the plasmid genome.

Preparative electrophoresis was used to separate the Sma I digestion products of labeled (bottom slot) and unlabeled (top slot) A277 plasmid (Figure 2). An autoradiogram of the labeled plasmid is shown in Figure 2 beside the photograph of the preparative gel. Faint bands are visible in the labeled plasmid digest. Both the autoradiogram and the photograph show that bands 1 and 2 are not present in stoichiometric amounts in the labeled digest, presumably due to greater radiation damage and/or nuclease damage to these large targets during nick translation. Furthermore, the autoradiogram shows slight interband film blackening beyond band 1 of the digest. This presumably indicates that damaged fragments were running ahead of each band. The resolution of bands 1 and 2, as well as 6–9, is not clearly shown by this autoradiogram of the 7 mm thick gel, owing to the finite distance between the radiation source and the film. Autoradiograms of much higher resolution are obtained when such gels are dehydrated (data not shown). The sharpness of the bands in the labeled digest is better indicated by the faint bands visible in the photograph of the gel (Figure 2). Labeled bands were excised and isolated from the preparative gel as detailed in Experimental Procedures.

Table 1. Molecular Weights of Sma I Digestion Products of A277 Plasmid[a]

Band Number	Fragment Molecular Weight + 10^{-6}	Multiplicity
1	14.9	1
2	14.0	1
3	9.2 × 2 = 18.4	2
4	7.5	1
5	6.6	1
6	6.05	1
7	6.0	1
8	5.7	1
9	5.5	1
10	4.0 × 3 = 12.0	3
11	3.78	1
12	3.22	1
13	2.65	1
14	2.5 × 2 = 5.0	2
15	2.28	1
16	1.58 × 2 = 3.16	2
17	1.23	1
18	1.17	1
19	0.88	1
Total	120.02	

[a] Molecular weight calibration standards were bacteriophage λ DNA (Helling, Goodman and Boyer, 1974) and E. coli F plasmid (E. Ohtsubo, personal communication), digested with restriction endonuclease Eco RI.

Molecular weight of A277 plasmid from contour length measurement: 120 × 10^6 ± 2 × 10^6 daltons (measured as described by Currier and Nester, 1976a).

Hybridization of Tumor DNA with Sma I-Digested Plasmid Bands 1, 2 and 3

The labeled (probe) DNA from bands 1 and 2 reassociated at the same rate in the presence of tumor DNA as control DNA (Figures 3 and 4). Thus band 1 and band 2 DNA sequences are not present in tumor DNA in detectable concentration. The "low reconstruction" reaction (see Experimental Procedures) in each case showed a distinct acceleration (Figures 3 and 4); hence 0.8 copies of band 1 or band 2 per tumor cell would have been detected.

In contrast, the DNA from band 3 showed a marked change in reassociation kinetics in the presence of tumor DNA but not normal tobacco callus DNA (Figure 5). These results have been confirmed in three experiments with different isolates of tumor DNA. The observation that reassociation of DNA from bands 1 and 2 was *not* accelerated by tumor DNA shows that the effect of tumor DNA on band 3 DNA reassociation is highly specific. Furthermore, tumor DNA did not influence the rate of reassociation of external standard probe, ^{3}H-labeled S. marcescens DNA (data not shown). Treatment of tumor DNA with DNAase

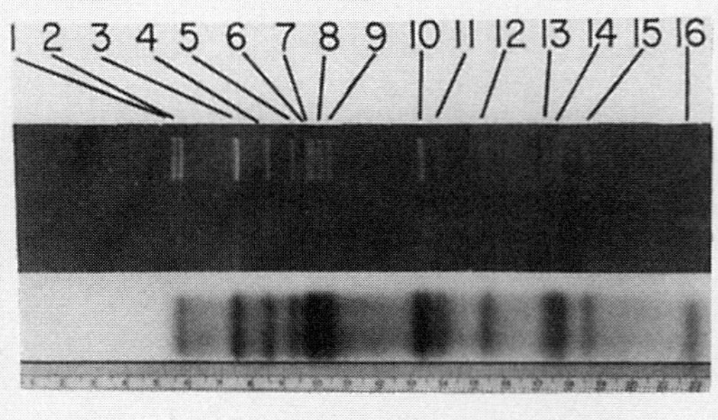

Figure 2. Preparative Horizontal Agarose Slab Gel Electrophoretic Separation of ^{32}P-Labeled A277 Plasmid Fragments

A277 plasmid (spec. act. 9 × 10^6 cpm/μg) labeled as described in Experimental Procedures with ^{32}P-thymidine triphosphate was cleaved with restriction endonuclease Sma I and separated by horizontal slab gel electrophoresis. The origin is at the left, and bands are numbered from largest to smallest. Fluorescent bands of an unlabeled A277 plasmid digest used as marker are visible in the upper slot. Below it, faint fluorescent bands are visible in the labeled DNA digest. The lower pattern is an autoradiogram of this labeled digest, reduced to the same scale photographically. Bands from this labeled digest were used for the hybridization studies presented in Figures 3–5.

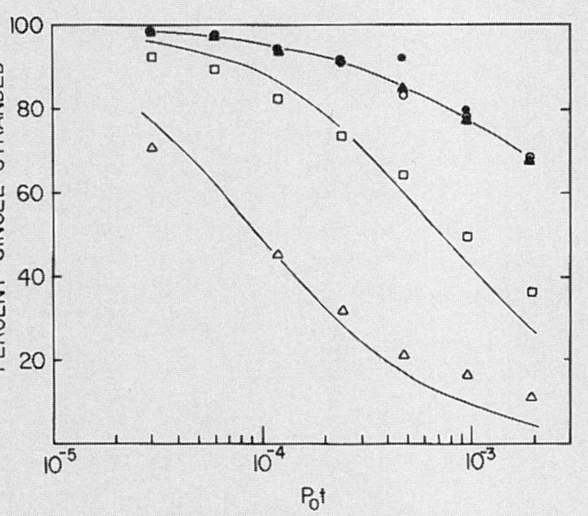

Figure 3. Renaturation Kinetics of A. tumefaciens A277 Plasmid Sma I Digest Band 1 DNA in the Presence of Tumor DNA and Control DNAs

^{32}P-labeled DNA (8.2×10^6 cpm/μg, 3.7×10^{-3} μg/ml) was allowed to reassociate in the presence of various sheared DNA samples (symbols as in Figure 1).

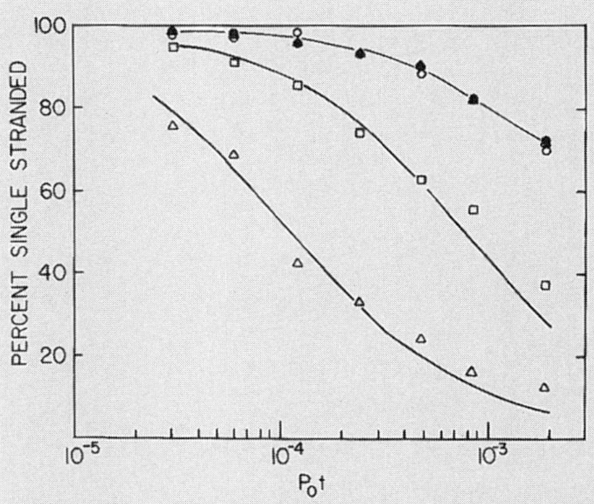

Figure 4. Renaturation Kinetics of A. tumefaciens A277 Plasmid Sma I Digest Band 2 DNA in the Presence of Tumor DNA and Control DNAs

Details are exactly as for Figure 3.

prior to hybridization eliminated the acceleration (Figure 5).

The fidelity of base pairing between band 3 DNA and tumor DNA was demonstrated by a thermal denaturation study (Figure 6). Aliquots were taken early in the reassociation reaction to minimize the contribution of duplexes formed by probe DNA reassociating with itself. To demonstrate the melting behavior of authentic plasmid duplexes under these conditions, aliquots of the "high reconstruction" reaction (see Experimental Procedures) were similarly subjected to thermal dissociation. The T_m (midpoint of the thermal transition) for the tumor DNA/plasmid DNA duplexes was 87°C, very close to that of authentic plasmid duplexes, 88°C (Figure 6).

The kinetics of reassociation of band 3 DNA with tumor DNA clearly deviate from those of a single component. By inspection of Figure 5, we estimate that the driven (rapid) component of the reaction constitutes 14–18% of the total. Scatchard analysis (Marsh and McCarthy, 1974) of the data (not shown) also gives a value in this range. At 7–9% reassociation (Pot = 3.7–4.7×10^{-5}), the driven component is 50% reassociated. To estimate the number of copies of foreign DNA present per tumor cell, we compare the $Pot_{1/2}$ of the driven component with the $Pot_{1/2}$ for the 10 copy reconstruction mixture, 10^{-4} (Figure 5). The driven component has a rate about 2.1–2.7 times faster than that of the 10 copy reconstruction reaction and therefore amounts to 21–27 copies per diploid cell. Estimates of 25–30 and 50 copies were obtained in two other experiments (data not shown).

Because band 3 is a doublet (Figure 2, Table 1), labeled band 3 DNA is a mixture of two probes, of which only one presumably bears homology to the tumor DNA. If pure, this member of the doublet should exhibit twice the percentage of homology with tumor DNA found for total band 3 DNA. Prolonged preparative electrophoresis (8 days) of plasmid digests in 0.7% agarose failed to resolve the band 3 doublet; however, 4 days of electrophoresis in 0.5% agarose produced a small separation of bands 3a and 3b (Figure 7). From bands 1, 2, 3a and 3b, DNA was recovered and labeled in vitro as described in Experimental Procedures. As expected, the reassociation of labeled DNA from bands 1 and 2 was not accelerated in the presence of tumor DNA (data not shown). Labeled band 3a DNA reassociation kinetics exhibited a slight rate increase in the presence of tumor DNA (Figure 8), while labeled band 3b DNA renaturation was dramatically accelerated (Figure 9). Because the resolution of bands 3a and 3b was imperfect (Figure 7), cross-contamination of 3a DNA with 3b DNA is the most reasonable explanation of the small rate change of 3a DNA reassociation in the presence of tumor DNA. However, the posssibility that a small amount of DNA homologous to band 3a DNA is present in the tumor DNA is not ruled out.

By inspection of Figure 9, we estimate that approximately 40% of band 3b DNA exhibits an accelerated rate, about twice the percentage estimated above for total band 3 DNA. A similar value was obtained by Scatchard analysis (Marsh and McCarthy, 1974) (data not shown). When the tumor (driver) DNA concentration was doubled, the rate of the rapid component of the reaction increased by about 2 fold (Figure 9), and a plateau between

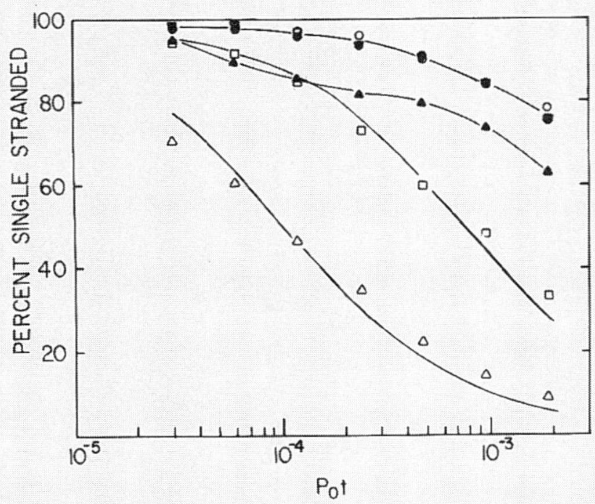

Figure 5. Renaturation Kinetics of A. tumefaciens A277 Plasmid Sma I Digest Band 3 DNA in the Presence of Tumor DNA and Control DNAs

Details are exactly as for Figure 3.

kinetic components was more clearly discernible. The foreign genetic information detected in the tumor thus amounts to 40% of that in band 3b or 3.7×10^6 daltons. At 20% reassociation, the driven component is 50% renatured; its $Pot_{1/2}$ is therefore 6×10^{-5} (2.2 mg/ml tumor DNA). The $Pot_{1/2}$ for the high reconstruction mixture (a 9.6 copy model with respect to 2.2 mg/ml tumor DNA) is 1.15×10^{-4}. The driven component thus has a $Pot_{1/2}$ 1/1.9 times that of the 9.6 copy model reaction, consistent with the presence of $1.9 \times 9.6 = 18$ copies of foreign DNA per diploid tumor cell.

Hybridization of Tumor DNA with Labeled DNA from Other Sma I Digest Bands

Labeled DNAs from bands 4–16 from the gel of Figure 2 were also used as probes to test for homologous DNA sequences in tumor DNA (data not shown). The results are equivocal because of the possibility that these smaller bands may be contaminated with band 3b DNA breakdown products. Negative results are therefore more significant than marginally positive results. We found small accelerations of reassociation using labeled probe DNA from bands 4 and (to a lesser extent) 5 and 6, the bands closest to band 3 on the gel and thus most susceptible to cross-contamination by breakdown products. DNA sequences homologous to bands 7, 8 and 9 were not detected in tumor DNA. Labeled probe DNA from bands 10 and 11 (but the latter may be contaminated with 10) were also driven to a limited extent by tumor DNA. No DNA sequences homologous to bands 12 and 14 were detected in tumor DNA, while the DNA probes from bands 13, 15 and 16 exhibited only a small rate increase in the presence of tumor DNA. Bands 17–19 were not

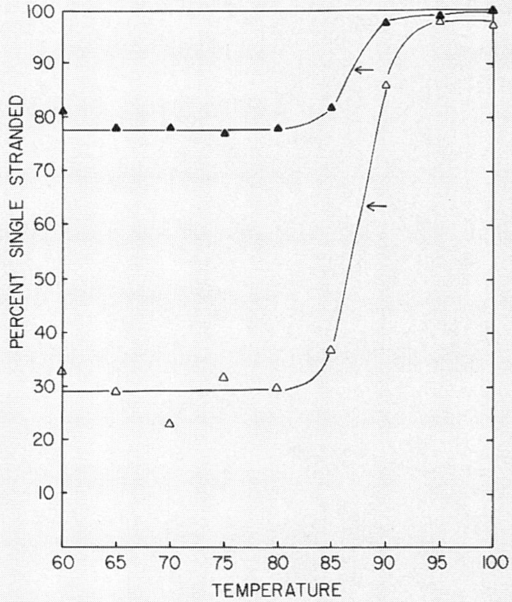

Figure 6. Thermal Dissociation Profile of Tumor DNA/Plasmid DNA Duplexes

Band 3 labeled DNA was used as probe in tumor DNA-driven and 10 copy reconstruction reactions similar to those of Figure 5. Ten replicate 10 λ aliquots were stopped after 3 hr of incubation. The former samples were 24% double-stranded, and the latter were 72% double-stranded. The control reaction (probe plus salmon DNA) was 9% double-stranded at this time (data not shown). A 10 λ aliquot was incubated for 5 min at the indicated temperatures, cooled to 0°C, diluted and assayed for duplexes in the usual way. Note that the data are *not* corrected for the single-stranded DNA present due to incomplete reassociation. (▲) tumor DNA/plasmid band 3 DNA duplexes (T_m = 87°C); (△) plasmid DNA/plasmid band 3 DNA duplexes (T_m = 88°C).

investigated. It is our view that band 10 may be the only one besides band 3b whose DNA is actually homologous to the tumor DNA. Further studies are required, however, before the details of this question can be completely clarified.

Discussion

Our results suggest that the tumor-inducing principle first proposed by Braun (1947) is indeed DNA, as many investigators have long suspected. Our data demonstrate that a small, presumably contiguous region within band 3b of the Sma I-digested virulence plasmid is maintained in tumor tissue at a level of about 20 copies per cell. If all foreign DNA in this tumor line is present at this level, the data of Figure 1 (whole plasmid probe DNA) would imply that about 5% of the entire plasmid is present. This amounts to about 6×10^6 daltons of foreign genetic

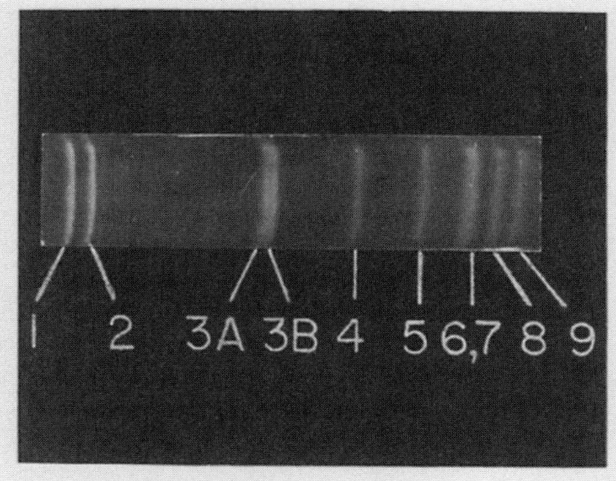

Figure 7. Preparative Horizontal Agarose Slab Gel Electrophoretic Separation of A277 Plasmid Fragments

A277 plasmid was cleaved with Sma I as described in Experimental Procedures. The band 3 doublet was visibly resolved after 4 days of electrophoresis of the digest in 0.5% agarose. Bands from this gel were excised, labeled by nick translation and used as probe DNA in the experiments of Figures 8 and 9.

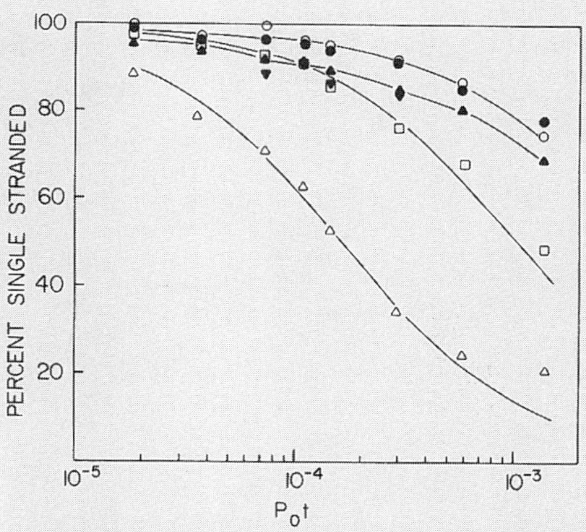

Figure 8. Renaturation Kinetics of A. tumefaciens A277 Plasmid Sma I Digest Band 3a in the Presence of Tumor and Control DNAs

Band 3a DNA (4.6×10^6 cpm/μg, 6.52×10^{-3} μg/ml) was allowed to reassociate in the presence of the following sheared DNA samples: (○) 2.2 mg/ml salmon DNA; (●) 1.9 mg/ml normal tobacco callus DNA; (▲) 2.2 mg/ml crown gall tumor E9 DNA; (▼) 4.4 mg/ml crown gall tumor E9 DNA; (□) 1.77 mg/ml salmon DNA + 0.035 μg/ml A277 plasmid DNA (0.8 copy model); (△) 2.22 mg/ml salmon DNA + 0.44 μg/ml A277 plasmid DNA (10 copy model). Ideal second-order renaturation curves are drawn for the two reconstruction reactions and the control reactions. Data have been corrected for 18% foldback observed upon denaturing and quick cooling of labeled probe DNA.

information, somewhat more than the amount accounted for in band 3b. On the basis of this simplest model, we would expect to find about 20 copies of at least a part of another fragment of the virulence plasmid. The most promising candidate thus far appears to be a member of triplet band 10.

Alternatively, it is possible that variable small numbers of copies of various portions of the plasmid genome may be incorporated into crown gall tumor cells. Although our data do not rule out this possibility, they clearly rule out the presence of 20 copies of a large fraction of the plasmid genome in tumor cells. Since the finding of plasmid DNA sequences stably maintained in a eucaryotic cell is unprecedented, we have no basis for predicting whether the pattern of incorporation will be simple or complex. The available data are thus far reconcilable with a simple pattern.

Our experiments do not bear directly upon the questions of whether foreign DNA in the tumor cell is covalently joined to plant DNA, either nuclear or plastid, or whether the multiple copies are tandemly or dispersively arranged. If the plasmid DNA sequences are integrated into plant nuclear DNA, crown gall tumors violate the posited barrier to recombination between procaryotic and eucaryotic DNA.

Crown gall tumors can, in light of our findings, be viewed as a feat of genetic engineering on the part of A. tumefaciens. It appears probable that production of the novel amino acid derivatives, octopine and nopaline, by tumor cells, now known to be specified by plasmid genes (Bomhoff et al., 1976; Gordon et al., 1976; Schell et al., 1976; Montoya et al., 1977), may be determined by the transferred DNA as first suggested by Morel and his colleagues (Goldman et al., 1968). The virulent bacterium may thus subvert the arginine supply of the plant by transmitting to the plant a gene which brings about conversion of this metabolite to octopine or nopaline, nutrients which only the bacterium can metabolize. If the foreign DNA in the tumor cell does function in this way, the potential applications of this transmissible element of the virulence plasmid as a vector in future genetic engineering studies in higher plants are apparent. As far as we are aware, this system is the only well documented case of the stable incorporation of bacterial genes into a plant.

Our data reveal an important etiological similarity between this plant cancer and certain virally induced animal cancers (Dulbecco, 1976; Temin, 1976; Baltimore, 1976). Although the hosts are in different taxonomic kingdoms, and the inciting agents are a bacterium in one case and a DNA or RNA virus in the other, in both cases, DNA homologous to the nucleic acid of the causative agent is present in the transformed cell. The potential of the crown gall system as a model for understanding neoplastic transformation in animal cells thus appears to be greatly enhanced. Furthermore, while the possible role of plasmid-bearing bacteria in

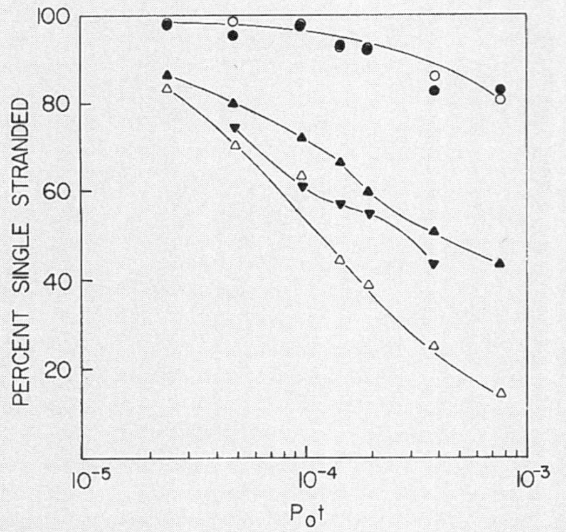

Figure 9. Renaturation Kinetics of A. tumefaciens A277 Plasmid Sma I Digest Band 3b in the Presence of Tumor and Control DNAs

Band 3b DNA (3.5×10^6 cpm/μg, 8.57×10^{-3} μg/ml) was allowed to reassociate in the presence of sheared DNAs described for Figure 8. Data for the 0.8 copy model reaction are not shown. Ideal second-order curves are drawn through all except the tumor DNA-driven reactions. Data have been corrected for 15% foldback observed upon denaturing and quick cooling of labeled probe DNA. Symbols are as in Fig. 8.

animal cancer systems has not to our knowledge been explored, it now seems reasonable to expect that some animal tumors may contain stably incorporated bacterial genes. We anticipate new insight into the general phenomenon of uncontrolled cell division from future studies of the location, replication and transcription of plasmid genes in the transformed plant cell.

Experimental Procedures

Reagents

α-^{32}P-thymidine triphosphate ($\geq$100 Ci/mmol) was purchased from New England Nuclear, and α-^{32}P-deoxycytidine triphosphate (150 Ci/mmol) was purchased from Amersham-Searle. Unlabeled deoxyribonucleotide triphosphates were products of P. L. Biochemicals. DNAase (DPFF, electrophoretically pure) was obtained from Worthington Biochemicals, and DNA polymerase I from Boehringer. Restriction endonuclease Sma I was isolated from Serratia marcescens by the procedure of Dr. Lyle Brown (Oregon State University) (personal communication).

Bacterial Strains

A. tumefaciens strain A277, an exconjugant containing only the virulence plasmid of B6-806 (Gordon et al., 1976), was isolated following conjugation in planta of B6-806 donor bacteria with A136, a plasmidless (avirulent) derivative of A. tumefaciens C58 (Watson et al., 1975). Serratia marcescens was obtained from Dr. Lyle Brown.

Tumor Strain

A. tumefaciens strain B6-806 was used to incite a tumor on Nicotiana tabacum var. Xanthi Nc., and a tissue culture line was derived from it. This tumor line was cloned by Dr. M. D. Sacristan-Allaily and Dr. G. Melchers (Max-Planck Institut für Biologie, Tübingen, Germany). DNA from one of the clones, E9, was used in the present experiments.

DNA Isolation

Plamid DNA was isolated by the procedure of Currier and Nester (1976b). Serratia marcescens DNA was isolated by Marmur's (1961) procedure, except that a phenol extraction was performed immediately after lysis of the bacteria in the presence of 1% (w/v) sodium dodecylsulfate (SDS) and pronase (0.5 mg/ml).

All glassware and implements used for isolation of tumor and normal callus DNA were cleaned with dichromate cleaning solution or were fired in a self-cleaning oven. To eliminate accidental contamination of tumor DNA with A. tumefaciens DNA, a separate laboratory was used for all plant DNA isolation. To test for possible bacterial contamination of tumor tissue cultures, immediately after harvest, the tissue culture agar was rinsed with nutrient broth which was then streaked onto nutrient agar. Any contaminated tissues were discarded.

DNA was isolated from crown gall and normal tobacco callus tissues by a modification of the technique of Heyn, Hermans and Schilperoort (1973). DNA voided from a Sepharose 4B column was precipitated with 2 vol of ethanol. Following treatment with RNAase (20 μg/ml, 2 $\times$ SSC, 37°C, 30 min), the DNA was extracted with phenol, spooled from solution, redissolved and dialyzed against 0.1 $\times$ SSC (1 $\times$ SSC is 0.15 M NaCl, 0.015 M trisodium citrate). After shearing (12,000 psi, French pressure cell), the DNA was dialyzed against 0.15 M PB (an equimolar mixture of NaH_2PO_4 and Na_2HPO_4), adsorbed to a hydroxylapatite column (BioRad HTP, DNA grade, 60°C, 0.15 M PB, 1 g packing per 1 mg DNA) and eluted with 0.3 M PB. The DNA was dialyzed exhaustively against deionized water, concentrated by evaporation and used as "driver" DNA in the hybridization experiments. The yield of DNA was 2–5 mg from 28 g of dry tissue. The A_{260}/A_{280} ratios were $\geq$1.9, and the A_{260}/A_{230} ratios were $\geq$2.4. The sheared DNA was not viscous at a concentration of 2 mg/ml.

Labeling of Probe DNA

A slight modification (Kelly et al., 1970) of the "nick translation" procedure of Maniatis, Jeffrey and Kleid (1975) was used to incorporate a labeled nucleotide triphosphate into 0.4–2 μg samples of DNA. Specific activities of up to 40×10^6 cpm/μg were obtained routinely by this procedure. Alkaline sucrose gradient centrifugation of labeled DNA immediately after preparation showed it to be similar in single-strand molecular weight to sheared unlabeled salmon DNA (approximately 5×10^5) used as marker.

Separation of Specific Fragments of Labeled Plasmid DNA

^{32}P-plasmid DNA from strain A277 was cleaved by Sma I as follows. Plasmid DNA (0.5–2 μg) in 50 μl of 20 mM Tris (pH 9.0), 15 mM KCl, 6 mM $MgCl_2$ was incubated at 30°C for 30 min with enough Sma I to give a complete digest. After addition of 5 μl of a solution of 0.07% bromophenol blue, 7% SDS in 33% glycerol, the sample was heated to 50°C, and 20 μl of 0.7% agarose (SeaKem ME) in electrophoresis buffer [0.08 M Tris, 0.04 M sodium acetate, 4 mM Versene (EDTA) adjusted to pH 8.0 with glacial acetic acid] were added. The digest was loaded into the bottom third of a well (7 mm deep, 1 mm wide, 2 cm long) in a 0.7% agarose horizontal slab gel (26 cm long) and overlaid with 0.7% agarose. The sample was electrophoresed at 50 V for 24–36 hr at room temperature. Autoradiography of the labeled DNA (30 min exposure on the underside of the gel, Kodak RP/R2 X-Omat film) revealed bands of labeled fragments migrating at the same position as unlabeled marker digests in flanking wells (Figure 2). The unlabeled fragments were visualized by soaking the gel for 30 min in 1–3 μg/ml ethidium bromide and illuminating it on a Blak-Ray Transilluminator (Ultraviolet Products).

Isolation of Radioactive DNA from Bands on Gel

Using flanking marker bands as guides, labeled bands were excised from the gel with a scalpel. Each gel slice was transferred to a 3 ml syringe, and under- and overlaid with 0.5% agarose in half-strength electrophoresis buffer. The tip of the syringe was fitted with a small dialysis bag containing 1 ml of half-strength electrophoresis buffer and 50 μg of sheared salmon DNA as a carrier. DNA was electrophoresed out of the agar and into dialysis bags (half-strength electrophoresis buffer, 110 V, 12–18 hr, 4°C), and its concentration was determined from its radioactivity.

Resolution of the Band 3 Doublet by Electrophoresis

Unlabeled plasmid DNA from strain A277 (20 μg) in 100 mM KCl, 6 mM $MgCl_2$, 15 mM Tris (pH 7.5) was digested for 90 min at 30°C with Sma I. The digest was loaded into three 3.5 cm × 1 mm wells in a 26 cm long × 0.8 cm thick slab gel of 0.5% agarose in electrophoresis buffer. After 4 days of electrophoresis at 50 V (room temperature, electrode buffers mixed once daily), bands 3a and 3b were visibly resolved (Figure 7).

Extraction of Unlabeled DNA from Agarose

Slices of agarose containing individual DNA bands (Figure 7) were cut from the gel, macerated with a metal spatula in an equal volume of extraction buffer [0.5 M ammonium acetate, (pH 7.6), 0.01 M magnesium acetate, 0.1 mM trisodium EDTA, 0.1% SDS] (Maxam and Gilbert, 1977) and frozen in a dry ice-ethanol bath. After thawing at 37°C, this process was repeated. Agarose was pelleted by centrifugation at 27,000 × g for 15 min at 4°C, and the pellet was again twice extracted as above. Ethidium bromide was removed from the pooled supernates by two extractions with equal volumes of n-butanol equilibrated with extraction buffer; residual agarose was removed by two extractions with equal volumes of a 1:1 (v:v) mixture of chloroform-phenol equilibrated with 1 mM Tris (pH 7.5), 0.5 mM NaCl. Following two extractions with ether, the DNA was dialyzed exhaustively against 1 mM Tris (pH 7.5) and concentrated by evaporation.

DNA recovery was determined by a modification of the ethidium bromide fluorescence assay of Le Pecq (1971). The yield from each band was 78–93% of the theoretical yield. A 0.4 μg portion of each DNA sample was labeled by nick translation and used in hybridization studies (see text).

Hybridization Reactions

Labeled "probe" DNA (whole plasmid or Sma I digest bands) at a concentration of 30,000 cpm/ml in 2 M $NaClO_4$, 0.03 M PB was added to tubes in which unlabeled "driver" DNA had been evaporated to dryness. Driver DNA (usually approximately 200 μg) was dissolved in 0.1 ml of probe DNA solution, and aliquots (10 μl) were sealed in capillary tubes. After 5 min of thermal denaturation at 105°C, the DNA was allowed to reassociate at 63°C. At various times, aliquots were cooled to 0°C and diluted into 1 ml of 0.15 M PB. Duplexes were assayed on hydroxylapatite as described previously (Chilton et al., 1974b).

Reconstruction Mixtures

To calibrate the hybridization reactions, reassociation of each probe DNA was measured in the presence of two known mixtures of salmon DNA and unlabeled plasmid DNA. One mixture ("low reconstruction") demonstrated the sensitivity of the experiment. This mixture contained 0.038 μg/ml A277 plasmid DNA. If tumor DNA contained 1 copy of the entire plasmid (1.2×10^8 daltons) per diploid tobacco DNA equivalent (6×10^{12} daltons) (Siegel et al., 1973), then 2.3 mg/ml of tumor DNA would contain $(1.2 \times 10^8 \div 6 \times 10^{12}) \times 2.3 \times 10^3 = 0.046$ μg/ml plasmid DNA. The "low reconstruction" therefore contained $0.038 \div 0.046 = 0.83$ times the concentration of plasmid DNA expected for 1 copy per tumor cell, and was thus designated "0.8 copy model."

The "high reconstruction" mixture contained 0.46 μg/ml plasmid DNA, exactly 10 times the concentration expected for 1 copy per tumor cell. The rate of the reaction of labeled probe DNAs with this "10 copy model" was used to calculate the concentration of homologous DNA sequences in tumor DNA (see Results).

Acknowledgments

We thank Dr. Arnold J. Bendich and Dr. Walter Halperin for illuminating discussions. We thank Dr. Lyle Brown for the procedure for isolation and use of Sma I. The expert technical assitance of Ross Bowman and James Martenson is gratefully acknowledged. We thank Marietta Ryba for electron microscopy and contour length measurements. This research was supported by grants from the NIH and the American Cancer Society. D.J.M. was supported by a USPHS postdoctoral research fellowship from the National Cancer Institute.

Portions of this work were presented at the Tenth Annual Miles Symposium June 1976, and the Annual Meeting of the American Phytopathological Society, July 1976.

Received January 31, 1977

References

Baltimore, D. (1976). Viruses, polymerases, and cancer. Science *192*, 632–636.

Bomhoff, G. (1974). Studies on crown gall—a plant tumor. Investigations on protein composition and on the use of guanidine compounds as a marker for transformed cells. Ph.D. Thesis, University of Leiden, The Netherlands.

Bomhoff, G., Klapwijk, P. M., Kester, H. C. M., Shilperoort, R. A., Hernalsteens, J. P. and Schell, J. (1976). Octopine and nopaline synthesis and breakdown genetically controlled by a plasmid of *Agrobacterium tumefaciens*. Mol. Gen. Genet. *145*, 177–181.

Braun, A. C. (1947). Thermal studies on the factors responsible for tumor initiation in crown gall. Am. J. Bot. *34*, 234–240.

Chilton, M.-D., Farrand, S. K., Eden, F., Currier, T. C., Bendich, A. J., Gordon, M. P. and Nester, E. W. (1974a). Is there foreign DNA in crown gall tumors? In Modification of the Information Content of Plant Cells, R. Markham, D. R. Davies, D. A. Hopwood and R. W. Horne, eds. (New York: Elsevier), pp. 297–311.

Chilton, M.-D., Currier, T. C., Farrand, S. K., Bendich, A. J., Gordon, M. P. and Nester, E. W. (1974b). *Agrobacterium tumefaciens* DNA and PS8 bacteriophage DNA not detected in crown gall tumors. Proc. Nat. Acad. Sci. USA *71*, 3672–3676.

Chilton, M.-D., Farrand, S. K., Levin, R. L. and Nester, E. W. (1976). RP4 promotion of transfer of a large *Agrobacterium* plasmid which confers virulence. Genetics *83*, 609–618.

Currier, T. C. and Nester, E. W. (1976a). Evidence for diverse types of large plasmids in tumor inducing strains of *Agrobacterium*. J. Bacteriol. *126*, 157–165.

Currier, T. C. and Nester, E. W. (1976b). Isolation of covalently closed circular DNA of high molecular weight from bacteria. Anal. Biochem. *76*, 431–441.

Dons, J. J. M. (1975). Crown gall—a plant tumor. Investigation on the nuclear DNA content and on the presence of *Agrobacterium tumefaciens* DNA and phage PS8 DNA in crown gall cells. Ph.D. Thesis, University of Leiden, The Netherlands.

Dulbecco, R. (1976). From the molecular biology of oncogenic DNA viruses to cancer. Science *192*, 437–440.

Goldman, A., Tempé, J. and Morel, G. (1968). Quelques particularités de diverses souches *d'Agrobacterium tumefaciens*. CR Acad. Sci. *162*, 630–631.

Goldman, A. Thomas, D. W. and Morel, G. (1969). Sur la structure de la nopaline métabolite anormal de certaines tumerus de crown-gall. CRH Acad. Sci. *268*, 852–854.

Gordon, M. P., Farrand, S. K., Sciaky, D., Montoya, A. L., Chilton, M.-D., Merlo, D. J. and Nester, E. W. (1976). The crown gall problem. In A Symposium on the Molecular Biology of Plants, I. Rubenstein, ed. (Minneapolis: University of Minnesota Press), in press.

Helling, R. B., Goodman, H. M. and Boyer, H. W. (1974). Analysis of endonuclease R EcoRI fragments of DNA from lambdoid bacteriophages and other viruses by agarose gel electrophoresis. J. Virol. *14,* 1235–1244.

Heyn, R. F., Hermans, A. K. and Schilperoort, R. A. (1973). Rapid and efficient isolation of highly polymerized plant DNA. Plant. Sci. Letters *2,* 73–78.

Holderbach, E. and Beiderbeck, R. (1976). Octopingehalt in normal und tumorgeweben einiger hoherer pflanzen. Phytochemistry *15,* 955–956.

Kelly, R. .B, Cozzarelli, N. R., Deutscher, M. P., Lehman, I. R. and Kornberg, A. (1970). Enzymatic synthesis of deoxyribonucleic acid: XXXII. Replication of duplex deoxyribonucleic acid by polymerase at a single strand break. J. Biol. Chem. *245,* 39–45.

Kemp, J. D. (1976). Octopine as a marker for the induction of tumorous growth by *Agrobacterium tumefaciens* strain B6. Biochem. Biophys. Res. Commun. *69,* 816–822.

Le Pecq, J.-B. (1971). Use of ethidium bromide for separation and determination of nucleic acids of various conformational forms and measurement of their associated enzymes. In Methods of Biochemical Analysis, *20,* D. Glick, ed. (New York: Interscience Publishers), pp. 41–86.

Maniatis, T., Jeffrey, A. and Kleid, D. G. (1975). Nucleotide sequence of the rightward operator of phage λ. Proc. Nat. Acad. Sci. USA *72,* 1184–1188.

Marmur, J. (1961). A procedure for isolation of deoxyribonucleic acid from microorganisms. J. Mol. Biol. *12,* 468–487.

Marsh, J. L. and McCarthy, B. J. (1974). Effect of reaction conditions on the reassociation of divergent deoxyribonucleic acid sequences. Biochemistry *13,* 3382–3388.

Matthysse, A. G. and Stump, A. J. (1976). The presence of *Agrobacterium tumefaciens* plasmid DNA in crown gall tumor cells. J. Gen. Microbiol. *95,* 9–16.

Maxam, A. and Gilbert, W. (1977). A New Method for Sequencing DNA. Proc. Nat. Acad. Sci. USA, in press.

Menagé, A. and Morel, G. (1964). Sur la présence d'octopine dans les tissue de crown-gall. CR Acad. Sci. *259,* 4795–4796.

Merlo, D. J. and Kemp, J. D. (1976). Attempts to detect Agrobacterium tumefaciens DNA in crown gall tumor tissue. Plant Physiol. *58,* 100–106.

Montoya, A. L., Chilton, M.-D., Gordon, M. P., Sciaky, D. and Nester, E. W. (1977). Octopine and nopaline metabolism in *Agrobacterium tumefaciens* and crown gall tumor cells: role of plasmid genes. J. Bacteriol. *129,* 101–107.

Nester, E. W., Chilton, M.-D., Drummond, M., Merlo, D., Montoya, A., Sciaky, D. and Gordon, M. P. (1977). Search for bacterial DNA in crown gall tumors. In Tenth Annual Miles Symposium, Impact of Recombinant Molecules on Science and Society, June 1976, F. Young, ed. (New York: Raven Press), in press.

Petit, A., Delhaye, S., Tempé, J. and Morel, G. (1970). Recherches sur les guanidines des tissus de crown gall. Mise en évidence d'une relation biochimique spécifique entre les souches d'*Agrobacterium tumefaciens* et les tumeurs qu'elles induisent. Physiol. Vég. *8,* 205–213.

Schell, J., Van Montagu, M., De Picker, A., DeWaele, D., Engler, G., Genetello, C., Hernalsteens, J. P., Holsters, M., Messens, E., Silva, B., Van den Elsacker, S., Van Larebeke, N. and Zaenen, I. (1976). Crown-gall: bacterial plasmids as oncogenic elements for eucaryotic cells. In A Symposium on the Molecular Biology of Plants, I. Rubenstein, ed. (Minneapolis: University of Minnesota Press), in press.

Schilperoort, R. A. and Bomhoff, G. H. (1975). Crown gall: a model for tumor research and genetic engineering. In Genetic Manipulations with Plant Material, L. Ledoux, ed. (New York: Plenum Press), pp. 141–162.

Sciaky, D. (1977). Plasmids of *Agrobacterium tumefaciens* and their role in crown gall tumorigenesis. Ph.D. Thesis, Washington State University, Pullman, Washington.

Sharp, P. A., Pettersson, U. and Sambrook, J. (1974). Viral DNA in transformed cells: I. A study of the sequences of adenovirus 2 DNA in a line of transformed rat cells using specific fragments of the viral genome. J. Mol. Biol. *86,* 709–726.

Siegel, A., Lightfoot, D., Ward, O. G. and Keener, S. (1973). DNA complementary to ribosomal RNA: relation between genomic proportion and ploidy. Science *179,* 682–683.

Temin, H. M. (1976). The DNA provirus hypothesis. The establishment and implications of RNA-directed DNA synthesis. Science *192,* 1075–1080.

Van Larebeke, N., Engler, G., Holsters, M., Van den Elsacker, S., Zaenen , I., Schilperoort, R. A. and Schell, J. (1974). Large plasmid in *Agrobacterium tumefaciens* essential for crown gall-inducing activity. Nature *252,* 169–170.

Van Larebeke, N., Genetello, C., Schell, J., Schilperoort, R. A., Hermans, A. K., Hernalsteens, J. P. and Van Montagu, M. (1975). Acquisition of tumor-inducing ability by non-oncogenic agrobacteria as a result of plasmid transfer. Nature *255,* 742–743.

Watson, B., Currier, T. C., Gordon, M. P., Chilton, M.-D. and Nester, E. W. (1975). Plasmid required for virulence of *Agrobacterium tumefaciens*. J. Bacteriol. *123,* 255–264.

Zaenen, I., Van Larebeke, N., Teuchy, H., Van Montagu, M. and Schell, J. (1974). Supercoiled circular DNA in crown-gall inducing *Agrobacterium* strains. J. Mol. Biol. *86,* 109–127.

I'M CERTAINLY GLAD I WAS THERE

Martin Drummond
Becketswell House, Wymondham, Norfolk, United Kingdom
E-mail: mh_drummond@hotmail.com

Commentary on the paper:

Drummond, M. H., Gordon, M. P., Nester, E. W., and Chilton, M.-D. 1977. Foreign DNA of bacterial plasmid origin is transcribed in crown gall tumors. Nature 269:535–536.

When I arrived in Gene Nester's lab in 1975, almost the entire group was focused on the problem of demonstrating the existence of bacterial DNA sequences in crown gall tissue. For decades a relatively small community of plant physiologists interested in the crown gall problem had suspected that the 'tumor inducing principle' was nucleic acid, maybe RNA, maybe a phage, and had produced a mass of unconvincing data to support the idea. By 1975, however, there was excellent evidence that the large plasmids discovered by Schell's group were involved and the race was on to detect plasmid sequences in plant tissue (Zaenen et al. 1974; Van Larebeke et al. 1975). We were well organised, specialised even; for months I did nothing but prepare radiolabeled probes using the nick-translation reaction, and we held regular councils of war at Gene Nester's house. When we were finally successful and the resulting *Cell* paper (Chilton et al. 1977) published, this collective obsession dissolved and we were faced with a gamut of secondary questions. People mulled them over, picked one that seemed attractive and went off to solve it in relative isolation.

I started looking for T-DNA transcripts. This was not entirely straightforward; cultured tumor tissue appeared to be full of nucleases that were remarkably difficult to destroy and would reduce mRNA to oligonucleotides within seconds unless properly dealt with. Fortunately Francine Eden had already solved that problem. No quarter was given. Callus tissue was subjected to a treatment that no self-respecting macromolecule could be expected to survive intact, being first dropped into liquid nitrogen, then quickly pulverised in a blender and finally sheared in hot phenol. Labelling was simplicity itself. ^{32}P was all the rage, having displaced tritium in the affection of hybridizers, and its short half life added to the general sense of urgency. Tumor tissue was cultured on a medium low in phosphate and then simply impregnated with $^{32}PO_4$, more or less straight out of the vial, which passed into the interstices of the friable callus. The quantities involved were impressive. Milt Gordon got hold of 40 mCi of the stuff, and on delivery day we could hear him coming down the corridor ordering people aside as though driving a plague cart. The vial sat in the corner of the lab, well swathed in perspex. In retrospect my handling of it appears a little nonchalant, but then, we now live in an age when one is required to don gloves and safety spectacles when weighing out table salt. For the *Cell* paper, we had relied on hybridization in solution, filter hybridization being out of favor because of the noise caused by non-specific binding of the radioactive probe. In 1975, however, Ed Southern had published his ingenious technique for transferring restriction fragments straight from the agarose gel in which they had been separated onto cellulose nitrate sheets (Southern 1975). Non-specific binding to these filters then appeared as irregular blobs, easily distinguishable from the signal of interest. The animal virologists in whose wake we tended to sail had already demonstrated the power and sensitivity of the method, and Mary-Dell decided that this was the way forward. I would have found it difficult to argue even if I had wished to—I had no vision of my own in those days, and we shared rather a small lab. Unlike those techniques that look great on paper but only actually deliver results in the hands of their inventors, causing deep anguish in everyone else, blotting worked like a charm. Well, perhaps not immediately, but it did work. Suddenly the huge effort the group had recently put into isolating restriction fragments of the octopine plasmid for solution hybridization seemed outdated. Just as agarose gel electrophoresis, simplicity itself, had made analytical ultracentrifugation look absurdly expensive, cumbersome and ineffective as a means of separating DNA molecules, so Southern's simple method opened all kinds of doors.

I forget how many times I did the experiment before everything came right, but I seem to remember that there wasn't much X-ray film left in the box when I'd finished. The nearest facility for developing autoradiograms was in the Radiology Department of the hospital, which was in the same building. That is not to say that it was close by. Construction of the Medical School and Hospital at the University site had proceeded by simple concatenation, each additional block being designated by a new letter. Getting to the hospital involved a ten minute walk through the first third of the alphabet down what was reputed to be the longest corridor in the United States or possibly the Northern hemisphere. I quite enjoyed my daily strolls through the various medical departments. There was one apparently uninhabited section of corridor with resonant qualities that made it difficult to resist humming, and on one occasion my rendering of the slow movement of Mozart's clarinet concerto for human nasal passages drew a round of applause from a small group of maintenance men who emerged from an adjoining room.

The final result was delightfully unequivocal, no suggestive smudge but a bold image corresponding exactly to what we already knew about T-DNA. My return from the hospital was less of a stroll and more of a strut, and when I showed the autoradiogram to Mary-Dell, her celebratory yelp echoed down the hall.

Struggling with manuscripts in later years, I have often wondered at the speed with which I wrote the first draft of the paper. Looking at it again now provides part of the answer; it is very short. The other part was that I had Mary-Dell breathing down my neck. "Bring it up to my house Sunday evening" she said one Friday afternoon, and I spent the weekend massaging it into shape, a particularly fine weekend, I recall. In a sense I needn't have bothered. She immediately set about dismembering my matchless prose, ultimately retaining only the final sentence, albeit with an important caveat inserted. I suppose it had a nice ring to it. Unfortunately it was also false, since even if the same transcripts had been found in crown gall tissue and in *Agrobacterium,* the possibility of overlapping eukaryotic and prokaryotic promoters could have accounted for this observation.

It is difficult to remember now how ignorant we then were about transcriptional processes in prokaryotes and eukaryotes. I find it even more difficult to reconcile the stunning explosion of knowledge produced by molecular genetics over the past thirty years with my own experience of work at the bench, where things always seem to be going appallingly slowly, and success bobs precariously in a sea of failure. Very odd! But however modest my part, I'm certainly glad I was there.

References

Chilton, M.-D., Drummond, M. H., Merlo, D. J., Sciaky, D., Montoya, A. L., Gordon, M. P., and Nester, E. W. 1977. Stable incorporation of plasmid DNA into higher plant cells: The molecular basis of crown gall tumorigenesis. Cell 11:263–271.

Southern, E. M. 1975. Detection of specific sequences among DNA fragments separated by gel electrophoresis. J. Mol. Biol. 98:503-517.

Van Larebeke, N., Genetello, C., Schell, J., Schilperoort, R. A., Hermans, A. K., Hernalsteens, J.-P., and Van Montagu, M. 1975. Acquisition of tumor-inducing ability by non-oncogenic agrobacteria as a result of plasmid transfer. Nature 255:742–743.

Zaenen, I., Van Larebeke, N., Teuchy, H., Van Montagu, M., and Schell, J. 1974. Supercoiled circular DNA in crown-gall-inducing *Agrobacterium* strains. J. Mol. Biol. 86:109–127.

MARTIN DRUMMOND: Growing up in the 1950s I was attracted by the prestige of science and its apparent promise of sure knowledge about life. I graduated from the University of East Anglia, Norwich, U.K. with a degree in Biological Sciences in 1967, and after a brief foray into teaching, embarked on a Ph.D. project on crown gall at the University of Leeds, U.K., under the tutelage of Dr. G. R. K. Sastry. I then joined Gene Nester's group at the University of Washington and found myself in Mary-Dell Chilton's lab. I learned more about the practice of science from Mary-Dell than from any other single person, and without her would not have had a research career at all. In 1978 I returned to London, and after a brief stint with Peter Tattersall at the Imperial Cancer Research Fund labs in Mill Hill, took up a permanent position in the (then) ARC Unit of Nitrogen Fixation at the University of Sussex. This somewhat anarchic multidisciplinary group of chemists, biochemists and geneticists provided a most stimulating intellectual environment, and from that point I worked on prokaryotic transcriptional regulation, with a minor interest in electron transport.

(Reprinted from Nature, Vol. 269, No. 5628, pp. 535–536, October 6, 1977)

Foreign DNA of bacterial plasmid origin is transcribed in crown gall tumours

Agrobacterium tumefaciens incites cancerous growths in dicotyledonous plants called crown gall tumours. Large plasmids in oncogenic *Agrobacterium* strains[1] carry genetic information which is essential for tumour induction[2,3]. We recently reported that axenic crown gall tumour tissue contains multiple copies of a small part of the oncogenic (Ti) plasmid of the inciting bacterial strain[4]. The mechanism of induction of crown gall tumours thus seems to resemble that of some virally induced animal tumours: the eukaryotic cell is transformed by addition of new genetic information which is stably maintained through subsequent cell divisions[5–7]. The means by which foreign DNA brings about the tumorous growth pattern of crown gall cells is unknown. The first step in elucidating its mode of action is reported here: we have detected RNA transcripts of the foreign genetic information in the tumour cell.

^{32}P-pulse-labelled RNA was isolated from tumour and normal tobacco callus tissues as described in Fig. 1. To test these RNA preparations for the presence of RNA transcripts of Ti plasmid DNA, we used a filter hybridisation technique developed by Southern[8]. Ti plasmid DNA from the inciting bacterial strain was first cleaved into 25 specific fragments with the restriction endonuclease *Sma* I. These were resolved into 21 bands by Agarose gel electrophoresis[4]. Plasmid DNA fragments were alkali-denatured in the gel, neutralised and transferred directly to a sheet of cellulose nitrate[8,9]. The pattern of bands obtained by electrophoresis was thereby reproduced on the sheet, which was then cut into longitudinal strips for separate hybridisation experiments.

Strips were incubated with labelled RNA isolated from clone E9 of a tobacco tumour incited by strain B6-806, and from normal tobacco callus tissue. In addition, a strip was incubated with ^{32}P-cRNA synthesised[9] from the whole Ti plasmid genome by *Escherichia coli* RNA polymerase. Autoradiograms of the hybridised filter strips are shown in Fig. 1 together with a photograph of the bands of DNA in the gel used to prepare the cellulose nitrate transfers.

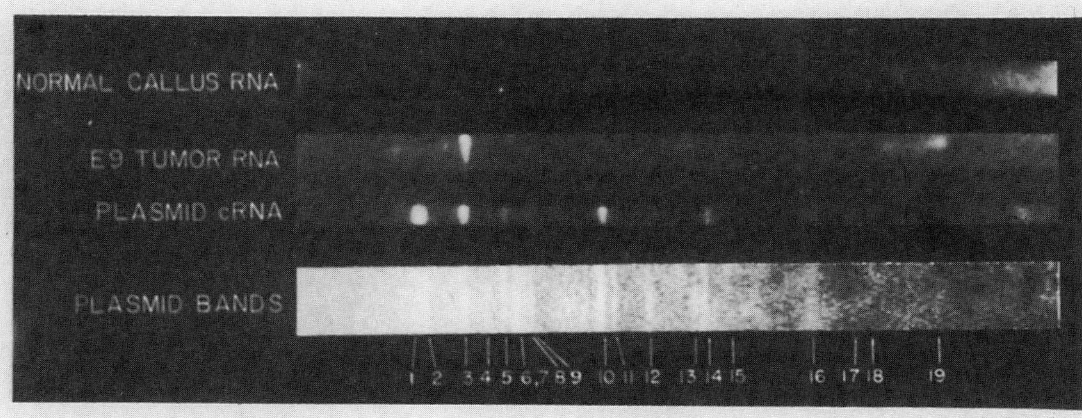

Fig. 1 Hybridisation of labelled E9 tumour and control RNAs to strips bearing *Sma* I digest bands. Tumour and normal callus tissues, tested for sterility at the time of labelling, were incubated in Petri dishes with approximately 1 mCi ^{32}P (neutralised) per g of tissue for 4 h. RNA was isolated by hot phenol extraction, DEAE–cellulose chromatography, DNase digestion and gel filtration on Sephadex G-75 (ref. 12). Specific activities of RNAs were 127,000 c.p.m. per μg (tumour) and 34,000 c.p.m. per μg (normal callus). For cRNA synthesis ^{32}P-UTP (ICN, 75 Ci mMol^{-1}) was used as labelled precursor, and plasmid from strain A277 served as template. Pure virulence plasmid of *A. tumefaciens* B6-806 was isolated from exconjugant strain A277 and digested with *Sma* I as described previously[4]. After electrophoretic separation in 0.5% Agarose, the bands were stained and photographed[4] before denaturation and transfer of bands to cellulose nitrate sheets (Schleicher and Schuell, B6 membrane). Strips were soaked in Denhardt's[13] pre-incubation mixture and baked at 55 °C overnight before use. Hybridisation was carried out with 5×10^6 c.p.m. ml^{-1} labelled RNA and 200 μg of unlabelled carrier RNA in 0.3 M NaCl, 0.03 M tri-sodium citrate (2×SSC) at 67 °C for 20 h. Filter strips were wetted with 2×SSC before addition of labelled RNA. Unlabelled carrier RNA was isolated from strain A114, an avirulent plasmidless derivative of strain C58 (ref. 3). After incubation, strips were washed (2×SSC 67 °C), treated with RNase (20 μg ml^{-1}, 1 h, 22 °C), rinsed, dried and autoradiographed for 36 h to 2 weeks. All radioactive label was entirely acid precipitable at the end of the incubation. Autoradiograms of filter strips hybridised with E9 tumour RNA, normal callus RNA and cRNA are presented beside a photograph of the original gel, adjusted to the same scale photographically.

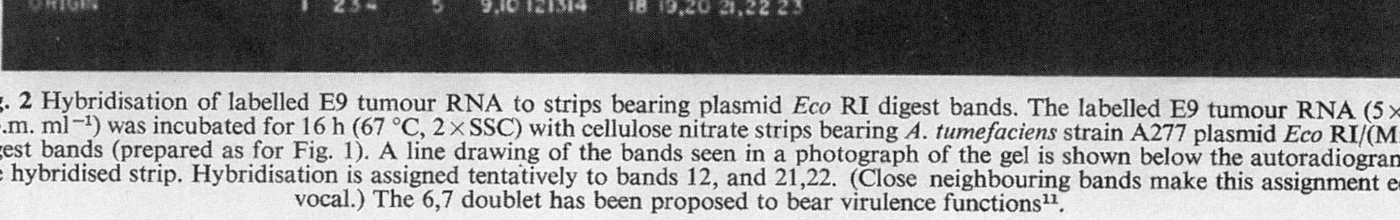

Fig. 2 Hybridisation of labelled E9 tumour RNA to strips bearing plasmid *Eco* RI digest bands. The labelled E9 tumour RNA (5×10^6 c.p.m. ml^{-1}) was incubated for 16 h (67 °C, 2×SSC) with cellulose nitrate strips bearing *A. tumefaciens* strain A277 plasmid *Eco* RI/(Miles) digest bands (prepared as for Fig. 1). A line drawing of the bands seen in a photograph of the gel is shown below the autoradiogram of the hybridised strip. Hybridisation is assigned tentatively to bands 12, and 21,22. (Close neighbouring bands make this assignment equivocal.) The 6,7 doublet has been proposed to bear virulence functions[11].

Fig. 3 Hybridisation of labelled E1 tumour RNA to strips bearing plasmid *Hind*III digest bands. Labelled E1 tumour RNA (3 h pulse, 46,000 c.p.m. per μg) was isolated as described in Fig. 1. Cellulose nitrate strips bearing *Hind*III (New England Biolabs) digest bands of strain A277 plasmid were prepared, and labelled RNA (4×10^6 c.p.m. ml^{-1}) was hybridised with the strips for 16 h (2 × SSC, 67 °C), washed and RNase treated as for Fig. 1. A line drawing of the bands, prepared from a photograph of the gel, is shown below the autoradiogram of the strip. Labelled E1 tumour RNA hybridisation is assigned to *Hind*III bands 1 and 5.

Labelled tumour RNA hybridised to a single band of the *Sma* I digest, identified as band 3 by comparison with the photograph of the gel and the autoradiogram of the cRNA hybrid strip. We have shown that this tumour clone contains about 18 copies, per diploid tobacco cell DNA equivalent, of part of *Sma* I digest band 3, by direct DNA–DNA hybridisation technique[4]. Our findings show that at least a part of this genetic information is transcribed in the tumour cell. Recent studies have shown that this tumour clone also contains a part of *Sma* I digest band 10 (ref. 10). In three separate experiments using different lots of pulse-labelled tumour RNA from this clone, no hybridisation was noted to band 10 or any other *Sma* I digest bands; only band 3 transcripts were detected. Normal callus tissue RNA was never observed to bind to any fragment of the plasmid, even after fivefold longer exposure of the autoradiograms.

Further experiments were carried out using ^{32}P-RNA form clone E9 and filter strips carrying bands of *Eco* RI digest of the Ti plasmid. The numerous fragments in this digest (Fig. 2) are not well separated by electrophoresis; thus the identification of bands is imprecise. Tumour RNA hybridised well to a band in the region of band 12, and weakly to a band in the region of the band 21,22 doublet. The latter band, while clearly visible in the autoradiogram, is too faint for photographic reproduction in Fig. 2. Doublet *Eco* RI digest band 6,7 has been proposed to be important in tumour initiation because it seems to be common to distantly related virulence plasmids[11]. We detected no transcript of this band in this tumour clone. Labelled RNA from normal callus tissue did not bind to any bands on filter strips bearing *Eco* RI digest bands (data not shown).

Labelled RNA from tumour clone E1, a single-cell isolate from the same tumour as E9, hybridised to *Sma* I digest band 3 (data not shown). When this RNA was hybridised to cellulose nitrate strips bearing *Hind*III digest fragments of the Ti plasmid, bands 1 and 5 gave positive results (Fig. 3). Thus these singlet bands appear to contain the same genetic information as the relevant member of the *Sma* I band 3 doublet, which we have identified as band 3b as described elsewhere[4].

Results presented here show that pulse-labelled crown gall tumour RNA can be used to identify DNA fragments, within any restriction endonuclease digest, which contain Ti plasmid DNA base sequences transferred to the tumour cell. Although not all the foreign DNA is transcribed at detectable levels in the tumour, at least part of the transferred DNA can be detected in this manner. DNA fragments shown to be contiguous by conventional mapping procedures may then be tested for homology with tumour DNA.

Our data confirm, by an independent experimental approach, the presence of a specific part of the Ti plasmid in crown gall tumour cells. Also, these data constitute the first convincing evidence that DNA of bacterial origin is transcribed in demonstrably axenic plant tissue. The possibility must be considered, however, that the transcribed genes in question may ultimately be of plant origin, having been picked up by the *A. tumefaciens* Ti plasmid in the course of its evolution. Should this be the case, one might expect these genes to remain untranscribed in the bacterium. Experiments are in progress to determine whether this is in fact the case. If the same RNA transcripts prove to be present in both the bacterial and the crown gall tumour cell, an unsuspected degree of compatibility between transcriptional processes in eukaryotic and prokaryotic systems will be revealed.

This research was supported by grant CA13015 from the National Cancer Institute and grant number NP194 from the American Cancer Society. We thank Mr Chips Armstrong for prints of autoradiograms. We thank Dr Daniela Sciaky for useful discussions and Dr Benjamin D. Hall for a gift of *E. coli* RNA polymerase.

MARTIN H. DRUMMOND
MILTON P. GORDON
EUGENE W. NESTER
MARY-DELL CHILTON

Department of Microbiology and Immunology,
University of Washington,
Seattle, Washington 98195

Received 14 June; accepted 15 August 1977.

1. Zaenen, I. Van Larebeke, N., Teuchy, H., van Montagu M. & Schell, J. *J. molec. Biol.* **86**, 109–127 (1974).
2. Van Larebeke, N. *et al. Nature* **252**, 169–170 (1974).
3. Watson, B., Currier, T. C., Gordon, M. P., Chilton, M. D. & Nester, E. W. *J. Bact.* **123**, 255–264 (1975).
4. Chilton, M. D. *et al. Cell* **11**, 263–271 (1977).
5. Baltimore, D. *Science* **192**, 632–636 (1976).
6. Dulbecco, R. *Science* **192**, 437–440 (1976).
7. Temin, H. M. *Science* **192**, 1075–1080 (1976).
8. Southern, E. M. *J. molec. Biol.* **98**, 503–517 (1975).
9. Ketner, G. & Kelley, J. T. *Proc. natn. Acad. Sci. U.S.A.* **73**, 1102–1106 (1976).
10. Chilton, M. D. *et al. Micro biology* (in the press).
11. Schell, J. *et al. Molecular Biology of Plants* (ed. Rubinstein, I.) (Academic, New York, in the press).
12. Eden, F. C. thesis, Univ. Washington (1974).
13. Denhardt, D. T. *Biochem. biophys. Res. Commun.* **23**, 641–646 (1966).

Printed in Great Britain by Henry Ling Ltd., at the Dorset Press, Dorchester, Dorset

AN ADVENTURE IN AGROBACTERIOLOGY

Marcelle Holsters
Department of Plant Systems Biology
Flanders Interuniversity Institute for Biotechnology
Ghent University, Gent, Belgium
E-mail: marcelle.holsters@psb.ugent.be

Commentary on the paper:

Holsters, M., Silva, B., Van Vliet, F., Genetello, C., De Block, M., Dhaese, P., Depicker, A., Inzé, D., Engler, G., Villarroel, R., Van Montagu, M., and Schell, J. 1980. The functional organization of the nopaline *A. tumefaciens* plasmid pTiC58. Plasmid 3:212–230.

A good traveler has no fixed plans
And is not intent upon arriving.
A good artist lets his intuition
Lead him wherever it wants.
A good scientist has freed himself of concepts
and keeps his mind open to what is

Tao Te Ching

This quotation from the Tao is dedicated to Jef Schell, a good scientist and "a thinker; that means he knew how to make things simpler than they are" (Nietzsche, *The Gay Scientist*). Jeff started up *Agrobacterium* research at the University of Ghent, Belgium in the eventful year 1968. He was a microbiologist and biochemist by training, an "offspring" from the famous microbiology tradition built up in Gent by Professor De Ley (renowned for *Agrobacterium* and *Rhizobium* taxonomy studies), but became fascinated by the awakening world of bacterial genetics. His contribution to the field of "agrobacteriology" (Zupan et al. 2000) has been creative and huge and he walked us through the adventure as through a thrilling detective story.

How did we get into agrobacteriology? I can recall the beginnings because I was there from the start as an undergraduate student, and participated in many stages of the evolving field. In 1967–1968, Jeff, freshly back from a stay in the States, was setting up a university laboratory, while starting to teach a course in genetics and recruiting students. He offered interested candidates the choice between two subjects: one he was familiar with, namely the study of restriction modification phenomena in *Escherichia coli* and its phage lambda; the other, novel and uncertain, namely the search for the tumor-inducing principle (TIP) of *A. tumefaciens.* I chose the challenge and was assigned the task of investigating the hypothesis: "is the TIP a temperate phage?" That phage was the favorite culprit was not surprising, given the established involvement of animal viruses in oncogenesis at that time. A few years of fervent quest followed to associate agrobacterial virulence and crown galls with prophages. I found no such correlation and, despite occasional disbelief from my colleagues, stubbornly stuck to this conclusion. In retrospect, perhaps a heuristic contribution because it turned attention to the wider concept of episomes, as defined by circular replicons that can be detected by physical separation in centrifugation gradients, a technology that was within our reach. Luckily, the tumor-inducing (Ti) plasmid was not a linear molecule; the future of the Gent laboratory might have been totally different. As is the case for a great many success stories, in science and in life in general, a good share of luck is important.

After we had established and demonstrated that the TIP is linked to the Ti plasmid, we entered the era of research that culminated in the following paper and in numerous associated manuscripts. A lot of people (12 authors!) with very complementary skills contributed to this project. It was a real group effort, involving large-scale plasmid purifications, construction of physical maps using restriction enzymes and cloning, bacterial genetics, and beautiful electron microscopic heteroduplex analysis. In 1977, an important book was published entitled *DNA Insertions, Plasmids and Episomes* (Bukhari et al. 1977). These authors assembled a lot of basic knowledge and applications following the discovery of DNA insertion elements. We applied these tools with enthusiasm to the Ti plasmid. To create mutations in the Ti plasmid a complete assortment of tags was tried including RP4 co-integration, different transposons, and phage Mu. Because the Ti plasmid was not exactly the same as RP4, and *A. tumefaciens* was different from *E. coli,* the task was not that easy. But eventually, we became experts in transformation, conjugation, and antibiotic selection and got acquainted with the specific whims and charms of different insertion elements in an *Agrobacterium*-genomic background. Several side branches of the Ti plasmid project produced elegant suicide plasmid vectors based on phage Mu genetics and insights into the mechanisms of nonhomologous recombination. While we were actively playing with insertion element tools for the functional analysis of the Ti plasmid, the T-DNA was doing something very similar (we realize this only in retrospect, of course), each time we performed oncogenicity assays on plant hosts. Indeed, we now know that the T-DNA is also a transposable genetic element that affects gene activity and chromosome structure and that it is integrated into the plant nuclear DNA via non-homologous recombination.

In the functional analysis study of the nopaline plasmid pTiC58, the different phenotypes known to be associated with the Ti plasmid were located on the rough physical map, thus providing an outline for further detailed studies and for applications. In parallel, work was going on with octopine plasmids (De Greve et al. 1981). The comparison of these two types of Ti plasmids provided interesting information on regions of homology as indicators of essential segments for particular functions (Depicker et al. 1978). For example, mutations that affected oncogenicity were located in regions of homology between octopine and nopaline plasmids. Importantly, what we now call the virulence (*vir*) region was identified as separate from the T-DNA; both regions are essential for oncogenesis, but in a different manner. This distinction had great consequences for later vector design. Also important was the observation that Tra functions, required for interbacterial plasmid transfer, were different from Vir functions, required for transfer of T-DNA to the plant. The Vir functions are still continuously under study and they have become the prototype of type IV secretion systems, which are widespread in animal pathogens.

By 1980, the base had been laid for all of the beautiful future work to explore the molecular mechanisms of the inter-kingdom DNA transfer, and also for the powerful applications of this extraordinary plant-microbe interaction. Plant molecular biology was beginning to flourish (Fox 1981). Thus, the seventies were the years in which crown gall was rediscovered and fantastic new tools for plant biology were created. This work was done in an atmosphere of "gay science" and collaboration, in a nurturing climate for raising young scientists, while the political and social context allowed for free fundamental research at the university, funded by academic sources.

References

Bukhari, A. I., Shapiro, J. A., and Adhya, S. L., eds. 1977. DNA Insertion Elements, Plasmids, and Episomes. Cold Spring Harbor Laboratory Press, Cold Spring Harbor, NY, U.S.A.

De Greve, H., Decraemer, H., Seurinck, J., Van Montagu, M., and Schell, J. 1981. The functional organization of the octopine *Agrobacterium tumefaciens* plasmid pTiB6S3. Plasmid 6:235–248.

Depicker, A., Van Montagu, M., and Schell, J. 1978. Homologous DNA sequences in different Ti-plasmids are essential for oncogenicity. Nature 275:150–153.

Fox, J. L. 1981. Plant molecular biology beginning to flourish. Chem. Eng. News 59:33–44.

Zupan, J., Muth, T. R., Draper, O., and Zambryski, P. 2000. The transfer of DNA from *Agrobacterium tumefaciens* into plants: A feast of fundamental insights. Plant J. 23:11–28.

MARCELLE HOLSTERS: In 1970, I obtained my licentiate in Zoology from Ghent State University, Belgium and my Ph.D. from the same university in 1976.

Looking back, I realize I had several mentors. First, Jeff Schell, who was my tutor for undergraduate and Ph.D. theses. Jeff's charismatic demeanor, his brightness and kindness had a major impact. Jeff counselled me in the ways of science.

Marc Van Montagu, Jeff's companion professor, added his wise and generous intelligence and energy and opened doors to the wide world of cultural values and events far beyond basic research. My own companions and colleagues, Suzanne Van den Elsacker, Dani De Waele and Patti Zambryski were and are my guides in science and life, by their wisdom, creativity and extraordinary strength and independence.

PLASMID 3, 212–230 (1980)

The Functional Organization of the Nopaline *A. tumefaciens* Plasmid pTiC58

M. HOLSTERS, B. SILVA, F. VAN VLIET, C. GENETELLO, M. DE BLOCK, P. DHAESE, A. DEPICKER, D. INZÉ, G. ENGLER, R. VILLARROEL, M. VAN MONTAGU, AND J. SCHELL

Laboratorium voor Genetika en Histologie, Rijksuniversiteit Gent, Ledeganckstraat 35, B-9000 Gent, Belgium

Received October 30, 1979

We have employed the P type plasmid RP4 and the transposons Tn*1* and Tn*7* to isolate insertion and deletion mutations in the nopaline Ti-plasmid pTiC58. Mutations that inactivate all known Ti phenotypes have been located on the physical map. Most importantly, we have positioned several regions involved in the determination of oncogenicity. They correspond to regions of homology between octopine and nopaline plasmids. One of these regions is part of the T-DNA, the Ti-plasmid DNA present in transformed plant cells. There are also segments of the T-DNA that are not essential for oncogenicity. One of these determines the biosynthesis of nopaline in tumors. The latter regions might allow insertion of foreign DNA that can then be introduced into plant cells.

Agrobacterium tumefaciens is a plant pathogenic soil bacterium that induces crown gall tumors upon infection of wounded plant tissues. Strains of this species contain one of a group of large DNA plasmids, the Ti-plasmids (Zaenen *et al.*, 1974), that are essential for the tumor-inducing capacity of the *tumefaciens* strains (Van Larebeke *et al.*, 1974; Watson *et al.*, 1975).

It has been demonstrated that part of the Ti-plasmid, the so-called T-DNA, is stably present in transformed plant cells (Chilton *et al.*, 1977; Schell *et al.*, 1979) where it is at least partially transcribed into RNA (Drummond *et al.*, 1977; Gurley *et al.*, 1979). Several types of Ti-plasmids exist. They can be classified according to their capacity (i) to determine the biosynthesis in crown gall tumors of a particular amino acid derivative ("opine") such as octopine or nopaline and (ii) to determine functions for the bacterial catabolism of the same compound (Petit *et al.*, 1970; Lippincott *et al.*, 1973; Watson *et al.*, 1975; Bomhoff *et al.*, 1976; Genetello *et al.*, 1977; Van Larebeke *et al.*, 1977; Montoya *et al.*, 1977; Schell and Van Montagu, 1977).

In our laboratory one particular octopine plasmid, pTiAch5, and one particular nopaline plasmid, pTiC58, have been extensively studied. For both of these plasmids physical maps of DNA fragments generated by digestion with several restriction endonucleases have been constructed (Depicker *et al.*, 1980; De Vos *et al.*, in preparation). Hybridizations with the Southern blotting method (Southern, 1975) between plant DNA (isolated from crown gall tumors induced with both of these plasmids) and cloned fragments of the respective plasmids have allowed a precise definition of the T-DNA region for both of the plasmids pTiAch5 and pTiC58 (De Beuckeleer *et al.*, in preparation). Furthermore the extent of DNA homology between the octopine and the nopaline plasmids has been determined both by the Southern blotting hybridization technique and by electron microscopic analysis of heteroduplex molecules (Engler *et al.*, in preparation). This knowledge

0147-619X/80/020212-19$02.00/0

provided an excellent background for genetic studies that attempted to localize known pTi phenotypes on the physical map and to understand the functional organization of these large plasmids.

Here we report on the isolation and characterization of mutants of the nopaline plasmid pTiC58. In order to locate mutations easily, we have isolated insertions and deletions that provided physical markers on the Ti-plasmid. For this purpose we have employed the transposons Tn*1* (Hedges and Jacob, 1974) and Tn*7* (Barth *et al.*, 1976), and the P-type plasmid RP4 (Datta *et al.*, 1971) which inserts into the Ti-plasmids (Holsters *et al.*, 1978). Transposons can exert a mutagenic effect either by direct or polar insertion, and can promote deletions (Kleckner, 1977). Previously, we reported that the streptomycin–trimethoprim resistance transposon Tn*7* can insert in several sites of the octopine plasmid pTiB6S3 (Hernalsteens *et al.*, 1978). To isolate these insertion derivatives we used a cointegrated pTiB6S3::RP4 plasmid. Concomitantly, techniques were developed to isolate Tn*1* and Tn*7* transpositions directly in Ti-plasmids. Using the latter approach we have now isolated a large set of transposon insertions in the pTiC58 plasmid which has enabled us to locate most of the known Ti phenotypes on the physical map. We obtained evidence for the existence of several distinct regions that are required for the full expression of the oncogenic phenotype and most importantly we have obtained genetic evidence for the role of the T-DNA in oncogenicity and in the determination of nopaline biosynthesis in plant tumor cells.

MATERIALS AND METHODS

Bacterial strains, plasmids, and transposons. The bacterial strains, plasmids, and transposons used in this work are listed in Table 1.

Media. YEB medium and minimal nopaline or octopine medium were as described previously (Van Larebeke *et al.*, 1977).

Minimal A medium was as described by Miller (1972).

Antibacterial drugs. Antibacterial drugs were added to the media at the following concentrations: rifampicin 50 μg/ml; erythromycin 100 μg/ml; chloramphenicol 50 μg/ml; carbenicillin 200 μg/ml; streptomycin and spectinomycin 100 μg/ml in rich medium; tetracycline 2.5 μg/ml.

Restriction endonucleases. The origin of the restriction endonucleases and the reaction conditions used were described by Depicker *et al.* (1980).

Oncogenicity. Oncogenicity tests were performed on pea seedlings according to Manigault and Kurkdjian (1967), on potato tuber slices according to Anand and Heberlein (1977), and on young *Kalanchoë* plants as described by Bomhoff *et al.* (1976).

Detection of nopaline in tumor tissues. Crude extracts of tumor tissues were analyzed for the presence of nopaline with a rapid electrophoretic method, as described by Aerts *et al.* (1979). Tissues that contained no detectable amounts of nopaline were thoroughly reexamined with the more elaborate technique described in Van Larebeke *et al.* (1977).

Exclusion of phage AP1 and sensitivity to the agrocin 84. Exclusion of phage AP1 (Schell, 1975) and sensitivity to agrocin 84 were determined as described by Van Larebeke *et al.* (1977).

Assay for the catabolic Ti-functions. Utilization of nopaline or octopine as a nitrogen (N) source was assayed by inspecting the growth on minimal nopaline or octopine agar medium (Van Larebeke *et al.*, 1977) with 0.2% sucrose as a carbon (C) source and 100 μg nopaline (or octopine)/ml as sole N source. Known positive and negative strains were included on each plate as controls.

Utilization of arginine or ornithine as C source was determined by growth of strains on minimal nopaline agar medium without sucrose but with 0.2% arginine or ornithine

TABLE 1

BACTERIAL STRAINS, PLASMIDS, AND TRANSPOSONS

Strain		Origin
A. tumefaciens strains		
GV3100	C58-C1, a derivative cured of pTiC58	Van Larebeke *et al.* (1974)
GV3101	C58-C1 Rif^R	Our isolate
GV3102	C58-C1 Rif^R Ery^R	Our isolate
GV3103	C58-C1 Str^R Spc^R	Our isolate
GV3105	C58-C1 Ery^R Cm^R	Our isolate
GV3145	C58 (RP4)	Our isolate
GV3159	C58-C1 Str^R Spc^R (pGV3100)	This work
GV3166	C58-C1::Tn7	Hernalsteens *et al.* (in preparation)
GV3129	C58-C1 Rif^R (pGV5007)	Hernalsteens *et al.* (in preparation)
GV3170	C58-C1::Tn7 (pGV3100) constructed by mating strain GV3159 with GV3166	This work
GV3152	C58-C1 Rif^R (pGV5007) (pGV3100) constructed by mating strain GV3159 with GV3129	This work
Plasmids		
pTiC58	nopaline Ti plasmid from strain C58	Van Larebeke *et al.* (1974)
pGV3100	pTiC58 Tra^C, a spontaneous mutant of pTiC58, derepressed for autotransfer	This work
pAtC58	272 M*d* cryptic plasmid present in strains C58 and C58-C1	Cassé *et al.* (1979) Villarroel *et al.* (in preparation)
RP4	P-type plasmid; Km^R Tc^R Cb^R	Datta *et al.* (1971)
pGV4000	pTiC580cc⁺::RP4	Holsters *et al.* (1978)
pGV5007	deletion mutant of RP4::Mucts61; Mu^- Km^S Tra^- Cb^R Tc^R	Hernalsteens *et al.* (in preparation)
pGV0329	Hybrid plasmid between pBR322 and the pTiC58 *Hin*dIII fragments 31, 23, 33, 2, 30, 43 and 26	Depicker *et al.* (1980)
Transposons		
Tn*1*	carbenicillin resistance transposon	Hedges and Jacob, (1974)
Tn*7*	streptomycin, spectinomycin, and trimethoprim resistance transposon	Barth *et al.* (1976)

Abbreviations for the pTi encoded phenotypes[a]

Noc, nopaline catabolism	Agr^S, agrocin 84 sensitivity	Occ, octopine catabolism	Agr^R, agrocin 84 resistance
Orc, ornithine catabolism	Ape, phage AP1 exclusion	Arc, arginine catabolism	Onc, oncogenicity
Nos, nopaline synthesis	Tra, autotransfer		

[a] As agreed in the IVth International Conference on Plant Pathogenic Bacteria (Angers, France, August 27 to September 1, 1978).

as C source. When mutants constitutive for arginine or ornithine catabolism were selected, nopaline was omitted from the latter medium.

Autotransfer of pTiC58 Tra^c: Filter conjugation. Donor and recipient strains were grown to exponential growth phase in minimal A medium and mixed in a proportion of 5:1 donors to recipients. One milliliter of the mixture was placed on the surface of a Millipore filter (pore size of 0.45 μm) and incubated for 24 h at 28°C on minimal A agar medium. The cells were resuspended in buffer and suitable dilutions were plated on media selective for recipients and transconjugants.

Replica conjugation. Potential donor colonies were inoculated in 0.5 ml minimal A medium in the wells of a replica device that contained 25 subdivisions. With a block of 25 needles, 25 donors could be replica plated simultaneously. The (rifampicin-sensitive) donor colonies were spotted onto the surface of a minimal nopaline agar plate with 50 μg rifampicin/ml that had been seeded with a Rif^R C58-C1 recipient strain. Plates were scored for nopaline consuming transconjugants after 4 days of incubation. As a control the "donor" cultures were spotted on nopaline rifampicin plates without the recipient strain. No growth occurred on these plates.

Transposition of Tn1 to pTiC58 Tra^c. A filter conjugation was carried out between the donor strain GV3152 (Table 1) and an $Ery^R Cm^R$ C58-C1 recipient strain (GV3105). The donor strain harbored both a Tra^c pTiC58 derivative (pGV3100) and, as a source of Tn*1*, a Km^S Tra^- Cb^R Tc^R derivative of the RP4 plasmid (pGV5007; Hernalsteens *et al.*, in preparation). After incubation the undiluted conjugation mixture was plated on YEB agar plates with 100 μg erythromycin/ml, 50 μg chloramphenicol/ml, and 200 μg carbenicillin/ml. Also, in order to determine the efficiency of the pTiC58 Tra^c transfer, suitable dilutions were plated on minimal nopaline agar plates with 100 μg erythromycin/ml and 50 μg chloramphenicol/ml.

Transposition of Tn7 to pTiC58 Tra^c. A donor strain, GV3170 (Table 1), was mated with an $Ery^R Cm^R$ C58-C1 recipient (GV-3105). The donor strain harbored the Tra^c TiC58 plasmid and a chromosomally located Tn7 insertion. After incubation the conjugation mixture was plated undiluted on YEB agar medium with 100 μg erythromycin/ml, 50 μg chloramphenicol/ml, and 100 μg streptomycin and 100 μg spectinomycin/ml to select for Tn7 containing transconjugants. Also, for a determination of the total pTiC58 Tra^c transfer frequency, suitable dilutions were plated on minimal nopaline medium with erythromycin and chloramphenicol.

Isolation of pTiC58::RP4 cointegrates. The strain GV3145, harboring separate pTiC58 and RP4 replicons, was grown for several days at 37°C in minimal medium with nopaline as the sole source of N. Subsequently the total bacterial population was used as the donor in a RP4-directed conjugation as described by Holsters *et al.* (1978) with strain GV3102 as recipient. The conjugation mixture was plated on minimal nopaline plates containing 50 μg rifampicin/ml and 100 μg erythromycin/ml. Noc^+ colonies that appeared on these plates were further analyzed for the presence of cointegrated plasmids as described by Holsters *et al.* (1978).

Fingerprint analysis of restriction endonuclease digests. Ti-plasmid fingerprints were obtained by digesting 1 μg of purified Ti-plasmid DNA with a restriction endonuclease followed by electrophoresis of the digest in a 0.8% agarose gel as described by Depicker *et al.* (1980).

Alternatively pTi fingerprints were obtained by a quick method described by Dhaese *et al.* (1979): total bacterial DNA was digested with a restriction endonuclease and subjected to gel electrophoresis in 0.8% agarose. The gel was blotted onto a nitrocellulose filter (Southern, 1975) and the filter hybridized against a plasmid probe (pTi or a subclone in pBR322), ^{32}P-labeled by nick translation (Rigby *et al.*, 1977). The hybridization conditions and the preparation of plasmid DNAs were described by Depicker *et al.* (1980).

Electron microscopy. The electron microscope techniques used were as described by Kleinschmidt (1968). Heteroduplex analysis was performed according to Davis *et al.* (1971).

RESULTS

Isolation of a Tra^c pTiC58 Mutant

In order to isolate transposon insertions in pTiC58 we made use of the conjugative properties of the Ti plasmids. Since the conjugational transfer of the wild type TiC58 plasmid is repressed (Kerr *et al.*, 1977; Genetello *et al.*, 1977; Petit *et al.*, 1978), we have isolated a derepressed mutant, pTiC58 Tra^c, by mating a wild type C58 donor strain with a streptomycin- and spectinomycin-resistant C58-C1 recipient

(GV3103) (see Materials and Methods). Transconjugants that had acquired the Ti-linked capacity to utilize nopaline as N source (Noc$^+$) were secondarily scored for ability to donate the same capacity at an elevated frequency. Among a total of over 10^9 cells in the original C58 starting culture, a single exceptional colony was found that in subsequent steps transferred its Ti plasmid with efficiencies of up to 30% (transconjugants per recipient). This mutant plasmid was used for the isolation of all transposon insertions. The rarity of this type of mutant suggests that more than one mutational step may be required.

Isolation of Transposon Insertions in pTiC58 Tra^c

Hernalsteens *et al.* (in preparation) describe the construction of Ti-plasmid-free *Agrobacterium* strains with a Tn*1* or Tn*7* transposon inserted in a nontransferrable replicon. After introducing the TiC58 Tra^c plasmid into these strains, transpositions to the conjugative plasmid could be selected by cotransfer of the transposon-coded antibiotic resistances with the Ti plasmid. For the isolation of Tn*7* insertions, the TiC58 Tra^c plasmid was transferred from a donor-strain GV3170 (Table 1) that harbored a chromosomally located Tn*7* transposon, to an erythromycin- and chloramphenicol-resistant C58-C1 recipient. The total Ti transfer frequency was determined by the number of Noc$^+$ transconjugants. On average a fraction of 10^{-5} of these had acquired the Tn*7*-coded resistance to streptomycin and spectinomycin. Two hundred Ery^R Cm^R Sp^R Sm^R colonies derived from several independent crosses were purified and further analyzed.

As a source of Tn*1* transpositions we used the P-type plasmid RP4 since it harbors this carbenicillin-resistance (Cb^R) transposon. A donor strain GV3152 with a Tra^c Ti-plasmid and a nontransferrable RP4 derivative was mated with an Ery^{R}Cm^R C58-C1 recipient. The transconjugant population contained on average 10^{-6} or fewer Cb^R colonies. These Cb^R colonies were of four different types: (i) the majority were Tc^S and expressed pTi markers; (ii) 5% were Tc^R (an RP4-determined feature) and also expressed pTi markers; (iii) 3% were Tc^R and expressed no pTi markers; (iv) 3% were Tc^S and expressed no pTi markers.

Colonies of type (i) were likely candidates for harboring Tn*1* insertions in the TiC58 plasmid and were purified and further analyzed.

Colonies of types (ii) and (iii) were not studied further. By some mobilization mechanism they had probably received the Tra$^-$ RP4 plasmid either alone or together with the Ti plasmid. Colonies of type (iv) were interesting because it was shown that they harbored a Tn*1* transposon inserted in the large 272 Md cryptic pAtC58 plasmid that is present in C58 and in C58-C1 (Cassé *et al.*, 1979; Villarroel *et al.*, in preparation).

Strategy for the Characterization of Transposon Insertions and Transposon Promoted Deletions

Cb^R and Sp^R Sm^R colonies obtained as described above were assayed for the Ti determined phenotypes of oncogenicity (Onc), nopaline synthesis in tumors (Nos), capacity to catabolize nopaline (Noc), exclusion of phage AP1 (Ape), sensitivity to the agrocin 84 (Agr) (Engler *et al.*, 1975) and autotransfer (Tra). Mutations for all phenotypes were found. In all mutants studied, the presence of a transposon or a transposon-promoted deletion in the Ti plasmid could be demonstrated. Mutant Ti plasmids were compared to wild-type by fingerprinting with several restriction endonucleases. Insertion of a transposon in a particular restriction fragment causes the disappearance of this fragment and the appearance, depending on the transposon and on the enzyme, of one or more new bands. Similarly deletion mutants were mapped by alterations in fingerprint patterns and by sizes of new composite frag-

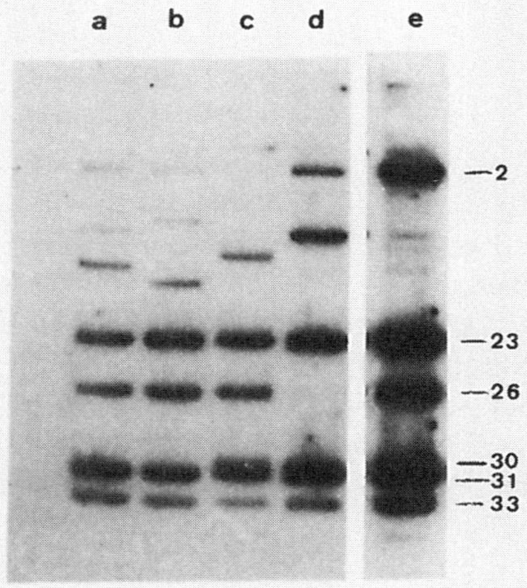

FIG. 1. Analysis of Noc⁻Arc⁻Orc⁻ Tn7 insertion mutations in pTiC58. Hybridization of 1–3 μg of *Hin*dIII digested total bacterial DNA with ^{32}P-labeled pGV0329 (see Table 1; 5 × 10^{6} cpm/ml). Bacteria harbored the following plasmids: track a, pGV3129; b, pGV3109; c, pGV3131; d, pGV3130; e, pTiC58. Note that the insertions in pGV3129 and 3109 are at different locations in the same fragment.

ments. In all cases missing fragments constituted a continuous segment of DNA on the physical map of pTiC58. Since this restriction map was obtained independently (Depicker *et al.*, 1980), both sets of results were mutually confirmative.

The fingerprints were obtained either directly, by digestion of purified plasmid DNA followed by gel electrophoresis, or indirectly (Dhaese *et al.*, 1979). With this latter method total bacterial DNA was digested with a restriction endonuclease and the Ti-plasmid composition was visualized after gel electrophoresis, Southern blotting, and hybridization with radioactively labeled, purified pTiC58 as a probe. After the locaion of a mutation was approximated, mapping was refined using a cloned subregion of the TiC58 plasmid as probe (see Figs. 1 and 2). Inasmuch as Tn7 contains three internal *Hin*dIII sites, insertions of the transposon generate four new *Hin*dIII fragments, two involving the transposon junctions and two containing only transposon sequences. Use of pTiC58 DNA as a probe resulted in the visualization of only the junction fragments. Internal transposon fragments were visualized only with transposon-containing probes. We examined 42 independent insertion mutations and 9 independent deletions, but we shall present here only an illustration of mapping techniques applied to Tn7 induced mutations (Figs. 1 and 2).

To confirm or refine the mapping obtained by analysis of restriction endonuclease fragment patterns, several mutant plasmids were also studied by electron microscopic examination of heteroduplex molecules (Fig. 3).

In subsequent sections features of several transposon-induced plasmid mutations will be described. These results are summarized

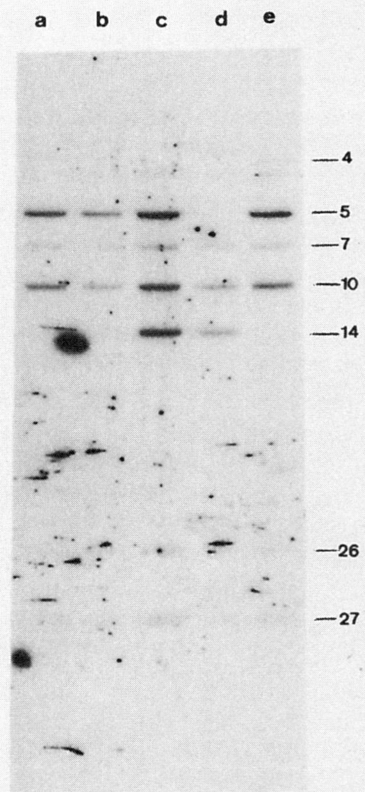

FIG. 2. Analysis of Noc⁻Arc⁻Orc⁻ Tn7 insertions in pTiC58. Hybridization of 1–3 μg of *Sma*I digested total bacteria DNA with ^{32}P-labeled pGV0329 (Table 1; 5 × 10^{6} cpm/ml). Bacteria harbored the following plasmids: track a, pGV3129; b, pGV3109; c, pTiC58; d, pGV3131; e, pGV3130.

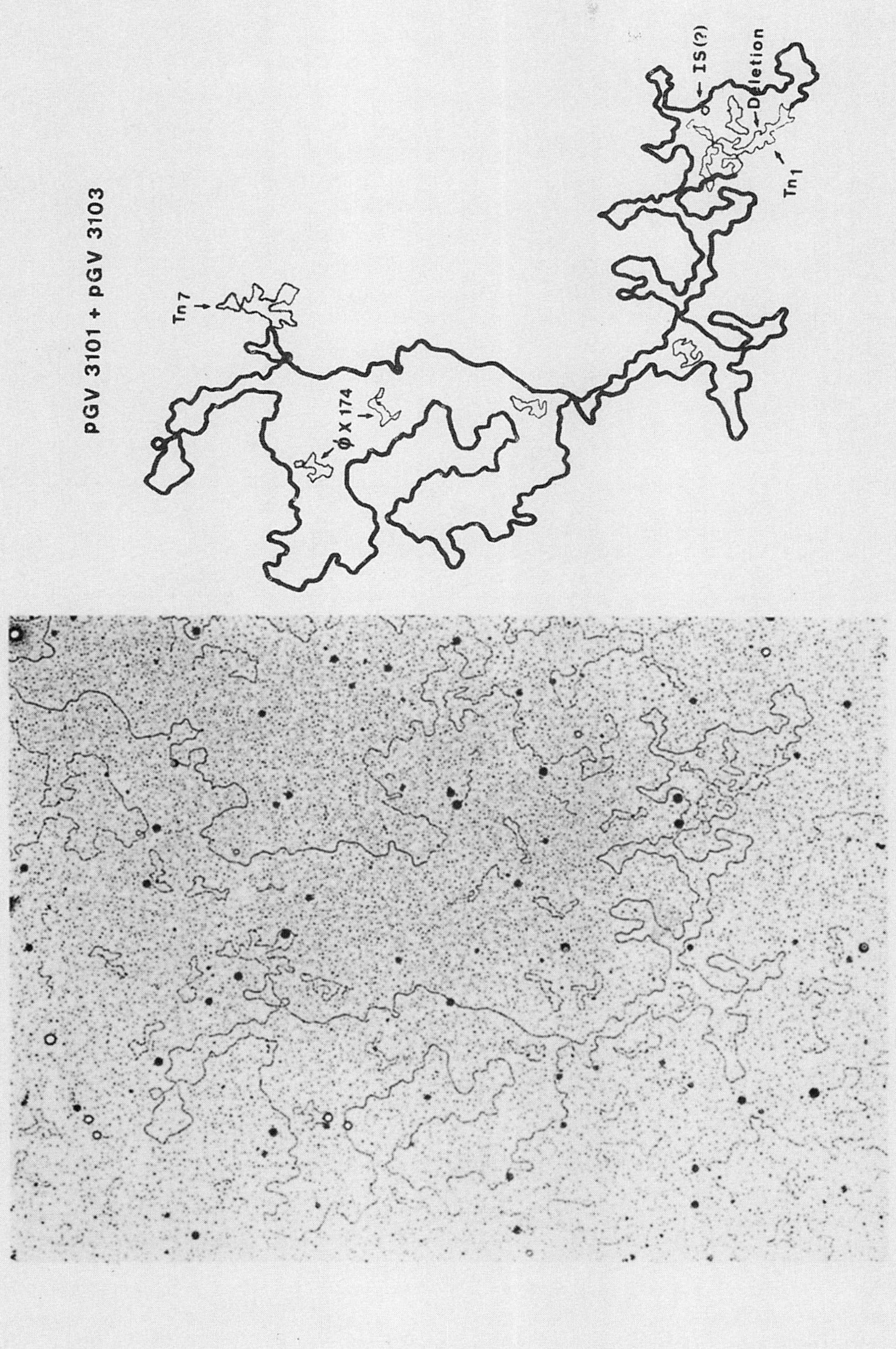

FIG. 3. Electron micrograph and tracing of a heteroduplex molecule formed by renaturation between single-stranded pGV3101 (Table 3) and single-stranded pGV3103 DNA (Table 2). The pGV3101 plasmid harbors a Tn*1* insertion (3.2 Md) and an adjacent deletion of 11.8 Md. The pGV3103 plasmid has a Tn7 insertion (9.5 Md). ×10,000.

TABLE 2

CHARACTERISTICS OF 42 INDEPENDENT INSERTION MUTATIONS IN pTiC58

A. Insertions affecting oncogenicity

Plasmid number	Insertion of	Restriction fragments modified				Approx. map position (Fig. 4) in Md	Oncogenicity on		
		*Hin*dIII	*Sma*I	*Hpa*I	*Kpn*I		Peas	Potatoes	Kalanchoë
3139	Tn7	4		1a		Between 70.4 and 76.4	−	−	
4009	RP4	4	3a	1a		75	−	−	
3185	Tn1	9		15	5	98	−	−	−
3138	Tn7	9		11		99	−	−	
3121	Tn7	9	16	9	5	100	−	−	−
3165	Tn1	14a		5		105	−	−	−
3133	Tn7	27		8		107	−	−	−
3134	Tn7	27		8		107	−	−	−
3135	Tn7	27		8		107	−	−	−
3136	Tn7	27		8		107	−	−	−
3137	Tn7	27		8		107	−	−	−
3187	Tn1	7		2	9	Between 113 and 116	−	−	−
3177	Tn1	7		2		Between 113 and 116	−	−	+
4012	RP4	10	1a	2		117	+	−	
3132	Tn7	19	7	6		128	−	−	+
3166	Tn1	41		6		129	−	−	+
4000	RP4	22	7	12		130.4	−	−	
4007	RP4		5		3	000.4	−	−	

B. Other insertions

Plasmid number	Insertion of	Restriction fragments modified			Approx. map position (see Fig. 4) in Md[a]	Phenotypes
		*Hin*dIII	*Sma*I	*Kpn*I		
3189	Tn1	23[b]	5		000.8	Nos⁻
3143	Tn7	4			Between 70.4 and 76.4	Tra⁻
3103	Tn7	11	1b		87	Agr^R
3108	Tn7	11	1b		87	Agr^R
3102	Tn1				a)[c] 89.6	None affected
	Tn1				b)[c] 91.6	None affected
3140	Tn7	34		10	92	None affected
3181	Tn1	10	1a	13	119.2	None affected
3182	Tn1	14	1a		123.6	None affected
3188	Tn1	13			Between 19 and 22	None affected
3123	Tn7	2	5	12	4.4	Noc^- Arc^- Orc^-; the bacteria harbor-
3131	Tn7	2	5		4.4	ing these mutant plasmids can no
3127	Tn7		26		5.6	longer utilize nopaline as N source,
3124	Tn7		11a	12	6.2	arginine and nopaline as C source
3161	Tn1	30	14		11.4	
3176	Tn1	30	14		11.4	
3129	Tn7	30	14		11.4	
3128	Tn7	30	14		11.4	
3109	Tn7	30	14		11.4	
3126	Tn7	30	14		11.4	
3144	Tn7	43			12.0	
3130	Tn7	26	14		12.2	$Noc^\pm$ $Occ^\pm$ Arc^- Orc^-; bacteria with
3162	Tn1	26	14		12.2	these mutants plasmids can to some
3164	Tn1	26	14		12.2	extent utilize both octopine and nopaline as N source

[a] The listing of several insertions at the same map position is not intended to imply that they are at the same site but rather indicates their relative proximity on the map (see Fig. 1 tracks a and b for example).

[b] The Tn1 insertion in pGV3189 was allocated to the left part of *Hin*dIII 23 as indicated in Fig. 4 by analysis with the restriction endonuclease *Sac*II (data not shown).

[c] Spontaneous isolation of two Tn1 insertions in the same plasmid. They were mapped by electron microscopic technique and are present in inverted orientation.

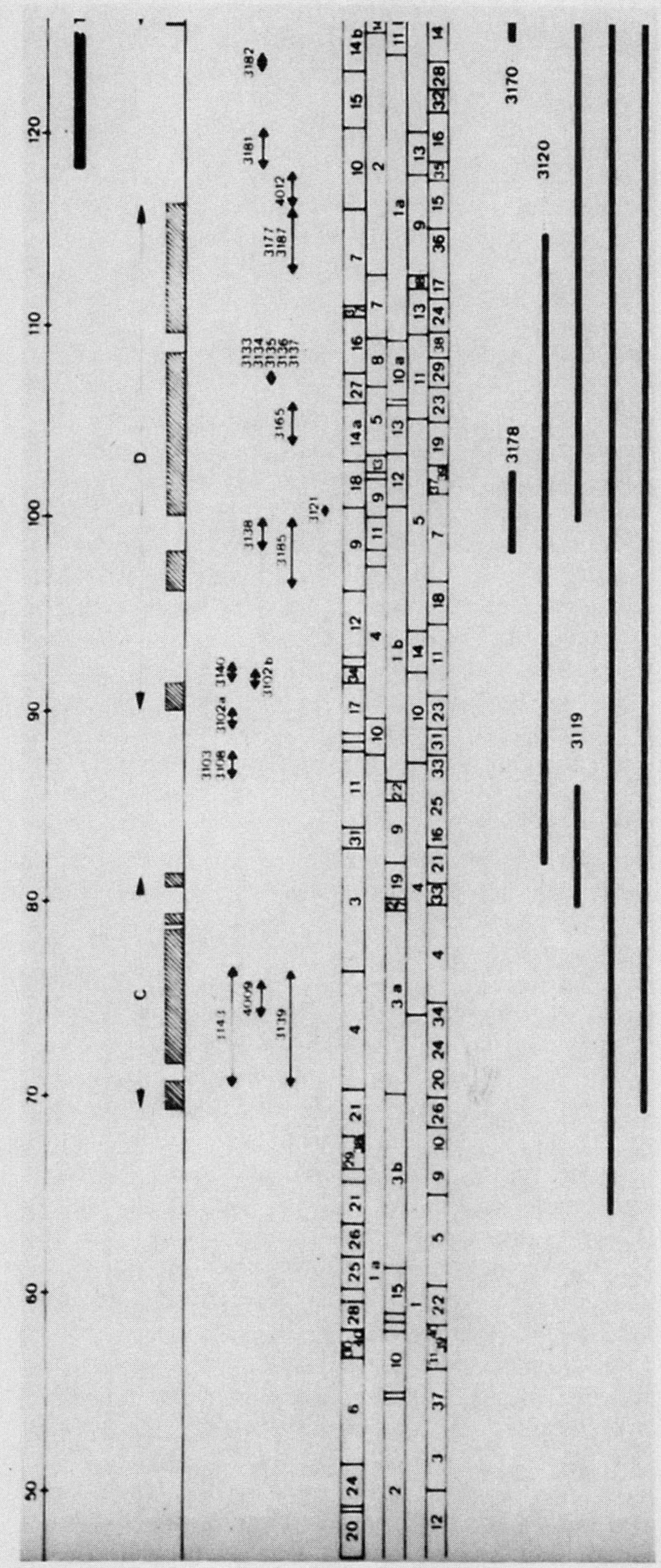
50
60
70
80
90
100
110
120
C
D
3170
3120
3178
3119

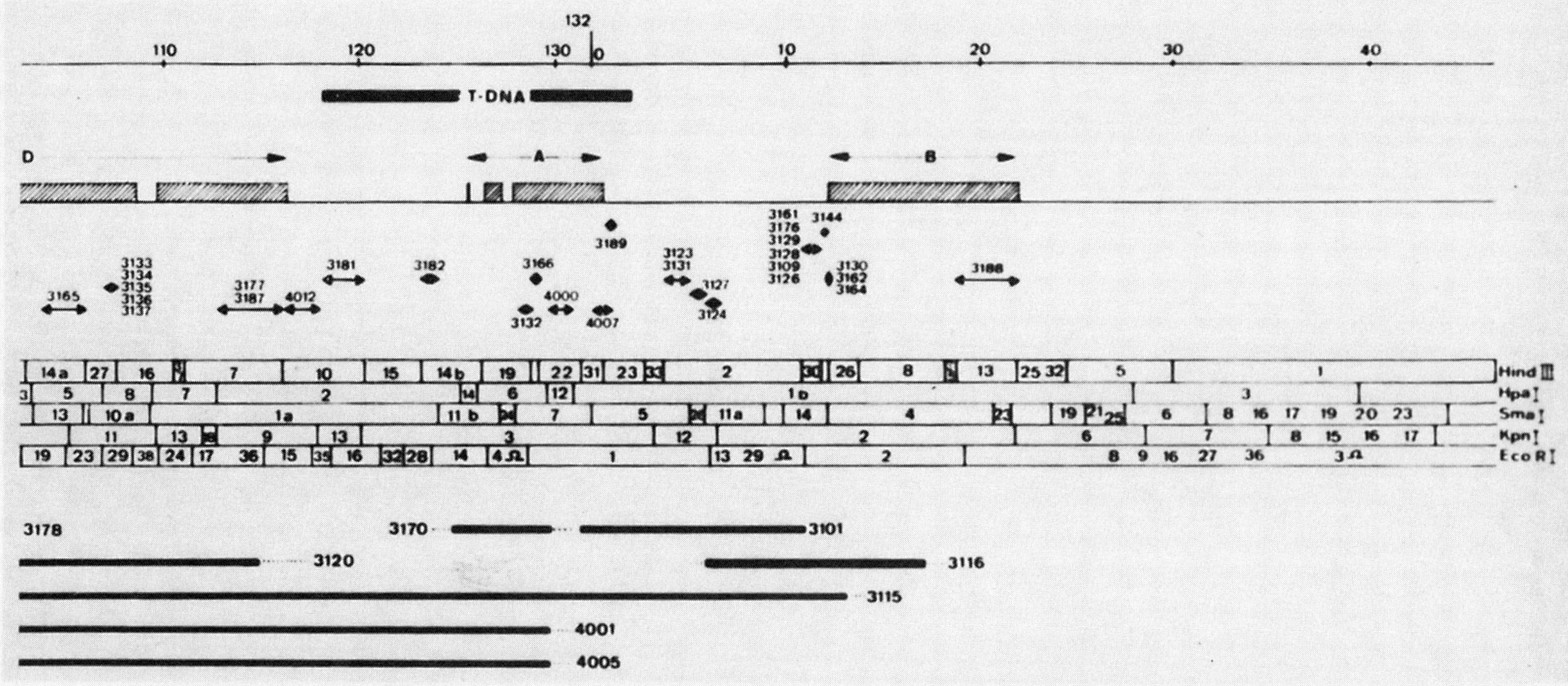

FIG. 4. The location of independent insertion and deletion mutations in the physical map of pTiC58. The physical map of pTiC58 and the numbering of the restriction fragments are according to Depicker *et al.* (1980). The *Eco*RI map depicted here however is a preliminary one; we refer to Depicker *et al.* for the updated *Eco*RI map. The T-DNA region was determined by De Beuckeleer *et al.* (in preparation). The dashed areas (A, B, C, and D) are the regions of homology of pTiC58 with the octopine plasmid pTiAch5 as determined by Engler *et al.* (in preparation). Deletions are represented by full black lines (——). The arrowed lines (↔) delimit regions wherein particular insertions are integrated. The distance between the arrows thus gives an estimation of the accuracy of mapping of these insertions. The features of the mutations represented in this figure are listed in Tables 2 and 3.

TABLE 3

CHARACTERISTICS OF DELETIONS IN pTiC58

Plasmid number	Deletion generated by	Approx. length in Md	Approx. map position in Md (see Fig. 4)	Phenotype				
				Oncogenicity on			Other phenotypes	Remarks
				Peas	Potatoes	Kalanchoë		
4001	RP4	66	From 64 to 130	–	–	–	Agr^R Ape^-	RP4 intact
4005	RP4	61	From 69 to 130	–	–	–	Agr^R Ape^-	RP4 intact
3178	Tn*1*	4.2	From 98.2 to 102.4	–	–	–	None	Tn*1* intact
3170	Tn*1*	5.0	From 125 to 130	–	–	+	None	Tn*1* intact
3101	Tn*1*	11.8	From 131.2 to 11	+	+	+	Noc^- Arc^- Orc^- Nos^-	Tn*1* intact
				Slow growing tumors				
3119	Tn*7*	8.0	From 80 to 88	+	+	+	Agr^R Tra^-	Part Tn*7* deleted[a]
3120	Tn*7*	33.0	From 82 to 115	–	–	–	Agr^R Tra^-	Original Tn*7* gone[a]
3115	Tn*7*	46.2	From 99 to 13.2	–	–	–	Noc^- Arc^- Orc^- Tra?	Original Tn*7* gone[a]
3116	Tn*7*	11.4	From 6 to 17.4	+	+	+	Noc^- Arc^- Orc^- Tra^- (leaky)	Tn*7* intact

[a] In all cases when the original Tn*7* transposon was shown to be partly or completely deleted, the cells were still resistant to streptomycin and spectinomycin, probably due to a second copy of Tn*7*, inserted in the chromosome or in the pAtC58 plasmid.

(Tables 2 and 3) and the positions of the mutations mapped (Fig. 4).

Transpositions Affecting Oncogenicity

A number of Cb^R and Sm^RSp^R colonies could no longer induce tumors on peas and potatoes. We localized four independent Tn*1* insertions and seven independent Tn*7* insertions that caused this phenotype (see Table 2 and Fig. 4). Two of the Tn*1* insertions were associated with a deletion that was adjacent to such insertion (mutant plasmids 3170 and 3178, Table 3 and Fig. 4).

A different phenotype was caused by the Tn*1* deletion-associated insertion in plasmid 3101 (Table 3 and Fig. 4). Bacteria harboring the mutant 3101 plasmid could still induce tumors, but the tumors grew very slowly, remained small, and did not contain any detectable nopaline. This 3101 deletion is very interesting because it extends into the T-DNA of pTiC58 and thus defines a segment of the T-DNA that is not essential for oncogenicity (see Discussion). A similar result has been obtained with an octopine Ti-plasmid by Koekman *et al.* (1979). The 3101 deletion also affects nopaline catabolism (Noc) as described below. The other known Ti-plasmid functions are normally expressed (Ape^+ Agr^S Tra^+).

Mutant strains unable to produce tumors on peas and potatoes were subsequently tested on *Kalanchoë*, to see whether some of the genes involved in tumor induction in one plant would also be required for tumor induction and/or maintenance in other plants. Bacteria harboring the mutant Ti-plasmids 3187, 3135, 3134, 3133, 3165, 3121, 3185, or 3178 (see Fig. 4) proved nononcogenic on *Kalanchoë*. However bacteria harboring the mutant plasmids 3170, 3166, or 3132 were able to induce tumors on *Kalanchoë*. Interestingly the mutations in plasmids 3132, 3166, and 3170 map within the T-DNA!

Transposon Insertions Affecting the Catabolic Ti-Functions

The Ti-plasmid determined capacity to utilize nopaline as N and C source is normally repressed and is induced by nopaline but not by octopine (Petit and Tempé, 1975, 1978). However, Ti-plasmid mutations that derepress nopaline catabolism allow the utilization of octopine as N and C source. Thus octopine catabolism (Occ) measures constitutivity of the nopaline operon of pTiC58. Recently Ellis *et al.* (1979) have shown that the pTiC58 plasmid also determines the capacity to utilize arginine and ornithine as C source. These functions are also repressed and similarly are induced by nopaline but not by arginine or ornithine. We have screened our collection of Ti-mutant strains for both the inactivation of and the constitutivity of the capacity to utilize nopaline as N source (Noc) and arginine or ornithine as C source (Arc, Orc). We found two phenotypically distinct types of mutants. The first type consisted of Tn*1* and Tn*7* insertions that inactivated the Noc as well as the Arc and Orc phenotypes (Noc^- Arc^- Orc^-). Mutants of the type Noc^- Arc^+ or Noc^- Orc^+ were not found; moreover, we have recently been able to isolate from such insertions secondary deletions that were Noc^- Arc^+ (data not shown). These observations suggest that mutations in genes for earlier functions in the pathway for nopaline catabolism might exert a direct or indirect polar effect on Arc and Orc functions and support a common regulation mechanism for the expression of these catabolic functions. Amazingly the Noc^- Arc^- Orc^- mutations are located in two separate regions of the physical map of pTiC58, one of them around map position 5 Md, the other around map position 12 Md (see Fig. 4). In between is a stretch of at least 4 Md without known insertion mutations. Some subtle differences in growth on minimal nopaline agar media (more or less mucoid) were observed between mutants representative of the two Noc^- regions. An explanation of these differences is not yet known.

A second series of mutants show a rather complicated phenotype. They behave to some extent as constitutive for nopaline catabolism. They can utilize both octopine and nopaline as N-source but require longer to display the full positive phenotype on minimal nopaline or octopine medium. Also they cannot utilize arginine or ornithine. Thus their behavior cannot be explained by inactivation of a repressor gene that might negatively control the nopaline catabolism operon. All mutations of this type map at the same position, 12 Md (see Fig. 4). Presently we have not isolated any transposon-induced mutation that would render the catabolic functions constitutive without also affecting other phenotypes. We are studying the partially constitutive mutants and representatives of the two classes of Noc^- mutants in the hope of identifying the impaired steps in the catabolic pathway.

*Other Transposon Insertions in pTiC58 Tra^c^*

We have also mapped some Tn*1* and Tn*7* insertions that affect none of the known pTiC58 functions (see Table 2). These mutants define pTiC58 regions that are nonessential for oncogenicity.

The two independently isolated Tn*7* insertions in plasmids 3103 and 3108 that inactivate the Agr^s phenotype have been allocated to position 87 Md on the pTiC58 map (see Fig. 4).

Out of a total of 400 transposon insertions in pTiC58 Tra^c that have been studied only one was found that caused the loss of the transfer ability (Tra^-). It is located in *Hin*d-III fragment 4 (see Fig. 4). This Tra^- mutant plasmid is nevertheless Onc^+. Tra^- insertions are extremely infrequent (compared to Onc^- or Noc^- mutations) probably because these transpositions in pTiC58 were isolated by an initial transfer selection.

Finally, one Tn*1* insertion in pTiC58Tra^c is located in the extreme right (Figs. 4 and 6) of the T-region and inactivates the Nos phenotype. That is, bacteria with mutant plasmid 3189 induce quasi-normal tumors but such tumors contain no nopaline. A mutation with similar phenotype and location is caused by a Tn*7* transposition in the nopaline plasmid pTi T37. In the left non-conserved part of the T-DNA two independent Tn*1* insertions, in plasmids 3181 and 3182, were found that do not affect either oncogenicity or nopaline biosynthesis. A more detailed report on the features of these particular transposon insertions in the T-DNA and on their relevance for genetic engineering of plants will be published separately. We note here that we have recently succeeded in demonstrating by the

Southern blotting technique (Southern, 1975) that the Tn*1* DNA has been incorporated into the plant genome along with the T-DNA following the induction of tumors with these two plasmids (De Beuckeleer *et al.*, in preparation).

Deletions Derived from Tn7 Insertions

Some Tn*7*-induced mutations reverted to the wild-type phenotype with a high frequency and were a ready source for the isolation of deletions in pTiC58. Such deletions produced several different phenotypes. Individual colonies of such strains were screened for reversion. Nonreverting derivatives were analyzed and were found to harbor deletions. Two deletions, in plasmids 3119 and 3120 (Table 3), were isolated from the Tn*7* insertion, in plasmid 3103, which has an extremely unstable Agr^R phenotype. Both deletions showed a stable Agr^R phenotype and were Tra^-. In addition, the deletion in plasmid 3120 affected oncogenicity. The deletion in 3119 is interesting because it comprises an 8 Md region and has a Tra^- but Onc^+ phenotype. Also from the difference between the 3119 deletion (Ape^+) and that in plasmid 4005 (Ape^-) one can deduce that a 9 Md region must be correlated with phage AP1 exclusion. Two deletions, in plasmids 3115 and 3116, were derived from an unstable Noc^- Tn*7* insertion 3109 (Table 3). The smallest deletion in 3116, affects autotransfer but not oncogenicity. The deleted plasmid still harbors the original Tn*7* insertion but the conjugative transfer of the Sp^RSm^R markers is reduced by a factor of 10^3. Exceptional Sp^RSm^R transconjugants were isolated and showed an equally reduced transfer efficiency as donors in subsequent crosses. Thus the deletion in 3116 is a leaky Tra^-. As neither the previously described deletion in 3101 nor the Noc^- insertions near map position 12 Md affect transfer, it follows from Fig. 4 that the Tra functions affected in the 3116 deletion must lie in a 6 Md region to the right of the Noc region.

Isolation of New pTiC58::RP4 Cointegrates

It has been described previously (Holsters *et al.*, 1978) that the P-type plasmid RP4 can insert in the TiC58 plasmid with formation of a large cointegrated replicon. We have isolated and characterized three new independent pTiC58::RP4 cointegrates: pGV4007, pGV4009, and pGV4012. Like the original cointegrate pGV4000 (Holsters *et al.*, 1978) they consist of a complete RP4 and a complete TiC58 plasmid. However the site of integration of RP4 in Ti is different in each isolate (see Fig. 4). The cointegration seems to be an RP4 directed event; it always occurs in the same region of RP4 and is accompanied by a 1.2 Md RP4 duplication in this region. The investigation of this particular cointegration mechanism will be described separately by Depicker *et al.* (in preparation).

Bacteria harboring the cointegrates pGV-4007 and pGV4009 cannot induce tumors on potatoes and peas. Interestingly the cointegrate pGV4012 allows tumor induction on peas but not on potatoes. This suggests a second pTiC58 region that might be required only for tumor induction on particular host plants.

Deletions Derived from the Cointegrate pGV4000

The cointegrated plasmid pGV4000 is less frequently cured by heat treatment than the TiC58 plasmid (Holsters *et al.*, 1978). Therefore it was used to select agrocin 84-resistant colonies and these were subsequently screened for the loss of a second Ti marker. Two large deletions of resp. 66 (4001) and 61 Md (4005) were isolated and both were $Agr^R \cdot Ape^-$. These deletions were presumably generated as a consequence of the presence of RP4 in pTiC58. Indeed, both start at one junction between RP4 and Ti and extend into the Ti DNA leaving RP4 intact. This was deduced from fingerprints of restriction endonuclease digests (data not shown) and from electron microscopic

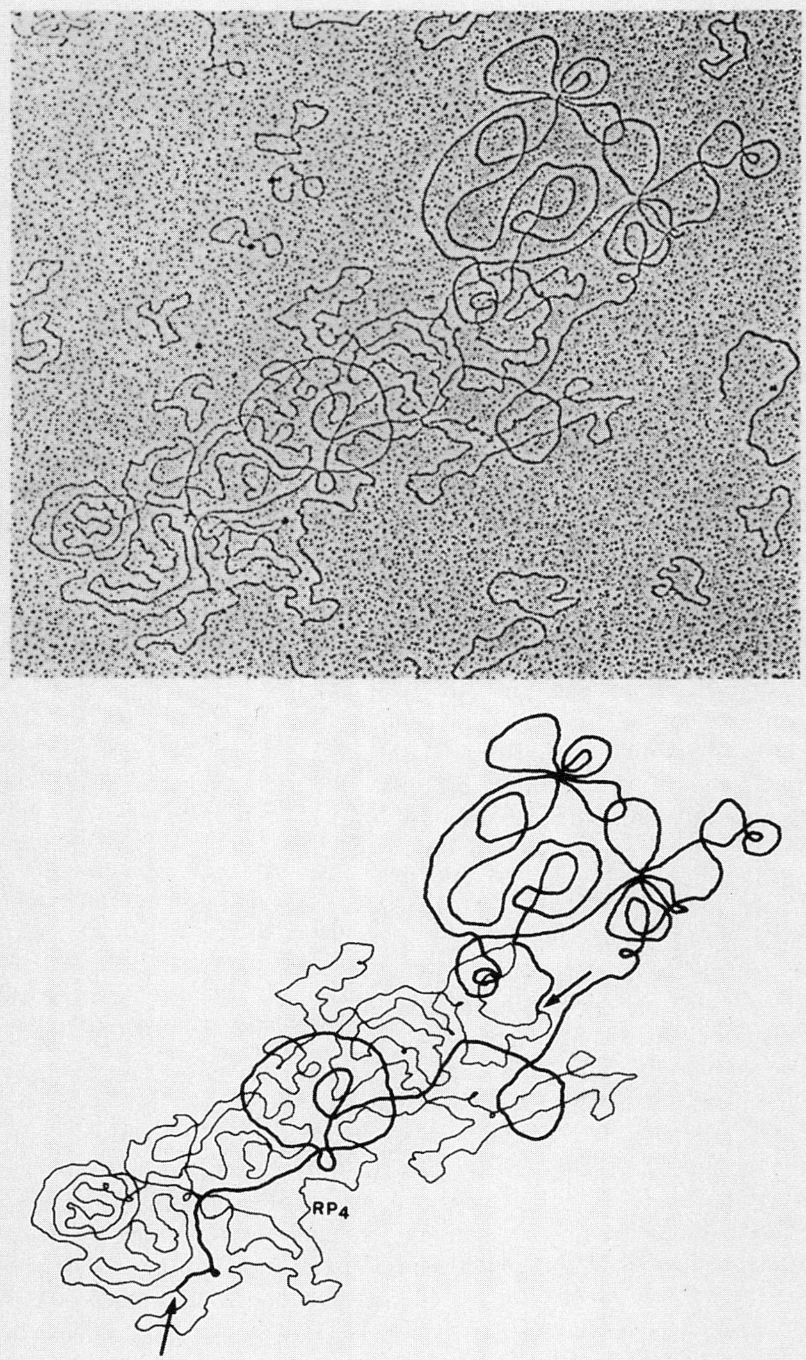

FIG. 5. Electron micrograph and tracing of a heteroduplex molecule formed by renaturation between single-stranded DNA of pGV4005 (Table 3) and single-stranded DNA of pTiC58. From the double-strand formed by renaturation between the pTiC58 strands two single-stranded loops protrude, one corresponding to the length of RP4 (36 Md), the other with an approximate length of 61 Md, corresponding to the pTiC58 DNA, the homologous part of which is deleted in pGV4005.

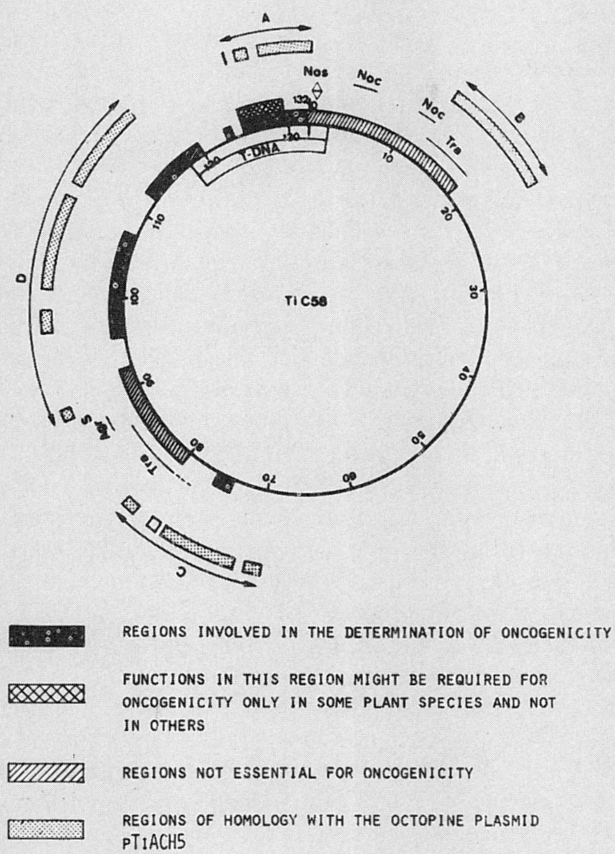

FIG. 6. Schematic representation of the functional organization of pTiC58.

analysis of heteroduplex molecules, an example of which is given in Fig. 5.

Like the parent cointegrate pGV4000, deleted cointegrates still dissociate and yield an intact RP4 and a deleted Ti plasmid replicon. Colonies harboring such dissociated plasmids were picked up by screening for the loss of cotransfer of the Noc^+ marker with RP4, exactly as described for pGV-4000 (Holsters *et al.*, 1978).

Bacteria with a pGV4000 plasmid were Onc^- but became Onc^+ after dissociation of the cointegrate. However even after dissociation the deleted Ti plasmids 4001 and 4005 could not induce tumors. This was expected since the deletions overlapped with several Onc^- transposon-insertion sites. After dissociation of deleted plasmids 4001 and 4005 into RP4 and Ti replicons, the Noc^+ Ti-marker remained stably present, indicating that neither large deletion affects the maintenance of the Ti replicon.

DISCUSSION

The approach used to study the organization of the TiC58 plasmid allowed approximate allocation of a large number of mutations to particular regions of the physical map. Figure 6 summarizes the conclusion of our study and represents a comprehensive summary of the relevant observations. Also indicated is the T-DNA region which is homologous to DNA in transformed plant cells (De Beuckeleer *et al.*, in preparation) and the regions of homology of pTiC58 with the octopine plasmid pTiAch5 (Engler *et al.*, in preparation).

We have thus located several pTiC58 regions that are essential and several that are nonessential for oncogenicity. A large region of about 18 Md extending from map position 131.6 Md to 18 Md can be deleted without loss of oncogenicity. It starts within the right end of the T-DNA and extends further to the right (as in Fig. 6). Thus the right end of the T-DNA, which is not conserved between octopine and nopaline plasmids, is nonessential for oncogenicity. It does however determine nopaline biosynthesis in tumors since Nos$^-$ insertions were located in this region. Located in the conserved T-DNA region of 5 Md adjacent to the nonconserved right end are several insertions and a deletion affecting oncogenicity. This provides the first genetic support that the T-DNA indeed plays a direct role in the establishment and/or maintenance of the tumoral transformation. Preliminary observations of oncogenicity tests on a wider range of host plants indicated that part of this Onc region might be essential for tumor induction on certain plant species but not on others.

Two insertions were isolated that mapped in the left, nonconserved part of the T-DNA without affecting oncogenicity. A large number of Onc$^-$ insertions and a small deletion mapped to the left of the T-DNA in the conserved region D. This region therefore must play an important role in oncogenicity. Two Onc$^-$ insertions have been located in the homology region C, but none has been ascribed so far to homology region B. However Onc$^-$ Tn7 insertions have been mapped in this region in the octopine plasmid pTiB6S3 (De Greve, personal communication).

In summary we can say that most if not all mutations that affect oncogenicity are located in regions of homology between octopine and nopaline plasmids. In the TiC58 plasmid there are at least three distinct Onc regions that are separated by regions that are nonessential for tumor induction. Only one of these Onc regions coincides in part with the T-DNA. The other Onc regions are probably involved in the mechanism of transfer of the T-DNA to the plant. However not all of the T-DNA is essential for oncogenicity. Both in right and left nonconserved T-DNA regions transposon insertions that do not affect tumor induction have been found. A detailed study of the T-DNA of tumors induced by these strains and a survey of prospects for engineering of plants through the use of these T-DNA sites will be published separately.

A large number of transposon insertions have been isolated that inactivate both the capacity to utilize nopaline as N source as well as arginine and ornithine as C source. These Noc$^-$ Arc$^-$ Orc$^-$ mutations map to the right of the T-DNA at two distinct positions, one at 5 Md, the other at 12 Md. None of these mutations affects the biosynthesis of nopaline in tumors and likewise the Nos$^-$ insertions do not affect the catabolic Ti-functions. From the different map positions of Nos$^-$ and Noc$^-$ mutations it is now convincingly proven that nopaline catabolism and biosynthesis are caused by separate functions. Preliminary results of a more detailed study of the Noc region to be published separately indicate that a function required for the catabolism of ornithine is located in the extreme left part of the Noc region (in fragment *Hin*dIII 33) and that a regulator function is located to the extreme right of this region (*Hin*dIII 26).

Some preliminary indications of Ti-plasmid regions involved in autotransfer have been obtained through comparison of Tra$^+$ and Tra$^-$ deletion mutations. Indeed, out of about 400 independent transposon insertions in pTiC58 only one was found to affect plasmid transfer. This Tra$^-$ insertion is approximately ascribed to homology region C. The method used for isolating the transpositions in pTiC58 was based on an initial transfer step which probably gave a counterselection for Tra⁻ insertions. Nevertheless using a similar approach Barth and Grinter (1977) have successfully isolated Tra⁻ mutants of RP4 and Hernalsteens *et al.* (1977) have isolated Tra⁻ mutants of R702. None of the Tra⁻ mutants we have studied affects oncogenicity, and vice versa. Tempé *et al.* (1977) have suggested that the bacterial conjugation machinery might also be responsible for the transfer of the T-DNA to the plant cells. Up to now we have obtained no evidence supporting this hypothesis. Tra functions exist that are not involved in oncogenicity but the possible existence of some functions common to both bacterial conjugation and to oncogenicity cannot be excluded. The analysis of more Ti mutations, isolated without counterselection for Tra⁻, should provide an answer to this problem.

The AP1 exclusion phenotype has been very roughly allocated to the region from 69–79 Md. Sensitivity to the agrocin 84 is allocated to approximately map position 87 Md. Finally, from the set of overlapping deletions shown in Fig. 4—none of which affects the stable maintenance of the Ti replicon—we can conclude that the origin of replication of the TiC58 plasmid must be located in the region between 18 and 64 Md.

In summary we can say that functions that are common between octopine and nopaline Ti-plasmids, such as oncogenicity and autotransfer, are determined by homologous DNA sequences but functions that are specific for either octopine or nopaline plasmids, such as the biosynthesis and catabolism of the opines and the sensitivity to the agrocin 84, are determined by heterogeneous sequences.

ACKNOWLEDGMENTS

The authors thank J. P. Hernalsteens for valuable help and advice and Dr. R. Ludwig for critical reading of the manuscript. This was supported by grants from the "Kankerfonds van de A.S.L.K" and from the "Fonds voor Kollektief Fundamenteel Onderzoek" (10.316) to J.S. and M.V.M. M.H., M.D.B., and P.D. are indebted for a fellowship to the N.F.W.O., and B.S. and R.V. to the A.S.L.K.

REFERENCES

Aerts, M., Jacobs, M., Hernalsteens, J. P., Van Montagu, M., and Schell, J. (1979). Induction and in vitro culture of *Arabidopsis thaliana* crown gall tumours. *Plant Sci. Lett.* **17**, 43–50.

Anand, V. K., and Heberlein, G. T. (1977) Crown gall tumorigenesis in potato tuber tissue. *Amer. J. Bot.* **64**, 153–158.

Barth, P. T., Datta, N., Hedges, R. W., and Grinter, N. J. (1976). Transposition of a deoxyribonucleic acid sequence encoding trimethoprim and streptomycin resistances from R483 to other replicons. *J. Bacteriol.* **125**, 800–810.

Barth, P. T., and Grinter, N. J. (1977). A Tn7 insertion map of RP4. *In* "DNA Insertion Elements, Plasmids, and Episomes" (A. I. Bukhari, J. A. Shapiro, and S. L. Adhya, eds.), pp. 675–676. Cold Spring Harbor Laboratory, New York.

Bomhoff, G., Klapwijk, P. M., Kester, H. C. M., Schilperoort, R. A., Hernalsteens, J. P., and Schell, J. (1976). Octopine and nopaline synthesis and breakdown genetically controlled by a plasmid of *Agrobacterium tumefaciens*. *Mol. Gen. Genet.* **145**, 177–181.

Cassé, F., Boucher, C., Julliot, J. S., Michel, M., and Dénarié, J. (1979) Identification and characterization of large plasmids in *Rhizobium meliloti* using agarose gel electrophoresis. *J. Gen. Microbiol.* **113**, 229–242.

Chilton, M. D., Drummond, M. H., Merlo, D. J., Sciaky, D., Montoya, A. L., Gordon, M. P., and Nester, E. W. (1977). Stable incorporation of plasmid DNA into higher plant cells: The molecular basis of crown gall tumorigenesis. *Cell* **11**, 263–271.

Datta, N., Hedges, R. W., Shaw, E. J., Sykes, R. B., and Richmond, M. H. (1971). Properties of an R-factor from *Pseudomonas aeruginosa*. *J. Bacteriol.* **108**, 1244–1249.

Davis, R. W., Simon, M., and Davidson, N. (1971). Electron microscope heteroduplex methods for mapping regions of base sequence homology in nucleic acids. *In* "Methods in Enzymology" (D. L. Grossmann and K. Moldave, eds.), Vol. XXI D, pp. 413–428. Academic Press, New York.

Depicker, A., De Wilde, M., De Vos, G., De Vos, R., Van Montagu, M., and Schell, J. (1980). Molecular cloning of overlapping segments of the nopaline Ti-plasmid pTiC58 as a means to restriction endonuclease mapping. *Plasmid* **3**, 193–211.

Dhaese, P., De Greve, H., Decraemer, H., Schell, J., and Van Montagu, M. (1979). Rapid mapping of a transposon insertion and deletion mutations in the large Ti-plasmids of *Agrobacterium tumefaciens*. *Nucl. Acids Res.* **7**, 1837–1849.

DRUMMOND, M. H., GORDON, M. P., NESTER, E. W., AND CHILTON, M.-D. (1977). Foreign DNA of bacterial plasmid origin is transcribed in crown gall tumours. *Nature (London)* **269,** 535–536.

ELLIS, J., KERR, A., TEMPÉ, J., AND PETIT, A. (1979). Arginine catabolism: A new function of both octopine and nopaline Ti-plasmids of *Agrobacterium. Molec. Gen. Genet.* **173,** 263–269.

ENGLER, G., HOLSTERS, M., VAN MONTAGU, M., SCHELL, J., HERNALSTEENS, J. P., AND SCHILPEROORT, R. A. (1975). Agrocin 84 sensitivity: A plasmid determined property in *Agrobacterium tumefaciens. Molec. Gen. Genet.* **138,** 345–349.

GENETELLO, CH., VAN LAREBEKE, N., HOLSTERS, M., DEPICKER, A., VAN MONTAGU, M., AND SCHELL, J. (1977). Ti-plasmids of *Agrobacterium* as conjugative plasmids. *Nature (London)* **265,** 561–563.

GURLEY, W. B., KEMP, J. D., ALBERT, M. J., SUTTON, D. W., AND CALLIS, J. (1979). Transcription of Ti plasmid-derived sequences in three octopine-type crown gall tumor lines. *Proc. Nat. Acad. Sci. U.S.A.* **76,** 2828–2832.

HEDGES, R. W., AND JACOB, A. E. (1974). Transposition of ampicillin resistance from RP4 to other replicons. *Mol. Gen. Genet.* **132,** 31–40.

HERNALSTEENS, J. P., VILLARROEL-MANDIOLA, R., VAN MONTAGU, M., AND SCHELL, J. (1977). Transposition of Tn*l* to a broad-host-range drug resistance plasmid. *In* "DNA Insertion Elements, Plasmids, and Episomes (A. I. Bukhari, J. A. Shapiro, and S. L. Adhya, eds.), pp. 179–185. Cold Spring Harbor Laboratory, New York.

HERNALSTEENS, J. P., DE GREVE, H., VAN MONTAGU, M., AND SCHELL, J. (1978). Mutagenesis by insertion of the drug resistance transposon Tn7 applied to the Ti plasmid of *Agrobacterium tumefaciens. Plasmid* **1,** 218–225.

HOLSTERS, M., SILVA, A., GENETELLO, C., ENGLER, G., VAN VLIET, F., DE BLOCK, M., VILLARROEL, R., VAN MONTAGU, M., AND SCHELL, J. (1978). Spontaneous formation of cointegrates of the oncogenic Ti-plasmid and the wide-host-range P-plasmid RP4. *Plasmid* **1,** 456–467.

KERR, A., MANIGAULT, P., AND TEMPÉ, J. (1977). Transfer of virulence in vivo and in vitro in *Agrobacterium. Nature (London)* **265,** 560–561.

KLECKNER, N. (1977). Tranlocatable elements in procaryotes. *Cell* **11,** 11–23.

KLEINSCHMIDT, A. K. (1968). Monolayer techniques in electron microscopy of nucleic acid molecules. *In* "Methods in Enzymology" (L. Grossmann and K. Moldave, eds.), Vol. XII B, pp. 361–377. Academic Press, New York.

KOEKMAN, B. T., OOMS, G., KLAPWIJK, P. M., AND SCHILPEROORT, R. A. (1979). Genetic map of an octopine Ti-plasmid. *Plasmid* **2,** 347–357.

LIPPINCOTT, J. A., BEIDERBECK, R., AND LIPPINCOTT, B. B. (1973). Utilization of octopine and nopaline by *Agrobacterium. J. Bacteriol.* **116,** 378–383.

MANIGAULT, P., AND KURKDJIAN, A. (1967). Influence de la lumière sur le développement des tumeurs de la plantule de pois, *Pisum sativum* L., induites par *Agrobacterium tumefaciens. C. R. M. Acad. Sci., Sér. D* **264,** 2304–2306.

MILLER, J. H. (1972). "Experiments in Molecular Genetics," p. 466. Cold Spring Harbor Laboratory, New York.

MONTOYA, A., CHILTON, M. D., GORDON, M. P., SCIAKY, D., AND NESTER, E. W. (1977). Octopine and nopaline metabolism in *Agrobacterium tumefaciens* and crown-gall tumor cells: Role of plasmid genes. *J. Bacteriol.* **129,** 101–107.

PETIT, A., DELHAYE, S., TEMPÉ, J., AND MOREL, G. (1970). Recherches sur les guanidines des tissus de crown gall. Mise en évidence d'une relation biochimique spécifique entre les souches d' *Agrobacterium* et les tumeurs qu'elles induisent. *Physiol. vég.* **8,** 205–213.

PETIT, A., AND TEMPÉ, J. (1975). Etude du métabolisme des guanidines des tissus de crown-gall par la souche T37 *d'Agrobacterium tumefaciens. C. R. Acad. Sci. (Paris)* **281,** 467–469.

PETIT, A., AND TEMPÉ, J. (1978). Isolation of *Agrobacterium* Ti-plasmid regulatory mutants. *Molec. Gen. Genet.* **167,** 147–155.

RIGBY, P. W. W., DIECKMANN, M., RHODES, C., AND BERG, P. (1977). Labeling deoxyribonucleic acid to high specific activity in vitro by nick translation with DNA polymerase I. *J. Mol. Biol.* **113,** 237–252.

SCHELL, J. (1975). The role of plasmids in crown gall formation by *A. tumefaciens. In* "Genetic Manipulations with Plant Materials (L. Ledoux, ed.), pp. 163–181. Plenum Press, New York.

SCHELL, J., AND VAN MONTAGU, M. (1977). On the transfer, maintenance and expression of bacterial Ti-plasmid DNA in plant cells transformed with *A. tumefaciens. Brookhaven Symp. Biol.* **29,** 36–49.

SCHELL, J., VAN MONTAGU, M., DE BEUCKELEER, M., DE BLOCK, M., DEPICKER, A., DE WILDE, M., ENGLER, G., GENETELLO, C., HERNALSTEENS, J. P., HOLSTERS, M., SEURINCK, J., SILVA, A., VAN VLIET, F., AND VILLARROEL, R. (1979). Interactions and DNA transfer between *Agrobacterium tumefaciens,* the Ti-plasmid and the plant host. *Proc. R. Soc. London B* **204,** 251–266.

SOUTHERN, E. M. (1975). Detection of specific sequences among DNA fragments separated by gel electrophoresis. *J. Mol. Biol.* **98,** 503–518.

TEMPÉ, J., PETIT, A., HOLSTERS, M., VAN MONTAGU, M., AND SCHELL, J. (1977). Thermosensitive step associated with transfer of the Ti-plasmid during conjugation: Possible relation to transformation in crown gall. *Proc. Nat. Acad. Sci. U.S.A.* **74,** 2848–2849.

Van Larebeke, N., Engler, G., Holsters, M., Van den Elsacker, S., Zaenen, I., Schilperoort, R. A., and Schell, J. (1974). Large plasmids in *Agrobacterium tumefaciens* essential for crown gall-inducing ability. *Nature (London)* **252**, 169–170.

Van Larebeke, N., Genetello, C., Hernalsteens, J. P., Depicker, A., Zaenen, I., Messens, E., Van Montagu, M., and Schell, J. (1977). Transfer of Ti-plasmids between *Agrobacterium* strains by mobilization with the conjugative plasmid RP4. *Molec. Gen. Genet.* **152**, 119–124.

Watson, B., Currier, T. C., Gordon, M. P., Chilton, M-D., and Nester, E. W. (1975). Plasmid required for virulence of *Agrobacterium tumefaciens*. *J. Bacteriol.* **123**, 255–264.

Zaenen, I., Van Larebeke, N., Teuchy, H., Van Montagu, M., and Schell, J. (1974). Supercoiled circular DNA in crown gall inducing *Agrobacterium* strains. *J. Mol. Biol.* **86**, 109–127.

I wanted the word to be "yow", but it was changed by an editor without my notice! I am still not happy about that!

WOW!!! THE T-DNA IS INTEGRATED

Michael F. Thomashow
MSU-DOE Plant Research Lab., Michigan State University
East Lansing, Michigan, United States
E-mail: thomash6@msu.edu

Commentary on the paper:

Thomashow, M. F., Nutter, R., Montoya, A. L., Gordon, M. P., and Nester, E. W. 1980. Integration and organization of Ti plasmid sequences in crown gall tumors. Cell 19:729–739.

It is hard to believe that it has been 23 years since our T-DNA integration paper was published. With just closing my eyes, I can transport myself back to the day that Bob Nutter and I developed the film of the Southern analysis that first indicated that the T-DNA was integrated into plant DNA. I can still feel our excitement as we virtually ran down the halls to the Nester labs and burst in to tell Frank White, Dave Garfinkel and the others "THE RESULT!!" After a few howls and cheers, and with big smiles across our faces, we headed over to the cafeteria to have lunch and celebrate. The only reason that we didn't head out to the tavern for some beers was that nick translations were on tap for the afternoon for the next experiment; there was simply no way that we would delay the experiment and lose time!!

The conversation that followed was remarkable. We quickly turned from the immediate result and discussion of follow-up experiments to the longer term potential practical applications of our results; specifically using the Ti plasmid as a vector to transform plants. We said "Hey, wouldn't it be wild to put *Bacillus thuringiensis* (BT) toxin genes into plants and make them resistant to insects!" This application came immediately to mind as Helen Whitely's lab was right down the hall from Gene's, and she and her colleagues had been having success isolating BT toxin genes. What is truly amazing to me now is that at that time, none of us thought seriously in terms of patenting these ideas, but instead, we simply basked in the beauty of biology, the fun of experimentation, and the good that could potentially come from the practical applications of our basic findings. Of course, much needed to be done before applications could be realized. There were no selectable marker genes for transformation at the time, it was uncertain whether the Ti-plasmid could be "disarmed" to introduce genes of choice without introducing genes that also resulted in tumor formation, and so on. But incredibly, within 7 years *Agrobacterium* was used to create transgenic tobacco (Barton et al. 1987) and tomato (Fischhoff et al. 1987) plants that expressed BT genes and displayed resistance to insects. I don't think that anyone at the table that day, certainly not I, would have bet that this could have been accomplished in such a short time. Simply staggering.

So, to put the present paper into an historical context, let me take you back to 1978 when I joined the Nester lab. (To give appropriate recognition and gratitude, I should point out that I was fortunate to be a Damon Runyon–Walter Winchell Cancer Research Fellow.) Not long before I had arrived, Mary-Dell Chilton and her colleagues in the Nester lab established the fact that Ti-plasmid sequences were present in crown gall tumor tissues (Chilton et al. 1977). This was accomplished by examining the reassociation kinetics of Ti-plasmid sequences in the presence of DNA, isolated from either crown gall or non-transformed tissues. It could be concluded from the rates of reassociation that a portion of the Ti-plasmid had been transferred from *Agrobacterium* to the plant cells where it became stably incorporated. This was a seminal finding in the course of crown gall research that was tremendously exciting.

As with all advances in science, however, the findings of Chilton and coworkers left a number of important questions unanswered. Arguably the most important was whether the portion of the Ti-plasmid DNA that was incorporated into the plant cell, the T-DNA, was integrated into the plant genome. I was drawn to this question, as my overall interest in crown gall was to understand the molecular basis for why tobacco crown gall tumor tissues could be grown in culture without hormones whereas non-transformed tissues required both an auxin and a cytokinin. Knowing exactly which sequences were transferred to the plant and knowing the "state" in which they existed was important to addressing this issue.

Upon arrival in the Nester lab, it was suggested that I try to conduct "DNA networking" experiments to establish whether the T-DNA was integrated into the plant DNA. In short, the approach is based on the fact that tobacco DNA contains highly repetitive DNA sequences that are dispersed throughout the genome. Thus, if total DNA is isolated by gentle procedures producing long stands, and these strands are denatured and renatured, the highly repetitive sequences from throughout the genome hybridize rapidly among themselves producing a network of DNA strands that can easily be collected by centrifugation. The idea was that if the T-DNA was integrated into the plant genome, the sequences should be present in the network fraction. If it existed as an independent replicon, it should be located in the supernatant fraction.

I conducted the network experiments, but soon decided to drop them and focus on what I thought was potentially a more powerful approach: genomic Southern analysis. The strategy had recently been used by Botchan, et al. (1976) to determine the arrangement of SV40 virus DNA sequences in the genomes of transformed cells. The approach was not only capable of providing evidence for T-DNA integration, but could also produce information about T-DNA organization, copy number and sites of integration. I was also drawn to the visceral reaction produced by blotting results in the early days; you could actually "see" the T-DNA on the filter and the physical presence of the DNA sequences of interest. There were, however, significant technical hurdles that had to be overcome before we could successfully conduct these experiments. The most important was producing nick-translated probes having a high specific activity; we needed to increase specific activities that we routinely obtained in the lab by about ten-fold. Also, the procedures required what, for the day, were large amounts of restriction enzymes. In certain cases, this necessitated either purifying restriction enzymes ourselves (Randy Saiki produced prodigious amounts of *Eco*RI) or obtaining them from others (Rich Meager was particularly generous with the amount of *Bst*I that he provided us).

There were also certain experimental idiosyncrasies that had to be dealt with. One was that radioactive nucleotides had to be used within a couple of days of arrival otherwise some ill-defined product seemed to form which inhibited the nick-translation reactions. Thus, when the "nukes" arrived, all of us went into action to prepare probes for our particular experiments producing traffic jams at the freezers, centrifuges and other critical common work areas in the lab. There was also one particularly irksome event that occurred as Bob Nutter and I were trying to finalize data for the T-DNA integration paper. All of a sudden we couldn't get the get nick-translations to work. After doing control upon control over the course of some three weeks or so, we finally figured out that the particular batch of size exclusion gel that we were using to purify our probes, physically bound double-stranded DNA. When I called the company (which will remain nameless) to tell them of this, venting a considerable amount of frustration and displeasure in the process, I got the response "Oh yes, that happens with some of our batches." The person said that given the situation, that they would be happy to send us a bottle of gel from a different batch, *for free.* "Wow! Thanks a lot!!"

Anyway, Bob and I got things back on track and finished up the experiments. The results allowed us to determine that each of the four independent tumor lines that we had analyzed contained a "core" T-DNA sequence that we proposed was responsible for maintaining the transformed state (phytohormone independence). This core T-DNA, which was co-linear with the Ti plasmid, included the sequences known as the "common DNA," DNA sequences that had been detected in all Ti plasmids that had been described to that time. Of course the most exciting results were those indicating that the majority, if not all, of the T-DNA was integrated into the plant genome, a result that we soon confirmed through cloning of the T-DNA. In addition, the results indicated that preferred regions of the Ti plasmid served as the junctions between the Ti plasmid and plant DNA and that the T-DNA could be integrated into multiple sites in the plant genome. When I finished writing-up the results, we all knew that we had a pretty exciting story. Indeed, the manuscript was sent off to *Cell* and the communication that came back from Benjamin Lewin was that the paper would appear, *as submitted,* in the March 1980 issue of the journal. "Champagne all around!!"

In closing, I must express how lucky I feel to have been a member of Gene's lab in those early years of bringing molecular genetic approaches to the study of crown gall. The excitement of crown gall research during the late 1970s was palpable; in part due to the neck-and-neck competition between Gene's lab and those of others including the groups headed by Jeff Schell, Marc Van Montagu and Rob Schilperoort. Also, Gene was a perfect mentor for me. He set the general course of direction, but fully supported my pursuing the lines of investigation that I thought would prove most fruitful. And finally, I must say that without Bob Nutter as a partner in the experiments, I am not sure how things would have turned out. He had his own line of investigation, but was willing to take out time to help me with the T-DNA integration story. We made a fabulous team that got the job done.

References

Barton, K. A., Whiteley, H. R., and Yang, N. S. 1987. *Bacillus thuringiensis* delta-endotoxin expressed in transgenic *Nicotiana tabacum* provides resistance to lepidopteran insects. Plant Physiol. 85:1103–1109.

Botchan, M., Topp, W., and Sambrook, J. 1976. The arrangement of simian virus 40 sequences in the DNA of transformed cells. Cell 9:269–287.

Chilton, M.-D., Drummond, M. H., Merlo, D. J., Sciaky, D., Montoya, A. L., Gordon, M. P., and Nester, E. W. 1977. Stable incorporation of plasmid DNA into higher plant cells: The molecular basis of crown gall tumorigenesis. Cell 11:263–271.

Fischhoff, D. A., Bowdish, K. S., Perlak, F. J., Marrone, P. G., McCormick, S. M., Niedermeyer, J. G., Dean, D. A., Kusanokretzmer, K., Mayer, E. J., Rochester, D. E., Rogers, S. G., and Fraley, R. T. 1987. Insect tolerant transgenic tomato plants. Bio/Technology 5:807–813.

Thomashow, M. F., Nutter, R., Postel, K., Chilton, M.-D., Blattner, F., Powell, A., Gordon, M. P., and Nester, E. W. 1980. Recombination between higher plant DNA and the Ti-plasmid of *Agrobacterium tumefaciens*. Proc. Natl. Acad. Sci. U.S.A. 77:6448–6452.

MICHAEL THOMASHOW: I studied as an undergraduate at the University of California, Los Angeles. I was pretty much muddling through college when I decided to take Dr. Sydney C. Rittenberg's class in introductory bacteriology. This decision set in motion the series of events that led me to become a scientist. "The Boss" as all in his lab called him, was an inspiring teacher (he was a recipient of the American Society for Microbiology, Carski Foundation Distinguished Teaching Award) and a superb scientist. He was a crystal clear thinker who immediately focused in on the core of whatever scientific issue he was confronted with. It is more than an understatement that I was extremely fortunate when Dr. Rittenberg invited me to conduct an undergraduate research project in his lab. I graduated with a B.A. in Bacteriology in 1972. Dr. Rittenberg then agreed to take me on as a graduate student and I completed my Ph.D. in 1978. Any success that I may achieve in science is due to the roots that were nourished and took hold under his guidance.

I went as a postdoc to Eugene W. Nester's lab in the Department of Microbiology, University of Washington, Seattle. As I mention in my remembrance of the T-DNA integration story, I could not have had a better postdoctoral adviser than Gene. (I would say this even if Gene weren't an editor of this book.) He afforded the people in his lab the opportunity to be successful by providing an appropriate amount of direction, a research environment that allowed one to conduct the experiments required to address a given question, and by giving individuals freedom to follow their own scientific instincts. In addition, he was encouraging and understanding at the personal level. How could one ask for more? Indeed, it is the same set of core principles that I now strive to bring to life in my own lab.

Cell, Vol. 19, 729–739, March 1980, Copyright © 1980 by MIT

Integration and Organization of Ti Plasmid Sequences in Crown Gall Tumors

Michael F. Thomashow,* Robert Nutter,† Alice L. Montoya,* Milton P. Gordon† and Eugene W. Nester*
* Department of Microbiology and Immunology
† Department of Biochemistry
University of Washington
Seattle, Washington 98195

Summary

The Ti plasmid sequences (T DNA) maintained in four unorganized crown gall tumor lines were defined by using restriction endonucleases and molecular hybridization techniques. Each tumor line contains a "core" T DNA which is apparently responsible for maintaining the transformed state; the core T DNA is colinear with the Ti plasmid and contains the Ti plasmid sequences referred to as the "common DNA"—sequences found in all Ti plasmids studied to date. A given Ti plasmid does not always give rise to the same T DNA complement. The data suggest that the majority if not all of the T DNA is integrated into plant DNA, that preferred regions of the Ti plasmid serve as the points of attachment to plant DNA and that the T DNA can be linked to more than one site in the plant genome.

Introduction

Crown gall is a neoplastic disease of dicotyledonous plants caused by virulent strains of Agrobacterium tumefaciens. Transformed plant tissue can be grown axenically in vitro without the addition of auxin or cytokinin, whereas normal plant tissues require these phytohormones for growth. The virulence trait of Agrobacterium is carried on one of a number of tumor-inducing (Ti) plasmids which range in size from approximately 90–150 × 10^6 daltons (Van Larebeke et al., 1974, 1975; Watson et al., 1975; Bomhoff et al., 1976). Most Ti plasmids fall into one of two groups depending on whether they code for octopine or nopaline catabolism (Watson et al., 1975; Bomhoff et al., 1976; Montoya et al., 1977). Strains of Agrobacterium harboring an octopine-type Ti plasmid generally induce unorganized tumors that produce octopine, while Agrobacterium with nopaline-type Ti plasmids usually induce teratomas [although they can form unorganized tumors (Gresshoff, Skotnicki and Rolfe, 1979)] that produce nopaline (Goldman, Tempé and Morel, 1968; Petit et al., 1970; Bomhoff et al., 1976; Montoya et al., 1977). These data suggested that crown gall might involve a transfer of Ti plasmid DNA from bacterium to plant. Chilton et al. (1977) first showed that this was true. Using DNA solution hybridization techniques, they found that tumor $B_6$806/E9 contained sequences common to approximately 5% of the Ti plasmid pTi-$B_6$806 at a level of some 20 copies per diploid cell. This transferred DNA, termed T DNA, was not found in normal tobacco cells. Merlo et al. (1980) subsequently confirmed and extended these results. They found T DNA in three independent tumor lines, $B_6$806/E9, 15955/1 and A6S/2, and concluded that T DNAs could vary and that within a given tumor line, the copy number of the right- and left-hand portion of the T DNA could differ. A teratoma tumor line that synthesizes nopaline has also been shown to contain plasmid DNA sequences (Yang et al., 1980).

The present study was undertaken to answer a number of basic questions concerning the T DNA in crown gall tissues incited by octopine-type Ti plasmids. Does T DNA become integrated into plant DNA? If it does, are the inserts stable, and is there specificity regarding the sequences that make up the plasmid DNA-plant DNA junctions? Is the T DNA colinear with Ti plasmid sequences? Does a given Ti plasmid always give rise to the same T DNA complement? The data obtained allow us to answer these questions to varying degrees and to reflect on the basic mechanisms of crown gall tumorigenesis.

Results

Description of Tumors Studied

The origins and phenotypes of the tumor tissues used in this study are given in Table 1. All tumors were incited on tobacco by A. tumefaciens strains harboring an octopine-type Ti plasmid, either pTi-$B_6$806, pTi-A6 or pTi-15955. These plasmids are very closely related. pTi-$B_6$806 and pTi-15955 are at least 90% homologous with pTi-A6 (Currier and Nester, 1976a). Eco RI, Bst I, Sma I and Hind III digestion of pTi-$B_6$806 and pTi-15955 yield fragments of identical molecular weights (Sciaky, Montoya and Chilton, 1978; M. Thomashow et al., 1980; M. Thomashow et al., unpublished results). The fragmentation patterns of pTi-A6 are consistent with its having suffered only a small deletion in the region of Sma I fragment 18. The Eco RI, Bst I (or the isoschizomer Bam HI), Sma I and Hind III restriction sites through the T DNA regions of the three plasmids are identical (Figure 1).

T DNA Maintained in Various Tumor Lines

The experimental approach taken to define the T DNA maintained in the various tumor lines was as follows. DNA from tobacco crown gall callus, normal tobacco callus (Nicotiana tabacum cv. Xanthi nc.) or Ti plasmid DNA was digested with various restriction endonucleases and the products were separated by electrophoresis in agarose gels. The fractionated DNAs were transferred to sheets of nitrocellulose and annealed with either whole Ti plasmid or cloned pieces of the Ti plasmid (Figure 1) that had been radioactively labeled

Table 1. Description of Tumor Tissues

Tumor Tissue	Cultivar of Nicotiana tabacum	Strain of A. tumefaciens	Ti Plasmid	Octopine Production	Date When Tissue Was Cloned
A6S/2[a]	White Burley	A6	pTi-A6	yes	August 1974
15955/1	Xanthi nc.	15955	pTi-15955	no	August 1974
$B_6$806/E9	Xanthi nc.	$B_6$806	pTi-$B_6$806	yes	Spring 1974
A277/5	Xanthi nc.	A277	pTi-$B_6$806	no	August 1978

[a] The uncloned tissue was provided by R. A. Schilperoort, University of Leiden, The Netherlands.

in vitro with ^{32}P. Plasmid sequences were then visualized by autoradiography. From the molecular weights and hybridization characteristics of the fragments produced, we deduced which portions of the Ti plasmid were present. [The basic strategy of this experimental approach was first outlined by Botchan, Topp and Sambrook (1976) in the study of SV40 integration into rodent DNA.]

It has been shown (Merlo et al., 1980) by DNA solution hybridization techniques that the DNA from A6S/2, $B_6$806/E9 and 15955/1 contains sequences homologous with pTi-$B_6$806 Hind III fragment 1; DNA from $B_6$806/E9 and 15955/1 also contains sequences homologous with Hind III fragment 5. To define further the region(s) of the Ti plasmid maintained in A6S/2, $B_6$806/E9 and 15955/1 as well as A277/5, we digested the tumor DNAs with Bst I and processed the products as outlined above. Fragments containing sequences homologous with the Ti plasmid were detected by hybridizing the nitrocellulose sheets with a mixture of ^{32}P–pNW31C-8,29-1 and ^{32}P–pNW31C-2,19-1 (Figures 2A and 2B) or ^{32}P–pTi-$B_6$806 (Figure 2C). Fragments with the same molecular weights and relative intensities were observed with either probe. These data indicate that the T DNA maintained in the tumors comes from the region of the Ti plasmid represented in the clones pNW31C-8,29-1 and pNW-31C-2,19-1 (Bst I fragments 2, 8, 19 and 29).

If a given tumor DNA contained an insert of T DNA that covered the entire region represented in pNW31C-8,29-1 and pNW31C-2,19-1, fragments migrating with the molecular weights of Bst I fragments 2, 8, 19 and 29 should have been observed, as was the case in the Ti plasmid reconstructions (Figure 2A, lanes 1, 2 and 3). Fragment 29 was present in all tumor lines (Figure 2B, lanes 5–8), and fragment 19 was also present in $B_6$806/E9 and A6S/2. Neither fragment 2 nor 8 was detected in any of the tumor DNAs. These data demonstrate the presence of the Bst I 8/29, 29/19 and 19/2 recognition sites in A6S/2 and $B_6$806/E9. We conclude that each of these tumor DNAs have a T DNA insert(s) that represents, at a minimum, the Ti plasmid sequences between a point somewhere in Bst I fragment 8 and a point in Bst I fragment 2. A277/5 and 15955/1 have T DNAs that represent, at a minimum, Ti plasmid sequences

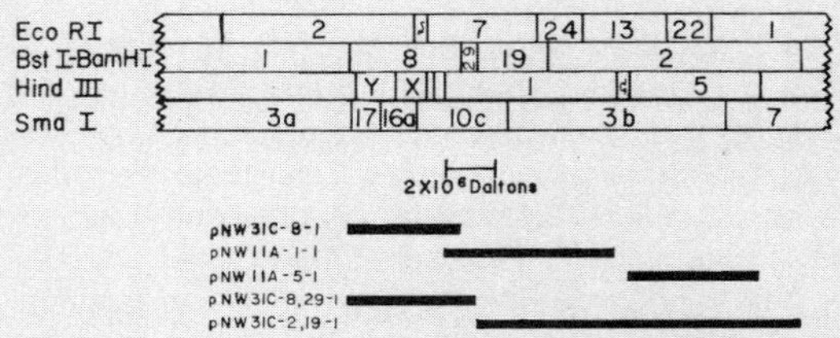

Figure 1. Restriction Maps of the T Region of pTi-$B_6$806, pTi-A6 and pTi-15955 and a Graphic Representation of the Cloned Ti Plasmid Sequences Used in the Study

The portion of the Ti plasmid contained in each clone is represented by a solid bar. Bst I and Bam HI are isoschizomers and thus they generate identical maps. For details concerning mapping and cloning see Experimental Procedures.

between a point in Bst I fragment 8 and a point in Bst I fragment 19.

To define the left extremes of the T DNAs more precisely, the tumor DNAs were digested with Hind III and the resulting nitrocellulose sheet was hybridized with ^{32}P–pNW31C-8-1. The data show (Figure 3A) that Hind III fragment X was present in all tumor lines whereas fragment Y was absent. The fragments that hybridized only to pNW31C-8-1, having molecular weights different than Hind III fragments X and Y, presumably contain a portion of the sequences represented in Hind III fragment Y. These data suggest that the left extremes of the T DNAs are within fragment Y. The observation that Sma I fragments 10c and 16a were present in all the tumor lines but a complete Sma I fragment 17 was not (data not shown) was consistent with this conclusion.

To determine the right extremes of the T DNAs, the tumor DNAs were digested with Eco RI and the resulting nitrocellulose sheets were hybridized with ^{32}P–pNW11A-5-1 (Figure 4B). On the basis of previous work (Merlo et al., 1980) we expected that $B_6$806/E9 and 15955/1 would contain sequences homologous with this region (Hind III fragment 5) of the plasmid but that A6S/2 would not. This was the case. The data suggest that the right extremes of the T DNAs in both $B_6$806/E9 and 15955/1 (Figure 4B, lanes 2 and 4, respectively) are in Eco RI fragment 22; in both cases fragment 13 was present but a complete fragment 22 was not. We could not detect sequences complementary to pNW11A-5-1 in A6S/2 digested with Eco RI

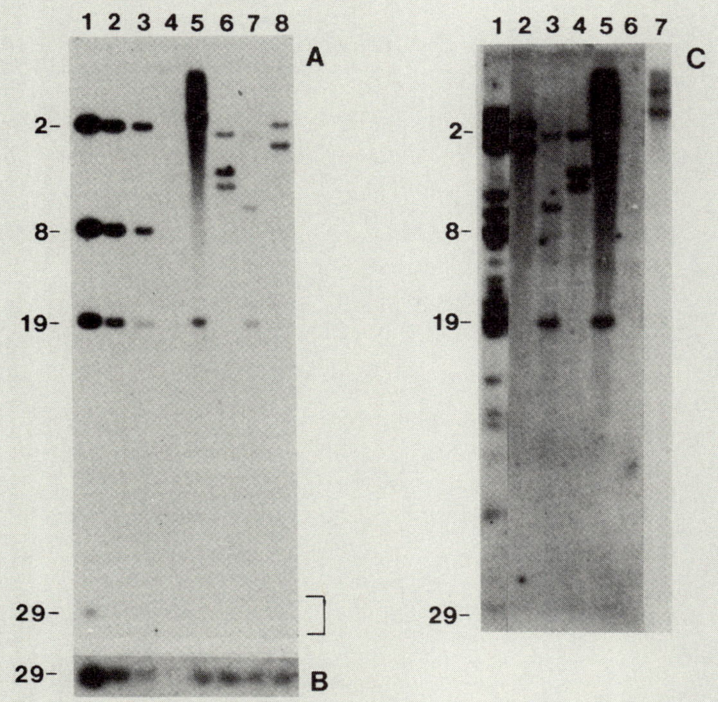

Figure 2. Detection of DNA Fragments Containing Ti Plasmid Sequences Produced upon Bst I Digestion of Tumor and Normal Tobacco DNA

The products of Bst I digestion were separated by agarose gel electrophoresis, transferred to nitrocellulose filters and hybridized with ^{32}P-labeled probes. Ti plasmid sequences were located by autoradiography (see Experimental Procedures for details).

(A) Hybridization of the nitrocellulose filter with a mixture of ^{32}P–pNW31C-8,29-1 and ^{32}P–pNW31C-2,19-1. Lanes (1, 2 and 3) contain 5, 2.5 and 1 copy reconstructions of the Ti plasmids (see Experimental Procedures); (4) normal tobacco callus DNA; (5) $B_6$806/E9 tumor DNA; (6) 15955/1 tumor DNA; (7) A6S/2 tumor DNA and (8) A277/5 tumor DNA.

(B) A longer exposure of the bracketed section of the nitrocellulose filter shown in (A).

(C) Hybridization of the nitrocellulose filter with ^{32}P-labeled p-Ti-$B_6$806. Lane (1) contains a 2.5 copy Ti plasmid reconstruction; (2) A277/5 tumor DNA; (3) A6S/2 tumor DNA; (4) 15955/1 tumor DNA; (5) $B_6$806/E9 tumor DNA; (6) normal tobacco callus DNA and (7) a shorter exposure of lane (5).

The positions of Bst I fragments 2, 8, 19 and 29 (10.7, 4.7, 2.2 and 0.6 $\times$ 10^6 daltons, respectively) are indicated at the sides of the autoradiographs.

(Figure 4B, lane 5), Bst I (Figure 6C, lane 2) or Hind III (data not shown). The data also show that A277/5 contains sequences complementary to Hind III fragment 5. When A277/5 was digested with Eco RI, a fragment was produced that hybridized with pNW11A-5-1 (Figure 4B, lane 6). Although these data do not allow us to specify which portion of Hind III fragment 5 was present, they show that this piece of T DNA extends only a short way, if at all, into Hind III fragment 1: unlike $B_6$806/E9 and 15955/1, A277/5 does not contain a complete copy of Eco RI fragment 13, and the fragment that hybridized to pNW11A-5-1 did not hybridize to pNW11A-1-1 (compare Figure 4B, lane 6, with Figure 4A, lane 8).

The Eco RI digests of the tumor DNAs were also useful for examining the T DNA from the middle of the T region. After hybridization with ^{32}P–pNW11A-1-1, the banding patterns of plasmid sequences shown in Figure 4A were observed. As discussed above, Eco RI fragment 13 was present in $B_6$806/E9 and 15955/1 (lanes 4 and 6, respectively). Eco RI fragment 7 was present in all tumor lines, but there was not a complete copy of fragment 24. The absence of fragment 24 in 15955/1 and $B_6$806/E9 was surprising, since the fragments to the left (fragment 7) and the right (fragment 13) of it were present. The simplest interpretation is that there are no continuous pieces of T DNA that include the entire sequence of Eco RI fragment 24. The finding that Hind III digestion did not produce a fragment with a molecular weight of Hind III fragment 1 in any tumor line supports this interpretation (Figure 3B).

From the data presented we conclude that all the tumor lines examined contain a "core" T DNA sequence representing the region between Hind III fragment Y and Eco RI fragment 24. This is illustrated in Figure 5. The right-hand ends of the core T DNAs, however, are slightly different. In A6S/2 and $B_6$806/E9 they end to the right of the Bst I recognition site in Eco RI fragment 24, while in 15955/1 and A277/5 they end to the left of this Bst I recognition site. The observation that Sma I fragments 16a and 10c (data not shown), Hind III fragment X (Figure 3A), Eco RI fragments δ (data not shown) and 7 (Figure 4A) and Bst I fragment 29 (Figure 2B) are present in all the tumor lines demonstrates that this region of T DNA is colinear with Ti plasmid DNA. In addition, the intensity of the hybridization observed with these fragments as compared with the Ti plasmid reconstructions suggests that these T DNAs are present in low copy—perhaps no more than one or two per diploid cell. This conclusion is consistent with the data presented below and the DNA solution hybridization data of Merlo et al. (1980).

In addition to the core T DNA, the data show that $B_6$806/E9, 15955/1 and A277/5 contain T DNA from a second adjacent region of the Ti plasmid (Figure 5). Tumor lines $B_6$806/E9 and 15955/1 contain T DNA representing the sequences between Eco RI fragments 24 and 22. The number of copies of this T DNA

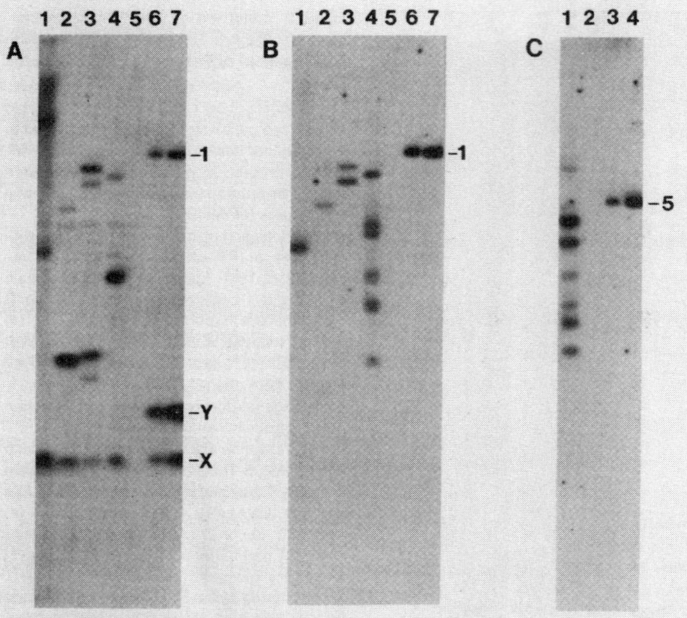

Figure 3. Detection of DNA Fragments Containing Ti Plasmid Sequences Produced upon Hind III Digestion of Tumor and Normal Tobacco DNA

Digestion products were fractionated and transferred to nitrocellulose as described in Experimental Procedures. In (A) and (B) the nitrocellulose filters were prepared from the same agarose gel and hybridized with ^{32}P–pNW31C-8-1 and ^{32}P–pNW11A-1-1, respectively. Lane (1) contains A277/5 tumor DNA; (2) A6S/2 tumor DNA; (3) 15955/1 tumor DNA; (4) B_6806/E9 tumor DNA; (5) normal tobacco callus DNA and (6 and 7) 1 and 2.5 copy Ti plasmid reconstructions, respectively. (C) The nitrocellulose filter was hybridized with ^{32}P–pNW11A-5-1. Lane (1) contains B_6806/E9 tumor DNA; (2) normal tobacco DNA and (3 and 4) 1 and 2.5 copy Ti plasmid reconstructions, respectively. The fragments shown in (C) can be compared directly to those in (A) and (B) since the standard molecular weight markers (Eco RI fragments of λ phage DNA) migrated to the same positions in each gel. The positions of Hind III fragments 1, 5, Y and X (7.3, 5.1, 1.8 and 1.2 × 10^6 daltons, respectively) are indicated at the sides of the autoradiograms.

differs greatly in the two tumor lines. The intensity of hybridization of Eco RI fragment 13 in 15955/1 (Figure 4A, lane 6) suggests that the sequence is present in only one or two copies per diploid cell. The intensity of hybridization of Eco RI fragment 13 in B_6806/E9 suggests that this T DNA is present at a level of approximately 15–30 copies per diploid genome. These estimations of T DNA copy number are consistent with the DNA solution hybridization data of Merlo et al. (1980). The data also show that A277/5 contains T DNA complementary to Hind III fragment 5; however, a more precise definition of the sequences represented was not determined.

We also conclude that a given Ti plasmid does not always give rise to the same T DNA complement. The B_6806/E9 and A277/5 tumors were both incited on Nicotiana tabacum cv. Xanthi nc. by strains of Agrobacterium harboring the Ti plasmid pTi-B_6806. It is clear (Figure 5) that a different T DNA is maintained in each tumor. It is not known whether this difference results from different initial integration events or identical integration events followed by rearrangement of the T DNA.

Integration of T DNA

The T DNA in plant cells could be present as free replicons and/or as integrated segments of DNA. To date, however, we have been unable to demonstrate the existence of T DNA replicons. When undigested tumor DNA was subjected to electrophoresis through agarose and transferred to nitrocellulose sheets, we did not observe any discrete bands after hybridizing the sheets with ^{32}P–pTi-B_6806. Hybridization was to the broad region of high molecular weight DNA resulting in an autoradiogram which resembled the original ethidium bromide-stained gel (our unpublished results). On the other hand, the data presented below suggest that most if not all of the T DNA is integrated into plant DNA. To facilitate the presentation of data we refer to the T DNA representing the Ti plasmid sequences to the left and right of Eco RI fragment 24 as the T_L DNA and T_R DNA, respectively (see Figure 5).

The simplest way to determine whether T DNA is integrated into plant DNA, and if so into how many sites, would involve the use of a restriction endonuclease that does not cleave in the T DNA sequence (Botchan et al., 1976). Since we are unaware of such an enzyme, a slightly more complex method of analysis was necessary. For example, digestion of A6S/2 tumor DNA with Bst I cuts the T DNA at the Bst I 8/29 recognition site. The next cut to the left should be in plant DNA if the T DNA is integrated (see Figure 5) and the number of "junction fragments" (plant DNA plus T DNA) produced represents the minimum number of T DNA inserts. The molecular weight of the junction fragment(s) could be smaller, the same as or larger than that of Bst I fragment 8, depending on the distance between the Bst I 8/29 recognition site in the T DNA and the nearest Bst I recognition site in the plant DNA. If the T DNA were present as a free circular replicon, its left and right ends should be joined and a "fusion" fragment produced. A fusion fragment would also be produced if the T DNA were integrated and present in tandem repeats. If the T DNA were present as a free linear DNA fragment, Bst I digestion

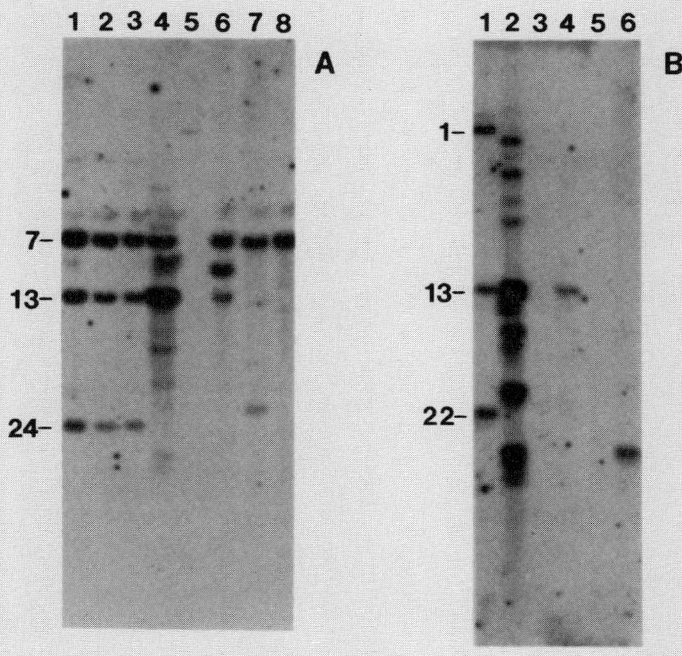

Figure 4. Detection of DNA Fragments Containing Ti Plasmid Sequences Produced upon Eco RI Digestion of Tumor and Normal Tobacco DNA

Digestion products were fractionated and transferred to nitrocellulose as described in Experimental Procedures. The nitrocellulose filters displayed in (A) and (B) were prepared from the same agarose gel.

(A) The nitrocellulose filter was hybridized with ^{32}P-pNW11A-1-1. Lanes (1, 2 and 3) contain 5, 2.5 and 1 copy Ti plasmid reconstructions, respectively; (4) $B_6$806/E9 tumor DNA; (5) normal tobacco callus DNA; (6) 15955/1 tumor DNA; (7) A6S/2 tumor DNA and (8) A277/5 tumor DNA.

(B) The nitrocellulose filter was hybridized with ^{32}P-pNW11A-5-1. Lane (1) contains a 1 copy Ti plasmid reconstruction; (2) $B_6$806/E9 tumor DNA; (3) normal tobacco callus DNA; (4) 15955/1 tumor DNA; (5) A6S/2 tumor DNA and (6) A277/5 tumor DNA.

The positions of Eco RI fragments 1, 7, 13, 22 and 24 (9.0, 4.7, 3.5, 2.7 and 2.6 × 10^6 daltons, respectively) are indicated at the sides of the autoradiograms.

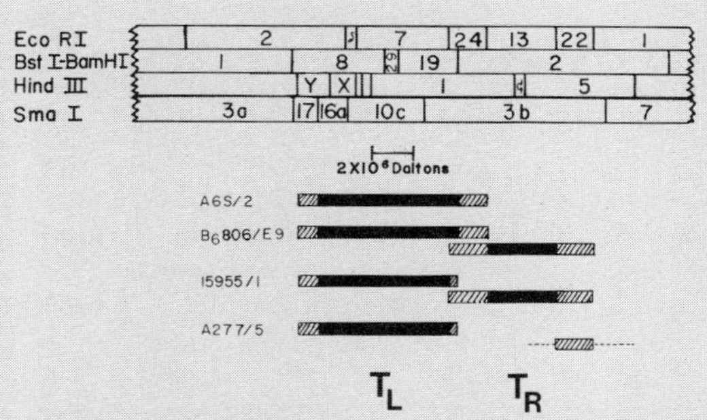

Figure 5. T DNA Maintained in Tumor Lines A6S/2, $B_6$806/E9, 15955/1 and A277/5

Ti plasmid sequences found in the various tumor DNAs are indicated by the solid bar. The left- and right-hand extremes of the T DNAs fall within the crosshatched areas of the bars. In A277/5, the region denoted by (---▨---) indicates that an unknown portion of Hind III fragment 5 is present (see text for details). The regions between Hind III fragment Y and Eco RI fragment 24 and between Eco RI fragments 24 and 22 are referred to in the text as the T_L DNA and T_R DNA, respectively.

should produce a fragment smaller than, but with sequences homologous to, Bst I fragment 8.

When A6S/2 tumor DNA was digested with Bst I, two fragments (9.3 and 5.5 × 10^6 daltons) were produced which hybridized with pNW31C-8-1 (Figure 6A, lane 5). Both fragments were larger than Bst I fragment 8 (4.7 × 10^6 daltons). To account for these molecular weights, the T DNA must have been either integrated into plant DNA or fused to itself. These fragments, however, did not hybridize to pNW31C-2,19-1 as would be expected for a fusion fragment (Figure 6D, lane 4). This result suggests strongly that the T DNA is integrated into plant DNA and that tandem repeats do no exist. The detection of two putative left end junction fragments suggests that there are two T DNA inserts.

Similar analysis of A277/5, 15955/1 and $B_6$806/E9 also suggests integration of T_L DNA. When A277/5 tumor DNA was digested with Bst I, a fragment (10.8 × 10^6 daltons) larger than Bst I fragment 8 that hybridized to pNW31C-8-1 (Bst I fragment 8) was produced (Figure 6A, lane 6). Two lines of reasoning argue against a fusion event and in favor of integration. First, if the left and right ends of the T_L DNA were joined together, a fragment with a maximum molecular weight of 7.7 × 10^6 daltons should have been produced (the sum of the molecular weights of Bst I fragments 19 and 8) and it should have hybridized strongly with pNW11A-1-1. In fact, the fragment was too large, and it hybridized only slightly with pNW11A-1-1 (Figure 6B, lane 1) which can be attributed to the small overlap of Hind III fragment 1 and Bst I fragment 8 (see Figure 1). Second, if the left end of the T_L DNA had become joined to the T_R DNA, the 10.8 × 10^6 dalton fragment produced by Bst I digestion should have, but did not, hybridize with pNW11A-5-1 (Figure 6C, lane 1). The detection of only one left end junction fragment suggests that there may be only one T_L DNA insert in A277/5.

In the case of 15955/1, Bst I digestion produced a fragment (9.3 × 10^6 daltons) larger than Bst I frag-

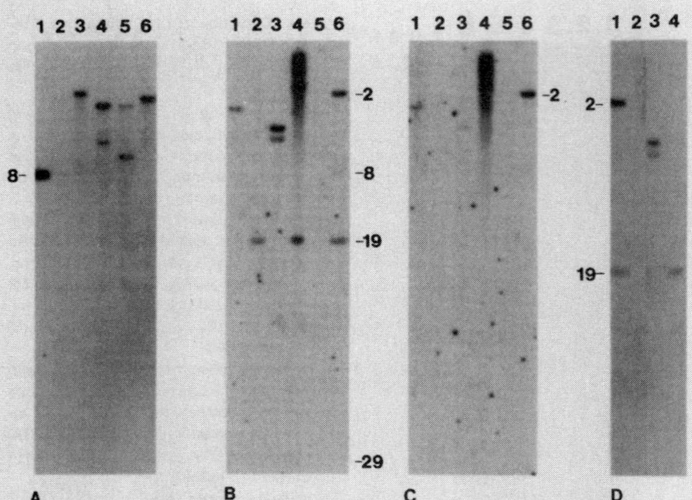

Figure 6. Detection of DNA Fragments Containing Ti Plasmid Sequences Produced upon Bst I Digestion of Tumor and Normal Tobacco DNA

Digestion products were fractionated and transferred to nitrocellulose as described in Experimental Procedures. The filters displayed in (A), (B) and (C) were prepared from the same agarose gel.

(A) The nitrocellulose filter was hybridized with ^{32}P–pNW31C-8-1. Lane (1) contains a 1 copy Ti plasmid reconstruction; (2) normal tobacco callus DNA; (3) $B_6$806/E9 tumor DNA; (4) 15955/1 tumor DNA; (5) A6S/2 tumor DNA and (6) A277/5 tumor DNA.

(B and C) The nitrocellulose filters were hybridized with ^{32}P–pNW11A-1-1 and ^{32}P–pNW11A-5-1, respectively. Lane (1) contains A277/5 tumor DNA; (2) A6S/2 tumor DNA; (3) 15955/1 tumor DNA; (4) $B_6$806/E9 tumor DNA; (5) normal tobacco callus DNA and (6) a 1 copy Ti plasmid reconstruction. These filters had previously been hybridized with either a mixture of ^{32}P–pNW31C-8,29-1 and ^{32}P–pNW31C-2,19-1 (B) or ^{32}P–pNW31C-8-1 (C). They were washed free of probe as described in Experimental Procedures.

(D) The nitrocellulose filter was hybridized with ^{32}P–pNW31C-2,19-1. Lane (1) contains a 1 copy Ti plasmid reconstruction; (2) normal tobacco callus DNA; (3) 15955/1 tumor DNA and (4) A6S/2 tumor DNA.

The positions of Bst I fragments 2, 8, 19 and 29 (10.7, 4.7, 3.0 and 0.7 × 10^6 daltons, respectively) are indicated at the sides of the autoradiograms.

ment 8 which hybridized with pNW31C-8-1 (Figure 6A, lane 4) but not with pNW31C-2,19-1 (Figure 6D, lane 3) or pNW11A-5-1 (Figure 6C, lane 3), thus arguing against fusion and suggesting integration of T_L DNA. Two additional fragments (6.6 and 5.9 × 10^6 daltons) which hybridized to pNW31C-8-1 (Figure 6A, lane 4) appear to be the products of a fusion event(s); these are discussed below.

Finally, Bst I digestion of $B_6$806/E9 produced one fragment (12 × 10^6 daltons) larger than Bst I fragment 8 that hybridized with pNW31C-8-1 (Figure 6A, lane 3). Fusion of the left and right ends of the T_L DNA could not account for a fragment of such large molecular weight without the T_L DNA having been integrated. If the left end of the T_L DNA had become fused to the T_R DNA and the structure existed as a free circular replicon, Bst I digestion should have produced a fragment with a maximum molecular weight of only 11 × 10^6 daltons, the sum of Bst I fragment 8 plus a T_R DNA segment containing Eco RI fragments 24, 13 and 22 (the data presented below suggest that T_R DNA is not tandemly repeated). The simplest interpretation of these data is that the T_L DNA is integrated. The situation regarding the presence or absence of tandemly repeated T_L DNA and/or T_L DNA fused to T_R DNA is equivocal at this time.

The data also suggest that the T_R DNA of $B_6$806/E9 is integrated and that tandem repeats do not exist. If the $B_6$806/E9 T_R DNA existed as nonintegrated linear DNA molecules, the fragments produced by Bst I digestion should have had a molecular weight of less than 8 × 10^6 daltons (the sum of Eco RI fragments 22, 13 and 24). When $B_6$806/E9 tumor DNA was digested with Bst I, however, and the T_R DNA fragments were located by hybridization with ^{32}P–pNW11A-5-1 (Figure 6C, lane 4), the fragments produced were larger than Bst I fragment 2, which is 10.7 × 10^6 daltons (Figure 6C, lane 6). This result suggested that a fusion or integration event had occurred. A fusion of the right and left ends of the T_R DNA should, upon Hind III digestion, produce fragments that hybridize with both pNW11A-5-1 and pNW11A-1-1. However, most if not all of the fragments produced upon Hind III digestion hybridized with either one or the other probe but not with both probes (compare Figure 3B, lane 4 with Figure 3C, lane 1).

Organization of T DNA

The organization of the T DNA in A6S/2 seems relatively simple. The number of T DNA inserts is low, only T_L DNA is present and there is no evidence for tandem repetition of T DNA. The organization of the T DNA in the other tumor lines, however, is more complex.

In the case of A277/5, the data suggest that the right-hand end of the T_L DNA is fused to the T_R DNA. After digestion with Bst I, the right-hand portion of the T_L DNA should hybridize with pNW11A-1-1 while the T_R DNA should hybridize with pNW11A-5-1. We found, however, that a single fragment (9.0 × 10^6 daltons) hybridized to both probes (compare Figure 6B, lane 1 with Figure 6C, lane 1). This suggests the possibility that the initial T DNA insert was a continuous piece of DNA with its left and right extremes composed of sequences from Hind III fragments Y and

5, respectively, and that the region between the Eco RI fragment 7/24 recognition site and some point in Hind III fragment 5 was subsequently deleted.

Bst I digestion of 15955/1 tumor DNA produced four fragments (9.3, 6.6, 5.9 and 0.6 $\times 10^6$ daltons) that contained DNA sequences homologous to the Ti plasmid (Figure 2A, lane 6). The largest fragment is the left end junction fragment described above and the smallest fragment corresponds to Bst I fragment 29. The 6.6 $\times 10^6$ dalton fragment hybridized with pNW31C-8-1 (Figure 6A, lane 4), pNW31C-2,19-1 (Figure 6D, lane 3), pNW11A-1-1 (Figure 6B, lane 3) and pNW11A-5-1 (Figure 6C, lane 3), while the 5.9 $\times 10^6$ dalton fragment hybridized to pNW31C-8-1 (Figure 6A, lane 4), pNW31C-2,19-1 (Figure 6D, lane 3) and pNW11A-1-1 (Figure 6B, lane 3). These results suggest that noncontiguous portions of the Ti plasmid are fused in 15955/1 T DNA. The formation of a fragment (the 6.3 $\times 10^6$ dalton fragment) that hybridized to both pNW11A-1-1 and pNW11A-5-1 suggests that as with A277/5, the T_L DNA and T_R DNA of 15955/1 are fused. The precise nature and significance of these fusion bands are as yet unknown.

The organization of the T_R DNA in B_6806/E9 is also complex. As described above, the data suggest that the T_R DNA copies are integrated and that tandem repeats are not present. However, the relatively large number of putative junction fragments of different sizes produced by both Hind III (Figure 3C, lane 1) and Eco RI (Figure 4B, lane 2) suggests that the T_R DNA is integrated into numerous sites. In addition, the intensity of the signal observed with a number of the fragments suggests that they are present in multiple copies. Two of the simpler interpretations are that the T_R DNA originates from multiple integration events into a few families of repetitive DNA and/or from single inserts into a few sites that have subsequently become amplified.

T DNA Is Integrated into Multiple Sites

Tumor lines B_6806/E9, A277/5 and 15955/1 were all incited in Nicotiana tabacum cv. Xanthi nc. The putative left end junction fragments of the T_L DNAs produced by Bst I digestion, however, had different molecular weights: for B_6806/E9, 12 $\times 10^6$ daltons; for 15955/1, 9.3 $\times 10^6$ daltons; and for A277/5, 10.8 $\times 10^6$ daltons (Figure 6A, lanes 3, 4 and 6, respectively). If the left-hand extremes of the T_L DNAs are the same in each tumor line, the distance to the nearest Bst I cut site in the plant must vary for each tumor DNA; that is, the surrounding plant DNA must be different for each insert. Even if the left ends of the T_L DNAs do vary, they differ by less than 1 $\times 10^6$ daltons (Figure 5), which is not enough to account for the difference in sizes observed among the left T_L DNA junction fragments. As discussed above, the T_R DNA inserts in B_6806/E9 also seem to be integrated into multiple sites.

Stability of T DNA Inserts

The tumors lines studied are cloned tissues that have been in culture for varying numbers of years (Table 1). If the T DNA inserts move from site to site, a new junction fragment should be produced upon each transposition event. If such transpositions occurred at a significant rate, each junction fragment would be present in only a small percentage of the cells and probably would not be observed. If the junction fragments could be detected, their numbers should exceed the number predicted by the copy number of the T DNA. In the experiments presented above, discrete left-hand junction fragments of the T_L DNAs were observed (Figure 6A, lanes 3–6) and the number of junction fragments agrees with the T DNA copy number. Thus the T DNAs in the cloned tumor lines are stable. This, however, does not preclude rearrangement of T DNAs early in tumor formation, as might have occurred, for example, in the case of the B_6806/E9 tumor line.

Homology between Ti Plasmid and Normal Tobacco DNA

It was previously reported that no homology exists between normal tobacco DNA and the octopine-type Ti plasmid pTi-B_6806 (Chilton et al., 1977). Using more sensitive techniques, however, we have occasionally observed fragments in the digests of normal tobacco DNA from axenic tissue culture that hybridized very weakly with the Ti plasmid clones used in this study. For example, in an Eco RI digest of normal tobacco DNA we detected a fragment of approximately 13 $\times 10^6$ daltons that hybridized weakly to pNW11A-1-1 (Figure 4A, lane 5) but did not hybridize to pNW11A-5-1 (Figure 4B, lane 3) or pBR322 (data not shown). These data suggest that it is premature to conclude firmly that the Ti plasmid and normal tobacco DNA do not share regions of homology.

Discussion

We have defined the T DNA sequences present in four crown gall tumor lines and have begun to describe their organization in plant DNA. Three general conclusions may be drawn:

—Each tumor line contains a "core" T DNA which is colinear with the Ti plasmid, includes the Ti plasmid sequences referred to as "common DNA" and is apparently responsible for maintaining the transformed state.

—A given Ti plasmid does not always give rise to the same T DNA complement.

—Most if not all T DNA is integrated into plant DNA.

Each of the four tumor lines studied contains a core T DNA insert(s) representing the Ti plasmid sequences between Hind III fragment Y and Eco RI fragment 24 (see Figure 5). The restriction endonucleases Eco RI, Sma I, Bst I and Hind III recognize a total of 12 sites in this region of the Ti plasmid. Our experimental approach allowed us to determine that 11 of these sites are maintained in the T DNA of each tumor line. We can therefore conclude that these T DNA sequences are colinear with Ti plasmid sequences. Since only these plasmid sequences could be detected in the A6S/2 tumor DNA, we must also conclude that these T DNA sequences contain information required for the maintenance of the transformed state. This is of particular significance since this T DNA includes Ti plasmid sequences (Sma I fragments 17, 16a and most of 10c) that have been found in all octopine- and nopaline-type Ti plasmids reported to date (Depicker, Van Montagu and Schell, 1978; Chilton et al., 1978). When the conjugative plasmid RP4 was inserted into this region of a nopaline-type Ti plasmid, the bacteria were avirulent (Depicker et al., 1978). From these data it was concluded that the "common DNA" has some central role in oncogenesis.

The mechanism of crown gall tumorigenesis is unknown. It was recently shown (Yang et al., 1980), however, that common DNA is present in the teratoma tissue incited by A. tumefaciens strain T-37. This strain harbors the nopaline-type pTi-T37 plasmid. This observation, along with our finding the common DNA in four independent unorganized tumor lines, suggests that common DNA has a direct role in tumor maintenance and perhaps tumor induction (see below). The same conclusion was recently reached by Gurley et al. (1979) on the basis of the finding that total RNA isolated from three tumor lines (a sibling clone of $B_6806/E9$ was included in this study) contained sequences complementary to common DNA.

In addition to maintaining the transformed state, the Ti plasmid sequences between Hind III fragment Y and Eco RI fragment 24 (T_L DNA) must contain information for the production of octopine, since A6S/2, an octopine-producing tumor, contains only these plasmid sequences. More precisely, the data suggest that the right-hand end of this T DNA is important for octopine production. The right-hand extremes of the A6S/2 and $B_6806/E9$ T_L DNA lie to the right of the Bst I recognition site in Eco RI fragment 24 whereas in 15955/1 and A277/5 they lie slightly to the left of this recognition site (Figure 5). Both A6S/2 and $B_6806/E9$ produce octopine whereas A277/5 and 15955/1 do not. The genetic studies performed by Klapwijk, Scheulderman and Schilperoort (1978) also suggest that this region of the Ti plasmid has a role in octopine production. It remains to be seen whether this region of the Ti plasmid codes directly for the octopine-synthesizing enzyme lysopine dehydrogenase.

Another fundamental conclusion that may be drawn from this work is that the T DNA is integrated into plant DNA. It also appears that integration can take place at multiple sites. In this respect the crown gall tumor system resembles simian virus 40 (Botchan et al., 1976) and the retroviruses avian sarcoma virus (Hughes et al., 1978; Sabran et al., 1979), murine leukemia virus (Steffen and Weinberg, 1978) and spleen necrosis virus (Keshet and Temin, 1978). Our data also suggest that the multicopy T_R DNA of $B_6806/E9$ is either integrated into a few families of repetitive DNA or integrated into a few unique sites that have subsequently become amplified. A similar situation has been postulated to occur for adenovirus type 12 (Sutter, Westphal and Doerfler, 1978).

Since we have analyzed only three tumors incited on the same host plant, we cannot yet conclude that T DNA integration is a random event. It is perhaps more significant that the data suggest that the core T DNA does not have to be integrated into a unique site to maintain the transformed state. This implies that the T DNA has an active role in tumorigenesis, perhaps coding for a transformation-inducing protein(s). Alternatively, there may be a number of plant genes important in the regulation of cell growth, and integration of T DNA into any one of these genes causes a mutation resulting in cell transformation.

The mechanism by which T DNA becomes integrated into plant DNA is unknown. Our data, however, suggest that the sequences contained in Hind III fragment Y and Eco RI fragments 24 and 22 serve as the preferred points of attachment of Ti plasmid DNA to plant DNA. In this respect, Ti plasmid integration at least superficially resembles that the Mu (Howe and Bade, 1975) and the retroviruses (Hughes et al., 1978; Keshet, O'Rear and Temin, 1979). In these systems a particular site of the phage or virus serves as the point of fusion into multiple cellular sites. Integration of simian virus 40 appears to be fundamentally different in that specific viral sequences do not serve as the junction points between virus and cellular DNA (Botchan et al., 1976). Further insight into the mechanism of T DNA integration should be gained by cloning and sequencing the Ti plasmid plant DNA junction fragments.

Experimental Procedures

Tumor Lines

The derivation of tumor lines $B_6806/E9$, A6S/2 and 15955/1 has been described elsewhere (Merlo et al., 1980). Tumor line A277 was incited by inoculating A. tumefaciens strain A277 into Nicotiana tabacum cv. Xanthi nc. Axenic cultures were established from the resultant tumor on standard phytohormone-free media (Johnson et al., 1974). A277/5 was derived from a single cell of the axenic culture. Octopine was not detected in A277/5 by the standard assay (Montoya et al., 1977), nor could lysopine dehydrogenase activity be demonstrated (L. Otten, personal communication).

Bacterial Strains and Plasmids

A detailed description of the plasmids pTi-B_6806, pTi-A6 and pTi-15955 and the history of the host bacteria have been published previously (Sciaky et al., 1978).

Isolation of Ti Plasmid DNA

Ti plasmids from the appropriate strains were isolated essentially as described previously (Currier and Nester, 1976b), except that the

DNA was precipitated after the addition of sodium acetate (0.3 M final concentration) instead of magnesium phosphate.

Isolation of Tumor DNA
High molecular weight tumor DNA was prepared as described by Chilton et al. (1977, 1978). The DNA was purified further by equilibrium centrifugation in cesium chloride-ethidium bromide gradients at a refractive index of 1.3870. The dye was extracted with isopropanol equilibrated against 20 × SSC (SSC is 0.15 M NaCl, 0.015 M sodium citrate) and the DNA was dialyzed against 1 mM Tris-HCl (pH 8.0).

Molecular Cloning and Mapping of the T Region of the Ti Plasmid
The Sma I and Hpa I restriction maps of the entire pTi-$B_6$806 plasmid have been published elsewhere (Chilton et al., 1978). Details of the mapping and cloning of Hind III fragments 1 and 5 of pTi-$B_6$806 have been described by Merlo et al. (1980). The map positions of Hind III fragments X and Y have been established by Gurley et al. (1979) for the pTi-15955 plasmid. The pTi-15955, pTi-A6 and pTi-$B_6$806 plasmids have been shown to be colinear through the T region (Sciaky et al., 1978; our unpublished results).

Clones containing Bam HI fragments were generated as follows. The plasmids pBR322 and pTi-A6 were digested with restriction enzyme Bam HI and the fragments were ligated with T4 ligase (Dugaiczyk et al., 1974). E. coli strain χ1776 was transformed with the recombinant molecules as previously described (Norgaard, Keem and Monahan, 1978). Ampicillin-resistant, tetracycline-sensitive transformants were screened by the colony filter hybridization technique (Grunstein and Hogness, 1975) for inserts containing sequences from the T region. The positions of the cloned fragments relative to the existing Sma I, Hpa I and Hind III maps were determined by digesting the recombinant plasmids with various restriction enzymes and comparing the molecular weights of the resultant fragments with the existing maps. The Eco RI restriction sites through the T region were determined by digesting the Hind III or Bam HI clones with Eco RI and correlating the molecular weights of resultant fragments with the existing map. A detailed physical map of the T region and a graphic representation of the fragments contained in the various clones used in this study are shown in Figure 1. All experiments involving recombinant DNA molecules were conducted under P2/EK2 containment as described in the NIH Guidelines (1976), or under P2/EK1 containment as described by the revised Guidelines (1978).

Isolation of E. coli Recombinant Plasmids
E. coli plasmids containing recombinant DNA were isolated essentially as outlined by Blair et al. (1972) after amplification with chloramphenicol. The plasmids were subjected to two cycles of equilibrium centrifugation through cesium chloride-ethidium bromide. After extracting the ethidium bromide with isopropanol saturated with 20 × SSC, the DNA solution was diluted with 2 vol of 10 mM Tris (pH 8), 1 mM EDTA and 0.3 vol of 3 M sodium acetate, and the DNA was held at −20°C for a minimum of 2 hr following the addition of 6 vol of absolute ethanol. The precipitate was collected by centrifugation, redissolved in 1 ml of buffer [10 mM Tris (pH 8), 1 mM EDTA] extracted twice with an equal volume of 1:1 neutralized phenol: chloroform (equilibrated against 5 mM NaCl), and then extracted twice with an equal volume of ether. The sodium concentration was adjusted to 100 mM by the addition of 3 M sodium acetate and the DNA was again precipitated at −20°C for at least 2 hr by the addition of 2 vol of ethanol. After centrifugation, the DNA was resuspended in buffer [10 mM Tris (pH 8), 1 mM EDTA] and its concentration was determined by the spectrofluorometric method of LePecq (1971).

In Vitro Labeling of DNA
Whole Ti plasmid and recombinant plasmids were labeled by a slight modification of the nick translation procedure described by Maniatis, Jeffrey and Kleid (1975). Plasmid DNA was suspended to a concentration of 10 μg/ml at 0°C in reaction buffer [50 mM Tris (pH 7.8), 5 mM $MgCl_2$, 50 μg/ml bovine serum albumin and 6 mM 2-mercaptoethanol] containing the ^{32}P-deoxyribonucleotides and DNAase, and was then placed at 15°C. After 30 min, DNA polymerase I (Worthington) was added (4 U/μg DNA) and the reaction was allowed to continue for an additional 90 min at 15°C. DNA preparations with specific activities of 5×10^7 to 1.5×10^8 cpm/μg DNA were routinely achieved.

Restriction Enzymes
Restriction endonucleases Hind III and Bam HI were used under the conditions recommended by the suppliers. Restriction enzyme Bst I, which has the same recognition sequence as Bam HI, was a gift from R. Meagher and was used with a reaction buffer composed of 100 mM Tris (pH 7.6), 6 mM $MgCl_2$ and 100 mM NaCl. Restriction enzyme Eco RI was purified in our laboratory by the procedure of Bickle, Pirrotta and Imber (1977). Reaction buffer consisted of 10 mM Tris (pH 7.6), 6 mM $MgCl_2$, 100 mM NaCl. Normal plant and tumor DNAs were digested with at least twice the enzyme for at least twice the time required to obtain a complete digest of an equivalent quantity of whole Ti plasmid. Digests were judged to be complete if the fragments containing Ti plasmid sequences did not change molecular weight when the amount of enzyme and time of digestion were increased. Reactions were terminated with 0.1 vol of a mixture containing 50% glycerol, 7% SDS and 0.7% bromophenol blue. Eco RI and Hind III digests of lambda DNA were run in parallel lanes as molecular weight standards.

Gel Electrophoresis and Preparation of Nitrocellulose Filters
7 μg of digested DNA were loaded into 1 × 5 × 10 mm wells in a horizontal 0.7% agarose gel (1 cm thick) prepared in a Tris-acetate buffer as described (Chilton et al., 1977). The samples were subjected to electrophoresis at 1.0–1.5 V/cm until the dye front had migrated approximately 18 cm. The gel was stained with 1 μg/ml ethidium bromide in water for 1 hr with gentle rocking, and the restriction pattern was photographed during illumination with ultraviolet light (Blak-Ray Transilluminator). The gels were then treated for 2 hr in 0.5 M NaOH and 0.8 M NaCl followed by 4 hr in 0.5 M Tris-HCl (pH 7), 1.5 M NaCl with gentle rocking at all times. The DNA was transferred from the gels to nitrocellulose sheets that had been wetted in water, followed by 10 × SSC, and blotted according to the method of Southern (1975) using 10 × SSC for 48 hr. Restaining the gel with ethidium bromide after transfer showed that only a trace of the highest molecular weight DNA was not transferred. The filters were then marked, cut to size, rinsed with 2 × SSC, air-dried at room temperature, baked overnight at 68°C and finally baked at 80°C for 2 hr in vacuo.

Ti Plasmid Reconstruction Mixtures
Reconstruction mixtures consisted of Eco RI-digested salmon sperm DNA (7 μg) mixed with either pTi-A6, pTi-15955 or pTi-$B_6$806 DNA (for the 1, 2.5 and 5 copy reconstructions, respectively) that had previously been digested with either Bst I, Hind III, Eco RI or Sma I. Since the molecular weights of the Ti plasmids are 1.2×10^8 daltons and the DNA of a diploid tobacco cell is 6×10^{12} daltons, it follows that if the entire Ti plasmid was present at one copy per dipoid genome, 140 pg of Ti plasmid would be present in 7 μg of plant DNA (the amount used in each experiment).

Hybridization
The prepared filters were first wetted in 2 × SSC then immersed in a prehybridization mix consisting of 6 × SSC and 10X Denhardt's solution [1X Denhardt's is 0.02% bovine serum albumin, 0.02% Ficoll and 0.02% polyvinylpyrrolidone (Denhardt, 1966)] at 68°C for 12–16 hr. The filters were then transferred to polyethylene bags containing hybridization solution with a final composition of 3 × SSC, 5X Denhardt's, 20 mM Tris (pH 7.6), 0.1% SDS, 20 μg/ml denatured sheared salmon sperm or steelhead trout sperm DNA and 2 mM EDTA. The filters were incubated at 68°C for 4 hr, after which time the denatured probe (100°C for 5 min) was added at a final concentration of 25 ng/ml and the mixture was incubated at 68°C for approximately 40 hr. The blots were rinsed briefly in 2 × SSC at room temperature followed by four washes for 30 min each in 3 × SSC, 0.2% SDS, 5 mM EDTA at 68°C, and by one wash for 30 min in 0.3 × SSC, 0.2% SDS, 5 mM EDTA at 68°C. The SDS was removed by two brief rinses at room temperature with 2 × SSC and

the filters were allowed to dry at room temperature. They were then mounted on cellulose acetate sheets, sandwiched between Dupont Lightning Plus intensification screens with Kodak RP-X-omat film and placed at −70°C for 1–14 days. If a filter was to be annealed with a second probe, the first annealing reagent was removed by heating the filter to 68°C for 30 min in hybridization mixture supplemented with formamide (50% final concentration). The filters were washed twice with 2 × SSC at room temperature, transferred to fresh hybridization buffer and processed further in the usual manner.

Acknowledgments

We wish to thank M.-D. Chilton for helpful discussions; Bob Simpson and Arnie Bendich for critical readings of the manuscript; Rich Meagher for providing a generous amount of Bst I; Stan Gelvin for leading the way in the cloning of the Bam HI fragments generated from pTi-A6; Ross Bowman and Rob Jensen for helping prepare the tumor DNAs; Randy Saiki for preparing the Eco RI and helping develop the final blotting and hybridization conditions; Fritz Lang for his advice concerning the lowering of background on nitrocellulose; Kathy Spangler for helping prepare the manuscript; Dr. L. Otten for assaying lysopine dehydrogenase activity; and Dr. Schilperoort for generously providing the uncloned A6S tumor tissue.

This work was funded by a grant from the National Cancer Institute. M. F. T. was a Fellow in Cancer Research supported by a grant from the Damon Runyon-Walter Winchell Cancer Fund. R.N. held a postdoctoral fellowship from the NCI.

The costs of publication of this article were defrayed in part by the payment of page charges. This article must therefore be hereby marked "*advertisement*" in accordance with 18 U.S.C. Section 1734 solely to indicate this fact.

Received November 16, 1979

References

Bickle, T. A., Pirrotta, V. and Imber, R. (1977). A simple, general procedure for purifying restriction endonucleases. Nucl. Acids Res. *4*, 2561–2572.

Blair, D. G., Sherratt, D. J., Clewell, D. B. and Helinski, D. R. (1972). Isolation of supercoiled colicinogenic factor E1 DNA sensitive to ribonuclease and alkali. Proc. Nat. Acad. Sci. USA *69*, 2518–2522.

Bomhoff, G., Klapwijk, P. M., Kester, H. C. M., Schilperoort, R. A., Hernalsteens, J. P. and Schell, J. (1976). Octopine and nopaline synthesis and breakdown genetically controlled by a plasmid of *Agrobacterium tumefaciens*. Mol. Gen. Genet. *145*, 177–181.

Botchan, M., Topp, W. and Sambrook, J. (1976). The arrangement of simian virus 40 sequences in the DNA of transformed cells. Cell *9*, 269–287.

Chilton, M.-D., Drummond, M. H., Merlo, D. J., Sciaky, D., Montoya, A. L., Gordon, M. P. and Nester, E. W. (1977). Stable incorporation of plasmid DNA into higher plant cells: the molecular basis of crown gall tumorigenesis. Cell *11*, 263–271.

Chilton, M.-D., Drummond, M. H., Merlo, D. J. and Sciaky, D. (1978). Highly conserved DNA of Ti-plasmids overlaps T-DNA maintained in plant tumors. Nature *275*, 147–149.

Currier, T. C. and Nester, E. W. (1976a). Evidence for diverse types of large plasmids in tumor-inducing strains of *Agrobacterium*. J. Bacteriol. *126*, 157–165.

Currier, T. C. and Nester, E. W. (1976b). Isolation of covalently closed circular DNA of high molecular weight from bacteria. Anal. Biochem. *76*, 431–441.

Denhardt, D. T. (1966). A membrane-filter technique for the detection of complementary DNA. Biochem. Biophys. Res. Commun. *23*, 641–646.

Depicker, A., Van Montagu, M. and Schell, J. (1978). Homologous DNA sequences in different Ti-plasmids are essential for oncogenicity. Nature *275*, 150–153.

Dugaiczyk, A., Hedgepeth, J., Boyer, H. W. and Goodman, H. M. (1974). Physical indentity of the SV40 deoxyribonucleic acid sequence recognized by the Eco RI restriction endonuclease and modification methylase. Biochemistry *13*, 503–512.

Goldman, A, Tempé, J. and Morel, G. (1968). Quelques particularités de diverses souches *d'Agrobacterium tumefaciens*. CR Acad. Sci. *162*, 630–631.

Gresshoff, P. M., Skotnicki, M. L. and Rolfe, B. G. (1979). Crown Gall teratoma formation is plasmid and plant controlled. J. Bacteriol. *137*, 1020–1021.

Grunstein, M. and Hogness, D. S. (1975). Colony hybridization: a method for the isolation of cloned DNAs that contain a specific gene. Proc. Nat. Acad. Sci. USA *72*, 3961–3965.

Gurley, W. B., Kemp, J. D., Albert, M. J., Sutton, D. W. and Collins, J. (1979). Transcription of Ti-plasmid derived sequences in three octopine-type crown gall tumor lines. Proc. Nat. Acad. Sci. USA *76*, 2828–2832.

Howe, M. M. and Bade, E. G. (1975). Molecular biology of bacteriophage Mu. Science *190*, 624–632.

Hughes, S. H., Shank, P. R., Spector, D. H., Kung, H.-J., Bishop, J. M., Varmus, H. E., Vogt, P. K. and Breitman, M. L. (1978). Proviruses of avian sarcoma virus are terminally redundant, co-extensive with unintegrated linear DNA and integrated at many sites. Cell *15*, 1397–1410.

Johnson, R., Guderian, R. H., Eden, F., Chilton, M.-D., Gordon, M. P. and Nester, E. W. (1974). Detection and quantitation of octopine in normal plant tissue and in crown gall tumors. Proc. Nat. Acad. Sci. USA *71*, 536–539.

Keshet, E. and Temin, H. M. (1978). Sites of integration of reticuloendotheliosis virus DNA in chicken DNA. Proc. Nat. Acad. Sci. USA *75*, 3372–3376.

Keshet, E., O'Rear, J. J. and Temin, H. M. (1979). DNA of noninfectious and infectious integrated spleen necrosis virus (SNV) is colinear with unintegrated SNV DNA and not grossly abnormal. Cell *16*, 51–61.

Klapwijk, P. M., Scheulderman, T. and Schilperoort, R. A. (1978). Coordinated regulation of octopine degradation and conjugative transfer of Ti-plasmids in *Agrobacterium tumefaciens*: evidence for a common regulatory gene and separate operons. J. Bacteriol. *136*, 775–785.

LePecq, J.-B. (1971). Use of ethidium bromide for separation and determination of nucleic acids of various conformational forms and measurement of their associated enzymes. In Methods of Biochemical Analysis, *20*, D. Glick, ed. (New York: Interscience Publishers), pp. 41–86.

Maniatis, T., Jeffrey, A. and Kleid, D. G. (1975). Nucleotide sequence of the rightward operator of phage λ. Proc. Nat. Acad. Sci. USA *72*, 1184–1188.

Merlo, D. J., Nutter, R. C., Montoya, A. L., Garfinkel, D. J., Drummond, M. H., Chilton, M.-D., Gordon, M. P. and Nester, E. W. (1980). The boundaries and copy numbers of Ti plasmid T-DNA vary in crown gall tumors. Mol. Gen. Genet., in press.

Montoya, A. L., Chilton, M.-D., Gordon, M. P., Sciaky, D. and Nester, E. W. (1977). Octopine and nopaline metabolism in *Agrobacterium tumefaciens* and crown gall tumor cells: role of plasmid genes. J. Bacteriol. *129*, 101–107.

Norgaard, M. V., Keem, K. and Monahan, J. J. (1978). Factors affecting the transformation of *Escherichia coli* strain χ1776 by pBR322 plasmid DNA. Gene *3*, 279–292.

Petit, A., Delhaye, S., Tempé, J. and Morel, G. (1970). Recherches sur les quanidines des tissues de crown gall. Mise en évidence d'une relation biochimique spécifique entre les souches *d'Agrobacterium tumefaciens* et les tumeurs qu'elles induisent. Physiol. Vég. *8*, 205–213.

Sabran, J. L., Hsu, T. W., Yeater, C., Kaji, A., Mason, W. S. and Taylor, J. M. (1979). Analysis of integrated avian RNA tumor virus

DNA in transformed chicken, duck and quail fibroblasts. J. Virol. *29,* 170–178.

Sciaky, D., Montoya, A. L. and Chilton, M.-D. (1978). Fingerprints of *Agrobacterium* Ti-plasmids. Plasmid *1,* 238–253.

Southern, E. M. (1975). Detection of specific sequences among DNA fragments separated by gel electrophoresis. J. Mol. Biol. *98,* 503–517.

Steffen, D. and Weinberg, R. A. (1978). The integrated genome of murine leukemia virus. Cell *15,* 1003–1010.

Sutter, D., Westphal, M. and Doerfler, W. (1978). Patterns of integration of viral DNA sequences in the genomes of adenovirus type 12-transformed hamster cells. Cell *14,* 569–585.

Thomashow, M., Panagopoulos, C. G., Gordon, M. P. and Nester, E. W. (1980). Host range of Agrobacterium tumefaciens is determined by the Ti-plasmid. Nature, in press.

Van Larebeke, N., Engler, G., Holsters, M., Van den Elsecher, S., Zaenen, I., Schilperoort, R. A. and Schell, J. (1974). Large plasmid in *Agrobacterium tumefaciens* essential for crown gall-inducing activity. Nature *252,* 169–170.

Van Larebeke, N., Genetello, C., Schell, J., Schilperoort, R. A., Hermans, A. K., Hernalsteens, J. P. and Van Montagu, M. (1975). Aquisition of tumor-inducing ability by non-oncogenic agrobacteria as a result of plasmid transfer. Nature *255,* 742–743.

Watson, B., Currier, T. C., Gordon, M. P., Chilton, M.-D. and Nester, E. W. (1975). Plasmid required for virulence of *Agrobacterium tumefaciens.* J. Bacteriol. *123,* 255–264.

Yang, F., Montoya, A. L., Merlo, D. J., Drummond, M. H., Chilton, M.-D., Nester, E. W. and Gordon, M. P. (1980). Foreign DNA sequences in crown gall teratoma and their fate during the loss of the tumorous traits. Mol. Gen. Genet., in press.

Note Added in Proof

The T DNA from tumor line A6S/2 has been cloned using the phage vector Charon 4A (M. Thomashow et al., manuscript in preparation). Analysis of cloned DNA segments definitively demonstrates that the A6S/2 T DNA is covalently linked to plant DNA.

WHEN TOOTHPICKS MATTERED: USING MICROBIAL GENETICS AND TRANSPOSON TAGGING TO MAP THE T-DNA

David J. Garfinkel
National Cancer Institute–Frederick
Frederick, Maryland, United States
E-mail: garfinke@ncifcrf.gov

Commentary on the paper:

Garfinkel, D. J., Simpson, R. B., Ream, L. W., White, F. F., Gordon, M. P., Nester, E. W. 1981. Genetic analysis of crown gall: Fine structure map of the T-DNA by site-directed mutagenesis. Cell 27:143–153.

I became interested in crown gall from my undergraduate research on microbial nitrogen fixation in Winston Brill's lab at the University of Wisconsin. My initial exposure to crown gall occurred when I presented a paper by Bruce Watson and the Seattle Crown Gall Group on Ti plasmids for journal club. I was impressed with the system and the combination of molecular and genetic approaches used in the work. Crown gall seemed to have everything: cancerous transformation, plant-microbe interaction, and molecular access via the Ti plasmid. I also thought the Pacific Northwest would be a nice change in scenery and climate for a Wisconsin native. So I enrolled in the Microbiology and Immunology graduate program at the University of Washington in the fall of 1975, after Gene Nester and I had several interesting phone conversations, although I failed to wheedle a recruitment trip to Seattle from him. During my first year, I took the requisite lab rotations and was tempted to work on bacterial ecology, but came to my senses and entered the Nester lab at a very exciting time in the spring of 1976.

The microbiology component of the Seattle Crown Gall Group at that time included fellow graduate students Frank White and Daniela Sciaky, postdocs Don Merlo and Martin Drummond, technician Alice Montoya, and the incomparable Mary-Dell Chilton. The plant work was performed in Milt Gordon's lab in the Department of Biochemistry. The central question in crown gall research in 1976 was whether the Ti plasmid induced tumor formation by transfer of foreign DNA to plant cells, similar to virally mediated transformation of animal cells. Within a year after I joined the lab, members of the Seattle Crown Gall Group clearly showed that a segment of the large Ti plasmid, called "T-DNA", was stably incorporated in crown gall tumors. This stunning discovery and the heroic experiments leading up to it were heady stuff for a beginning graduate student to observe, and has had a lasting influence on me.

After floundering in the lab for awhile, I became interested in transposon tagging as a way to identify the *Agrobacterium* Ti plasmid genes required for tumor formation. The simple idea was to make avirulent mutants by transposable element insertion, and thereby then physically map the insertions using the restriction endonuclease map of the Ti plasmid to identify the genetic loci responsible for crown gall. Within a few years, several groups had developed techniques for transposon tagging in *Agrobacterium* and a variety of Ti plasmids were under investigation. We, along with others, identified the *vir* regulon and the T-DNA as being important for tumorigenesis, as well as chromosomal genes (Garfinkel and Nester 1980) that were important for binding *Agrobacterium* to plant cells (Douglas et al. 1982). We also obtained several T-DNA mutants that induced tumors with aberrant morphologies (Garfinkel and Nester 1980), which set the stage for defining the T-DNA "oncogenes." One class of mutants produced tumors with an overabundance of shoots while another produced tumors with many roots. The tumor morphology (tm) mutants suggested a phytohormone imbalance in the tumor, which strongly reinforced the straightforward hypothesis that the T-DNA encoded genes for auxin and cytokinin biosynthesis. Similar results were also obtained by Rob Schilperoort and colleagues (Ooms et al. 1981).

To saturate the T-DNA with transposon insertions in order to define the putative phytohormone genes, required a more "site-directed" method to enrich for T-DNA insertions in the Ti plasmid, since the T-DNA comprises only about 5% of the plasmid. Fortunately, Bob Simpson, one of several talented postdocs to join the lab during my stay, obtained a novel marker transfer technique from Gary Ruvkun in the Ausubel lab (Gary and Bob were friends from Harvard) well before publication (Ruvkun and Ausubel 1981). This technique was a spectacular success when applied to the T-DNA. It allowed individual Tn*5* insertions in a segment of T-DNA carried on a shuttle plasmid in *Escherichia coli* to be transferred back into the Ti plasmid in *Agrobacterium* by homologous recombination. Marker transfer worked so well that a team rapidly formed in the Nester lab to generate the necessary insertions in the minimal segment (TL) of T-DNA required for tumorigenesis in the octopine plasmid pTiA6NC. In addition to Bob, who helped me establish the technique in the lab, Frank White generated Tn*3* insertions in a region of the T-DNA that turned out to be a cold-spot for Tn*5* transposition. Walt Ream, another new postdoc, focused on the right end of the core T-DNA, where little was known except the rough position of the octopine synthesis gene, *ocs*.

We generated a congruent functional map of the TL T-DNA from the analysis of 75 Tn*5* insertiions and three Tn*3* insertions. Four genetic loci were identified; three of these corresponded to the location of genes present in octopine tumors. *Tms* mutants caused a proliferation of shoots from the tumor on a variety of plants. This locus is now known to contain two genes, *iaaH* and *iaaM,* that encode enzymes required for auxin synthesis from tryptophan. The *tmr* mutants caused a proliferation of roots from the tumor and is now called *ipt.* It encodes an enzyme required to form the cytokinin isopentenyl-AMP, from AMP and isopentylpyrophosphate. The *tml* mutants caused more massive tumors by fresh weight. However, this gene's biochemical function remains unknown. Of course, the *ocs* gene encoded octopine synthase. Another striking feature of the tumor morphology mutants was that no single T-DNA gene was required for virulence. Walt extended this observation in a subsequent study using T-DNA double and triple mutants. The TL T-DNA contains additional genes, including *ons* (opine export from plant cells), genes *5* (auxin antagonism) and *7* (unknown function) that may have been mutated in our study. However, these genes apparently do not affect tumorigenesis in the plant hosts we analyzed. Partially or completely crippled T-DNA mutants also aided the development of vectors for genetic engineering of plants. In addition, we showed that bacterial T-DNA Tn*5* sequences were present in the *tmr*-149 tumor.

This research significantly shaped my training and further kindled my interest in mobile DNA. This interest extends to my current work on retrotransposons. In retrospect, the technical aspects of the study would have been performed faster and easier if we had the modern-day tool kit, especially higher quality molecular biology reagents such as T4 DNA ligase, more convenient shuttle plasmids, PCR, DNA sequencing, and personal computers! The Ruvkun and Ausubel marker transfer technique, was crucial to the success of our project. I suspect it is still being used today. The thousands of sterile toothpicks used for picking bacterial colonies and inoculating plant hosts were also unavoidable, although I think the toothpicks were recycled a few times. Finally, the intellectual climate of the Seattle Crown Gall Group encouraged both imaginative thinking and rigorous experimentation, because crown gall itself seemed so unusual. Given our state of ignorance in 1981, it is remarkable that *Agrobacterium*-based gene transfer to plants is now commonly used in modern plant breeding throughout the world, and as a way to tag plant genes. Yet I remain intrigued by the consequences of genetic information transfer between unrelated organisms and believe that crown gall will continue to be the underpinning for understanding how DNA transfer can establish new ecosystems.

References

Douglas, C. J., Halperin, W., and Nester, E. W. 1982. *Agrobacterium tumefaciens* mutants affected in attachment to plant cells. J Bacteriol. 152:1265–1275.

Garfinkel, D. J., and Nester, E. W. 1980. *Agrobacterium tumefaciens* mutants affected in crown gall tumorigenesis and octopine catabolism. J. Bacteriol. 144:732–743.

Ooms, G., Hooykaas, P. J., Moolenaar, G., and Schilperoort, R. A. 1981. Grown gall plant tumors of abnormal morphology, induced by *Agrobacterium tumefaciens* carrying mutated octopine Ti plasmids; analysis of T-DNA functions. Gene 14:33–50.

Ruvkun, G. B., and Ausubel, F. M. 1981. A general method for site-directed mutagenesis in prokaryotes. Nature 289:85–88.

DAVID J. GARFINKEL: I majored in Bacteriology at the University of Wisconsin and graduated with my B.S. in 1975. I completed my Ph.D. at the University of Washington, Seattle, in 1981. Many people have helped me during the course of this work. Eugene Nester was a primary force in my graduate school experience, as both a mentor and friend. Gene seemed to know from the beginning what *Agrobacterium* was up to, and therefore, provided needed ballast when the seas were stormy. Don Merlo and Alice Montoya taught me how hard you have to work to reach a goal, while Frank White and Mike Thomashow showed me how to maintain focus and not yield to dogma. Bob Simpson was my most critical editor. Conrad Lichtenstein, Harry Klee, Joan McPherson, and Tsuni Kosuge provided many useful comments during the preparation of the manuscript. Milt Gordon taught me much about the history and mysteries of crown gall plant cell culture.

Cell, Vol. 27, 143–153, November 1981 (Part 2), Copyright © 1981 by MIT

Genetic Analysis of Crown Gall: Fine Structure Map of the T-DNA by Site-Directed Mutagenesis

David J. Garfinkel,* Robert B. Simpson,†
Lloyd W. Ream,* Frank F. White,*
Milton P. Gordon‡ and Eugene W. Nester*
* Department of Microbiology and Immunology
University of Washington
Seattle, Washington 98195
† ARCO Plant Cell Research Institute
6905 Sierra Court
Dublin, California 94566
‡ Department of Biochemistry
University of Washington
Seattle, Washington 98195

Summary

Seventy-five Tn5 and three Tn3 insertions were generated, characterized and mapped in the pTiA6NC plasmid sequences which are stably integrated in crown gall tumors (T-DNA). Four mutants containing Tn5 insertions in a specific region of the T-DNA incited tumors that no longer synthesized octopine. No single insertion resulted in complete loss of oncogenicity. Twenty-five transposon insertions defined three distinct loci affecting tumor morphology. The first group (*tml*), of seven mutants, contained Tn5 insertions within a 1.25 kilobase (kb) region and incited tumors two to three times larger than normal. The second group (*tmr*), of nine mutants, incited tumors with a massive amount of roots proliferating from the tumor callus and contained Tn5 insertions in a 1 kb cluster. The third group (*tms*), of nine mutants, incited tumors with shoots growing from the tumor callus and contained three Tn3 insertions and six Tn5 insertions distributed over a 3.1 kb region. Each of these loci was separated by Tn5 insertions that did not noticeably change tumor formation. We identified two more regions of the T-DNA where transposon insertions did not appear to affect tumor morphology: one contained 28 Tn5 insertions distributed over 3.2 kb; the other contained 17 Tn5 insertions distributed over 7 kb and the octopine synthesis locus. A detailed functional map of the T-DNA of pTiA6NC resulted from the characterization of these insertions. The incorporation of Tn5 sequences into the plant genome was also demonstrated. We discuss these results in relation to the map location of tumor-derived RNA transcribed from the T-DNA, the role of phytohormones in crown gall tumorigenesis and the eventual use of the Ti plasmid as a vehicle for introducing genes of choice into the genomes of higher plants.

Introduction

The ability of Agrobacterium tumefaciens to cause crown gall, a neoplastic disease of dicotyledonous plants, is encoded by tumor-inducing (Ti) plasmids (Van Larebeke et al., 1974; Watson et al., 1975). A portion of the Ti plasmid, the T-DNA, is stably transferred to the plant (Chilton et al., 1977), causing the plant cells to exhibit two new properties: the cells lose their requirement for exogenously added phytohormones (Braun, 1958); and crown gall tissues produce tumor-specific compounds or opines, such as octopine (Menage and Morel, 1964), nopaline (Goldman et al., 1968, 1969) and agropine (Firmin and Fenwick, 1978). Genes required for the synthesis and catabolism of opines are coded for by specific types of Ti plasmids (Bomhoff et al., 1976; Montoya et al., 1977).

The physical characterization of the T-DNA in several nopaline-type tumors (Lemmers et al., 1980; Zambryski et al., 1980; Yang and Simpson, 1981) and octopine-type tumors (Thomashow et al., 1980a) indicates that the T-DNA is covalently joined to plant nuclear DNA (Thomashow et al., 1980b; Yadav et al., 1980; Chilton et al., 1980; Willmitzer et al., 1980; Zambryski et al., 1980), and that the T-DNA from a tumor is generally colinear with the Ti plasmid. The T-DNA in nopaline tumors is present in a continuous segment, while the T-DNA in octopine tumors can be split between a core T-DNA segment present in all tumors and other segments present in some tumors.

The T-DNA of both octopine (pTiB$_6$806, pTiB$_6$S$_3$ and pTiACH5) and nopaline (pTiC58) Ti plasmids has been analyzed genetically by use of R factor–Ti plasmid cointegrates, transposon insertion mutants and deleted derivatives of the Ti plasmid. Several mutants containing transposon insertions, R-factor cointegrations or deletions in the T-DNA of the nopaline plasmid pTiC58 are avirulent on some plants (Holsters et al., 1980). Mutations caused by transposon insertions in the T-DNA of three highly related octopine-type Ti plasmids, pTiB$_6$806 (Garfinkel and Nester, 1980), pTiB$_6$S$_3$ (Ooms et al., 1981) and pTiACH5 (Ooms et al., 1980, 1981), affect the ability of strains harboring these plasmids to incite undifferentiated tumors. The tumors have either roots or shoots proliferating from the callus. Several large deletions and transposon insertions generated in pTiC58 (Holsters et al., 1980) and several deletions generated in pTiACH5 (Koekman et al., 1979) demonstrate that regions of the T-DNA in both pTiC58 and pTiACH5 are not essential for tumorigenesis. The nopaline synthesis locus of pTiC58 maps to a nonessential region of the T-DNA based on one Tn7 insertion (Holsters et al., 1980), while the precise location of the octopine synthesis locus is not known.

We sought to understand the detailed functional organization of the T-DNA of the octopine-type Ti plasmid pTiA6NC. Transposons Tn3 and Tn5 were used to mutagenize the T-DNA by insertional inactivation. The techniques used (Ruvkun and Ausubel, 1981) directed insertions to relatively small restriction fragments of the pTiA6NC T-DNA instead of to the whole Ti plasmid as in previous studies. The transposon insertions were mapped and introduced into pTiA6NC by homologous recombination, and recombinant strains were tested for altered tumor phenotypes. This approach enabled us to isolate and characterize a large collection of mutations within a relatively small portion of the Ti plasmid and resulted in a detailed functional map of the T-DNA.

Results

Experimental Approach

A method for site-directed mutagenesis (or marker transfer) in procaryotes (Ruvkun and Ausubel, 1981) was modified for use in Agrobacterium. This technique allows transposon insertions isolated in cloned DNA fragments to replace wild-type genomic DNA sequences by homologous recombination. Marker transfer consists of several steps as applied to the T-DNA (Figure 1). Overlapping restriction fragments of the T-DNA were cloned into either pRK290, a broad-host-range cloning vehicle derived from RK2 (Ditta et al., 1981), or pHK17, a broad-host-range cosmid vector derived from RK2 (H. Klee, M. P. Gordon and E. W. Nester, submitted manuscript). T-DNA clones harbored in Escherichia coli were mutagenized by Tn3 or Tn5. Hybrid plasmids carrying Tn3 or Tn5 insertions were isolated, and the locations of the transposon insertions in the T-DNA were determined by restriction enzyme analysis. After transformation of the mutagenized hybrid plasmids into an A. tumefaciens strain that harbors pTiA6NC, homologous recombination events resulted in the transfer of the selectable Tn3 or Tn5 insertions from the hybrid plasmid to pTiA6NC. Subsequent introduction of another broad-host-range plasmid pPH1J1, incompatible with pRK290 and pHK17, and simultaneous selection for kanamycin resistance (retention of Tn5) or carbenicillin resistance (retention of Tn3) as well as for gentamicin resistance (conferred by pPH1J1) enriched for those cells where the transposon had been introduced into pTiA6NC. Total DNA from the potential recombinants was cleaved with the appropriate restriction enzyme, and the products were separated by agarose gel electrophoresis. The fractionated DNAs were transferred to nitrocellulose filters (Southern, 1975) and annealed with in vitro ^{32}P-labeled T-DNA, and plasmid sequences were visualized by autoradiography. From molecular weights and hybridization characteristics of the fragments produced, we determined whether Tn3 or Tn5 recombined into the Ti plasmid by means of flanking T-DNA homology

Restriction Enzyme Analysis of Tn3 and Tn5 Inserts into the T-DNA of pTiA6NC

The T-DNA restriction fragments used for Tn5 mutagenesis were Bam HI fragments 8 and 2 and Eco RI fragments 7 and 24, cloned into pRK290 (Figure 2).

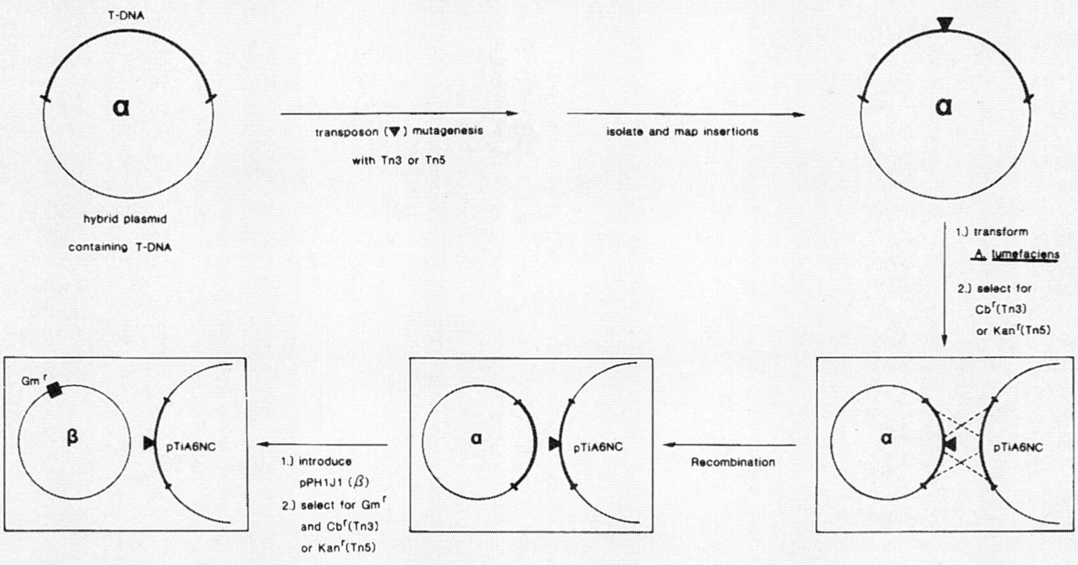

Figure 1. Experimental Strategy Used to Introduce Specific Tn3 or Tn5 Insertions into the T-DNA of pTiA6NC

This was a slight modification of a technique developed by Ruvkun and Ausubel (1981). Plasmids (α) containing different T-DNA fragments cloned into either pRK290 or pHK17 were mutagenized by either Tn3 or Tn5. The mutagenized plasmids were isolated, and the locations of the transposons were determined by restriction enzyme mapping. Hybrid plasmids containing insertions distributed throughout the T-DNA were transformed into A. tumefaciens, which harbors pTiA6NC. A. tumefaciens strains in which the T-DNA::Tn3 or Tn5 mutations had recombined into pTiA6NC were selected by mating pPH1J1 (β) into each strain. The resulting transconjugants were analyzed for the introduction of Tn3 or Tn5 into pTiA6NC by homologous recombination as described in the text and in Figure 4. T-DNA is shown in heavy lines. Boxes denote steps of the procedure that occur in A. tumefaciens, while the other steps occur in E. coli.

The cosmid clone pNW35D-210-1 used for Tn3 mutagenesis contains Eco RI fragment 7 and Bam HI fragment 8 of the T-DNA, as well as additional Ti plasmid sequences adjacent to the T-DNA (H. Klee, M. P. Gordon and E. W. Nester, submitted manuscript). Overlapping restriction fragments and the cosmid pNW35D-210-1 ensured that all sequences of the T-DNA would be included as potential targets for Tn3 or Tn5 insertional mutagenesis. However, no overlapping cloned fragment was obtained that covered the region between Eco RI fragments 7 and 24, resulting in the omission of two small Eco RI fragments totaling approximately 0.5 kb as targets for mutagenesis. Plasmids containing Tn3 or Tn5 insertions were isolated, and the transposons were located by restriction enzyme mapping (Figure 3) as outlined in Experimental Procedures.

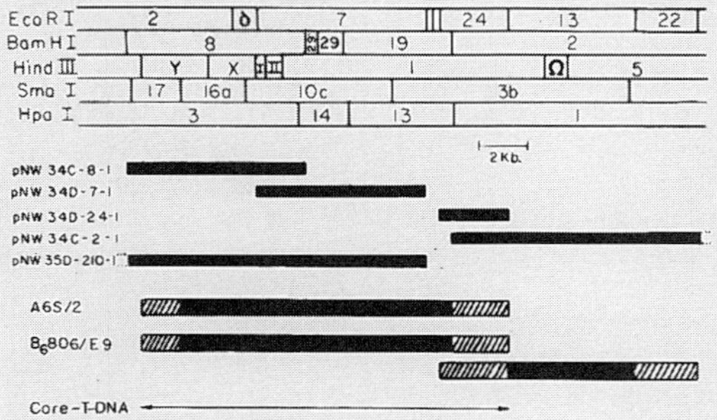

Figure 2. Sequence Organization of the pTiA6NC T-DNA Used for Transposon Mutagenesis

The restriction endonuclease map of the T-DNA is presented at the top. Solid bars: plasmid clones containing the T-DNA fragments used for transposon mutagenesis. Solid bars bracketed by crosshatched areas: the sequence organization of the T-DNA present in two octopine-type tobacco tumor lines, A6S/2 and E9. Crosshatched areas: junction regions where the T-DNA is attached to plant DNA (Thomashow et al., 1980a, 1980b). The core T-DNA of pTiA6NC is presented at the bottom.

Transfer of the Insertion Mutations of pTiA6NC

Seventy-five Tn5 insertions in pNW34C-8-1, pNW34D-7-1, pNW34D-24-1 and pNW34C-2-1 and three Tn3 insertions in pNW35D-210-1 were introduced into pTiA6NC. To detect each recombinant, total DNA from between two and twenty transconjugants (containing pPH1J1) was digested with Eco RI and in some cases with Bam HI or Hind III. The resulting nitrocellulose sheets were hybridized with the appropriate ^{32}P-labeled T-DNA probe. For example, ten transconjugants from a bacterial mating involving T-DNA::Tn5 insertion 147, which is located in Eco RI fragment 7 (see Figure 3 for map location), were analyzed for the introduction of Tn5 into pTiA6NC (Figure 4). If Tn5 were introduced into pTiA6NC by homologous recombination, the Eco RI fragment 7 of pTiA6NC should contain Tn5. Eco RI fragment 7 should be 5.7 kb larger than the parental fragment, since Tn5 contains no Eco RI sites (Jorgenson et al., 1979). For all transconjugants analyzed (lane b–k), only the Eco RI fragment 7::Tn5 hybridized with ^{32}P-labeled pNW33D-7-1, which contains Eco RI fragment 7, indicating that the desired recombination event had taken place. Approximately one quarter of the T-DNA insertions were also digested with Bam HI or Hind III and analyzed as described above. Since Bam HI cleaves Tn5 once approximately in the center of the element and Hind III cleaves Tn5 once in each inverted repeat (Jorgenson et al., 1979), predictable T-DNA::Tn5 junction fragments of different sizes were observed (data not shown). These banding patterns suggest that marker transfer occurs without any other DNA rearrangements.

The same experiment with a Tn5 insertion in

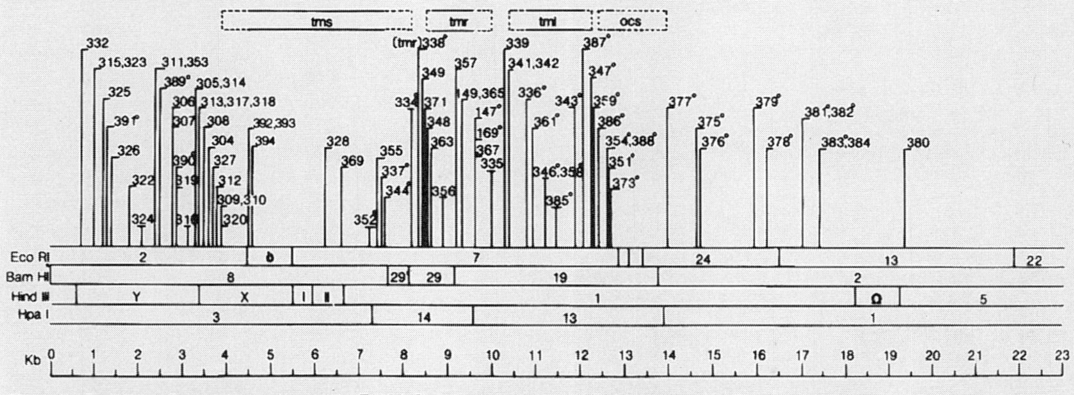

Figure 3. Restriction Endonuclease Map of the T-DNA Showing Location of 75 Tn5 Insertions and Three Tn3 Insertions (392, 393, and 394)

Map is from Thomashow et al., 1980a. Vertical lines: sites of insertion, accurate to ± 0.2 kb. Rectangular boxes above the map: the *tms*, *tmr*, *tml* and *ocs* loci. Solid lines of each box: regions where transposon insertions confer a change in tumor morphology or octopine synthesis. Dashed lines of each box: distance from the outermost transposon-induced mutation in a given locus to the nearest insertion that is phenotypically silent. Tumors incited by A. tumefaciens strain A136(pPH1J1) harboring the designated T-DNA insertions (○) in pTiA6NC were assayed for octopine synthesis. The scale is in kilobases (kb), with the left end of Bam HI fragment 8 arbitrarily chosen as zero (*).

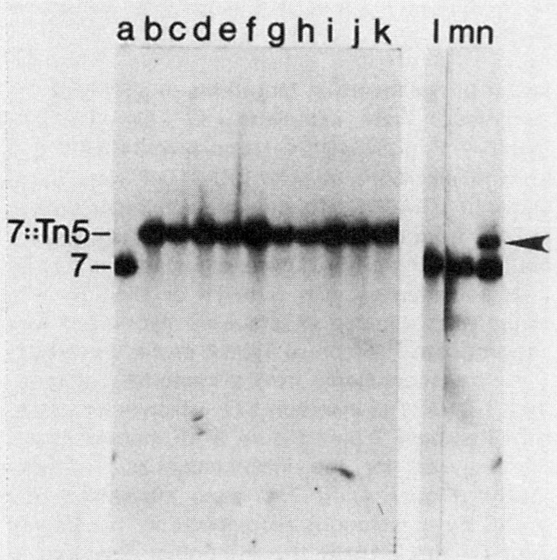

Figure 4. Analysis of T-DNA from Ten Stable Transconjugants That Originally Contained Eco RI Fragment 7::Tn5 Insertion 147 in pNW34D-7-1 and from Three That Originally Contained a Tn5 Insertion in pBR325 of pNW34D-7-1

Simultaneous selection for the retention of Tn5 and the introduction of pPH1J1, which is incompatible with pNW34D-7-1, enriched for two general classes of recombinants. These classes were distinguished by digesting total cellular DNA from the parental A. tumefaciens strain A348(pPH1J1) and from the 13 transconjugants with Eco RI, separating the fragments by electrophoresis and transferring the fragments to nitrocellulose according to Southern's procedure (1975). The filter-bound DNA was hybridized with in vitro ^{32}P-labeled pNW33D-7-1 (7×10^7 cpm/μg) (Maniatis et al., 1975), which contains Eco RI fragment 7 joined to pBR325, and the fragments possessing complementary sequences were detected by autoradiography. Strain A348(pPH1J1) (lane a) possesses Eco RI fragment 7. The first class of transconjugants had Eco RI fragment 7::Tn5 insertion 147 transferred from the hybrid plasmid to pTiA6NC by homologous recombination. For the ten transconjugants analyzed (lanes b–k), only Eco RI fragment 7::Tn5 remained, indicating that the desired recombination had occurred. The second class of transconjugants resulted when a Tn5 insertion in pBR325 was used instead of an insertion in DNA homologous with pTiA6NC. Two transconjugants retained Eco RI fragment 7 (lanes l and m), while the third transconjugant also retained the Eco RI–cleaved pBR325::Tn5 fragment (lane n). The positions of Eco RI fragments 7 (7.25 kb), 7::Tn5 (13 kb) and pBR325::Tn5 (◄; 12 kb) are indicated at the sides of the autoradiogram.

pBR325 instead of in Eco RI fragment 7 provides a measure of other recombinational events that could result in stable transconjugants (see Table 2; note that plasmid pNW34D-7-1 contains the original cloning vehicle pBR325 as well as Eco RI fragment 7 and pRK290). Stable transconjugants appeared after the introduction of the plasmid pPH1J1, but at frequencies lower by a factor of approximately 100 (1×10^{-8} per recipient plated) than if the insertion were present in the middle of Eco RI fragment 7, and lower by a factor of 2–3 than if the insertion were present within 1 kb from the end of Eco RI fragment 7. Hybridization of DNA from three representative transconjugants with ^{32}P-labeled pNW33D-7-1 indicated that each transconjugant retained the pTiA6NC Eco RI fragment 7 (Figure 4, lanes l, m and n). The results for two of these transconjugants (lanes l and m) suggest that Tn5 transposed to a new site prior to the exclusion of the mutagenized hybrid plasmid by pPH1J1. The third transconjugant (lane n) contained pBR325::Tn5, which hybridized with the cloning vehicle (pBR325) present in pNW33D-7-1, in addition to Eco RI fragment 7. Although this strain was not characterized further, the presence of fragment 7 and pBR325::Tn5 is consistent with the formation of a cointegrate between the hybrid plasmid and pTiA6NC by means of T-DNA homology. These transconjugants were also fully oncogenic when used as inoculum on Kalanchoe.

Functional Organization of the T-DNA of pTiA6NC

Each verified pTiA6NC::Tn3 or Tn5 recombinant strain was used as an inoculum on Kalanchoe daigremontiana leaf and stem wounds and tobacco (Nicotiana tobacum var white burley or xanthi) stem wounds. Plants inoculated with the virulent parental strain A348(PH1J1) and with the Ti-plasmidless strain A136(pPH1J1) were used as controls. Other than the presence of pTiA6NC, the strains are isogenic. Strain A348(pPH1J1) induces an undifferentiated tumor on Kalanchoe leaves and tobacco stems and induces root proliferation from the tumor periphery on Kalanchoe stems. Strain A136(pPH1J1) is completely avirulent.

A large number of strains harbored Ti plasmid insertions that apparently did not affect virulence. These insertions were located in two regions (Figure 3). The first region (3.2 kb) is located 0.7–3.9 kb from a Bam HI restriction site (designated as zero) near the left T-DNA–plant DNA junction (Figures 2 and 3). The second region (7 kb), located between 12.3 and 19.3 kb, extends through the right T-DNA–plant DNA junction, as found in the tumor line A6S/2, and into an adjacent segment of the Ti plasmid that is only found in some tumor lines.

Twenty-five T-DNA insertions defined three distinct loci affecting tumor morphology. These regions probably represent distinct genetic loci, since Tn5 insertions were isolated that conferred no noticeable change in tumorigenicity, yet mapped between clusters of insertions that affect tumor morphology.

The first region encodes the *tms* locus (Figure 3). The Tms$^-$ phenotype was characterized previously (Garfinkel and Nester, 1980; Ooms et al., 1981) by a lack of root proliferation on Kalanchoe stems, by an atrophied callus formation on both Kalanchoe leaves and stems and by shoot proliferation from an unorganized tumor callus on tobacco stems. The Tms$^-$ phenotype was not observed with the parental strain A348 (with or without pPH1J1). The insertions, located between 4.5 and 7.6 kb, form a cluster of approximately 3.1 kb.

The second region encodes the *tmr* locus (Figure 3). As described previously (Garfinkel and Nester, 1980; Ooms et al., 1980, 1981), *tmr* mutants were avirulent on Kalanchoe leaves, yet incited tumors on Kalanchoe stems that produced a massive amount of roots. On tobacco, roots also developed from the tumor callus. The Tmr^- phenotype was not observed when parental strains A348 or A348(pH1J1) were used. Eight of the nine *tmr* mutants contained Tn5 insertions that form a 1 kb cluster between 8.65 and 9.65 kb. However, one insertion, pTiA6NC*tmr*-338::Tn5, was separated from the *tmr* cluster by three other Tn5 insertions that did not confer any noticeable change in tumorigenicity. When total DNA from this *tmr* mutant was initially digested with Eco RI and the resulting nitrocellulose filter hybridized with ^{32}P-labeled pNW33D-7-1, a fragment 7::Tn5 band of the expected size appeared. Furthermore, Tn5 insertion 338 was reintroduced into pTiA6NC from the original mutagenized hybrid plasmid. Kalanchoe leaves and stems were inoculated with 32 of the resulting transconjugants, and 31 of these acquired the Tmr^- phenotype. One transconjugant, which apparently did not undergo marker transfer, induced a morphologically unaltered tumor. The implications of these results are unclear because only one *tmr* mutant separated from the *tmr* cluster was isolated. One possibility is that the Tmr^- phenotype can result from mutations in either of two separate loci.

The third region of the T-DNA encodes the *tml* locus (Figure 3). This region is located between 10.8 and 12.05 kb and forms a 1.25 kb cluster. Tumors incited on Kalanchoe leaves or stems by *tml* mutants were two to three fold larger than tumors incited by the parental strain A348 or A348(pPH1J1). The size differences of tumors incited by a representative *tml* mutant, by three strains that carry phenotypically silent T-DNA::Tn5 insertions and by the parental strain were quantified by comparing the fresh weight of the tumors excised from Kalanchoe leaves (Table 1). However, this dramatic difference in tumor size was not observed on tobacco.

Mapping of the Locus Specifying Octopine Synthesis

Previous results (Koekman et al., 1979; Thomashow et al., 1980a) suggested that a region of the T-DNA located between 12 and 15 kb coded for octopine synthesis (*ocs*) in the tumor (Figure 3). Tumors incited on Kalanchoe leaves or stems by strains containing Tn5 insertions in this region, as well as by strains containing insertions in other regions of the T-DNA, were assayed for octopine synthesis (Figures 3 and 5). Strains containing Ti plasmid pTiA6NC*ocs*-351, -354, -373 and -388::Tn5 incited tumors that did not synthesize detectable levels of octopine. The T-DNA::Tn5 insertions that define the maximum size of the *ocs* locus were located at 12.4 kb (pTiA6NC::Tn5 number 386) and at 14 kb (pTiA6NC::Tn5 number 377). Tumors incited by representative *tms*, *tmr* and *tml* mutants as well as by strains carrying phenotypically silent T-DNA::Tn5 insertions synthesized octopine (Figure 3). Uninoculated Kalanchoe did not synthesize octopine whereas tumors incited by the parental strain A348(pPH1J1) always synthesized octopine.

Incorporation of Tn5 Sequences into the Plant Genome

A strain containing pTiA6NC*tmr*-149::Tn5 (Figure 3) was inoculated onto tobacco (N. tabacum var white burley). An axenic tumor line (149/UN) was established from the resulting tumor and passaged for several months on phytohormone-free medium before analysis. Total genomic DNAs from the 149/UN and A6S/2 tumor lines and from normal callus (WBSR) were digested with Hind III, fractionated by agarose gel electrophoresis and transferred to a nitrocellulose filter. Hind III fragments containing sequences homologous with Tn5 were detected by hybridizing the nitrocellulose filter with ^{32}P-labeled ColE1::Tn5 (Figure 6). The Tn5 insertion present in pTiA6NC*tmr*-149::Tn5 is located in a region of the T-DNA that is colinear with the core T-DNA maintained in octopine tumors (Figure 2). If the Tn5 sequences incorporated into the 149/UN genome as part of the T-DNA are not considerably rearranged, a 3 kb and a 4 kb Hind III fragment should hybridize with ^{32}P-labeled ColE1::Tn5 (Figure

Table 1. Fresh Weights of Tumor Tissue[a]

pTiA6NC::Tn5[b] Insertion	Inoculum Size (Cells × 10^{10})	Weight of Tumor (mg) Incited by Mutant Strains	Weight of Adjacent Tumor (mg) Incited by A348(pPH1J1)[c]	Normalized Tumor Weight (% of A348(pPH1J1))
pTiA6NC::Tn5 no. 334	0.95	114	79	144
pTiA6NC::Tn5 no. 335	1.90	79	81	98
pTiA6NC::Tn5 no. 339	2.21	94	114	82
pTiA6NC*tml*-336::Tn5	0.79	293	79	371

[a] The weight of the tumor is an average from four individually incited tumors.
[b] A. tumefaciens strain A136(pPH1J1) harbored each pTiA6NC::Tn5 plasmid.
[c] The inoculum size of the parental strain A348(pPH1J1) was 1.7 × 10^{10} cells.

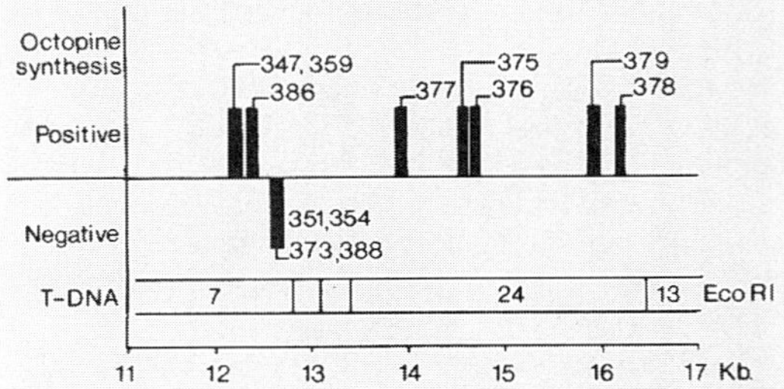

Figure 5. Restriction Map of the *ocs* Region of the T-DNA

Tumors incited by strains containing Tn5 insertions located in this region (Otten and Schilperoort, 1978) were assayed for octopine synthesis. Vertical solid bars: locations of insertions 351, 354, 373 and 388 that affect octopine synthesis and insertions 347, 359, 375, 376, 377, 378, 379 and 386 that do not affect octopine synthesis.

6). A 3 kb fragment was detected in 149/UN, and it comigrated with the expected 3 kb internal Hind III fragment of Tn5 (Jorgenson et al., 1979) present in the ColE1::Tn5 reconstruction mixture. A 4 kb Hind III fragment was also detected, which probably corresponds to the 4 kb T-DNA::Tn5 junction fragment of pTiA6NC*tmr*-149::Tn5 generated by cleavage at Hind III fragment 1 and at a Hind III site in the inverted repeat of Tn5. The organization of the 3.4 kb and 8.5 kb fragments that hybridize with ColE1::Tn5 is not known because the complete physical organization of the 149/UN T-DNA has not been determined. No homology existed between ColE1::Tn5 and white burley tobacco DNA derived from either transformed (A6S/2) or untransformed (WBSR) tissues. These results indicate that Tn5 inserted in the T-DNA can be incorporated into the plant genome and suggest that the Tn5 insertion in at least one copy of the 149/UN T-DNA is colinear with the Tn5 insertion in pTiA6NC.

Discussion

The characterization of transposon insertions generated in this study defines several features of the T-DNA. Four apparently distinct genetic loci (*tms*, *tmr*, *tml* and *ocs*) are located in the T-DNA. Three loci (*tms*, *tmr* and *tml*) affect tumor morphology and are located in the core T-DNA segment present in all octopine-type tumors. The fourth locus (*ocs*) controls the synthesis of octopine in the tumor. Each of these loci is separated by regions of the T-DNA that appear phenotypically silent.

The gene products encoded by the *tms*, *tmr* and *tml* loci are unknown, but they could influence phytohormone levels in the developing tumor, resulting in aberrant tumor morphologies. This idea is supported by tissue culture experiments (Skoog and Miller, 1957) demonstrating that development of roots or shoots from normal tobacco callus is regulated by the relative ratio of auxin to cytokinin in the culture medium. In addition, Ooms et al. (1981) demonstrated that supplying naphthalene acetic acid (an auxin) to developing tumors incited by a *tms* mutant or supplying kinetin (a cytokinin) to developing tumors incited by a *tmr* mutant stimulates tumor formation on tomato stems. The following comparisons indicate that the mutants reported by these investigators are equivalent to the *tms* and *tmr* mutant classes in our analysis. In the absence of phytohormones, their mutants, as well as *tmr* and *tms* mutants characterized previously (Garfinkel and Nester, 1980; Ooms et al., 1980), incited small tumors on tomato stems. The *tms* and *tmr* mutants reported by Ooms et al. (1981) also incited tumors on Kalanchoe and tobacco morphologically indistinguishable from the *tms* and *tmr* mutant classes. These mutations mapped in regions of the T-DNA of pTiB$_6$S$_3$ and pTiACH5 that correspond to the *tms* and *tmr* loci of pTiA6NC. All three Ti plasmids are highly related (Currier and Nester, 1976; Sciaky et al., 1978; Koekman et al., 1979; Ooms et al., 1980) and possess identical restriction endonuclease maps through the T-DNA (Thomashow et al., 1980a; Ooms et al., 1981). Their results suggest that gene products encoded by the *tms* locus influence auxin levels and gene products encoded by the *tmr* locus influence cytokinin levels during tumorigenesis. The tumor morphology loci of the T-DNA may also contribute to the ability of crown gall cells to grow under tissue culture conditions in the absence of exogenously added phytohormones.

A striking feature of the transposon-induced mutations located in the T-DNA is that no single insertion causes complete loss of oncogenicity, although the T-DNA is essential for tumor formation (Holsters et al., 1980) and for tumor maintenance (Lemmers et al., 1980; Yang and Simpson, 1981). Thus transformation apparently does not depend solely on the expression of a single T-DNA-encoded oncogene conceptually similar to the *src* gene of avian sarcoma virus (Hanafusa, 1977). Mutations in the *tmr*, *tms* and *tml* loci might shift the phytohormone balance in the developing tumor, resulting in aberrant tumor morphologies but not necessarily in loss of oncogenicity. It will be interesting to learn the effects different combinations of mutated loci have on tumorigenesis.

The aberrant tumor morphologies are also influenced by both the plant species and the plant tissue that is inoculated. For example, mutating the *tmr* locus results in loss of oncogenicity on Kalanchoe leaves, but results in the development of morphologically altered tumors on Kalanchoe and tobacco stems. Perhaps differences in the endogenous phytohormone levels of the target tissue cause these variations in the tumorous phenotype. The tumor callus that appears, however, is transformed since octopine is synthesized. In addition, the tumor line 149/UN, originally established from a tumor incited by a *tmr* mutant, grows on phytohormone-free medium and has Tn5 incorporated into the plant genome as part of the T-

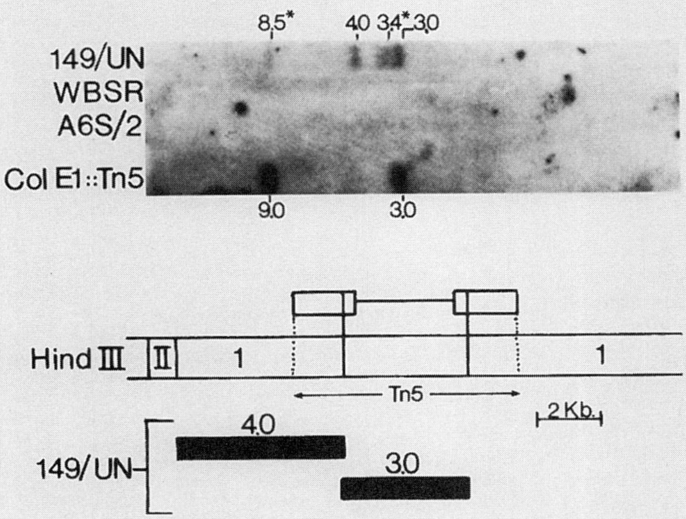

Figure 6. Detection of Tn5 Sequences in Hind III–Digested 149/UN Tumor DNA

Hind III–cleaved ColE1::Tn5 DNA (65 picograms per lane; five-copy reconstruction), 149/UN and A6S/2 tumor DNA (10 micrograms per lane) and white burley tobacco DNA (WBSR, 10 micrograms per lane) was fractionated by agarose gel electrophoresis and transferred to a nitrocellulose filter according to Southern's procedure (1975). The filter-bound DNA was hybridized with in vitro ^{32}P-labeled ColE1::Tn5 (1.2×10^8 cpm/μg) (Maniatis et al., 1975), and the fragments containing complementary sequences were detected after autoradiography. Dotted lines in the T-DNA restriction endonuclease map: location of the Tn5 insertion from pTiA6NC*tmr*-149::Tn5. Open boxes: the 1.45 kb inverted repeats of Tn5. Solid bars: organization of the 3 kb and 4 kb Hind III fragments present in 149/UN. Organization of the 3.4 kb and 8 kb Hind III fragments (*) have not seen determined (see text for details).

DNA. Whether the roots or shoots proliferating from the mutant tumors are transformed has not been determined.

From the data presented, we conclude that the maximum size of the *ocs* locus is approximately 1.6 kb and that it is located between 12.4 and 14 kb of the T-DNA. This result defines more precisely the *ocs* region outlined previously (Koekman et al., 1979; Thomashow et al., 1980a). Schroder et al. (1981) have shown that tumor-derived RNA complementary to a T-DNA restriction fragment (Hind III fragment 1) containing the genetically defined *ocs* locus directs the in vitro synthesis of a protein that is immunoprecipitated by antisera raised against purified octopine synthase. The size of the precipitated protein (molecular weight 39,000) corresponded to that of the purified enzyme (Hack and Kemp, 1980). Thus the *ocs* locus probably codes for octopine synthase.

The approximate locations of six discrete polyadenylated RNAs from the core T-DNA of octopine-producing tumors have been determined by "Northern" blot hybridization (S. B. Gelvin, M. F. Thomashow, J. C. McPherson, M. P. Gordon and E. W. Nester, submitted manuscript). The location of at least four transcripts can be correlated with the present fine structure genetic map of the pTiA6NC T-DNA. A 1.2 kb transcript and a 0.73 kb transcript hybridize with Hind III fragment Y, a region of the T-DNA that can be mutagenized without affecting tumorigenesis or octopine synthesis. A 1.25 kb transcript hybridizes with sequences in the *tmr* region. The transcript probably crosses the Bam HI site separating Bam HI fragments 29 and 19 because it is homologous with both of these contiguous Bam HI fragments. This Bam HI site is also located in the middle of the *tmr* locus, which suggests that the 1.25 kb transcript codes for some function affecting tumor morphology. A 1.45 kb RNA maps in the vicinity of the genetically defined *ocs* locus and is absent from tumors that do not synthesize detectable levels of octopine. It hybridizes with a DNA fragment located within 1 kb of the right end of Eco RI fragment 7, with Eco RI fragment 24 and with the two small Eco RI fragments located between Eco RI fragments 7 and 24. These results suggest that the 1.45 kb transcript codes for octopine synthase. A 1.75 kb RNA may overlap part of the *tms* locus because this RNA hybridizes with only the 2 kb Hind III fragment X probe. However, using the "Northern" transfer technique, we detect no transcripts that map completely within the *tml* or *tms* loci. Although it is possible that the *tml* and *tms* loci are not transcribed in plant cells, it is also possible that these plant transcripts are present at concentrations below the limits of detection of the "Northern" transfer technique, or that they are not polyadenylated.

Another conclusion that can be drawn from the 70 insertions located in the 13 kb core T-DNA segment (Figures 2 and 3) concerns saturation of the genetic map with insertions. To address this point we make two assumptions: that the genetic loci considered code for protein; and that insertion of Tn3 or Tn5 into the gene destroys the activity of the protein product (Berg, 1977; Kleckner, 1977; Calos and Miller, 1980), regardless of whether the gene is expressed in Agrobacterium or in the plant cell. In 9 kb of the core T-DNA, the largest interval without an insertion is 0.4 kb. This interval size occurs once in a 3.2 kb region (0.7–3.9 kb), once within the *tml* locus and again between the *tml* and *tmr* loci. The largest continuous coding sequence therefore could theoretically produce a polypeptide of approximately 16,000 daltons. Larger continuous coding sequences would probably be mutated by insertional inactivation. The defined *tms* locus is located within a 3.1 kb region where nine transposon insertions were isolated. Three intervals that occur within this region could theoretically code for proteins of 24,000 to 50,000 daltons. Thus this region is not saturated to the same extent as the rest of the core T-DNA and could easily contain additional loci.

A Tn7 insertion in the nopaline synthesis locus of pTiC58 (Hernalsteens et al., 1980) and an IS60 insertion in the *tms* locus of pTiACH5 (Ooms et al., 1981) are stably incorporated into the plant genome as part of the T-DNA. In the experiment we describe, a Tn5 insertion in the *tmr* locus of pTiA6NC can also be transferred and incorporated into the plant genome. These results indicate that non-T-DNA sequences can be incorporated into the plant genome with pTiC58, pTiACH5 or pTiA6NC used as the delivery vehicle. Other regions of the pTiA6NC T-DNA suitable for the introduction of specific genes can be identified if Tn5 inserted at various locations in the T-DNA are incorporated without rearrangement into the plant genome. The transmission of Tn5 into the plant cell demonstrates another application of the marker transfer technique. By slightly modifying the procedure (Matzke and Chilton, 1981), genes linked to Tn5 sequences can be introduced, along with the T-DNA, into the genomes of higher plants.

Experimental Procedures

Bacterial Strains, Plasmids and Bacteriophage

Agrobacterium tumefaciens strain A136 (Watson et al., 1975) harbored the mutated Ti plasmids. A. tumefaciens strain A348 was constructed by transforming pTiA6NC into strain A136. These derivatives, as well as the E. coli strains, plasmids and bacteriophage used are presented in the text or in Table 2.

Tumor lines

An axenic uncloned tumor line (149/UN) was established from a tumor incited on tobacco (N. tabacum var white burley) by A. tumefaciens strain A136 (pTiA6NC*tmr*-149::Tn5)(pPH1J1). Nutrient broth (Difco) inoculated with 149/UN tissue was incubated for 3 days at 28°C with aeration and gave no indication of growth, either by turbidity or by plating on nutrient agar at 28°C for 5 days. A6S/2 (Thomashow et al., 1980a) is a cloned octopine-type tumor line originally established from a white burley tobacco tumor, and WBSR (Thomashow et al., 1980b) is a normal callus line derived from a white burley root cell. Tumor tissues were propagated on standard phytohormone-free media, while WBSR required the addition of auxin and cytokinin (Johnson et al., 1974).

Selective Media

The concentrations of antibiotics in micrograms per milliliter and types of media used were as follows. For E. coli strains, L agar (Miller et al., 1972) containing 10 μg tetracycline (Sigma), 50 μg kanamycin (Sigma) or 100 μg ampicillin (Wyeth) was used. For A. tumefaciens strain A348 and its derivatives, AB minimal medium (Chilton et al., 1974), containing 100 μg kanamycin or 200–300 μg carbenicillin (Roerig) was used. The selective medium (AB/KGN) used for E. coli–A. tumefaciens conjugations was AB medium containing either 100 μg kanamycin or 200–300 μg carbenicillin, 100 μg gentamicin (Sigma) and 50 μg nalidixic acid (Sigma). In some cases, only gentamicin and nalidixic acid were added (AB/GN).

DNA Isolation

The rapid plasmid purification technique of Birnboim and Doly (1979) was used for E. coli. In some cases, plasmid DNA was further purified by cesium chloride–ethidium bromide density gradient centrifugation. Total DNA isolated from A. tumefaciens was prepared from 2 ml late-logarithmic-phase cultures by a modification of Marmur's procedure (Currier and Nester, 1976). High molecular weight DNA from tobacco callus or tobacco leaves was prepared according to established procedures (Chilton et al., 1977; Thomashow et al., 1980a).

Transformation

Standard techniques were used to transform A. tumefaciens (Holsters et al., 1978) and E. coli (Cohen et al., 1972) with plasmid DNA.

Bacterial Matings

Overnight filter matings between E. coli strain 2174 (pPH1J1) and A. tumefaciens strains A348(pRK290-T-DNA::Tn5), A348(pHK17-T-DNA::Tn3) or A348 were performed as described previously (Garfinkel and Nester, 1980). Transconjugant colonies, with A348(pRK290-T-DNA::Tn5) as the recipient, appeared on AB/KGN with a frequency of between 2.7×10^{-6} and 3.5×10^{-8} per recipient. The 75-fold variation in conjugation frequency, obtained when the matings were spread on AB/KGN plates, depends on the position of Tn5 in the T-DNA. For example, when Tn5 was located within 1 kb of the end of a T-DNA fragment, the number of stable transconjugants decreased dramatically. On AB/GN plates, transconjugant colonies appeared with a frequency of approximately 5×10^{-4} per recipient. Transconjugant colonies with strain A348 as the recipient appeared on AB/GN with a frequency of 1×10^{-2} per recipient.

Restriction Endonuclease Reactions

Restriction endonucleases Bam HI (Boehringer Mannheim or Bethesda Research Laboratories) or Bgl II, Eco RI, Hind III, Hpa I, Sst I and Taq I were used according to supplier's (Bethesda Research Laboratories) recommendations. Normal plant and tumor DNAs were digested with three times the normal amount of enzyme for at least twice the time required to obtain a complete digest of an equivalent quantity of Ti plasmid.

Gel Electrophoresis and Recovery of DNA from Agarose

Vertical agarose (Bethesda Research Laboratories) slab gels (Garfinkel and Nester, 1980) of varying concentrations (0.7% to 1.5% w/v) were used to locate the transposon insertions. Horizontal agarose gels were constructed as described previously for preparative fractionation of plasmid DNA fragments (Chilton et al., 1977) and for fractionation of genomic digests of tobacco DNA (Thomashow et al., 1980a). DNA was recovered from agarose by electroelution (Tabak and Flavell, 1978). Staining of gels and photographic conditions were described previously (Garfinkel and Nester, 1980).

DNA Filter Hybridization

After gel electrophoresis, DNA was transferred to nitrocellulose filters (0.45 μm, Schleicher and Schuell) by the method of Southern (1975). The nitrocellulose filters were used in filter hybridizations as described previously (Thomashow et al., 1980a). Radioactive probes were labeled in vitro to 70×10^6 to 150×10^6 cpm/μg of DNA with α-^{32}P-dATP, α-^{32}P-dCTP and α-^{32}P-TTP (300–700 Ci/mmole) (New England Nuclear) by nick translation (Maniatis et al., 1975). Autoradiograms were prepared as previously described (Thomashow et al., 1980a).

Tn5 Reconstruction Mixture and Determination of Molecular Weights

Reconstruction mixtures (five copies per diploid tobacco genome) for ColE1::Tn5 cleaved with Hind III were prepared as described previously (Thomashow et al., 1980a). Lambda DNA cleaved with Hind III was used as a molecular weight standard to determine the sizes of tumor DNA fragments that hybridized with ^{32}P-labeled ColE1::Tn5. As an internal control, the set of Hind III–digested λ DNA fragments correctly estimated the sizes of the previously known Hind III–digested ColE1::Tn5 DNA fragments of 3 kb and 9 kb present in the five-copy reconstruction.

Molecular Cloning

Purified T-DNA Bam HI fragments 2 and 8 were ligated to pRK290 linearized after cleavage with Bgl II and Eco RI fragments 7 and 24

Table 2. Bacterial Strains, Plasmids and Phage

Strain	Relevant Phenotype[a]	Source or Relevant Comments
HB101	E. coli, Thr Leu Thi RecA HsdR HsdM Pro Str^r	Boyer and Roulland-Dussoix, 1969
2174	E. coli, Met Pro	Obtained from J. Beringer
A136	A. tumefaciens, Nal^r, Rif^r	Ti plasmid cured derivative of strain C58 (Watson et al., 1975)
Plasmids		
pTiA6NC	Onc$^+$, Whr, Occ$^+$	Ti plasmid isolated from nonclumping derivative of strain A6 (Sciaky et al., 1978)
pPH1J1	IncP Gm^r, Cam^r, Str^r, Spc^r	Obtained from J. Beringer
pRK290	IncP, Tet^r	Derivative of RK2 (Ditta et al., 1981)
pHK17	IncP, Kan^r, Tet^r	Derivative of pRK2501 containing the λcos site (H. Klee, M. P. Gordon and E. W. Nester, submitted manuscript)
pNW31C-8, -29′, -29-1	Amp^r	pBR322–pTiA6NC Bam HI fragments 8, 29′ and 29 (Thomashow et al., 1980a)
pNW31C-2, -19-1	Amp^r	pBR322–pTiA6NC Bam HI fragments 2 and 19 (Thomashow et al., 1980a)
pNW33D-7-1	Amp^r, Tet^r	pBR325–pTiA6NC Eco RI fragment 7 (S. Gelvin and R. Nutter, unpublished results)
ColE1::Tn5	Kan^r	Obtained from D. Berg
pNW34C-8-1	Tet^r	pRK290–pTiA6NC Bam HI fragment 8 (this study)
pNW34D-7-1	Tet^r, Amp^r	pRK290–pBR325–pTiA6NC Eco RI fragment 7 (this study)
pNW34D-24-1	Tet^r	pRK290–pTiA6NC Eco RI fragment 24 (this study)
pNW34C-2-1	Tet^r	pR290–pTiA6NC Bam HI fragment 2 (this study)
pNW35D-210-1	Kan^r, Tet^r	pHK17–pTiA6NC cosmid clone no. 210 (H. Klee, M. P. Gordon and E. W. Nester, submitted manuscript)
Bacteriophage		
λ::Tn5	λb221 rex::Tn5 cI857 $O_{am8}P_{am29}$	Ruvkun and Ausubel, 1981

[a] Amp^r: ampicillin resistance. Cam^r: chloramphenicol resistance. Gm^r: gentamicin resistance. IncP: incompatibility group P. Kan^r: kanamycin resistance. Nal^r: nalidixic acid resistance. Occ: octopine catabolism. Onc: tumorigenesis. Rif^r: rifampicin resistance. Spc^r: spectinomycin resistance. Str^r: streptomycin resistance. Tet^r: tetracycline resistance. Whr: wide-host-range.

were ligated to pRK290 after cleavage with Eco RI, using T4 ligase (Bethesda Research Laboratories). E. coli strain HB101 (Boyer and Roulland-Dussoix, 1969) was transformed by recombinant molecules, and colonies resistant to tetracycline were screened by colony hybridization (Grunstein and Hogness, 1975) for the presence of T-DNA inserts, with ^{32}P-labeled pTiA6NC as the hybridization probe. The construction of pHK17 and the cosmid clone pNW35D-210-1 will be described separately (H. Klee, M. P. Gordon and E. W. Nester, submitted manuscript).

Mutagenesis of the pRK290-T-DNA Clones or pNW35D-210-1 with Tn5 or Tn3

Tn5 insertions were generated by infecting E. coli strain HB101 harboring the different pRK290-T-DNA plasmids with λ::Tn5 as previously described (Ruvkun and Ausubel, 1981). The technique used to generate Tn3 insertions in the T-DNA of pNW35D-210-1 will be described separately (H. Klee, F. White and E. W. Nester, manuscript in preparation).

Location of Tn3 and Tn5 Insertions in the T-DNA of pTiA6NC

We located transposon insertions relative to restriction sites in the T-DNA (Thomashow et al., 1980a) using the restriction sites introduced by Tn3 (Heffron et al., 1979) or Tn5 (Jorgenson et al., 1979). Plasmid DNA was digested with Bam HI, Eco RI, Hind III, Hpa I or Sst I or with combinations of these enzymes and subjected to vertical agarose gel electrophoresis. Various concentrations of agarose were used (0.7% to 1.5%, w/v) depending on the size of the digestion products. Molecular weight standards used were pBR322 cleaved with Taq I (Sutcliffe, 1978) and λ DNA cleaved with Eco RI or Hind III (Szybalski and Szybalski, 1979). After the T-DNA recombinants were analyzed for virulence and octopine synthesis, Tn5 and Tn3 insertions that initially mapped close together yet conferred different phenotypes were mapped again on the same gel. The details of the mapping strategy are as follows (also see Figure 2).

Bam HI Fragment 8 (pNW34C-8-1)

This fragment was ligated to the Bgl II site of pRK290. The cloned insert was subsequently unable to be regenerated by digestion with either Bam HI or Bgl II. Plasmid DNA was digested with Hpa I, and the sizes of the T-DNA::Tn5 junction fragments were determined. This predicted two possible insertion sites for Tn5 in the plasmid based on the presence of one Hpa I site in pNW34C-8-1, approximately 0.3 kb from the end of Bam HI fragment 8, and the presence of a Hpa I site in each of the inverted repeats of Tn5, approximately 0.12 kb from either end of the element. DNA was then digested with Hind III, which also cleaves in the inverted repeats of Tn5 (approximately 1.25 kb from the ends of the element), to confirm the location of Tn5.

Eco RI Fragment 7 (pNW34D-7-1)

This fragment, isolated from pNW33D-7-1, was ligated into the Eco RI site of pRK290. The recombinant plasmid obtained, however, contained pBR325 (from pNW33D-7-1) as well as pRK290 and Eco RI fragment 7. DNA was initially cleaved with Hind III, and the size of the T-DNA::Tn5 junction fragment was determined. There are two Hind III sites in Eco RI fragment 7, both located at the same end of the fragment. The junction fragment generated should give the distance from the Hind III site in the inverted repeat of the Tn5 insertion to the Hind III sites in Eco RI fragment 7. The restriction digest pattern was then analyzed after cleavage with Hind III plus Eco RI (Eco RI does not cleave Tn5) or Bam HI (Bam HI cleaves Tn5 approximately in half) plus Eco RI to determine the transposon's location unambiguously.

Eco RI Fragment 24 (pNW34D-24-1)

This fragment was cloned into the Eco RI site of pRK290. We initially located insertions by using Eco RI. We then mapped the Tn5 insertions present in Eco RI fragment 24 by using Hpa I and Hpa I plus Eco RI double digests.

Bam HI Fragment 2 (pNW34C-2-1)

This fragment was cloned into the Bgl II site of pRK290. Tn5 insertions were mapped in this fragment by the same strategy as was used for Bam HI fragment 8 insertions because Bam HI fragment 2 also contains only one Hpa I site, located approximately 0.25 kb from one end of the fragment. After locating the insertion using Hpa I, we cleaved the clones with Hpa I plus Eco RI or with Hind III to determine the location of Tn5 unambiguously.

The Cosmid Clone pNW35D-210-1

This clone was constructed with Eco RI partial digests of pTiA6NC and contains Eco RI fragments 7, δ and 2 of T-DNA (H. Klee, M. P. Gordon and E. W. Nester, submitted manuscript). Tn3 insertions in pNW35D-210-1 were initially located by cleaving the DNA with Hind III or Eco RI (neither enzyme cleaves Tn3). The insertions were then mapped by cleaving the DNA with Bam HI, Bam HI plus Eco RI and Bam HI plus Sst I. Bam HI cleaves Tn3 once, 1.4 kb from the end of the element, while Sst I cleaves Tn3 once, 0.27 kb from the opposite end of the element.

Oncogenicity

We inoculated Kalanchoe daigremontiana leaves or stems or tobacco (N. tabacum var xanthi or white burley) stems with bacteria as previously described (Garfinkel and Nester, 1980).

Octopine Production

Tumor tissue was excised from tobacco or Kalanchoe after one month of incubation and assayed for octopine synthesis (Otten and Schilperoort, 1978).

Determination of Tumor Size

Bacterial cultures were grown in yeast extract peptone broth (YEP) with shaking at 28°C for 60 hr. Immediately before they were used to inoculate plants, the cultures were titered for viable cells on nutrient agar plates. Kalanchoe leaves were wounded with a sterile No. 1 cork borer, and the center of the circular wound was abraded with a sterile toothpick. Adjacent wounds were made on opposite sides of and equidistant from the midrib of the leaf. In each wound pair, one wound was inoculated with the parental strain A348(pPH1J1) and the other was inoculated with a mutant strain. Wounds were inoculated by centrifuging 1 ml of culture, resuspending the pellet in 0.05 ml of YEP broth and placing the entire suspension onto the wound by pipette. Tumor tissue was excised and weighed 37 days after inoculation.

Acknowledgments

We wish to thank Amanda Garfinkel, Conrad Lichtenstein, Joan McPherson, Tsuni Kosuge and Harry Klee for critical readings of the manuscript; Gary Ruvkun and Fred Ausubel for valuable discussion and communication prior to publication of their marker transfer technique; Brian Watson and Gina Ghidossi for help in the plant room; Jim Champoux for the use of his high-voltage paper electrophoresis apparatus; Kathy Spangler for helping prepare the manuscript; Nancy Shaw for drafting Figure 1; and Alice Montoya for drafting Figure 2.

This work was funded by grants from the National Cancer Institute and American Cancer Society. R. B. S. was a fellow in cancer research supported by a grant from the Damon Runyon-Walter Winchell Cancer Fund. L. W. R. holds a postdoctoral fellowship from the American Cancer Society. D. J. G. and F. F. W. were supported by a training grant from the National Institutes of Health.

The costs of publication of this article were defrayed in part by the payment of page charges. This article must therefore be hereby marked "*advertisement*" in accordance with 18 U.S.C. Section 1734 solely to indicate this fact.

Received August 5, 1981; revised September 14, 1981

References

Berg, D. E. (1977). Insertion and excision of the transposable kanamycin resistance determinant Tn5. In DNA Insertion Elements, Plasmids and Episomes. A. I. Bukhari, J. A. Shapiro and S. L. Adhya (eds.) (Cold Spring Harbor New York: Cold Spring Harbor Laboratory), pp. 205–212.

Birnboim, H. C. and Doly, J. (1979). A rapid extraction procedure for screening recombinant plasmid DNA. Nucl. Acids Res. *7*, 1513–1525.

Bomhoff, G., Klapwijk, P. M., Kester, H. C. M., Schilperoort, R. A., Hernalsteens, J. P. and Schell, J. (1976). Octopine and nopaline synthesis and breakdown genetically controlled by a plasmid of *Agrobacterium tumefaciens*. Mol. Gen. Genet. *145*, 177–181.

Boyer, H. W. and Roulland-Dussoix, D. (1969). A complementation analysis of the restriction and modification of DNA in *Escherichia coli*. J. Mol. Biol. *41*, 459–472.

Braun, A. C. (1958). A physiological basis for autonomous growth of the crown gall tumor cell. Proc. Nat. Acad. Sci. USA *44*, 344–349.

Calos, M. P. and Miller, J. H. (1980). Transposable elements. Cell *20*, 579–595.

Chilton, M.-D., Currier, T. C., Farrand, S. K., Bendich, A. J., Gordon, M. P. and Nester, E. W. (1974). *Agrobacterium tumefaciens* DNA and PS8 bacteriophage DNA not detected in crown gall tumors. Proc. Nat. Acad. Sci. USA *71*, 3672–3676.

Chilton, M.-D., Drummond, M. H., Merlo, D. J., Sciaky, D., Montoya, A. L., Gordon, M. P. and Nester, E. W. (1977). Stable incorporation of plasmid DNA into higher plant cells: the molecular basis of crown gall tumorigenesis. Cell *11*, 263–271.

Chilton, M.-D., Saiki, R. K., Yadav, N., Gordon, M. P. and Quetier, F. (1980). T-DNA from *Agrobacterium* Ti-plasmid is in the nuclear DNA of crown gall tumor cells. Proc. Nat. Acad. Sci. USA *77*, 4060–4064.

Cohen, S. N., Chang, A. C. Y. and Hsu, L. (1972). Nonchromosomal antibiotic resistance in bacteria: genetic transformation of *Escherichia coli* by R-factor DNA. Proc. Nat. Acad. Sci. USA *69*, 2110–2114.

Currier, T. C. and Nester, E. W. (1976). Evidence for diverse types of large plasmids in tumor-inducing strains of *Agrobacterium*. J. Bacteriol. *126*, 157–165.

Ditta, G., Stanfield, S., Corbin, D. and Helinski, D. R. (1981). Broad host range DNA cloning system for gram-negative bacteria: construction of a gene bank of *Rhizobium meliloti*. Proc. Nat. Acad. Sci. USA *77*, 7347–7351.

Firmin, J. L. and Fenwick, R. G. (1978). Agropine: a major new plasmid determined metabolite in crown gall tumors. Nature *276*, 842–844.

Garfinkel, D. J. and Nester, E. W. (1980). *Agrobacterium tumefaciens* mutants affected in crown gall tumorigenesis and octopine catabolism. J. Bacteriol. *144*, 732–743.

Goldman, A., Tempe, J. and Morel, G. (1968). Quelques particularites de diverses souches d'*Agrobacterium tumefaciens*. C. R. Acad. Sci. *162*, 630–631.

Goldman, A., Thomas, D. W. and Morel, G. (1969). Sur le structure de la nopaline, metabolite anormale de certaines tumeurs de crown gall. C. R. Acad. Sci. *268*, 852–854.

Grunstein, M. and Hogness, D. S. (1975). Colony hybridization: a method for the isolation of cloned DNAs that contain a specific gene. Proc. Nat. Acad. Sci. USA *72*, 3961–3965.

Hack, E. and Kemp, J. D. (1980). Purification and characterization of the crown gall-specific enzyme, octopine synthase. Plant Physiol. *65*, 949–955.

Hanafusa, H. (1977). Cell transformation by RNA tumor viruses. In Comprehensive Virology *10*. H. Fraenkel-Conrat and R. R. Wagner, eds. (New York: Plenum), pp. 401–482.

Heffron, F., McCarthy, B. J., Ohtsubo, H. and Ohtsubo, E. (1979). DNA sequence analysis of the transposon Tn3: three genes and three sites involved in transposition of Tn3. Cell *18*, 1153–1163.

Hernalsteens, J.-P., Van Vliet, F., De Beuckeleer, M., Depicker, A., Engler, G., Lemmers, M., Holsters, M., Van Montagu, M. and Schell, J. (1980). The *Agrobacterium tumefaciens* Ti-plasmid as a host vector for introducing foreign DNA in plant cells. Nature *287*, 654–656.

Holsters, M., de Waele, D., Depicker, A., Messens, E., Van Montagu, M. and Schell, J. (1978). Transfection and transformation of *Agrobacterium tumefaciens*. Mol. Gen. Genet. *163*, 181–187.

Holsters, M., Silva, B., Van Vliet, F., Genetello, C., De Block, M., Dhaese, P., Depicker, A., Inze, D., Engler, G., Villarroel, R., Van Montagu, M. and Schell, J. (1980). The functional organization of the nopaline *A. tumefaciens* plasmid pTiC58. Plasmid *3*, 212–230.

Johnson, R., Guderian, R. H., Eden, F., Chilton, M.-D., Gordon, M. P. and Nester, E. W. (1974). Detection and quantitation of octopine in normal plant tissues and crown gall tumors. Proc. Nat. Acad. Sci. USA *71*, 536–539.

Jorgenson, R. A., Rothstein, S. J. and Reznikoff, W. S. (1979). A restriction enzyme cleavage map of Tn5 and location of a region encoding neomycin resistance. Mol. Gen. Genet. *177*, 65–72.

Kleckner, N. (1977). Translocatable elements in procaryotes. Cell *11*, 11–23.

Koekman, B. P., Ooms, G., Klapwijk, P. M. and Schilperoort, R. A. (1979). Genetic map of an octopine Ti-plasmid. Plasmid *2*, 347–357.

Lemmers, M., De Beuckeleer, M., Holsters, M., Zambryski, P., Depicker, A., Hernalsteens, J. P., Van Montagu, M. and Schell, J. (1980). Internal organization, boundaries and integration of Ti-plasmid DNA in nopaline crown gall tumors. J. Mol. Biol. *144*, 355–378.

Maniatis, T., Jeffrey, A. and Kleid, D. G. (1975). Nucleotide sequence of the rightward operator of phage λ. Proc. Nat. Acad. Sci. USA *72*, 1184–1188.

Matzke, A. J. M. and Chilton, M.-D. (1981). Site-specific insertion of genes into the T-DNA of the *Agrobacterium* tumor-inducing plasmid: an approach to genetic engineering of higher plant cells. J. Mol. Appl. Genet. *1*, 39–49.

Menage, A. and Morel, G. (1964). Sur la presence d'octopine dans les tissue de crown-gall. C. R. Acad. Sci. *259*, 4795–4796.

Miller, J. H. (1972). Experiments in Molecular Genetics. (Cold Spring Harbor New York: Cold Spring Harbor Laboratory), p. 21.

Montoya, A. L., Chilton, M.-D., Gordon, M. P., Sciaky, D. and Nester, E. W. (1977). Octopine and nopaline metabolism in *Agrobacterium tumefaciens* and crown gall tumor cells: role of plasmid genes. J. Bacteriol. *129*, 101–107.

Ooms, G., Hooykaas, P. J. J., Moolenaar, G. and Schilperoort, R. A. (1981). Crown gall plant tumors of abnormal morphology, induced by *Agrobacterium tumefaciens* carrying mutated octopine Ti-plasmids; analysis of T-DNA functions. Gene *14*, 33–50.

Ooms, G., Klapwijk, P., Poulis, J. A. and Schilperoort, R. A. (1980). Characterization of Tn904 insertions in octopine Ti-plasmid mutants of *Agrobacterium tumefaciens*. J. Bacteriol. *14*, 82–91.

Otten, L. A. B. M. and Schilperoort, R. A. (1978). A rapid microscale method for the detection of lysopine and nopaline dehydrogenase activities. Biochem. Biophys. Acta. *527*, 497–500.

Ruvkun, G. B. and Ausubel, F. M. (1981). A general method for site-directed mutagenesis in prokaryotes. Nature *289*, 85–88.

Schroder, J., Schroder, G., Huisman, H., Schilperoort, R. A. and Schell, J. (1981). The mRNA for lysopine dehydrogenase in plant tumor cells is complementary to a Ti-plasmid fragment. FEBS Letters *129*, 166–168.

Sciaky, D., Montoya, A. L. and Chilton, M.-D. (1978). Fingerprints of *Agrobacterium* Ti-plasmids. Plasmid *1*, 238–253.

Skoog, F. and Miller, C. O. (1957). Chemical regulation of growth and organ formation in plant tissue cultures *in vitro*. Symp. Soc. Exp. Biol. *11*, 118–131.

Southern, E. M. (1975). Detection of specific sequences among DNA fragments separated by gel electrophoresis. J. Mol. Biol. *98*, 503–517.

Sutcliffe, J. G. (1978). Complete nucleotide sequence of the *Escherichia coli* plasmid pBR322. Cold Spring Harbor Symp. Quant. Biol. *43*, 77–90.

Szybalski, E. H. and Szybalski, W. (1979). A comprehensive molecular map of bacteriophage lambda. Gene *7*, 217–270.

Tabak, H. F. and Flavell, R. A. (1978). A method for the recovery of DNA from agarose gels. Nucl. Acids Res. *5*, 2321–2332.

Thomashow, M. F., Nutter, R., Montoya, A. L., Gordon, M. P. and Nester, E. W. (1980a). Integration and organization of Ti plasmid sequences in crown gall tumors. Cell *19*, 729–739.

Thomashow, M. F., Nutter, R., Postle, K., Chilton, M.-D., Blattner, F. R., Powell, A., Gordon, M. P. and Nester, E. W. (1980b). Recombination between higher plant DNA and the Ti-plasmid of *Agrobacterium tumefaciens*. Proc. Nat. Acad. Sci. USA *77*, 6448–6452.

Van Larebeke, N., Engler, G., Holsters, M., Van den Elsacker, S., Zaenen, I., Schilperoort, R. A. and Schell, J. (1974). Large plasmid in *Agrobacterium tumefaciens* is essential for crown gall-inducing ability. Nature *252*, 169–170.

Watson, B., Currier, T. C., Gordon, M. P. Chilton, M.-D. and Nester, E. W. (1975). Plasmid required for virulence of *Agrobacterium tumefaciens*. J. Bacteriol. *123*, 255–264.

Willmitzer, L., DeBeukeleer, M., Lemmers, M., Van Montagu, M. and Schell, J. (1980). DNA from Ti-plasmid is present in the nucleus and absent from the plastids of crown gall plant cells. Nature *287*, 359–361.

Yadav, N. S., Postle, K., Saiki, R., Thomashow, M. F. and Chilton, M.-D. (1980). T-DNA of a crown gall teratoma is covalently joined to host plant DNA. Nature *287*, 458–461.

Yang, F.-M. and Simpson, R. B. (1981). Revertant seedlings from crown gall tumors retain a portion of the bacterial Ti-plasmid DNA sequences. Proc. Nat. Acad. Sci. USA, *78*, 4151–4155.

Zambryski, P., Holsters, M., Kruger, K., Schell, J., Van Montagu, M. and Goodman, H. (1980). Tumor DNA structure in plant cells transformed by *A. tumefaciens*. Science *209*, 1385–1391.

CROSS, COUNT AND CALCULATE! T-DNA SATISFIES MENDELIAN RULES

Léon Otten
Institut de Biologie Moléculaire des Plantes, Strasbourg, France
E-mail: leon.otten@ibmp-ulp.u-stasbg.fr

Commentary on the paper:

Otten, L., De Greve, H., Hernalsteens, J.-P., Van Montagu, M., Schieder, O., Straub, J., and Schell, J. 1981. Mendelian transmission of genes introduced into plants by the Ti plasmids of *Agrobacterium tumefaciens.* Mol. Gen. Genet. 183:209–213.

In 1979, the late Jeff Schell started his new *Agrobacterium* group at the Max-Planck-Institut in Cologne, and I was lucky enough to join him as a postdoc at the end of that year. These were exciting times; the revolutionary hypothesis of DNA transfer from a soil bacterium to a eukaryote had just been definitely proven by the technically masterful experiments of Mary-Dell Chilton. The long-suspected potential of the system for plant genetic engineering was now obvious to everyone.

However, there were still doubts about T-DNA stability and seed transmissibility. It had been shown that T-DNA transfer was not always complete. George Wullems, Laszlo Marton and Gert Ooms in Leiden regenerated morphologically aberrant tobacco lines from protoplasts, transformed with different *Agrobacterium* strains, and showed that these contained truncated T-DNAs (Wullems et al. 1981). Some lines formed teratomata (aberrant shoots) that had been grafted on to normal rootstocks and produced seeds. But, in the case of octopine teratomata, the pollen was sterile. Among the progeny of nopaline plants only one in forty contained nopaline synthase activity. Grafting experiments with T37 crown gall tissues had been pioneered as early as 1959 by Armin Braun. However, analysis of T37-derived teratoma plants had to await the discovery of the opines that could then be used as markers for transformation (although some reports claimed the presence of opines in untransformed plants, this could not be confirmed). Opine and T-DNA analysis demonstrated that one of the T37 regenerants was a *bona fide* transformant but its first (S1) generation was morphologically normal and lacked the T-DNA (Yang et al. 1980). This of course caused some worry among the advocates of *Agrobacterium*-based genetic engineering. Later experiments showed that in some cases, seed transmission of part of the T37 T-DNA was possible.

I arrived in Cologne having just completed my Ph.D. dissertation, in Rob Schilperoort's lab in Leiden, on the properties of the octopine and nopaline synthesis enzymes. I had developed a simple enzymatic microassay to track the enzymes during the different purification steps; the test required only a few microliters of crude plant extract and incubation with substrates led to the accumulation of high concentrations of the arginine derivatives octopine or nopaline (Otten and Schilperoort 1978). Substrates and products were separated by high-voltage paper electrophoresis, and revealed with the guanidine-specific phenanthrenequinone staining. The electrophoresis method was used routinely by "opinologists" Jacques Tempé and Annik Petit at Versailles and had been introduced to Leiden. The 4000 Volt tension required an impressive apparatus using a decane coolant. Sometimes during electrophoresis the one meter long paper would dry up and partially burn. I therefore prudently scaled down the dimensions, using 500 V and a normal electrophoresis tank. Our studies showed that contrary to the octopine dehydrogenase of *Octopus octopodia,* the octopine enzyme from crown gall preferred lysine (yielding lysopine) over arginine (yielding octopine). Since lysopine was the first opine to be discovered and is the main opine in some crown gall tumors, Bernard Lejeune had proposed the name lysopine dehydrogenase (LpDH, EC 1.5.1.16). However, because octopine was easier to detect, the name octopine synthase (Ocs) finally stuck among plant scientists. Recent work has shown that LpDH/Ocs-like enzymes also occur in *Sinorhizobium, Arthrobacter* and other bacteria, and even have some homology to animal octopine dehydrogenase sequences.

In Cologne I could soon put my experience with opine enzymes to good use. Marc Van Montagu and Jan Leemans in Gent and Jean-Pierre Hernalsteens and Henri de Greve in Brussels had produced *Agrobacterium* pTiB6S3 mutants, using the Tn*7* transposon. One mutant (pGV2100) induced shooty tumours from which plants could be regenerated. The Ti plasmid maps (established in several laboratories) and the first T-DNA transcript maps showed the insertion to be in T-DNA gene *2* essential for auxin synthesis. The induction of shooty tumors by pGV2100 was due to T-DNA gene *4,* encoding cytokinin synthesis. The auxin/cytokinin biosynthesis system of the T-DNA is a most remarkable natural example of the famous Skoog and Miller hormone experiments, that showed the role of auxins and cytokinins in root, shoot and callus induction. One of the pGV2100 regenerants developed into a morphologically normal plant, rGV-1, that remained LpDH/Ocs-positive. Thus, *ocs* became one of the first marker genes for transgenic plants, a primitive forerunner of the *nptII, hptII, uidA, gfp* or *bar* marker genes; its presence was revealed with the LpDH microassay. Crude extracts of rGV-1 contained no octopine, but incubation with arginine, pyruvate and NADH led to easily detectable quantities. LpDH analysis of protoplast-derived clones showed that all rGV-1 were transformed, and similar analysis of the progeny demonstrated that rGV-1 contained a single T-DNA locus in the heterozygous state. The T-DNA was transmitted according to the rules of Gregor Mendel, enabling us to recover homozygous lines in the F1.

Over twenty years later, progeny analysis of *Agrobacterium* transformants all over the world probably provides one of the most common occasions for students to test their working knowledge of the Mendelian rules. Regeneration of haploid plants from rGV-1 anther cultures by Otto Schieder and crosses with wild-type plants in both directions confirmed that the T-DNA was transmitted by pollen and egg cells. Later, DNA analysis showed that rGV-1 lacked the major part of the pGV2100 T-DNA leaving only *ocs.* Our results were rapidly confirmed. In Cologne, other regenerants carrying *ocs* and the cytokinin gene provided one of the first examples of transformants with phenotypic markers (rootless seedlings). Regenerants were also obtained from *A. rhizogenes*–induced hairy roots in Versailles. It is worth noting that already in 1977, Claudius Ackermann in Heidelberg regenerated morphologically abnormal tobacco plants from hairy root tissues and obtained progeny (Ackermann 1977), but no opine or DNA analysis was done, and his studies were not pursued.

Apart from proving the possibility of seed transmission, our rGV-1 results had some other interesting implications. They indicated that *Agrobacterium* could transfer non-tumorigenic T-DNA fragments under natural infection conditions, allowing regeneration and lead to naturally transgenic plants. Amazingly, Frank White discovered *A. rhizogenes* T-DNA sequences (*rol* genes) in wild-type *N. glauca* and other species (White et al. 1983); certainly one of the most astonishing discoveries in *Agrobacterium* research. The *rol* gene distribution patterns suggested transfer from the bacterium to the plant. Thus, plant species with Mendelian transmission of T-DNA genes existed well before 1981! It would also seem logical that people strictly opposed to the use of transgenic plants stop smoking.

Even after 30 years of research with transgenic plants, we do not quite know whether and how phenomena like post-transcriptional gene silencing and other epigenetic effects affect T-DNA expression during seed transmission. In spite of this, experiments with thousands of transgenic plant lines show that phenotypes and expression patterns are more often than not stably and predictably inherited.

Thus, our early hopes to develop a reliable gene transfer system for plants have been amply fulfilled.

References

Ackermann, C. 1977. Pflanzen aus *Agrobacterium rhizogenes*-Tumoren an *Nicotiana tabacum.* Plant Sci. Lett. 8:23–30.

Otten, L., and Schilperoort, A. 1978. A rapid micro scale method for the detection of lysopine and nopaline dehydrogenase activities. Biochim. Biophys. Acta 527:497–500.

White, F. F., Garfinkel, D. J., Huffmann, G. A., Gordon, M. P., and Nester, E. W. 1983. Sequences homologous to *Agrobacterium rhizogenes* T-DNA in the genomes of uninfected plants. Nature 301:348–350.

Wullems, G. J., Molendijk, L., Ooms, G., and Schilperoort, R. A. 1981. Retention of tumor markers in F1 progeny plants from in vitro induced octopine and nopaline tumor tissues. Cell 24:719–727.

Yang, F.-M., Montoya, A., Merlo, D. J., Drummond, M. H., Chilton, M.-D., Nester, E. W., and Gordon, M. P. 1980. Foreign DNA sequences in crown gall teratomas and their fate during the loss of the tumorous traits. Mol. Gen. Genet. 177, 707–714.

LEON OTTEN: I did my Ph.D. thesis in Rob Schilperoort's lab in Leiden. Rob was a very inspiring mentor and always had plenty of ideas. He combined chemistry with biology in a way that was still rare at that time, and managed to attract many excellent people. I then continued my work in the lab of Jeff Schell who was to become my second mentor. Jeff had an enormous amount of important responsibilities but that did not hinder him from discussing our results in detail and suggesting a way to proceed. He always seemed to make the right decision. His last years were very sad but his many contributions to science, his charismatic personality and his great courage to face adversity will not be forgotten.

Mol Gen Genet (1981) 183:209–213

Mendelian Transmission of Genes Introduced into Plants by the Ti Plasmids of *Agrobacterium tumefaciens**

L. Otten[1], H. De Greve[2], J.P. Hernalsteens[2], M. Van Montagu[2,3], O. Schieder[1], J. Straub[1], and J. Schell[1,3]

[1] Max-Planck-Institut für Züchtungsforschung, D-5000 Köln 30, Federal Republic of Germany
[2] Laboratorium Genetische Virologie, Vrije Universiteit Brussel (Belgium)
[3] Laboratorium voor Genetica, Rijksuniversiteit Gent (Belgium)

Summary. Insertion of the bacterial transposon Tn*7* was used to obtain mutants of an octopine Ti plasmid. Crown gall tumours induced on tobacco by an *Agrobacterium tumefaciens* strain carrying a particular mutant Ti plasmid (pGV2100) were found to give rise to shoots. These shoots were grown in vitro and one of them (rGV-1) was found to contain the T-DNA specific enzyme lysopine dehydrogenase (LpDH) and to form roots. After transfer to soil, rGV-1 developed into a morphologically and functionally normal tobacco plant. All cells of the regenerant and of vegetatively produced offspring were shown, by cloning of leaf protoplasts, to contain T-DNA and LpDH activity. rGV-1 and vegetatively produced offspring flowered normally. Plantlets obtained from haploid anther cultures were tested for LpDH activity. Forty-one percent of these plantlets were LpDH positive. Moreover, both self-pollination of rGV-1 and crosses between rGV-1 and normal tobacco plants showed that the LpDH character was transmitted both through the pollen and through the eggs of rGV-1 as a single dominant factor with Mendelian segregation ratios typical for monohybrid crosses. By repeated selfing, homozygous plants were obtained which bred true with respect to LpDH. The importance of these findings with respect to the use of *Agrobacterium tumefaciens* and Ti plasmids for genetic engineering in plants is discussed.

Introduction

Crown gall is a neoplastic disease of most dicotyledonous plants caused by the soil bacterium *Agrobacterium tumefaciens*. A large extrachromosomal plasmid in these bacteria was found to be responsible for its tumour-inducing capacity and was, therefore, called Ti plasmid (Van Larebeke et al. 1974). Bacteria-free crown gall cells can be cultured in vitro in the absence of phytohormones (Braun 1956). Such sterile tumour tissues have been shown to contain a DNA segment (called T-DNA) which is homologous and colinear with a defined fragment of the Ti plasmid and is covalently linked to plant DNA (Chilton et al. 1977; Schell et al. 1979; Lemmers et al. 1980; Zambryski et al. 1980; Yadav et al. 1980; Thomashow et al. 1980a, 1980b). The T-DNA has been localised in the nucleus (Chilton et al. 1980; Willmitzer et al. 1980). The T-DNA sequences in plants contain active genes. Indeed, T-DNA sequences were found to be transcribed according to a reproducible pattern (Drummond et al. 1977; Gurley et al. 1979; Willmitzer et al. 1981) and in an octopine tumour line (A6), at least seven independent transcripts were identified (Willmitzer et al. 1981). Recently, it was demonstrated (Schröder et al. 1981) that one of these transcripts can be translated in vitro to form the crown gall specific LpDH enzyme. This enzyme produces, amongst other compounds, N^2-[D-1-carboxyethyl]-L-arginine (or octopine) (Kemp 1976). Because the formation of crown gall tumours involves the transfer of genetic information from a bacterium to a higher plant, it can be described as a natural instance of genetic engineering (Schell and Van Montagu 1977; Schilperoort et al. 1978; Schell et al. 1979).

Foreign DNA can be inserted into the T-region of Ti plasmids and thus be cotransferred to the plant nucleus (Hernalsteens et al. 1980; Leemans et al. 1981). Since one of the goals of such experiments is the production of plants with interesting new properties, it is important to establish whether cells transformed with T-DNA can be regenerated into plants and whether the extra information is maintained during the sexual reproduction of the regenerated plants. Up to now, in spite of a number of attempts (Braun 1959; Braun and Wood 1976; Turgeon et al. 1976; Wood et al. 1978; Yang et al. 1980), these two requirements for successful use of the Ti plasmid in plant improvement were not met, probably because the tumorous properties of the crown gall cells could not be entirely suppressed. An alternative approach would be to alter Ti plasmids genetically so as to inactivate their tumour controlling properties without interference with the transfer of the T-DNA to the plant cell.

Such a pTi mutant has recently been obtained by Tn*7* mutagenesis (De Greve et al. 1981). One of these mutants (pGV2100) produced tumours from which shoots proliferate. One of these shoots contained LpDH activity, and was easily regenerated to a normal flowering plant. It was demonstrated that the nuclei of cells of this plant contained a T-DNA with an internal deletion eliminating the Tn*7* insertion and adjacent T-DNA sequences. The T-DNA sequences determining LpDH activity were present and actively transcribed (De Greve et al., submitted for publication).

Materials and Methods

1. Bacterial Strains and Inoculation of Tobacco Plants. The octopine Ti plasmid pGV2100 with a Tn*7* transposon inserted in the *Eco*RI fragment 32 of the B6S3 plasmid (De Greve et al. 1981) was used to induce tumours on *Nicotiana tabacum* cv.

* This paper is dedicated to Prof. Georg Melchers on the occasion of his 75th birthday, in recognition and gratitude for his relentless and enthousiastic pioneering efforts in the field of experimental plant genetics and cell biology

Offprint requests to: Prof. J. Schell, Max-Planck-Institut für Züchtungsforschung, D-5000 Köln 30 (Vogelsang), Federal Republic of Germany

0026-8925/81/0183/0209/$01.00

Wisconsin 38 (W38) (De Greve et al. submitted for publication). For the test inoculation of various plants, including the rGV-1 regenerant, C58C1 Ery^R Cm^R(pGV2100) (De Greve et al. 1981), B6S3 (an octopine strain) and C58 (a nopaline strain) were used. Bacterial strains were grown overnight at 27° C in 0.5% (w/v) beef extract (Difco), 0.5% (w/v) Peptone, 0.5% (w/v) saccharose and 2 mM Mg_2SO_4 in a rotary shaker. The stems of fully grown tobacco plants were wounded with a needle 48 h before inoculation with bacteria.

2. Shoot cultures. Sterile shoot cultures of rGV-1 were cultivated axenically on Murashige and Skoog medium (Murashige and Skoog 1962). Shoot cultures were sometimes maintained on a medium with cytokinin to suppress rapid growth, according to Gamborg (Gamborg et al. 1968) with 30 g/l sucrose instead of 20 g/l and 1 mg/l 6-benzylaminopurine. Roots were induced by transferring these shoots to the same medium without hormones.

3. Tobacco plants. Both rGV-1 tobacco plants and control tobacco plants (*N. tabacum*, cv. JWB and W38) were grown in a greenhouse in soil. When they had reached a height of about 30 cm, they were transferred to soil-filled buckets where they reached a height of about 2 m, until they flowered and set seed. The original rGV-1 tobacco plant was vegetatively multiplied by cutting.

4. Anther cultures. Anther cultures were performed as described by Nitsch and Nitsch (1969). Medium H was used with the addition of 0.1 mg/l of IAA. Cultures were kept in a culture room at 26±1° C with a 16 h light cycle (2,000–4,000 lux). For root induction, plantlets with a size of 3–4 mm were transferred to medium with an IAA concentration of 0.2 mg/l.

5. Plant Crosses. Plants which were to be self-pollinated were separated from other tobacco plants before they started to produce pollen. Because at the time these experiments were carried out, no flowering W38 tobacco plants were available, the tobacco cultivar JWB was used when normal tobacco material was needed for crosses.

6. Protoplast Isolation and Regeneration of Plants from Calli. The isolation and culturing of tobacco leaf protoplasts was carried out according to Schieder (1978). Callus and regenerating plantlets were kept on the modified Gamborg medium with cytokinin, as described under 2.

7. LpDH Assays. To test the presence of LpDH, in protoplast-derived tissues and in seedlings, an enzymatic micro-assay for LpDH was carried out essentially as described (Otten and Schilperoort 1978). To test seedlings, one-day old plantlets (leaf size 2 mm, weight about 1 mg) were separated from their roots and crushed with a match stick on the bottom of an Eppendorf tube with 20 µl of a 1:1 (v/v) mixture of incubation mixture and extraction buffer which was freshly prepared and kept on ice. The assay was incubated for 60 min at room temperature and further analysed as described. Semi-quantitative LpDH assays were carried out by adding 3 ml/g fresh weight of extraction buffer to fresh leaf pieces, and incubating the extract with substrates for 60 min. The intensity of the octopine spots resolved by electrophoresis and stained by phenanthrenequinone was estimated by visual inspection.

8. DNA Isolation and Southern Gel Electrophoresis. DNA isolation from tobacco material and gel electrophoresis of this DNA, followed by Southern hybridisation with labelled probes of cloned Ti plasmid DNA was carried out as described by Lemmers et al. (1980).

Results

1. Clonal Nature of the $LpDH^+$ rGV-1 Regenerant

To prove the clonal nature of the $LpDH^+$ rGV-1 regenerant with respect to the presence of LpDH, leaf protoplasts isolated from sterile shoot cultures derived from rGV-1 shoots were cultured until they formed small plantlets. Fifty-two of these plantlets were tested and all were found to be $LpDH^+$. Moreover, leaves, shoots and roots of these plants were all found to be $LpDH^+$, although there was significantly less LpDH activity in the shoots when compared to roots or leaves (based on fresh weight, and on polyA-containing polysomal RNA content, results not shown).

2. Anther Cultures

After the formation of roots, the rGV-1 tobacco shoot was brought in the greenhouse and developed into a normal tobacco plant which flowered. From the anthers, anther cultures were started which after callus formation gave rise to shoots. Of 102 shoots tested 42 (41%) were $LpDH^+$. $LpDH^+$ and $LpDH^-$ shoots were put on medium with increased IAA content to induce root formation, after which they were brought into the greenhouse and further developed into characteristic haploid tobacco plants. No morphological differences between $LpDH^+$ and $LpDH^-$ haploid plants could be detected.

3. Self-Pollination and Crosses with Normal Tobacco

After flowering of the rGV-1 plant, it was self-pollinated and set seed. Of 303 germinating S1 plants 214 (70%) were found to be LpDH positive, whereas 89 (30%) were LpDH negative. Of 137 positive plants, 42 (31%) showed a higher LpDH activity when compared to the remaining 95 (69%). (For an example of this difference, see Fig. 1). In about 10% of the cases it was not clear to which activity class a particular plant belonged. The plants with a high and a low LpDH activity were designated as "presumably homozygous" and "presumably hemizygous" for LpDH. These plants were allowed to develop flowers and by the time the flowers started to ripen five plants of each

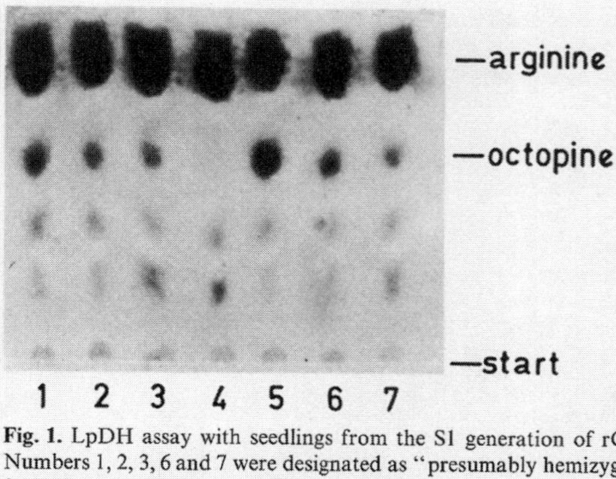

Fig. 1. LpDH assay with seedlings from the S1 generation of rGV-1. Numbers 1, 2, 3, 6 and 7 were designated as "presumably hemizygous" for LpDH, number 4 as negative and number 5 as "presumably homozygous" for LpDH

Table 1. Enzymatic LpDH analysis of S2 seedlings of rGV-1. After measuring the LpDH activity of rGV-1 derived S1 plants five plants expected to be homozygous for LpDH (1–5) and five expected to be hemizygous for LpDH (6–10) were self-pollinated and, from each S2, 52 seedlings were tested for the presence of LpDH activity

	Expected to be LpDH-positive	Found to be LpDH-positive
1	52	52
2	52	52
3	52	52
4	52	40
5	52	39
6	39	42
7	39	37
8	39	35
9	39	42
10	39	43

class were put separately in greenhous cabins, self-pollinated, and the resulting offspring were tested for LpDH activity. The results are shown in Table 1. Of five presumably homozygous plants three produced an offspring which was 100% LpDH$^+$ whereas the two others gave a segregation of about 3:1, indicating that they were in fact hemizygous. On the other hand, of five presumably hemizygous plants all produced an offspring which was about 75% LpDH$^+$ (mean value of all plants tested). The results of these tests showed conclusively that plants homozygous for LpDH had been obtained at the first selfing of the original rGV-1 plant.

The original rGV-1 plant was also crossed with normal tobacco plants. In one cross, the regenerant was pollinated, after castration, with pollen from a normal tobacco plant. Of 187 F1 plants, 81 (43%) were LpDH$^+$. In a second cross, pollen of rGV-1 was used to fertilise previously castrated normal tobacco plants. Of 248 plants tested, 124 (50%) were LpDH$^+$.

4. Morphological and Growth Properties of Transformed Plants

Plants homozygous for LpDH and control W38 tobacco plants sown at the same time are shown at three different stages of development in Fig. 2. No external morphological differences could be observed between these plants, nor was there any difference in their rate of development. The morphology of the flowers was also compared and found to be identical.

5. Sensitivity of Regenerants towards Agrobacterium Infection

Five plants presumably homozygous and five plants presumably hemizygous for LpDH (as determined by their LpDH activities), obtained from the S1 of the original self-pollinated rGV-1 plant, as well as a normal W38 tobacco plant, were inoculated with the *Agrobacterium* strains C58C1 Ery^R Cm^R(pGV2100), C58 and B6S3. All plants developed tumours; those induced with C58 were found to synthesise both octopine and nopaline.

6. Localization of LpDH in Different Parts of the Regenerants

LpDH tests were carried out with different tissues from an S1 plant homozygous for LpDH. The following tissues were tested: leaves, stem tissue (separated into the outer green part of the stem and the inner part or parenchyma), roots and seeds. In all tissues, except the parenchyma, we were able to demonstrate the presence of LpDH activity (results not shown).

Fig. 2a–c. Three different developmental stages of tobacco plants homozygous for LpDH (S2 generation of rGV-1) compared with normal tobacco plants sown at the same time. **a** Two days after germination. **b** Two weeks after germination. **c** Nine weeks after germination. The plants homozygous for LpDH are shown on the left. The normal tobacco is shown on the right

7. Southern Analysis of the Regenerant DNA

The rGV-1 plant and its offspring in various crosses were analysed for the presence of T-DNA by isolation of DNA, Southern gel electrophoresis, blotting and hybridisation with labelled T-DNA probes. This analysis confirmed that all these plants contained a T-DNA segment similar to the one observed in rGV-1 (data not shown) (De Greve et al., submitted for publication).

Discussion

It was recently established that morphologically and functionally normal plants can be obtained from plant cells transformed by *Agrobacterium tumefaciens* strains harbouring a mutant Ti plasmid (De Greve et al., submitted for publication). The results described in this paper demonstrate that a T-DNA linked gene (LpDH) in such a plant is transmitted by sexual crosses as a normal chromosomal locus. The main importance of these observations is that they prove that the stable incorporation of a well defined set of genes into higher plants is possible. In all the breeding experiments involving the regenerated LpDH-positive tobacco plant rGV-1, the LpDH gene was transmitted as a single dominant factor in a normal Mendelian way. (i) About 40% of the pollen, when cultured in vitro, gave rise to $LpDH^+$ plantlets. Moreover, when the pollen of rGV-1 was used to fertilise normal tobacco plants, 50% of the F1 seedlings were found to be $LpDH^+$; (ii) 43% of the F1 seedlings obtained by fertilisation of castrated rGV-1 plants with normal tobacco pollen were $LpDH^+$; (iii) from the F1 seedlings obtained by selfing of rGV-1, 71% were $LpDH^+$. This result is consistent with the results of (i) and (ii). Moreover, a semi-quantitative LpDH test was used to distinguish between plants homozygous for LpDH (31%) and those hemizygous for LpDH (69%) as shown by subsequent selfing of such plants and LpDH analysis of the S2 generation. Earlier experiments have already shown that normal plants can be regenerated from sterile crown gall tissue (Sacristán and Melchers 1977). It was not established, however, whether these plants contained T-DNA sequences and active T-DNA linked genes. Braun and collaborators (Braun 1959; Turgeon et al. 1976) have demonstrated that cloned tobacco crown gall cells of the nopaline type retain some morphogenetic potential and form aberrant shoots called teratomas. Such teratoma shoots can be forced to flower and to set seed through self-pollination, by repeated grafting onto normal tobacco plants. The regenerants contained nopaline, but the S1 obtained by selfing was nopaline-negative. Analysis of tissue culture material of leaves, flowers and filaments of such regenerants showed that they still contained T-DNA, whereas T-DNA was absent from tissue cultures of the anthers and the S1 generation (Yang et al. 1980; Lemmers et al. 1980). It was concluded that the T-DNA was either lost during meiosis or that it was lost in some somatic cells which would selectively give rise to gametes. In fact, recent observations (Y. Wu and O. Schieder, personal communication) indicate that the BT37 nopaline tissues originally cloned by Braun, again contain or produce nopaline-negative cells, because protoplasts derived from these teratomas gave rise to nopaline negative shoots and plants. These results indicate that deletions involving at least part of the T-DNA can occur in tissue cultures.

More recently, Wullems et al. (Wullems, personal communication) have produced regenerants of both octopine and nopaline tumours by using regenerating protoplasts of tobacco transformed in vitro with wild-type *Agrobacterium* (Martón et al. 1979) and also (for the octopine type) by the formation of somatic hybrids between octopine crown gall cells and normal tobacco cells (Wullems et al. 1980). All three types of regenerants were shown to produce opines in every cell as evidenced by leaf protoplast cloning. The leaf petals of such plants were shorter than normal, whereas the pistil remained of normal size. The pollen were shown to be sterile. Seedlings obtained from the octopine regenerants did not contain LpDH, whereas from 40 F1 seedlings obtained from the nopaline regenerant only one contained NpDH.

It is not clear why the LpDH gene and, in most cases, the NpDH gene was not transmitted in these experiments. In the experiments reported in this paper the LpDH gene originally introduced in the rGV-1 plant was stable for at least three generations. Apart from the LpDH gene, other T-DNA sequences are still present in rGV-1 plants (De Greve et al., submitted for publication).

One interesting property of the rGV-1 plant and of both its hemizygous and homozygous progeny is that they are fully sensitive to infection with a variety of *Agrobacterium* strains, including the C58C1 Ery^R Cm^R(pGV2100) strain originally used to produce the rGV-1 plant. If fertile plants can be regenerated from such doubly transformed tumours, it will be possible to establish by selfing of such plants whether the T-DNA introduced by the second transformation integrates in or near the T-DNA sequence already present in rGV-1.

An important difference between our approach and those reported previously, is the use of an *Agrobacterium* strain harbouring a mutant Ti plasmid with a Tn7 transposon inserted at a defined locus of the T-region (De Greve et al., submitted for publication). This approach is currently being extended to a number of other plants and preliminary observations (A. Wöstemeyer, personal communication) indicate that results similar to those reported here can be obtained with potatoes.

In view of the fact that the genes of the T-region coding for opine synthesis can be replaced by foreign DNA sequences without affecting the transfer functions of the Ti plasmid (Hernalsteens et al. 1980; Leemans et al. 1981) and since the experiments reported here establish that this segment of the T-DNA can be stably maintained, expressed and sexually transmitted by transformed plants, it follows that the LpDH gene of the mutant pTiGV2100 is a region of choice for the insertion of foreign genes to be transferred to higher plants.

Acknowledgements. We thank Mr. E. Czerny, Mrs A. Kaufmann and Mrs M.-T. Loogen for technical assistance, and Mrs M. Kalda for the photografic work. We are grateful to Dr. G. Wullems for the communication of his data prior to publication. L.O. was supported by a long-term EMBO fellowship and J.P.H. is a Research Associate of the Belgian National Fund for Scientific Research (N.F.W.O.). This work was supported by the Max-Planck-Gesellschaft, München and by grants from the "Onderling Overlegde Akties" (O.O.A./12052179) to J.S. and M.V.M.

References

Braun AC (1956) The activation of two growth substance systems accompanying the conversion of normal to tumor cells in crown gall. Cancer Res 16:53–56

Braun AC (1959) A demonstration of the recovery of the crown gall tumor cell with the use of complex tumors of single cell origin. Proc Natl Acad Sci USA 45:932–938

Braun AC, Wood HN (1976) Suppression of the neoplastic state with the acquisition of specilized functions in cells, tissues, and organs of crown gall teratomas of tobacco. Proc Natl Acad Sci USA 73:496–500

Chilton M-D, Drummond HJ, Merlo DJ, Sciaky D, Montoya AL, Gordon MP, Nester EW (1977) Stable incorporation of plasmid DNA into higher plant cells: the molecular basis of crown gall tumorigenesis. Cell 11:263–271

Chilton M-D, Saiki RK, Yadav N, Gordon MP, Quetier F (1980) T-DNA from *Agrobacterium* Ti plasmid is in the nuclear DNA fraction of crown gall tumor cells. Proc Natl Acad Sci USA 77:4060–4064

De Greve H, Decraemer H, Seurinck J, Van Montagu M, Schell J (1981) The functional organization of the octopine *Agrobacterium tumefaciens* plasmid pTiB6S3. Plasmid 6 (in press)

Drummond MH, Gordon MP, Nester EW, Chilton M-D (1977) Foreign DNA of bacterial plasmid origin is transcribed in crown gall tumours. Nature 269:535–536

Gamborg OL, Miller RA, Ojima K (1968) Nutrient requirements of suspension cultures of soybean root cells. Exp Cell Res 50:151–158

Gurley WB, Kemp JD, Albert MJ, Sutton DW, Callis J (1979) Transcription of Ti plasmid-derived sequences in three octopine-type crown gall tumor lines. Proc Natl Acad Sci USA 76:2828–2832

Hernalsteens JP, Van Vliet F, De Beuckeleer M, Depicker A, Engler G, Lemmers M, Holsters M, Van Montagu M, Schell J (1980) The *Agrobacterium tumefaciens* Ti plasmid as a host vector system for introducing foreign DNA in plant cells. Nature 287:654–656

Kemp JD (1976) Octopine as a marker for the induction of tumorous growth by *Agrobacterium tumefaciens* strain B6. Biochem Biophys Res Commun 69:816–822

Leemans J, Shaw Ch, De Blaere R, De Greve H, Hernalsteens JP, Van Montagu M, Schell J (1981) Intermediate Ti-vectors for site specific mutagenesis of T-DNA and for transfer of genes to plant cells. J Mol Appl Genet (in press)

Lemmers M, De Beuckeleer M, Holsters M Zambryski P, Depicker A, Hernalsteens JP, Van Montagu M, Schell J (1980) Internal organization, boundaries and integration of Ti-plasmid DNA in nopaline crown gall tumours. J Mol Biol 144:355–378

Martón L, Wullems GJ, Molendijk L, Schilperoort RA (1979) In vitro transformation of cultured cells from *Nicotiana tabacum* by *Agrobacterium tumefaciens*. Nature 277:129–130

Murashige T, Skoog S (1962) A revised medium for rapid growth and bioassays with tobacco tissue cultures. Physiol Plantarum 15:473–497

Nitsch JP, Nitsch C (1969) Haploid plants from pollen grains. Science 163:85–87

Otten LA, Schilperoort RA (1978) A rapid microscale method for the detection of lysopine- and nopaline dehydrogenase activities. Biochim Biophys Acta 527:497–500

Sacristán MD, Melchers G (1977) Regeneration of plants from "habituated" and "*Agrobacterium*-transformed" single cell clones of tobacco. Mol Gen Genet 152:111–117

Schell J, Van Montagu M (1977) The Ti plasmid of *Agrobacterium tumefaciens*, a natural vector for the introduction of *NIF* genes in plants? In: Hollaender A (ed) Genetic engineering for nitrogen fixation. Plenum Press, New York, p 159–179

Schell J, Van Montagu M, De Beuckeleer M, De Block M, Depicker A, De Wilde M, Engler G, Genetello C, Hernalsteens JP, Holsters M, Seurinck J, Silva B, Van Vliet F, Villarroel R (1979) Interactions and DNA transfer between *Agrobacterium tumefaciens*, the Ti-plasmid and the plant host. Proc R Soc Lond B 204:251–266

Schieder O (1978) Somatic hybrids of *Datura innoxia* Mill. + *Datura discolor* Bernh. and of *Datura innoxia* Mill. + *Datura stramonium* L. var. *tatula* L. I. Selection and characterization. Mol Gen Genet 162:113–119

Schilperoort RA, Klapwijk PM, Hooykaas PJJ, Koekman BP, Ooms G, Otten LABM, Wurzer-Figurelli EM, Wullems G, Rörsch A (1978) *A. tumefaciens* plasmids as vectors for genetic transformation of plant cells. In: Thorpe TA (ed) Frontiers of plant tissue culture, Proc. 4th Int. Congress of plant tissue and cell culture. University of Calgary, Alberta, p 85–94

Schröder J, Schröder G, Huisman H, Schilperoort RA, Schell J (1981) The mRNA for lysopine dehydrogenase in plant tumor cells is complementary to a Ti-plasmid fragment. FEBS Lett 129:166–168

Thomashow MF, Nutter R, Montoya AL, Gordon MP, Nester EW (1980a) Integration and organization of Ti plasmid sequences in crown gall tumors. Cell 19:729–739

Thomashow MF, Nutter R, Postle K, Chilton M-D, Blattner FR, Powell A, Gordon MP, Nester EW (1980b) Recombination between higher plant DNA and the Ti plasmid of *Agrobacterium tumefaciens*. Proc Natl Acad Sci USA 77:6448–6452

Turgeon R, Wood MN, Braun AC (1976) Studies on the recovery of crown gall tumor cells. Proc Natl Acad Sci USA 73:3562–3564

Van Larebeke N, Engler G, Holsters M, Van den Elsacker S, Zaenen I, Schilperoort RA, Schell J (1974) Large plasmid in *Agrobacterium tumefaciens* essential for crown gall-inducing ability. Nature 252:169–170

Willmitzer L, Otten L, Simons G, Schmalenbach W, Schröder J, Schröder G, Van Montagu M, De Vos G, Schell J (1981) Nuclear and polysomal transcripts of T-DNA in octopine crown gall suspension and callus cultures. Mol Gen Genet 182:255–262

Willmitzer L, De Beuckeleer M, Lemmers M, Van Montagu M, Schell J (1980) The Ti-plasmid derived T-DNA is present in the nucleus and absent from plastids of plant crown-gall cells. Nature 287:359–361

Wood HN, Binns AN, Braun AC (1978) Differential expression of oncogenicity and nopaline synthesis in intact leaves derived from crown gall teratomas of tobacco. Differentiation 11:175–180

Wullems GJ, Molendijk L, Schilperoort RA (1980). The expression of tumour markers in intraspecific somatic hybrids of normal and crown gall cells from *Nicotiana tabacum*. Theor Appl Genet 56:203–208

Yadav NS, Postle K, Saiki RK, Thomashow MF, Chilton M-D (1980) T-DNA of a crown gall teratoma is covalently joined to host plant DNA. Nature 287:458–461

Yang F, Montoya AL, Merlo DJ, Drummond MH, Chilton M-D, Nester EW, Gordon MP (1980) Foreign DNA sequences in crown gall teratomas and their fate during the loss of the tumorous traits. Mol Gen Genet 177:707–714

Zambryski P, Holsters M, Kruger K, Depicker A, Schell J, Van Montagu M, Goodman HM (1980) Tumor DNA structure in plant cells transformed by *A. tumefaciens*. Science 209:1385–1391

Communicated by W. Arber

Received June 6, 1981

HOW DOES *AGROBACTERIUM* "KNOW" WHAT IS T-DNA ON THE Ti PLASMID?

Mary-Dell Chilton
Novartis Seeds, Inc., Research Triangle Park, North Carolina, United States
E-mail: marydell.chilton@syngenta.com

Commentary on the paper:

Yadav, N. S., Vanderleyden, J., Bennett, D. R., Barnes, W. M., and Chilton, M.-D. 1982. Short direct repeats flank the T-DNA on a nopaline Ti plasmid. Proc. Natl. Acad. Sci. U.S.A. 79:6322–6326.

From the time we first found T-DNA in crown gall tumor cells, we wondered whether it was discrete, defined part of the Ti plasmid, or whether random pieces of plasmid were incorporated and the cells with a certain part of the plasmid grew out as a gall. Once the Southern blotting technique was perfected to genomic sensitivity, we could identify in an autoradiogram from a transformed plant certain T-DNA fragments whose size matched their counterpart fragments on the Ti plasmid. Different tumor lines initiated by the same Ti plasmid had these same internal T-DNA fragments. In addition there were T-DNA fragments that varied from one tumor line to another, and we thought they were border fragments, junctions of plant DNA and T-DNA. In the case of nopaline Ti plasmid pTiT37, the border seemed to map in *Eco*RI fragment 29. The fact that adjacent fragment 13 was an internal T-DNA fragment but 29 was not, suggested that the border was discrete. We believed that there must be some signal sequence that defined the borders of T-DNA on the Ti plasmid.

In order to identify such a border sequence, it was necessary to isolate a T-DNA/plant DNA junction by molecular cloning in order to study its sequence and look for the signal. At that time, in the late 1970s, it was feasible to clone a gene from a genome the size of tobacco, our host plant of choice, but the large number of clones that must be screened, dictated that cloning must be done in bacteriophage lambda. The elegantly designed lambda Charon phage vectors had just been developed by Fred Blattner et al. (1978) but exploiting their efficient delivery of DNA into *E. coli* required the tricky process of packaging the ligated DNA into phage coats *in vitro.* Kotty Postle, a new postdoc from the University of Wisconsin, spearheaded our effort. Ultimately Kotty, Mike Thomashow and I flew to Madison, Wisconsin, in the wake of a paralyzing blizzard and worked under Blattner's supervision for several days. We had brought both octopine and nopaline type tobacco tumor DNA, from which we prepared cloning target DNAs that were a mix of partial and complete *Eco*RI digest fragments. Blattner kindly provided phage vector arms and packaging components that worked, and we went home triumphantly with literally millions of recombinant plaque-forming units. Mike screened the octopine-line library, while Kotty and I screened the nopaline one, hybridizing labeled T-DNA probe to plaque "lifts", screening hundreds of thousands of plaques grown on cafeteria trays that served as mega-petri dishes. We found a wealth of positives, of which most turned out to be uninteresting pieces from the middle of the T-DNA. Mike Thomashow identified amongst his positive plaques the left and right junctions of octopine tumor T-DNA as evidenced by hybridization with Ti plasmid and with tobacco genomic DNA (Thomashow et al. 1980).

The positive nopaline T-DNA clones moved with me to Washington University in St. Louis in 1979, where a new postdoc, Narendra Yadav, carried on with their analysis. He found that one lambda clone contained eight small *Eco*RI fragment inserts, of which seven matched the left-most seven *Eco*RI fragments of T-DNA. The eighth fragment hybridized to *Eco*RI fragment 29 and to repeated DNA in the tobacco genome (Yadav et al. 1980). Patti Zambryski, together with collaborators in the Schell/Van Montagu laboratory, also performed lambda cloning on the T-DNA of a nopaline tumor. This group cloned a fragment that represented a fusion between tandem T-DNA inserts and in addition, a right border fragment that hybridized at low stringency to highly repeated tobacco DNA (Zambryski et al. 1980). They published a description of their right border fragment and show in one of their figures the 40 bp of Ti plasmid DNA sequence preceding the right T-DNA/plant DNA junction.

Yadav, having identified the fortunately very small left border fragments in Ti plasmid and tumor, set about their analysis at the DNA sequence level. Like Charon phage cloning, DNA sequencing at that time was still an arcane procedure. We were fortunate to collaborate with Wayne Barnes, who was the local master of this art at Washington University. We were hoping to get the sequence of the border fragment in time for the upcoming Gordon Conference on Plant Molecular Biology. Strips of paper with As, Ts, Gs and Cs fluttered from the bulletin board in my office as we compared, base by base, the DNA sequence of tumor DNA and Ti plasmid DNA looking for something interesting—we didn't know exactly what. Yadav toiled long hours, sometimes overnight, toward his goal. Finally, after his third all-night session, he had the border identified. "Aren't you tired after staying up all night?" I asked him. He shook his head and chuckled: "I'm getting used to it now!" We were able to report the border sequence at the Gordon Conference. And we could see already the future implication: if you want to trick the bacteria into putting a gene of your choice into the plant cell, you use the Trojan horse approach: flank the DNA by 25-basepair border repeats and *Agrobacterium* will—perhaps—think it is T-DNA!

References

Blattner, F. R., Blechl, A. E., Denniston-Thompson, K., Faber, H. E., Richards, J. E., Slightom, J. L., Tucker, P. W., and Smithies, O. 1978. Cloning fetal γ globin and mouse α type globin DNA: Preparation and screening of shotgun collections. Science 202: 1279–1284.

Thomashow, M. F., Nutter, R., Postle, K., Chilton, M.-D., Blattner, F. R., Powell, A., Gordon, M. P., and Nester, E. W. 1980. Recombination between higher plant DNA and the Ti plasmid of *Agrobacterium tumefaciens*. Proc. Natl. Acad. Sci. U.S.A. 77:6448–6452.

Yadav, N. S., Postle, K., Saiki, R. K., Thomashow, M. F., and Chilton, M.-D. 1980. T-DNA of a crown gall teratoma is covalently joined to host plant DNA. Nature 287:458–461.

Zambryski, P., Holsters, M., Kruger, K., Depicker, A., Schell, J., Van Montagu, M., and Goodman, H. M. 1980. Tumor DNA structure in plant cells transformed by *A. tumefaciens*. Science 209:1385–1391.

MARY-DELL CHILTON: In 1960, I completed a B.S. in chemistry at the University of Illinois, Urbana, and obtained a Ph.D. in chemistry from the same university in 1967.

My Ph.D. mentor was Ben Hall, in whose laboratory I was privileged to work at the time he developed the technique of measuring DNA/RNA hybridization by binding to nitrocellulose filters. I discovered that *Bacillus subtilis* could be transformed by single-stranded DNA, and after physically separating the "Watson and Crick" strands of transforming DNA, showed that either strand of the double helix could transform the bacteria.

My post-doctoral mentor was Brian McCarthy, under whose direction I refined my technical expertise in DNA hybridization, renaturation kinetics analysis and work with the complex genomes of higher organisms (mouse and maize). For one of my projects, I measured the degree of homology required for transformation of *B. subtilis* by DNA from related *Bacillus* species. The lesson I took from this early work was that DNA transformation required extensive homology between donor and recipient DNA. Based on this experience, I was very sceptical of early claims in the crown gall literature that *A. tumefaciens* could transfer bacterial or bacteriophage DNA to plant cells. It was about the same time that the earliest reports were published claiming animal cell transformation by naked DNA. My dogmatic belief that transforming DNA must match the genome of the recipient was soon to be shaken.

Proc. Natl. Acad. Sci. USA
Vol. 79, pp. 6322–6326, October 1982
Genetics

Short direct repeats flank the T-DNA on a nopaline Ti plasmid

(transposable elements/site-specific recombination/plant tumors/transformation)

NARENDRA S. YADAV*†, JOS VANDERLEYDEN*, DONALD R. BENNETT*, WAYNE M. BARNES‡, AND MARY-DELL CHILTON*§

Departments of *Biology and ‡Biological Chemistry, Washington University, St. Louis, Missouri 63130

Communicated by Arthur Kelman, July 12, 1982

ABSTRACT Crown gall disease results from the insertion of a segment of the *Agrobacterium* Ti plasmid, called T-DNA, into host plant nuclear DNA. We have subjected to sequence analysis the border regions of pTi T37 (ends of T-DNA) and one left T-DNA/plant DNA border fragment isolated from BT37 tobacco teratoma by molecular cloning. These sequence studies, taken together with published sequence of a right T-DNA/plant DNA border fragment, allowed us to identify the positions of left and right borders at the DNA sequence level. Comparison of left and right border regions of the Ti plasmid revealed a "core" direct repeat of 13 of 14 bases (12 contiguous) precisely at the borders of T-DNA. An extended repeat of 21 of 25 bases overlaps this core repeat. T-DNA on the Ti plasmid exhibits no longer direct or inverted repeats in the border regions, based on Southern hybridization studies. The physical structure of T-DNA differs from that of known prokaryotic and eukaryotic transposable elements but bears a structural resemblance to the prophage of bacteriophage λ.

Crown gall is a neoplastic disease of higher plants caused by oncogenic strains of *Agrobacterium tumefaciens* (1). Tumor cells are stably transformed in that they exhibit hormone autotrophy *in vitro* in the absence of the inciting bacterium (2). Axenic tumor cells synthesize one or more novel metabolites called opines (3–7) that are specific catabolic substrates for the inciting *Agrobacterium* strain (5, 7–9). The pathogen harbors large plasmids called Ti (tumor inducing) plasmids (10) that code for oncogenicity and specify which opines the tumor will synthesize as well as which opines the bacteria can catabolize (8, 9, 11, 12). A part of the Ti plasmid called T-DNA is stably maintained in the tumor cells (13–17) and is transcribed into polyadenylylated RNAs (14, 18–20). T-DNA transcripts are found on polysomes in the tumor cell (20) and can be translated into proteins *in vitro* (21). Insertion of transposons into the T-DNA region of the Ti plasmid can affect tumor morphology (22–24) or eliminate the synthesis of opines by the tumor (24, 25). T-DNA is in the nuclear fraction of the plant tumor cell (26, 27) covalently joined to host plant DNA (28–30). T-DNA insertion into plant DNA can be viewed as an example of genetic engineering that occurs in nature, for it diverts plant metabolites into opines useful only to the pathogen.

The mechanism by which T-DNA integrates into the plant genome is unknown. The T-DNA region of the Ti plasmid, when compared with T-DNA in the tumor cell by Southern hybridization, indicates that several octopine-type tumors contain two Ti plasmid segments (T_L and T_R) inserted into plant DNA independently (16). T_R has been found at 0–20 copies per tumor cell; T_L has been found at one to several copies, and its edges are variable in different tumor lines (16). A right border fragment of T_L joined to plant DNA, isolated by molecular cloning, hybridizes to a noncontiguous region of the Ti plasmid that maps in the center of T-DNA (29). In contrast to the complex T-DNA inserts found in the octopine tumors, T-DNA of the nopaline tumors has a more simple arrangement. It appears colinear with T-DNA of the Ti plasmid, with borders that appear fixed, within the resolution of the technique of Southern blot analysis. The host plant DNA sites at which it inserts appear to be variable (17). In addition, T-DNA in nopaline tumors includes "fusion" DNA fragments homologous to the left and right edges of T-DNA, presumed to derive from either tandem copies or circular forms of T-DNA (30).

If T-DNA is a discrete physical and genetic element, the DNA sequences at its edges might structurally define it. We have subjected to sequence analysis DNA from regions of the nopaline Ti plasmid pTi T37 that contain the left and right edges of T-DNA. The flanking Ti plasmid DNA exhibits short direct repeats that we propose to be border signals.

MATERIALS AND METHODS

Recombinant Plasmids. *Eco*RI fragments G and 29 (Fig. 1) were subcloned into pBR325 (31) from recombinant phage clone PC1 (28) and recombinant plasmid pBR322 (*Bam*HI-6) (28), respectively. *Hin*dIII fragment 23 (Fig. 1) was subcloned into pBR325 from *Eco*RI fragment 1 of pTi T37. *Eco*RI fragment 1 had previously been cloned into Charon 4A bacteriophage (32, 33).

M13 Cloning Vectors. M13mp7 vector (34) was obtained as *Eco*RI linear DNA from New England BioLabs. M13 phage derivative mWB2341 carries *lac* operon DNA that has been modified to contain one *Eco*RI site and one *Hin*dIII site adjacent to the commercially available primer (arrow on the vector in Fig. 2) for dideoxy sequence analysis (unpublished data). The *lac* DNA was cloned out of mWJ22 (two *Eco*RI sites, one *Hin*dIII site) (35) into the unique *Sau* 96 site of wild-type M13 DNA to obtain mWB2341. Details of construction of this and other vectors in this series will be published elsewhere.

Cloning of T-DNA Fragments into M13 Vectors. The restriction enzymes employed were from New England BioLabs. Sticky ends of restriction fragments were rendered blunt by treatment for 20 min at 37°C with the large fragment of *Escherichia coli* DNA polymerase I (36–38) (New England BioLabs) in 10 mM Tris·HCl, pH 7.9/10 mM $MgCl_2$/50 mM NaCl/10 mM 2-mercaptoethanol and 100 μM each of dATP, dCTP, dGTP, and dTTP at a DNA concentration of 200 μg/ml. All other enzymatic reactions were performed as recommended by the manufacturer and terminated by phenol extraction and ethanol precipitation of the DNA. T4 ligase, from cells infected by λ T4 lig (39), was a gift from M. Bittner. Ligation of sticky ends was carried out in 30 mM Tris·HCl, pH 7.9/4 mM $MgCl_2$/1 mM

The publication costs of this article were defrayed in part by page charge payment. This article must therefore be hereby marked "*advertisement*" in accordance with 18 U. S. C. §1734 solely to indicate this fact.

Abbreviation: bp, base pair(s).
† Present address: Central Research and Development Dept., E. I. Du Pont de Nemours & Co., Wilmington, DE 19898.
§ To whom reprint requests should be addressed.

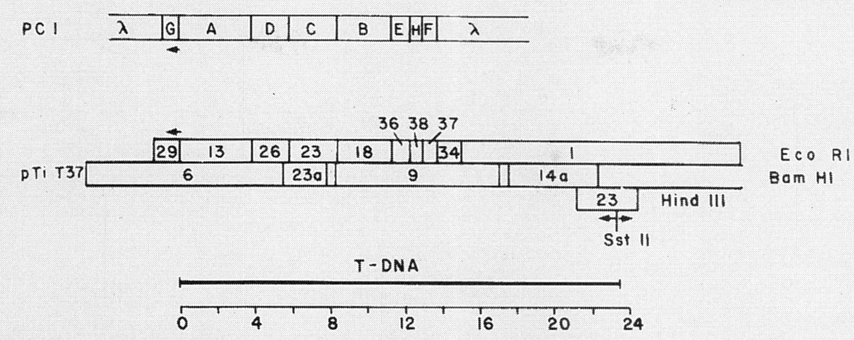

FIG. 1. Restriction maps of recombinant phage clone PC1 and the T-DNA region of pTi T37. PC1 is a recombinant phage containing T-DNA fragments from BT37 tumor tissue obtained as an *Eco*RI partial digest product (28); its *Eco*RI map is shown above. The map of the T-DNA region of pTi T37 shown (M. W. Bevan, personal communication) corrects details of a previous version (28). Phage inserts A, D, C, B, E, H, and F correspond to Ti plasmid *Eco*RI fragments 13, 26, 23, 18, 36, 38, and 37, respectively. Phage insert G is a border fragment containing T-DNA joined to tobacco DNA (28). The arrows indicate the direction of DNA sequence analysis of the fragments at the left and right edges of T-DNA. The scale indicates the size of T-DNA in kilobases.

EDTA/40 μM ATP/10 mM dithiothreitol at a DNA concentration of 16 μg/ml. Blunt-end ligations employed the same buffer with 1 mM ATP/1 mM hexamine cobalt chloride (British Drug House, Poole, England) (unpublished data) at a DNA concentration of 100 μg/ml. All ligations were performed at 4°C for 15 hr.

*Eco*RI fragments 29 and G were cloned into the *Eco*RI site of M13mp7 as described (34). The *Sst* II/*Hin*dIII subfragments of *Hin*dIII fragment 23 were subcloned out of pBR325 into mWB2341 replicative form by the strategy outlined in Fig. 2, which directs the blunted *Sst* II site adjacent to the *lac* primer (New England BioLabs) used for dideoxy sequence analysis. Phage clones containing the large and small (40) *Sst* II/*Hin*dIII subfragments were designated mTi 35 and mTi 40, respectively.

DNA Sequence Analysis. DNA sequence was obtained from recombinant M13 phage DNA by the method of Sanger *et al.* (41) as adapted by Barnes (40).

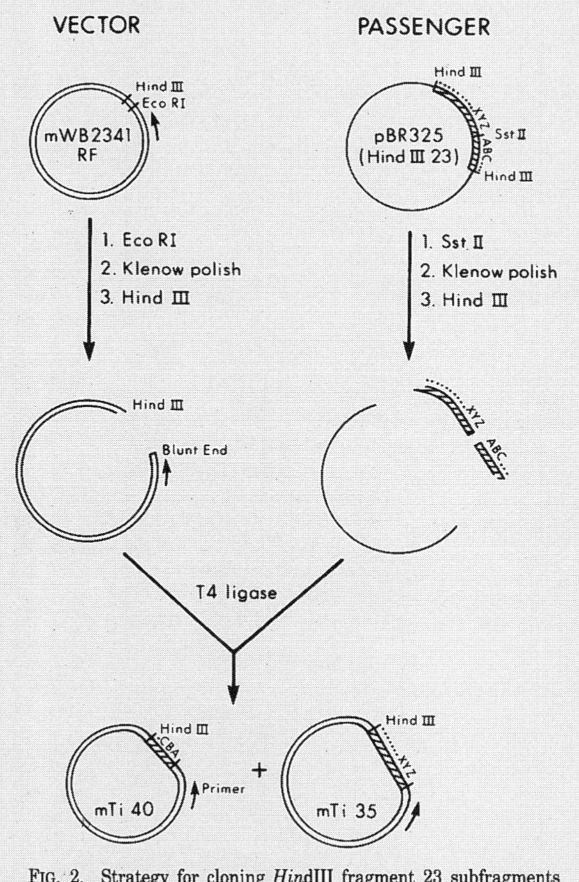

FIG. 2. Strategy for cloning *Hin*dIII fragment 23 subfragments into M13 phage vector. The replicative form (RF) of the M13 phage vector mWB2341 (double line) and the passenger DNA, consisting of *Hin*dIII/blunted *Sst* II subfragments of *Hin*dIII fragment 23, were ligated together as indicated. Recombinant phage clones contained the large (. . . XYZ) or small (ABC . . .) subfragment with the *Sst* II end adjacent to the primer site for DNA sequence determination (indicated by an arrow on the vector). mTi 40 and mTi 35 are not drawn to scale.

RESULTS

Left Boundary. To determine at the nucleotide sequence level the position of a left border of T-DNA in the nopaline crown gall tumor, we subjected to sequence analysis a cloned left T-DNA/plant DNA border fragment (fragment G) (28) and the corresponding region of pTi T37 (*Eco*RI fragment 29) (Fig. 1) leftward from their common *Eco*RI site. The sequence from fragment G perfectly matched the sequence from fragment 29 for the first 68 bases to the left of the *Eco*RI site and then diverged, presumably into plant DNA (Fig. 3*A*). Computer dot matrix analysis of 153 bases of fragment G sequence vs. fragment 29 sequence revealed no other direct or inverted repeats.

Computer analysis of the sequence of *Eco*RI fragment 29 revealed a Chi site, 5′-G-C-T-G-G-T-G-G-3′ (43), and four Chi-like octamer sequences, one differing from Chi by one base and the others by two bases (Fig. 3*A*). These all occur in the same orientation within a 34-base region that lies 12 bases outside the left T-DNA border. We found no such Chi-related elements in either orientation in the region of pTi T37 that contains the right border of T-DNA (see below).

Right Border. The BT37 tumor line contains several T-DNA copies; thus, the single right T-DNA/plant DNA border fragment that was subjected to sequence analysis by Zambryski *et al.* (30) may not derive from the insert whose left border our fragment G represents. Nevertheless, any right border fragment should serve to determine the location of the border, if T-DNA is a discrete element.

*Hin*dIII fragment 23, which contains the right border (17, 30), was subcloned as *Hin*dIII/*Sst* II fragments into M13 vector mWB2341 to yield mTi 35 and mTi 40 (Fig. 2). By Sanger (32) sequence analysis of mTi 35, 310 bases were determined to the left (M. W. Bevan, personal communication) of the *Sst* II site and from mTi 40, 270 bases to the right (Fig. 3*B*) (see arrows in

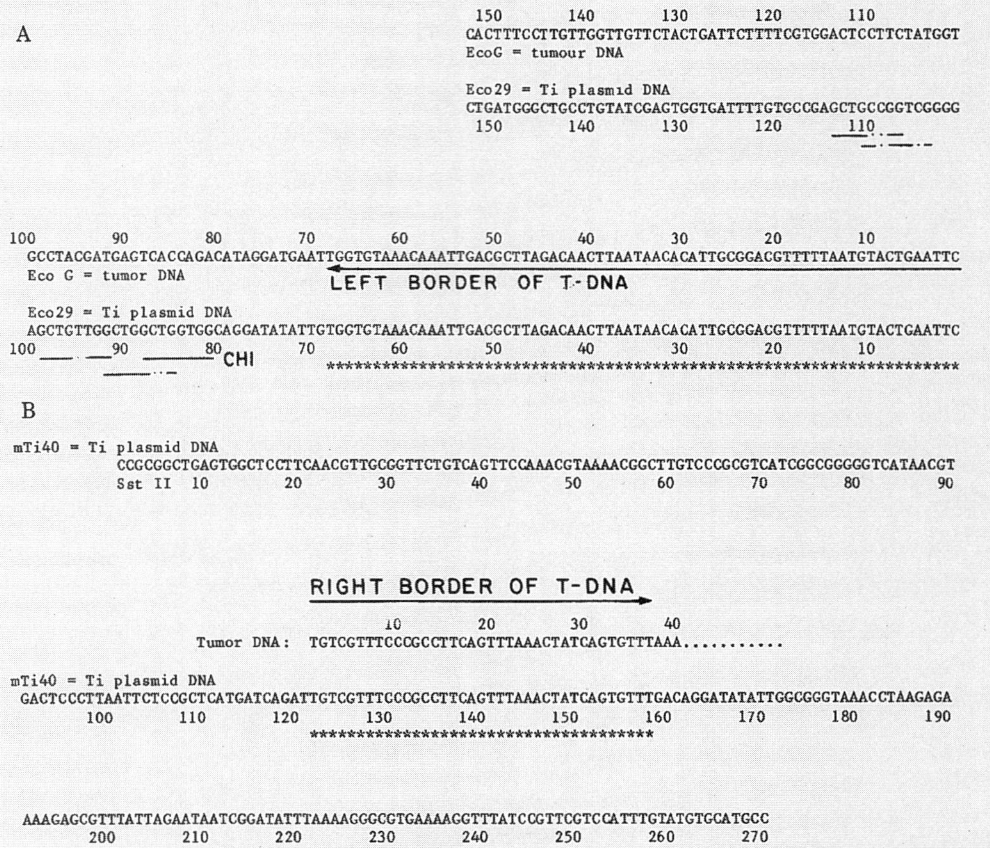

FIG. 3. DNA sequence of left and right border regions of pTi T37 vs. left and right border clones from BT37 tumor DNA. (*A*) Sequence of *Eco*RI fragments G (from tumor) and 29 (from Ti plasmid). Stars and arrow mark the 68 base pairs (bp) of matching sequence to the left of the *Eco*RI site (numbered 1–6). The figure presents the complement of the strands actually subjected to sequence analysis, so that the orientation of this sequence corresponds to the orientation of fragments G and 29 on the map in Fig. 1. A Chi site (G-C-T-G-G-T-G-G) (42) and four Chi-like octamers are underlined in the sequence of fragment 29. (*B*) Sequence of mTi 40 (from Ti plasmid) is compared with the published BT37 tumor border fragment sequence (30). Stars and arrow mark matching sequences that define the point of divergence into plant DNA 158 bp to the right of the *Sst* II site (numbered 1–6).

Fig. 1). Comparison with the partial sequence of the right border fragment of Zambryski *et al.* (30) (reproduced in Fig. 3*B*) shows the position of the right border to be 158 bases to the right of the *Sst* II site.

Direct Repeats Flank T-DNA on the Ti Plasmid. Comparison of Ti plasmid DNA sequence in the left and right border regions revealed that the first 14 bases outside T-DNA on the Ti plasmid constitute a direct repeat of 13 of 14 bases, 12 contiguous. This 14-mer forms a part of an extended repeat of 21 of 25 bases (Fig. 4). The Chi sequence near the left border overlaps the extended repeat by two bases. Computer dot matrix analysis revealed no other significant direct or inverted repeats in the 153 bases sequenced in *Eco*RI fragment 29 vs. 580 bases sequenced around the *Sst* II site in *Hin*dIII fragment 23.

Absence of Longer Repeats. To test whether there was a much longer repeated sequence at the borders of T-DNA in pTi

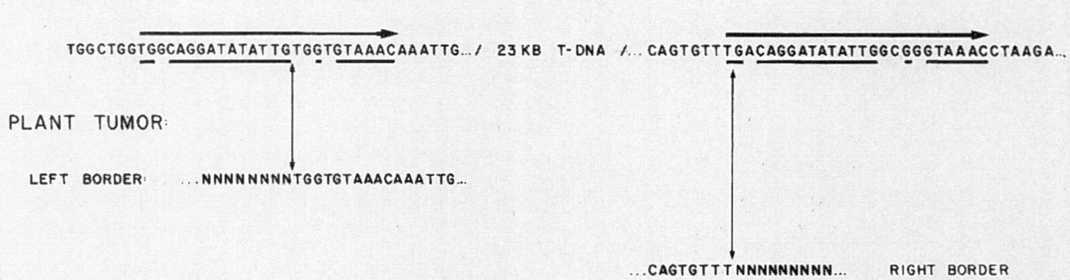

FIG. 4. DNA sequence of border repeats. Sequence data from the border regions (Fig. 3) are presented with the 25-bp direct repeats marked by horizontal arrows and bases within the repeat that match are marked by underlining. The border fragments from tumor DNA are abstracted, with NNNNN . . . representing presumed plant DNA to emphasize the fate of the 25-bp repeat upon T-DNA integration into plant DNA.

T37 that escaped detection by our limited sequence study, we looked for homology between regions of pTi T37 that contain the left and right borders of T-DNA by DNA hybridization. For this purpose, the recombinant clone of pBR325 containing *Hin*dIII fragment 23 was digested with *Hin*dIII, *Sst* II, and *Bam*HI. The three resulting subfragments (designated 1, 2, and 3 in Fig. 5) were excised and collected by electroelution. Similarly, *Eco*RI fragments 29 and G were isolated after digestion of the appropriate recombinant plasmids with *Eco*RI. Isolated fragments were subjected to gel electrophoresis and Southern blotting (42) and were hybridized to labeled *Eco*RI fragment 29 (Fig. 5) and labeled *Eco*RI fragment 13 (recombinant plasmid) (data not shown). These probes from the left border of T-DNA failed to hybridize to fragments 1, 2, and 3 from the right border region but hybridized strongly to homologous DNA. We conclude that there is no extended homology between the left and right border regions of pTi T37.

DISCUSSION

Conservation of 25-bp Repeat on Octopine Ti Plasmid. The data presented here suggest that the borders of T-DNA in nopaline-type Ti plasmid pTi T37 are determined by specific DNA sequences. T-DNA on the Ti plasmid is flanked by direct repeats of 13 of 14 bp (12 contiguous). Overlapping these are 25-bp imperfect repeats. The significance of these repeats is underscored by the finding of similar sequence at the left border of an octopine Ti plasmid (44):

Octopine Ti left repeat:

C ↓ **G-G-C-A-G-G-A-T-A-T-A-T-T**-C - A-A-T- T- **G-T-A-A-A**-T

Nopaline Ti left repeat:

T - **G-G-C-A-G-G-A-T-A-T-A-T-T-G** ↓ T-G-G-T -**G-T-A-A-A**-C

Nopaline Ti right repeat:

T ↓ **G**-A-**C-A-G-G-A-T-A-T-A-T-T-G** - G-C-G-G-**G-T-A-A-A**-C

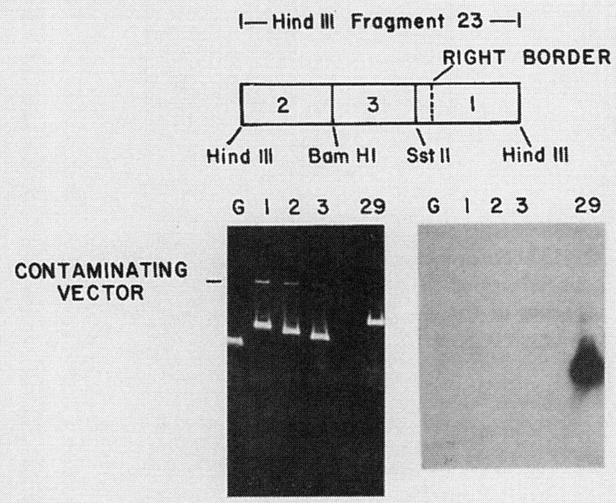

FIG. 5. Hybridization of labeled left border probe with Southern blots of right border region fragments. (*Upper*) Map of *Hin*dIII fragment 23. (*Lower Left*) Photograph of ethidium bromide-stained agarose gel containing 1 μg of the following DNA fragments, previously isolated from preparative agarose gels by electroelution: G (the left T-DNA/plant DNA border fragment); 1, 2, and 3, the subfragments of *Hin*dIII fragment 23; and 29, *Eco*RI fragment 29. (*Lower Right*) Autoradiogram of Southern blot made from the agarose gel after hybridization with *Eco*RI fragment 29. No homology is detected between left and right border regions. Similar results were obtained with *Eco*RI fragment 13 (data not shown). DNA fragments are numbered as in *Lower Left*.

Arrows in the above sequence indicate the ends of colinear T-DNA sequence found in border fragment clones isolated from plant tumors; bases shown in **boldface** are those conserved on both ends of nopaline T-DNA and their counterparts in the octopine Ti repeat. Arrows (points of discontinuity) occur at the left and right ends of the imperfect 14-mer G-N-C-A-G-G-A-T-A-T-A-T-T-N (N is any nucleoside), consistent with a site-specific recombination mechanism.

The DNA sequence surrounding the 25-bp repeat on the octopine Ti plasmid is unrelated to the context sequence of the nopaline Ti plasmid repeats, based on our dot matrix analysis (data not shown) of the sequence of the left border region of the octopine Ti plasmid (44). We infer that this repeat *per se* signals T-DNA borders and is of functional significance rather than structural significance because it is preserved on the octopine Ti plasmid whereas surrounding DNA is not.

T-DNA Structure Differs from That of Transposons. No known prokaryotic transposon or eukaryotic transposable element has border structures resembling those of T-DNA reported here. Prokaryotic transposons possess long direct or inverted repeats that stay joined to the transposon when it moves to a new location (45). Eukaryotic transposable elements characterized thus far have long direct repeats that appear to move with the element (46–49). Retrovirus proviruses structurally resemble the eukaryotic transposable elements, with direct repeats hundreds of bp in length (50). Thus, none of these elements affords a clear precedent for the structure and recombination pattern of T-DNA: short imperfect direct repeats that do not become inserted intact into the plant genome.

All of the known transposable elements generate short direct repeats of host DNA flanking each insert (45–50). These are diverse for a given element and appear to be a structural by-product of the transposition process (45). We have not determined whether the T-DNA insert in the plant genome is flanked by variable direct repeats of plant DNA (a study requiring isolation of both borders of a single T-DNA insert). T-DNA *on the Ti plasmid* (as defined by the border positions identified here) is flanked by imperfect 14-mers.

The closest parallel that we can find between T-DNA and a transposable element is a certain resemblance to the prophage of bacteriophage λ. Although present information is insufficient to allow any firm conclusion, the following parallels with λ exist: The λ prophage is flanked by functional 15-bp repeats, which are signals for site-specific recombination (51), resulting in excision of λ as a circle carrying one copy of the repeat. A similar model for T-DNA excision would predict the net preservation of one copy of the extended 25-bp repeat in each T-DNA insert, possibly split between the two borders where T-DNA adjoins plant DNA. Our data are consistent with such a model.

Further, a λ-like mechanism would predict that circular T-DNA would insert at plant DNA sites somewhat homologous to the 25-bp repeat. Consistent with this idea, we find homology between the presumed plant DNA component of border fragment G and the "missing" end of the 25-bp repeat. From the sequences of Fig. 3*A* we see the following matching bases (shown in **boldface**) on the plant DNA side of the border, 2 bp out of register:

Border fragment G:

A-G-A-C-A-T-**A-G-G-A-T**-G-A-**A-T** ↓ T-G-G-T-G-T-A-A-A-C

Left Ti plasmid repeat:

T-G-G-C-**A-G-G-A-T**-A-T-**A-T**-T-G ↑ T-G-G-T-G-T-A-A-A-C

Border

Insufficient data have been published for the right border fragment to allow similar analysis for it.

Chi Site. The role of the Chi element at the left end of T-

DNA on pTi T37 is not clear. Chi augments recombination in λ over a broad region; its greatest effect extends leftward as Chi is written here (52). Site-directed mutagenesis of Chi on the Ti plasmid will test whether it affects the process of T-DNA excision. We note that Chi is not transferred intact to the plant genome with T-DNA; thus, any effect it might exert would presumably be in the bacterium.

Genetic Analysis of Role of Repeats. The data presented here suggest that the border repeats define what will be T-DNA. Deletion of the right border repeat by Tn*1* has been found to attenuate but not abolish tumor induction (17, 24). The Tn*1* DNA in this case formed the border of T-DNA (17). This interesting result implies that alternative borders can substitute when a normal border is deleted. λ phage can utilize secondary attachment sites when the primary site is deleted, even when the homology is as low as 9 of 15, although the efficiency is diminished (53, 54).

The unusual nopaline Ti plasmid pTi AT181, which is closely related to pTi T37, has a large deletion at the left end of its T-DNA (compared to pTi T37); however, DNA sequence analysis has revealed that the left border repeat survives intact in this virulent strain (unpublished data).

Deletion of border repeats followed by their reintroduction at new positions can be achieved by site-directed insertions into the Ti plasmid (55–57). Such studies will rigorously test whether the border repeats signal the functional ends of T-DNA. The presence of the 25-bp direct repeats at the edges of T-DNA provides the first clue to the mechanism of illegitimate recombination between prokaryotic Ti plasmid DNA and the eukaryotic host plant genome.

We thank Dr. P. Zambryski and Dr. A. Depicker for discussions of their unpublished data and Drs. R. Simpson, P. O'Hara, E. W. Nester, and M. P. Gordon for access to their unpublished sequence data. This research was supported by Grants NP194 to M.-D.C. and JFRA-1 to W.M.B. from the American Cancer Society and Grant GM 24956 to W.M.B. from the National Institute of General Medical Sciences.

1. Smith, E. F. & Townsend, C. O. (1907) *Science* **25**, 671–673.
2. White, P. R. & Braun, A. C. (1942) *Cancer Res.* **2**, 597–617.
3. Menage, A. & Morel, G. (1964) *C. R. Hebd. Seances Acad. Sci. Ser. C* **259**, 4795–4796.
4. Menage, A. & Morel, G. (1965) *C. R. Hebd. Seances Acad. Sci. Ser. C* **261**, 2001–2002.
5. Petit, A., Delhaye, S., Tempé, J. & Morel, G. (1970) *Physiol. Vég.* **8**, 205–213.
6. Firmin, J. L. & Fenwick, G. R. (1978) *Nature (London)* **276**, 842–844.
7. Guyon, P., Chilton, M.-D., Petit, A. & Tempe, J. (1980) *Proc. Natl. Acad. Sci. USA* **77**, 2693–2697.
8. Bomhoff, G., Klapwijk, P. M., Kester, H. C., Schilperoort, R. A., Hernalsteens, J. P. & Schell, J. (1976) *Mol. Gen. Genet.* **145**, 171–181.
9. Montoya, A., Chilton, M.-D., Gordon, M. P., Sciaky, D. & Nester, E. W. (1977) *J. Bacteriol.* **129**, 101–107.
10. Zaenen, I., Van Larebeke, N., Teuchy, H., Van Montagu, M. & Schell, J. (1974) *J. Mol. Biol.* **86**, 109–127.
11. Van Larebeke, N., Engler, G., Holsters, M., Van den Elsacker, S., Zaenen, I., Schilperoort, R. A. & Schell, J. (1974) *Nature (London)* **252**, 169–170.
12. Watson, B., Currier, T. C., Gordon, M. P., Chilton, M.-D. & Nester, E. W. (1975) *J. Bacteriol.* **123**, 255–264.
13. Chilton, M.-D., Drummond, M. H., Merlo, D. J., Sciaky, D., Montoya, A., Gordon, M. P. & Nester, E. W. (1977) *Cell* **11**, 263–271.
14. Yang, F., Montoya, A. L., Merlo, D. J., Drummond, M. H., Chilton, M.-D., Nester, E. W. & Gordon, M. P. (1980) *Mol. Gen. Genet.* **177**, 707–714.
15. Merlo, D. J., Nutter, R. C., Montoya, A. L., Garfinkel, D. J., Drummond, M. H., Chilton, M.-D., Gordon, M. P. & Nester, E. W. (1980) *Mol. Gen. Genet.* **177**, 637–643.
16. Thomashow, M. F., Nutter, R. C., Montoya, A. L., Gordon, M. P. & Nester, E. W. (1980) *Cell* **19**, 729–739.
17. Lemmers, M., De Beuckeleer, M., Holsters, M., Zambryski, P., Depicker, A., Hernalsteens, J. P., Van Montagu, M. & Schell, J. (1980) *J. Mol. Biol.* **144**, 353–376.
18. Drummond, M. H., Gordon, M. P., Nester, E. W. & Chilton, M.-D. (1977) *Nature (London)* **269**, 535–536.
19. Gurley, W. B., Kemp, J. D., Albert, M. J., Sutton, D. W. & Callis, J. (1979) *Proc. Natl. Acad. Sci. USA* **76**, 2828–2832.
20. Yang, F., McPherson, J. C., Gordon, M. P. & Nester, E. W. (1980) *Biochem. Biophys. Res. Commun.* **92**, 1273–1277.
21. McPherson, J. C., Nester, E. W. & Gordon, M. P. (1980) *Proc. Natl. Acad. Sci. USA* **77**, 2666–2670.
22. Garfinkel, D. J. & Nester, E. W. (1980) *J. Bacteriol.* **144**, 732–743.
23. Ooms, G., Klapwijk, P. M., Poulis, J. A. & Schilperoort, R. A. (1980) *J. Bacteriol.* **144**, 82–91.
24. Holsters, M., Silva, B., Van Vliet, F., Genetello, C., De Block, M., Dhaese, P., Depicker, A., Inze, D., Engler, G., Villarroel, R., Van Montagu, M. & Schell, J. (1980) *Plasmid* **3**, 193–211.
25. Hernalsteens, J. P., Van Vliet, F., De Beuckeleer, M., Depicker, A., Engler, G., Lemmers, M., Holsters, M., Van Montagu, M. & Schell, J. (1980) *Nature (London)* **287**, 654–656.
26. Chilton, M.-D., Saiki, R. K., Yadav, N., Gordon, M. P. & Quétier, F. (1980) *Proc. Natl. Acad. Sci. USA* **77**, 4060–4064.
27. Willmitzer, L., De Beuckeleer, M., Lemmers, M., Van Montagu, M. & Schell, J. (1980) *Nature (London)* **287**, 359–361.
28. Yadav, N. S., Postle, K., Saiki, R. K., Thomashow, M. F. & Chilton, M.-D. (1980) *Nature (London)* **287**, 458–461.
29. Thomashow, M. F., Nutter, R. C., Postle, K., Chilton, M.-D., Blattner, F. R., Powell, A., Gordon, M. P. & Nester, E. W. (1980) *Proc. Natl. Acad. Sci. USA* **77**, 6448–6452.
30. Zambryski, P., Holsters, M., Kruger, K., Depicker, A., Schell, J., Van Montagu, M. & Goodman, H. M. (1980) *Science* **209**, 1385–1391.
31. Bolivar, F. (1978) *Gene* **4**, 121–136.
32. Blattner, F. R., Blechl, A. E., Denniston-Thompson, K., Faber, H. E., Richards, J. E., Slightom, J. L., Tucker, P. W. & Smithies, O. (1978) *Science* **202**, 1279–1284.
33. Becker, A. & Gold, M. (1975) *Proc. Natl. Acad. Sci. USA* **72**, 581–585.
34. Messing, J., Crea, R. & Seeburg, P. H. (1981) *Nucleic Acids Res.* **9**, 309–321.
35. Rothstein, R. J., Lau, L. F., Bahl, C. P., Narang, S. A. & Wu, R. (1979) *Methods Enzymol.* **68**, 98–109.
36. Klenow, H., Overgaard-Hansen, K. & Patkar, F. A. (1971) *Eur. J. Biochem.* **22**, 371–381.
37. Setlow, P., Brutlag, D. & Kornberg, A. (1972) *J. Biol. Chem.* **247**, 224–231.
38. Kelly, W. S. & Stumpf, K. H. (1979) *J. Biol. Chem.* **254**, 3206–3210.
39. Murray, N. E., Bruce, S. A. & Murray, K. (1979) *J. Mol. Biol.* **132**, 493–495.
40. Barnes, W. M. (1977) *Science* **195**, 393–394.
41. Sanger, F., Nicklen, S. & Coulson, A. R. (1977) *Proc. Natl. Acad. Sci. USA* **74**, 5463–5467.
42. Southern, E. M. (1975) *J. Mol. Biol.* **98**, 503–517.
43. Smith, G. R., Kunes, S. M., Schultz, D. W., Taylor, A. & Triman, K. L. (1981) *Cell* **24**, 429–436.
44. Simpson, R., O'Hara, P., Montoya, A. L., Lichtenstein, C., Gordon, M. P. & Nester, E. W. (1982) *Cell* **29**, 1005–1014.
45. Calos, M. P. & Miller, J. H. (1980) *Cell* **20**, 579–595.
46. Farabaugh, P. J. & Fink, G. R. (1980) *Nature (London)* **286**, 352–356.
47. Gafner, J. & Philippsen, P. (1980) *Nature (London)* **286**, 414–418.
48. Levis, R., Dunsmuir, P. & Rubin, G. M. (1980) *Cell* **21**, 581–588.
49. Dunsmuir, P., Brorein, W. J., Simon, M. A. & Rubin, G. M. (1980) *Cell* **21**, 575–579.
50. Temin, H. M. (1980) *Cell* **21**, 599–600.
51. Landy, A. & Ross, W. (1977) *Science* **197**, 1147–1160.
52. Stahl, F. W., Stahl, M. M., Malone, R. E. & Crasemann, J. M. (1980) *Genetics* **94**, 235–248.
53. Nash, H. A. (1981) *Annu. Rev. Genet.* **15**, 143–167.
54. Chapman, J. & Gardner, J. F. (1981) *J. Bacteriol.* **146**, 1046–1054.
55. Matzke, A. J. M. & Chilton, M.-D. (1981) *J. Mol. Appl. Genet.* **1**, 39–49.
56. Garfinkel, D. J., Simpson, R. B., Ream, L. W., White, F. F., Gordon, M. P. & Nester, E. W. (1981) *Cell* **27**, 143–153.
57. Leemans, J., Shaw, C., Deblaere, R., De Greve, H., Hernalsteens, J. P., Maes, M., Van Montagu, M. & Schell, J. (1981) *J. Mol. Appl. Genet.* **1**, 149–164.

SUCH A CLEVER BUG

Scott E. Stachel
2120 South Park Ranch Road
Jackson, Wyoming, United States
E-mail: stachel@molbio.mgh.harvard.edu

Commentary on the papers:

Stachel, S. E., Messens, E., Van Montagu, M., and Zambryski, P. 1985. Identification of the signal molecules produced by wounded plant cells that activate T-DNA transfer in *Agrobacterium tumefaciens.* Nature 318:624–629.

Stachel, S. E., Timmerman, B., and Zambryski, P. 1986. Generation of single-stranded T-DNA molecules during the initial stages of T-DNA transfer from *Agrobacterium tumefaciens* to plant cells. Nature 322:706–712.

I'm intrigued and awed by stories that reveal nature's cleverness. My favorite stories involve the seemingly devious manipulation of one organism by another, so that the manipulator thrives in a unique niche—such as *Wolbachia* bacteria causing insect parthenogenesis, to be propagated ever onward via female hosts; or trematodes triggering the fish they parasitize to swim to light, to be eaten by unsuspecting birds, their subsequent hosts; or *Nosema* protozoa synthesizing excess juvenile hormone to halt beetle metamorphosis. An ever growing grub makes an ideal home for *Nosema.* But my all-time favorite tale of organismal trickery is *Agrobacterium tumefaciens*' genetic parasitism of plants, as it is such a complete solution of niche creation, and elegant example of evolution's ingenuity.

I first was told *Agrobacterium*'s story as a beginning graduate student at UC San Francisco. It was my turn to lead journal club, and I wanted to present something other than gene regulation or the latest cloning technique. Pat O'Farrell told me that a postdoc in Howard Goodman's lab, Patti Zambryski, studied plants. At a medical school this was different, so I wandered past the centrifuges to find her. Howard's lab was intimidating—filled with assertive postdocs poised to pioneer the biotech industry. Patti, herself had just published the *Science* article concerning the T-DNA borders, and she introduced me to the Ti-plasmid. I asked how T-DNA transfer worked, and was quietly told it was not yet known. I subsequently gave a T-DNA journal club talk, and with graduate student enthusiasm, we hypothesized transfer mechanisms—perhaps via a prophage, or an SV40-like particle, or subverted conjugation, or transposition of a retrovirus—and how we might distinguish between them. Here was a mystery I wanted to solve, and I subsequently joined Howard's lab as a Ph.D. student. It was early days of plant genetic engineering, and I was first assigned to identify the *nos* gene's promoter and terminator, though my real goal was to decipher *Agrobacterium*'s special trick.

The Setup

Several seminal events led to my two *Nature* articles reproduced in this anthology. The first was a nice seminar by Cynthia Kenyon, where she described her use of random *Mud*::*lacZ* mutagenesis to first isolate *E. coli* DNA repair genes, and then to determine their transcriptional organization and regulation. This talk provided the starting point for my thesis proposal. Using a *lacZ* transposon in *E. coli,* I would randomly mutagenize the Ti-plasmid's 40 Kbp virulence (*vir*) region, which seemed to encode T-DNA transfer functions. I would then use these presumptive gene fusions in *Agrobacterium* to investigate the transfer process. Unfortunately, the *lacZ* elements then available were unsuitable for plasmid mutagenesis. Fred Heffron suggested Tn*3* as a platform and helped me design a Tn*3*::*lacZ* system. The rest of my proposal was decidedly optimistic, especially to my committee—using hundreds of *vir*::*lacZ* insertions I would sort out *vir* genetic organization, find conditions in a plant elixir that activate *vir* expression, establish the transcriptional organization and regulatory control of the (hypothesized) *vir* regulon, isolate and identify the plant-produced signal molecule, use this inducer to activate bacteria (in the absence of plant cells) in order to identify a T-DNA intermediate, and ultimately, the transfer mechanism. Surprisingly, all this came to pass.

While I was writing my proposal, Howard moved the lab to Harvard, and Patti joined our collaborators in Gent. I could go to Boston, Belgium, or change projects and stay at UCSF. So I found an alternative: Gene Nester's lab at the University of Washington was a mecca for *Agrobacterium* research. One fall weekend I visited Seattle, where I found several wonderful people: Dave Garfinkel, Frank White, Conrad Lichtenstein, and Alice Montoya. Gene welcomed me to join them. I don't recommend this path, graduate students need mentors, but Harold Varmus agreed to be my adviser *in absentia* at UCSF, and I headed north. The incomparable Gynheung An initially provided me with daily guidance and sandwiches, and following months of DNA manipulation agony (this was pre-PCR), I produced the Tn*3*::*lacZ* and transposase-helper constructs. Then, using Vic Knauf's pTiA6 cosmid clones, and with Carlos Flores' and Corty Thiens' able assistance, I assembled a library of 143 Tn*3*::*lacZ* insertions in both orientations every 50–100 bases across the entire *vir* region. (Stachel et al. 1985, Stachel and Nester 1986) I also began a daily routine of β-galactosidase assays, of which I carried out more than 25,000 over the next four years. But would my Tn*3* element produce functional *vir*::*lacZ* fusions? And could plants induce *vir* gene expression? One afternoon I cocultivated several presumptive *vir*::*lacZ* strains with regenerating tobacco mesophyll protoplasts. The following morning when I added the β-gal substrate to the separated bacteria, several samples immediately turned blue. Plant-mediated induction of *Agrobacterium vir*-gene expression existed (Stachel et al. 1986).

This proved the seminal event that led to my identification of the *vir* regulon and its regulation by the *virA*/*virG* two-component system, as well, as my discovery of the inducing factor and the T-DNA transfer mechanism. (Stachel and Zambryski 1986a,b) Another key observation was that callus and root suspension cultures also produced, in limited amounts, a *vir*-inducing activity. Freed from protoplasts, I found that the factor was small, soluble, stable, and dependent on plant metabolism. But I was no chemist, and required help to proceed further. While in Europe for a meeting, I visited Marc Van Montagu, and was chased and toyed by his beautiful and wild marguay cat. Don't wear sandals at Marc's house. Marc suggested I join his lab in Gent, to purify the factor with Eric Messens, who had isolated opines, and had his own LPLC and ready access to a GC/MS. UCSF gave me permission to further my wandering student act. And during my last week at UW I worked almost 100 hours—to finish the *virA*/*virG* data, teach induction assays, and organize strains.

The Signal

While I was in Seattle, Patti Zambryski in Gent had significantly contributed to the foundation of plant genetic engineering via pGV3850 and produced insights into T-DNA transfer via the 25 bp border repeat. Marc had me work with Patti, and I initiated her into the joys of β-gal assays. Another extraordinary event then occurred—Max Tate, a chemist at the Waite Institute in Australia, came to visit. In just a few days, Max helped us determine that the inducer is hydrophobic and partitions into chloroform. Its subsequent purification proved straightforward. From several litres of conditioned root media, Eric performed two differential separations on silica C-2/C-18 to isolate two active compounds. Subsequent GC/MS showed the most abundant molecule had the composition $C_{10}H_{12}O_4$ and the structure acetyl-dimethoxyphenol. This structure has 15 possible isomers, only one of which was commercially available. It arrived in two days. When Marc came by the following morning to ask about synthesizing all of the isomers, he received my boldest grin. We had it—the major inducer was 4-acetyl-2,6-dimethoxyphenol (AS, acetosyringone)—three years after that first protoplast assay.

I subsequently quantified the *vir*-inducing activities of some 200 related phenolics, and found that AS and OH-AS were the best inducers (much of these data were never published). For me, the most intriguing part of this story came to be (and remains) why these molecules? Why did *Agrobacterium* evolve to activate its *vir*-regulon in their presence, versus other plant metabolites? To this end, the second half of our plant-signal article was directed at asking whether AS makes biological sense, that is, does AS possess the properties to signal in nature the presence of plant cells most susceptible to *Agrobacterium* transformation. We conjectured such a molecule should be exudate specific, produced only by active cells, produced in quantities sufficient to induce *vir* expression, and associated with wounded versus intact plant tissues. Rather remarkably, we found that AS and OH-AS possess all these properties. A question we didn't address is what is acetosyringone's role in the normal physiology of a plant, although we speculated it is related to flavonoid and lignin metabolism, and might function in plant defense.

I've anticipated further studies into AS biology. While a substantial literature now exists on the use of AS to encourage transformation of organisms not ordinarily susceptible to *Agrobacterium*'s charms (why didn't we patent this process!), as well as reports that it stimulates hair growth, and inhibits polymorphonuclear granulocytes, my curiosity remains unsated. Does AS have other rhizosphere (perhaps antimicrobial) activities? Does it affect gene expression (now, a simple transcriptome experiment) in other organisms? Why do certain plants actively produce AS, and via what pathway (and what is the mutant phenotype)?

Finally, does AS also function as a plant cell signal (what happens if plants are grown in it)? Interesting questions, still awaiting answers.

T-Strand Tales,
Trans-Kingdom Conjugation

Patti and I shared a fascination with T-DNA transfer, and with AS in hand, we focused on identifying the transfer mechanism. By now it was known that any DNA located between the 25 bp direct border repeats could be transferred and integrated, with an apparent rightward to leftward polarity. Also, Patti had just published a clever paper with Zdena Koukolikova-Nicola and Barbara Hohn, which suggested that the transfer intermediate is a circular double stranded T-DNA joined at its ends by a single border element (Koukolikova-Nicola et al. 1985). But I was skeptical, as this structure had been indirectly recovered from *E. coli* transformed with DNA from plant-induced *Agrobacterium,* and at a frequency of less than 1 copy per 100 induced cells.

AS allowed us to directly look for free T-DNA structures in total DNA from highly induced *Agrobacterium* cultures. We initially failed to detect such changes—until we used protease digestion to isolate the DNA, and a radioactive probe for analysis. I still recall my excitement when I developed that first film and saw a novel T-DNA homologous signal migrating at half the size of a double-stranded T-DNA. This signal was a revelation. With no evidence beyond expectation, I knew it had to correspond to single stranded T-DNA; and that it announced (if only then to me) that T-DNA transfer is trans-kingdom conjugation. We called it the T-strand.

Some results, such as those concerning *vir* induction, are long interminable roads that require countless numbers of bench hours to produce a finished story. Luckily, the results concerning the T-strand were clear paths that ask for just a few direct experiments to complete. But UCSF was demanding my thesis, and Marc had ordered me to my desk to write backlogged manuscripts. I was released from the bench for only a few hours each day. Fortunately, the delightful Ben Timmerman joined Patti and me to help characterize the novel T-DNA signal. These experiments relied on many of the factoids I had learned as a beginning student; for example, the fact that nitrocellulose only binds single stranded DNA provided the basis for the nondenatured transfer assay—which we initially used to convince ourselves that the T-strand is single stranded. In a few exciting weeks, our experiments further revealed that it was a linear unipolar copy of the T-DNA region's bottom strand between the Ti-plasmid border sites. These characteristics were strikingly similar to those of the molecule mobilized by bacterial conjugation. Moreover, since we found the T-strand at a frequency of about one copy per induced cell, this suggested that as for conjugation, its synthesis is conservative, and its presumptive mode of transfer to the recipient cell is direct and efficient.

Additional results solidified this connection. For example, we found induction generates S_1 nuclease sensitive sites in the Ti-plasmid border repeats. These border nicks, which Kan Wang later nicely showed correspond to a site-specific endonuclease cleavage in the bottom strand of each border, implied T-strand synthesis initiates and terminates at the right and left nicks, respectively, and evoked the nicked *oriT* site that initiates conjugal transfer. Furthermore, they provided an explanation for the T-DNA circles recovered in *E. coli*—as artifacts of nick-induced homologous recombination. We also found that induction generates Ti-plasmid T-region structures having both single-stranded and double-stranded DNA. These structures suggested that T-strands are produced via displacement strand synthesis, and evoked the donor intermediates generated during conjugation. Finally, that we found the T-strand at all suggested its synthesis and transfer are not coupled. That it was observed only after proteolysis indicated it exists in the induced cell as a nucleoprotein T-complex, the actual transfer intermediate. Subsequent studies have shown that this complex consists of the T-strand covalently bound at its 5′-terminus to the *vir*D2 relaxase/pilot protein, and coated with virE2 SSB sub-units.

Fortunately, we initially chose pGV3850 to look for T-DNA intermediates, as its T-region carries a pair of borders and generates a single T-strand. But we also concurrently characterized the T-strands generated from pTiA6, whose complex T-region harbors four border repeats. These results further validated our T-DNA conjugation model. My *vir::lacZ* mutants were constructed in pTiA6, and we wanted to identify which genes direct T-strand synthesis—but this is another story, and Marty Yanofsky similarly identified these genes using the same mutant set. These studies are reported in this Anthology. (See commentary by M. Yanofsky) We surprisingly found that pTiA6 generates, in limited quantities, six distinct T-strands that correspond to all the possible border-to-border combinations. This showed that each border can function to initiate and to terminate T-strand synthesis, or be skipped altogether. In hindsight this border promiscuity explains why plant transformants can carry unwanted sequences 3′ of the left border; and suggests construction of a left border deficient in T-strand initiation and highly efficient in termination would benefit plant genetic engineering. Indeed, *oriT* mutation studies imply these functions are separable.

We proposed that T-DNA transfer is bacterial conjugation applied to plant cells. Everything fits—conjugation is a conservative and direct mechanism, as T-strand transfer seemed to be; and the border nicks, T-region intermediates, and T-strands appeared equivalent, respectively, to conjugation's *oriT,* displacement strand donor intermediates, and transfer strand (and all subsequent results have reinforced this correspondence). On the other hand the striking disparity between the bacterial versus plant recipients of these two processes, suggested to me that conjugation had probably required considerable tinkering to target plants. But the pTiA6 T-strands marked my final experiments in Gent, whose labyrinth of corridors and incredible characters and dedicated beer refrigerator I still miss. And soon, Patti was in Berkeley starting her lab, and I had moved on to become a developmental neurobiology postdoc.

Evolutionary Antecedents, Present Day Confederates

Although I left *Agrobacterium* long ago, I remain awed by its ingenuity, and I'd like to finish here with two T-DNA transfer questions that still intrigue me: Where did it come from? Does it have functional cohorts? Two notable sets of post T-strand results speak to these questions. First, when stimulated with AS, *Agrobacterium* becomes a remarkable generalist that can use T-DNA transfer to transform any cell that is presented—including dicots, monocots, gymnosperms, bacteria, several fungi, and even human cells. Moreover, this amazing capability is not unique to T-DNA transfer, as broad-host-range plasmids also can transport conjugal DNA from *E. coli* to yeast and animal cells. Thus, at least in the lab, conjugation doesn't require specific cell recognition or binding. Second, T-complex transport occurs via a Type IV secretion system (T4SS). The T4SS also mediates bacterial conjugation. They are found in many diverse bacterial pathogens, where they move a multiplicity of macromolecule effectors into eukaryotic cells. T4SSs encode a translocation apparatus and corresponding substrate coupling protein; and all T4SSs function as protein delivery devices. Indeed, the principle substrate of conjugation (and T-DNA transfer) is the pilot protein, which leads its attached conjugal DNA through the translocation apparatus. Thus, the progenitor T4SS could have been a protein delivery device, and conjugation arose, perhaps via association with a rolling circle replication system, as an adaptation for DNA transport. Furthermore, since different conjugation systems can be more closely related to specific T4SSs than to each other, conjugation has possibly evolved several times.

These observations suggest that getting from T4SS protein secretion, to conjugation, to T-DNA transfer might actually not be so difficult. There is also an implied possible model for this system's origin. *Agrobacterium*'s ancestor expressed genes for opine synthesis and catabolism, and phytohormone synthesis. This ancestor also encoded an ancestral *virB* T4SS that it used to translocate opine synthesis enzymes into adjacent plant cells, whose growth it promoted by secreting phytohormones. Indeed, *Agobacterium* does translocate proteins, independent of the T-complex, during transformation. Think of this ancestor as cheap crown gall. Its opine and phytohormone genes then came to be linked to an element that used rolling circle replication, whose pilot protein then came to be recognized by the T4SS *vir*D4 coupling protein, and voila—transkingdom conjugation. This model is partially testable. For instance, do T-DNA encoded proteins (if engineered to be expressed in *Agrobacterium*) still harbor sequences that direct their transport into plant cells via the *virB/virD4* T4SS? If they do, can they provide selective advantage to bacteria expressing opine catabolism?

When I first heard *Agrobacterium*'s story, I believed it unique. But genetic parasitism is a great gig for a bacterium—a home of one's own with a perpetually full refrigerator. Also, all of the pieces mediating T-DNA transfer occur throughout the bacterial pangenome; and many pathogenic bacteria use T4SSs to transfer macromolecules into eukayotes. Thus, evolution has seemingly had both the motive and means to produce systems of genetic these systems? Crown gall is common, dramatic, and well studied. Perhaps, transformants of other genetic parasitism systems have much subtler phenotypes. Recent results do suggest potential candidates. Genomic sequencing shows several bacteria (including the fascinating *Wolbachia*) encode *virB, virD2,* and *virD4* relatives, hallmarks of T-DNA transfer, and several diseases involving aberrant growths. Included among these diseases are atherosclerotic plaques, ulcers, and various polyps—all associated with T4SS expressing pathogens. Additionally, bacteria which encode pathogenic functions, that when mutated can't be *cis*-complemented, also represent possible contenders. While I now find it a scary notion that I'm under constant genetic attack from bacteria (and that I might make opines, after working with *Agrobacterium*), I look forward with great interest to someday meeting *Agrobacterium*'s confederates, and learning about their own clever manipulations.

References

Koukolikova-Nicola, S., Shillito, R. D., Hohn, B. Wang, K., Van Montagu, M., and Zambryski, P. 1985. Involvement of circular intermediates in the transfer of T-DNA from *Agrobacterium* to plant cells. Nature 313:191–196.

Stachel, S. E., An, G., Flores, C., and Nester, E. W. 1985. A Tn3 *lacZ* trans-poson for the random generation of β-galactosidase gene fusions: Application to the analysis of gene expression in *Agrobacterium tumefaciens.* EMBO J. 4:891–898.

Stachel, S. E., and Nester, E. W. 1986. The genetic and transcriptional organization of the *vir* region of the A6 Ti-plasmid of *Agrobacterium tumefaciens.* EMBO J. 5:1445–1454.

Stachel, S. E., Nester, E. W., and Zambryski, P. C., 1986. A plant cell factor induces *Agrobacterium tumefaciens vir* gene expression. Proc. Natl. Acad. Sci. U.S.A. 83:379–383.

Stachel, S. E., and Zambryski, P. C. 1986a. *vir*A and *vir*G control the plant-induced activation of the T-DNA transfer process in *Agrobacterium tumefaciens.* Cell 46:325–333.

Stachel, S. E., and Zambryski, P. C. 1986b. *Agrobacterium* and the susceptible plant cell: A novel adaptation of extracellular recognition and DNA conjugation. Cell 47:155–157.

SCOTT E. STACHEL: As a kid I enjoyed examining insects and plants and animals, and learning their book names and pondering all their differences. Like many, I had that one wonderful high school science teacher, who always taught up to us. I first experienced lab research while an undergraduate student at UC Santa Barbara. I worked with Ed Tripplett, a keen and boisterous developmental biologist who was intent on solving the mysteries of Speeman's Organizer and frog gastrulation twenty years too early. I surfed, while my protein columns dripped and my embryo explants grew. I then attended graduate school at UC San Francisco, in the Department of Biochemistry. UCSF was such a creative environment, with an amazing faculty, including three Nobel laureates-in-waiting, and all the energy that was generated from the initial application of molecular biology tools to long-standing biology questions. Although I received my Ph.D. from UCSF, I improbably pursued my thesis research and fascination with *Agrobacterium tumefaciens* through three highly productive labs directed by Howard Goodman, Gene Nester, and Marc Van Montagu. Since I was allowed, and took, far too much independence, I lacked the usual thesis adviser, and found my mentors in the people I worked with—especially Gynheung An and Pat Zambryski; and the many scientists whose methods and insights inspired me—from Hans Speeman to Barbara McClintock to Jeff Schell. My goal was to be a scientist and teacher of developmental biology, but I never obtained a permanent position. I still maintain my joy for science, and treasure thinking about biology and following new discoveries.

Reprinted from Nature, Vol. 318, No. 6047, pp. 624-628, 19 December 1985

Identification of the signal molecules produced by wounded plant cells that activate T-DNA transfer in *Agrobacterium tumefaciens*

Scott E. Stachel, Eric Messens*, Marc Van Montagu & Patricia Zambryski

Laboratorium voor Genetica and * Histologisch Instituut, Rijksuniversiteit Gent, B-9000 Gent, Belgium

We show here that Agrobacterium tumefaciens *virulence* (Vir) *gene expression is activated specifically by the plant molecules acetosyringone (AS) and α-hydroxyacetosyringone (OH-AS). These molecules induce the entire* vir *regulon in* Agrobacterium *as well as the formation of T-DNA intermediate molecules. AS and OH-AS occur specifically in exudates of wounded and metabolically active plant cells and probably allow* Agrobacterium *to recognize susceptible cells in nature.*

SOIL bacteria often form specialized interactions with plant cells. Usually a particular bacterium interacts with only a few species and/or types of plant cells[1,2]. A primary step in the formation of a bacterial/plant interaction is the detection by the bacterium of the appropriate susceptible plant cell. This recognition then triggers the activation of the bacterial genes whose products direct the development and/or maintenance of the interaction. The soil is a complex biological and chemical environment, and the signals that mediate the detection of a specific plant cell by a bacterium are unknown.

A. tumefaciens, a soil phytopathogen, genetically transforms dicotyledonous plant cells to cause the neoplastic disease crown gall[3,4]. Only cells that have been wounded are seen to be susceptible[5,6]. The bacterium transfers a specific segment of DNA, the T-DNA, from its large (>200 kilobases, kb) tumour-inducing (Ti) plasmid to the susceptible plant cell, where it becomes integrated into the nuclear genome[7,8]. In the Ti plasmid, the T-DNA is defined and bounded by identical 25-base pair (bp) direct repeats; only DNA between these T-DNA borders is transferred to the plant genome[9-12]. During co-cultivation with plant cells, independent T-DNA circles are formed in *Agrobacterium*; these molecules arise by a specific recombination between the 25-bp sequences at the ends of the T-DNA and are potential intermediates in the transfer of the T-DNA from *Agrobacterium* to the plant cell[13].

The Ti plasmid genes required for plant transformation are not contained in the T-DNA but are located in the ~40-kb *vir* region[14-16]. Genetic analysis of the *vir* region of the A6 Ti plasmid has shown that it encodes at least six separate complementation groups, *virA, B, C, D, E* and *G*, and *pinF* that are organized as a single regulon (ref. 17; S.E.S., in preparation). In the vegetative bacterium only *virA* and *virG*, the *vir* regulatory genes, are significantly expressed; however, when *Agrobacterium* is co-cultivated with plant cells the expression of *virB, C, D, E, G* and *pinF* is induced to highlevels[17,18]. This activation of *vir* expression by plant cells probably initiates the steps of T-DNA transfer and integration into the plant cell genome. We have shown that *vir* induction and the production of T-DNA circles is mediated by a small deffusible factor produced by actively growing plant cells[17]. Here, we purify and establish the chemical identity of this factor, and demonstrate that its production is related to plant cell wounding and that its recognition by the bacterium is a highly specific process. This factor is likely to be the signal that allows *Agrobacterium* to recognize in nature a plant cell susceptible to transformation.

Purification of signal molecules

To purify the plant molecule(s) that specifically signals *Agrobacterium* to initiate its interaction with plant cells requires first, an efficient and quantitative bioassay for a primary event in this transformation process, specifically the induction of *vir* gene expression; and second, a starting source of the signal molecule(s). We have previously described an assay for *vir* induction in *Agrobacterium* that used gene fusions between the pTiA6 *vir* loci and the *Escherichia coli lacZ* gene[17,18]. Briefly, in a bacterium carrying a *vir*::*lac* gene fusion, the production of β-galactosidase (the *lacZ* gene product) is controlled by the *vir* locus to which *lacZ* has been fused. Thus, the state of expression of the locus can be monitored by measuring the β-galactosidase activity present in the bacterium; increased activity indicates increased *vir* expression. (The relative amount of induced activity reflects the relative amount of *vir*-inducing activity to which the cell has been exposed.) Here we use *Agrobacterium* strain A348(pSM30) (ref. 17) to detect and measure *vir*-inducing activity. This strain contains wild-type pTiA6 and pSM30, a *virB*::*lac* fusion plasmid, and gives high levels (up to ~100-fold of basal activity) of induced β-galactosidase activity[17].

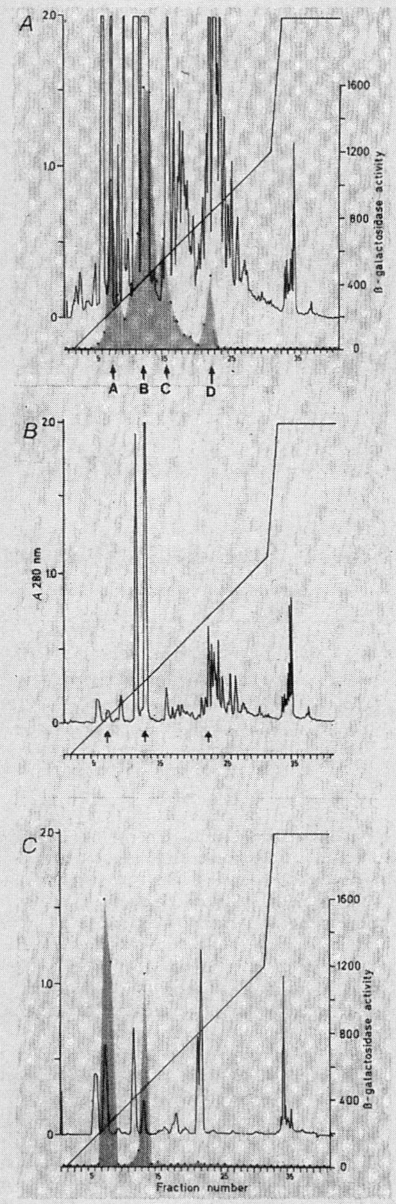

Fig. 1 RPC/FPLC fractionation of the *vir*-inducing activity in plant-cell exudates. The organic compounds present in plant cell medium conditioned with root culture (*cmr*; *A*, *B*), or with leaf disks (*cml*; *C*), were prepared by chloroform extraction and analysed by RPC/FPLC. The material analysed in *A*, *B* and *C* was prepared from 2 litres *cmr*, 40 ml *cmr* and 50 ml *cml*, respectively. Each sample was fractionated on a C-2/C-18 RPC column eluted with a linear methanol gradient. In *A* and *C*, column fractions were bioassayed for *vir*-inducing activity in *Agrobacterium*. The diagonal line represents the elution gradient and the solid line indicates the ultraviolet absorbance measured at 280 nm of the column fractions; shaded curves indicate relative specific units of β-galactosidase activity induced in the *virB*::*lacZ* tester strain A348(pM30) by the column fractions. The major peaks of *vir*-inducing activity in *A* are indicated by arrows labelled A, B, C and D. *B*, Arrows indicate peaks corresponding to peaks A, B and D of *A*.

Methods. Root culture of *N. tabacum* transformed with *Agrobacterium rhizogenes* A15834 was grown and maintained as described previously[17]. Transformed roots were used because they are easy to propagate. Every 72 h the conditioned medium (*cmr*) was removed and stored at −20 °C. Leaf disks were prepared from 6-week-old untransformed *N. tabacum* SR1 plants, and 2-g samples of 1.5-cm-diameter disks were incubated in 50 ml MS medium (Murashige and Skoog[28] salts, 3% sucrose, supplemented with 0.018% K_2HPO_4, 0.01% inositol, 0.0001% biotin, *p*H 5.5) in 150-mm Petri dishes. After 72 h, the conditioned medium (*cml*) was removed and stored at −20 °C. The conditioned medium (*cmr* and *cml*) was filtered through 0.22 μm nitrocellulose and the filtrate extracted twice with a 25% volume of chloroform and the pooled chloroform phase was back-extracted with 1 vol. MS medium, rotary-evaporated to dryness and resuspended in 500 μl 20% CH_3OH, 0.1% CH_3COOH for analysis by RPC/FPLC. (Note that initial solvent extraction experiments were performed to determine the solvent partitioning character of the *cmr* and *cml vir*-inducing activities. In these experiments 50 ml of the respective conditioned medium was extracted with chloroform; the interphase and chloroform phase were lyophilized and each resuspended in 2 ml MSSP (MS medium supplemented with 12.5 mm sodium phosphate, pH 5.5[17]), and the aqueous phase was blown with a stream of air to remove traces of chloroform. Each of these samples was bioassayed for *vir*-inducing activity; only the chloroform phase material contained this activity.) A Pep RPC pre-packed 5 mm × 50 mm column (HR5/5) containing 6-μm silica particles with C-2 and C-18 alkyl side chains (Pharmacia) was pre-equilibrated with 10% CH_3OH, 0.1% CH_3COOH. The column was run at a flow rate of 0.7 ml min^{-1} using a Pharmacia FPLC system equipped with the LCC500 chromatographic programmer. A single-path ultraviolet monitor was used to monitor absorbance at 280 nm. The column was eluted with a linear gradient of 10–60% $CH_3OH:H_2O$ (v/v), 0.1% CH_3COOH. We collected 40 1.4-ml (2-min) fractions; 140-μl aliquots (*A*) or the entire fractions (*C*) were lyophilized, resuspended in 1.5 ml MSSP and bioassayed for *vir*-inducing activity. Overnight cultures of strain A348(pSM30) were centrifuged and resuspended in MSSP medium. Material to be tested for *vir*-inducing activity was inoculated with bacteria at 0.1 absorbance unit ml^{-1} at 600 nm cm^{-1}. Incubations were for 10 h at 28 °C and 200 r.p.m. Specific units of β-galactosidase activity were determined as described previously[17,18] and are expressed as U per bacterial cell.

Induction of *vir* expression occurs during co-cultivation of *Agrobacterium* with plant cells, and during incubation of bacteria in plant-cell exudates. We have shown that the medium in which *Nicotiana tabacum* root culture has grown (designated *cmr* for conditioned medium roots) contains substantial amounts of a *vir*-inducing activity[17]. This activity stimulates the expression of each of the inducble pTiA6 *vir* loci, and also the formation of T-DNA circular intermediates, indicating that it triggers in *Agrobacterium* the initiation of plant cell transformation. This activity has relative molecluar mass less than 1,000; is stable to boiling, freezing, lyophilization, and high and low *p*H; and is partially hydrophobic, as it is retained by silica C-18 and completely elutes from this matrix with 40% CH_3OH (ref. 17). Here we purify and identify this *cmr* activity.

The above properties suggest that the *cmr vir*-inducing activity is composed of one or more small organic molecules; that it completely partitions into the organic solvent chloroform (Fig. 1) confirms this identity, and provided a basis for its purification. The *cmr vir*-inducing activity was fractionated by reverse-phase chromatography (RPC) using a high-resolution fast-performance liquid chromatography (FPLC) system. Activity was localized to two major and two minor peaks that had eluted, respectively, with 18 (peak A), 27.5 (peak B), 32 (peak C) and 42% (peak D) CH_3OH (Fig. 1*A*). Thus, the *cmr vir*-inducing activity is represented by at least four distinct compounds. When less material is loaded onto the FPLC column (Fig. 1*B*), peak B is the major component of the organic fraction of *cmr*. Because of their low abundance, the peak C and peak D activities were not analysed further. The signal molecules in peaks A and B were purified to homogeneity by further RPC fractionation and analysed by gas chromatography/mass spectroscopy (GC-MS) (Fig. 2) and ultraviolet absorption spectroscopy (Fig. 3).

Identification of *vir* inducers

The peak B compound, the major *cmr vir*-inducing molecule, is 4-acetyl-2,6-dimethoxyphenol, based on several experimental observations: (1) The GC/MS spectrum of the peak B compound

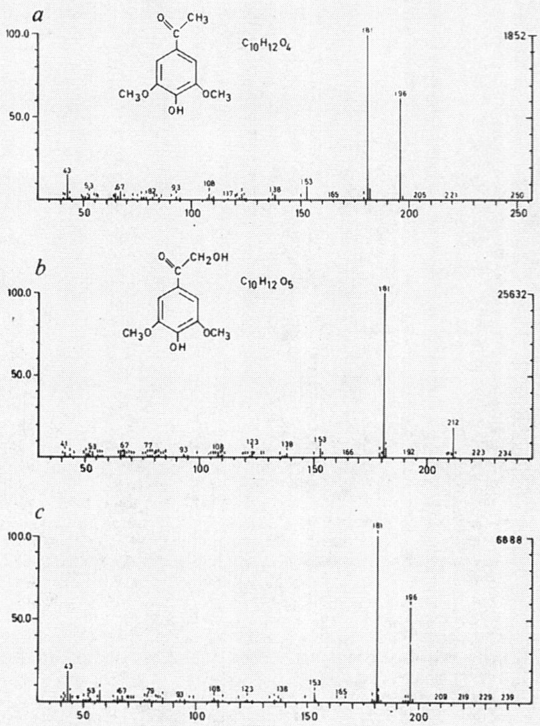

Fig. 2 Mass spectra of *vir*-inducing molecules. The mass spectra of purified peak B compound (Fig. 1), purified peak A compound (Fig. 1) and authentic acetosyringone (Janssen Chimica), are shown in *a*, *b* and *c*, respectively. Vertical axes, relative intensity of the fragment ions; horizontal axes, the mass/charge ratio of the fragment ions. The determined chemical formulae and structures of peak B and peak A (from Fig. 1) are indicated in *a* and *b*, respectively.

Methods. The biologically active compounds present in the peak A and B fractions were purified to homogeneity by RPC/FPLC; fractionation conditions were identical to those described in Fig. 1, except for the column solvent. The fractions containing peak A or peak B (from Fig. 1*A*) were pooled, lyophilized, resuspended in 5% acetonitrile, 0.1% trifluoroacetic acid (TFA) and injected onto the C-2/C-18 column equilibrated with this buffer. The column was eluted with a linear gradient of 5–30% acetonitrile, 0.1% TFA, and monitored for ultraviolet absorbance at 280 nm and for biological activity. For each sample a single peak was resolved from other minor ultraviolet-absorbing compounds, and a portion of the peak fraction was lyophilized, dissolved in 500 µl chloroform and evaporated to 40 µl under a nitrogen stream for GC/MS analysis. Separate portions of each sample were dissolved in CH_3OH for ultraviolet absorption analysis (Fig. 3) and in MSSP for analysis of biological activity. We injected 0.5 µl of sample into a 0.32 mm × 25 m Carlo Erba (HRGC) gas chromatography column SF-52, directly coupled to a Finnigan 4000 mass spectrometer. GC fractionation was with a 50–250° gradient, 5° min^{-1}. Data were collected and processed on a Nova 3 computer (Data General).

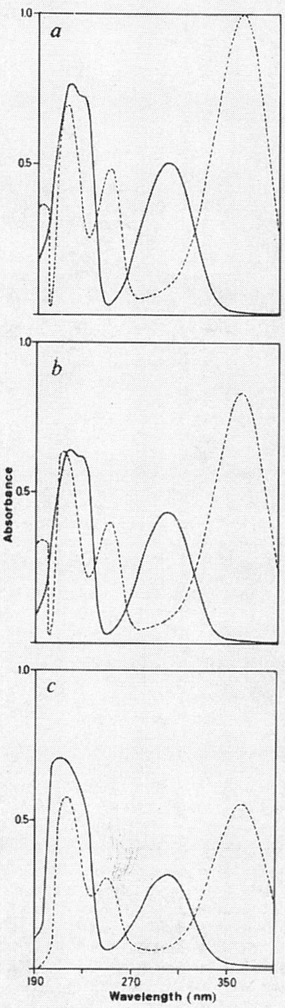

Fig. 3 Ultraviolet absorption spectroscopy of *vir*-inducing compounds purified from *cmr*. Spectra were determined in both CH_3OH (solid line) and $CH_3OH/NaOH$ (dashed line). *a*, Spectra of peak B compound; *b*, spectra of commercially obtained acetosyringone; *c*, spectra of peak A compound.

Methods. Absorption spectra were measured in a Beckman DU-6 spectrophotometer scanning at 60 nm min^{-1}. For each compound dried material was resuspended in CH_3OH (HPLC grade, Rathburn Chemicals) to an approximate concentration of 50 µM, and scanned against CH_3OH as a blank solution. The *p*H of the sample was subsequently adjusted to basic *p*H by the addition of 1 M NaOH to 20 mM and the sample scanned against CH_3OH/20 mM NaOH. (Note that when the base-adjusted sample of each compound is readjusted to acid *p*H by adding 1 M HCl to 40 mM, the determined spectrum is equivalent to the original CH_3OH spectrum of the compound (data not shown). Thus, the base-induced redshift of each compound is fully reversible in the conditions used.)

(Fig. 2*a*) indicates a molecule of relative molecular mass (M_r) 196, chemical composition $C_{10}H_{12}O_4$ and general structure acetyl-dimethoxyphenol. (2) Comparison of the ultraviolet absorption spectrum of the peak B compound in CH_3OH with its spectrum in $CH_3OH/NaOH$ (Fig. 3*a*) indicates that the molecule exhibits a strong absorption shift to longer wavelengths in the presence of NaOH; this base-induced redshift is diagnostic of phenolic compounds in which the phenolic hydroxyl group is *para* (but not *ortho* or *meta*) to a conjugated ring substituent, such as a ketone or allyl group[19]. (3) The *p*-hydroxyl derivative, 4-acetyl-2,6-dimethoxyphenol, commonly termed acetosyringone (AS), is commercially available and greatly stimulates *vir* expression in *Agrobacterium* (see below). AS and the purified peak *B* compound co-elute on RPC/FPLC (data not shown) and their GC/MS spectra are indistinguishable (Fig. 2). Furthermore, their respective ultraviolet spectra, both in CH_3OH and in $CH_3OH/NaOH$, are identical (Fig. 3*a*, *b*). (4) The quantitative *vir*-inducing activities of authentic AS and the peak *B* compound are equivalent. In the experiment described in Fig. 4, the units of induced β-galactosidase activity as a function of concentration of inducing compound were determined for each compound at several concentrations ; the respective activity/concentration curves are identical.

The peak *A* compound, the second major *cmr vir*-inducer, is 4-(2-hydroxyacetyl)-2,6-dimethoxyphenol, termed α-hydroxy acetosyringone (OH-AS), an analogue of AS. This identification is based on the following observations: (1) The GC/MS spectrum of the peak A compound (Fig. 2*c*) indicates a molecule of M_r 212 whose only significant fragmentation product has M_r 181. The GC/MS spectrum of compound A following trimethylsilyation indicates a mass increase corresponding to two trimethylsilyl groups (data not shown). Thus, compound A contains two hydroxyl groups and has general structure of either (3-propanol)dimethoxyphenol ($C_{11}H_{14}O_4$) or (2-hydroxyacetyl)dimethoxyphenol ($C_{10}H_{12}O_5$). Only the latter structure is consistent with the observation that the peak A compound is more polar than AS, as it elutes before AS on RPC/FPLC

Table 1 Biological activity of derivatives of acetosyringone

a acetosyringone	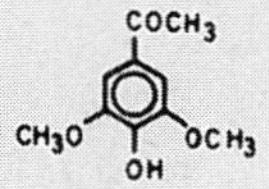	110	109	93	29
b acetovanillone	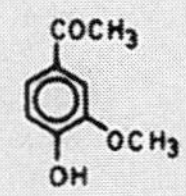	80	33	1.9	1.6
c 3,5-dimethoxy acetophenone	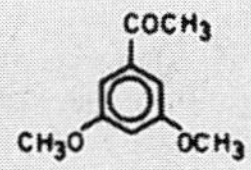	ND	1.0	ND	ND
d 4-hydroxy acetophenone		ND	1.1	ND	ND
e syringaldehyde	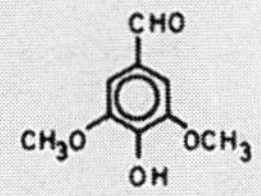	86	51	2.2	ND
f syringic acid	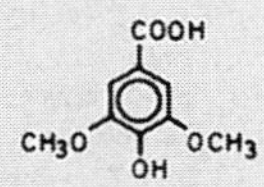	21	8.6	1.7	ND
g sinapinic acid	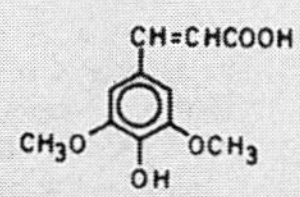	104	98	68	18

Acetysyringone (*a*) and six related compounds; acetovanillone (*b*); 3,5-dimethoxy acetophenone (*c*); 4-hydroxyacetophenone (*d*); syringaldehyde (*e*); syringic acid (*f*); and sinapinic acid (*g*), were tested for their ability to induce β-galactosidase in the *Agrobacterium virB*:: *lacZ* strain A348(pSM30). All compounds were purchased from Janssen Chemica, and prepared as 0.1 M solutions in dimethyl sulphoxide. Each compound was serially diluted into MSSP (Fig. 1, legend) medium to 200, 50, 5 and 0.5 μM (*a, b, e, f, g*) or 50 μM (*c, d*). These solutions were inoculated with bacteria at 0.05 absorbance unit ml^{-1} at 600 nm cm^{-1} and incubated at 28 °C, 200 r.p.m. After 12 h, the bacterial β-galactosidase activity in each sample was determined. The data are expressed as activity of the bacteria incubated in the presence of a compound relative to the basal activity present in bacteria incubated without added compound. The basal activity for the pSM30 strain is 10 U. ND, not determined.

(Fig. 1). (2) Comparison of the ultraviolet adsorption spectrum of the peak A compound in CH_3OH with its spectrum in CH_3OH/NaOH indicates that, similarly to AS, this molecule exhibits a strong base-induced redshift (Fig. 3*c*). Thus, compound A contains a phenolic hydroxyl group *para* to a conjugated ring substituent. Furthermore, the absorption spectra of compound A are almost identical to the equivalent spectra of AS; the respective absorption maxima occur at identical wavelengths (the largest maxima occur at 298 nm in CH_3OH and 355 nm in CH_3OH/NaOH), although their relative extinction coefficients are different. These similarities strongly indicate that the peak *A* compound is very closely related to AS. (3) The quantitative *vir*-inducing activity of the peak A compound is approximately equivalent to that of AS (Fig. 1). As described below, the structure/activity specificity of *vir*-induction indicates that only molecules with similar structure to AS induce *vir* expression to high levels (Table 1).

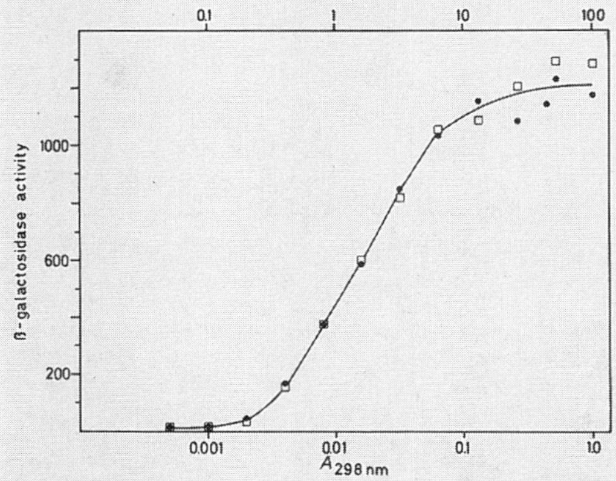

Fig. 4 Comparison of *vir*-inducing activity of purified peak *B* compound (Fig. 2) and commercially obtained acetosyringone. Vertical axis, β-galactosidase activity (U $bacterium^{-1}$) in the *Agrobacterium virB*:: *lacZ* strain A348(pSM30); lower horizontal axis, relative concentration of purified peak B compound (Fig. 2) or authentic AS (Janssen Chimica) in the inducing medium, measured as absorbance at 298 nm (cm^{-1}). The extinction coefficient of AS at 298 nm is 10,300 (ref. 29); this value was used to calculate the concentration of inducer compound in the inducing medium, indicated by the upper horizontal axis.
Methods. Purified peak B compound and authentic AS were separately dissolved in MSSP to 1.0 absorbance unit ml^{-1} at 298 nm (cm^{-1}), measured against fresh MSSP. Each solution was serially diluted and the resultant samples were inoculated with bacterium at 0.09 absorbance unit ml^{-1} at 600 nm (cm^{-1}). Samples were incubated for 14 h at 28 °C, 200 r.p.m. and the β-galactosidase activity of the bacteria in each sample determined.

Thus, we have identified plant molecules that induce *vir* expression. Our next experiments seek to provide insight into the relationship between these molecules and *Agrobacterium* in nature. For the *Agrobacterium*/plant transformation system to be most efficient, its activation should be limited to the presence of susceptible plant cells. This could be achieved if activation is signalled only by molecules specific to these cells. For instance, such moleclues should first, be produced by different types of plant cells, as *Agobacterium* can transform several different cell types[20]; second, be specifically synthesized by susceptible plant cells, such as wounded cells; and third, be available to the bacterium in quantities sufficient for efficient activation. The following experiments demonstrate that AS has these properties and that activation of *Agrobacterium* is a highly specific process.

Different plant tissues produce AS/OH-AS

To determine whether AS and OH-AS are specific to roots or whether they are also produced by other plant tissues, we purified and identified the *vir*-inducing activity produced by leaf cells (Fig. 1). Medium in which *N. tabacum* leaves that have been cut into disks have been cultured (*cml*; conditioned medium leaves) contains substantial amounts of inducing activity that completely partitions into chloroform. The RPC/FPLC profile and corresponding bioassay profile of this material (Fig. 1*C*) show that the *cml* activity fractionates into two peaks that co-elute with OH-AS and AS. Subsequent purification and GC/MS and ultraviolet spectrophotometric analysis of the active molecules in these peaks confirmed that the major *vir*-inducing activity in *cml* is composed of OH-AS and AS (data not shown). Thus, these two compounds are present in the exudates of at least two different plant tissues.

AS and OH-AS are exudate-specific

In the soil, *Agrobacterium* probably detects plant cells through their exudates. To determine whether AS and OH-AS are exudate-specific we assessed their relative concentrations within the *N. tabacum* leaf disks of Fig. 1*C*. These disks were extracted with chloroform and the extract was analysed by RPC/FPLC fractionation and corresponding bioassay. The relative concentrations of AS and OH-AS are <0.5% of the organic compounds present in the total leaf disks (data not shown). In comparison, the relative concentrations of these compounds in the *cml* extract is ~25% and greater (Figs 1*C*, 5*A*). Thus, AS and OH-AS probably do not leak out of damaged plant cells and are exudate-specific compounds.

Concentration of *vir* inducers in *cmr*/*cml*

Figure 4 shows the relationship between the concentration of AS (both commercial and *cmr*-purified) and induction of *virB* expression as measured in units of β-galactosidase activity. Under the conditions used, AS at $\geq$10 μM, and AS at 1.5 μM, induce maximal (1,200 U) and half-maximal (600 U) expression, respectively. Note that AS induction does not significantly affect cell viability (that is, induction is a non-lethal event), and that concentrations of AS >200 μM are not significantly toxic to the bacteria. Using the dose–response curve of Fig. 4, the concentration of *vir*-inducing compounds in plant-cell exudates relative to AS can be estimated. Our starting *cmr* and *cml* preparations typically induce between 250 and 500 U β-galactosidase activity in A348 (pSM30); this activity approximately corresponds to between 0.5 and 1 μM AS.

Biological activity of acetyosyringone

The primary molecular signal for the initiation in *Agobacterium* of the events of plant cell transformation should induce in the bacterium both the entire *vir* regulon and the initial steps of T-DNA transfer. We determined that AS has these activities. Several *vir*::*lac* and *pin*::*lac Agrobacterium* strains were incubated in 20 μM AS (both commerical and *cmr*-purified) and assayed for β-galactosidase activity. Inductions of all the previously identified inducible *vir* loci (*B, C, D, E, G*) and *pinF* was obtained, and the levels of induction were at least 20–50% higher than those stimulated during co-cultivation with plant cell cultures or incubation with *cmr*[17] (data not shown). AS also induces the production of T-DNA intermediates: when bacteria are incubated with 20 μM AS for 12–18 h, T-DNA intermediates[13] are found at a frequency two- to fivefold greater than that obtained following 48-h co-cultivation with prrotoplasts[17] (data not shown). We have also seen that AS induces the appearance and/or disappearance of several major proteins in *Agrobacterium* (P. Engström, P.Z. and S.E.S., in preparation). Thus, a single compound is sufficient to trigger the complete activation of *vir* and the initial events of T-DNA transfer; and AS (probably OH-AS also) is a primary signal for plant cell transformation by *Agrobacterium*.

Molecular specificity of *vir* induction

Table 1 shows the respective *vir*-inducing activities of AS (*a*) and analogues of AS (*b*–*g*). These results indicate that *Agrobacterium vir* expression is efficiently induced by molecules that conform to a structure best represented by AS itself, and suggest that *Agrobacterium* has evolved to recognize and respond to AS as a specific signal for the activation of plant cell transformation. For instance, AS without one methoxy group (*b*) has greatly attenuated *vir*-inducing activity, whereas AS without both methoxy groups (*d*) or its hydroxyl group (*c*) is inactive. The acetyl substituent of AS is important for activity. The formyl (*e*) and carboxylic acid (*f*) analogues of AS have attenuated activity, whereas the cinnamic acid analogue of AS, sinapinic acid (*g*), has approximately equivalent activity to AS. This latter result is interesting in that sinapinic acid is a precusor of lignin, an integral cell-wall constituent of all vascular plants (see below).

Wounding stimulates AS/OH-AS production

Although many dicotyledonous plant cells can be transformed by *Agrobacterium*, only cells that have been wounded or traumatized are susceptible. Therefore, we tested whether the production of AS and OH-AS is stimulated by plant cell wounding. Figure 5 shows the RPC/FPLC profiles of the organic extracts of *cml* produced by equivalent amounts of *N. tabacum* leaf disks (wounded cells; Fig. 5*a*) or intact leaves (unwounded; Fig. 5*b*). Comparison of these profiles indicates that the 'wounded' *cml* contains >10-fold more AS and OH-AS than the 'unwounded' *cml*, demonstrating that wounding stimulates the appearance of these molecules in the cell exudate. The low levels of AS and OH-AS in the 'unwounded' leaf exudate could be caused by the cut stem surfaces of the leaves.

These results do not define which cells of the wounded tissue produce AS and OH-AS. For example, damaged or dead cells could release AS, although such cells are not good targets for

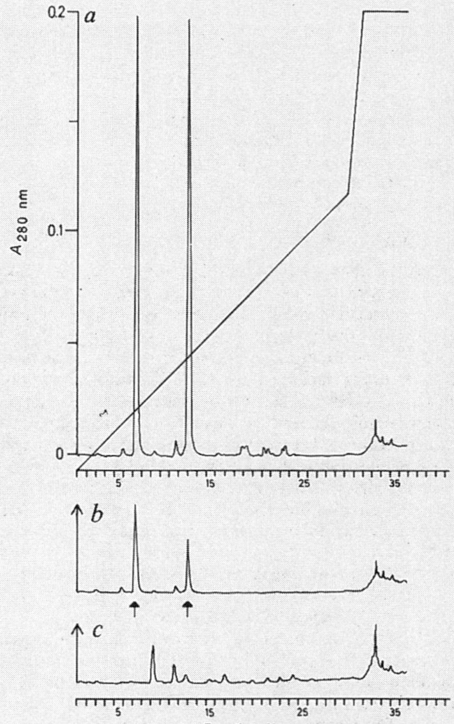

Fig. 5 Effect of wounding and inhibition of plant cell metabolism on production of AS and OH-AS by plant cells. RPC/FPLC fractionations of equivalent amounts of medium conditioned with: *a*, 1.5-cm-diameter leaf disks; *b*, intact leaves; *c*, leaf disks in the presence of cycloheximide. The vertical axes are drawn to the same scale in *a*–*c*. Arrows, elution positions of OH-AS (fraction 7), and AS (fraction 13).

Methods. Leaves (4 g) excised from 6-week-old *N. tabacum* SR1 plants were treated as described below and incubated in 50 ml MS medium in a 150-ml Petri dish. Care was taken to use visually equivalent leaves in each experiment. *a*, 5-6 1-cm^2 disks were obtained per leaf; *b*, intact leaves were used; *c*, same as *a* but with 5 p.p.m. (12.8 μM) cycloheximide (Actidion, 99%; Aldrich Chemicals) added to the medium. After a 72-h incubation, 40 ml of each respective conditioned medium was recovered, filtered through 0.22 μm nitrocellulose and 35 ml extracted with chloroform (Fig. 1), and the organic pellet was resuspended in 400 μl 10% CH_3OH, 0.1% CH_3COOH. We analysed 100 μl of the sample by RPC/FPLC using the conditions described in Fig. 1. The remaining 5 ml of conditioned medium was bioassayed for *vir*-inducing activity; the units of β-galactosidase activity induced by the conditioned media of *a*, *b* and *c* were 565, 135 and 20, respectively.

Agrobacterium infection. We tested whether active plant cell metabolism is required for the production of AS by analysing the *cml* produced by leaf disks incubated in cycloheximide. This material does not contain AS and OH-AS (Fig. 5*c*). These results confirm and extend our previous observation that in plant/*Agrobacterium* co-cultivations only actively growing plant cell cultures are able to stimulate efficient *vir* gene expression[17]. Our data concerning the production of AS and OH-AS in total suggest that these compounds are specifically synthesized and exuded by metabolically active wounded cells.

Discussion

The molecules that we have purified signal *Agrobacterium* to activate the expression of its virulence genes, setting in motion a series of molecular events ultimately resulting in the transfer of T-DNA from the bacterium to the plant cell.

Wounded cells are known to be susceptible to *Agrobacterium* infection, perhaps because the intact cell walls of undamaged cells restrict T-DNA transfer, or because T-DNA integration is dependent on host-cell replication and wounding stimulates this replication. Because *Agrobacterium* responds to AS and OH-AS to initiate plant cell transformation, these molecules potentially represent the signal that *Agrobacterium* detects and recognizes in the soil as susceptible plant cells. The presence of AS and OH-AS specifically in the exudates of wounded but actively metabolizing plant tissues supports our hypothesis.

The activation of *Agrobacterium vir* expression by plant signal molecules probably involves at least two steps: extracellular recognition and intracellular response. The first step could depend on the signal molecule acting as a chemical attractant and/or nutritive source for the bacterium. The latter must depend on the ability of the bacterium to convey the information of the signal from outside to inside the cell and to activate *vir* expression. The mechanism of these events is unknown. Induction of the pTiA6 *vir* region is attentuated in *virA* and does not occur in *virG* mutant bacteria (S.E.S., in preparation). Also, the amino-acid sequence of the *virG* gene product is highly homologous with several positive regulatory proteins of *E. coli*[21]. Because AS can cause the induction of the complete *vir* regulon, we suggest that this compound acts to activate allosterically the *virG* protein, which then activates *vir* transcription by directly interacting with *vir* gene promoter sequences. The *virA* protein potentially functions in the initial extracellular/intracellular recognition and/or intracellular transport of the signal molecule.

AS and OH-AS have not previously been identified as natural component of plant cells, suggesting that the appearance of these molecules in nature is not widespread. Thus, these molecules potentially represent to *Agrobacterium* only those cells which are its desired targets. In addition, these or related molecules might also serve to initiate other bacterial/plant interactions in the soil. The observed resistance of most monocotyledonous plants to *Agrobacterium* could result because these plants do not produce, or only produce in low quantities, *vir* signal molecules. AS and OH-AS could be useful for obtaining *Agrobacterium* transformation of plant species previously seen to be resistant to *Agrobacterium*, and also for the analysis of the initial molecular steps of the T-DNA transfer process.

Although it is important for *Agrobacterium* to detect and respond to such molecules, it is equally important that it does not respond to closely related but functionally different molecules; we have shown that *vir* induction is most efficiently stimulated by AS and by the lignin precursor sinapinic acid. This latter compound is not present in detectable quantities in our exudate preparations; however, sinapinic acid could be present in the soil in exudates of wounded cells in the process of cell-wall rebuilding.

It is interesting to speculate on the function of AS and OH-AS in plants. These compounds are likely products of the shikimic acid biosynthetic pathway that provides the plant cell with the precursors to a broad spectrum of molecules, including the flavonoids, and lignins[22–24]. These classes of compounds are important to a plant subjected to stress or injury. For example, many flavonoid-derived phytoalexins are potent inhibitors of the growth of invading pathogens[25], whereas lignin, a major component of the cell wall, provides a physical barrier to invasion[26]. Thus, AS and OH-AS could be part of the wound response[27] of plant cells. These compounds are potentially toxic to other soil pathogens, and *Agrobacterium* has evolved to be resistant to AS and OH-AS and to use these chemicals to recognize wounded cells. Alternatively, the compounds could be products related to lignin repair in damaged cells.

We propose that there is a continual excretion of wound-related phenolics during growth, caused by abrasion from the soil. *Agrobacterium* may be attracted to plants by recognition of these compounds. However, significant levels of *vir* induction and the events of T-DNA transfer will only occur if signal molecules are present in sufficient concentrations. As the highest concentrations are found at wound sites, *A. tumefaciens* effectively infects only these sites.

We thank Marc Schelfaut and Luc Van Royen (Laboratory of Organic Chemistry, Rijksuniversiteit, Gent) for help with the GC/MS analyses, and Max Tate (Waite Agricultural Research Institute, Adelaide) for suggestions concerning solvent partitioning. We also thank Klaus Hahlbrock (Max-Planck Institute, Köln) for discussion, Ernest T. Wombatt III, Martine De Cock for preparation of the manuscript and Albert Verstraete and Karel Spruyt for preparation of the figures. This work was supported by grants to M.V.M. and J. Schell from the ASLK-Kankerfonds, the Fonds voor Wetenschappelijk Geneeskundig Onderzoek (FGWO 3.0001.82), the Services of the Prime Minister (OOA 12.0561.84), and was carried out under research contract no. GBI-4-017-B (RS) of the Biomolecular Engineering Programme of the Commission of the European Communities. S.E.S. was funded in part by an EMBO short-term fellowship.

Received 4 September; accepted 11 October 1985.

1. Verma, D. P. S. & Hohn, Th. *Genes Involved in Microbe-Plant Interactions* (Springer, Wien, 1984).
2. Rhodes-Roberts, M. & Skinner, F. A. *Bacteria and Plants* (Academic, New York, 1982).
3. Kahl, G. & Schell, J. *Molecular Biology of Plant Tumours* (Academic, New York, 1982).
4. Nester, E. W., Gordon, M. P., Amasino, R. M. & Yanofsky, M. F. *A. Rev. Pl. Physiol.* **35,** 387–413 (1984).
5. Smith, E. F., Brown, N. A. & Townsend, C. O. *US Dept Agric. Bur. Pl. Ind. Bull.* **213,** 1–201 (1911).
6. Braun, A. C. *Biochim. biophys. Acta* **516,** 167–191 (1978).
7. Caplan, A. *et al. Science* **222,** 815–821 (1983).
8. Gheysen, G., Dhaese, P., Van Montagu, M. & Schell, J. in *Genetic Flux in Plants* (eds Hohn, B. & Dennis, E. S.) 11–17 (Springer, Wien, 1985).
9. Yadav, N. S., Vanderleyden, J., Bennett, D. R., Barnes, W. M. & Chilton, M.-D. *Proc. natn. Acad. Sci. U.S.A.* **79,** 6322–6326 (1982).
10. Simpson, R. B. *et al. Cell* **29,** 1005–1014 (1982).
11. Holsters, M. *et al. Molec. gen. Genet.* **190,** 35–41 (1983).
12. Wang, K., Herrera-Estrella, L., Van Montagu, M. & Zambryski, P. *Cell* **38,** 455–462 (1984).
13. Koukolíková-Nicola, Z. *et al. Nature* **313,** 191–196 (1985).
14. Garfinkel, D. J. & Nester, E. W. *J. Bact.* **144,** 732–743 (1980).
15. Ooms, G., Klapwijk, P. M., Poulis, J. A. & Schilperoort, R. A. *J. Bact.* **144,** 82–91 (1980).
16. Klee, H. J., White, F. F., Iyer, V. N., Gordon, M. P. & Nester, E. W. *J. Bact.* **153,** 878–883 (1983).
17. Stachel, S., Nester, E. W. & Zambryski, P. *Proc. natn. Acad. Sci. U.S.A.* (in the press).
18. Stachel, S., An, G., Flores, C. & Nester, E. W. *EMBO J.* **4,** 891–898 (1985).
19. Scott, A. I. in *Interpretation of the Ultraviolet Spectra of Natural Products* (eds Barton, D. H. R. & Doering, W.) 89–134 (Pergamon, London, 1964).
20. De Cleene, M. & De Ley, J. *Bot. Rev.* **42,** 389–466 (1976).
21. Ebert, P., Stachel, S. E., Winans, S. & Nester, E. W. *Science* (submitted).
22. Halhbrock, K. & Grisebach, H. *A. Rev. Pl. Physiol.* **30,** 105–130 (1979).
23. Halhbrock, K. in *The Biochemistry of Plants, a Comprehensive Treatise* Vol 7 (ed. Conn, E. E.) 425–456 (Academic, London, 1981).
24. Grisebach, H. in *The Biochemistry of Plants, a Comprehensive Treatise* Vol. 7 (ed. Conn, E. E.) 457–478 (Academic, London, 1981).
25. Darvill, A. G. & Albersheim, P. *A. Rev. Pl. Physiol.* **35,** 243–275 (1984).
26. Bell, A. A. *A. Rev. Pl. Physiol.* **32,** 21–81 (1981).
27. Kahl, G. *Biochemistry of Wounded Plant Tissues* (de Gruyter, Berlin, 1978).
28. Murashige, T. & Skoog, F. *Physiologia Pl.* **15,** 473–497 (1962).
29. Grasselli, J. G. & Ritchey, W. M. *CRC Atlas of Spectral Data and Physical Constants for Organic Compounds* 2nd edn (CRC, Boca Raton, 1975).

Printed in Great Britain by Macmillan Production Ltd., Basingstoke, Hampshire

Reprinted from Nature, Vol. 322, No. 6081, pp. 706-712, 21 August 1986

Generation of single-stranded T-DNA molecules during the initial stages of T-DNA transfer from *Agrobacterium tumefaciens* to plant cells

Scott E. Stachel*, Benedikt Timmerman & Patricia Zambryski*

Laboratorium voor Genetica, Rijksuniversiteit Gent, B-9000 Gent, Belgium

Activation of the T-DNA transfer process of Agrobacterium *by the plant signal molecule acetosyringone generates a single-stranded, unipolar, linear T-DNA molecule (T-strand)—a potential conjugative intermediate in the transfer of the T-DNA to plant cells. Acetosyringone induction also leads to S_1 nuclease-sensitive sites at the Ti plasmid T-DNA borders, and other molecular changes associated with the Ti plasmid T-DNA sequences, which may correspond to specific steps of T-strand synthesis.*

DURING the genetic transformation of plant cells by the soil pathogen *Agrobacterium tumefaciens* (reviewed in ref. 1), a specific segment of DNA, the T-DNA, is recognized in and mobilized from the large (>200 kilobase pairs; kbp) Ti plasmid of the bacterium, transferred across the cell walls of the bacterium and plant cell, and integrated as an unaltered fragment into the plant nuclear genome. Analysis of the transfer process has focused on what defines the T-DNA; the genetic requirements for transfer other than the T-DNA; and the mechanism of transfer.

In the Ti plasmid the T-DNA is bounded by essentially identical 25-base pair (bp) direct repeats[2-5]. These sequences define the T-DNA, for any DNA, and only DNA, located between T-DNA borders is efficiently transferred and integrated[6-9]. The T-DNA transfer process is directed by the products of the Ti plasmid virulence (*vir*) and chromosomal virulence (*chv*) loci (reviewed in refs 10, 11). Whereas *chv* expression is constitutive, *vir* expression is tightly regulated[12,13], and its activation initiates the transfer process. This activation is mediated by specific phenolic compounds present in the exudates of wounded and actively metabolizing plant cells. One such compound is acetosyringone (AS; 4-acetyl-2,6-dimethoxyphenol)[14].

Genetic analyses of the 25-bp sequences have indicated that they are polar in function[15,16]. While deletion of the left border repeat has no significant effect on pathogenicity[17], deletion of the right repeat totally abolishes it[15,16,18]. Furthermore, when the orientation of the right border is reversed with regard to its natural orientation on the Ti plasmid, the efficient transfer and/or integration of the T-DNA sequences is greatly attenuated. These results indicate that T-DNA transfer may occur in a rightward to leftward fashion, determined by the orientation of the 25-bp border repeats, and suggest that transfer might be via a conjugative mechanism[15].

Here we use hybridization techniques to directly identify and characterize novel structures associated with the T-DNA and its border sequences in *Agrobacterium* following induction of *vir* gene expression with AS. A variation of the Southern blotting transfer procedure is used to distinguish between single-stranded (ss) and double-stranded (ds) DNA molecules present in total DNA prepared from these cells. Evidence is presented for (1) free ss T-DNA molecules (T-strands) whose polarity corresponds to that of the T-DNA borders; (2) ss endonuclease-sensitive structures associated with the Ti plasmid T-DNA borders (S_1 border sites); and (3) molecular alterations associated with the internal sequences of the Ti plasmid T-region (T-region structures). We discuss the potential role of each of these AS-induced T-DNA homologous molecules in the transfer of the T-DNA to the plant cell.

* Present address: Institute of Cancer Research and Howard Hughes Medical Institute, College of Physicians and Surgeons, Colombia University, New York 10032, USA (S.E.S.); Division of Molecular Plant Biology, Hilgard Hall, University of California, Berkeley, California 94720, USA (P.Z.).

Free T-DNAs in AS-induced cells

For these studies we used *Agrobacterium* carrying Ti plasmid pGV3850 (ref. 6). pGV3850 has been derived from the nopaline C58 Ti plasmid, and the structure of its minimal T-DNA region is shown in Fig. 1*A*. Strain 3850 has been used previously to isolate and identify a 9.5 kbp ds T-DNA circle molecule (Fig. 1*B*) following transformation of *Escherichia coli* with total DNA prepared from plant[19] or AS-induced *Agrobacteria*[14]; this T-DNA circle has been proposed as a candidate for the T-DNA molecule that is transferred to the plant cell. Our initial experiments aimed to identify the presence of this molecule directly in *Agrobacterium*.

Total DNA was prepared from AS-induced cells and found to produce ds T-DNA circles in *E. coli*; this AS-induced DNA, along with total DNA prepared from uninduced cells, was fractionated by agarose gel electrophoresis and transferred to a nitrocellulose filter following gel denaturation (normal Southern transfer procedure, see below). Two T-DNA homologous signals are observed in Fig. 1*C*, lane 1. Since the DNA is undigested, the upper signal represents Ti-linked T-DNA sequences, while the lower signal represents free T-DNA sequences. This novel signal is specific to the AS-induced DNA, and thus corresponds to a Ti-independent T-DNA molecule whose synthesis is the result of the AS-induced activation of the pGV3850 *vir* loci.

The free T-DNA signal does not migrate as a 9.5-kbp ds DNA molecule (supercoiled, relaxed-circular or linear), but instead migrates at a size corresponding to a ds linear molecule of 4.4 kbp. Thus, by hybridization we find no evidence for ds T-DNA circles in AS-induced bacteria. This result is not totally unexpected because the frequency of recovery of these molecules in *E. coli* is low. We obtain on average 50 ds circle transformants per μg AS-induced DNA (at a transformation efficiency of 10^7 transformants per μg supercoiled pBR322 plasmid DNA). Since 1 μg of total *Agrobacterium* DNA contains ~1.7 ng of pGV3850 T-DNA[19], at most only one in every 100 AS-induced cells harbours a ds T-DNA circle. That these molecules are not detected by hybridization indicates that the transformation results are representative of their actual concentration in the AS-induced DNA. We discuss below a model for how ds circle molecules might be generated at a low frequency as a result of AS induction (Fig. 6).

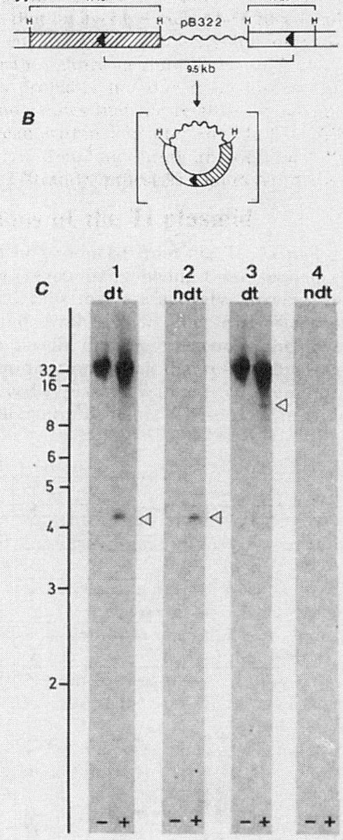

Fig. 1 Hybridization analysis of total DNA prepared from AS-induced *Agrobacterium*. *A*, T-DNA region of the pGV3850 Ti plasmid[6]. The interior portion of the T-DNA of pTiC58 has been replaced with PBR322 (wavy line). *Hin*dIII fragments 10 (hatched) and 23 (white) from pTiC58 carrying the right and left 25-bp T-DNA border sequences respectively (black arrows); H, *Hin*dIII restriction site. *B*, Structure of the 9.5-kbp ds T-DNA circle molecule isolated after transformation of *E. coli* with undigested total DNA from AS-induced pGV3850 *Agrobacteria*. *C*, Southern blot hybridization analyses of *Agrobacterium* DNA. Total DNA prepared from ininduced (−) and AS-induced (+) 3850 *agrobacteria* was analysed by hybridization against nick-translated ^{32}P-labelled pBR322 probe. Lane 1, denatured transfer (dt) of untreated DNA; lane 2, non-denatured transfer (ndt) of untreated DNA; lane 3, denatured transfer of total DNA digested with S_1 nuclease; lane 4, non-denatured transfer of S_1-treated DNA. Scaling on left is in kbp. Arrows indicate novel signals observed in the AS-induced lanes. Note that all of these novel signals are also observed, albeit at a lower level, with DNA prepared from *Agrobacterium* co-cultivated with *Nicotiana tabacum* protoplast cells (data not shown).

Methods. An overnight culture of 3850 *Agrobacteria* grown in YEB liquid medium was resuspended in MSSP medium[14] at 0.05 A_{600} units ml^{-1} and grown at 28 °C with high aeration. After 5 h of preincubation growth, acetosyringone (Janssen) was added to half of the MSSP culture at 100 μM, a concentration that is non-limiting for induction of the Ti-plasmid *vir* genes[14]. The uninduced and AS-induced cultures were grown for another 12–18 h and a parallel culture of the *vir*B :: *lac* strain A348(pSM30) (refs 10, 14) was used to monitor *vir* induction; under these conditions the cultures undergo two to three doublings, and >100-fold increases in β-galactosidase activity are observed in A348(pSM30). The bacteria were collected by centrifugation and total DNA prepared as described previously[33]. Briefly, the pellet from 5 ml of cells is lysed in 200 μl TE (50 mM Tris, 20 mM Na_2-EDTA, *p*H 8.0), 100 μl 5% sodium sarkosyl and 100 μl pronase (10 mg ml^{-1}) for 45 min at 37 °C. The lysate is then vortexed for 15 s (light-shear), extracted twice with phenol and twice with chloroform, and the DNA recovered by EtOH precipitation. *E. coli* was transformed with uninduced and AS-induced DNA as described previously[19], and the AS-induced preparation was determined to give transformants carrying the ds T-DNA circle molecule diagrammed in *B*. Aliquots (1 μg) of total *Agrobacterium* DNA (either untreated or S_1-digested) were then electrophoresed in 0.9% TBE agarose gels containing 0.5 μg ml^{-1} EtBr, transferred to nitrocellulose in 10× SSC, and analysed by Southern blot filter hybridization. Two different transfer condtions have been used, denatured and non-denatured. For denatured transfer, the agarose gel is soaked in denaturing solution for 60 min followed by neutralizing solution for 60 min before capillary blotting. For non-denatured transfer, the gel is soaked in H_2O for 10 min, then 10× SSC for 10 min, before blotting. For the S_1 nuclease digestions, 1 μg total DNA in 200 μl S_1 digestion buffer was incubated with 50 U S_1 nuclease (Boehringer Mannheim) for 30 min at 20 °C. The reaction is terminated by adding 20 μl 10× S_1 termination buffer, followed by phenol extraction and EtOH precipitation. Unless otherwise specified, all buffers and conditions used here and in Figs 2–5 are according to Maniatis *et al.*[34].

Free T-DNAs are single stranded

Since the size of the AS-induced free T-DNA molecule is about half that of the T-region of pGV3850 (Fig. 1*A*), it may be a single-strand copy of the pGV3850 T-DNA. The following experiments demonstrate that the free T-DNA molecule has properties characteristic of ss DNA. Evidence is also given for other novel T-DNA-homologous molecules present in the AS-induced bacteria; in contrast to the free T-DNA, these structures are still linked to the Ti plasmid, as they are only detected after enzymatic digestion.

Transfer assay for ss T-DNAs. DNA binds to nitrocellulose only if it is single-stranded[20] or associated with protein[21]. In the Southern blotting technique, ds DNA fragments fractionated in an agarose gel must be denatured before being transferred to nitrocellulose[20]. Thus, ss (and protein-associated) DNA molecules can be easily distinguished from ds molecules by comparing their transfer to nitrocellulose from agarose gels treated with or without NaOH before transfer. Figure 1*C*, lanes 1 and 2, show the denatured and non-denatured transfers of uninduced and AS-induced total DNA. Of the two T-DNA-homologous signals present in the AS-induced DNA sample, only the lower signal, which represents the free T-DNA molecule, transfers without gel denaturation (Fig. 1*C*, lane 2). This free T-DNA must be ss DNA because the AS-induced total DNA has been fully deproteinated during its preparation (Fig. 1 legend) and RNase digestion of the DNA before hybridization analysis does not affect the ss signal (data not shown).

Nuclease sensitivity of ss T-DNAs. S_1 nuclease is a ss-specific endonuclease[22]. In Fig. 1*C*, the denatured (lane 3) and non-denatured (lane 4) transfers of S_1 nuclease-digested samples of uninduced and AS-induced total DNA demonstrate that the free T-DNA molecule is fully degraded by S_1 nuclease. Interestingly, S_1 treatment also appears to release the T-region from the AS-induced Tl plasmid. A novel T-DNA homologous signal is observed in the S_1-treated AS-induced sample in the denatured transfer (Fig. 1*C*, lane 3), and this signal corresponds to a ds molecule, which migrates as a linear fragment of 9.5 kbp, the precise size of the pGV3850 T-DNA from its left to right T-DNA borders (Fig. 1*A*). Thus, AS induction also results in a Ti plasmid whose T-DNA borders are sensitive to cleavage by S_1 nuclease (see Figs 4 and 5).

The activity of type II restriction endonucleases is limited primarily to duplex DNA. Aliquots of AS-induced DNA were digested to completion with *Bam*HI, *Eco*RI, *Hin*dIII, *Pst*I and *Sal*I (each of these enzymes cleaves within the T-DNA region of pGV3850; Fig. 2*A*), and hybridized to the *Hin*dIII fragment 10/pBR probe following non-denatured transfer (Fig. 2*B*). Because the ss signal is present in all the digests (except for *Pst*I), the free T-DNA is generally resistant to restriction digestion.

The *Hin*dIII and *Pst*I results are more complex. The ss signal is weak in the *Hin*dIII lane and absent in the *Pst*I lane; also, several novel signals are observed in these lanes. Several restriction enzymes have been shown to restrict ss DNA[23], and the sensitivity of the free T-DNA to *Hin*dIII and *Pst*I may be an example of this effect. Interestingly, fragments both smaller and larger than the ss T-DNA molecule are produced by *Hin*dIII and *Pst*I digestion (Fig. 2*B*). While the smaller fragments can be cleavage products of the ss T-DNA, the larger fragments must be derived from a large molecule, presumably the AS-induced Ti plasmid. Thus, AS induction also results in a Ti plasmid molecule whose T-region carries ss sequences (see Fig. 5).

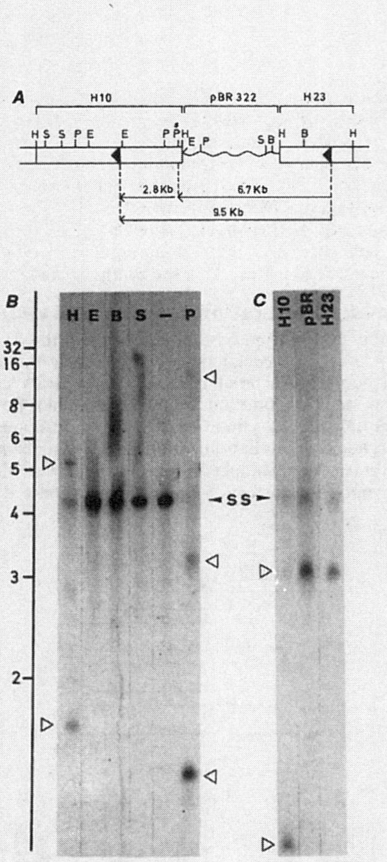

Fig. 2 Restriction endonuclease treatment of AS-induced DNA. *A*, Restriction map of the T-DNA region of pGV3850. Black arrows, T-DNA border sequences; the 9.5-kb arrow designates the ss T-DNA molecule, and the 2.8-kb and 6.7-kb arrows correspond to the ss fragments that would be produced by *Pst*I cleavage of the ss T-DNA molecule at the *Pst*I* site. *B*, Aliquots (1 μg) of AS-induced DNA were digested to completion with five restriction endonucleases, fractionated on 0.9% agarose, transferred under non-denaturing conditions to nitrocellulose, and analysed by hybridization against *Hin*dIII fragment 10/pBR322 probe. H, *Hin*dIII; E. *Eco*RI; B, *Bam*HI; S, *Sal*I; P, *Pst*I; –, undigested; ss, free T-DNA molecule; open arrows, unexpected fragments produced by *Hin*dIII or *Pst*I digestion of the AS-induced DNA. *C*, Three equivalent non-denatured transfers of *Pst*I-digested AS-induced DNA were hybridized against *Hin*dIII fragment 10 probe (H10); pBR322 probe (pBR); and *Hin*dIII fragment 23 probe (H23). Scaling is different from that in *B*, such that the signals marked with triangles correspond to ss DNA molecules of ~6.6 and 2.8 kb, respectively. Note that different preparations of AS-induced DNA were used in the experiments of *B* and *C*. The ~9.2-kbp signal observed in the *Pst*I digest in *B* is not observed in *C*; this difference may reflect the detection of different levels of AS-induced events in the two preparations.

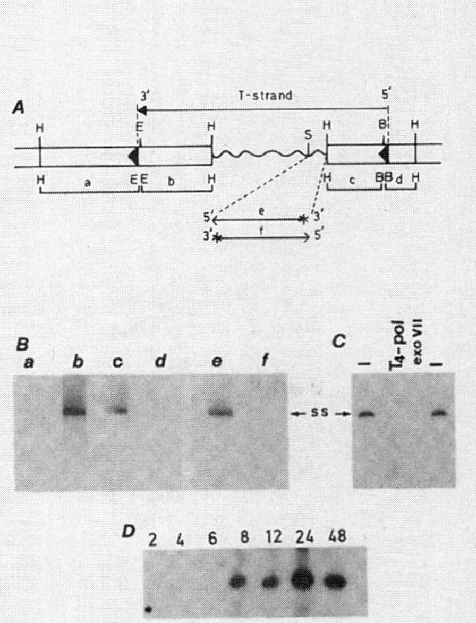

Fig. 3 Molecular characterization of ss T-DNA molecules. *A*, Six probes used to analyse the sequence content and strandedness of the ss T-DNA molecule. *B*, *Bcl*I; *E*, *Eco*RI; *H*, *Hin*dIII. The *Eco*RI site of probes *a* and *b* falls 55 bases inside the left T-DNA border, and the *Bcl*I site of probes *c* and *d* falls 42 bases inside the right border.

Methods. Fragments *a*, *b*, *c* and *d* were gel-purified twice and labelled by nick translation. Probes *e* and *f* are 3'-end-labelled probes where the position of the label is marked by an asterisk, and correspond to the upper and lower strands of the pGV3850 T-DNA, respectively. For probe *e*, pBR322 was digested with *Hin*dIII, 3'-end-labelled at the *Hin*dIII site by Klenow fill-in, recut with *Sal*I, and the resultant 622-bp *Hin*dIII/*Sal*I fragment was gel-purified. For probe *f*, the order of the restriction digests was reversed to give a 622-bp *Sal*I/*Hin*dIII fragment, 3'-end-labelled at the *Sal*I site. The polarity of the ss T-DNA molecule, or T-strand, is shown at the top of the figure. ***B***, Sequence content and strandedness of the ss T-DNA. Six equivalent strips of a non-denatured transfer of undigested AS-induced DNA hybridized against the six probes shown in *A*. ss, Single-stranded T-DNA signal. ***C***, Exonuclease sensitivity of the ss T-DNA. The non-denatured transfer of untreated (–), T4 polymerase-digested (T4-pol), and exonuclease VII-digested (ExoVII) AS-induced DNA was hybridized against nick-translated pBR322 probe. For exonuclease digestion, 1-μg aliquots of AS-induced DNA were incubated with 2.5 U T4 polymerase (Anglian Biotechnology Ltd) or 1.1 U exonuclease VII (Gibco-BRL), respectively, for 30 min at 37 °C. Each reaction was carried out in 15 μl in the enzyme buffer recommended by the manufacturer. Note that under identical conditions these treatments have no effect on M13 ss circular phage DNA; also, the ss T-DNA is fully degraded after 90 min digestion with T4 polymerase (data not shown). ***D***, Kinetics of synthesis of the ss T-DNA molecule. A 50-ml culture of pGV3850 *Agrobacteria* pregrown for 5 h was then induced with AS at 100 μM. At 2, 4, 6, 8, 12, 24 and 48 h after the start of induction, 5-ml aliquots were removed, frozen at –70 °C, and total DNA was then prepared. These DNA samples were fractionated on agarose and transferred to nitrocellulose without gel denaturation, and the amount of ss T-DNA molecules present in the AS-induced cells for each time point was assessed by hybridization against nick-translated pBR322 probe. Note that by EtBR staining the amount of DNA in the 24-h lane was about double that for the other lanes.

Characterization of ss T-DNAs

ss T-DNAs are full-length and unipolar. Six identical strips of a non-denatured transfer of undigested AS-induced DNA were hybridized separately with the six probes labelled *a* to *f* in Fig. 3*A*. Probes *a* and *d* correspond to pGV3850 sequences just outside the left and right T-DNA border repeats; probes *b* and *c* correspond to T-region sequences just within these repeats; and probes *e* and *f* correspond to the upper and lower strands of the pGV3850 T-DNA region, respectively. Figure 3*b* shows that the ss T-DNA molecule hybridizes only to probes *b*, *c* and *e*. Thus, it is composed only of sequences located internal to the T-DNA borders, and which correspond to the lower strand of the pGV3850 T-region. We designate this unipolar ss T-DNA molecule as the T-strand.

T-strands are linear. The T-strand is sensitive to both exonuclease VII (3'→5' and 5'→3' activities specific for ss DNA[24]) and T4 polymerase (3'→5' exonuclease that accepts both ss and ds DNA as substrate[25]) (Fig. 3*C*). Thus, the T-strand, as isolated, must be a linear ss DNA molecule whose 3' terminus (and perhaps also 5' terminus) is available to exonuclease digestion.

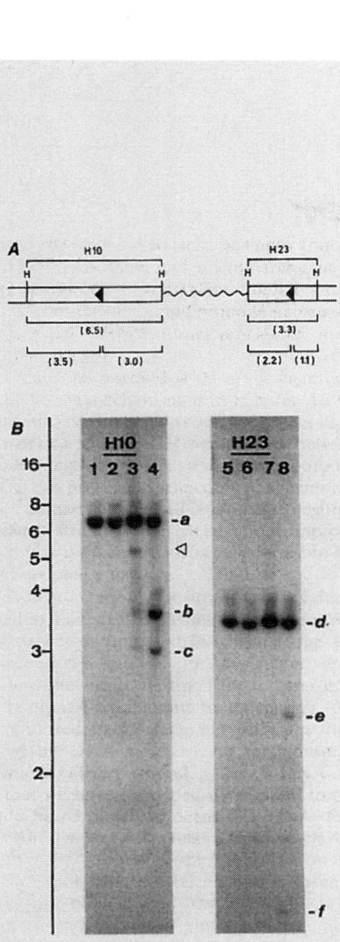

Fig. 4 Identification of AS-induced S_1-sensitive T-DNA border structures. *A*, *Hin*dIII fragments produced by ds cleavages of the pGV3850 T-DNA borders. Fragments *b* and *c* result from cleavage of the left border repeat carried by *Hin*dIII fragment 10 (*a*). Fragments *e* and *f* result from cleavage of the right T-DNA border carried by *Hin*dIII fragment 23 (*d*). Fragment sizes are in kbp. Black arrows, T-DNA border sequences; H, *Hin*dIII. *B*, Uninduced and AS-induced DNA was digested with *Hin*dIII, and half of each sample was treated further with S_1 nuclease. The four samples were analysed on two denatured transfers hybridized against nick-translated *Hin*dIII fragment 10 (lanes 1–4) or *Hin*dIII fragment 23 probe (lanes 5–8). Lanes 1, 5, uninduced DNA; lanes 2, 6, uninduced DNA plus S_1 treatment; lanes 3, 7, AS-induced DNA; lanes 4, 8, AS-induced DNA plus S_1 treatment. *a–f* refer to the fragments shown in *A*. The open arrow indicates a novel S_1-sensitive fragment of ~5.2 kbp. Scaling is in kbp. Note that the low level of the border cleavage fragments (*b*, *c*, *e*, *f*) seen in the absence of S_1 treatment (lane 3, and lane 7 after long exposure) may reflect mechanical breakage of the AS-induced S_1-sensitive border structures during sample preparation.

The *Pst*I cleavage products of the T-strand indicate that its 5′ and 3′ ends map to the right and left T-DNA borders, respectively. Two fragments of ~3.3 and 1.4 kbp (~ss lengths 6.7 and 2.8 kbp) are detected. The 6.6-kbp fragment is homologous to pBR322 and *Hin*dIII fragment 23, while the 2.8-kbp fragment is homologous only to *Hin*dIIII fragment 10 (Fig. 2*C*). Cleavage of a circular T-strand molecule at any (or all) of its three *Pst*I sites would yield fragments of different sizes and sequence content from those observed. The simplest explanation for the results of Fig. 2*C* is that the T-strand is a linear ss molecule, cleaved by *Pst*I principally at its middle *Pst*I site (just within the right end of its *Hin*dIII fragment 10 sequences), to produce a 6.7-kbp 5′ and a 2.8-kbp 3′ fragment.

T-strand synthesis is limited. Using a probe homologus to both strands of the T-DNA, the intensity of the T-strand signal is ~5–10-fold less than the Ti-plasmid T-region signal (Fig. 1*C*, lane 1; also Fig. 5). As the T-strand corresponds to one strand of the T-region, and the copy number of the Ti plasmid is about two (and assuming no specific loss of T-strands during DNA preparation), we estimate that on average each AS-induced cell carries 0.4–0.8 T-strand molecules. As T-DNA homologous molecules are not found in the culture medium (data not shown), this low copy number is probably not due to export of the T-strand. Figure 3*D* demonstrates that the relative amount of T-strands begins to plateau within 8 h after the start of AS induction, and shows only a doubling during an additional 40 hours of induction. Thus, T-strand synthesis is a limited process.

AS-induced alterations of the Ti plasmid

Since the T-strand must be produced from the Ti-plasmid T-region, several molecular reactions involving these sequences should occur in bacterial cells that are actively carrying out different steps of T-strand synthesis, or other early steps of T-DNA transfer. To gain insight into these reactions, the novel T-DNA homologous hybridization signals that result from fragmentation of the AS-induced Ti plasmid with restriction enzymes and S_1 nuclease are characterized below. These data provide

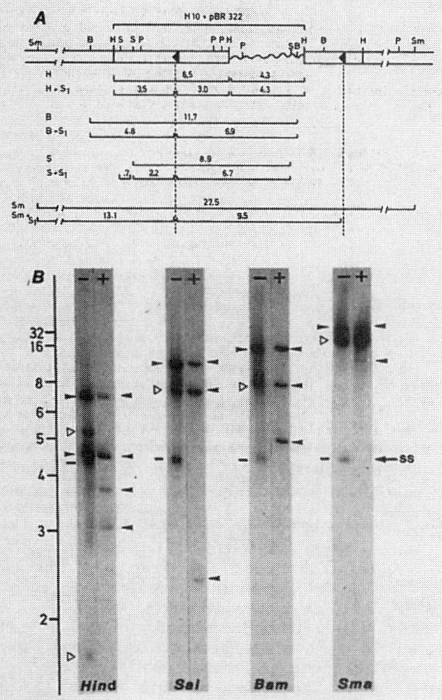

Fig. 5 Hybridization analysis of T-region intermediate structures. *A*, Restriction map of the left portion of the pGV3850 T-DNA region and border cleavage fragments predicted to arise after S_1 treatment of AS-induced DNA. Sizes are given in kbp. B, *Bam*HI; H, *Hin*dIII; P, *Pst*I; S, *Sal*I; Sm, *Sma*I; S_1, S_1 nuclease. The black arrows and dotted lines indicate the positions of the T-DNA borders. *B*, Denatured transfers of *Hin*dIII, *Bam*HI, *Sal*I and *Sma*I digests of AS-induced DNA treated (+) and untreated (−) with S_1 nuclease were hybridized against nick-translated *Hin*dIII fragment 10/pBR322 probe (*A*). Scaling on left is in kbp. The single-stranded T-strand is indicated by ss and the bar adjacent to the digests. Solid arrows indicate the expected fragments before and after S_1 treatment and correspond in size to the fragments shown in *A*. The open arrows indicate unexpected AS-induced fragments which correspond to novel T-DNA homologous structures linked to pGV3850. Note that other unexpected signals are observed when the filter is reprobed with *Hin*dIII fragment 23 sequences, and that all of the unexpected fragments are specific to AS-induced DNA and have been observed with DNA prepared from at least two independent AS-induced bacterial cultures (data not shown).

evidence for specific reactions associated with the Ti-plasmid T-DNA border repeats and with the internal sequences of the T-region.

S_1-sensitive T-DNA border sites. Figure 4 demonstrates that AS induction leads to the generation of Ti-plasmid T-DNA border structures which are sensitive to cleavage by S_1 nuclease. Uninduced and AS-induced DNA was restricted with *Hin*dIII, half of each sample was treated further with S_1 nuclease, and the fragments homologous to the left and right border regions of pGV3850 were detected by hybridization. The AS-induced DNA lanes contain novel signals whose sizes correspond exactly to the fragments that would be produced by ds cleavages at the left and right T-DNA border sequences of pGV3850 (Fig. 4*A*), and the intensities of these signals are greatly increased by S_1 digestion of the DNA. We designate these S_1-sensitive structures S_1 border sites.

The relative intensities of the signals which correspond to the S_1 border cleavage fragments show that ~30% of the left T-DNA borders and ~10% of the right T-DNA borders of the AS-induced Ti plasmid population are S_1-sensitive. Thus, the generation of S_1 border sites is a frequent event. Furthermore, since a 9.5-kbp ds T-DNA fragment is produced by S_1 digestion of AS-induced DNA that is unrestricted (Fig. 1*C*, lane 3), or restricted with *Sma*I (which does not cleave within the pGV3850 T-region (Fig. 5)), S_1 border sites can simultaneosly occur at both the right and left border repeats on a single Ti plasmid. We note that incubation of the uncut AS-induced DNA at 65 °C does not result in the release of the 9.5-kbp T-DNA fragment from the AS-induced Ti plasmid (data not shown); that is, the S_1 border site does not correspond to a ds staggered cleavage of the T-DNA border sequence. Other experiments have shown that the S_1-sensitive structure corresponds to an AS-induced ss endonucleolytic cleavage in the bottom strand of the 25-bp border sequence (K. Wang and M. Van Montagu, in preparation); S_1 is known to cleave opposite such structures[26,27].

Other AS-induced T-region structures. Figure 5 demonstrates that other novel Ti-plasmid-linked T-region structures are also present in the AS-induced cell population. Three different classes of T-DNA-homologous signals are observed when AS-induced DNA is digested with various restriction enzymes and S_1 nuclease (Fig. 5*B*): those that correspond to the ss T-DNA molecule (black bars, $-S_1$ lanes); those that correspond to the predicted border cleavage fragments (Fig. 5*A*) produced by S_1 digestion of the AS-induced S_1 border sites (solid arrows, $+S_1$ lanes); and those that correspond to novel unexpected fragments (open triangles) derived from pGV3850 Ti plasmids whose internal T-region sequences have been altered as a result of AS induction.

For example, the 5.2-kbp fragment present in the $-S_1$ *Hin*dIII sample (Fig. 5*B*; also Figs 2*B* and 4*B*) is S_1-sensitive, binds to nitrocellulose, specifically hybridizes to *Hin*dIII fragment 10 (and not to pBR322 or *Hin*dIII fragment 23; data not shown), and is ~1.3 kbp smaller than *Hin*dIII fragment 10 (Fig. 5*A*). Thus, this novel *Hin*dIII fragment must be partially single-stranded and derived from an AS-induced Ti plasmid whose T-region is partially single-stranded or associated with ss T-DNA sequences, as in a D-loop structure. More perplexing fragments are also detected: the ~15-kbp S_1-insensitive *Sma*I signal corresponds to a ds DNA fragment which is ~12 kbp smaller than the *Sma*I fragment that covers the T-region of the uninduced Ti plasmid; also, in the $-S_1$ *Bam*HI and *Sal*I lanes, signals are observed whose sizes correspond to the S_1 cleavage fragments of these digests (Fig. 5*A*) which are internal, but not external, to the left T-DNA border. While the present data do not allow the precise identification of the T-region structure(s) to which these unexpected fragments correspond, they illustrate the complexity of the molecular reactions associated with the Ti-plasmid T-DNA sequences which occur in AS-induced *Agrobacteria*.

Discussion

Agrobacterium tumefaciens transfers its Ti-plasmid T-DNA to plant cells, and this process is activated by the induction of the Ti-plasmid virulence genes with the plant phenolic compound AS. We show that AS induction results in the generation of several novel T-DNA homologous molecules in *Agrobacterium*: a linear ss molecule, the T-strand, which corresponds to the lower strand of the Ti-plasmid T-region; and Ti-plasmid molecules whose T-DNA border repeats are sensitive to cleavage by S_1 nuclease (S_1 border sites), and whose internal sequences have been altered (T-region structures). The specific properties of these AS-induced molecules allows the formulation of mechanistic models for the generation of the T-strand, and for its potential transfer to the plant cell.

Figure 6*A* presents a model for the generation of the T-strand molecule. First, ss endonucleolytic cleavages occur within the left and right T-DNA border sequences on the Ti plasmid. These cleavages correspond to S_1 border sites and provide free 3′ OH groups from which DNA synthesis can be primed. Second, using the top strand of the T-region as template, DNA synthesis initiates at the right border cleavage site, and proceeds unidirectionally across the T-region. This synthesis displaces the bottom strand of the T-region and produces a transitory triple-stranded structure which may correspond to one of the AS-induced T-region structures that we have detected. Third, DNA synthesis terminates when it encounters the left border cleavage site, and the displaced bottom strand is released from the Ti plasmid, as the free T-strand ss molecule. Alternatively, a $5' \rightarrow 3'$ helicase activity unwinds the T-region of the nicked Ti plasmid to free the T-strand molecule and produce a Ti plasmid whose T-region is momentarily ss prior to replacement strand synthesis. Since we observe at most one T-strand molecule per AS-induced cell, T-strand production must be tightly regulated.

The sequences internal to the T-DNA of the wild-type Ti plasmid encode genes whose expression in the transformed plant cell result in the tumorous phenotype, crown gall[1], and when the orientation of the right border on the Ti plasmid is flipped (by *in vitro* manipulation), phenotypically transformed plant cells are not obtained[15,16]. The model of Fig. 6*A* explains this functional polarity of the right T-DNA border; that is, the T-strand corresponds to the bottom strand of the T-region and must be generated in a right-to-left (5′ to 3′) direction. Thus, if the orientation of the right border is reversed in the Ti plasmid, ss molecules will be generated away from (rightward of) the T-DNA tumour genes. This model may also explain how the ds

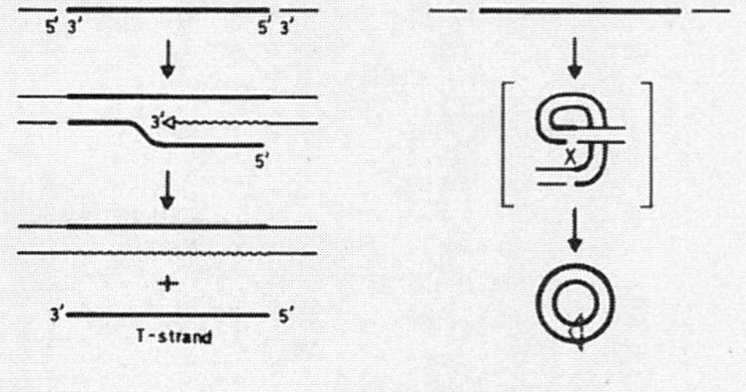

Fig. 6 Proposed reactions associated with the Ti-plasmid T-region in AS-induced cells. *A*, Model for the generation of the T-strand; *B*, model for the generation of a ds T-DNA circle molecule. Thick lines represent T-region sequences, thin lines represent adjoining Ti-plasmid sequences external to the 25-bp T-DNA border repeats, indicated by the dashed triangles. ss breaks in the bottom strand of the border repeats correspond to S_1 border sites, and the wavy line corresponds to a newly synthesized bottom strand of the T-region. This model implies that ss molecules are not generated leftward of the left T-DNA border; while we have not observed such molecules, they might not have been detected if they are heterogeneous in length.

T-DNA circle molecules, recovered in *E. coli* transformed with AS-induced DNA, are generated at a low frequency in response to AS induction. Since nicked DNA stimulates recombination events[28], the S_1 border sites could promote pairing and recombination between the T-DNA border repeats of the AS-induced Ti plasmid (Fig. 6*B*).

Assuming that the T-strand is the transfer intermediate, we can speculate on the mechanism of its transfer. This mechanism might share features with known bacterial processes which mediate the transfer of specific DNA molecules between bacteria, such as phage infection or conjugation. An important distinction between these two processes is that only phage infection involves the synthesis of many copies of the molecule destined for transfer. Since the T-strand is present at about one copy per AS-induced cell, it is unlikely that it is transferred to the plant cell as an infectious phage particle.

In bacterial conjugation, one strand of a ds donor molecule is transferred as a linear ss molecule from the donor to recipient cell[29]. This process is initiated by nicking one strand of the donor molecule at a specific site (designated *ori*T, origin of transfer)[30], and the strand destined for transfer is unwound in a $5' \rightarrow 3'$ direction. Concomitant to unwinding, the unwound strand is mobilized to the recipient cell, and this transfer is accompanied by DNA synthesis on the donor molecule to replace the mobilized strand. The T-DNA homologous molecules that we have described correspond to the structures which would be predicted to occur if T-DNA transfer occurs through a conjugative mechanism. The S_1 border sites are analogous to nicked *ori*T sites, the T-strand is analogous to the linear ss DNA molecule transferred during bacterial conjugation, and the internal T-region structures are analogous to the replacement strand synthesis intermediates of the donor molecule. Furthermore, bacterial conjugation requires direct contact between donor and recipient cells[31], and the same requirement is observed for the T-DNA transfer process[32].

If the T-strand is transferred by conjugation to the plant cell, it is still not known how it ultimately finds its way into the plant cell nucleus and becomes integrated into the nuclear genome. While the deproteinized T-strand that we have described is a naked linear molecule, presumably it is transferred as a complex that carries proteins which in part mediate the post-transfer events. These proteins, as well as the proteins involved in the generation and transfer of the T-strand, are probably encoded by the plant-inducible Ti-plasmid *vir* loci[10].

We thank Marc Van Montagu for his encouragement and suggestions for the exonuclease studies, Kan Wang for helpful discussion, Jeff Schell for support, and Martine De Cock, Karel Spruyt and Albert Verstraete for preparation of this manuscript and figures. This work was supported by grants from the 'ASLK-Kankerfonds', the 'Fonds voor Geneeskundig Wetenschappelijk Onderzoek' (FGWO 3.001.82), and the Services of the Prime Minister (OOA 12.0561.84) (to Jeff Schell and Marc Van Montagu).

Note added in proof: The nondenatured transfer assay for single-stranded DNAs has also been recently described by Reile, H. T., Michel, B. and Ehrlich, S. D. *Proc. natn Acad. Sci. U.S.A.* **83,** 2541–2545 (1986).

Received 24 April; accepted 19 June 1986.

1. Gheysen, G., Dhaese, P., Van Montagu, M. & Schell, J. in *Advances in Plant Gene Research* Vol. 2 (eds Hohn, B. & Dennis, E. S.) 11–47 (Springer, Vienna, 1985).
2. Zambryski, P., Depicker, A., Kruger, K. & Goodman, H. *J. molec. appl. Genet.* **1,** 361–370 (1982).
3. Yadav, N. S., Vanderleyden, J., Bennett, D. R., Barnes, W. M. & Chilton, M.-D. *Proc. natn. Acad. Sci. U.S.A.* **79,** 6322–6326 (1982).
4. Simpson, R. B. *et al. Cell* **29,** 1005–1014 (1982).
5. Holsters, M. *et al. Molec. gen. Genet.* **190,** 35–41 (1983).
6. Zambryski, P. *et al. EMBO J.* **2,** 2143–2150 (1983).
7. de Framond, A. J., Barton, K. A. & Chilton, M.-D. *Bio/technology* **1,** 262–269 (1983).
8. Bevan, M. *Nucleic Acids Res.* **12,** 8177–8721 (1984).
9. An, G., Watson, B. D., Stachel, S., Gordon, M. P. & Nester, E. W. *EMBO J.* **4,** 277–284 (1985).
10. Stachel, S. E. & Nester, E. W. *EMBO J.* **5,** 1445–1454 (1986).
11. Douglas, C. J., Staneloni, R. J., Rubin, R. A. & Nester, E. W. *J. Bact.* **161,** 850–860 (1985).
12. Stachel, S. E., Nester, E. W. & Zambryski, P. C. *Proc. natn. Acad. Sci. U.S.A.* **83,** 379–383 (1986).
13. Stachel, S. E. & Zambryski, P. *Cell* (in the press).
14. Stachel, S. E., Messens, E., Van Montagu, M. & Zambryski, P. *Nature* **318,** 624–629 (1985).
15. Wang, K., Herrera-Estrella, L., Van Montagu, M. & Zambryski, P. *Cell* **38,** 455–462 (1984).
16. Peralta, E. G. & Ream, L. W. *Proc. natn. Acad. Sci. U.S.A.* **82,** 5112–5116 (1985).
17. Joos, H. *et al. Cell* **32,** 1057–1067 (1983).
18. Shaw, C. H., Watson, M. D., Carter, G. H. & Shaw, C. H. *Nucleic Acids Res.* **12,** 6031–6041 (1984).
19. Koukoliková-Nicola, Z. *et al. Nature* **313,** 191–196 (1985).
20. Southern, E. M. *J. molec. Biol.* **98,** 503–518 (1975).
21. Bourgeois, S. *Meth. Enzym.* **21,** 491–500 (1971).
22. Vogt, V. *Eur. J. Biochem.* **33,** 192–200 (1973).
23. Nishigaki, K., Kaneko, Y., Wakuda, H., Husimi, Y. & Tanaka, T. *Nucleic Acids Res.* **13,** 5747–5760 (1985).
24. Huang, W. M. & Lehman, I. R. *J. biol. Chem.* **247,** 3139–3146 (1972).
25. Chase, J. W. & Richardson, C. C. *J. biol. Chem.* **249,** 4545–4561 (1974).
26. Lilley, D. M. J. *Nucleic Acids Res.* **9,** 1271–1289 (1981).
27. Panayotatos, N. & Wells, R. D. *Nature* **289,** 466–470 (1981).
28. Dressler, D. & Potter, H. *A. Rev. Biochem.* **51,** 727–761 (1982).
29. Willetts, N. & Wilkins, B. *Microbiol. Rev.* **48,** 24–41 (1984).
30. Thompson, R., Taylor, L., Kelly, K., Everett, R. & Willetts, N. *EMBO J.* **3,** 1175–1180 (1984).
31. Manning, P. A. & Achtman, M. in *Bacterial Outer Membranes* (ed. Inouye, M.) 409–447 (Wiley, New York, 1979).
32. Douglas, C. J., Halperin, W. & Nester, E. W. *J. Bact.* **152,** 1265–1275 (1982).
33. Dhaese, P., De Greve, H., Decraemer, H., Schell, J. & Van Montagu, M. *Nucleic Acids Res.* **7,** 1837–1849 (1979).
34. Maniatis, T., Fritsch, E. F. & Sambrook, J. *Molecular Cloning, a Laboratory Manual* (Cold Spring Harbor Laboratories, New York, 1982).

Printed in Great Britain by Turnergraphic Limited, Basingstoke, Hampshire

EARLY EVENTS IN THE T-DNA TRANSFER PROCESS

Martin Yanofsky
Section of Cell and Developmental Biology, University of California at San Diego
La Jolla, California, United States
E-mail: marty@ucsd.edu

Commentary on the paper:

Yanofsky, M. F., Porter, S. G., Young, C., Albright, L. M., Gordon, M. P., and Nester, E. W. 1986. The *virD* operon of *Agrobacterium tumefaciens* encodes a site-specific endonuclease. Cell 47:471–477.

My interest in *Agrobacterium* can be traced to my undergraduate days when I worked for several years in the lab of Don Helinski at the University of California at San Diego on the *Rhizobium*-plant interaction. Similar techniques that were being used in both *Rhizobium* and *Agrobacterium* studies made it rather simple for me to enter into the *Agrobacterium* field. I recall hearing a seminar by Gene Nester on the current state of *Agrobacterium* research and I found it to be so stimulating that I immediately knew that I wanted to study *Agrobacterium* for my graduate school research. As luck would have it, I followed this lead and entered into the Microbiology department at the University of Washington, where I was welcomed, by Gene, into his laboratory.

Having spent a significant amount of time reading the recent literature, I recall being somewhat intimidated when I first entered Gene's lab. First of all, the lab was quite large and consisted almost entirely of postdocs, and second, these postdocs were all people whose papers I had read, including Frank White, Harry Klee, David Garfinkel and others, each of whom had already made major contributions to the *Agrobacterium* field. It didn't take long, however, for me to settle into the lab and before long I became good friends with Harry, Frank, and others. These friendships continue to this day.

During my stay in Gene's lab, I developed a very close working relationship with two postdocs, Lisa Albright and Bernard Leroux. It was this relationship that blossomed into a successful collaboration, where the three of us focused our efforts on what was at the time a major question in the *Agrobacterium* field. What is the mechanism of T-DNA transfer? What are the initial steps that occur in *Agrobacterium* and which genes are responsible for carrying out these reactions?

It was those thoughts that seemed to captivate our attention as we tried to devise strategies for gaining insights into the T-DNA transfer problem. I recall that we were particularly excited about a recent finding by Zdena Koukolikova-Nicola (1985) from the Barbara Hohn lab. She had used an elegant selection strategy that seemed to identify a double stranded circular T-DNA molecule that at the time was speculated to represent an intermediate in the transfer process. Of course, we were also stimulated by a wealth of additional information from prior investigators who had established the role of the T-DNA border sequences. We were also aided by the knowledge from the work of Scott Stachel, Patti Zambryski and others who showed that the *vir* genes were induced upon exposure to plant cells (Stachel and Zambryski, 1986). Since everyone believed that the *vir* genes likely held the secret to the T-DNA transfer problem, it made sense that we would not only take advantage of this ability to control *vir* gene expression, but also look at the possible functions of the *vir* genes in terms of their likely roles in mediating T-DNA transfer.

In the ensuing months, we developed several strategies to look at molecular events that occurred at the T-DNA borders in *Agrobacterium,* upon induction by exposure to plant cells. We also began to look more closely at the sequences and possible functions of the *vir* genes (Albright et al. 1987). One of the strategies we developed was a simple DNA blotting (Southern) assay. We reasoned that, if the first step in the T-DNA transfer process involved cutting at the T-DNA borders, we should be able to detect such cleavage events upon induction of *Agrobacterium* by exposure to plant cells. Although this assay provided some success, it proved ultimately to be unreliable. We didn't realize at the time that the reason it was unreliable was due to the fact that it relied on double strand T-DNA breaks, something that we had assumed was occurring.

We subsequently turned to more sensitive approaches. The first of these was a relatively straightforward RNase protection assay to look for molecular changes at the T-DNA borders. This procedure turned out to offer a number of advantages. First, it was very sensitive, allowing the ready detection of cleavage events. Second, it was a strand-specific assay, allowing us individually to monitor changes on the "top" and "bottom" strands. This fact was particularly important, because it didn't take long for us to notice that we consistently observed T-DNA cleavage on what we referred to as the "bottom" strand, whereas T-DNA cleavage on the "top" strand was largely undetected. This strand-specific cleavage explained why it was so difficult to detect these border-cleavage events by our DNA-blotting assay. The RNase protection assay also allowed us to map, with near precision, the specific site of border cleavage. As expected, the assay detected specific cleavage within the previously defined T-DNA border sequences, which were loosely defined as approximately 25 base pair imperfect direct repeats.

In order precisely to map the site of border cleavage, we next turned to a simple primer extension assay. Although less sensitive than the RNase protection assay, the primer extension results were very clear. We detected cleavage only on the bottom strand at the left and right T-DNA borders. Moreover, we could map the exact nucleotide of border cleavage and found that it occurred at the same relative position in both T-DNA borders. Furthermore, the fact that the border cleavage could only be detected after inducing the *Agrobacterium* by exposure to plant cells suggested that the enzyme responsible for the border cleavage was likely to be *vir* encoded.

While these exciting experiments were unfolding, we were also delving into the functions of the *vir* genes. With the exception of the *virA* and *virG* genes, whose roles in regulating *vir* gene expression were elegantly worked out by Stachel and Zambryski (1986), none of the functions of the *vir* genes involved in T-DNA transfer had been elucidated. Since we now had a functional assay for one of the first steps in T-DNA transfer, border cleavage, we were poised to examine the possible contribution of the *vir* genes to border cleavage.

Our first experiment, using the RNase protection assay, worked like a charm. We simply took *Agrobacterium* mutants for all of the *vir* genes and used this assay to look for border cleavage before and after *vir* gene induction. The results were very clear. We detected T-DNA border cleavage in the *virB, virC,* and *virE* mutants, but not in the *virA, virG* and *virD* mutants. Since we already knew that *virA* and *virG* were regulators of *vir* gene induction, it was logical to suspect that the border-specific endonuclease activity was in fact encoded by *virD.*

Fortunately, since we were already investigating *vir* gene function, we had by now sequenced the *virD* operon and had constructs in hand that over-expressed *virD* in *E. coli.* Moreover, we had a deletion series of the *virD* operon that might allow us to further dissect the region of *virD* that encoded the T-DNA border-specific endonuclease. We quickly developed a simple *E. coli* system in which we could induce *virD* expression from one plasmid, and on a separate plasmid, we had the T-DNA border sequences. Although nearly twenty years have passed, I still remember these experiments as if they just happened yesterday. As I recall, we did this key experiment to look for *virD* cleavage of the T-DNA borders in *E. coli* along with a number of additional experiments. As usual, we exposed the radioactive gel to film and went home for the night. The next morning, Lisa was the one who developed the films while I waited in the lab. At the time, the nearest film-developing room was located in the hospital some distance away, so it was quite some time before she returned with the film in her hands. The results were as clear as one could have hoped. In a single experiment we were able to show that *virD* encoded a site specific endonuclease that cut only the bottom strand of the T-DNA borders at exactly the same site that we had observed cleavage in *Agrobacterium.* Moreover, using a deletion series, this same experiment defined a truncated portion of the *virD* operon that was sufficient for T-DNA border cleavage.

I recall quite vividly attending an *Agrobacterium* conference at Fallen Leaf Lake many months later. Although I remember being quite sick at the time, I presented the findings that Lisa, Bernie and I had made toward understanding the first step in T-DNA transfer. We presented a model in which the T-DNA transfer process involved a linear single-stranded molecule, rather than the double-stranded circular intermediate that had been previously postulated. I believe that Stachel et al. (1986), at around the same time, had come to the same conclusion using different approaches. However, others were not quite ready to give up on the idea of a double stranded intermediate and I remember one of *Agrobacterium*'s pioneers, Mary-Dell Chilton, following my presentation, gave a talk in which she gave a list of reasons why the T-DNA intermediate must be double stranded, and not single stranded as we had suggested. Certainly her presentation served as a great catalyst for debate.

I look back on these early days of my scientific career with great fondness, partly for the long-lasting friendships that started during these days and partly because of the great excitement that we all felt about those experiments. It has been my experience that significant leaps in understanding are relatively rare, and the sense of satisfaction that I felt as a young graduate student has continued to be a motivating factor throughout my career.

References

Albright, L. M., Yanofsky, M. F., Leroux, B., Ma, D., and Nester, E. W. 1987. Processing of the T-DNA of *Agrobacterium tumefaciens* generates border nicks and linear, single-stranded T-DNA. J. Bacteriol. 169: 1046–1055.

Koukolikova-Nicola, Z., Shillito, R. D., Hohn, B., Wang, K., Van Montagu, M., and Zambryski, P. 1985. Involvement of circular intermediates in the transfer of T-DNA from *Agrobacterium tumefaciens* to plant cells. Nature 313:191–196.

Stachel, S. E., Timmerman, B., and Zambryski, P. C. 1986. Generation of single-stranded T-DNA molecules during the initial stages of T-DNA transfer from *Agrobacterium tumefaciens* to plant cells. Nature 322: 706–712.

Stachel, S. E., and Zambryski, P. C. 1986. *VirA* and *virG* control the plant-induced activation of the T-DNA transfer process of *Agrobacterium tumefaciens.* Cell 46:325–333.

MARTIN YANOFSKY: As a teenager, I worked in my father's laboratory, and there is no doubt that his passion for science was a major factor in my choosing Biology as a career. As an undergraduate at the University of California at San Diego, I worked for a number or years with Gary Ditta in Don Helinski's lab. Both Gary and Don had a strong influence on me and helped to spark my interest in plant biology. I followed their lead and was mentored by Gene Nester, together with Milt Gordon, as a graduate student at the University of Washington, where I was also stimulated by many students and postdocs, including Harry Klee, Frank White, and many others.

Cell, Vol. 47, 471–477, November 7, 1986, Copyright © 1986 by Cell Press

The *virD* Operon of Agrobacterium tumefaciens Encodes a Site-Specific Endonuclease

Martin F. Yanofsky, Sandra G. Porter, Calvin Young, Lisa M. Albright,* Milton P. Gordon, and Eugene W. Nester
Department of Microbiology and Immunology
University of Washington
Seattle, Washington 98195

Summary

Tumor formation by Agrobacterium tumefaciens involves the transfer and integration of a defined segment (T-DNA) of tumor-inducing (Ti) plasmid DNA into the plant nuclear genome. A set of plasmid genes outside the T-DNA, the *vir* genes, are thought to mediate the transfer process. We report here that the *virD* operon encodes a site-specific endonuclease that cleaves at a unique site within each of the 24 bp direct repeats that flank the T-DNA. The endonuclease function was further localized to the 5′ end of this operon by demonstrating that cleavage does not occur in *virD* mutant strains of Agrobacterium and that the 5′ end of the *virD* operon is sufficient to direct cleavage in E. coli. Analysis of nucleotide sequence and protein data indicate that two proteins of 16.2 and 47.4 kd are involved.

Introduction

The process of tumor formation by Agrobacterium tumefaciens involves the transfer and integration of a specific segment of plasmid DNA (T-DNA) into the plant nuclear genome (for reviews see Bevan et al., 1982; Depicker et al., 1983; Hooykaas et al., 1984; Nester et al., 1984). The T-DNA encodes gene products that catalyze the constitutive synthesis of the phytohormones auxin and cytokinin (Akiyoshi et al., 1984; Barry et al., 1984; Schröder et al., 1983; Thomashow et al., 1984, 1986), resulting in the uncontrolled cell division characteristic of a tumor. The T-DNA is flanked by 24 bp imperfect direct repeats (Yadav et al., 1982; Barker et al., 1983) which are required in *cis* for transfer (Wang et al., 1984). None of the genes within the T-DNA are required for transfer or integration.

A second plasmid region, designated the virulence (*vir*) region, is thought to mediate T-DNA transfer. The *vir* region spans 35 kb of plasmid DNA and encodes at least six transcriptional loci (Klee et al., 1983; Stachel et al., 1986). Mutations in these loci result in either an avirulent (*virA*, *B*, *D*, and *G*) or attenuated (*virC* and *E*) phenotype (Hille et al., 1984; Stachel et al., 1986; Yanofsky et al., 1985). Expression of *vir* genes is specifically induced by plant factors (Stachel et al., 1985) and *virA* and *virG* were recently shown to be required for induction of the other *vir* loci (Stachel et al., 1986; Winans et al., 1986). However, the precise function of these or any other *vir* locus has yet to be determined.

We have recently shown that cocultivation of Agrobacterium with plant cells leads to cleavage at a unique site within the 24 bp direct repeats (borders A and B; Barker et al., 1983) that flank the T-DNA (Albright et al., submitted) and that a single-stranded molecule is a major product of T-DNA processing. Similar conclusions were recently reached by Stachel et al. (1986c). Here we show that a specific region of the *virD* operon encodes this site-specific endonuclease and we characterize the proteins involved. This represents the first specific function assigned to a *vir* locus involved in T-DNA transfer.

Results

After cocultivation of Agrobacterium with plant cells, a nick is introduced at unique and identical locations within the 24 bp direct repeats flanking the T-DNA (Albright et al., submitted). We used an RNAase protection assay (Figure 1) to identify the *vir* locus that encodes this border-specific endonuclease. RNA probes were hybridized in solution to total Agrobacterium or E. coli DNA. The hybrids were digested with RNAase, and the protected RNA was visualized by gel electrophoresis and autoradiography. An interruption in the DNA complementary to the RNA probe would result in protection of RNA fragments smaller than full-length, and the size of the resulting fragments localizes the cleavage site. The advantages of these assays are that the probe is of defined length, uniformly labeled, and strand-specific.

vir Genes in Multiple Copies Enhance Cutting

In each of the experiments shown here, we used an RNA probe specific for the bottom strand of the right T-DNA border (border B) (Figure 1) and homologous to Agrobacterium DNA from base pairs 13774 to 14276 (Barker et al., 1983). This RNA probe was hybridized to DNA isolated from Agrobacterium that had been cultured in either the presence or absence of plant cells. Hybridization to DNA from bacteria cultured in the absence of plant cells protected a 502 nucleotide fragment that corresponds to the full-length of homology between the RNA probe and the Agrobacterium DNA. Hybridization to DNA from bacteria cocultivated with plant cells revealed two new fragments of 285 and 215 nucleotides, in addition to the 502 nucleotide fragment. These two new fragments result from cleavage at border B.

The intensities of the 285 and 215 nucleotide long bands, relative to the 502 nucleotide long band, are weak for the Agrobacterium strain A348 (Figure 2, lanes 6 and 7) and for the derivative of A348 that harbors the vector pVK102 (Figure 2, lanes 10 and 11). These data indicate that only a small fraction of T-DNA molecules are cleaved at border B in these strains. The relative proportion of the 285 and 215 nucleotide signals is greatly enhanced for A348(pVK225) (Figure 2, lanes 14 and 15), indicating that

* Present address: Department of Molecular Biology, Massachusetts General Hospital, Boston, Massachusetts 02114.

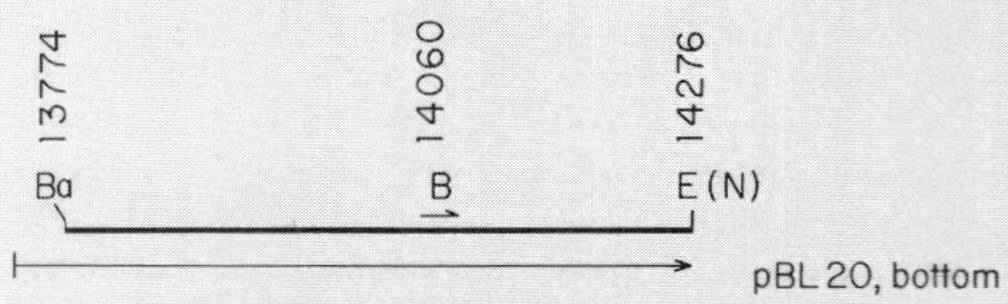

Figure 1. Diagram of RNA Probe

The RNA probe used throughout this paper was synthesized from pBL20 and is complementary to the bottom strand at border B. This plasmid contains sequences from bp 13,774 to bp 14,277 (Barker et al., 1983) inserted into the vector GEM2. This plasmid was linearized with EcoRI and the resulting 544 nucleotide transcript shares 502 nucleotides of homology with the border B region. The position of border B is at bp 14,060.

the additional copies of one or more of the *vir* loci present on this clone (see Figure 3A) increase the amount of cleavage at border B.

We had previously shown that the overlapping cosmid clones pVK257, pVK242, and pVK225 (Figure 3A) enhanced double-stranded cleavage at border B (L. A., data not shown). These clones share a common region spanning the *virG* and *virC* operons, and the 5′ end of the *virD* operon. The formation of double-strand cleavages appears to be linked to border nicking; we therefore tested nonoverlapping subclones from this common region for their ability to enhance strand-specific cleavage at border B. The plasmids pLA110 and pLA111, which contain the SalI 13B and SalI 7B restriction fragments, respectively (see Figure 3A), each increased the cleavage over that observed for A348 alone (Figure 2, lanes 18, 19, 22, 23), indicating that at least two separate *vir* operons are involved. The Sal 13B fragment spans the *virG* locus, which has been shown to be involved in the transcriptional activation of the *vir* region. SalI 7B spans the entire *virC* operon and the 5′ end (0.8 kb) of the *virD* operon. Taken together, these data indicate that *virG* and either *virC* or a portion of *virD* are involved in enhancing the cutting at border B.

virD Is Required for Cutting in Agrobacterium

We analyzed a number of *vir* mutant strains to identify the *vir* loci required for cutting at the T-DNA borders. Since strain A348 yielded only a relatively small amount of cutting, we introduced the plasmid containing the SalI 13B fragment into each of the *vir* mutant strains in order to increase cutting. Since *virG* is directly involved in the transcriptional activation of the *vir* region (Stachel et al., 1986; Winans et al., 1986), a direct involvement in cutting seemed unlikely. A representative set of *vir* mutants is shown in Figure 3B, while the results of RNAase protection assays are summarized in Figure 3A. Cutting was observed for mutations in the *virB*, *C*, and *E* operons (Figure 3B, lanes 7, 9, and 13). However, two mutations in *virA* and three mutations at the 5′ end of *virD* eliminated cutting (Figure 3A; Figure 3B, lanes 5 and 11). Since *virA* is also involved in the transcriptional activation of the *vir* region (Stachel and Zambryski, 1986; Winans et al., 1986), it is probably not responsible for cutting. Thus these data indicate that *virD* is required for the endonuclease activity. Furthermore, since mutations toward the 3′ end of *virD* still allowed cutting, the nicking activity was localized to the 5′ end of this operon.

Gene Products of *virD* Are Sufficient for Cleavage in E. coli

The analysis of *vir* region mutants indicated that the *virD* operon was required for the border-sequence-specific endonuclease activity. We used a binary vector system (Figure 4A) to determine if *virD* was sufficient to specify this activity in E. coli. Various lengths of this operon were transcriptionally fused to the strong E. coli "Tac" promoter (Amann et al., 1983) (Figure 4B). This permitted the regulated expression of the *virD* proteins and the detection of the native protein products (see protein analysis below). These *virD*-containing plasmids were placed in *trans* to the binary vector pE2 (Klee et al., 1985), which contains fragments spanning the T-DNA borders A and B. E. coli strain JM101 (Messing et al., 1981) harboring these plasmids was grown in the presence of IPTG to induce the "Tac" promoter and DNA was isolated and analyzed for border cleavage.

Three *virD*-containing plasmids (pMY1125, pMY1151, and pMY1153), when induced with IPTG, resulted in protection of the 285 and 215 nucleotide long fragments,

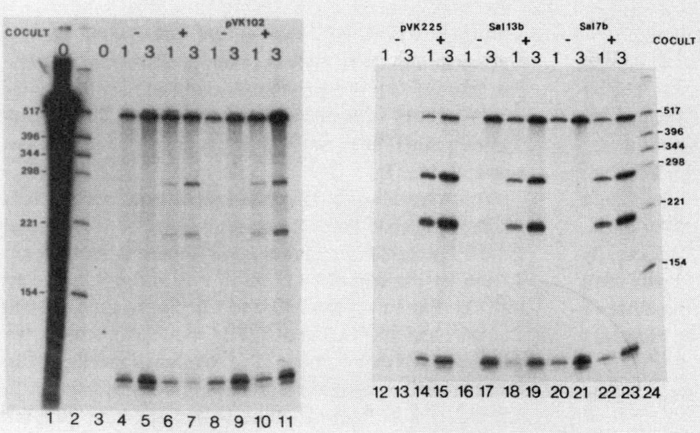

Figure 2. Enhancement of Cleavage by *vir* Region Clones

Protection of an RNA probe synthesized from pBL20 was performed for DNA isolated from Agrobacterium cultured in the absence (−) or presence (+) of plant cells over a 3-fold range of DNA concentration. Agrobacterium strains include A348 (lanes 4–7), A348 (pVK102) (lanes 8–11), A348 (pVK225) (lanes 12–15), A348 (pLA110) (lanes 16–19), and A348 (pLA111) (lanes 20–23). No DNA was added to samples for lanes 1 and 3; no RNAase was added to the sample in lane 1. Lanes 2 and 24 are markers from pBR322 digested with HinfI.

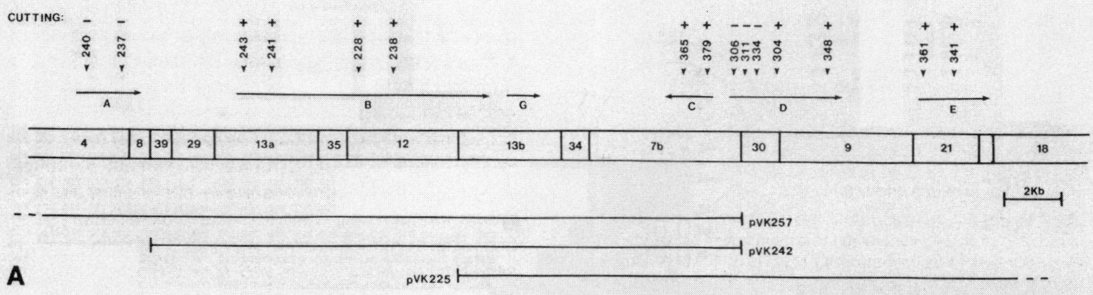

A

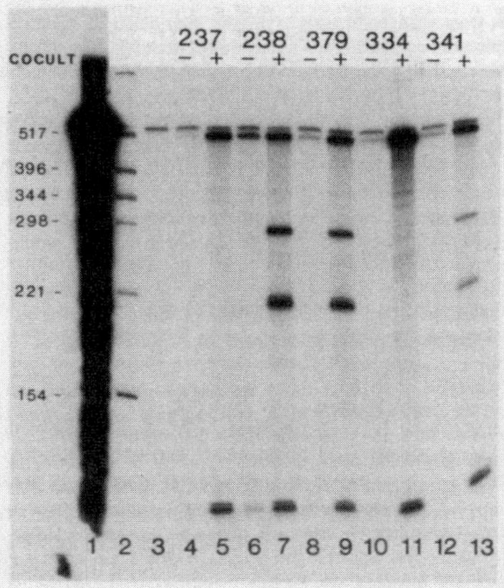

B

Figure 3. RNAase Protection Analysis of *vir* Mutants

(A) A SalI restriction map of the *vir* region together with overlapping cosmid clones pVK257, pVK242, and pVK225 from this region are shown. Also shown in the transcriptional organization of the *vir* loci and the position of transposon insertion mutations. The ability of each mutant strain to induce border-directed cleavage at border B is also shown (+ = cleavage; − = no cleavage).

(B) Protection of an RNA probe synthesized from pBL20 was performed for DNA isolated from Agrobacterium cultured in the absence (−) or presence (+) of plant cells. Agrobacterium strains harboring mutant Ti-plasmids include pTi237 (lanes 4 and 5), pTi238 (lanes 6 and 7), pTi379 (lanes 8 and 9), pTi334 (lanes 10 and 11), and pTi341 (lanes 12 and 13). These include mutations in the *virA, B, C, D,* and *E* loci respectively. No DNA was added to samples for lanes 1 and 3; no RNAase was added to the sample in lane 1. Lane 2 is a marker from pBR322 digested with HinfI.

indicating cleavage of the bottom strand at border B (Figure 4C, lanes 8–10). These data demonstrate that the endonuclease activity is encoded within the 5′ end of the *virD* operon. Three smaller derivatives (pMY1172, pMY1174, and pMY1176), or the "Tac" promoter vector alone (pMY1133), did not yield the 285 and 215 nucleotide long fragments, indicating that these plasmids do not contain sufficient information to specify the endonuclease activity. We also analyzed each of these same DNA preparations for cutting on the top strand at border B, and for cutting on both strands at border A. In each case, the DNA prepared from E. coli protected RNA fragments that comigrated with those protected by Agrobacterium DNA (data not shown). Cleavage by *virD* in Agrobacterium and E. coli resulted largely in bottom strand nicks, although a small fraction of top strand cleavage was also observed (data not shown; Albright et al.). We also used the technique of primer extension to map precisely the site of bottom strand cleavage in E. coli (data not shown). These data demonstrate that cleavage by *virD* occurs after the third nucleotide of the 24 bp repeat in both E. coli and Agrobacterium.

Nucleotide Sequence Analysis

We determined the nucleotide sequence of the portion of *virD* required for endonuclease activity. The 5′ region of *virD* was subcloned into the vectors pUC118 and pUC119 (J. Vierra, personal communication) and the nucleotide sequence determined by the chain termination method (Sanger et al., 1977) on both strands over its entire length (Figure 5). This sequence analysis identified two open reading frames, designated *virD1* and *virD2*, whose molecular weights are 16.2 and 47.4 kd, respectively. Each of these reading frames is preceded by a sequence homologous to the E. coli consensus ribosome binding sequence of TAAGGAGGTG . . . 5–9 bp . . . ATG (Shine and Dalgarno, 1974). *VirD1* is preceded by the sequence of GGAGG . . . 5 bp . . . ATG (bp 147–159) and *virD2* is preceded by the sequence GGAG . . . 8 bp . . . ATG (bp 621–636). The start of the third open reading frame of *virD* is also shown and is preceded by the sequence AAGGA . . . 7 bp . . . ATG (bp 1916–1930). The transcription initiation site was previously mapped to the T residue at bp 96, and the promoter for *virD* has been described (Das et al., 1986).

Protein Analysis

A coupled in vitro transcription–translation system (Collins et al., 1979) was used to visualize the *virD* protein products. Two distinct protein products of approximately 16 kd and 56 kd were expressed from the two larger *virD* constructs, pMY1151 and pMY1153 (Figure 6, lanes 4 and 5). These two polypeptides agree with the nucleotide sequence data (see above) which predicted the molecular

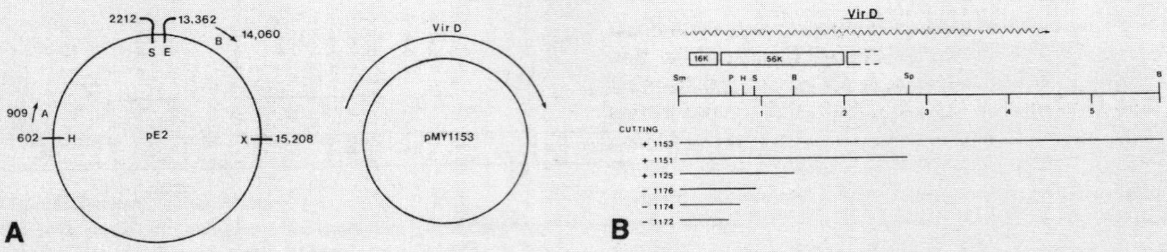

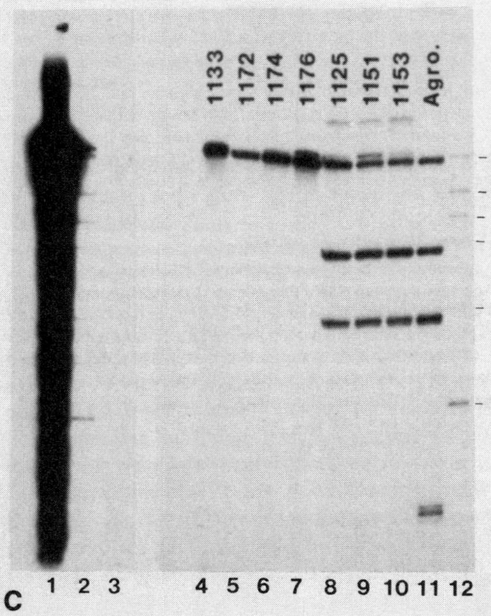

Figure 4. Analysis of Nicking in E. coli

(A) Plasmids containing various lengths of *virD*, fused to the "Tac" promoter, were assayed in *trans* to pE2 for their ability to induce border-specific cleavage. pE2 contains fragments from bp 602 to 2212, which contains border A (bp 909), and bp 13,362 to 15,208, which contains border B (bp 14,060) (Barker et al., 1983).

(B) The *virD* operon, defined by transposon mutagenesis, is shown by a wavy line; also shown is a partial restriction map encompassing this locus: Sm, SmaI; P, PstI; H, HindIII; B, BamHI; Sp, SphI. The locations of the first two open reading frames that code for proteins that migrate in SDS-polyacrylamide gels with apparent molecular weights of 16 and 56 kd are given. Restriction fragments fused to the "Tac" promoter include SmaI–BamHI–BamHI (pMY1153), SmaI–SphI (pMY1151), SmaI–BamHI (pMY1125), SmaI–SalI (pMY1176), SmaI–HindIII (pMY1174), and SmaI–PstI (pMY1172). The ability of these plasmids to direct border-specific cleavage is indicated by a "+" (cutting) or "−" (no cutting).

(C) Protection of an RNA probe synthesized from pBL20 was performed for DNA isolated from E. coli as described in Experimental Procedures. All E. coli strains harbor the plasmid pE2 and the following plasmids: pMY1133, lane 4; pMY1172, lane 5; pMY1174, lane 6; pMY1176, lane 7; pMY1125, lane 8; pMY1151, lane 9; and pMY1153, lane 10. An Agrobacterium control is shown in lane 11 for DNA isolated from A348 (pLA110). No DNA was added for samples in lanes 1 and 3; no RNAase was added to the sample in lane 1. Lane 2 is a marker with pBR322 digested with HinfI.

weights for the first two proteins of *virD* to be 16.2 kd and 47.4 kd. pMY1125, which contains only the first 1.4 kb of the *virD* locus, resulted in the expression of polypeptides of approximately 16 kd and 38 kd (Figure 6, lane 3). The 38 kd protein is most likely a truncated version of the 56 kd protein, and was not observed when smaller subclones of *virD* (pMY1172, pMY1174, and pMY1176) were used (data not shown). Our nucleotide sequence data indicate that this 38 kd polypeptide is a fusion of the amino-terminal 31.5 kd of *virD2* and vector sequences. We note that this truncated polypeptide, together with the 16 kd protein, retains full activity and specificity as a site-specific endonuclease.

Discussion

We have recently shown that after cocultivation of Agrobacterium with plant cells, cleavage occurs at unique and identical sites within each of the 24 bp direct repeats flanking the T-DNA (Albright et al., submitted). A strand-specific RNAase protection assay showed that cleavage does not occur in *virD* mutant strains of Agrobacterium (Figures 3A and 3B) and that the *virD* locus is sufficient to direct cleavage in E. coli (Figure 4). Primer extension demonstrated that cleavage by *virD* in E. coli and Agrobacterium occurs at identical locations. Therefore the *virD* operon encodes this border-specific endonuclease. This activity was further localized to the 5′ end of this operon for the following reasons. First, only mutations of the 5′ end of *virD* abolished cleavage, whereas cleavage was still observed for mutations toward the 3′ end of this operon (Figure 2A). Second, deletions in *virD* demonstrated that only the first 1.4 kb of this operon were required for cleavage in E. coli (Figure 4).

Nucleotide sequence analysis of the 5′ end of the *virD* operon identified two open reading frames of 16.2 and 47.4 kd (Figure 5). A coupled transcription–translation system synthesized these protein products (Figure 6). The second (47.4 kd) *virD* protein, which is characterized by its high degree of hydrophilicity, migrated in SDS-polyacrylamide gels with an apparent molecular weight of approximately 56 kd (Figure 6, lanes 4 and 5). This protein was shown definitively to be encoded within this region by demonstrating that a deletion within this open reading frame yielded the predicted truncated protein (Figure 6, lane 3). The reason for the discrepancy in predicted and apparent molecular weights is unclear, but has been observed in other proteins. For example, the first protein of the *virC* operon is predicted to be 25.7 kd and migrates with an apparent molecular weight of 31 kd (Yanofsky and

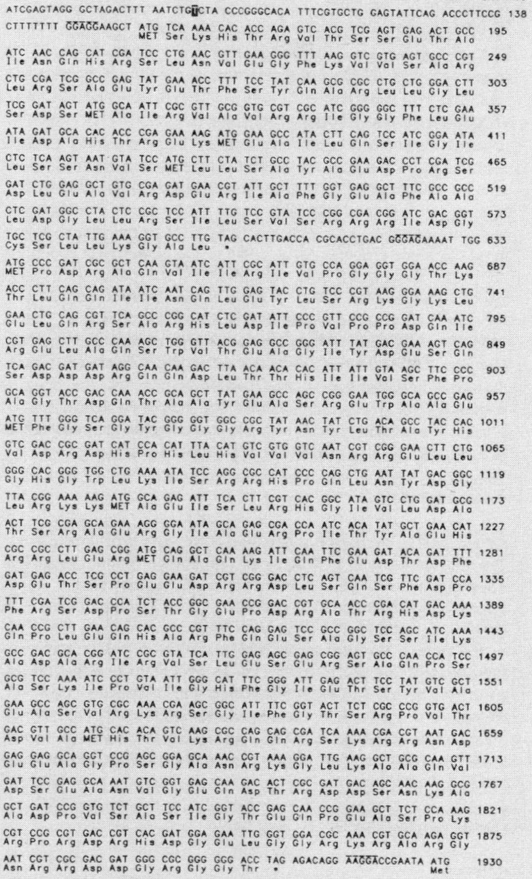

Figure 5. Nucleotide Sequence Analysis

The nucleotide sequence for the 5′ end of the *virD* locus is shown, together with the predicted amino acid sequences of the first two open reading frames. Presumed ribosome binding sites are overlined and the transcription initiation site is shown for the T residue at position 96.

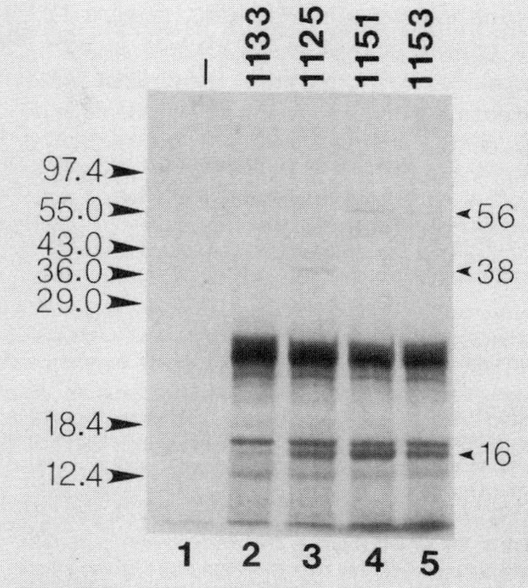

Figure 6. Analysis of Protein Products

A coupled in vitro transcription–translation system was used to express the *virD* protein products as described in Experimental Procedures. Plasmids include pMY1133 (lane 2), pMY1125 (lane 3), pMY1151 (lane 4), and pMY1153 (lane 5). No plasmid DNA was used for lane 1. The position of the *virD*-specific proteins of approximately 16, 38, and 56 kd are shown. Molecular weight markers (in kilodaltons) include phosphorylase B (97.4), glutamate dehydrogenase (55.0), ovalbumin (43.0), lactate dehydrogenase (36.0), carbonic anhydrase (29.0), lactoglobulin (18.4), and cytochrome c (12.4).

Nester, 1986). Similarly, a *virE* protein has a predicted size of 60 kd but migrates as a 66 kd protein (Winans et al., submitted). Similar aberrant mobilities on gels have been observed for the products of *rpoD*, *nusA*, and *ntrA* (Merrick et al., 1985). The 16.2 kd protein shares a region of homology with a previously characterized DNA-binding domain (Drummond et al., 1986), suggesting that it may also bind DNA. A computer search of the Protein Identification Resource (PIR) database, using the program of FASTP (Lipman and Pearson, 1985), failed to identify any significant sequence homology to either of the *virD* gene products. Neither of the two open reading frames for *virD* presented here are consistent with the previously reported sequence from the C58 Ti-plasmid (Hagiya et al., 1985).

Deletions that abolish expression of the 47.4 kd protein also eliminate endonuclease activity (Figure 4C, lanes 5–7), demonstrating that this protein is required for cleavage. These deletions still express the 16.2 kd protein, indicating that this protein is not sufficient for cutting. Thus the 47.4 kd protein, perhaps together with the 16.2 kd protein, encodes the endonuclease activity. It has been suggested that the 24 bp direct repeats flanking the T-DNA are similar to the *cis*-acting conjugal transfer origins (oriT) (Wang et al., 1984; Albright et al., submitted). It is tempting to speculate that these *virD* proteins are functionally analogous to the endonuclease activities responsible for nicking at oriT since each of these sets of proteins generates single-stranded cleavages, an early step in the transfer of DNA from donor to recipient. For the F plasmid, two proteins make up the *traYZ* endonuclease, and for the plasmid ColE1, proteins of 11, 16, and 60 kd are involved in nicking at oriT (for review see Willetts and Wilkins, 1984).

Several groups of investigators have isolated circular T-DNA molecules using strong genetic selection pressures (Koukolikova-Nicola et al., 1985; Machida et al., 1986). These molecules contain a single T-DNA border which is a fusion between the left and right borders, the formal equivalent of a site-specific recombinational event. More recently, Alt-Moerbe and co-workers (1986) have shown that proteins in the *virC-D* region are sufficient to direct the formation of T-DNA circles in E. coli. They identified protein products of 56, 43, 31, and 15 kd involved in this process. The 15 and 56 kd proteins correspond to the first two open reading frames of *virD* reported here, and the 31 kd protein is the first protein of the *virC* operon (Yanofsky and Nester, 1986). It is perhaps significant that T-DNA circle formation in E. coli occurs at a very low frequency (approximately 10^{-5}), whereas the endonuclease activity described here results in the cleavage of more than half of the T-DNA molecules (compare the intensities of the 502 nucleotide fragment to those of the 285 and 215 nucleotide fragments in Figure 4C, lanes 8–11). Since similar proteins appear to be involved in both border cleavage and circle formation, it seems likely that the endonuclease activity described here is involved in creating molecules that can be recovered as T-DNA circles. Whether T-DNA circles are an artifact of the powerful selection used to recover them or whether they represent true intermediates in the transfer process remains to be determined.

Experimental Procedures

Bacterial Strains and Plant Cell Culture

A348 is Agrobacterium tumefaciens strain A136 containing pTiA6NC (Garfinkel et al., 1981). pVK225, pVK242, and pVK257 are cosmid clones from the *vir* region of pTiA6NC (Knauf and Nester, 1982). Mutations in the *vir* region result from insertion of the transposon Tn3HoHo1 into each of the *vir* loci (Stachel and Nester, 1986). pE2 (Klee et al., 1985) is a binary vector that contains fragments spanning the left and right border regions (borders A and B, Barker et al., 1983) inserted into the broad host range replicon pTJS75 (Schmidhauser et al., 1986). GEM transcription vectors were from Promega Biotec. E. coli hosts included JM101 and MV1193 and were maintained on either L agar or M9 minimal agar (Miller, 1972) supplemented with the appropriate antibiotic (kanamycin, 50 μg/ml; carbenicillin, 100 μg/ml; tetracycline, 10 μg/ml). Agrobacterium strains were maintained on AB minimal agar (Chilton et al., 1974) supplemented with the appropriate antibiotic (kanamycin, 100 μg/ml; carbenicillin, 100 μg/ml). N. tabacum suspension culture cells were maintained in liquid Murashige and Skoog (1962) medium supplemented with 0.2 mg/1 2,4-dichlorophenoxy acetic acid, as described by An (1985).

Enzymes and Reagents

Restriction endonucleases were purchased from Bethesda Research Laboratories or New England Biolabs and used according to the supplier's recommendations. DNA polymerase I Klenow fragment was purchased from Boehringer-Mannheim. The M13 sequencing primer and sequencing mixes were from Pharmacia. ^{35}S-dATP and ^{32}P-CTP were purchased from New England Nuclear. Ribonuclease A, type III-A, was from Sigma; exonuclease III, ribonuclease T1, and T7 RNA polymerase were from Bethesda Research Laboratories. RNAase-free DNAase and RNAasin ribonuclease inhibitor were from Promega Biotec. Unlabeled rNTPs were from P-L Biochemicals.

Plasmid Constructions

The SmaI–BamHI fragment carrying the 5′ end of the *virD* locus was inserted into the vector pUC18CM (Schmidhauser et al., 1986) to yield pMY1122. An EcoRI–PvuII fragment, carrying the "Tac" promoter (Amann et al., 1983) was inserted into the EcoRI–SmaI sites of PMY1122 to yield the plasmid pMY1125. This placed the *virD* sequences under "Tac" promotor control. This same "Tac" promoter fragment was also inserted into the pUC18CM vector to create pMY1133, a vector-only control. Digestion of pMY1125 with PstI, HindIII, or SalI, followed by ligation and transformation, resulted in deletions that retained only the SmaI–PstI (pMY1172), SmaI–HindIII (pMY1174), or SmaI–SalI (pMY1176) fragments of *virD* (Figure 4B). Insertion of the BamHI–SphI fragment of *virD* into pMY1125 resulted in pMY1151; insertion of the entire BamHI fragment resulted in pMY1153, which contains the entire *virD* locus under "Tac" promoter control (Figure 4B). The SalI 13b fragment, carrying the *virG* locus, was inserted into pVK102 (Knauf and Nester, 1982) to yield pLA110; SalI 7b was also inserted into pVK102 yielding pLA111.

DNA Isolation and RNAase Protection

Plasmid DNA was isolated by the method of Birnboim and Doly (1979). Agrobacterium cocultivations were performed exactly as described (Albright et al., submitted). Briefly, log phase Agrobacterium strains were cultivated in the presence or absence of Nicotiana tabacum suspension cells for approximately 16 h at 28°C in a rotary shaker. The bacteria were separated from tobacco cells by filtration through Miracloth (Calbiochem) and DNA was isolated by methods previously described (Albright et al., submitted). E. coli strains, harboring various *virD* constructions and the binary vector pE2 (Klee et al., 1985), were grown at 37°C in L broth to an A_{600} of 0.5 before adding IPTG to a final concentration of 1 mM. After 2 hr of further growth, DNA was extracted as described for Agrobacterium above.

DNA Sequencing and Protein Analysis

Restriction fragments of *virD* were inserted into the vectors pUC118 and pUC119 (J. Vieira, personal communication) and overlapping deletions were constructed by exonuclease III and nuclease S1 digestion as previously described (Henikoff et al., 1984). Single-stranded templates were prepared by infecting with M13K07 helper phage essentially as described (Zagursky et al., 1984; J. Vieira, personal communication) and sequencing was performed by the chain termination method of Sanger et al. (1977). In some cases, synthetic oligonucleotides were used to prime sequencing reactions (Strauss et al., 1986). Nucleotide sequencing was performed on both strands over their entire lengths. Protein analyses were carried out with a coupled transcription translation system, purchased from Amersham and described by Collins et al. (1979). For each reaction, 2 μg of purified plasmid DNA was used. The labeled proteins were separated on SDS-polyacrylamide gels (Laemmli, 1970) and visualized by autoradiography.

Acknowledgments

We thank Drs. J. Ward, Srinivasan, and S. Winans for critical reading of the manuscript, and S. Bradley for preparation of the manuscript. We also thank L. Wong for excellent technical assistance. M. F. Y. was supported by PHS NRSA 5 T32 GM07270, C. Y. was supported by PHS NRSA 1 F32 GM11082-01, and L. M. A. was supported by a fellowship from the Damon Runyon–Walter Winchell Cancer Fund (DRG 819). This work was supported by 5 R01 GM32618-14 from the National Institutes of Health and 84-CRCR-1-1438 from the USDA.

The costs of publication of this article were defrayed in part by the payment of page charges. This article must therefore be hereby marked "*advertisement*" in accordance with 18 U.S.C. Section 1734 solely to indicate this fact.

Received August 18, 1986; revised September 24, 1986.

References

Akiyoshi, D. E., Klee, H., Amasino, R. M., Nester, E. W., and Gordon, M. P. (1984). T-DNA of Agrobacterium tumefaciens encodes an enzyme of cytokinin biosynthesis. Proc. Natl. Acad. Sci. USA *81*, 5994–5998.

Alt-Moerbe, B. Rak, and Schröder, J. (1986). A 3.6 kbp segment from the *vir* region of Ti plasmids contains genes responsible for border sequence-directed production of T region circles in E. coli. EMBO J. *5*, 1129–1135.

Amann, E., Brosius, J., and Ptashne, M. (1983). Vectors bearing a hybrid trp-lac promoter useful for regulated expression of cloned genes in Escherichia coli. Gene *25*, 167–178.

An, G. (1985). High efficiency transformation of cultured tobacco cells. Plant Physiol. *79*, 568–570.

Barker, R. F., Idler, K. B., Thompson, D. V., and Kemp, J. D. (1983). Nucleotide sequence of the T-DNA region from the Agrobacterium tumefaciens octopine Ti plasmid pTi15955. Plant Mol. Biol. *2*, 335–350.

Barry, G. F., Rogers, S. G., Fraley, R. T., and Brand, L. (1984). Identification of a cloned cytokinin biosynthetic gene. Proc. Natl. Acad. Sci. USA *81*, 4776–4780.

Bevan, M. W., and Chilton, M.-D. (1982). T-DNA of the Agrobacterium Ti and Ti plasmids. Ann. Rev. Genet. *16*, 357–384.

Birnboim, H. C., and Doly, J. (1979). A rapid alkaline extraction procedure for screening recombinant plasmid DNA. Nucl. Acids Res. *7*, 1513–1523.

Chilton, M.-D., Currier, T. C., Farrand, S. K., Bendich, A. J., Gordon, M. P., and Nester, E. W. (1974). Agrobacterium DNA and PS8 bacteriophage DNA not detected in crown gall tumors. Proc. Natl. Acad. Sci. USA *71*, 3672–3676.

Collins, J. (1979). Cell-free synthesis of proteins coding for metabolism functions of ColE1 and transposition functions of Tn3. Gene *6*, 29–42.

Das, A., Stachel, S., Ebert, P., Allenza, P., Montoya, A., and Nester, E. (1986). Promoters of Agrobacterium tumefaciens Ti-plasmid virulence genes. Nucl. Acids Res. *14*, 1335–1364.

Drummond, M., Whitty, P., and Wootton, J. (1986). Sequence and domain relationships of *ntrC* and *nifA* from Klebsiella pneumoniae: homologies to other regulatory proteins. EMBO J. *5*, 441–447.

Hagiya, M., Close, T. J., Tait, R. C., and Kado, C. I. (1985). Identification of pTiC58 plasmid-encoded proteins for virulence in Agrobacterium tumefaciens. Proc. Natl. Acad. Sci. USA *82*, 2669–2673.

Henikoff, S. (1984). Unidirectional digestion with exonuclease III creates targeted breakpoints for DNA sequencing. Gene *28*, 351–359.

Hille, J., van Kan, J., and Schilperoort, R. (1984). *trans*-acting virulence functions of the octopine Ti plasmid from Agrobacterium tumefaciens. J. Bacteriol. *158*, 754–756.

Hooykaas, P. J., and Schilperoort, R. A. (1984). The molecular genetics of crown gall tumorigenesis. Adv. Genet. *22*, 210–283.

Klee, H. J., White, F. F., Iyer, V. N., Gordon, M. P., and Nester, E. W. (1983). Mutational analysis of the virulence region of an Agrobacterium tumefaciens Ti-plasmid. J. Bacteriol. *153*, 878–883.

Knauf, V. C., and Nester, E. W. (1982). Wide host range cloning vectors: a cosmid clone bank of an Agrobacterium Ti plasmid. Plasmid *8*, 45–54.

Koukoliková-Nicola, Z., Shillito, R. D., Hohn, B., Wang, K., Van Montagu, M., and Zambryski, P. (1985). Involvement of circular intermediates in the transfer of T-DNA from Agrobacterium tumefaciens to plant cells. Nature *313*, 191–196.

Laemmli, U. K. (1970). Cleavage of structural proteins during the assembly of the head of bacteriophage T4. Nature *277*, 680–685.

Lipman, D. J., and Pearson, W. R. (1985). Rapid and sensitive protein similarity searches. Science *227*, 1435–1441.

Machida, Y., Usami, S., Yamamoto, A., and Takebe, I. (1986). Plant-inducible recombination between the 25-base-pair border sequences of T-DNA in Agrobacterium tumefaciens. Mol. Gen. Genet., in press.

Merrick, M. J., and Gibbins, J. R. (1985). The nucleotide sequence of the nitrogen-regulation gene *ntrA* of Klebsiella pneumoniae and comparison with conserved features in bacterial RNA polymerase sigma factors. Nucl. Acids Res. *13*, 7607–7620.

Messing, J., Crea, R., and Seeburg, P. H. (1981). A system for shotgun DNA sequencing. Nucl. Acids Res. *9*, 309–314.

Miller, J. H. (1972). Experiments in Molecular Genetics (Cold Spring Harbor, New York: Cold Spring Harbor Laboratory), p. 21.

Murashige, T., and Skoog, F. (1962). A revised medium for rapid growth and bio assays with tobacco tissue cultures. Physiol. Plant *15*, 473–497.

Nester, E. W., Gordon, M. P., Amasino, R. M., and Yanofsky, M. F. (1984). Crown gall: a molecular and physiological analysis. Ann. Rev. Plant Physiol. *35*, 387–413.

Sanger, F., Nicklen, S., and Coulson, A. R. (1977). DNA sequencing with chain terminating inhibitors. Proc. Natl. Acad. Sci. USA *74*, 5463–5467.

Schmidhauser, T. J., and Helinski, D. R. (1985). Regions of broad-host-range plasmid RK2 involved in replication and stable maintenance in nine species of gram-negative bacteria. J. Bacteriol. *164*, 446–455.

Schröder, G., Waffenschmidt, S., Weiler, E. W., and Schröder, J. (1983). The T-region of Ti-plasmids codes for an enzyme synthesizing indole-3-acetic acide. EMBO J. *2*, 403–409.

Shine, J., and Dalgarno, L. (1974). The 3′-terminal sequence of Escherichia coli 16S ribosomal RNA: complementarity of nonsense triplets and ribosome binding sites. Proc. Natl. Acad. Sci. USA *71*, 1342–1346.

Stachel, S. E., and Nester, E. W. (1986). The genetic and transcriptional organization of the *vir* region of the A6 Ti-plasmid of Agrobacterium tumefaciens. EMBO J. *5*, 1445–1454.

Stachel, S. E., and Zambryski, P. C. (1986). *VirA* and *virG* control the plant-induced activation of the T-DNA transfer process of Agrobacterium tumefaciens. Cell *46*, 325–333.

Stachel, S. E., Messens, E., Van Montagu, M., and Zambryski, P. (1985). Identification of the signal molecules produced by wounded plant cells that activate T-DNA transfer in Agrobacterium tumefaciens. Nature *318*, 624–629.

Stachel, S. E., Timmerman, B., and Zambryski, P. (1986). Generation of single-stranded T-DNA molecules during the initial stages of T-DNA transfer from Agrobacterium tumefaciens to plant cells. Nature *322*, 706–712.

Strauss, E. C., Kobori, J. A., Su, G., and Hood, L. (1986). Specific-primer-directed DNA sequencing. Anal. Biochem. *154*, 353–360.

Thomashow, L., Reeves, S., and Thomashow, M. F. (1984). Crown gall oncogenesis: evidence that a T-DNA gene from the Agrobacterium Ti-plasmid pTiA6 encodes an enzyme that catalyzes synthesis of indoleacetic acid. Proc. Natl. Acad. Sci. USA *81*, 5071–5075.

Thomashow, M. F., Hugly, S., Buchholz, W. G., and Thomashow, L. S. (1986). Molecular basis for the auxin-independent phenotype of crown gall tumor tissues. Science *231*, 616–618.

Wang, K., Herrera-Estrella, L., Van Montagu, M., and Zambryski, P. (1984). Right 25 bp terminus sequence of the nopaline T-DNA is essential for and determines direction of DNA transfer from Agrobacterium to the plant genome. Cell *38*, 455–462.

Willetts, N., and Wilkins, B. (1984). Processing of plasmid DNA during bacterial conjugation. Microbiol. Rev. *48*, 24–41.

Winans, S. C., Ebert, P. R., Stachel, S. E., Gordon, M. P., and Nester, E. W. (1986). A gene essential for Agrobacterium virulence is homologous to a family of positive regulatory loci. Proc. Natl. Acad. Sci. USA, in press.

Yadav, N. S., Vanderleyden, J., Bennett, D. R., Barnes, W. M., and Chilton, M.-D. (1982). Short direct repeats flank the T-DNA on a nopaline Ti-plasmid. Proc. Natl. Acad. Sci. USA *79*, 6322–6326.

Yanofsky, M., Lowe, B., Montoya, A., Rubin, R., Krul, W., Gordon, M., and Nester, E. (1985). Molecular and genetic analysis of factors controlling host range in Agrobacterium tumefaciens. Mol. Gen. Genet. *201*, 237–246.

Yanofsky, M. F., and Nester, E. W. (1986). Molecular characterization of a host range determining locus from Agrobacterium tumefaciens. J. Bacteriol., in press.

Zagursky, R. J., and Berman, M. L. (1984). Cloning vectors that yield high levels of single-stranded DNA for rapid DNA sequencing. Gene *27*, 183–191.

VirD TARGETS T-DNA INTO THE PLANT NUCLEUS

Alfredo Herrera-Estrella
Departamento de Ingeniería Genética, Unidad de Biotecnología Agrícola
Centro de Investigacion y Estudios Avanzados Irapuato, Guanajuato, Mexico
E-mail: aherrera@ira.cinvestav.mx

Commentary on the paper:

Herrera-Estrella, A., Van Montagu, M., and Wang, K. 1990. A bacterial peptide acting as a plant nuclear targeting signal: The amino-terminal portion of *Agrobacterium* VirD2 protein directs a β-galactosidase fusion protein into tobacco nuclei. Proc. Natl. Acad. Sci. U.S.A. 87:9534–537.

My dreams of working with genetic engineering became reality while visiting my brother Luis at one of the leading laboratories in plant genetic engineering in Gent, Belgium, where I briefly met Marc Van Montagu. Marc said I would be welcome to join their team. I arrived to work in Gent in April 1986. During a short meeting with Marc, it was decided that I would work on two subjects during my doctorate. The first subject would be the *Agrobacterium* virulence genes and the second the mycoparasite *Trichoderma*—with which I still work. Right after the meeting I went to meet Scott Stachel who, I was told, had all the information I needed regarding *Agrobacterium.* Scott was packing because he and Patricia Zambryski were leaving the lab in Gent. After an unforgettable encounter with Scott I had some of the information I needed.

The information, which I hardly understood, indicated that it was already known that the 25-bp direct repeats at the ends of the T-DNA were the only factors required *in cis* for its mobilization to the plant cell. It was also known that T-DNA transfer is polar, which together with the functionality of the T-DNA borders reflected their role in the generation of a free transferable T-DNA copy following the activation of *vir* gene expression. Other *vir*-induced changes observed that occurred at the T-DNA borders included the single-stranded endonucleolytic cleavages of the 25-bp border repeat (Stachel et al. 1986; Wang et al. 1987). Only two proteins encoded by the 5′-half of the *virD* locus, namely the VirD1 and VirD2 polypeptides, were known to directly participate in the generation of T-strand molecules, these proteins being responsible for the T-DNA border specific endonuclease activity (Yamamoto et al. 1987). Resultant nicks at the right and left borders were presumably then used as initiation and termination sites for T-strand production.

However, as a rookie, I hardly got the picture and began looking at the expression of *vir* genes when transferred to other bacterial systems. Nothing came of these experiments. Luckily, lunchtime was full of surprises. One surprise was to meet Kan Wang, a nice Chinese doctoral student. She used to come to my lab and say a single word: "eat". Her English was extremely difficult for me to follow. As foreigners we had little to do besides work, so partying with other friends was our major distraction. One of those party nights (at around 3 am) Kan asked me to go back to the lab and help her load a gel. That night she began talking to me about *Agrobacterium* and teaching me the secrets of the system. After about a year of working together on the induction of the *vir* genes it was clear to me that we were not getting anything exciting and that the *virD* genes had to play another important role in the T-DNA transfer process besides nicking the DNA. Kan had the same idea. However, for one reason or other, it seemed that we were not supposed to work on VirD. After working hard on convincing her, I managed to get her to lead our work on VirD using the following argument: the physical structure and proposed (nicked) intermediates in the synthesis of the T-strand suggested a possible mechanism for T-DNA transfer to plants. By further assuming that biological processes are not evolutionary unique, it was likely that *Agrobacterium* had modified a pre-existing process to suit its ends. In fact, there were many aspects of T-DNA transfer that were clearly analogous to bacterial conjugation (Stachel et al. 1986). T-DNA border nicks were analogous to nicks at the origin of conjugal DNA transfer, and the T-strand was analogous to the linear donor ssDNA molecule produced during bacterial conjugation. Furthermore, bacterial conjugation requires direct contact between donor and recipient cells, and the same requirement held for *Agrobacterium*-mediated T-DNA transfer. Our work on *vir*D allowed us to catch up with other groups in studies that indicated that the *vir*D2 product was tightly associated with the 5′-termini of T-strands and nicked molecules (Herrera-Estrella et al. 1988). Those findings led us to the idea that this protein could potentially serve as a pilot and/or nuclear localization signal that would facilitate translocation of the transfer complex through bacterial and plant membranes.

The production of a T-complex molecule consisting of T-strand and covalently bound protein (Herrera-Estrella 1988) was similar to DNA-protein complexes demonstrated for virus, phage, and plasmid relaxation complexes. In all cases the protein-DNA interaction took place at the 5′-terminus of the DNA molecule. Strong support for the model, in which T-DNA transfer is analogous to bacterial conjugation but applied to plants, derived from experiments showing that the origin of transfer (oriT) from a conjugative *E. coli* plasmid, pRSF-1010, can substitute for T-DNA borders in directing DNA transfer to plant cells from *Agrobacterium* (Buchanan-Wollaston et al. 1987). So a conjugation model was sufficient to explain the entire process of *Agrobacterium* infection of plant cells. Unlike DNA transferred between bacteria during conjugation, T-DNA should be capable of penetrating the plant cell wall, localizing and penetrating the plant cell membrane, and integrating efficiently into the genome, which is encased in a complex nuclear membrane. These latter features were more reminiscent of virus/cell interactions. Thus, T-DNA transfer should involve at least two phases: initial T-DNA transfer to the plant cytoplasm by a conjugative process, and later steps by more virus-like processes.

Based on the available information it was possible to suggest a sequence of events leading to T-DNA transfer from bacteria to plants. The process was initiated with the T-DNA border cleavage by the *virD* encoded endonuclease activity (Wang et al. 1987). The second step would be the 5′ to 3′ replication of the bottom strand from the nicked border, which produces a single-stranded DNA molecule (Stachel et al. 1986) with VirD proteins associated at the 5′-terminus (Herrera-Estrella et al. 1988). This T-complex subsequently would associate with non-specific single-stranded DNA binding proteins, such as VirE2, resulting in the formation of a coated intermediate molecule. These proteins would protect the T-strand from attack by nucleases, and could also participate in active transport through the cytoplasmic membrane in conjunction with membrane-associated proteins such as those encoded by the *vir*B locus. However, there was still the issue of the nuclear membrane that had to be solved.

Because we were heavily involved in our work with *vir*D and at that time sequencing was still not a simple task and the data could have mistakes, we decided to re-sequence the *vir*D operon. When analyzing the sequence data we realized, with the help of our favorite in house paper reviewer, Allan Caplan, that there was an obvious nuclear localization sequence in the N-terminal region of *vir*D2. It seemed to us that to prove that *vir*D2 helped the T-DNA get into the nucleus was a very difficult task. Nevertheless we tried to purify the T-complex to prove that introduction of the complex by any mechanical means or electroporation into plant cells would result in more efficient integration of the T-complex into the plant genome than the non-protein linked DNA. However, all efforts to purify sufficient amounts of the T-complex were unsuccessful. An alternative approach to proving that VirD2 directs nuclear targeting was to fuse ths protein to a reporter enzyme, and prove that it led the reporter into the plant nucleus. Fortunately, our friend Teemu Teeri in Finland had just developed a series of vectors to generate fusions with *lacZ* and had all the protocols for its use as a reporter in plants. We received all of the necessary materials and information from Teemu and made the constructs to transform plants. Once we had obtained transformed plants it took us a month to obtain what we thought were convincing data proving that VirD2 targeted lacZ into the plant nucleus. Obviously we were very excited but Kan had already left the lab and we had to write the manuscript exchanging comments via long telephone calls and fax (no e-mail). After finishing the manuscript, we thought it was important enough to publish in a major journal. We sent the manuscript to our preferred journal; we received very positive comments from one reviewer but the second reviewer was not satisfied and the paper was rejected. We solved all problems encountered and sent the manuscript to our second choice journal, again one positive review and one with objections, apparently by the same reviewer who had previously rejected it. Once again we solved all problems and sent out the manuscript. This time the manuscript was finally accepted. Our work was later confirmed and extended by other groups, who in addition found a function for the C-terminal nuclear localization signals that we had ignored.

References

Buchanan-Wollaston, V., Passiatore, J. E., and Cannon, F. 1987. The mob and oriT mobilization functions of a bacterial plasmid promote its transfer to plants. Nature 328:172–175.

Herrera-Estrella, A., Chen, Z.-M., Van Montagu, M., and Wang, K. 1988. VirD proteins of *Agrobacterium tumefaciens* are required for the formation of a covalent DNA-protein complex at the 5′ terminus of T-strand molecules. EMBO J. 7:4055–4062.

Stachel, S. E., Timmerman, B., and Zambryski, P. 1986. Generation of single-stranded T-DNA molecules during the initial stages of T-DNA transfer from *Agrobacterium tumefaciens* to plant cells. Nature 322: 706–712.

Wang, K., Stachel, S., Timmerman, B., Van Montagu, M., and Zambryski, P. 1987. Site-specific nick in the T-DNA border sequence following *vir* gene expression in *Agrobacterium.* Science 235:587–591.

Yamamoto, A., Iwahashi, M., Yanofsky, M. F., Nester, E. W., Takebe, I., and Machida, Y. 1987. The promoter proximal region in the virD locus of *Agrobacterium tumefaciens* is necessary for the plant-inducible circularization of T-DNA. Mol. Gen. Genet. 206: 174–177.

ALFREDO HERRERA-ESTRELLA: In 1985, I obtained my B.S. in Biochemical Engineering from Escuela Nacional de Ciencias Biologicas, Instituto Politécnico Nacional, Mexico. I did my graduate studies at State University of Ghent and graduated with a Ph.D. in 1990.

My official mentor was Marc Van Montagu, although the person who actually introduced me to science and led me in the art, was Kan Wang, who was then an experienced doctorate student.

Proc. Natl. Acad. Sci. USA
Vol. 87, pp. 9534–9537, December 1990
Cell Biology

A bacterial peptide acting as a plant nuclear targeting signal: The amino-terminal portion of *Agrobacterium* VirD2 protein directs a β-galactosidase fusion protein into tobacco nuclei

(protein transport/T-complex/transformation/virulence genes)

A. HERRERA-ESTRELLA, M. VAN MONTAGU, AND K. WANG*

Laboratorium voor Genetica, Rijksuniversiteit Gent, B-9000 Gent, Belgium

Contributed by M. Van Montagu, September 10, 1990

ABSTRACT ***Agrobacterium tumefaciens*** **is a soil bacterium capable of transferring DNA to the genome of higher plants. Of the virulence region-encoded proteins of the tumor-inducing (Ti) plasmid of *A. tumefaciens*, the VirD1 and VirD2 proteins are essential for T-DNA transfer to plant cells. These two proteins have been shown to be directly responsible for the formation of T-strands. VirD2 was also shown to be firmly attached to the 5′ termini of T-strands; these facts have led to its postulation as a pilot protein in the T-DNA transfer process and as a nucleus-targeting signal in plants. We have constructed a chimeric gene by fusing the *virD2* gene and the *Escherichia coli lacZ* gene. Cell fractionation and electron microscopy studies with transgenic tobacco plants containing the VirD2–LacZ fusion protein indicate that the first 292 amino acids of VirD2 are able to direct the cytoplasmic protein β-galactosidase to the plant nucleus. This provides an example of cross-kingdom nuclear localization between two free-living organisms: a bacterial peptide is capable of acting as a eukaryotic (plant) nuclear targeting signal.**

Agrobacterium tumefaciens is a soil bacterium capable of transferring DNA to the genome of higher plants (1). Among the virulence region-encoded proteins of its tumor-inducing (Ti) plasmid, only two polypeptides, VirD1 and VirD2, encoded by the 5′ half of the *Agrobacterium virD* locus, are known to participate directly in the generation of T-strand DNA molecules (1–4). Further insight into these polypeptides has shown (5, 6) that VirD1 possesses a DNA-relaxing activity and the VirD2 is a strand-specific and sequence-specific endonuclease. Recent studies (7–11) indicate that the VirD2 product becomes tightly associated with the 5′ termini of T-strand and nicked molecules. All of the processes summarized above have evolved to produce a structure, the T-DNA transfer complex, capable of penetrating bacterial and recipient plant cell plasma membranes and possibly cell walls to introduce new genetic information into the plant genome. Major functions that may be attributed to the proteins of the transfer complex (encoded by the *virD* and *virE* operons) include protection of the T-strand molecule (8), recognition of transmembrane routes from the bacterium to the plant cell (1), targeting of the transfer complex to the plant nucleus, and efficient integration of the transferred DNA into the plant genome.

MATERIALS AND METHODS

Plasmid Construction and Plant Transformation. Plasmid pTR-lacZ was constructed as follows. A 3.1-kilobase *Nco* I–*Bam*HI fragment containing the *Escherichia coli lacZ* gene (a gift from T. Teeri, University of Helsinki, Finland) was cloned behind the plant promoter T_R2' (12) at the *Xba* I site in vector pAH1. pAH1 is a derivative of pC27-3 (13) in which an ATG codon within polylinkers was deleted. The selectable marker gene *nptII*, encoding neomycin phosphotransferase II, is driven by the other half of this dual promoter (T_R1'). To construct pTR-vd-lacZ, the same *lacZ* fragment was first translationally fused to an 881-base-pair *Xmn* I fragment carrying 65% of the 5′ portion of the *virD2* gene. This chimeric gene was then cloned at the *Xba* I site in pAH1 as in pTR-lacZ.

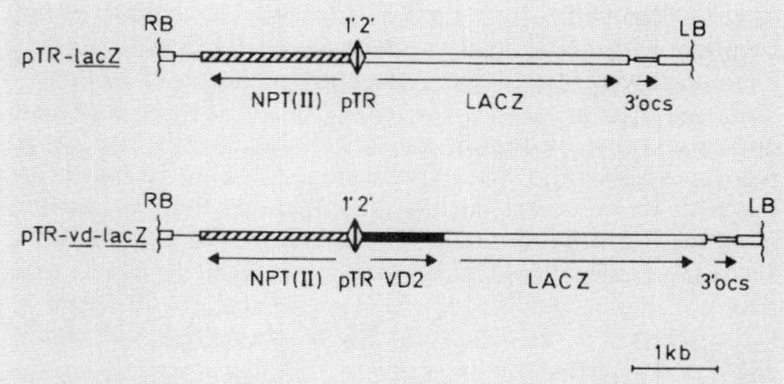

FIG. 1. Constructs used to determine the function of the VirD2 polypeptide in nuclear localization. LB and RB, left and right borders of the T-DNA; 3′ocs, 3′ terminator of the octopine synthase gene; kb, kilobase.

Plasmids pTR-lacZ and pTR-vd-lacZ (see Fig. 1) were mobilized into *A. tumefaciens* harboring pGV2260 (14) by triparental crosses using pRK2013 (15) as a helper plasmid. Leaf disc of *Nicotiana tabacum* (W38) was used for plant transformation as described (16).

Isolation of Nuclei and Assay of β-Galactosidase Activity in Subcellular Fractions. Nuclei were isolated according to Mazzolini *et al.* (17). Samples were treated with 2% glutaraldehyde prior to the assay. Activity was measured using *o*-nitrophenyl β-D-galactoside as substrate, as described by Miller (18) except that the pH of the buffer was 7.4.

For the electrophoretic analysis, equivalent amounts of protein from the three subcellular fractions (nuclear, organellar, and cytoplasmic) of the TR-VD-LACZ and TR-LACZ transgenic plants were subjected to SDS/PAGE (19). The gel was then used to assay for β-galactosidase activity as described (20).

Protease Protection. Nuclei of the TR-VD-LACZ and TR-LACZ transgenic plants were incubated with Pronase (125 μg/ml; Calbiochem-Behring) in 10 mM Tris·HCl, pH 7.6/2 mM $CaCl_2$ either with 0.2 M sucrose (intact nuclei) or without sucrose (and vigorously mixed with a Vortex to break the

The publication costs of this article were defrayed in part by page charge payment. This article must therefore be hereby marked "*advertisement*" in accordance with 18 U.S.C. §1734 solely to indicate this fact.

Abbreviation: X-Gal, 5-bromo-4-chloro-3-indolyl β-D-galactopyranoside.

*Present address: Garst Seed Company, Research Department, Box 500, Slater, IA 50224.

nuclei) for 3 hr at 37°C. Untreated nuclei were incubated in the same conditions but no Pronase was added.

Electron Microscopy. Tissue sections and nuclei were reacted with the chromogenic substrate 5-bromo-4-chloro-3-indolyl β-D-galactopyranoside (X-Gal) after fixation with 2% glutaraldehyde (20) and were embedded in LR white acrylic resin (London Resin, London) (T. Hürek, unpublished work). β-Galactosidase activity is detected by the formation of electron-dense crystals of indigo when X-Gal is used as substrate.

RESULTS

A major obstacle in studying the role of the VirD2 protein at late stages in the T-DNA transfer process is the very limited amount of this protein associated with the T-strand (T-complex) which is theoretically transferred to the plant cell. We approached this problem by introducing and expressing the *virD2* gene in the plant. Fig. 1 illustrates the plasmid constructs used for plant transformation. A *virD2–lacZ* chimeric gene was cloned behind the wound-inducible plant promoter T_R2' (12, 20) to generate plasmid pTR-vd-lacZ. The control construct pTR-lacZ shares all the features of pTR-vd-lacZ except for the *virD2* gene fused to the *lacZ* gene. The *lacZ* gene (encoding β-galactosidase) was chosen for this purpose due to its versatility as a reporter gene (20) and its proven value in studies of protein accumulation in nuclei (21, 22).

Transgenic plants carrying *lacZ* or the *virD2–lacZ* fusion were induced to express the corresponding proteins by preparing protoplasts. Nuclei were isolated from the protoplasts. The various cellular fractions were assayed for β-galactosidase activity. As shown in Table 1, equivalent amounts of β-galactosidase activity were found in the cytosolic fraction of plants transformed with pTR-vd-lacZ and pTR-lacZ. However, a striking difference was observed when the amounts of activity accumulated in the nuclear fraction of the two plants were compared. The specific activity accumulated in the nuclei of the TR-VD-LACZ plants was ≈3-fold higher than that found in the cytosol, whereas in the TR-LACZ plant the specific activity in the nuclei was about 5-fold lower than that of the cytosolic fraction. Almost no activity was found in the organellar fractions (chloroplasts, mitochondria, peroxisomes, etc.) of both plants. These data strongly indicate that the 292 amino-terminal amino acids from VirD2 promote the localization of β-galactosidase activity in the nucleus and not in any other organelle.

Protein samples from the different cellular fractions were used to analyze the pattern of β-galactosidase activity after separation by SDS/PAGE. A very distinct band of enzymatic activity was detected in the nuclear fraction of the TR-VD-LACZ plant (Fig. 2, lane 2). This band, corresponding to the size predicted for the VirD2–LacZ fusion protein, was still present after the nuclei were washed with 1 M NaCl (data not shown), indicating its strong association with them. A weakly active band was also observed in the organellar fraction (lane 4). This was very likely due to contamination with VirD2–LacZ released from broken nuclei during isolation. As expected, no detectable activity was found in the nuclear (lane 1) and organellar (lane 3) fractions of the TR-LACZ transgenic plant. Both plants showed active bands in their cytosolic fraction (lanes 5 and 6). The β-galactosidase-active band detected in the TR-LACZ plant corresponded to that of the bacterial enzyme (lane 5 vs. lane 8). Interestingly, a smear of activity was detected in the cytosolic fraction of the TR-VD-LACZ plant (lane 6); its highest band corresponded with the position of the VirD–LacZ fusion protein whereas the lowest had an electrophoretic mobility similar to that of the bacterial LacZ. This could be explained if the VirD2 portion of the protein is especially sensitive to degradation by protease when present in the cytosol and is protected when it migrates into the nucleus. This hypothesis would also explain the high level of activity detected in the cytosol of the TR-VD-LACZ plant (Table 1). That no endogenous LacZ activity could be detected in any of the fractions is due to inactivation during the fractionation procedure (data not shown).

Protease protection is widely used to demonstrate that proteins have fully traversed membranes from one cellular compartment into another. Using this approach, we found that the β-galactosidase activity associated with nuclei was partially protected from degradation by Pronase (Table 2). When isolated nuclei of the TR-VD-LACZ plant were Pronase-treated in a buffer containing an osmoprotectant (intact nuclei), only 36% of the original activity was lost after the

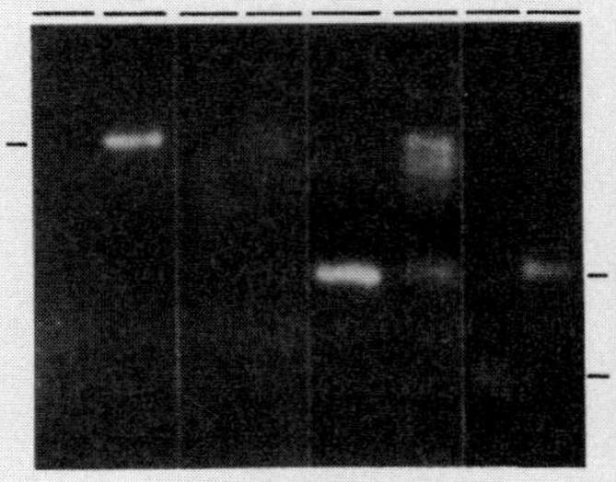

FIG. 2. Electrophoretic analysis of β-galactosidase activity in subcellular fractions. Lanes 1 and 2, nuclear fractions; lanes 3 and 4, organellar fractions; lanes 5 and 6, cytosolic fractions. Even numbers represent the fractions obtained from TR-VD-LACZ plants, the odd numbers those of the TR-LACZ plants. Total protein of an untransformed tobacco plant (lane 7) and of an *E. coli* strain carrying the bacterial β-galactosidase gene (lane 8) was used as reference. VD-LACZ, the VirD2–LacZ hybrid protein; LACZ, the bacterial β-galactosidase; Endo, the endogenous β-galactosidase in total extracts of a nontransformed tobacco plant.

Table 1. β-Galactosidase activity in subcellular fractions of TR-LACZ and TR-VD-LACZ plants

	Activity, units	
Fraction	TR-LACZ	TR-VD-LACZ
Cytosolic	526.98	377.98
Organellar	34.85	72.62
Nuclear	112.00	1036.41

Activity was measured as described in *Materials and Methods*; 1 unit produces 1 nmol of nitrophenol produced per mg of protein per min. Organellar fraction included mitochondria, chloroplasts, peroxisomes, etc.

Table 2. β-Galactosidase activity of protease-treated nuclei

	Activity	
Nuclei	Units	% retained
TR-VD-LACZ		
Untreated	424.33	100.00
Intact	270.89	63.83
Broken	153.43	36.15
TR-LACZ		
Untreated	63.80	100.00
Intact	22.33	35.00
Broken	23.92	37.50

Nuclei of the TR-VD-LACZ and TR-LACZ transgenic plants either were treated with protease directly after isolation (intact), or had their membranes disrupted prior to treatment (broken), or did not receive the protease treatment (untreated) (see *Materials and Methods*). β-Galactosidase activity was assayed and is expressed as described in Table 1.

treatment. Nuclei resuspended in a buffer without osmoprotectant and broken by Vortex mixing lost 64% of the original activity. The incomplete protection of the β-galactosidase activity observed with intact nuclei may have been due to penetration of the nuclear envelope by Pronase, a mixture of proteases of undetermined mass. Another possibility is that part of the fusion protein might have been attached to the nuclear membrane and thus have remained exposed to the protease. On the other hand, the incomplete loss of activity in broken nuclei may have been due to the reported tolerance of LacZ to proteolytic attack (23). This hypothesis is supported by data obtained with isolated nuclei of TR-LACZ plants (Table 2), where the low activity associated with this fraction was reduced to 35% and 37.5% of the original when treated with Pronase before and after breakage, respectively.

Electron microscopy studies of tissue sections of the transgenic plants showed that the fusion protein from the TR-VD-LACZ plant was located in the nucleus (Fig. 3*A*), as suggested by the protease protection and cell fractionation experiments. However, an important part of the β-galactosidase activity appeared to have accumulated in the nuclear membrane of both the control TR-LACZ plant and the TR-VD-LACZ plant (Fig. 3 *A* and *B*). The nonspecific accumulation of indigo precipitate at the nuclear envelope has been observed in other cases (T. Teeri, personal communication). It is not known whether this phenomenon is due to accumulation of β-galactosidase at the nuclear membrane or to accumulation of the reaction product formed in the cytoplasm. Since the cell fractionation experiments had shown 10-fold higher β-galactosidase activity in nuclear fractions of the TR-VD-LACZ plant than in nuclear fractions of the control TR-LACZ plant, further electron microscopy studies were carried out using isolated nuclei. In nuclei from the TR-VD-LACZ plant, the indigo precipitate was found both associated with the nuclear envelope and in the nucleoplasm (Fig. 3*C*). In contrast, no reaction product was found in nuclei of the control plant, either in the membrane or in the nucleoplasm (Fig. 3*D*). These data suggest that the association of β-galactosidase activity with the nuclear envelope observed in complete cells (Fig. 3 *A* and *B*) was most likely nonspecific and due to manipulation. Although a considerable amount of precipitate was detected in the cytosol of both plants, no specific accumulation of activity was observed in any other organelle from either transgenic plant. No indigo precipitate was detected in untransformed control plants (data not shown).

DISCUSSION

The experiments reported here demonstrate the ability of a truncated VirD2 protein from *A. tumefaciens* to direct nuclear localization of the bacterial protein β-galactosidase in plants. The ability of peptides to target proteins into the nucleus has been extensively studied in animal cells and yeast (24–26). Recent studies have led to a proposed consensus signal sequence for nuclear targeting, Lys-(Lys or Arg)-Xaa-(Lys or Arg) (27). Analysis of the VirD2 protein sequence revealed the presence of three conserved peptides in *A. tumefaciens* and *A. rhizogenes* that resemble the proposed signal (27). Two of these peptides are located at the carboxyl-terminal part of the nopaline (pTiC58) VirD2 protein, at positions 417–420 (Lys-Arg-Pro-Arg) and 431–434 (Lys-Arg-Glu-Arg). The third potential signal is located at positions 31–34 (Arg-Lys-Gly-Arg). The truncated protein used in this work contained only the putative signal from the amino-terminal part of VirD2 (positions 31–34). This peptide alone is capable of promoting accumulation of β-galactosidase activity in the nuclei. Although the putative signal sequence (Arg-Lys-Gly-Arg) contained in this peptide has an arginine at the first position instead of the lysine proposed for the

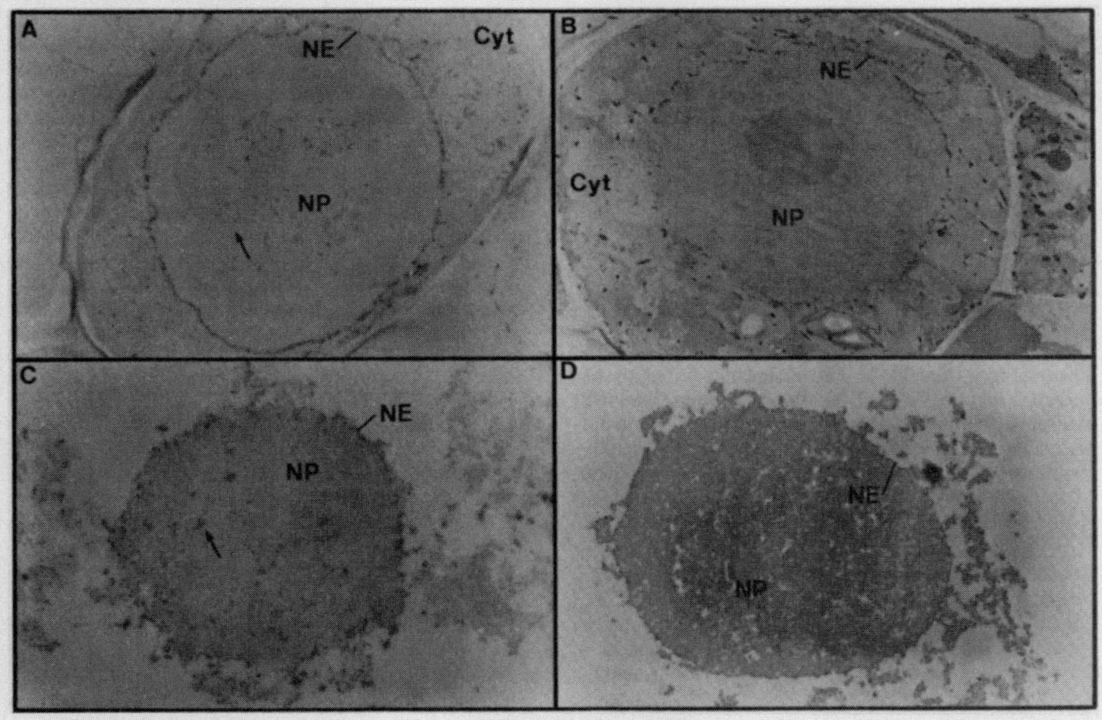

FIG. 3. Electron microscopic analysis of β-galactosidase activity in tissue sections of TR-VD-LACZ (*A*) and TR-LACZ (*B*) transgenic plants and in isolated nuclei from TR-VD-LACZ (*C*) and TR-LACZ (*D*) plants. Examples of the indigo-precipitated crystals formed by hydrolysis of the β-galactosidase substrate X-Gal are indicated by arrows. Cyt, cytoplasm; NE, nuclear envelope; NP, nucleoplasm.

consensus signal described for animal cells, it has been shown for the simian virus 40 targeting signal that such a substitution yields a peptide that is active as a nuclear targeting signal (28). Our data do not eliminate the possibility that the remaining two postulated signals at the carboxyl terminus could act as auxiliary nuclear targeting signals. In a similar way, the polyoma virus large tumor antigen employs two different peptides to direct nuclear accumulation (29). Moreover, increasing the number of signals per molecule increases the relative uptake by nuclei of albumin–peptide conjugates and the effective size of the channels available for translocation in animal cells (30). Therefore, the presence of three putative nuclear targeting signals in the VirD2 protein from *A. tumefaciens* might lead to a higher efficiency of transport through the nuclear envelope, which presumably would be required for the transport of the T-strand.

Our data, together with other recent observations, allow us to suggest a sequence of events leading to T-DNA transfer from bacteria to plants. It is initiated by the T-DNA border cleavage by the *virD*-encoded endonuclease activity (3, 4, 31). The second step is most probably the replication of the bottom strand in a polar way, 5′ to 3′ from the nicked border, thereby producing a single-stranded molecule (2) with VirD proteins associated at the 5′ terminus (7–11). This T-complex subsequently associates with nonspecific single-stranded DNA-binding proteins, such as VirE2 (32, 33), resulting in the formation of an intermediate form. These proteins would protect the T-strand from attack by nucleases (11) and might participate in active transport through the cytoplasmic membrane in conjunction with membrane-associated proteins such as those encoded by the *virB* loci (34, 35). The presence of the VirD2 protein at the 5′ terminus of the T-complex molecules, its function as a nuclear targeting signal, and its endonuclease activity strongly implicate its involvement in subsequent T-DNA integration into the plant genome. These data are supported by the fact that after integration, the T-DNA molecules have well-defined borders at the right-hand end of the T-DNA but not at the left (36–38).

Further investigation is required to determine whether the postulated signal peptide (Arg-Lys-Gly-Arg) is actually responsible for the nuclear accumulation that has been observed and whether the carboxyl-terminal part of the VirD2 protein plays a role in targeting to nuclei. A question of major interest is whether the VirD2 protein is responsible for promoting T-DNA transport into the plant nucleus and whether it plays a role in integration of the T-DNA into the plant nuclear genome.

We thank Drs. A. Caplan, C. Castresana, and L. Herrera-Estrella for critical reading of the manuscript and M. De Cock, V. Vermaercke, S. Van Gijsegem, and K. Spruyt for preparation of the manuscript and figures. We are grateful to R. De Rijcke for the excellent electron microscopy work and to Dr. T. Hürek for invaluable advice. A.H.-E. is indebted to the Commission of the European Communities for a training grant.

1. Zambryski, P. (1988) *Annu. Rev. Genet.* **22,** 1–30.
2. Stachel, S. E., Timmerman, B. & Zambryski, P. (1987) *EMBO J.* **6,** 857–863.
3. Yanofsky, M. F., Porter, S. G., Young, C., Albright, L. M., Gordon, M. P. & Nester, E. W. (1986) *Cell* **47,** 471–477.
4. Albright, L. M., Yanofsky, M. F., Leroux, B., Ma, S. & Nester, E. W. (1987) *J. Bacteriol.* **169,** 1046–1055.
5. Ghai, J. & Das, A. (1989) *Proc. Natl. Acad. Sci. USA* **86,** 3109–3113.
6. Soliman, M. & Das, A. (1990) *J. Bacteriol.*, in press.
7. Herrera-Estrella, A., Chen, Z.-M., Van Montagu, M. & Wang, K. (1988) *EMBO J.* **7,** 4055–4062.
8. Ward, E. R. & Barnes, W. M. (1988) *Science* **242,** 927–930.
9. Young, C. & Nester, E. W. (1988) *J. Bacteriol.* **170,** 3367–3374.
10. Howard, E. A., Winsor, B. A., De Vos, G. & Zambryski, P. (1989) *Proc. Natl. Acad. Sci. USA* **86,** 4017–4021.
11. Dürrenberger, F., Crameri, A., Hohn, B. & Koukolíková-Nicola, Z. (1989) *Proc. Natl. Acad. Sci. USA* **86,** 9154–9158.
12. Velten, J., Velten, L., Hain, R. & Schell, J. (1984) *EMBO J.* **3,** 2723–2730.
13. Alliotte, T., Zhu, L. H., Van Montagu, M. & Inzé, D. (1988) *Plasmid* **19,** 251–254.
14. Deblaere, R., Bytebier, B., De Greve, H., Deboeck, F., Schell, J., Van Montagu, M. & Leemans, J. (1985) *Nucleic Acids Res.* **13,** 4777–4788.
15. Van Haute, E., Joos, H., Maes, M., Warren, G., Van Montagu, M. & Schell, J. (1983) *EMBO J.* **2,** 411–418.
16. Budar, F., Deboeck, F., Van Montagu, M. & Hernalsteens, J.-P. (1986) *Plant Science (Lucknow, India)* **46,** 195–206.
17. Mazzolini, L., Vaeck, M. & Van Montagu, M. (1989) *Eur. J. Biochem.* **178,** 779–787.
18. Miller, J. H. (1972) *Experiments in Molecular Genetics* (Cold Spring Harbor Lab., Cold Spring Harbor, NY).
19. Laemmli, U. K. (1970) *Nature (London)* **227,** 680–685.
20. Teeri, T. H., Lehväshlaiho, H., Franck, M., Uotila, J., Heino, P., Palva, E. T., Van Montagu, M. & Herrera-Estrella, L. (1989) *EMBO J.* **8,** 343–350.
21. Kalderon, D., Roberts, B. L., Richardson, W. D. & Smith, A. E. (1984) *Cell* **39,** 499–509.
22. Hall, M. N., Hereford, L. & Herskowitz, I. (1984) *Cell* **36,** 1057–1065.
23. Ullman, A. & Perrin, D. (1970) in *The Lactose Operon*, eds. Beckwith, J. R. & Zipser, D. (Cold Spring Harbor Lab., Cold Spring Harbor, NY), pp. 143–172.
24. Chelsky, D., Ralph, R. & Jonak, G. (1989) *Mol. Cell. Biol.* **9,** 2487–2492.
25. Richardson, W. D., Mills, A. D., Dilworth, S. M., Laskey, R. A. & Dingwall, C. (1988) *Cell* **52,** 655–664.
26. Silver, P. A., Keegan, L. P. & Ptashne, M. (1984) *Proc. Natl. Acad. Sci. USA* **81,** 5951–5955.
27. Wang, K., Herrera-Estrella, A. & Van Montagu, M. (1990) *J. Bacteriol.* **172,** 4432–4440.
28. Kalderon, D., Richardson, W. D., Markham, A. T. & Smith, A. E. (1984) *Nature (London)* **311,** 33–38.
29. Richardson, W. D., Roberts, B. L. & Smith, A. E. (1986) *Cell* **44,** 77–85.
30. Dworetzky, S. I., Lanford, R. E. & Feldherr, C. M. (1988) *J. Cell Biol.* **107,** 1279–1287.
31. Wang, K., Stachel, S., Timmerman, B., Van Montagu, M. & Zambryski, P. (1987) *Science* **235,** 587–591.
32. Gietl, C., Koukolíková-Nicola, Z. & Hohn, B. (1987) *Proc. Natl. Acad. Sci. USA* **84,** 9006–9010.
33. Citovsky, V., De Vos, G. & Zambryski, P. (1988) *Science* **240,** 501–504.
34. Thompson, D. V., Melchers, L. S., Idler, K. B., Schilperoort, R. A. & Hooykaas, P. J. J. (1988) *Nucleic Acids Res.* **16,** 4621–4636.
35. Ward, J. E., Akiyoshi, D. E., Regier, D., Datta, A., Gordon, M. P. & Nester, E. W. (1988) *J. Biol. Chem.* **263,** 5804–5814.
36. Zambryski, P., Depicker, A., Kruger, K. & Goodman, H. (1982) *J. Mol. Appl. Genet.* **1,** 361–370.
37. Holsters, M., Villarroel, R., Gielen, J., Seurinck, J., De Greve, H., Van Montagu, M. & Schell, J. (1983) *Mol. Gen. Genet.* **190,** 35–41.
38. Slightom, J. L., Jouanin, L., Leach, F., Drong, R. F. & Tepfer, D. (1985) *EMBO J.* **4,** 3069–3077.

FROM PLANT PATHOGEN TO VECTOR FOR PLANTS, YEASTS AND FUNGI

Paul J. J. Hooykaas
Institute of Molecular Plantsciences, Clusius Laboratory
Universiteit Leiden, Leiden, The Netherlands
E-mail: genetica@rulbim.leidenuniv.nl

Commentary on the paper:

Bundock, P., den Dulk-Ras, A., Beijersbergen, A., and Hooykaas, P. J. J. 1995. Transkingdom T-DNA transfer from *Agrobacterium tumefaciens* to *Saccharomyces cerevisiae.* EMBO J. 14:3206–3214.

I came into the crown gall field in 1974 when I entered the department of Biochemistry of Arthur Rörsch as an undergraduate to perform research in the Crown Gall group (project leader Rob Schilperoort) together with Ph.D. student Piet Klapwijk. At that time the important role of the Ti plasmid in the virulence properties of *Agrobacterium* had just been discovered. The objective of my project was to try to mobilize the Ti plasmid from one *Agrobacterium* strain to another *ex planta.* I succeeded by using the R-plasmids RP4 and R702. A great difficulty was the selection of transconjugants. There were already indications that the Ti plasmid encoded the ability to degrade opines and thus that mobilization of the Ti plasmid from an "octopine" strain would confer the ability to degrade octopine on the Ti plasmid cured nopaline strain C58-C9. Thus we tried to select for recipients that could utilize octopine as a sole nitrogen source. This worked out quite well, but only after selecting an agar source that contained few impurities that *Agrobacterium* could use as an alternative N-source and by the use of auxotrophic donor strains. These results led to a key publication by Bomhoff et al. (1976). A year later, I published a paper on transfer of the Ti plasmid to *R. trifolii.* Following transfer, *R. trifolii* was converted into a bacterium that not only could still nodulate clover but also induce crown gall tumors (Hooykaas et al. 1977). Several years later as a postdoc, I performed the reverse experiment showing that the transfer of a large plasmid from *Rhizobium trifolii* into *Agrobacterium* converted *Agrobacterium* into a nodulating bacterium, after which we proposed the name Sym-plasmid for this class of *Rhizobium* plasmid. These results stressed the importance of plasmid genes as key determinants of the symbiotic and phytopathogenic properties of members of the Rhizobiaceae family.

The Ti mobilization experiments by R plasmids, which were helped later on by the development of the Bromthymol Blue indicator medium, led us to recognize that mobilization occurred by cointegrate formation of Ti and R plasmids. Transposition of then unknown transposable elements such a Tn*904,* Tn*1831* and IS*70* were responsible for this cointegrate formation. Thus a collection of transposon insertion mutants was built up that was used to characterize the genes on the Ti plasmid. Selecting for loss of the ability to catabolize octopine enabled us to obtain deletions that had these transposon-insertions as starting points. These were mapped physically by my fellow Ph.D. student Gert Ooms. One of these, LBA4404, is still used worldwide now as the helper plasmid in the binary vector system that was developed in our lab by Ph.D. student André Hoekema, postdoc Penny Hirsch, Rob Schilperoort and me. All of these results were published in my Ph.D. thesis entitled "The role of plasmid determined functions in the interactions of Rhizobiaceae with plant cells. A genetic approach." After my graduation, I stayed on first as a postdoc and from 1983 as a staff member in the same department which was now headed by Rob Schilperoort, who succeeded Arthur Rörsch.

Among the transposon insertion mutants were some that gave phenotypically abnormal tumors on plants; these mutants were called shooters and rooters. Some of the tobacco shoots, derived from such shooty tumors were found to contain octopine and thus to consist of tumor cells. These could be propagated *in vitro,* but, as they did not form roots, had to be grafted onto normal tobacco rootstocks for flowering and seed set. Interestingly, stem inoculation with *Agrobacterium* of these already transformed shoots did not result in tumor formation (Hooykaas-Van Slogteren et al. 1984). The hypothesis was that "tumor" cells had a cell wall that was not suitable for an interaction with *Agrobacterium.* My wife (Truus van Slogteren) and I did think of another possibility: that transformation did occur, but did not result in crown gall formation due to an altered hormonal balance as a consequence of the presence of the first T-DNA. When we infected such octopine producing shoots with a nopaline strain we could indeed detect nopaline as well as octopine at the infection site! Then it was a small step to apply the same concept to monocots. As there were no monocots available in the lab, we sacrificed some of the ornamental monocots (*Chlorophytum, Narcissus*) in our home, and were very excited indeed to find the first evidence for T-DNA transfer to monocots (Hooykaas-Van Slogteren et al. 1984).

The 10 years or so that followed were used in our lab and that of our colleagues abroad for a detailed analysis of the virulence system and the regulation of its expression. This led eventually to the hypothesis that the virulence system encoded a bacterial conjugation system that had been adapted for plants. We hypothesized that if this were true, the system might still mediate classic conjugation to other bacteria. My Ph.D. student Alice Beijersbergen and technician Amke den Dulk-Ras put this to the test and could show, indeed, that under the proper inducing conditions *Agrobacterium* could use its virulence system to mobilize a non-conjugative incQ plasmid to a recipient bacterial cell (Beijersbergen et al. 1992). This result also indicated to us that the "conjugative" system of *Agrobacterium* was not very selective in the choice of the target host cell and apparently could interact not only with different types of dicotyledonous and monocotyledonous plant cells, but also with totally different cells, such as those of bacteria. This led us to think that other cell types, such as yeast and fungal cells, might also be targets for *Agrobacterium.*

The unicellular yeast *Saccharomyces cerevisiae* is a model eukaryote for studies of many important processes such as DNA recombination and cell cycle regulation. If transfer to this host were possible, this would open a new avenue in which the well-developed yeast genetic technology could be used for the analysis of *Agrobacterium* T-DNA transfer. First, we developed T-DNA vectors in which the yeast URA3 gene was present as a selective marker together with the replicon of the yeast 2μ plasmid, as vectors based on this replicon are the most efficient for yeast transformation. Co-cultivation of yeast with *Agrobacterium* strains containing such vectors, performed by Amke den Dulk-Ras and Ph.D. students Alice Beijersbergen and Paul Bundock, led to high frequency transformation. Subsequently, other T-DNA vectors that were meant for genomic integration either by homologous or nonhomologous recombination were tested (Bundock and Hooykaas 1996). As T-DNA integrated efficiently into the yeast genome by homologous recombination, this demonstrated that the low frequency of gene targeting which was observed in plants by another Ph.D. student, Remko Offringa, was not due to peculiarities of the T-DNA and associated virulence proteins, but to the constitution of the plant host enzymatic machinery for DNA recombination. We found also that T-DNA vectors lacking a replicon or homology with the host genome integrated, albeit with low frequency, in the yeast genome by a process of nonhomologous recombination similar to that used by plants for T-DNA integration (Bundock and Hooykaas 1996). This allowed us to use the available yeast mutant collections to define the genes coding for the enzymatic machinery involved in T-DNA integration. This led to the discovery by Ph.D. student Haico van Attikum that the enzymes involved in DNA double strand repair by non-homologous end-joining were involved in T-DNA integration (Van Attikum et al. 2001). These enzymes are conserved from yeast to man, which explains the apparent similarity of T-DNA integration in yeast and plants. The results obtained with yeast initiated the idea that *Agrobacterium* might be useful also as a DNA vector for the transformation of biotechnologically important fungal species, which are often difficult to transform. Together with previous Ph.D. students Marcel de Groot and Alice Beijersbergen, who had moved to the Unilever Research lab, Paul Bundock who stayed on in the lab as a postdoc, and I embarked on a study of fungal transformation by *Agrobacterium.* Attempts to transform fungi by *Agrobacterium* were immediately successful, and even species such as the mushroom *Agaricus bisporus,* which are recalcitrant to transfomation by other means, could be successfully transformed by *Agrobacterium.* More recently my postdoc Annette Vergunst, Ph.D. student Barbara Schrammeijer, Amke and I, found that the practical use of *Agrobacterium* is not limited to being a DNA vector, but that the bacterium can also be used as a protein donor to yeast and plant cells and thus may become an agent for "protein therapy."

References

Beijersbergen, A., Den Dulk-Ras, A., Schilperoort, R. A., and Hooykaas, P. J. J. 1992. Conjugative transfer by the virulence system of *Agrobacterium tumefaciens.* Science 256:1324–1327.

Bomhoff, G., Klapwijk, P. M., Kester, H. C. M., Schilperoort, R. A., Hernalsteens, J. P., and Schell, J. 1976. Octopine and nopaline synthesis and breakdown genetically controlled by a plasmid of *Agrobacterium tumefaciens.* Mol. Gen. Genet. 145:177–181.

Bundock, P., and Hooykaas, P. J. J. 1996. Integration of *Agrobacterium tumefaciens* T-DNA in the *Saccharomyces cerevisiae* genome by illegitimate recombination. Proc. Natl. Acad Sci. U.S.A. 93:15272–15275.

Hooykaas, P. J. J., Klapwijk, P. M., Nuti, M. P., Schilperoort, R. A., and Rorsch, A. 1977. Transfer of the *Agrobacterium tumefaciens* Ti plasmid to avirulent agrobacteria and to *Rhizobium ex planta.* J. Gen. Microbiol. 98:477–484.

Hooykaas-Van Slogteren, G. M., Hooykaas, P. J. J., and Schilperoort, R. A. 1984. Expression of Ti plasmid genes in monocotyledonous plants infected with *Agrobacterium tumefaciens.* Nature 311:763–764.

Van Attikum, H., Bundock, P., and Hooykaas, P. J. J. 2001. Non-homologous end-joining proteins are required for *Agrobacterium* T-DNA integration. EMBO J. 20:6550–6558.

PAUL HOOYKAAS: In 1979, I obtained my Ph.D. from the University of Leiden, The Netherlands.

The person who has been my mentor for many years is Rob Schilperoort. Thanks to his inspiring personality, the department has been an extremely stimulating environment for research into *Agrobacterium.* Thanks also are due to Piet Klapwijk, Arthur Rörsch and Piet van de Putte, who taught me bacterial genetics and to Marco Nuti (Universita di Pisa) for introducing me into the world of *Rhizobium.* Furthermore, thanks to all these colleagues and students with whom I had the pleasure to share the lab and exchange ideas.

The EMBO Journal vol.14 no.13 pp.3206–3214, 1995

Trans-kingdom T-DNA transfer from *Agrobacterium tumefaciens* to *Saccharomyces cerevisiae*

Paul Bundock, Amke den Dulk-Ras, Alice Beijersbergen[1] and Paul J.J.Hooykaas[2]

Institute of Molecular Plant Sciences, Clusius Laboratory, Leiden University, Wassenaarseweg 64, 2333 AL Leiden, The Netherlands
[1]Present address: Unilever Research Laboratorium Vlaardingen, PO Box 114, 3130 AC, Vlaardingen, The Netherlands

[2]Corresponding author

P.Bundock, A.den Dulk-Ras and A.Beijersbergen have contributed equally to this article

Agrobacterium tumefaciens **transfers part of its tumour-inducing (Ti) plasmid, the transferred or T-DNA, to plants during tumourigenesis. This represents the only example of naturally occurring trans-kingdom transfer of genetic material. Here we report that *A.tumefaciens* can transfer its T-DNA not only to plant cells, but also to another eukaryote, namely the yeast *Saccharomyces cerevisiae*. The Ti plasmid virulence (*vir*) genes that mediate T-DNA transfer to plants were found to be essential for transfer to yeast as well. Transgenic *S.cerevisiae* strains were analysed for their T-DNA content. Results showed that T-DNA circles were formed in yeast with precise fusions between the left and right borders. Such T-DNA circles were stably maintained by the yeast if the replicator from the yeast 2μ plasmid was present in the T-DNA. Integration of T-DNA in the *S.cerevisiae* genome was found to occur via homologous recombination. This contrasts with integration in the plant genome, where T-DNA integrates preferentially via illegitimate recombination. Our results thus suggest that the process of T-DNA integration is predominantly determined by host factors.**

Key words: *Agrobacterium tumefaciens*/homologous recombination/T-DNA transfer/virulence genes/yeast

Introduction

Agrobacterium tumefaciens is a Gram-negative soil bacterium able to induce tumours, or crown galls, at plant wound sites. During tumourigenesis part of its tumour-inducing (Ti) plasmid, the T-DNA, can be mobilized from the bacteria into the plant cell by the virulence (*vir*) genes located on the Ti plasmid. Expression of *onc* genes on the T-DNA leads to plant cell proliferation and formation of a tumour. *vir*-mediated plant cell transformation represents the only example of naturally occurring trans-kingdom DNA transfer.

vir genes involved in T-DNA transfer have been extensively studied (for reviews see Hooykaas and Schilperoort, 1992; Winans, 1992; Zambryski, 1992; Hooykaas and Beijersbergen, 1994). The transfer system is activated when VirA senses inducing compounds produced at the plant wound sites, such as acetosyringone (AS), and activates the remaining *vir* loci via the transcriptional activator VirG. Upon activation of *vir* gene expression a linear single-stranded DNA, the T-strand, is generated following nicking of the border repeats surrounding the T region (Albright *et al.*, 1987; Wang *et al.*, 1987). The border repeats are nicked by the VirD2 protein with help from the VirD1 protein (Lessl and Lanka, 1994). The VirC proteins increase the efficiency of the right border (RB) nicking reaction by binding to an enhancer sequence located next to the RB (Toro *et al.*, 1988) and in this way can affect host range. Nicking leads to covalent attachment of the VirD2 protein to the 5′-end of the T-DNA. The C-terminal part of VirD2 contains nuclear localization sequences (NLSs) for transport of the T-DNA to the nucleus (Tinland *et al.*, 1992; Rossi *et al.*, 1993). The T-DNA is transferred to the plant cell as a single-stranded molecule (Chaudhury *et al.*, 1994; Tinland *et al.*, 1994; Yusibov *et al.*, 1994). Also transferred to the plant cell is the VirE2 protein, a single-stranded DNA binding protein which may coat the T-DNA along its length to form long nucleoprotein filaments (Citovsky *et al.*, 1989). It has been proposed that transferred *vir* proteins may promote T-DNA integration in the genome of the plant cell. T-DNA is thought to leave the *Agrobacterium* cell through a transmembrane structure consisting of products of the *virB* operon. Most of the 11 VirB proteins are located in the membrane (Beijersbergen *et al.*, 1994) and, with the exception of the VirB1 protein, are all essential for tumourigenesis (Berger and Christie, 1994). T-DNA transfer to certain plant species also requires the host range proteins VirF and VirH. The precise function of these is unknown. The VirH protein shares some homology with the P450 cytochrome class of enzymes and therefore may play a role in detoxification of plant exudate products (Kanemoto *et al.*, 1989). VirF is probably transported to the plant cell and is necessary for T-DNA transport to some plant species (Regensburg-Tuink and Hooykaas, 1993).

Agrobacterium tumefaciens induces tumours in a wide range of dicotyledonous plant species, but not in monocotyledonous plants. Nevertheless, T-DNA transfer to monocots could be demonstrated indirectly by the use of sensitive reporter systems, such as agroinfection (Grimsley *et al.*, 1987). Recently Hiei *et al.* (1994) used the *Agrobacterium* vector system to obtain transgenic rice plants. Subsequently it was found that T-DNA integrated into the genome of rice in the same way as in dicots.

We were interested to see whether T-DNA transfer and integration would be possible to species belonging to another kingdom than that of plants. We chose the yeast *Saccharomyces cerevisiae* (kingdom Fungi) because of its

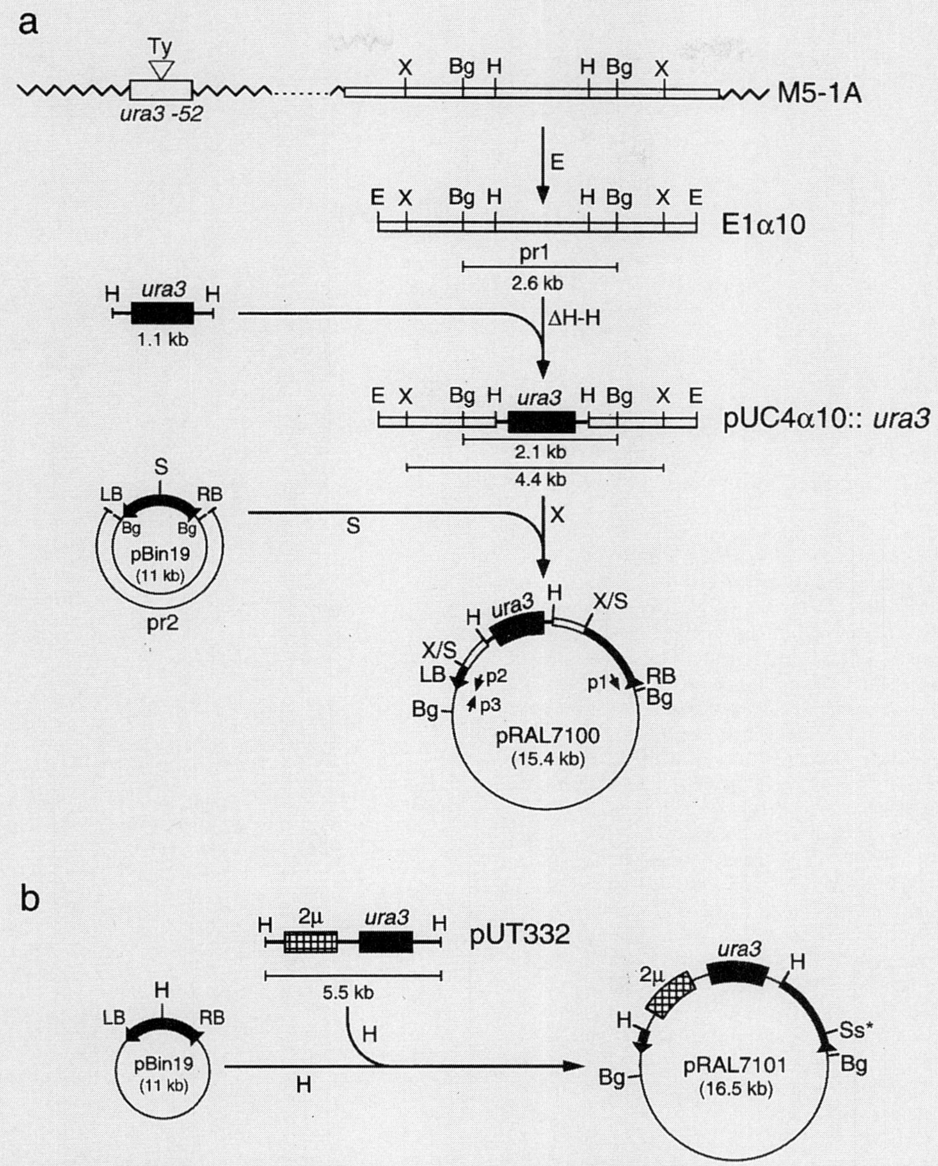

Fig. 1. Construction of the 'integrative' plasmid pRAL7100 (**a**) and 'replicative' plasmid pRAL7101 (**b**). The wild-type chromosomal *PDA1* locus of M5-1a is located between the *Hin*dIII sites. Restriction enzymes: Bg, *Bgl*II; E, *Eco*RI; H, *Hin*dIII; S, *Sal*I; Ss, *Sst*II; X, *Xho*I. ΔH-H, deletion of *Hin*dIII–*Hin*dIII *PDA1* fragment from clone E1α10 (Steensma *et al.*, 1990). Ty, transposon Ty; pr1, probe 1; pr2, probe 2; p1–p3, primers 1–3; LB, left border repeat; RB, right border repeat; *, no unique restriction site (not drawn to scale).

ease of handling, the availability of suitable vectors and its fast growth rate. The results of these studies are described in this article. We discovered that T-DNA transfer to *S.cerevisiae* is possible, suggesting that in nature the host range of *Agrobacterium* may be even broader than so far anticipated. We found that *vir* genes required for T-DNA transfer to plants were also necessary for T-DNA transfer to yeast. T-DNA was integrated via homologous recombination in the yeast genome. Therefore the process of T-DNA integration seems to be largely determined by host factors.

Results

T-DNA transfer from A.tumefaciens to S.cerevisiae

Transformation of yeast with a replicative vector is much more efficient than with an integrative vector. Therefore, in order to establish whether *A.tumefaciens* could transfer T-DNA to the yeast *S.cerevisiae*, we constructed the binary vector pRAL7101. This plasmid contains the yeast *URA3* selection gene (Rose *et al.*, 1984) and the yeast 2μ origin of replication between the left border (LB) and RB repeats of the binary vector pBIN19 (Figure 1b). To be able to

Table I. T-DNA transfer to *Saccharomyces cerevisiae* strain M5-1a

Plasmid present in donor *Agrobacterium* strain LBA1100	Medium	M5-1a colonies on medium without uracil	M5-1a colonies (10^8/ml) on medium with uracil	Frequency of Ura^+ colonies per output recipient
pRAL7100	+AS	272	1.6	1.7×10^{-6}
	−AS	0	0.2	$<5\ \times10^{-8}$
pRAL7101	+AS	200	0.6	3.3×10^{-6}
	−AS	0	2.2	$<4.5\times10^{-9}$

test for transfer, pRAL7101 was electroporated into the *A.tumefaciens* helper strain LBA1100, which carries the *vir* genes that code for the T-DNA transfer system (Beijersbergen *et al.*, 1992). Activation of the *vir* genes can be accomplished *in vitro* by incubating the strain in a low pH medium containing the phenolic inducer AS (Scheeren-Groot *et al.*, 1994). Incubation of *S.cerevisiae* in this medium did not inhibit growth or lead to lethality and mixtures of agrobacteria and yeast cells also survived provided that the glucose concentration was <5 mM. After these control experiments, co-cultivations between *A.tumefaciens* and *S.cerevisiae* strain M5-1a, which is a haploid Ura^- strain, were carried out to test for T-DNA transfer. After incubation on low pH medium with or without AS (as a control), cells were resuspended in 0.9% NaCl and plated onto selective media. As can be seen in Table I, URA3 transfer from *A.tumefaciens* to *S.cerevisiae* was indeed observed and was dependent on the presence of the *vir* inducer AS in the medium. The Ura^+ yeast strains thus obtained were purified and then characterized.

The Ura^+ phenotype turned out to be stable. There was no loss of marker after overnight growth in non-selective medium. To assay for the presence of T-DNA, a simple DNA preparation was carried out on these Ura^+ strains. Since we suspected that a plasmid replicating via the 2μ replicator was present, the isolated DNA was used for transformation of *Escherichia coli*, from which DNA can be purified more easily than from yeast. Selection was for carbenicillin resistance (Cb^r), a selective marker that was present in the T-region of pRAL7101, together with the ColE1 origin of replication and the *URA3* gene. DNA preparations from all of the purified Ura^+ yeast strains resulted in Cb^r *E.coli* transformants. Forty-eight *S.cerevisiae* strains were analysed for their T-DNA content in this way.

Three of the 48 strains contained a complete circular T-DNA and their restriction patterns confirmed that the two *Bgl*II sites present on either side of the borders of pRAL7101 had been lost, while the *Sst*II site (Figure 1b) next to the RB was still present (data not shown). We concluded that in these co-cultivations transfer of a T-strand to M5-1a had occurred. T-DNA is transferred as a linear single-stranded molecule (Chaudhury *et al.*, 1994; Tinland *et al.*, 1994; Yusibov *et al.*, 1994) and therefore a ligation step is required for the formation of a plasmid in yeast. Plasmids from these strains were tested with primers p1 and p2, which flank the border repeats of the T region. All three strains amplified a 703 bp fragment consistent with a border fusion (Figure 2a, lane 4). The PCR product was sequenced and a comparison of sequences surrounding the T-region revealed that the T-circles of these three strains each contained one intact border repeat. The sequence of this border repeat was

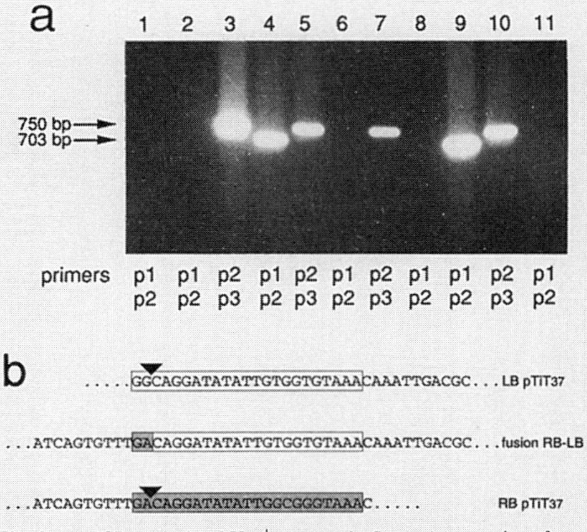

Fig. 2. (**a**) PCR analysis of Ura^+ transgenic *S.cerevisiae* strains after T-DNA transfer from LBA1100 (pRAL7101) (lanes 2–5) or LBA1100 (pRAL7100) (lanes 6–10). Primers used are indicated as p1–p3 (see also Figure 1). The size of the products (in base pairs) is indicated by arrows. Lane 1, M5-1a; lane 2, pRAL7101 (plasmid control, 100 pg); lane 3, pRAL7101 (plasmid control, 100 pg); lane 4, pRAL7101 T-region transferred; lane 5, entire pRAL7101 transferred; lane 6, pRAL7100 (plasmid control, 100 pg); lane 7, pRAL7100 (plasmid control, 100 pg); lane 8, double cross-over between the T-region of pRAL7100 and the yeast *PDA1* locus; lane 9, M5-1a containing a circularized T-strand derived from pRAL7100; lane 10, integration of entire pRAL7100 plasmid; lane 11, H_2O control. (**b**) Comparison of the nucleotide sequences of the LB and RB repeats with the fused border sequence found in the transferred T-DNAs. pBIN19 border sequences are derived from the wild-type nopaline plasmid pTiT37 (Bevan, 1984). The boxed areas indicate the border repeats; ▼, nick site, established *in vivo*, *in vitro* and corresponding to homology with the RP4 nick site (Lessl and Lanka, 1991).

identical for all three strains and was compatible with a precise fusion between processed RB and LB repeats, i.e. the first 2 nt derived from the RB and the remaining 22 nt from the LB (Figures 2b and 4a).

Sixteen of the Ura^+ M5-1a colonies contained a plasmid smaller than expected if T-strand transfer and circularization had occurred and did not amplify a fragment during PCR with primers p1 and p2. In order to characterize these plasmids DNA sequencing was performed. In these 16 strains the plasmid formed was the result of homologous recombination between a direct repeat consisting of part (150 bp) of the bleomycin gene. This 150 bp sequence turned out to be accidently present between the borders of the original pBIN19 vector and also on the introduced pUT332 fragment (Mazodier *et al.*, 1985). Therefore, these small T-circles were formed by intra-T-DNA recombination after transfer, resulting in deletion of the border sequences.

Table II. No transformation of M5-1a after addition of plasmid DNA to the donor–recipient mixtures

Plasmid DNA added	No. of M5-1a colonies on medium without uracil	No. of M5-1a colonies (10^7/ml) on medium with uracil	Transformation frequency
pRAL7100[a]	0	4.2	$<2.4\times10^{-8}$
pRAL7100[b]	0	6.4	$<1.6\times10^{-8}$
pRAL7101[a]	0	8.4	$<1.2\times10^{-8}$
pRAL7101[b]	0	4.2	$<2.4\times10^{-8}$

[a]Addition of 1 μg DNA to the co-cultivation medium.
[b]Addition of 5 μg DNA to the co-cultivation medium.
Strain LBA1100 (lacking the T-DNA *in trans*) was used as the donor.

Surprisingly, plasmids from the remaining 29 Ura$^+$ yeast strains showed restriction patterns which indicated that they contained the entire pRAL7101 plasmid. It can be argued that in these cases T-DNA transfer may not have occurred, but rather that these strains may have been generated by uptake of intact plasmid DNA present in the medium. However, in a model transfer experiment (Table II), no Ura$^+$ *S.cerevisiae* strains were found when 1 or 5 μg plasmid DNA (pRAL7101 or pRAL7100) was added to a mixture of donor (LBA1100 lacking the T-region *in trans*) and acceptor cells (M5-1a). Also, transfer from *Agrobacterium* was unaffected by addition of 1 μg/ml DNase I to the incubation medium, showing that the T-DNA is protected against this enzyme during transfer. Therefore, the presence of the whole pRAL7101 plasmid in a large number of the yeast Ura$^+$ strains can be best explained by missing of the LB, which was not nicked during T-DNA processing in the bacterial cell. This has been shown to occur during *in vitro* T-DNA processing experiments (Stachel *et al.*, 1987), as well as in T-DNA transfer to plants, where 20% of transformed plants were found to contain DNA sequences present outside the T-DNA borders (Martineau *et al.*, 1994). It was confirmed that the Ura$^+$ yeast strains which showed the entire restriction pattern of pRAL7101 contained an intact LB, as shown by amplification of a 750 bp fragment using primers p2 and p3, which were located on either side of the LB (Figure 1 and Figure 2a, lane 5). Missing of the LB, transfer of the entire plasmid and ligation at the RB would be expected to re-create the original binary vector in *S.cerevisiae*.

Evidence for T-DNA integration into the yeast genome

In plant species (T-)DNA integration occurs via illegitimate recombination (Offringa *et al.*, 1990), but DNA introduced into *S.cerevisiae* integrates into the genome predominantly via homologous recombination. We were interested in whether T-DNA carrying extensive homology with the *S.cerevisiae* genome would integrate via homologous recombination or illegitimate recombination. Therefore, an 'integrative' binary vector, pRAL7100, was constructed with the *URA3* selection gene surrounded by DNA derived from the flanking regions of the *S.cerevisiae PDA1* gene (Figure 1a), which could promote homologous recombination between the introduced T-DNA and the *S.cerevisiae PDA1* locus on chromosome V. The T-DNA of pRAL7100 carries no yeast origin of replication. Therefore, growth of

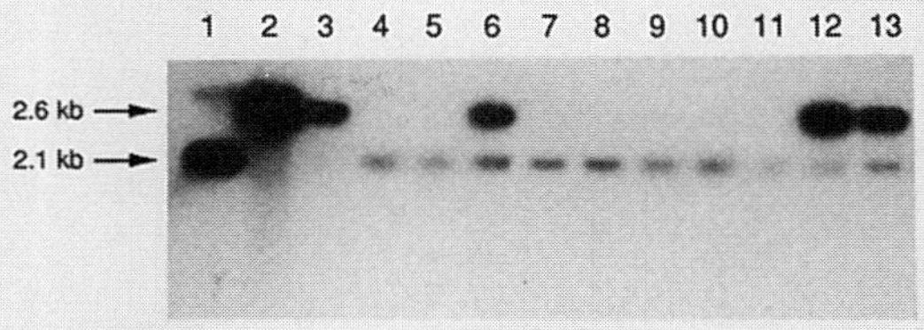

Fig. 3. Autoradiograph of a Southern blot of Ura$^+$ transgenic *S.cerevisiae* strains obtained after transfer of DNA from LBA1100 (pRAL7100) (see Materials and methods). Total DNA digested with *Bgl*II was hybridized with the 2.6 kb *Bgl*II fragment from clone E1α10 (probe pr1, Figure 1a). The fragment sizes (in kilobase pairs) are indicted by arrows. Lane 1, pRAL7100 (0.1 μg); lane 2, E1α10 (0.1 μg); lane 3, M5-1a; lanes 4–13, Ura$^+$ yeast strains. Lanes 4, 5 and 7–11 lack the 2.6 kb band, indicating a gene replacement event had occurred. Lanes 6, 12 and 13 represent Ura$^+$ M5-1a strains in which the whole pRAL7100 plasmid had integrated into the genome via a single cross-over.

S.cerevisiae on medium lacking uracil after co-cultivation with *A.tumefaciens* should be due to integration of the T-DNA carrying the *URA3* marker into the yeast genome.

Ura$^+$ M5-1a colonies were indeed found after co-cultivation with LBA1100 (pRAL7100) on plates containing AS (Table I) and, surprisingly, at approximately the same frequency as after co-cultivation using LBA1100 (pRAL7101) as the donor. This result was unexpected, because the replicating vector pRAL7101 gave an ~100-fold higher frequency of Ura$^+$ M5-1a transformants than the integrative vector pRAL7100 after electroporation (data not shown).

Twenty of the Ura$^+$ yeast strains generated after co-cultivation with *A.tumefaciens* containing pRAL7100 were purified. Total DNA was isolated from these and a Southern blot was carried out to detect the presence of integrated T-DNA. Using a 2.6 kb *Bgl*II fragment from the *PDA1* gene as a probe (Figure 1a, probe pr1), the same 2.6 kb *Bgl*II fragment was expected in the wild-type and in strains still containing an unmodified *PDA1* gene. However, a 2.1 kb *Bgl*II fragment was expected for strains containing the modified *PDA1* gene with an integrated *URA3* gene. In the DNA of 12 of the Ura$^+$ yeast strains only a 2.1 kb homologous fragment was detected (Figure 3). In these strains, apparently, the incoming T-DNA with the modified *PDA1* gene had recombined with the homologous *PDA1* locus and introduced the *URA3* gene via a double cross-over or gene conversion (Figure 4c). The remaining eight strains still contained the 2.6 kb wild-type *PDA1* fragment, in addition to the 2.1 kb fragment. This is consistent with a single cross-over event having occurred between one of the *PDA1* flanking sequences present on the incoming T-DNA and the homologous sequence in the yeast genome. To find further evidence for this, DNA from the eight Ura$^+$ strains were digested with *Eco*RI and probed with pr3 (a 0.5 kb *Eco*RI–*Xho*I fragment of the *PDA1* gene). A single cross-over on either side of the *PDA1* locus would be expected to give a band of 16.5 kb with this probe and indeed in three of the strains the 16.5 kb band was detected (data not shown). All three strains showing this 16.5 kb band also amplifed the 750 bp fragment with primers p2 and p3 (Figure 2a, lane 10), showing that they all contain an integrated complete binary vector. Together, these data suggest that these strains were formed after

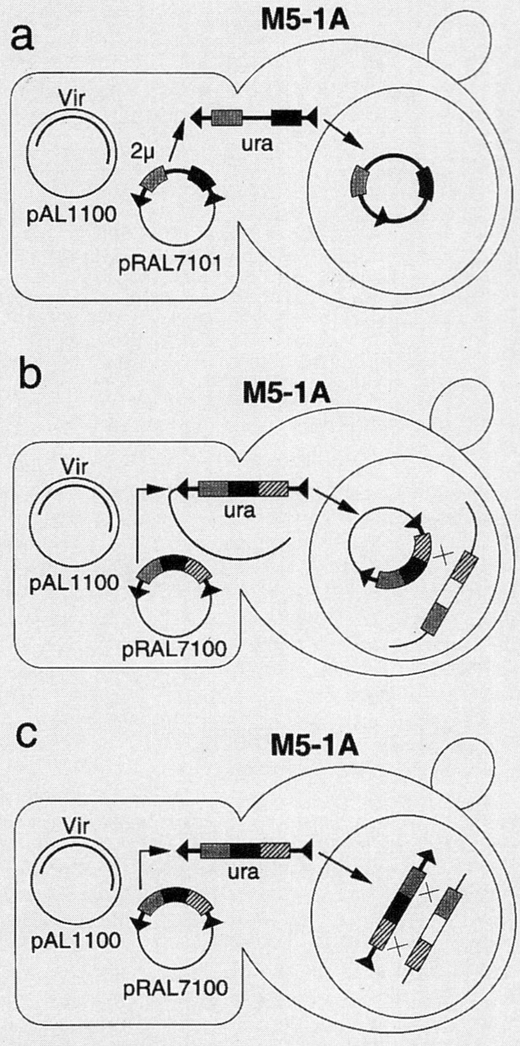

Fig. 4. Representation of the different T-DNA transfer events from *A.tumefaciens* to *S.cerevisiae* using plasmid pRAL7101 (**a**) or pRAL7100 (**b** and **c**). An *Agrobacterium* cell is shown on the left and a *Saccharomyces* cell on the right. Transgenic *S.cerevisiae* strains were formed by (a) autonomous replication of the transferred T-DNA, (b) integration of pRAL7101 via a single cross-over and (c) integration via a double cross-over. In (a) only T-DNA is transferred; in 60% of the cases we found transfer of the entire pRAL7101.

missing of pRAL7100 LB during T-DNA processing, transfer of the whole pRAL7100 plasmid and subsequent circularization and integration of the whole pRAL7100 via a single cross-over event at the *PDA1* locus.

In the remaining five strains no evidence of homologous recombination between the *PDA1* locus and the T-DNA could be seen. We therefore attempted to localize the T-DNA insertions in these strains by separating the yeast chromosomes on a CHEF gel and blotting with a 300 bp *PstI–SphI* fragment of the T-DNA as a probe. No signal on any chromosome could be detected, whereas in controls the T-DNA integrated by a single cross-over could be localized to chromosome V, which carries the *PDA1* locus.

S.cerevisiae itself is able to ligate introduced DNA to mitochondrial sequences (Schiestl *et al.*, 1994) and this could also account for the as yet unidentified forms of T-DNA in these strains.

Effect of vir mutations on the transfer of T-DNA to S.cerevisiae

To define the Vir functions necessary for T-DNA transfer to *S.cerevisiae*, the binary vectors pRAL7100 and pRAL7101 were electroporated into a series of *A.tumefaciens* strains containing mutations in the *vir* genes. Co-cultivations with the *A.tumefaciens* mutants and *S.cerevisiae* strain M5-1a were performed as in the previous experiments and the results are summarized in Table III.

All mutants tested showed decreased or zero T-DNA transfer as compared with the wild-type, except for the *virF* mutant. Mutations in the *virA* and *virG* genes prevented transfer as expected, as these two Vir proteins are essential for induction of the remaining *vir* genes. No Ura^+ M5-1a strains were produced by co-cultivation with the *virD4* or *virB* mutants, supporting the proposed role of these proteins in transfer of the T-strand. No Ura^+ M5-1a colonies were produced during co-cultivation with a strain carrying a mutation in the 3′-region of *virD2*. This particular mutation does not affect the nicking function or *vir*-mediated bacterial conjugation (Beijersbergen *et al.*, 1992), but deletes the NLS sequences and the proposed omega sequence, which are important for T-DNA transfer to plants (Shurvinton *et al.*, 1992). Apparently this 3′-region of VirD2 is also important for transfer to yeast. The *virE2* mutant showed a much reduced transfer frequency. The VirE2 protein is important, but not absolutely essential, for T-DNA transfer to plants (Yusibov *et al.*, 1994), however, it plays no role in *vir*-mediated bacterial conjugation (Beijersbergen *et al.*, 1992). The VirE2 protein was reported to be able to coat T-strands and thus be able to protect the T-DNA against nucleases present in eukaryotic cytosol. Since the absence of VirE2 might lead to the generation of deletions in the T-DNA, autonomously replicating plasmids were isolated from Ura^+ M5-1a strains obtained after co-cultivation with the *virE2* mutant containing pRAL7101. The same three plasmid types were isolated as were previously obtained when wild-type LBA1100 (pRAL7101) was used in co-cultivations (see above). Therefore, the lack of VirE2 protein did not lead to any apparent deletions during transfer of the T-DNA to yeast. However, it is possible that in the absence of VirE2 nuclease action digests 3′ sequences necessary for T-DNA circularization in yeast. This could account for the observed drop in frequency for all three plasmid types. Alternatively, the reduction in frequency seen after co-cultivation of yeast with the *virE2* mutant may be due to the loss of the nuclear targeting function mediated by the NLS present on the VirE2 protein (Citovsky *et al.*, 1992). Also, in the absence of VirE2, the T-DNA may be unable to form a long unfolded T-complex, which seems necessary for efficient T-DNA transport (Citovsky *et al.*, 1989).

The *virF* mutant was the only *A.tumefaciens* strain which transferred T-DNA at a frequency equivalent to the wild-type strain during co-cultivations. It is thought that the *virF* protein influences the host range of *A.tumefaciens* by an interaction with a structure or protein in the plant cell (Jarchow *et al.*, 1991; Regensburg-Tuink and

Table III. T-DNA transfer from *A.tumefaciens* donor strains to the *S.cerevisiae* recipient haploid strain M5-1a

Mutant donor	*vir* mutation	Plasmid	Titre input (10^7 cells/ml)		Titre output (10^7 cells/ml)		No. of M5-1a colonies on medium without uracil	Frequency of Ura^+ colonies per output recipient
			Donor	Recipient	Donor	Recipient		
LBA1100		pRAL7101	1.8	2.0	24.0	6.0	200	3.3×10^{-6}
		pRAL7100	1.9	2.0	34.0	16.0	272	1.7×10^{-6}
LBA1142	*vir*A	pRAL7101	5.0	2.3	2.0	9.0	0	$<1.1\times10^{-8}$
		pRAL7100	3.0	1.5	7.0	8.0	0	$<1.2\times10^{-8}$
LBA1143	*vir*B4	pRAL7101	2.2	1.9	1.0	31.0	0	$<3.2\times10^{-9}$
		pRAL7100	2.4	1.6	3.0	5.0	0	$<2\times10^{-8}$
LBA1144	*vir*B7	pRAL7101	4.6	1.9	1.5	7.0	0	$<1\times10^{-8}$
		pRAL7100	4.4	2.0	1.0	11.0	0	$<9\times10^{-9}$
LBA1145	*vir*G	pRAL7101	5.6	1.9	2.0	15.0	0	$<6.6\times10^{-9}$
		pRAL7100	5.6	1.5	8.0	9.0	0	$<1\times10^{-8}$
LBA1147	3′*vir*D2	pRAL7101	4.1	2.1	4.5	13.0	0	$<7.7\times10^{-9}$
		pRAL7100	4.8	1.5	6.0	13.0	0	$<7.7\times10^{-9}$
LBA1148	*vir*D4	pRAL7101	2.0	2.1	3.2	16.0	0	$<6.3\times10^{-9}$
		pRAL7100	3.0	1.5	6.5	7.0	0	$<1.4\times10^{-8}$
LBA1149	*vir*E2	pRAL7101	3.0	2.0	24.0	6.0	18	3×10^{-7}
		pRAL7100	2.4	2.0	6.0	14.0	24	1.7×10^{-7}
LBA1517	*vir*F	pRAL7101	3.7	2.0	3.0	6.0	231	3.9×10^{-6}
		pRAL7100	2.7	2.0	4.5	11.0	233	2.1×10^{-6}

Each *A.tumefaciens* mutant carried either pRAL7101 (replicative vector) or pRAL7100 (integrative vector). AS (200 μM) was included in the co-cultivation plates. Similar results were obtained from at least three independent experiments. The numbers of donor and recipient cells were determined at mixing (titre input) and after 3 days co-cultivation (titre output).

Hooykaas, 1993). Apparently, this interacting structure or protein is absent from the yeast cell. Plasmids isolated from Ura^+ M5-1a strains generated by co-cultivation with the *virF* mutant were also detected in the three forms described when LBA1100 (pRAL7101) was used as a donor.

Discussion

In this report we show for the first time that Ura^- *S.cerevisiae* cells can be converted to Ura^+ by incubation with an *Agrobacterium* donor containing a binary vector with the *URA3* gene. The transferred DNA was found to be protected against nucleases in the medium during transfer. In contrast, plasmid DNA added to an artificial co-cultivation mixture of agrobacteria and yeast cells was not taken up by *S.cerevisiae*. Transfer occurred via an active process mediated by the *Agrobacterium vir* system, as shown by the essential role of the *vir* genes, suggesting that T-DNA transfer to yeast and plants occurs via a common mechanism. Together, these data show that the host range for *Agrobacterium* T-DNA transfer includes yeasts, in addition to plants.

DNA transfer to *S.cerevisiae* has also been reported from *E.coli* containing a conjugative IncP plasmid (Heinemann and Sprague, 1989) or a mobilizable IncQ plasmid (Nishikawa *et al.*, 1992). It has been proposed that the *tra* system of certain conjugative plasmids and the *vir* system of Ti share common evolutionary origins. This idea has been supported by studies showing IncQ transfer from *A.tumefaciens* to plants (Buchanan-Wollaston *et al.*, 1987) and to other bacteria (Beijersbergen *et al.*, 1992) mediated by the *vir* system. Homology studies have found significant similarities between the processing and transport proteins of the *tra* and *vir* systems (reviewed in Lessl and Lanka, 1994) and between the nick regions of a variety of DNA transfer systems (Pansegrau and Lanka, 1991). Our experiments support the proposed link between the *vir* system and other DNA transfer systems, although *Agrobacterium* and its *vir* system seem optimized for DNA transfer to plant species.

Surprisingly, in our co-cultivation experiments with *Agrobacterium*, the frequencies with which Ura^+ *S.cerevisiae* strains were obtained were similar for the replicating vector (pRAL7101) and the integration vector (pRAL7100). In electroporation experiments, as expected, the replicating vector pRAL7101 was more efficient than the integration vector pRAL7100. This difference in results between co-cultivation and electroporation might be due to the different DNA structures that are introduced into yeast, single-stranded linear DNA during the T-DNA transfer process and double-stranded circular DNA during electroporation. However, when RSF1010-derived vectors containing the *URA3* gene were mobilized from *E.coli* into *S.cerevisiae* (Nishikawa *et al.*, 1992), Ura^+ *S.cerevisiae* colonies were obtained at a higher frequency with an ARS-type replicative vector compared with an integrative vector. In this latter case the DNA is probably introduced in a single-stranded linear form, like T-DNA. Since the yeast host factors in IncQ and T-DNA transfer are the same, the abnormal behaviour of the T-DNA transfer system in yeast must be due to the Vir proteins that accompany the T-DNA. It can be envisaged that these stimulate T-DNA integration, making integrative T-DNAs equally efficient as replicating T-DNAs. We think this unlikely, because in plants T-DNA is integrated in a manner strikingly different from that in yeast, i.e. illegitimate versus homologous recombination (see below). Alternatively, it can be argued that replicative T-DNAs have difficulty in establishing themselves in yeast. An interesting possibility could be that the VirD2 pilot protein, which shows *in vitro* nickase/ligase activity (Pansegrau *et al.*, 1993a, 1994), is inefficient in circularizing the T-strand. After all, in plants circularization does not normally occur and the ends of the T-DNA are used for integration (Mayerhofer *et al.*, 1991). The VirD2 protein may be different in its poor ligase activity from the functionally homologous nickase/ligase protein MobA encoded by the IncQ plasmid (Scherzinger *et al.*, 1992) and the TraI protein encoded by the IncP plasmid (Pansegrau *et al.*, 1993b). Work in our laboratory is in progress to find out whether *vir*-mediated IncQ plasmid transfer to yeast is different for integration and replication vectors.

We were surprised to find a large number of Ura^+ yeast strains containing all of the pRAL7101 binary vector DNA instead of only the T-DNA. Apparently, during T-DNA processing the LB is missed quite often. This seems to be typical for binary vectors and is in contrast to processing of the wild-type Ti plasmid. Unequivocal evidence for missing of the LB of the wild-type Ti has never been given, although many tumour lines were analysed for their T-DNA content. However, a large number of transgenic plants transformed with a binary vector were recently reported to contain sequences from outside the T-DNA borders (Martineau *et al.*, 1994). If transfer of the whole binary vector to yeast is common, it might be argued that the T-circles observed in yeast originate from these whole circles by homologous recombination on the 24 bp border repeats. There are three arguments against this reasoning. First, we have not observed instability in the pRAL7101 plasmid maintained by yeast, suggesting that recombination must take place early in establishment. Second, the frequency of recombination between direct repeats in yeast shows a linear dependence on the length of the DNA homology (Jinks-Robertson *et al.*, 1993), the minimal length necessary for homologous recombination being ~270 bp. This would make the 24 bp border repeats unlikely targets for homologous recombination. Third, if homologous recombination occurred between the border repeats then a cross-over may occur anywhere within this repeat. However, in all cases analysed by sequencing we found that the result was compatible with border ligation, whereas homologous recombination may equally well result in a product that is incompatible with border ligation. For these three reasons we favour the idea that the T-DNA circles found in yeast are formed via circularization of the T-DNA. Differences between recombination enzymes and nucleases in plants and yeast may explain why exact T-DNA border fusions were not recovered from plants, not even when these events were selected for (Bakkeren *et al.*, 1989). T-circles found in bacteria were the result of recombination between the border repeats (Koukolíková-Nicola *et al.*, 1985), but here this was induced by the nicking of these border repeats via the *vir* system (Timmerman *et al.*, 1988).

Plants integrate incoming (T-)DNA into the genome preferentially by illegitimate recombination; homologous recombination has only been shown to occur at low frequencies, when the T-DNA shares extensive homology with the target locus (Offringa *et al.*, 1990). However, in lower eukaryotes such as *S.cerevisiae*, integration of the introduced DNA occurs predominantly via homologous recombination (Hinnen *et al.*, 1978). We have demonstrated in this paper that T-DNA is also integrated in this latter way in *S.cerevisiae*, showing that the host proteins, rather than the *Agrobacterium vir* proteins, are the decisive factors in the integration process.

T-DNA transfer to *S.cerevisiae* can be used as a powerful tool to study the interaction of *A.tumefaciens* with eukaryotic recipient cells and to study the influence of different recombination systems on T-DNA integration. From our experiments we suggest that trans-kingdom DNA transfer may be more widespread than previously thought. When the Vir system is activated by plant wound products, *Agrobacterium*-mediated T-DNA transfer might be possible to a wide range of recipients, including fungal and maybe even certain animal cells. This system may therefore be useful in the genetic modification of species that are so far recalcitrant to transformation. However, in nature the correct conditions necessary for activation of the Vir system will only be available in the vicinity of plants, thus making natural T-DNA transfer to species other than plants most unlikely.

Materials and methods

Agrobacterium and Saccharomyces strains

The *Agrobacterium* strains used are listed in the Table IV (Beijersbergen *et al.*, 1992; Regensburg-Tuink and Hooykaas, 1992). *Saccharomyces* strain M5-1a (MAT**a** *trp1-92 leu2-3/112 ura3-52 his4*) was used as the recipient. Cloning was done in *E.coli* strain MH1.

Table IV. *Agrobacterium* strains used

Strain	Chromosomal background	Plasmid
LBA1100	C58	pAL1100ΔT-DNA, Δ*tra*, Δ*occ*
LBA1142	C58	pAL1100(*virA*::Tn*3*HoHo)
LBA1143	C58	pAL1100(*virB4*::Tn*3*HoHo)
LBA1144	C58	pAL1100(*virB7*::Tn*3*HoHo)
LBA1145	C58	pAL1100(*virG*::Tn*3*HoHo)
LBA1147	C58	pAL1100(3′*virD2*::Tn*3*HoHo)
LBA1148	C58	pAL1100(*virD4*::Tn*3*HoHo)
LBA1149	C58	pAL1100(*virE2*::Tn*3*HoHo)
LBA1517	C58	pTiB6(*virF*::Tn*1831*)

Plasmid constructions

Plasmid pRAL7100 was constructed by the ligation of a 4.4 kb *Xho*I–*Xho*I fragment of pUC4α10::*ura3* (Steensma *et al.*, 1990) to the *Sal*I-digested vector pBIN19 (Bevan, 1984). Construct pRAL7101 was made by ligation of *Hin*dIII-digested DNA of plasmid pUT332 (Gaitignol *et al.*, 1990) to pBIN19 digested with *Hin*dIII. *Agrobacterium* strains were electroporated with these constructs as described by Mozo and Hooykaas (1991).

T-DNA transfer experiments

Agrobacterium strains containing the binary vector pRAL7100 or pRAL7101 were grown at 29°C overnight in minimal medium (Hooykaas *et al.*, 1979) containing the appropriate antibiotics at the following concentrations: kanamycin, 100 μg/ml; streptomycin, 250 μg/ml; carbenicillin, 75 μg/ml. The *Saccharomyces* recipient strains were grown overnight at 30°C in CY medium (Sherman *et al.*, 1983). After dilution of the *Agrobacterium* cells to an $OD_{660\ nm} \approx 0.15$ in induction medium [IM; composed of MM salts and 40 mM 2-(*N*-morpholino)ethanesulfonic acid (MES), pH 5.3, 10 mM glucose, 0.5% (w/v) glycerol and 200 μM AS] the *Saccharomyces* cells were diluted 10 times in fresh CY medium and the cultures were subsequently grown for 6 h at 29 and 30°C respectively. The input number of *Agrobacterium* and *Saccharomyces* cells was accurately determined by plating out dilutions of the cultures: for *Agrobacterium* on LC containing rifampicin (20 μg/ml) and for *Saccharomyces* on MY containing the required amino acids. Subsequently, 50 μl of both the *Agrobacterium* and *Saccharomyces* cultures were mixed and placed on 0.45 μm cellulose nitrate filters on IM plates containing 5 mM glucose and the amino acids histidine, tryptophan

(20 µg/ml) and leucine and uracil (30 µg/ml). After incubation for 3 days at 29°C the co-cultivation mixture was then resuspended in 2 ml of a physiological salt solution (PZ; 9 g NaCl/l) and 100 µl aliquots of this mix were plated. Ura^+ M5-1a strains were selected for on MY medium (Zonneveld, 1986) containing 200 µM cefotaxim to kill the *Agrobacterium* cells. The number of surviving *Agrobacterium* cells was determined by plating on LC plus rifampicin and the number of M5-1a cells was determined by plating dilutions on MY containing the required amino acids, including uracil.

DNA isolation from *Saccharomyces* strains

Chromosomal DNA was isolated from a 100 ml culture using the method decribed by Holm *et al.* (1986). DNA (5 µg) was digested with *Bgl*II for 16 h and electrophoresed in a 0.7% (w/v) TBE gel for 16 h at 25 V. DNA was processed further for Southern blotting as described (Sambrook *et al.*, 1989). Plasmid DNA was isolated from M5-1a strains and then electroporated into *E.coli* strain MH1 and analysed by restriction analysis.

PCR amplification and nucleotide sequencing

One purifed colony of a Ura^+ M5-1a strain was suspended in 50 µl H_2O and heated for 5 min at 95°C. After cooling on ice the lysate was centrifuged for 5 min at 15 000 r.p.m. An aliquot (5 µl) of this DNA suspension was added to 45 µl of reaction mixture [50 mM KCl, 10 mM Tris–HCl, pH 8.3, 1.5 mM $MgCl_2$, 200 µM each nucleotide, 0.5 µM each primer, 0.1 U *Taq* polymerase (Super *Taq*; HT Biotechnology Ltd, UK)]. The PCR reaction cycles were: 2 min at 95°C; 1 min at the annealing temperature; 2 min at 72°C; 30 cycles of 1 min at 95°C, 1 min at the annealing temperature and 2 min at 72°C; 1 min at the annealing temperature; 10 min at 72°C. The primers used in the PCR reaction were: p1 5′-CGTTGCGGTTCTGTCAGTTCC-3′ (annealing temperature 59°C); p2 5′-CGCCTTGCAGCACATCCC-3′ (annealing temperature 60°C); p3 5′-TCAACATGCTACCCTCC-3′ (annealing temperature 45°C). The PCR product (15 µl, ~200 ng DNA) was purified on an agarose gel in 1× TBE for 90 min at 70 V. The amplified fragment was cut out and the gel slice heated at 65°C for 10 min. To 5 µl of this suspension was added 5 µl H_2O, 2 µl primers (0.5–2 pmol) and 2 µl annealing buffer. This mixture was incubated for 10 min at 65°C and then solidified by cooling on ice. Labelling mix (3.5 µl) containing 0.5 µl [α-^{35}S]dATP (<37 Tbq/nmol, 37 mBq/100 µl), 1.6 µl T7 dilution buffer and 3 U T7 DNA polymerase was added to the annealing mix and incubated for 5 min at 40°C. Aliquots (3 µl) of the labelling reaction were added to 2.5 µl of each termination mix and incubated for 5 min at 40°C. The reaction was stopped by addition of 5 µl stop buffer. The samples were heated for 5 min at 95°C before loading and then electrophoresed on a 6% polyacrylamide gel for 2.5 h in 1× TBE at 2000 V and 35 mA. The gel was fixed in 20% (v/v) methanol and 10% acetic acid for 1 h. After drying at 85°C, autoradiography was for 3 days (Kodak XAR film).

Acknowledgements

We would like to thank Y.Steensma and T.Wenzel for their generous gifts of yeast strains and helpful advice. This work was supported by the Netherlands Foundation for Chemical Research (SON) and the Biotechnological and Biological Sciences Research Council (BBSRC), UK. Thanks are also due to Remko Offringa for critical reading of the manuscript.

References

Albright,L.M., Yanofsky,M.F., Leroux,B., Ma,D. and Nester,E.W. (1987) Processing of the T-DNA of *Agrobacterium tumefaciens* generates border nicks and linear, single stranded T-DNA. *J. Bacteriol.*, **169**, 1046–1055.

Bakkeren,G., Koukolíková-Nicola,Z., Grimsley,N. and Hohn,B. (1989) Recovery of *Agrobacterium tumefaciens* T-DNA molecules from whole plants early after transfer. *Cell*, **57**, 847–857.

Beijersbergen,A., Den Dulk-Ras,A., Schilperoort,R.A. and Hooykaas, P.J.J. (1992) Conjugative transfer by the virulence system of *Agrobacterium tumefaciens*. *Science*, **256**, 1324–1327.

Beijersbergen,A., Smith,S.J. and Hooykaas,P.J.J. (1994) Localization and topology of VirB proteins of *Agrobacterium tumefaciens*. *Plasmid*, **32**, 212–218.

Berger,B.R. and Christie,P.J. (1994) Genetic complementation analysis of the *Agrobacterium tumefaciens* virB operon: virB2 through virB11 are essential virulence genes. *J. Bacteriol.*, **176**, 3646–3659.

Bevan,M. (1984) *Agrobacterium* vectors for plant transformation. *Nucleic Acids Res.*, **22**, 8711–8721.

Buchanan-Wollaston,V., Passiatore,J.E. and Cannon,F. (1987) The mob and oriT mobilization functions of a bacterial plasmid promote its transfer to plants. *Nature*, **328**, 172–175.

Chaudhury,A., Dennis,E. and Brettell,R. (1994) Gene-expression following T-DNA transfer into plant cells is aphidicolin-sensitive. *Aust. J. Plant. Physiol.*, **21**, 125–131.

Citovsky,V., Wong,M.L. and Zambryski,P. (1989) Cooperative interaction of *Agrobacterium* VirE2 protein with single-stranded DNA: implications for the T-DNA transfer process. *Proc. Natl Acad. Sci. USA*, **86**, 1193–1197.

Citovsky,V., Zupan,J., Warnick,D. and Zambryski,P. (1992) Nuclear localization of *Agrobacterium* VirE2 protein in plant cells. *Science*, **256**, 1802–1805.

Gatignol,A., Dassain,M. and Tiraby,G. (1990) Cloning of *Saccharomyces cerevisiae* promoters using a probe vector based on phleomycin resistance. *Gene*, **91**, 35–41.

Grimsley,N., Hohn,T., Davies,J.W., and Hohn,B. (1987) *Agrobacterium* mediated delivery of infectious maize streak virus into maize plants. *Nature*, **325**, 177-179.

Heinemann,J.A. and Sprague,G.F. (1989) Bacterial conjugative plasmids mobilize DNA transfer between bacteria and yeast. *Nature*, **340**, 205–209.

Hiei,Y., Ohta,S., Komari,T. and Kumashiro,T. (1994) Efficient transformation of rice (*Oryza sativa* L.) mediated by *Agrobacterium* and sequence analysis of the boundaries of the T-DNA. *Plant J.*, **6**, 271–282.

Hinnen,A., Hicks,J.B. and Fink,G.R. (1978) Transformation of yeast. *Proc. Natl Acad. Sci. USA*, **75**, 1929–1933.

Holm,C., Meeks-Wagner,D.W., Fangman,W.L. and Botstein,P. (1986) A rapid efficient method for isolating DNA from yeast. *Gene*, **42**, 169–173.

Hooykaas,P.J.J. and Beijersbergen,A. (1994) The virulence system of *Agrobacterium tumefaciens*. *Annu. Rev. Phytopathol.*, **32**, 157–79.

Hooykaas,P.J.J. and Schilperoort,R.A. (1992) *Agrobacterium* and plant genetic engineering. *Plant Mol. Biol.*, **19**, 15–38.

Hooykaas,P.J.J., Roobol,C. and Schilperoort,R.A. (1979) Regulation of the transfer of Ti-plasmids of *Agrobacterium tumefaciens*. *J. Gen. Microbiol.*, **110**, 99–109.

Jarchow,E., Grimsley,N.H. and Hohn,B. (1991) *virF*, the host range determining virulence gene of *Agrobacterium tumefaciens*, affects T-DNA transfer to *Zea mays*. *Proc. Natl Acad. Sci. USA*, **88**, 10426–10430.

Jinks-Robertson,S., Michelitch,M. and Ramcharan,S. (1993) Substrate length requirements for efficient mitotic recombination in *Saccharomyces cerevisiae*. *Mol. Cell. Biol.*, **13**, 3937–3950.

Kanemoto,R.H., Powell,A.T., Akiyoshi,D.E., Regier,D.A., Kerstetter, R.A., Nester, E.W., Hawes,M.C. and Gordon,M.P. (1989) Nucleotide sequence and analysis of the plant-inducible locus *pinF* from *Agrobacterium tumefaciens*. *J. Bacteriol.*, **171**, 2506–2512.

Koukolíková-Nicola,Z., Shillito,R.D., Hohn,B., Wang,K., Van Montagu,M. and Zambryski,P. (1985) Involvement of circular intermediates in the transfer of T-DNA from *Agrobacterium tumefaciens* to plant cells. *Nature*, **313**, 191–195.

Lessl,M. and Lanka,E. (1994) Common mechanisms in bacterial conjugation and Ti mediated T-DNA transfer to plant cells. *Cell*, **77**, 321–324.

Martineau,B., Voelker,T.A. and Sanders,R.A. (1994) On defining T-DNA. *Plant Cell*, **6**, 1032–1033.

Mayerhofer,R. *et al.* (1991) T-DNA integration: a mode of illegitimate recombination in plants. *EMBO J.*, **10**, 697–704.

Mazodier,P., Cossart,P., Giraud,E. and Gasser,F. (1985) Completion of the nucleotide sequence of the central region of Tn*5* confirms the presence of three resistance genes. *Nucleic Acids Res.*, **13**, 195–205.

Mozo,T. and Hooykaas,P.J.J. (1991) Electroporation of megaplasmids into *Agrobacterium*. *Plant Mol. Biol.*, **16**, 917–918.

Nisikawa,M., Suzuki,K. and Yoshida,K. (1992) DNA integration into recipient yeast chromosomes by trans-kingdom conjugation between *Escherichia coli* and *Saccharomyces cerevisiae*. *Curr. Genet.*, **21**, 101–108.

Offringa,R., de Groot,M.J.A., Haagsman,H.J., Does,M.P., van den Elzen,P.J.M. and Hooykaas,P.J.J. (1990) Extrachromosomal homologous recombination and gene targetting in plant cells after *Agrobacterium* mediated transformation. *EMBO J.*, **9**, 3077–3084.

Pansegrau,W. and Lanka,E. (1991) Common sequence motifs in DNA

relaxases and nick regions from a variety of DNA transfer systems. *Nucleic Acids Res.*, **19**, 3455.
Pansegrau,W., Schröder,W. and Lanka,E. (1993a) Relaxase (TraI) of IncPα plasmid RP4 catalyzes a site specific cleaving-joining reaction of single-stranded DNA. *Proc. Natl Acad. Sci. USA*, **90**, 2925–2929.
Pansegrau,W., Schoumacher,F., Hohn,B. and Lanka,E. (1993b) Site-specific cleavage and joining of single-stranded DNA by VirD2 protein of *Agrobacterium tumefaciens* Ti plasmids: analogy to bacterial conjugation. *Proc. Natl Acad. Sci. USA*, **90**, 11538–11542.
Pansegrau,W., Schröder,W. and Lanka,E. (1994) Concerted action of three distinct domains in the DNA cleaving-joining reaction catalyzed by relaxase (TraI) of conjugative plasmid RP4. *J. Biol. Chem.*, **269**, 2782–2789.
Regensburg-Tuink,A.J.G. and Hooykaas,P.J.J. (1993) Transgenic *N.glauca* plants expressing bacterial virulence gene *virF* are converted into hosts for nopaline strains of *A.tumefaciens*. *Nature*, **363**, 69–70.
Rose,M., Grisafi,P. and Botstein,D. (1984) Structure and function of the yeast URA3 gene: expression in *Escherichia coli*. *Gene*, **29**, 113–124.
Rossi,L., Hohn,B. and Tinland,B. (1993) The VirD2 protein of *Agrobacterium tumefaciens* carries nuclear localization signals important for transfer of T-DNA to plants. *Mol. Gen. Genet.*, **239**, 345-353.
Sambrook,J., Fritsch,E.F. and Maniatis,T. (1989) *Molecular Cloning: A Laboratory Manual*. Cold Spring Harbor Laboratory Press, Cold Spring Harbor, NY.
Scheeren-Groot,E.P., Rodenburg,K.W., Den Dulk-Ras,A., Turk,S.C.H.J. and Hooykaas,P.J.J. (1994) Mutational analysis of the transcriptional activator VirG of *Agrobacterium tumefaciens*. *J. Bacteriol.*, **176**, 6418–6426.
Scherzinger,E., Lurz,R., Otto,S. and Dobrinski,B. (1992) *In vitro* cleavage of double- and single-stranded DNA by plasmid RSF1010-encoded mobilization proteins. *Nucleic Acids Res.*, **20**, 41–48.
Schiestl,R.H., Zhu,J. and Petes,T.D. (1994) Effect of mutations in genes affecting homologous recombination on restriction enzyme-mediated and illigitimate recombination in *Saccharomyces cerevisiae*. *Mol. Cell. Biol.*, **14**, 4493–4500.
Sherman,F., Fink,G.R. and Lawrence,C.W. (1983) *Methods in Yeast Genetics*. 2nd Edn. Cold Spring Harbor Laboratory Press, Cold Spring Harbor, NY, p. 61.
Shurvinton,C.E., Hodges,L. and Ream,W. (1992) A nuclear localization signal and the C terminal omega sequence in *Agrobacterium tumefaciens* VirD2 endonuclease are important for tumor formation. *Proc. Natl Acad. Sci. USA*, **89**, 11837–11841.
Stachel,S.E., Timmerman,B. and Zambryski,P. (1987) Activation of *Agrobacterium tumefaciens vir* gene expression generates multiple single-strand T-strand molecules from the pTiA6 T-region: requirement for 5′ *virD* gene products. *EMBO J.*, **6**, 857–863.
Steensma,H.Y., Holterman,L., Dekker,I., Van Sluis,C.A. and Wenzel,T.J. (1990) Molecular cloning of the gene for the E1α subunit of the pyruvate dehydrogenase complex from *Saccharomyces cerevisiae*. *Eur. J. Biochem.*, **191**, 769–774.
Timmerman,B., Van Montagu,M. and Zambryski,P. (1988) Vir induced recombination in *Agrobacterium*. Physical characterization of precise and imprecise T-circle formation. *J. Mol. Biol.*, **203**, 373–384.
Tinland,B., Koukolíková-Nicola,Z., Hall,M.N. and Hohn,B. (1992) The T-DNA linked VirD2 protein contains two distinct functional nuclear localization signals. *Proc. Natl Acad. Sci. USA*, **89**, 7442–7446.
Tinland,B., Hohn,B. and Puchta,H. (1994) *Agrobacterium tumefaciens* transfers single-stranded transferred DNA (T-DNA) into the plant cell nucleus.. *Proc. Natl Acad. Sci. USA*, **91**, 8000–8004.
Toro,N., Datta,A., Yanofsky,M. and Nester,E. (1988) Role of the overdrive sequence in T-DNA border cleavage in *Agrobacterium*. *Proc. Natl Acad. Sci. USA*, **85**, 8558–8562.
Wang,K., Stachel,S.E., Timmerman,B., van Montagu,M. and Zambryski,P. (1987) Site specific nick in the T-DNA border sequence as a result of *Agrobacterium vir* gene expression. *Science*, **235**, 587–591.
Winans,S.C. (1992) Two-way chemical signalling in *Agrobacterium*–plant interactions. *Microbiol. Rev.*, **56**, 12–31.
Yusibov,V.M., Steck,T.R., Gupta,V. and Gelvin,S.B. (1994) Association of single-stranded transferred DNA from *Agrobacterium tumefaciens* with tobacco cells. *Proc. Natl Acad. Sci. USA*, **91**, 2994–2998.
Zambryski,P. (1992) Chronicles from the *Agrobacterium* plant cell DNA transfer story. *Annu. Rev. Plant Physiol. Plant Mol. Biol.*, **43**, 465–490.
Zonneveld,B.J.M. (1986) Cheap and simple yeast media. *J. Microbiol. Methods*, **4**, 287.

Received on January 16, 1995; revised on April 15, 1995

A RECKLESS DECISION?

Jean-Pierre Hernalsteens
Vrije Universiteit Brussel
Faculteit Wetenschappen Onderzoeksgroep Genetische Virologie
Pleinlaan 2, Brussel, Belgium
E-mail: jphernal@vub.ac.be

Commentary on the paper:

Hernalsteens, J.-P., Van Vliet, F., De Beuckeleer, M., Depicker, A., Engler, G., Lemmers, M., Holsters, M., Van Montagu, M., and Schell, J. 1980. The *Agrobacterium tumefaciens* Ti plasmid as a host vector system for introducing foreign DNA in plant cells. Nature 287:654–656.

I got involved in research on *Agrobacterium* as an undergraduate student in 1972, when I followed a course on advanced genetics by Jeff Schell. During a coffee break, Jeff spoke with enthusiasm about his studies on *Agrobacterium*. The aim of Jeff's group was to unravel the mechanism of tumor induction by *Agrobacterium* 'from the side of the bacterium'. A few months later, I joined Jeff's group to prepare a Master's thesis on the genetics of temperate bacteriophages of *Agrobacterium*. In the very confusing literature on *Agrobacterium,* several publications suggested that phages might carry the tumor inducing principle. However, Jeff's people observed that large plasmids were present in all oncogenic wild type *A. tumefaciens* isolates and absent from related non-oncogenic *A. radiobacter* isolates (Zaenen et al. 1974). This was an excellent argument to believe that these plasmids carry the tumor inducing principle.

For the subject of my Ph.D. thesis, Jeff gave me the choice between continuing the work on the *Agrobacterium* bacteriophages and starting a new project on the genetics of the Ti-plasmids. In retrospect, my decision to work on Ti-plasmids seems reckless, because in 1973 there were very few tools available to tackle this project efficiently. Luckily enough, in the group of Jeff Schell and Marc Van Montagu, there were several young people working on different aspects of the subject. None of us had a lot of experience, but the friendly atmosphere and enthusiastic collaboration in the group helped a lot to get the essential techniques started.

My task was to develop a system to map the genes of Ti-plasmids. It was obvious that to establish such a genetic system totally in *Agrobacterium* would be extremely complex, if not impossible. Therefore we decided to find out whether the genetic tools that were available in *Escherichia coli* could be transferred to *Agrobacterium*. Most plasmids known at that time, such as the F-type plasmids, could not be transferred to *Agrobacterium*. However, the group of Naomi Datta had reported that the plasmids of the incompatibility group P could be introduced into *Agrobacterium* by conjugation (Datta et al. 1971). These were therefore good candidates for the genetic shuttle that we needed. Naomi provided us with several P-type plasmids and we indeed confirmed that these could be transferred to different *Agrobacterium* strains. The best-known of these plasmids is RP4, that later became an excellent tool for the genetics of a whole range of Gram-negative bacteria.

Around that time, the observation was made that well-defined DNA sequences carrying resistance genes could transpose and thereby insert into another replicon in the same cell, such as the chromosome or a plasmid. It was attractive to test whether these so-called transposons could insert into the *Agrobacterium* plasmids and inactivate genes. We received two transposons from Naomi Datta and Robert Hedges for our studies, the ampicillin resistance transposon Tn*1* and the streptomycin and trimethoprim resistance transposon Tn*7* (Hernalsteens et al. 1978). Initially, we made transposon Tn*7* insertions in a RP4-Ti plasmid cointegrate in *E coli*. The insertions in the Ti-part of the plasmid were subsequently identified using the instability of the cointegrate during conjugation. This was a very labor-intensive procedure and more efficient methods were required for detailed mapping of the Ti-plasmid.

The simplest strategy to obtain transposon insertions in the Ti-plasmid was to construct an *Agrobacterium* strain with a transposon insertion in its chromosome and subsequently to use the conjugative properties of the Ti-plasmid to select for transposon insertions in the plasmid. Because at that time no suicide plasmid systems were available, obtaining the insertion of the transposon in the *Agrobacterium* genome was not easy. This was performed by introducing a P-type plasmid carrying the transposon in *Agrobacterium* and subsequently selecting for clones that had lost the plasmid, while still retaining the transposon. Different methods to cure *Agrobacterium* of the P-plasmid were tested. Selection using the bacteriophage GU5 that we isolated from river water turned out to be the most reliable way to obtain P-plasmid free clones. This bacteriophage uses the P-type pili as a receptor and can grow on numerous Gram-negative bacterial species if they carry a P-type plasmid.

When the work on *Agrobacterium* plasmids started, the degree of similarity between the octopine and nopaline plasmids was not known. Therefore we decided to work on octopine and nopaline plasmids in parallel. As a nopaline strain we used T37, provided by Armin Braun, which has the intriguing property of inducing teratoma tumors. It seemed likely that these could somehow provide information about cell differentiation in plants. At that time, it was not at all clear how the transferred DNA was organized and how many changes the transfer and integration system could tolerate. A study on this subject using transposon mutagenesis is reported in this Anthology (see commentary by D. Garfinkle). As a first step, we obtained a transposon insertion mutation in a gene encoding an enzyme involved in opine synthesis. These genes would most likely not be required for tumor induction. A chromosomal insertion of Tn*7* in *Agrobacterium* strain T37Noc^{c}1 (a mutant derived from strain T37 that shows derepressed conjugative transfer of its Ti-plasmid) was first constructed. A set of insertions in its Ti-plasmid was subsequently obtained by selecting for the transfer of the resistance markers of Tn*7* after conjugation. To identify nopaline synthesis deficient mutants, we induced tumors on sterile sunflower hypocotyl segments cultured *in vitro* using the different mutant strains harbouring a Tn*7* insertion in the Ti-plasmid. Extracts of these tumors were analyzed to determine the presence of nopaline. The sixth candidate tested proved to be a nopaline synthesis defective (*nos*::Tn*7*) mutant. Both restriction enzyme analysis and heteroduplex mapping showed that in this mutant the Tn*7* is integrated near the right border of the T-DNA.

To study the fate and the expression of the transposon in plant cells, sterile cell cultures were initiated from tumors induced by *Agrobacterium* harboring the pTiT37*noc*^{c}1*nos*::Tn*7* mutant on tobacco. Analysis of DNA extracted from these cultures by Southern blot DNA hybridization confirmed that the same left and right T-DNA borders were used for the integration of the T-DNA containing Tn*7* and that the whole 14 kbp Tn*7* sequence was present at its original position near the right T-DNA border. Thus, the presence of a large foreign DNA insert does not interfere with the transfer and stable integration of the T-DNA. This was confirmed by cloning the T-DNA in a cosmid vector. It is interesting to note that no recombinant DNA manipulations were required for the transfer and integration of Tn*7* into the tobacco genome.

At that time, little was known about the differences in gene expression between prokaryotic and eukaryotic cells. Therefore we tested whether the presence of Tn*7* increases the resistance of the cells to streptomycin, spectinomycin, trimethoprim and methotrexate, drugs towards which Tn*7* confers resistance. No significant increase in resistance was observed in tobacco cells.

We did not observe any differentiation in the tobacco tumor lines that were obtained after infection of tobacco with this *nos*::Tn*7* mutant. However, Henri De Greve obtained a mutant of the octopine Ti-plasmid pTiB6S3 that, in contrast to the wild type, produced numerous shoots from the tumor tissue. This mutation inactivates the auxin synthesis system of the T-DNA and the profuse regeneration is caused by the cytokinin synthesis that is still active in the tumor tissue. Most of these shoots developed into morphologically normal plants but in some, the presence of octopine was detected. Southern blot analysis of DNA extracted from these octopine synthesizing plants showed large deletions in the T-DNA, that removed the oncogenic sequences (De Greve et al. 1982). These plants were fertile and segregated the octopine synthesis marker as a single dominant Mendelian gene. This was a significant observation, because in studies with shoots regenerated from teratoma lines, the T-DNA sequence was not transmitted sexually (Yang et al. 1980) and there was therefore some doubt whether transgenes introduced by *Agrobacterium* would be stable during meiosis. Our studies clearly established that they were.

References

Datta, N., Hedges, R. W., Shaw, E. J., Sykes, R. B., and Richmond, M. H. 1971. Properties of an R factor from *Pseudomonas aeruginosa*. J. Bacteriol. 108: 1244–1249.

De Greve, H., Leemans, J., Hernalsteens, J.-P., Thia-Toong, L., De Beuckeleer, M., Willmitzer, L., Otten, L., Van Montagu, M., and Schell, J. 1982. Regeneration of normal and fertile plants that express octopine synthase from tobacco crown galls after deletion of tumour-controlling functions. Nature 300:752–755.

Hernalsteens, J.-P., De Greve, H., Van Montagu, M., and Schell, J. 1978. Mutagenesis by insertion of the drug resistance transposon Tn*7* applied to the Ti plasmid of *Agrobacterium tumefaciens*. Plasmid 1:218–225.

Yang, F., Montoya, A. L., Merlo, D. J., Drummond, M. H., Chilton, M.-D., Nester, E. W., and Gordon, M. P. 1980. Foreign DNA sequences in crown gall teratomas and their fate during the loss of the tumorous traits. Mol. Gen. Genet. 177:707–714.

Zaenen, I., Van Larebeke, N., Van Montagu, M., and Schell, J. 1974. Supercoiled circular DNA in crown-gall inducing *Agrobacterium* strains. J. Mol. Biol. 86: 109–127.

JEAN-PIERRE HERNALSTEENS: I obtained a B.Sc. degree in Zoology from the University of Brussels (Vrije Universiteit Brussel) in 1973 and a Ph.D. from the same university in 1980. The research for my Ph.D., reported in the accompanying article, was performed in the new lab that Jeff Schell and Marc Van Montagu had started in Brussels and partly in Gent, in a nice informal collaboration between both labs.

Both Jeff and Marc had a strong influence on the course of my work. Both were always available when I needed help or advice and on the other hand, gave me sufficient freedom to develop my ideas independently. Marc often introduced new techniques right on time to allow the work to progress to the next step. Discussions with Jeff were often useful, because he frequently converted complex observations into simple hypotheses that could be easily tested.

Numerous other people also influenced the work significantly. In particular, Naomi Datta and Robert Hedges generously provided us with plasmids and transposons and valuable advice on their use. An essential contribution was that of Michel Jacobs and members of his group, who introduced me to the art of cell culture.

Reprinted from Nature, Vol. 287, No. 5783, pp. 654-656, October 16 1980

The *Agrobacterium tumefaciens* Ti plasmid as a host vector system for introducing foreign DNA in plant cells

Jean-Pierre Hernalsteens

Laboratorium voor Genetische Virologie, Vrije Universiteit Brussel, B-1640 Sint-Genesius-Rode, Belgium

Françoise Van Vliet, Marc De Beuckeleer, Ann Depicker, Gilbert Engler, Michel Lemmers, Marcelle Holsters, Marc Van Montagu & Jeff Schell*

Laboratoria voor Histologie en Algemene Genetika, Rijksuniversiteit Gent, B-9000 Gent, Belgium

The molecular basis of the neoplasmic transformation of plant cells by the crown gall bacterium *Agrobacterium tumefaciens* is the transfer to and stable maintenance of T DNA—a well defined segment of the bacterial Ti plasmid—in plant cells[1-4]. Transformed cells express new biosynthetic capacities, such as the synthesis of opines[5], of which octopine[6] and nopaline[7] are well known examples. Ti plasmids harbour genes enabling the bacteria to degrade opines and use them as carbon and nitrogen sources[8]. pTi-linked genes also determine the specificity of synthesis of opines in the plant tumour cells[8]. On this basis the Ti plasmid can be considered an unusual type of catabolic plasmid, which induces the synthesis of its substrate in transformed plant cells. This relationship, called genetic colonization[2], has opened prospects for the use of Ti plasmids as vectors for genetic manipulation of plants. To evaluate this possibility we inserted a well defined DNA segment, the bacterial transposon Tn7 (ref. 9), into the Ti-plasmid DNA sequence that determines nopaline synthesis in *A. tumefaciens* strain T37 Noc^{c}1 (ref. 10). As we report here, the inserted Tn7 DNA segment became part of the T DNA because the 9.6×10^6 molecular weight (MW) Tn7 DNA sequence was transferred to, and maintained in, the DNA of tumour tissue cultures induced by this mutant strain.

* Max Planck Institut für Züchtungsforschung, 30 Köln 5000, FRG.

Tn7, which encodes streptomycin, spectinomycin and trimethoprim resistance, was introduced into the chromosome of strain T37 Noc^{c}1. An RP4 :: Tn7 plasmid[9], pGV1 (our isolate) was transferred from the auxotrophic *Escherichia coli* strain GV1000 (ref. 11) to strain T37 Noc^{c}1 by plate conjugation[12]. A transconjugant subsequently cured of its pGV1 plasmid was obtained by selection with the pilus-specific bacteriophage GU5 (J. P. H., in preparation). A culture of a transconjugant colony, exponentially growing in LB medium, was infected with a multiplicity of 5 plaque-forming units of GU5 per bacterium and plated after overnight incubation at 32 °C. About 15% of surviving cells were shown to have lost the carbenicillin, kanamycin and tetracycline resistance markers of RP4 but retained the spectinomycin (100 μg ml^{-1}) and trimethoprim (1,000 μg ml^{-1}) resistance markers of the Tn7 transposon.

One such T37 Noc^{c}1 chr :: Tn7 strain, GV15, was used for the isolation of Tn7 insertions in the transfer-derepressed Ti plasmid of strain T37 Noc^{c}1 by selecting for the Ti-plasmid mediated conjugative transfer of the transposon resistance markers. GV15 was incubated for 24 h on minimal A[13] agar with the rifampicin- and erythromycin-resistant C58C1 derivative GV3102. Tn7-containing transconjugants were found at a frequency of 1.5×10^{-5} per recipient cell after plating on LB[13] medium supplemented with rifampicin (100 μg ml^{-1}), erythromycin (100 μg ml^{-1}) and spectinomycin (100 μg ml^{-1}). In 10% of these strains Ti plasmid could be demonstrated by its ability to degrade nopaline[8] and sensitivity to agrocin 84 (ref. 14). The remaining transconjugants expressed none of the Ti-plasmid markers and therefore probably harboured Tn7 insertions in the cryptic conjugative plasmids of strain

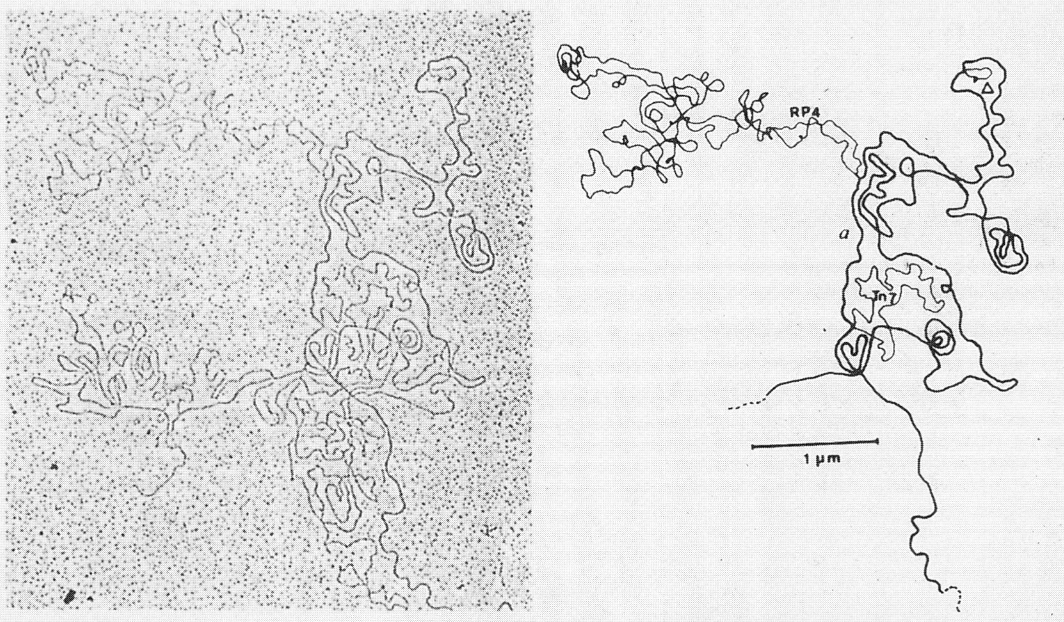

Fig. 1 Physical mapping of the Tn7 insertion in a nopaline-synthesis deficient (Nos$^-$) mutant of pTiT37. pGV3106 DNA was prepared from strain GV3171 as described previously[12]. The Tn7 insertion in pGV3106 was localized by electron microscopic analysis of heteroduplex molecules[20] formed by renaturation of single-stranded DNA molecules of pGV3106 and of pGV4000 (ref. 18), a co-integrate plasmid consisting of pTiC58 and the P plasmid RP4. The distance (*a*) between the Tn7 transposon in pGV3106 and the RP4 insertion (at map position 130.4)[18] in pGV4000 is 1.3 ± 0.1 μm. The 0.5-μm deletion loop (Δ) in pTiT37 (ref. 4) and present in pGV3106 (at map position 124.3–125.3) (G. E., in preparation) was used as a reference point for the Tn7 integration site. This mapping result agrees with the restriction endonuclease analysis (data not shown) of pGV3106.

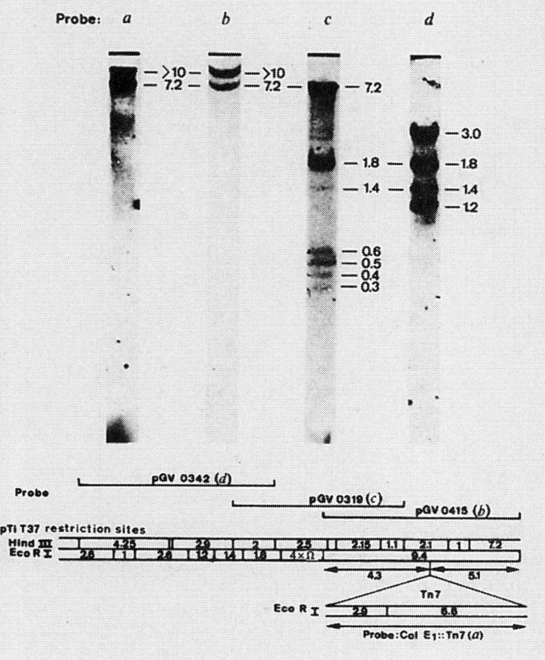

Fig. 2 Analysis of Tn*7* and Ti-plasmid related DNA fragments present in restriction endonuclease digests of tobacco tumour DNA. W38T37 :: Tn*7*-1, a bacterium-free tumour tissue culture, was started from a tumour induced by strain GV3171 on *N. tabacum* cv Wisconsin 38, essentially as described previously[16]. After 5 months of *in vitro* culture on hormone-free medium, samples of the tissue were extracted with 25 ml buffer (Tris 50mM *p*H 8.2, EDTA 100 mM, diethylammonium diethyldithiocarbamate (DIECA) 100mM, triisopropyl naphthalene sulphonic acid 2%, diethylpyrocarbonate 0.1%) per g lyophilized tissue. The mixture was brought to 0.3 M NaCl and extracted with an equal volume of phenol and chloroform. The aqueous phase was brought to 0.5 M NaCl and re-extracted with the same organic phase. After precipitation with two volumes of ethanol the crude DNA was further purified by CsCl gradient banding (starting density 1.55). After dialysis it was concentrated to about 1 mg ml^{-1} by ethanol precipitation. The DNA was digested to completion with restriction endonuclease *Eco*RI. Digested DNA (20 μg) was separated by electrophoresis through a 0.8% agarose slab gel and transferred to a nitrocellulose filter[19]. The filter was hybridized by Southern's[19] method to cloned Ti-plasmid fragments derived from the T region of pTiC58 (ref. 17) and to ColE1 :: Tn*7* plasmid DNA, labelled *in vitro* with ^{32}P by nick-translation[21]. ColE1 :: Tn*7* plasmid DNA was used as a probe in lane *a*. The extent of the cloned pTi fragments used as probes for the hybridizations of lanes *b*, *c* and *d* is indicated on top of the physical map of the T region of pTiT37. Lanes *a* and *b* demonstrate the presence in the tumour DNA of the complete Tn*7* sequence integrated in a sequence homologous to the 9.4×10^6 MW *Eco*RI fragment of pTiT37. The localization of the integration site of Tn*7* in the T-DNA segment based on these data and confirmed by similar hybridization experiments with *Hin*dIII digests of the tumour DNA (data not shown), corresponds to the integration site of Tn*7* in pGV3106 (Fig. 1). Lanes *b*, *c* and *d* demonstrate that the T-DNA segment in this tumour line is identical, except for the *Eco*RI fragment of MW 9.4×10^6, to the T-DNA segment independently shown to be present in the BT37 tobacco tumour line induced by wild-type pTiT37 (ref. 4). Numbers adjacent to bands and numbers in each segment indicate MW($\times 10^{-6}$).

T37 Noc^c1 (G. E., unpublished). Twelve transconjugants expressing both pTi and Tn*7* markers were inoculated on sunflower hypocotyl segments[15]. Nine were found to be pathogenic. Nopaline synthesis was demonstrated reproducibly by electrophoretic analysis[16] of the tumours induced by eight of these strains, but could not be detected in tumours induced by strain GV3171. The inability of strain GV3171 to induce nopaline synthesis was confirmed by analysis of tumours induced on potato, *Kalanchoë daigremontiana* and tobacco.

The Tn*7* insertion in strain GV3171 was localized on the mutant Ti plasmid (pGV3106) by restriction endonuclease analysis, and mapped in a DNA fragment homologous to the *Hin*dIII-23 fragment[17] of the standard nopaline Ti plasmid pTiC58 (data not shown). This localization was confirmed independently by electron microscopic heteroduplex analysis, which also showed that the Tn*7* inserted without detectable deletion (Fig. 1). This *Hin*dIII-23 fragment was subsequently shown to be homologous with the right-end border fragment of the T DNA[2,4] and thus to be part of the Ti-plasmid DNA segment which is normally transferred to plant cells. To confirm this result, several hundred Tn*1* and Tn*7* insertions in pTiC58 were screened for nopaline synthesis after induction of tumours. A single nopaline synthesis-deficient mutant was found and shown to be the result of a Tn*1* insertion in fragment *Hin*dIII-23 (ref. 18).

A bacterium-free tissue culture, W38T37 :: Tn*7*-1, was started from a tumour induced by strain GV3171 on *Nicotiana tabacum* cv Wisconsin 38. Analysis by Southern blot hybridizations[19] of a DNA sample isolated from this culture, using ColE1 :: Tn*7* (our isolate) DNA and various cloned segments of the nopaline plasmid pTiC58 as radioactive probes showed that the complete Tn*7* was present in the plant tumour genome,

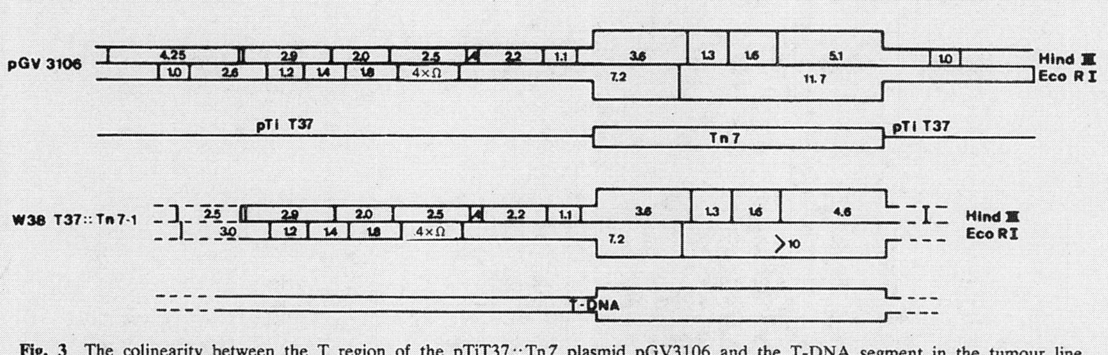

Fig. 3 The colinearity between the T region of the pTiT37 :: Tn*7* plasmid pGV3106 and the T-DNA segment in the tumour line W38T37 :: Tn*7*-1 induced by the *A. tumefaciens* strain GV3171 harbouring pGV3106. The figure is based on results obtained by Southern type DNA hybridizations as described in Fig. 2 legend.

integrated at its original site in a normal-size T DNA (Figs. 2, 3).

Our results show that the Ti plasmid can be used as a vector for the experimental introduction and stable maintenance of foreign genes in higher plants. Furthermore, the demonstration that the Ti-plasmid DNA segment specifying nopaline synthesis in transformed plant cells is part of the T DNA, but not functionally required for the induction or maintenance of the neoplasmic state, supports the genetic colonization model of crown gall[2].

We thank Ruth Trimarchi for help. This work was supported by grants from the Kankerfonds van de ASLK, Instituut tot Aanmoediging van het Wetenschappelijk Onderzoek in Nijverheid en Landbouw (IWONL) (2841A), the Fonds Wetenschappelijk Geneeskundig Onderzoek (3.0052.78) to J. S. and M. V. M., and Fonds voor Kollektief Fundamenteel Onderzoek (2.0007.77) to M. Jacobs and M. V. M. M. H. thanks the Belgian NFWO for a fellowship and M. L. and M. D. B. thank the IWONL for financial support.

Received 5 June; accepted 8 August 1980.

1. Chilton, M. D. *et al. Cell* **11,** 263–271 (1977).
2. Schell, J. *et al. Proc. R. Soc.* B**204,** 251–266 (1979).
3. Thomashow, M. F., Nutter, R., Montoya, A. L., Gordon, M. P. & Nester, E. W. *Cell* **19,** 729–739 (1980).
4. Lemmers, M. *et al. J. molec. Biol.* (in the press).
5. Petit, A., Dessaux, Y. & Tempé, J. *Proc. 4th int. Conf. Pl. Path. Bact.*, Angers, 143–152 (1978).
6. Ménagé, A. & Morel, G. *C. r. hebd. Séanc. Acad. Sci., Paris* **259,** 4795–4796 (1964).
7. Goldmann, A., Thomas, D. W. & Morel, G. *C. r. hebd. Séanc. Acad. Sci., Paris* **268,** 852–854 (1969).
8. Bomhoff, G. *et al. Molec. gen. Genet.* **145,** 177–181 (1976).
9. Barth, P. T., Datta, N., Hedges, R. W. & Grinter, N. J. *J. Bact.* **125,** 800–810 (1976).
10. Petit, A. *et al. Nature* **271,** 570–572 (1978).
11. Hernalsteens, J. P., De Greve, H., Van Montagu, M. & Schell, J. *Plasmid* **1,** 218–225 (1978).
12. Van Larebeke, N. *et al. Molec. gen. Genet.* **152,** 119–124 (1977).
13. Miller, J. H. *Experiments in Molecular Genetics*, 466 (Cold Spring Harbor Laboratory, New York, 1972).
14. Engler, G. *et al. Molec. gen. Genet.* **138,** 345–349 (1975).
15. Petit, A. & Tempé, J. *Molec. gen. Genet.* **167,** 147–155 (1978).
16. Aerts, M., Jacobs, M., Hernalsteens, J.-P., Van Montagu, M. & Schell, J. *Pl. Sci. Lett.* **17,** 43–50 (1979).
17. Depicker, A. *et al. Plasmid* **3,** 193–211 (1980).
18. Holsters, M. *et al. Plasmid* **3,** 212–230 (1980).
19. Southern, E. M. *J. molec. Biol.* **98,** 503–518 (1975).
20. Davis, R. W., Simon, M. & Davidson, N. *Meth. Enzym.* **21D,** 413–428 (1971).
21. Rigby, P. W. J., Dieckmann, M., Rhodes, C. & Berg, P. *J. molec. Biol.* **113,** 237–252 (1977).

SHRINKING THE Ti PLASMID

Mary-Dell Chilton
Novartis Seeds, Inc., Research Triangle Park, North Carolina, United States
E-mail: marydell.chilton@syngenta.com

Commentary on papers:

de Framond, A. J., Barton, K. A.,and Chilton, M.-D. 1983. Mini-Ti: A new vector strategy for plant genetic engineering. Bio/Technology 1:262–269.

Hoekema, A., Hirsch, P. R., Hooykaas, P. J. J., and Schilperoort, R. A. 1983. A binary plant vector strategy based on separation of vir- and T-region of the *Agrobacterium tumefaciens* Ti-plasmid. Nature 303:179–180.

The first obstacle in developing the Ti plasmid as a gene vector for plants was its enormous size, which rendered impossible any direct recombinant DNA approach to deletion or insertion of genes of interest. In the early 1980s, several indirect methods were developed for manipulating T-DNA. These were all based on the approach of "forcing" (that really means "selecting for" a hoped for event) recombination between an intermediate vector containing a piece of T-DNA and the corresponding piece of the Ti plasmid in *Agrobacterium.* One strategy involved a double recombination (replacing part of T-DNA by an engineered counterpart constructed *in vitro* (Matzke and Chilton 1981, Leemans et al. 1981). Another also utilized an intermediate vector but only one recombination with the Ti plasmid (Fraley et al. 1985) or a Ti plasmid mutant, in which, the oncogenic functions of the T-DNA had been deleted (Zambryski et al. 1983). But these methods were not easy or not yet developed at the time several groups began to investigate a zero recombination strategy: could T-DNA on a separate plasmid be transferred to plants by a Ti "helper" plasmid?

Anic de Framond, a French graduate student, came to my laboratory at Washington University in St. Louis to do her thesis research, and took up the project of testing this idea in the nopaline Ti plasmid system, where we had just found the border sequences. Her approach was to clone full-length T-DNA from pTiT37 onto an intermediate vector which could replicate in *Agrobacterium.* To test this T-DNA plasmid for tumor-inducing activity, she needed a helper Ti plasmid with no T-DNA. We found exactly what we needed in one of Rob Schilperoort's papers: the now famous strain LBA4404. When I called the University of Leiden to request this strain from Rob, the ensuing discussion brought to light that his group was engaged in a similar effort on the octopine T-DNA. Despite our competitive project goal, Paul Hooykaas, a member of Schilperoort's project team, honorably sent us LBA4404. Knowledge of each other's work no doubt increased the sense of urgency of both projects, which happily came to the same conclusion. T-DNA could be transferred to plants by the virulence functions on a separate plasmid, operating *in trans.* We called our shrunken Ti plasmid "Mini-Ti" and the Leiden authors coined the more widely adopted term "binary vector." As an interesting historical note, we learned later that Jeff Schell's group had also tested the binary vector concept and concluded that it did not work; we never heard any details.

Today binary vectors have been developed with chimeric selectable markers, scorable markers, convenient polylinker sites for introduction of passenger genes, and selectable markers of utility for selection in *Agrobacterium* (Hajdukiewicz et al. 1994 and references therein). This approach seems to be the most widely used for genetic engineering of crops.

References

Fraley, R. T., Rogers, S. G., Horsch, R. B., Eichholtz, D. A., Flick, J. S., Fink, C. L., Hoffmann, N. L., and Sanders, P. R. 1985. The SEV system: A new disarmed Ti plasmid vector system for plant transformation. Bio/Technology 3:629–635.

Hajdukiewicz, P., Svab, Z., and Maliga, P. 1994. The small, versatile pPZP family of *Agrobacterium* binary vectors for plant transformation. Plant Mol. Biol. 25: 989–994.

Leemans, J., Shaw, C., Deblaere, R., De Greve, H., Hernalsteens, J. P., Maes, M., Van Montagu, M., and

Schell, J. 1981. Site-specific mutagenesis of *Agrobacterium* Ti plasmids and transfer of genes to plant cells. J. Mol. Appl. Genet. 1:149–164.

Matzke, A. J. M., and Chilton, M.-D. 1981. Site-specific insertion of genes into T-DNA of the *Agrobacterium* tumor-inducing plasmid: An approach to genetic engineering of higher plant cells. J. Mol. Appl. Genet. 1:39–49.

Zambryski, P., Joos, H., Genetello, C., Leemans, J., Van Montagu, M., and Schell, J. 1983. Ti plasmid vector for the introduction of DNA into plant cells without alteration of their normal regeneration capacity. EMBO J. 2:2143–2150.

MARY-DELL CHILTON: See the commentary on page 139.

Reprinted from BIO/TECHNOLOGY, May 1983. Vol. 1, No. 3, © 1983 Nature Publishing Co.

RESEARCH PAPERS

MINI-Ti: A NEW VECTOR STRATEGY FOR PLANT GENETIC ENGINEERING

Annick J. de Framond, Kenneth A. Barton,* and Mary-Dell Chilton
Department of Biology, Washington University, St. Louis, Missouri 63130

Agrobacterium tumefaciens **has the natural ability to insert a specific part of its Ti (tumor-inducing) plasmid, called T-DNA, into the chromosomal DNA of host plant cells, causing tumorous growth. Foreign DNA artificially introduced into T-DNA is inserted into the plant genome by this natural vector system. Genes on the Ti plasmid required for T-DNA transfer, called VIR (virulence) genes, are situated outside T-DNA. We have separated T-DNA from the VIR functions onto a MINI-Ti plasmid that replicates in *E. coli* and contains full-length T-DNA (including the 25 basepair border signals) from pTi T37, a nopaline-type Ti plasmid. We present evidence that MINI-Ti, when transferred to *Agrobacterium* by cointegration with a shuttle vector, in the presence of helper plasmid pAL4404 containing VIR genes, can incite nopaline positive tumors, evidence of T-DNA transfer. It is not clear whether MINI-Ti functions in *trans* to the VIR genes, or whether spontaneous cointegration of pAL4404 with MINI-Ti/pRK (the shuttle form of MINI-Ti) accounts for the tumor-inducing ability of the bacteria. The MINI-Ti approach simplifies procedures for exploiting T-DNA as a gene vector for plants.**

Agrobacterium tumefaciens is the etiological agent of the plant cancer crown gall[1], which can affect most dicotyledonous plants[2]. Oncogenicity is encoded on large plasmids called Ti (tumor-inducing) plasmids[3, 4]. The mechanism by which the pathogen transforms host plants at wound sites is not known, but the end result is the integration of a specific part of the Ti plasmid, called T-DNA, into the nuclear genome of the transformed plant cells[5–7], where it is transcribed into mRNA[7–10]. The tumor tissue synthesizes specific metabolites, the opines[11, 12], never found in normal plant tissue. The pattern of opines produced by a tumor is determined by the T-DNA [11, 13, 14]. Ti plasmid genes outside T-DNA confer on the host bacterium the specific ability to catabolize the opines[11, 12, 15, 16, 17]. Another set of Ti plasmid genes, also located outside T-DNA, termed the VIR (virulence) region, is essential for oncogenicity[16, 17]. These functions presumably play a role in the transfer of the T-DNA from the bacterium to the plant cell.

The Ti plasmid is often referred to as a natural vector for the genetic engineering of plants because it can transfer its T-DNA from the prokaryotic pathogen to a eukaryotic plant cell. A foreign gene can be inserted into the T-DNA of a Ti plasmid and thus engineered into the genome of a tumor cell[18–20] or regenerated plant[21]. Because of the very large size of the Ti plasmids (*ca.* 200 kb), special methods have been devised for insertion of foreign DNA into specific sites in the T-DNA[19, 20, 22]. These techniques, while useful, are difficult and time-consuming. We report here that the Ti plasmid can be split into two independent replicons, one carrying the T-DNA, the other harboring the VIR region. An Agrobacterium strain harboring the pair of replicons is fully tumorigenic. T-DNA replicons can thus be used as smaller, more versatile vectors for genetic engineering of plants.

*Cetus Madison Corporation, 2208 Parview Rd., Middleton, WI 53562

RESULTS

In order to study nopaline T-DNA isolated from other functions of the Ti plasmid, we undertook the isolation of full length T-DNA of pTi T37 on a recombinant MINI-Ti plasmid. Because of its large size (23 kb), this T-DNA contains numerous recognition sites for the restriction endonucleases commonly used in cloning, *e.g.* Hind III, Eco RI or Bam HI. It would therefore be difficult to clone this entire T-DNA as a single contiguous fragment after partial digestion of pTi T37. In contrast, the enzyme Kpn I cuts this T-DNA only once, in Eco RI fragment 13 (Figure 1). Our strategy for construction of a MINI-Ti plasmid exploited this situation and allowed us to clone these two Kpn I fragments directly. We used a site-specific insertion technique[20] to introduce a Col El replicon into the Kpn I site of the T-DNA in a manner that destroyed this unique Kpn I site. Kpn I digestion of the resulting Ti plasmid produced, from the T-DNA sector, a single long Kpn I fragment harboring the Col El insert. After circularization with ligase, the resulting plasmid was rescue-cloned in *E. coli.*

Construction of pMoM

In order to insert a kanamycin resistant ColEl replicon, pAMl, into the unique Kpn I site of pTi T37 T-DNA, we used the approach developed by Ruvkun and Ausubel for

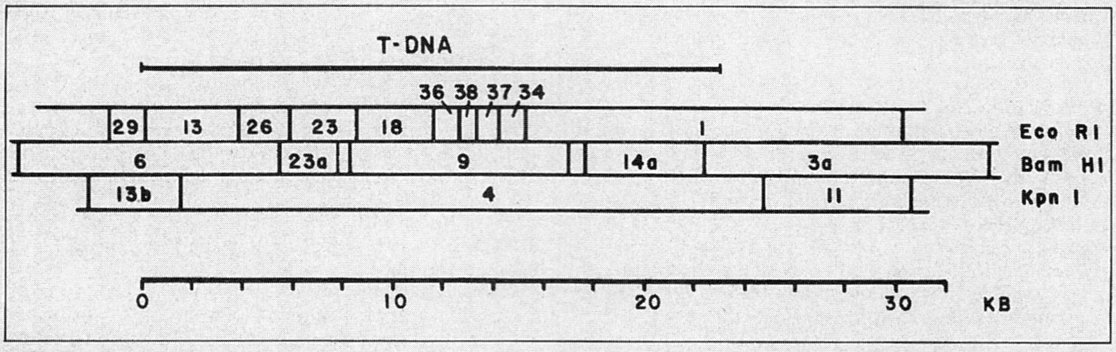

FIGURE 1 Map of the T-DNA region of pTi T37. The map shown above was adapted from Bevan and Chilton.[24] Eco RI, Bam HI and Kpn I restriction endonuclease cleavage sites were determined by single and multiple digest analysis of recombinant plasmids containing Eco RI or Bam HI fragments as inserts into pBR325 or pBR322.

similar constructions in Rhizobium[23] as applied to *A. tumefaciens* by A. Matzke in this laboratory[20]. Construction of pAMl and the successive steps of the insertion of pAMl into the Kpn I site of pTi T37 T-DNA are detailed in Figures 2 and 3. The first step was the insertion of pAMl into the Kpn I site of pTi T37 Eco RI fragment 13, subcloned into pBR325. The modified copy of Eco RI fragment 13 containing pAMl was then recloned into the Eco RI site of the wide host range plasmid pRK290. The resulting plasmid, pAM3, can thus replicate in *A. tumefaciens* (Fig. 2).

The second step, called homogenotization[20], selects for replacement of the wild-type Eco RI fragment 13 of pTi T37 by the engineered copy carried by pAM3 (Fig. 3). For this purpose, we first transformed pAM3 into *A. tumefaciens* strain A208, which harbors pTi T37. To select for the desired homogenotes, which occur spontaneously as a minor fraction of the bacterial population, we introduced into the same strain by conjugation a third plasmid, R751-pMG2, that is incompatible with pRK290 replicons such as pAM3 and confers resistance to gentamycin. We selected simultaneously for kanamycin and gentamycin resistance. Because the separate plasmids bearing these markers are incompatible, most of the bacterial population is killed. The surviving colonies carried a homogenote Ti plasmid in which double recombination between pTi T37 and pAM3 had replaced the wild-type Eco RI fragment 13 by the engineered copy. The structure of the homogenote was proven by Southern blotting analysis. The homogenote Ti plasmid DNA was purified, digested with Bam HI and the fragments separated electrophoretically and blotted. The blot was hybridized with labeled pAM2 DNA containing the engineered Eco RI fragment 13 as insert. Bam HI fragment 6 of pTi T37 contains Eco RI fragment 13[24] and should therefore be altered in the homogenote. Because no new Bam HI site was introduced in the engineering step, we expect a new fragment to appear in the Bam HI digest of the homogenote Ti plasmid, whose size is greater than fragment 6 by the size of the insert, pAMl. We observed indeed the absence of normal fragment 6 and the appearance of the expected band at higher molecular weight (Fig. 4). We named the homogenote Ti plasmid pMoM, an acronym for its role as "mother of MINI-Ti" as described below.

Construction of MINI-Ti

pMoM was digested to completion with Kpn I, which cuts it only outside T-DNA, and the resulting fragments were ligated at low DNA concentration, to favor circularization. The ligation mixture was transformed into *E. coli* strain LE392, selecting for the genetic markers carried by pAMl, kanamycin and ampicillin resistances. We obtained one transformant harboring a plasmid with a single Kpn I site that we named MINI-Ti (Fig. 5). It contains the entire T-DNA of pTi T37, including the 25 base pair direct repeats that flank the T-DNA[25]. We also obtained a second transformant containing a plasmid designated MINI-Ti:Kpn 11, which consists of the plasmid MINI-Ti plus the next contiguous Kpn I fragment on the right side of T-DNA, Kpn I fragment 11. We presume that it derived from a partial digest fragment of pTi T37 with

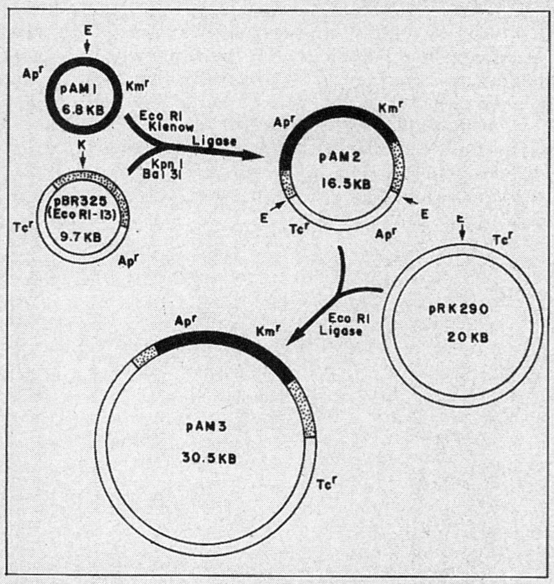

FIGURE 2 Insertion of a Col E1 replicon into pTi T37 Eco RI fragment 13 at its Kpn I site. Col E1 replicon pAM1 was constructed from pBR325 by substitution of its Hind III/Sal I fragment by the Hind III/Sal I fragment of Tn5, encoding kanamycin resistance (Km^r). pAM1 has the ampicillin resistance (Ap^r) and chloramphenicol resistance (Cm^r) traits of pBR325 but tetracycline resistance (Tc^r) is eliminated. pBR325 (Eco RI-13) was linearized at its unique Kpn I site and rendered blunt by brief treatment with BAL-31 (see Experimental Protocol). pAM1 was linearized with Eco RI and rendered blunt by "filling in" its ends with Klenow polymerase (see Experimental Protocol). The two blunt fragments were ligated together and transformed into *E. coli* LE392, selecting for Km and Tc resistances. The resulting plasmid, pAM2, has lost the Cm^r trait of pAM1. For introduction into *A. tumefaciens*, the engineered version of Eco RI fragment 13 was excised from pAM2 with Eco RI and recloned into the unique Eco RI site of wide host range plasmid pRK290,[37] yielding pAM3.

Kpn I. MINI-Ti:Kpn 11 contains all of the Eco RI fragment 1 (see Fig. 1), an attribute we exploited for the next construction.

MINI-Ti Does Not Replicate in *A. tumefaciens*

We made repeated attempts to transform the plasmid MINI-Ti into *A. tumefaciens* strain A114, containing no Ti plasmid, by selection for kanamycin resistance. While pMoM plasmid, a positive control, reproducibly yielded 25–50 transformants, MINI-Ti never gave any. MINI-Ti thus does not seem to contain an origin of replication that is functional in *A. tumefaciens*. This result is consistent with genetic evidence for a unique origin of replication on the Ti plasmid that maps outside of T-DNA[26].

Construction of MINI-Ti/pRK

In order to study the properties of MINI-Ti in *A. tumefaciens*, we isolated a cointegrate between MINI-Ti:Kpn 11 and a derivative of the wide host range Inc Pl plasmid pRK290, which does replicate in *A. tumefaciens*[20]. We used a "shuttle vector" method to force cointegration of the two plasmids[22]. The bacterial host strain was *E. coli* DS989, a mutant possessing a thermosensitive DNA polymerase I[33]. This enzymatic activity is essential for the replication of Col El replicons[27] such as MINI-Ti:Kpn 11. At permissive temperature, 28°C, Col El replicons are maintained in DS989, but at the restrictive temperature, 43°C, they are lost. The replication of Pl group wide host range plasmids such as pRK290 is not dependent on DNA polymerase I; therefore they are stable in DS989 at 43°C. If the Col El replicon and the wide host range plasmid share a region of homology, they can recombine to form a cointegrate. Selection at the restrictive temperature for maintenance of the Col El genetic markers allows isolation of cointegrate plasmids.

The shuttle vector that we used in our experiment is pRK290:Eco RI-1, which consists of pTi T37 Eco RI fragment 1 cloned into pRK290. Our Col El replicon was MINI-Ti:Kpn 11, which shares 15 kb of homology with the shuttle vector. The two plasmids were transformed into DS989 at 28°C; then, by growth of the transformants at 43°C and selection for genetic markers on the Col El replicon (kanamycin and ampicillin resistances), we obtained clones in which the two plasmids had cointegrated by homologous recombination through Eco RI fragment 1 (Fig. 6). In this paper we designated this cointegrate structure MINI-Ti/pRK.

MINI-Ti/pRK is Virulent only in the Presence of Helper pAL4404

When MINI-Ti/pRK was introduced into A114, an *A. tumefaciens* strain lacking any Ti plasmid, by transformation and selection for kanamycin resistance, the transformants obtained were avirulent. Analysis of plasmid DNA minipreps from the transformants confirmed that they indeed contained MINI-Ti/pRK plasmid. This result con-

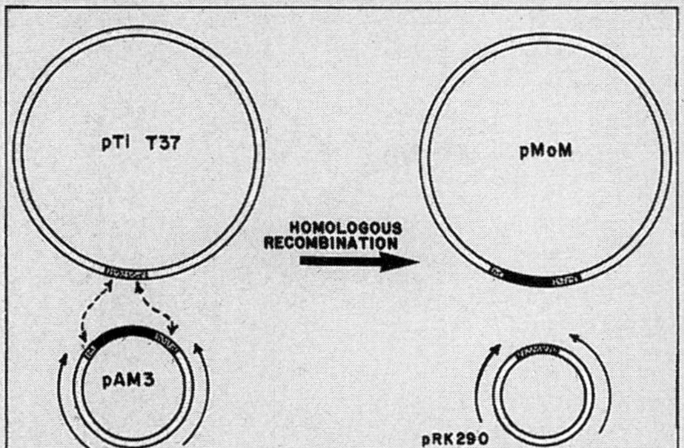

FIGURE 3 Homogenotization of engineered Eco RI fragment 13 in *A. tumefaciens* by double recombination with pTi T37. pAM3 was introduced into *A. tumefaciens* strain A208, carrying pTi T37, by transformation.[40] Homologous recombination in a minor fraction of the population brings about exchange of the engineered Eco RI fragment 13 with its wild type counterpart on the Ti plasmid. Stippled regions of pAM3 and pTi T37 indicate homologous DNA, and the arrows indicate potential regions of recombination. Selection for bacteria containing the homogenote Ti plasmid, pMoM, is accomplished by introducing R751-pMG2, a gentamycin resistant P1-type plasmid that is incompatible with pAM3, a pRK290-based replicon of the P1 incompatibility group. Selection for Km^r and gentamycin resistance yields colonies whose Ti plasmids have the structure of pMoM (see text).

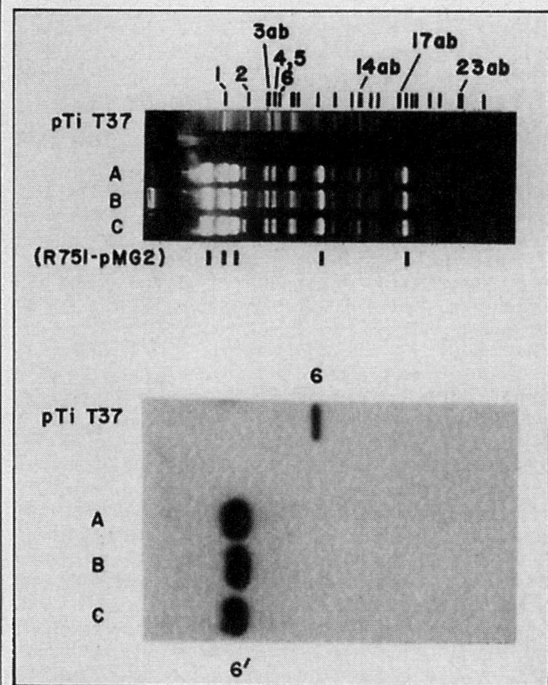

FIGURE 4 Confirmation of structure of Ti plasmid from homogenotes. **Upper:** Bam HI digests of 1.0 μg of pTi T37 and 3 μg of plasmid from candidate homogenotes (A, B and C) were separated by electrophoresis on a horizontal gel of 0.7% agarose. The gel was stained with ethidium bromide and illuminated with UV light to visualize fluorescent DNA bands. The bright pattern of five bands in A, B and C is due to the eviction plasmid R751-pMG2, whose fragment pattern is drawn below the photo. Fragment number assignments for pTi T37 are indicated above the photo. Note that in A, B and C, bands corresponding to 3ab and the 4,5 doublet are visible but band 6 is not detected. This gel was used to prepare the Southern blot shown below. **Lower:** Southern blot hybridization analysis of the fragments displayed above was performed with labeled pAM2 DNA as probe. The single band of hybridization in pTi T37 is at the position of fragment 6. In the digests of candidates A, B and C, no hybridization is detectable at the position of wild type fragment 6, and a new band (designated 6') has appeared at the position expected (*ca.* 17 kb) for engineered fragment 6 containing the pAM1 insert. More intense hybridization is seen in A, B and C than in pTi T37 principally because these engineered plasmids contain more DNA homologous to the labeled probe: in addition to the Ti plasmid homology (Eco RI fragment 13) they contain pAM1 insert which is also represented in the labeled probe.

firms the conclusion of earlier Ti plasmid mutagenesis studies, which demonstrated that the VIR region of the Ti plasmid located to the left of T-DNA is essential for tumor induction[16, 17].

The MINI-Ti/pRK plasmid was next transformed into an avirulent Agrobacterium strain, LBA4404, harboring pAL4404, an avirulent deletion mutant derivative of the octopine Ti plasmid pTi Ach5. This plasmid has a Tn904 insert in Sma I fragment 3a and has suffered a deletion extending from this point rightward to Sma I fragment 6 as indicated in Fig. 7. This "helper" plasmid thus lacks the entire octopine T-DNA and octopine catabolism functions. The VIR region, located on the left side of T-DNA, survives intact.

The transformant strain LBA4404(MINI-Ti/pRK) induces teratomatous tumors on *Nicotiana tabacum* stems and on *Kalanchoe daigremontiana* leaves (Fig. 8). Tobacco tumor tissue was excised and cultivated axenically on hormone-free Murashige and Skoog medium[48], where it synthesized nopaline, the diagnostic opine of pTi T37 T-DNA (Fig. 9).

Does MINI-Ti/pRK Cointegrate with pAL4404 Helper Plasmid?

The virulence of strain LBA4404(MINI-Ti/pRK) could reflect *trans* complementation between the two plasmids. Alternatively, they might form a cointegrate topologically equivalent to an intact Ti plasmid. To test for any DNA homology between the two plasmids, we hybridized labeled pAL4404 DNA to a Southern blot containing MINI-Ti/pRK and pAL4404 itself. Under non-stringent hybridization conditions (63°C, 3xSSC) we observed no hybridization to the MINI-Ti/pRK band even when the pAL4404 signal was grossly overexposed (data not shown).

Attempts to visualize a putative cointegrate of pAL4404 and MINI-Ti/pRK by gel electrophoresis of the intact plasmids of strain LBA4404(MINI-Ti/pRK) led to ambiguous results. Southern hybridization of the blotted plasmid bands using labeled pAL4404 as probe produced a strong signal at the position of pAL4404 that would obscure any signal from the putative cointegrate, which would migrate just above pAL4404. Probing similar Southern blots with labeled pAMI DNA to identify MINI-Ti containing bands produced a weak autoradiographic signal at the position of the putative cointegrate, but this was also seen in a control of MINI-Ti/pRK alone (data not shown). pRK290 replicons yield a ladder of multimeric bands, both in *E. coli* and in *Agrobacterium*, in our experience; one of these, visible as a faint band on the original gel, interferes in this analysis. Genetic studies will be required to test definitively for cointegration between pAL4404 and MINI-Ti/pRK.

DISCUSSION

The results presented here demonstrate that MINI-Ti/pRK, a replicon containing T-DNA but not the VIR functions of the Ti plasmid, can insert its T-DNA into plant cells when "helped" by pAL4404, a replicon containing VIR functions but no T-DNA. It is noteworthy that the VIR region derived from an octopine-type Ti plasmid (pTi Ach5) can bring about insertion of T-DNA derived from a nopaline-type Ti plasmid (pTi T37). This indicates that the homologous VIR regions of the octopine and nopaline Ti plasmids[28] do indeed code for equivalent functions. Signal repeats flanking octopine T-DNA are very similar to those flanking nopaline T-DNA[25, 29] (J. Schell, personal communication), consistent with the view that these are involved in the mechanism of T-DNA transfer.

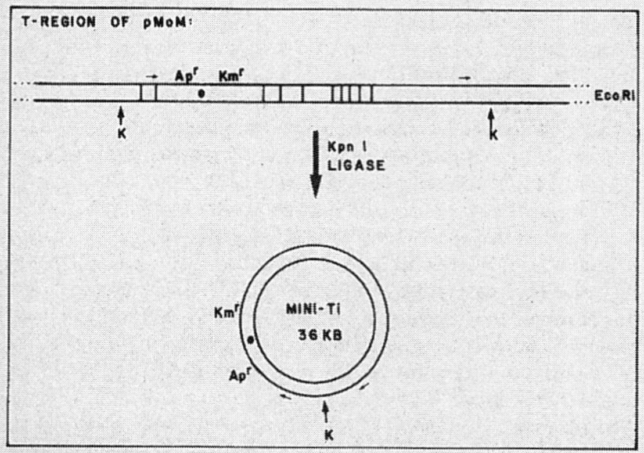

FIGURE 5 Formation of MINI-Ti from pMOM. pMOM was digested with Kpn I and the fragments ligated at low DNA concentration. Transformation of the mixture into *E. coli* LE392 selecting for Ap^r and Km^r allowed rescue cloning of the long Kpn I fragment containing all of T-DNA, including the left and right border 25 basepair repeats (small arrows) and the Col E1 origin of replication (black dot). Arrows labeled K indicate Kpn I sites. The linear map shows a small section of pMoM, with its Eco RI fragments indicated (vertical lines). The rescue-cloned T-DNA-containing plasmid was named MINI-Ti.

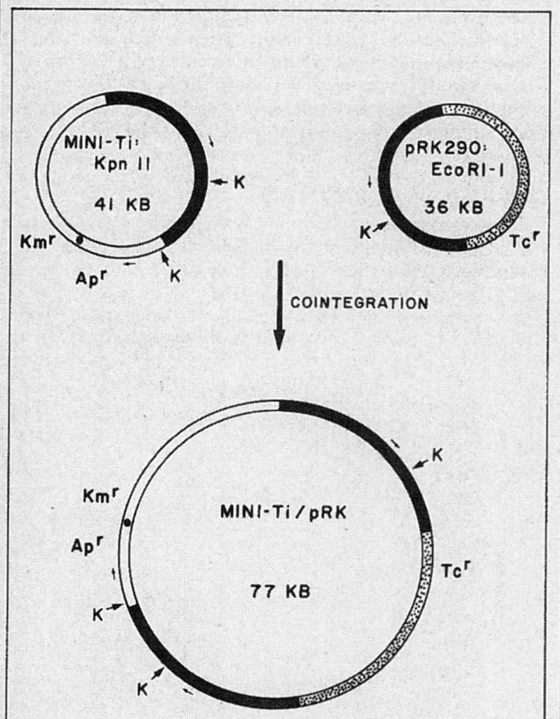

FIGURE 6 Formation of MINI-Ti/pRK by cointegration. For introduction into *Agrobacterium*, MINI-Ti:Kpn 11 (a homolog of MINI-Ti containing the contiguous Kpn I fragment 11, see text) was cointegrated into pRK290:Eco RI-1, with which it shares 15 kb of homology (black bar = Eco RI fragment 1). Selection for the cointegrate was in *E. coli* DS989, whose DNA polymerase I is temperature sensitive. At the restrictive temperature (43°), Col E1 plasmids such as MINI-Ti:Kpn 11 are lost; selection for the Km^r and Ap^r traits on this plasmid yields cointegrate plasmid MINI-Ti/pRK. Symbols are as described for Figure 4; pRK290 DNA is stippled.

The mechanism by which MINI-Ti/pRK is "helped" by pAL4404 is not clear. The two plasmids may interact in *trans*. We have not, however, ruled out the possibility of low frequency formation of cointegrate plasmid. Any such cointegration must presumably occur by a mechanism other than homologous recombination, since we observe no homology between the two plasmids.

The practical consequence of the results presented here is that the plasmid MINI-Ti/pRK, and presumably other similar T-DNA constructions, can be used directly for introducing foreign genes into plants. They need not be forced onto a Ti plasmid by single or double recombination[19, 20]. MINI-Ti-type replicons can replicate in *E. coli;* thus the recombinant DNA construction of genetically engineered T-DNA can be carried out readily in this host. The engineered MINI-Ti replicon can then be introduced into an Agrobacterium strain containing pAL4404, using the shuttle vector technique described here. The resulting Agrobacterium strain, with no further manipulation, can be used to introduce the engineered trait into host plant DNA.

Although substantially smaller than the Ti plasmid, MINI-Ti is still too large for easy manipulation by recombinant DNA techniques. It should be possible to delete from MINI-Ti all of the T-DNA functions except the 25 basepair border signals, replacing them with any desired genetic marker, preferably a selectable one. None of the genes in T-DNA appears to be essential for the transfer mechanism[24, 30, 31]. Thus the functions of T-DNA responsible for tumorous characteristics of the transformed cells can be eliminated, yielding a convenient disarmed vector based on the MINI-Ti strategy.

EXPERIMENTAL PROTOCOL

Bacterial strains:

E. coli LE392 (from Dr. R. Beachy of this department) was the recipient strain for routine transformation experiments. It is a derivative of ED8654 and is described by Murray *et al.*[32]. *E. coli* DS989 (from Dr. S. Rogers, Monsanto, St. Louis) is a mutant carrying temperature sensitive DNA polymerase I (*pol*AI)[33].

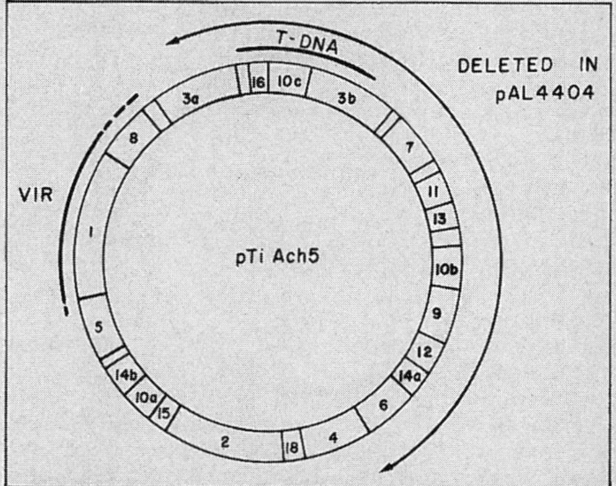

FIGURE 7 Sma I fragment map of pTi Ach5 indicating the part of the plasmid that is deleted in helper plasmid pAL4404. The fragment map of octopine Ti plasmid pTi Ach5 was adapted from Koekman et al.[50] pAL4404 derives from pTi Ach5 by deletion of the region indicated by the double arrow. It contains an insertion of Tn904 at the junction of the deletion ends (fragments 3a and 4). The VIR region is intact in pAL4404, while T-DNA[47] is deleted entirely. We observed no DNA homology between pAL4404 and our MINI-Ti/pRK plasmid (see text). Plasmid pAL4404 was kindly provided by Dr. Paul Hooykaas, University of Leiden, The Netherlands.

A. tumefaciens strain A114 is a derivative of the nopaline strain C58, cured of its Ti plasmid by growth at high temperature[4]. Strain A136 is a derivative of A114 bearing three chromosomal antibiotic resistance markers: resistance to rifampicin, nalidixic acid and chloramphenicol[4]. *A. tumefaciens* strain LBA4404, kindly provided to us by Dr. Paul Hooykaas (University of Leiden, The Netherlands), is an avirulent derivative of the octopine strain Ach5. It carries a deletion product of the wild-type Ti plasmid pTi Ach5, called pAL4404, that has lost the entire T-DNA and the octopine catabolism genes[34].

A. tumefaciens A208 is a transconjugant of strain A136 containing the nopaline Ti plasmid pTi T37[35].

Plasmids

pBR325(Eco RI-13) consists of Eco RI fragment 13 of pTi T37 T-DNA[36] cloned into the Eco RI site of pBR325. pRK290 was a gift of Dr. D. Helinski (University of California, San Diego). It is a derivative of the Inc Pl R-factor RK2, and confers resistance to tetracycline[37]. pRK290:EcoRI-1, constructed in our laboratory by Dr. A. Matzke, consists of Eco RI fragment 1 of pTi T37 cloned into the unique Eco RI site of pRK290. R751-pMG2 was provided by Dr. G. Jacoby (Massachusetts General Hospital, Boston) and is a recombinant plasmid derived from R751 and pMG2. It belongs to the Pl incompatibility group and confers resistance to gentamycin[38].

Media

A. tumefaciens cultures were grown at 29°C in YEP broth (1% yeast extract, 1% peptone, 0.5% NaCl) with vigorous agitation. Nutrient agar (Difco) plates served as solid medium for *A. tumefaciens* and were supplemented with kanamycin (50 μg/ml), gentamycin (50 μg/ml) or carbenicillin (100 μg/ml) as needed.

E. coli strains were cultivated in L-broth (1% tryptone, 1% NaCl, 0.5% yeast extract) or on L-agar plates (L-broth solidifed with 1.5% agar). The antibiotics ampicillin (100 μg/ml) and kanamycin (25 μg/ml) (Sigma) were added to agar immediately before pouring. Tetracycline (Sigma) (25 μg/ml) was dissolved in 95% ethanol before addition to melted agar.

Bacterial Transformation

E. coli transformation was performed according to the calcium chloride procedure of Cohen et al.[39]. The freeze-thaw method of Holsters et al.[40] was used to transform Agrobacterium strains.

Conjugal Crosses

Conjugation between *A. tumefaciens* recipient and *E. coli* strain J53-2(R751-pMG2) was performed as described by Barton and Chilton[22].

Plasmid DNA Isolation

E. coli plasmids were isolated from cleared lysates[41] and purified through isopycnic centrifugation in CsCl/ethidium bromide gradients. After removal of dye and CsCl, plasmid DNA was subjected to phenol deproteinization before dialysis against 0.01 M Tris, pH 8.0. Plasmid DNA was stored at 4°C. Recombinant plasmids in *E. coli* transformants were analyzed using a "mini-prep" procedure developed by Gill[42].

Ti plasmids were isolated from *A. tumefaciens* by the method of Currier and Nester[43], the lysate shearing step being omitted. Ti plasmid "mini-preps" were prepared using the alkaline lysis procedure of Casse et al.[44].

FIGURE 8 MINI-Ti/pRK is virulent when "helped" by pAL4404. Immature leaves of *Kalanchoe daigremontiana* were inoculated with LBA4404 (containing only the helper plasmid pAL4404) or LBA4404 (MINI-Ti/pRK). Only a scar persists on the leaf inoculated with LBA4404 (left); in contrast, LBA4404(MINI-Ti/pRK) induces a tumor at the site of inoculation (right). Similar results were observed with SR1 tobacco plants wounded by decapitation (see text), from which tumor tissue was excised and grown axenically in culture on hormone-free medium.

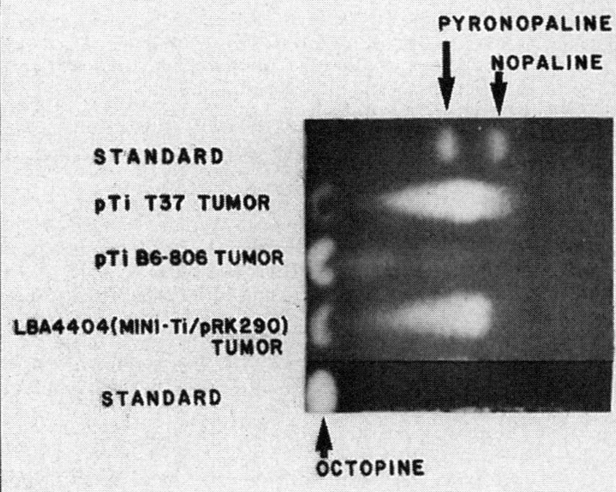

FIGURE 9 Tobacco tumor tissue incited by LBA4404(MINI-Ti/pRK) synthesizes nopaline. Tumor tissue extract (see Methods) corresponding to 50 mg of fresh tissue was electrophoresed at pH 4.0 on Whatman 3MM paper, dried and stained with the phenanthrene quinone reagent.[49] Interfering fluorescent material at the origin comigrates with the octopine standard at this pH. Tumor tissue incited by LBA4404(MINI-Ti/pRK) contains nopaline in amounts similar to tumor incited by a pTi T37-containing strain. The tumor tissue incited by an octopine-type *A. tumefaciens* strain containing pTi B6-806 produces no nopaline. Octopine production by this tumor was confirmed by paper chromatography in methyl ethyl ketone:2-butanol:t-butanol:diethyl amine:water (40:20:20:1:25), which resolves octopine from interfering fluorescent substances in the tumor extract.

Enzymatic Reactions

Restriction endonucleases were purchased from New England Biolabs or Bethesda Research Laboratories and were used as recommended by the supplier. For preparative digests, DNA was subjected to one phenol extraction and one chloroform extraction before ethanol precipitation. Ligations were carried out in 0.02 M Tris pH 7.5; 0.01 M $MgCl_2$; 0.05 M NaCl; 0.01 M DTT; 0.5 mM ATP; 100 μg/ml BSA and 0.05 unit of T4 DNA ligase per μg of DNA. For sticky end ligations, incubation was at 16°C for 12 hours with 100 μg/ml DNA. For blunt end ligations, incubation was at 4°C for 24–48 hours at a DNA concentration of 200–400 μg/ml. T4 DNA ligase was purchased from New England Biolabs.

Sticky ended DNA fragments were rendered blunt by addition of 1 unit of *E. coli* DNA polymerase I-large fragment (Klenow) per μg of DNA. Incubation was for 30 minutes at 37°C with four unlabeled deoxynucleotide triphosphates. BAl-31 resections employed 0.6 unit of exonuclease BAl-31 per μg of DNA, incubated at 30°C for 3 minutes at a DNA concentration of 34 μg/ml.

Nick translations employed a kit and α^{32}P-dCTP from Amersham.

Agarose Gel Electrophoresis

Vertical agarose (Sigma) gels (0.7% to 1%) in TB buffer (10.8 g Tris, 0.93 g disodium EDTA, 5.5 g boric acid per liter) were electrophoresed at 5 V/cm for 1–5 hours. Horizontal agarose gels were employed as described previously[35].

Southern Blots

Southern blots were prepared by the procedure of Southern[45] with the modifications of Wahl *et al.*[46] and were hybridized with labeled DNA as described by Thomashow *et al.*[47].

Virulence Assays

For *Kalanchoe daigremontiana* inoculations, bacterial colonies were picked from nutrient agar with a sterile toothpick and inoculated into immature leaves of young plants by wounding the leaf surface. Positive (virulent) and negative (avirulent) control strains were included in every assay series. A virulent strain induces a prominent cell proliferation (tumor) at the wound site after 10–15 days; an avirulent strain or uninoculated wound produces only a scar.

Nicotiana tabacum SRl plantlets (provided by Dr. Pàl Maliga) were grown in axenic conditions on Murashige and Skoog medium solidifed with 0.7% agar[48]. Plants were decapitated and inoculated with the bacterial strain at the wound site.

Nopaline and Octopine Assays

Opine was extracted from tumor tissue by macerating with an equal weight of 70% ethanol. Aliquots (100 μl) of the extract were applied to Whatman 3MM paper. The samples were subjected to electrophoresis at 45 volts per cm for 45 minutes in a 0.1 M acetate buffer (pH 4.0). Guanidines were stained with the phenanthrene quinone reagent of Yamada and Itano[49] and fluorescent nopaline and octopine spots were visualized under UV light. Intrinsically fluorescent compounds in the extract were noted before application of the stain.

ACKNOWLEDGMENTS

We thank Dr. Paul Hooykaas for providing strain LBA4404, without which this work would not have been possible. We thank Dr. Scott Chilton for assistance in the octopine and nopaline assays, Dr. A. J. M. Matzke for construction of the plasmid pRK290:Eco RI-1. We thank Dr. Ernest G. Jaworski, Monsanto Co., for suggesting this project. This research was supported by DOE contract AC02-81ER 10889. A. J. de Framond was supported by a French fellowship (D.G.R.S.T.) K. A. Barton was recipient of a fellowship from the Monsanto Company.

Received 15 March 1983; accepted 24 March 1983

This work was part of a doctoral thesis presented by AJdF at the National Institute of Agronomy of Paris-Grignon, January 1983.

REFERENCES

1. Smith, E. F. and Townsend, C. O. 1907. A plant tumor of bacterial origin. Science **25**:671–673.
2. De Cleene, M. and De Ley, J. 1976. The host range of crown gall. Bot. Rev. **42**:389–466.
3. Zaenen, I., Van Larebeke, N., Teuchy, H., Van Montagu, M., and Schell, J. 1974. Supercoiled circular DNA in crown gall-inducing *Agrobacterium* strains. J. Mol. Biol. **86**:109–127.
4. Watson, B., Currier, T. C., Gordon, M. P., Chilton, M.-D. and Nester, E. W. 1975. Plasmid required for virulence of *Agrobacterium tumefaciens*. J. Bacteriol. **123**:255–264.
5. Chilton, M.-D., Drummond, M. H., Merlo, D. J., Sciaky, D., Montoya, A. L., Gordon, M. P., and Nester, E. W. 1977. Stable incorporation of plasmid DNA into higher plant cells: the molecular basis of crown gall tumorigenesis. Cell **11**:263–271.
6. Chilton, M.-D., Saiki, R. K., Yadav, N., Gordon, M. P., and Quétier, F. 1980. T-DNA from *Agrobacterium* Ti plasmid is in the nuclear fraction of crown gall tumor cells. Proc. Natl. Acad. Sci. USA **77**:4060–4064.
7. Willmitzer, L., De Beuckeleer, M., Lemmers, M., Van Montagu, M., and Schell, J. 1980. DNA from Ti plasmid present in nucleus and absent from plastids of crown gall plant cells. Nature **287**:359–361.
8. Drummond, M. H., Gordon, M. P., Nester, E. W., and Chilton, M.-D. 1977. Foreign DNA of bacterial plasmid origin is transcribed in crown gall tumors. Nature **269**:535–536.
9. Gelvin, S. B., Gordon, M. P., Nester, E. W., and Aronson, A. A. 1981. Transcription of the *Agrobacterium* Ti plasmid in the bacterium and in crown gall tumors. Plasmid **6**:17–29.
10. Gurley, W. B., Kemp, J. D., Albert, M. J., Sutton, D. W., and Callis, J. 1979. Transcription of Ti plasmid-derived sequences in three octopine-type tumor lines. Proc. Natl. Acad. Sci. USA **76**:2828-2832.
11. Petit, A., Delhaye, S., Tempé, J., and Morel, G. 1970. Recherches sur les guanidines des tissues de crown gall. Mise en évidence d'une relation biochimique spécifique entre les souches d'*Agrobacterium tumefaciens* et les tumeurs qu'elles induisent. Physiol. Vég. **8**:205–213.
12. Bomhoff, G., Klapwijk, P. M., Kester, H. C. M., Schilperoort, R. A., Hernalsteens, J. P., and Schell, J. 1976. Octopine and nopaline synthesis and breakdown genetically controlled by a plasmid of *Agrobacterium tumefaciens*. Mol. Gen. Genet. **145**:177–181.
13. Schröder, J., Schröder, G., Huisman, H., Schilperoort, R. A., and Schell, J. 1981. The mRNA for lysopine dehydrogenase in plant tumor cells is complementary to a Ti plasmid fragment. FEBS. Lett. **129**:166–168.
14. Murai, N. and Kemp, J. D. 1982. Octopine synthase mRNA isolated from sunflower crown gall callus is homologous to the Ti plasmid of *Agrobacterium tumefaciens*. Proc. Natl. Acad. Sci. USA **79**:86–90.
15. Petit, A. and Tempé, J. 1978. Isolation of *Agrobacterium* Ti plasmid regulatory mutants. Mol. Gen. Genet. **167**:147–155.
16. Holsters, M., Silva, B., Van Vliet, F., Genetello, C., De Block, M., Dhaese, P., Depicker, A., Inze, D., Engler, G., Villarroel, R., Van Montagu, M., and Schell, J. 1980. The functional organization of the nopaline *A. tumefaciens* plasmid pTi C58. Plasmid **3**:212–230.
17. De Greve, H., Decraemer, H., Seurinck, J., Van Montagu, M., and Schell, J. 1981. The functional organization of the octopine *Agrobacterium tumefaciens* plasmid pTi B6S3. Plasmid **6**:235–248.
18. Hernalsteens, J. P., Van Vliet, F., De Beuckeleer, M., Depicker, A., Engler, G., Lemmers, M., Holsters, M., Van Montagu, M., and Schell, J. 1980. The *Agrobacterium tumefaciens* Ti plasmid as a host vector for introducing foreign DNA in plant cells. Nature **287**:654–656.
19. Leemans, J., Shaw, C., Deblaere, R., De Greve, H., Hernalsteens, J. P., Maes, M., Van Montagu, M., and Schell, J. 1981. Site-specific mutagenesis of *Agrobacterium* Ti plasmids and transfer of genes to plant cells. J. Mol. Appl. Genet. **1**:149–164.
20. Matzke, A. J. M. and Chilton, M.-D. 1981. Site-specific insertion of genes into T-DNA of the *Agrobacterium* Ti plasmid: an approach to genetic engineering of higher plant cells. J. Mol. Appl. Genet. **1**:39–49.
21. Barton, K. A., Binns, A. N., Matzke, A. J. M. and Chilton, M. D. 1983. Regeneration of intact tobacco plants containing full-length copies of genetically engineered T-DNA and transmission of T-DNA to R1 progeny. Cell, in press.
22. Barton, K. A. and Chilton, M.-D. 1982. *Agrobacterium* Ti plasmids as vectors for plant genetic engineering. Methods in Enzymology, in press.
23. Ruvkun, G. B. and Ausubel, F. M. 1981. A general method for site-directed mutagenesis in prokaryotes. Nature **289**:85–88.
24. Bevan, M. W. and Chilton, M.-D. 1982. T-DNA of the *Agrobacterium* Ti and Ri plasmids. Ann. Rev. Genet. **16**:357–384.
25. Yadav, N. S., Vanderleyden, J., Bennet, D., Barnes, W. M., and Chilton, M.-D. 1982. Short direct repeats flank the T-DNA on a nopaline Ti plasmid. Proc. Natl. Acad. Sci. USA **79**:6322–6326.
26. Koekman, B. P., Hooykaas, P. J. J., and Schilperoort, R. A. 1982. A functional map of the replicator region of the octopine Ti plasmid. Plasmid **7**:119–132.
27. Kingsbury, D. T. and Helinski, D. R. 1973. Temperature-sensitive mutants for the replication of plasmids in *Escherichia coli:* Requirement for deoxyribonucleic acid polymerase I in the replication of the plasmid Col E1. J. Bacteriol. **114**:1116–1124.
28. Engler, G., Depicker, A., Maenhaut, R., Villarroel-Mandiola, R., Van Montagu, M., and Schell, J. 1981. Physical mapping of DNA base sequence homologies between an octopine and a nopaline Ti plasmid of *Agrobacterium tumefaciens*. J. Mol. Biol. **152**:183–208.
29. Simpson, R., O'Hara, P., Lichtenstein, C., Montoya, A. L., Kwok, W., Gordon, M. P., and Nester, E. W. 1982. The DNA from A6S/2 tumor contains scrambled Ti plasmid sequence near its junction with plant DNA. Cell **29**:1005–1014.
30. Garfinkel, D. J., Simpson, R. B., Ream, L. W., White, F. F., Gordon, M. P., and Nester, E. W. 1981. Genetic analysis of crown gall: fine structure map of the T-DNA by site-directed mutagenesis. Cell **27**:143–153.
31. Leemans, J., Deblaere, R., Willmitzer, L., De Greve, H., Hernalsteens, J. P., Van Montagu, M. and Schell, J. 1982. Genetic identification of functions of T_L-DNA transcripts in octopine crown galls. The EMBO Journal **1**:147–152.
32. Murray, N. E., Brammer, W. J., and Murray, K. 1977. Lambdoid phages that simplify the recovery of *in vitro* recombinants. Mol. Gen. Genet. **150**:53–61.
33. Tacon, W. and Sherratt, D. 1976. Col E plasmid replication in DNA polymerase I deficient strains of *Escherichia coli*. Mol. Gen. Genet. **147**:331–335.
34. Ooms, G., Hooykaas, P. J. J., Van Veen, R. J. M., Van Beelen, P., Regensburg-Tuink, T. J. G. and Schilperoort, R. 1982. Octopine Ti plasmid deletion mutants of *Agrobacterium tumefaciens* with emphasis on the right side of the T-region. Plasmid **7**:15–29.
35. Sciaky, D., Montoya, A. L. and Chilton, M.-D. 1978. Fingerprints of *Agrobacterium* Ti plasmids. Plasmid **1**:238–253.
36. Yadav, N. S., Postle, K., Saiki, R. K., Thomashow, M. F. and Chilton, M.-D. 1980. T-DNA of a crown gall teratoma is covalently joined to host plant DNA. Nature **287**:458–461.
37. Ditta, G., Stanfield, S., Corbin, D., and Helinski, D. 1980. Broad host range DNA cloning system for gram-negative bacteria: construction of a gene bank of *Rhizobium meliloti*. Proc. Natl. Acad. Sci. USA **77**:7347–7351.
38. Jacoby, G. A., Jacob, A. E., and Hedges, R. W. 1976. Recombination between plasmids of incompatibility groups P-1 and P-2. J. Bacteriol. **127**:1278–1285.
39. Cohen, S. N., Chang, A. C. Y. and Hsu, L. 1972. Nonchromosomal antibiotic resistance in bacteria: genetic transformation of *Escherichia coli* by R-factor DNA. Proc. Natl.

Acad. Sci. USA **69**:2110–2114.
40. Holsters, M., de Waele, D., Depicker, A., Messens, E., Van Montagu, M., and Schell, J. 1978. Transfection and transformation of *Agrobacterium tumefaciens*. Mol. Gen. Genet. **163**:181–187.
41. Blair, D. G., Sherratt, D. J., Clewell, D. B., and Helinski, D. R. 1972. Isolation of supercoiled colicinogenic factor El DNA sensitive to ribonuclease and alkali. Proc. Natl. Acad. Sci. USA **69**:2518–2522.
42. Gill, R. Ph.D. Thesis, University of Washington, Seattle, Washington, 1980.
43. Currier, T. C. and Nester, E. W. 1976. Isolation of covalently closed circular DNA of high molecular weight from bacteria. Anal. Biochem. **76**:431–441.
44. Casse, F., Boucher, C., Julliot, J. S. Michel, M., and Dénarié, J. 1979. Identification and characterization of large plasmids in *Rhizobium meliloti* using agarose gel electrophoresis. J. Gen. Microbiol. **113**:229-242.
45. Southern, E. M. 1975. Detection of specific sequences among DNA fragments separated by gel electrophoresis. J. Mol. Biol. **98**:503–517.
46. Wahl, G. M., Stern, M., and Stark, G. R. 1979. Efficient transfer of large DNA fragments from agarose gels to diazobenzyloxymethyl paper and rapid hybridization by using dextran sulfate. Proc. Natl. Acad. Sci. USA **76**:3683–3687.
47. Thomashow, M. F., Nutter, R., Montoya, A. L., Gordon, M. P., and Nester, E. W. 1980. Integration and organization of Ti plasmid sequences in crown gall tumors. Cell **19**:729–739.
48. Murashige, T. and Skoog, F. 1962. A revised method for rapid growth and bioassays with tobacco tissue cultures. Physiol. Plant. **15**:473–497.
49. Yamada, S. and Itano, H. A. 1966. Phenanthrenequinone as an analytical reagent for arginine and other monosubstituted guanidines. Biochem. Biophys. Acta **130**:538–540.
50. Koekman, B. P., Ooms, G., Klapwijk, P. M., and Schilperoort, R. A. 1979. Genetic map of an octopine Ti plasmid. Plasmid **2**:347–357.

A binary plant vector strategy based on separation of *vir*- and T-region of the *Agrobacterium tumefaciens* Ti-plasmid

A. Hoekema, P. R. Hirsch*, P. J. J. Hooykaas & R. A. Schilperoort

Department of Biochemistry, University of Leiden, Wassenaarseweg 64, 2333 AL Leiden, The Netherlands

The soil bacterium *Agrobacterium tumefaciens* is a plant pathogen that causes crown-gall tumours after infection of wounded dicotyledonous plants. Large plasmids (Ti-plasmids) are responsible for the oncogenicity of the bacterium[1–3]. Crown-gall tumours contain a DNA segment, called the T-DNA, which is homologous with a defined part of the Ti-plasmid present in the tumour-inducing bacterium, and is stably integrated into the plant genome[4–7]. Apart from the T-DNA another region of the Ti-plasmid-called the *vir*-region, is essential for tumour induction[8–11]. We report here the interaction of two compatible plasmids, one containing the *vir*-region, the other carrying the T-DNA on a wide host-range replicon. An *A. tumefaciens* strain harbouring both plasmids has a normal tumour-inducing capacity, although neither plasmid is functional alone. With this approach, the T-DNA on one plasmid can, because of its size, be easily genetically manipulated using *Escherichia coli* as a host. Transfer of this plasmid into an *A. tumefaciens* strain harbouring the plasmid with the *vir*-region allows introduction of the manipulated T-DNA into plant cells. In this way, sophisticated binary vector systems for plant genetic engineering can be developed.

To obtain a plasmid that contains the entire T-DNA of the octopine Ti-plasmid pTiAch5 and that can replicate autonomously in *A. tumefaciens*, as well as in *E. coli*, we used the recombinant plasmid pOTY8. pOTY8 was constructed by inserting the T-DNA of pTiAch5 into the tetracycline resistance locus of pJDB207[12] (see Fig. 1), and also carries the ampicillin resistance gene of pAT153[13] and a *LEU2* gene as genetic markers. Details of the construction of pOTY8 will be described elsewhere (P. R. Hirsch, manuscript in preparation). As pOTY8 cannot replicate in *A. tumefaciens*, we converted it into a wide host-range plasmid by fusion with the IncP plasmid R772. To this end, R772 was introduced into *E. coli* strain HB101 (pOTY8) by conjugation, and transconjugants were used as donors in further crosses with the *A. tumefaciens* strain LBA202. Transconjugants of these crosses inheriting the ampicillin resistance marker of pOTY8 were expected to be carrying a co-integrate plasmid of pOTY8 and R772, as pOTY8 itself is non-conjugative and cannot replicate in *Agrobacterium*. The site of co-integration of R772 could be in either the vector or the T-DNA region of pOTY8.

To isolate a co-integrate with an intact T-DNA, we screened for a plasmid whose site of integration was the *LEU2* locus. Thirty of the co-integrate-carrying transconjugants were conjugated with the *E. coli* strain JA221[14] (C600 *trpE leuB*) and the offspring was tested for leucine auxotrophy. One of the 30 transconjugant strains failed to grow on a minimal medium without added leucine, and so presumably harboured an R772::pOTY8 cointegrate plasmid, whose site of co-integration has led to inactivation of the *LEU2* gene. Restriction endonuclease digestion patterns of this R772::pOTY8 plasmid (called pAL1050) were analysed in detail. Figure 1 presents evidence that plasmid pAL1050 had an insertion of R772 into the pJDB207 region of pOTY8, leaving the T-DNA region unaltered. The stucture of pAL1050 was further analysed by Southern hybridizations[15], using labelled plasmid DNAs of R772 and pOTY8 as probes (data not shown).

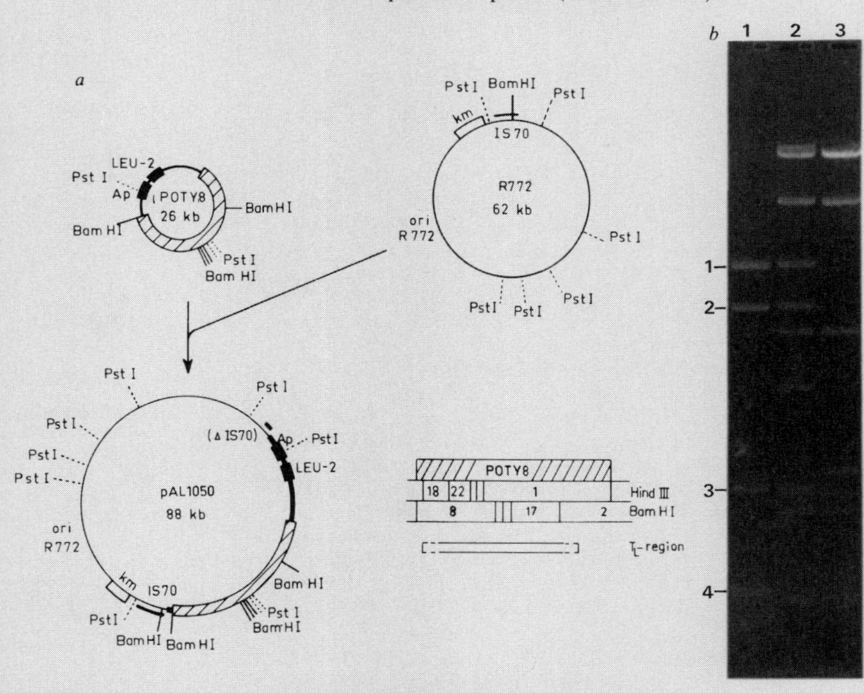

Fig. 1 *a*, The construction of pAL1050. The part of the T-DNA of pTiAch5 present on pOTY8 is shown in the bottom right-hand corner. pOTY8 was constructed from subclones of *Bam*HI fragment 8 and *Hind*III fragment 1, using pJDB 207 as a vector plasmid. For pAL1050, R772 and pOTY8, the restriction sites for *Bam*HI and *Pst*I are indicated. Analysis with many other restriction endonucleases revealed the presence of all T-DNA fragments of pOTY8 (data not shown). For R772 a physical map has been constructed with many enzymes[18]. As shown above, the insertion sequence present on R772, IS*70*, mediated co-integrate formation between pOTY8 and R772. At both junctions of R772 and pOTY8 a copy of IS*70* is present. Co-integrates formed between Ti-plasmids and R-plasmids usually dissociate after transfer from *A. tumefaciens* to *E. coli*[19]. This is due to recombination between the two IS elements bordering the fusion sites. Deletion of one of the copies of the IS generates stable co-integrates[18–20]. The co-integrate pAL1050 was shown to be perfectly stable in our experiments, which is explained by the fact that one of the copies of IS*70* was shown to be deleted (data not shown). *b*, An agarose gel showing the *Bam*HI+*Pst*I digest patterns of pOTY8 (lane 1), pAL1050 (lane 2) and R772 (lane 3). R772 has inserted into fragment 4 of pOTY8 to generate pAL1050. Clearly, fragment 1 of pOTY8, containing the *LEU2* locus itself, is unaltered, while the expression of *LEU2* is blocked on pAL1050. Presumably this inactivation is due to the polar effect of the integration of R772. The T-DNA part of this plasmid, however, is still intact.

*Present address: Cancer Research Campaign, Department of Biochemistry, Imperial College, London SW7 2AZ, UK.

0028-0836/83/190179—02$01.00

Table 1 Plant tumour-induction tests

Strain*	Plasmids	Tomato Tumour	Tomato OCS†	Kalanchoë Tumour	Kalanchoë OCS	Tobacco Tumour	Tobacco OCS	Pea Tumour	Pea OCS
LBA 4001	Cr, pTiAch5	+	+	+	+	+	+	+	+
LBA 4404	Cr, pAL 4404	–		–		–		–	
LBA 1050	Cr, pAL 1050	–		–		–		–	
LBA 4434	Cr, pAL 1050, pAL 4404	+	+	+	+	+	+	+	+

* The strains LBA 4001, 4404 and 4434 have Ach5 as chromosomal background. LBA 1050 has a C58-C9 chromosomal background. In all these strains a large cryptic plasmid (Cr) is present.
† OCS: synthesis of octopine in the tumour, detected as described by Otten and Schilperoort[23].

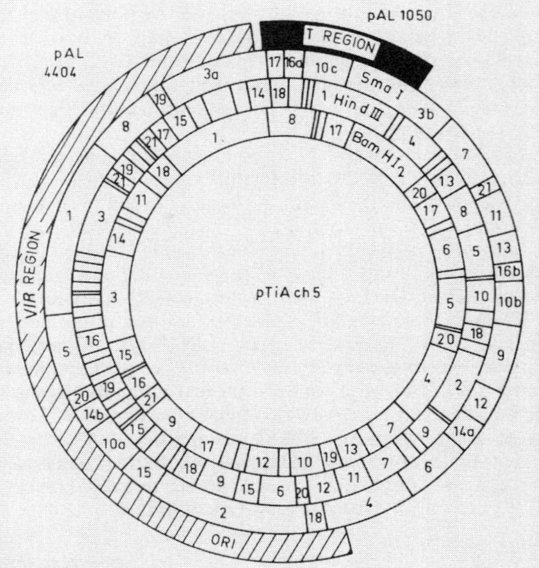

Fig. 2 The physical map of pTiAch5 is shown for the restriction endonucleases *Sma*I[21], *Bam*HI and *Hin*dIII[22]. In the hatched and shaded areas the Ti-plasmid segments are indicated that are present separately on pAL4404 and pAL1050.

The inability of pAL1050 to restore the tumour-inducing activity of a non-oncogenic *A. tumefaciens* (which lacks an endogenous Ti-plasmid) showed that the T-DNA region alone is insufficient for tumorigenicity (Table 1).

Subsequently, pAL1050 was conjugatively transferred into a different non-oncogenic, *A. tumefaciens* strain, LBA 4404[16] which contains a severely deleted Ti-plasmid, lacking the entire T-DNA region, but still with an intact *vir*-region (see Fig. 2). The tumorigenicity of the transconjugant strain LBA4434, carrying both the T-DNA-containing plasmid pAL1050 and the *vir*-region-containing plasmid pAL4404 as separate entities, was tested on several plant species. Strain LBA4434 induced normal tumours on all plants tested, and octopine was detected in these tumours (Table 1). These experiments show that the *vir*-region and the T-DNA of the octopine Ti-plasmid can be physically separated on different plasmids without affecting tumorigenicity of the bacterium. As no tumours were induced by *A. tumefaciens* carrying pAL1050 alone, genes on the *vir*-region must act in *trans* to effect transfer of the T-DNA to the plant cell.

One could argue that the tumorigenicity of the *A. tumefaciens* strain LBA4434 results from the formation of a co-integrate between pAL4404 and pAL1050 in a small part of the bacterial population. However, this is unlikely, for the following reasons. First, by Southern hybridization we have established that no homology exists between the two plasmids (data not shown), which excludes co-integrate formation by homologous recombination. Second, to study the possibility that co-integrates were formed by illegitimate recombination we crossed LBA4434 (pAL1050, pAL4404) with LBA4078, an erythromycin-resistant *A. tumefaciens* strain (no endogenous Ti-plasmid) as a recipient. Co-transfer of the non-conjugative plasmid pAL4404 with the IncP plasmid pAL1050 was not detected (frequency lower than 10^{-4}). Any co-integration between pAL1050 and pAL4404 must therefore be at a very low frequency, and so is unlikely to significantly contribute to tumour induction. In this regard, it should be kept in mind that infections of wound sites with mixtures of oncogenic and non-oncogenic *A. tumefaciens* strains in low ratios does not lead to tumour formation[17]. To verify this effect in our own tumour induction tests, we carried out a model experiment. We tested different mixtures of oncogenic and non-oncogenic bacteria, varying from a ratio of $1:1$ up to $1:10^4$. We found that in our system a ratio of $1:10^2$ resulted in a strongly attenuated tumour response, and at higher ratios tumour development was negligible. However, the size of the tumours induced by LBA4434 was the same as those induced by the wild-type strain Ach5. Thus it is very unlikely that tumour induction by LBA4434 is caused by a mixed population consisting mainly of non-oncogenic cells and a very limited number of cells containing a co-integrate plasmid.

Our results provide a new method for the introduction of genetic information into plant cells. One is no longer required to use an intact Ti-plasmid as a plant vector. A broad host-range plasmid containing T-DNA can be manipulated in *E. coli*, and subsequent introduction of such a plasmid into an *A. tumefaciens* strain already carrying a *vir*-region will bring about the transfer and integration of genetic material into plant cells.

This work was supported in part by the Netherlands Foundation of Chemical Research (SON) with financial aid from the Netherlands Organization for Advancement of Pure Scientific Research (ZWO). P.R.H. was sponsored by an EMBO short-term fellowship and the Cancer Research Campaign. We thank Drs J. Hille, R. J. M. van Veen and M. J. J. van Haaren for their critical reading of the manuscript, and Tonny Tuink for technical assistance during part of these studies.

Received 5 January; accepted 9 March 1983.

1. Van Larebeke, N. *et al. Nature* **252**, 169–170 (1974).
2. Zaenen, I., Van Larebeke, N., Teuchy, H., Van Montagu, M. & Schell, J. *J. molec. Biol.* **86**, 109–127 (1974).
3. Watson, B., Currier, T. C., Gordon, M. P., Chilton, M. D. & Nester, E. W. *J. Bact.* **123**, 255–264 (1975).
4. Chilton, M. D. *et al. Cell* **11**, 263–271 (1977).
5. Thomashow, M. F., Nutter, R., Montoya, A. L., Gordon, M. P. & Nester, E. W. *Cell* **19**, 729–739 (1980).
6. De Beuckeleer, M. *et al. Molec. gen. Genet.* **183**, 283–288 (1981).
7. Ooms, G. *et al. Cell* **30**, 589–597 (1982).
8. Ooms, G., Poulis, J. A. & Schilperoort, R. A. *J. Bact.* **144**, 82–91 (1980).
9. Garfinkel, D. J. & Nester, E. W. *J. Bact.* **144**, 732–743 (1980).
10. De Greve, H. *et al. Plasmid* **6**, 235–248 (1981).
11. Hille, J., Klasen, I. & Schilperoort, R. A. *Plasmid* **7**, 107–118 (1982).
12. Beggs, J. D. *Alfred Benson Symp.* **16**, 383–389 (1981).
13. Twigg, A. J. & Sheratt, D. *Nature* **283**, 216–218 (1980).
14. Beggs, J. D. *Nature* **275**, 104–109 (1978).
15. Southern, E. J. *J. molec. Biol.* **98**, 503–518 (1975).
16. Ooms, G., Hooykaas, P. J. J., Moolenaar, G. & Schilperoort, R. A. *Gene* **14**, 33–50 (1981).
17. Lippincott, B. B. & Lippincott, J. A. *J. Bact.* **97**, 620–628 (1969).
18. Hille, J., Van Kan, J., Klasen, I. & Schilperoort, R. A. *J. Bact.* (in the press).
19. De Picker, A., De Block, M., Inzé, D., Van Montagu, M. & Schell, J. *Gene* **10**, 329–338 (1980).
20. Hooykaas, P. J. J., Den Dulk-Ras, H. & Schilperoort, R. A. *Plasmid* **4**, 64–75 (1980).
21. Chilton, M. D. *et al. Plasmid* **1**, 254–259 (1978).
22. De Vos, G., De Beuckeleer, M., Van Montagu, M. & Schell, J. *Plasmid* **6**, 249–253 (1981).
23. Otten, L. A. B. M. & Schilperoort, R. A. *Biochim. biophys. Acta* **527**, 497–500 (1978).

THE FIRST TUMORLESS TRANSFORMED PLANT WAS TREATED WITH UTMOST CARE

Patricia C. Zambryski
Department of Plant and Microbial Biology
University of California, Berkeley, California, United States
E-mail: zambrysk@nature.berkeley.edu

Commentary on the paper:

Zambryski, P., Joos, H., Genetello, C., Leemans, J., Van Montagu, M., and Schell, J. 1983. Ti plasmid vector for the introduction of DNA into plant cells without alteration of their normal regeneration capacity. EMBO J. 2:2143–2150.

It was the summer of 1981. I had just left Howard Goodman's lab at UCSF, as Howard was moving his lab to Massachusetts General Hospital in Boston. Usually I spent summers in Gent to enable our UCSF-Gent collaboration. However, that summer I decided to stay long-term in Gent to immerse myself in the *Agrobacterium* system.

My research at that time predominantly revolved around defining and characterizing the boundaries of the T-DNA element in plants, a collaboration between myself, Kelly Kruger, and Howard Goodman at UCSF, with Marcelle Holsters, Ann Depicker, Marc Van Montagu and Jeff Schell of Gent. This work revealed that the boundaries of the T-DNA in plants were very precise. How this precision occurs was unveiled by comparing the nucleotide sequences of the ends of the T-DNA element on the Ti plasmid to the T-DNA ends in junctions with plant DNA. We were thrilled to find that the T-DNA element on the Ti plasmid was delimited by 25 bp direct repeats (Zambryski et al. 1980; Zambryski et al. 1982). At this time, it was known that transposons carried inverted DNA sequences at their ends critical for movement. Direct repeats at the ends of the T-DNA fit with the nature of a "mobile DNA element."

In 1981, while much of the Gent lab was abuzz with efforts to fine-map genes internal to the T-DNA, I still was focused on the T-DNA ends. In particular, I wished to test whether a Ti plasmid that was deleted for the internal region of the T-DNA, but that still carried the T-DNA borders, was capable of delivering the abbreviated T-DNA element to plants. This experiment would be a genetic test of the function of the T-DNA borders as well as whether the T-DNA encoded products were required for its own transfer.

Several students in Gent, including Dirk Inze, Henk Joos, and Eddy van Haute had developed genetic tools to enable mapping of genes on the Ti plasmid. Strains were created to facilitate conjugal transfer of plasmids from *E. coli* to *Agrobacterium:* once inside *Agrobacterium,* the plasmids were designed to force recombination (and insertion or deletion of DNA of interest) via homologous sequences, or to carry transposons to facilitate insertional mutagenesis. I was anxious to use this system to create a deletion of the entire internal T-DNA region.

How to make such a deletion? The basic idea was simple: design a plasmid to carry restriction fragments overlapping each of the T-DNA borders and then force a double recombination event to delete the internal T-DNA. Rather than create such an intermediate vector for this strategy, I looked over the T-DNA border clones that Marcelle Holsters and I had isolated from plant tumor DNA (Zambryski et al. 1980). One such "junction" clone did not contain adjacent plant DNA sequences; instead it contained a restriction fragment that bracketed the left and right ends of two directly repeated T-DNA elements in the plant genome. This "junction" clone carried 2 kbp just internal to the right T-DNA border immediately adjacent to 3 kbp just internal to the left T-DNA border. I was discussing using this clone with Henk Joos, and he eagerly demonstrated how the genetics would work. We became very excited, as this strategy would force a deletion larger than any attempted to date, and it would make a perfect test of T-DNA border function.

I cannot ignore that there was additional interest in such a T-DNA deletion mutant. Besides testing the functionality of the T-DNA borders, a Ti plasmid without internal T-DNA genes should not cause a tumorous phenotype. Thus, such a plasmid would be an excellent test of the idea to use *Agrobacterium* as a useful vector to transform plants without interfering with normal plant growth and development. Henk Joos and

I worked hard to make and confirm a double recombinant with the expected genetic organization of a deleted T-DNA region. We then asked Chris Genetello, the lab plant culture expert, to apply our T-DNA deletion strain to decapitated tobacco stems. Wild type *Agrobacterium* carrying a wild type Ti plasmid will form a robust tumor within a few weeks (as we now know the internal region of the T-DNA element carries genes to specify auxin and cytokinin synthesis that are expressed in transformed cells resulting in an undifferentiated, tumorous, phenotype). As expected, our strain did not induce any growth.

However, we reasoned that our strain, dubbed 3850 (the next number in the existing strain collection) would have still transformed the plant cells. Since the 2 kbp right border region used to force the right side recombination event carried the nopaline synthase gene, we could confirm that T-DNA transfer had occurred by a simple assay for nopaline in plant tissues. Next Chris Genetello removed a thin layer (less than 1 mm) of cells at the cut surface where we had applied our 3850 strain, and coaxed these cells to multiply using hormones in tissue culture. Finally, after much anxious waiting and constant looking into the plant growth chamber, we had enough tissue to do a nopaline assay. We took different parts of the tissue, and indeed found that some tissue contained nopaline! Chris then coaxed the nopaline containing callus tissue to regenerate, and we assayed regenerated shoots for nopaline. The nopaline positive shoots made roots and ultimately an entire plant. I recall that this first 3850 transformed plant was treated with utmost care. After removal from tissue culture, it sat attentively on my bench under a beaker where I scrupulously monitored that it did not wilt or dry up.

When this 3850 transformed plant had passed the test of time and health, Jeff Schell brought it to Cologne, where he filled several entire greenhouses with its progeny! These progeny were the final proof we needed to publish our results. The greenhouses were full of 3850 transformed progeny plants that were absolutely identical: each plant was entirely normal in phenotype, and each passed on the 3850 T-DNA element from generation to generation. Thus, *Agrobacterium* had evolved (via scientific intervention) from agent of tumorous growth, to a suitable vector capable of transferring DNA to plants without interfering with their growth and development.

During this time there was much excitement about this 3850 strain. Besides the Gent group, Jan Lemeens, of Marc Van Montagu's lab at the Vrije Universiteit Brussel, became the spokesperson for the potential applications of 3850. It may be hard to imagine this state given that genetic engineering of plants via *Agrobacterium* is now so routine, and one can order up numerous vectors for transformation according to one's specific needs.

In the above description, I neglected to mention one extremely useful attribute of the 3850 derived Ti plasmid. The "junction" clone that was used to force a deletion of the internal T-DNA was carried on an *E. coli* plasmid, pBR322. This plasma has an *E. coli* origin of replication, an ampicillin resistance gene, and a useful cloning site in the tetracycline resistance gene. That pBR322 could replicate in *E. coli* but not *Agrobacterium* was critical in forcing recombination of the junction clone with the Ti plasmid, i.e., selection of ampicillin resistant *Agrobacterium* implied that the junction clone had recombined at least once with the Ti plasmid. The double recombinant was selected by its sensitivity to a different antibiotic resistance gene present in the internal region of the T-DNA. So it came to be that the internal T-DNA region was replaced by plasmid sequences from pBR322. The presence of pBR322 sequences in between the T-DNA border regions provides two useful properties. First, ampicillin resistance provides a genetic marker for the presence of the pGV3850 plasmid. Even more beneficial was the fact that these pBR322 sequences would serve to insert any DNA contained on a related *E. coli* plasmid by simple homologous recombination!

At this time there were few plant genes available to test the vector potential of 3850 to deliver "any DNA of interest" to plant cells. However, in another part of the Gent lab, Luis Herrera-Estrella and colleagues used the transcriptional control sequences from the nopaline or octopine synthase genes (the 5′ promoter and 3′ regions) to test whether they enabled plant specific expression of bacterial kanamycin resistance (Km^R) coding sequences. Indeed, they successfully obtained Km^R calli (reviewed in this anthology). With this result in hand, the engineered plant specifc Km^R gene was cloned into a pBR322 plasmid derivative that also carried a bacterial Km^R gene. After mobilization of this plasmid from *E. coli* to *Agrobacterium* carrying pGV3850, a simple selection for bacterial Km^R was evidence for co-integration between the incoming plasmid and pGV3850. Growing *Agrobacterium* in the presence of kanamycin ensured that the "selectable marker gene of interest" was maintained as a cointegrate with pGV3850. This proof of principle that the Km^R plant specific marker gene could be transferred to plants via the pGV3850 vector was performed as a collaboration between myself, Luis Herrera-Estrella, and Marc de Block (De Block et al. 1984).

Note: This first pGV3850::NOS-Km co-integrate has had a long and useful life thanks to Ken Feldmann. Ken used *Agrobacterium* carrying this Ti plasmid derivative to generate thousands of independent T-DNA insertions in *Arabidopsis* that are a resource of T-DNA tagged mutant lines (reviewed in this anthology).

Thus a straightforward idea to perform a test for T-DNA border function led to the development of an easy to use and efficient *Agrobacterium* vector to transform plants with any gene of interest. The gene of interest was simply cloned in a pBR-derived plasmid that also carried a selectable marker other than ampicillin resistance. Selection for co-integration with 3850 then placed the gene of interest between the T-DNA borders and ready for transfer to plants!

The above historical account summarizes events in Gent related to the 3850 vector. However, in other *Agrobacterium* labs around the world, numerous other scientists were also fervently designing other useful vectors (reviewed in this anthology).

My interest in T-DNA borders has continued for many, many years, first in Gent with Ken Wang (showing right T-DNA border directed polar transfer) and during a period of remarkable discovery with Scott Stachel (such as T-strands and border nicks) and then in Berkeley with Vitaly Citovsky, Liz Howard and John Zupan. Now my interest lies in figuring out how the T-strand and its associated proteins exit the bacterial cell via a specific transport channel, called the type IV secretion apparatus. How the T-DNA moves has kept us busy for 25 years and I expect more surprises and fundamental insights are yet to come.

References

De Block, M., Herrera-Estrella, L., Van Montagu, M., Schell, J., and Zambryski, P. 1984. Expression of foreign genes in regenerated plants and in their progeny. EMBO J. 3:1681–1689.

Zambryski, P., Depicker, A., Kruger, K., and Goodman, H. 1982. Tumor induction by *Agrobacterium tumefaciens:* Analysis of the boundaries of T-DNA. J. Mol. Appl. Genet. 1:361–370.

Zambryski, P., Holsters, M., Kruger, K., Depicker, A., Schell, J., Van Montagu, M., and Goodman, H. M. 1980. Tumor DNA structure in plant cells transformed by *A. tumefaciens.* Science 209:1385–1391.

PATRICIA C. ZAMBRYSKI: In 1969, I graduated B.Sc. Honours Genetics from McGill University, Quebec, Canada and in 1974 I graduated Ph.D Molecular Biology from the University of Colorado, Boulder, Colorado, USA. From 1974–1981, I spent a Postdoc at the Department of Biochemistry and Molecular Biology, University of California, San Francisco, California, U.S.A., and from 1981–1986, I was a Senior Research Associate in the Department of Biochemistry, Genetics and Molecular Biology, University of Gent, Gent, Belgium.

I have been blessed with superb mentors. Etta Kafer, fungal geneticist at McGill University, introduced me to scientific research. Working in her lab in the summers and during the school years made me realize that I wanted to continue research as a graduate student.

Larry Gold, Ph. D. advisor, studying bacteriophage T4 development, at the University of Colorado. Larry was the most amazing scientist I have ever met, constantly throwing around ideas and insights for the next experiment. I learned a tremendous amount from him about how to think about biological questions, and also that one can do multiple experiments in a single day. It was Larry who suggested I work with Howard Goodman as a postdoctoral advisor.

Howard Goodman, postdoctoral advisor at UCSF, with too many interesting research topics to mention in a few words. I started working on SV40 virus as I was interested in how animal tumors form. I also worked for a time on yeast tRNA splicing, just after Howard discovered an intron in a tRNA gene. It is hard to remember the time when eukaryotic genes were thought to be like prokaryotic genes; however, introns were discovered in 1978! It was Howard who sent me to Gent when he realized I wanted to explore research in a new biological system, one that was just beginning to be studied.

Of course, Marc Van Montagu and Jeff Schell introduced me to *Agrobacterium,* and my scientific curiosity about this system has been unrelenting. I especially thank Marc for letting me spend so many years in his lab with its assortment of scientists with diverse expertise and ethnicity. It was during this time, that I was given the opportunity to mentor students and earn the confidence to apply for an independent position in the US. It was a privilege to be welcomed in their labs and families during my stay in Gent, and since then to have their continuous support and friendship.

The EMBO Journal Vol.2 No.12 pp.2143–2150, 1983

Ti plasmid vector for the introduction of DNA into plant cells without alteration of their normal regeneration capacity

P. Zambryski[1], H. Joos[1], C. Genetello[1], J. Leemans[2,4], M. Van Montagu[1,2]* and J. Schell[1,3]*

[1]Laboratorium voor Genetica, Rijksuniversiteit Gent, B-9000 Gent, [2]Laboratorium voor Genetische Virologie, Vrije Universiteit Brussel, B-1640 St.-Genesius-Rode, Belgium, and [3]Max-Planck-Institut für Züchtungsforschung, D-5000 Köln 30, FRG

Communicated by M. Van Montagu
Received on 22 July 1983

A Ti plasmid mutant was constructed in which all the oncogenic functions of the T-DNA have been deleted and replaced by pBR322. This Ti plasmid, pGV3850, still mediates efficient transfer and stabilization of its truncated T-DNA into infected plant cells. Moreover, integration and expression of this minimal T-DNA in plant cells does not interfere with normal plant cell differentiation. A DNA fragment cloned in a pBR vector can be inserted in the pGV3850 T-region upon a single recombination event through the pBR322 region of pGV3850 producing a co-integrate useful for the transformation of plant cells. Based upon these properties, pGV3850 is proposed as an extremely versatile vector for the introduction of any DNA of interest into plant cells.
Key words: Ti plasmid/vector/plant cell transformation

Introduction

The Ti plasmid of *Agrobacterium tumefaciens* has long been recognized as a natural vector for the transfer of DNA to plant cells. *Agrobacterium* harbouring a Ti plasmid causes neoplastic transformation, called crown gall, of the wounded tissue of a wide range of dicotyledonous plants (for reviews, see Nester and Kosuge, 1981; Bevan and Chilton, 1982; Kahl and Schell, 1982; Zambryski *et al.*, 1983). Crown gall tissues synthesize novel metabolites called opines, and the Ti plasmids are classified according to the type of opine they specify. The most commonly used Ti plasmids are those which code for the metabolism of nopaline or octopine. The molecular basis of crown gall transformation is the transfer and stable integration of a well-defined T-DNA (transferred DNA) fragment of the Ti plasmid into the plant cell genome. The mechanism of DNA transfer as well as the functions encoded by this DNA have only begun to be understood. However, we can already begin to take advantage of this system to modify the Ti plasmid for application to the genetic engineering of plants.

Knowledge of the T-DNA sequences which define the borders of the transferred DNA is a basic requirement for the use of the Ti plasmid as a vector for DNA transfer to plant cells. The borders of the T-DNAs in nopaline (Lemmers *et al.*, 1980; Yadav *et al.*, 1980,1982; Zambryski *et al.*, 1980,1982) and octopine (Thomashow *et al.*, 1980a,1980b; De Beuckeleer *et al.*, 1981; Simpson *et al.*, 1982; Holsters *et al.*, 1983) tumor lines have been accurately determined. In octopine tumors a 13.6-kb TL-DNA is always present, and an additional 6–7 kb TR-DNA derived from an adjacent region of the Ti plasmid is sometimes also present. The nopaline T-DNA is a continuous stretch, ~23 kb of the entire 200-kb plasmid. The ends of this T-DNA as found in the plant genome are very precise; for example, the right end of the T-DNA junction of four borders varies over only 1 bp (Zambryski *et al.*, 1982), and the left junction of four borders varies over 90 bp (Zambryski *et al.*, 1982; Yadav *et al.*, 1982). The precision of the T-DNA integration allows the insertion of foreign DNA within the borders of the T-DNA and ensures its transfer to the plant cell genome (Hernalsteens *et al.*, 1980; Holsters *et al.*, 1982).

The second most important criterion, if one expects to utilize the Ti plasmid as a vector system, is that the transformed plant cells can differentiate in a normal manner rather than grow in a tumorous fashion. To obtain normal differentiating cells after T-DNA transfer requires knowledge of the functions encoded by the T-DNA which prevent this normal differentiation. Thus, the genes coding for these functions have been localized to specific regions of the T-DNAs of both octopine and nopaline Ti plasmids by intense genetic analysis (Holsters *et al.*, 1980; Garfinkel and Nester, 1980; Garfinkel *et al.*, 1981; Ooms *et al.*, 1980,1981; De Greve *et al.*, 1981; Leemans *et al.*, 1982; Willmitzer *et al.*, 1982,1983; Joos *et al.*, 1983; Ream *et al.*, 1983).

One of the results of these genetic studies was the isolation of a class of mutants (*shi*) which induced shoots containing either untransformed plant cells or cells transformed with a shortened T-DNA segment which no longer contained any oncogenic functions. The latter transformed shoots regenerated into normal and fertile plants which even transmitted T-DNA-specific sequences through meiosis as a single dominant Mendelian factor (Otten *et al.*, 1981; De Greve *et al.*, 1982). While these results were important in establishing that normal and fertile plants could be produced following Ti plasmid-mediated transformation, there is an aspect of uncertainty in these studies as there is no information about how the 'transformed' shoots containing the shortened T-DNA occur. Either a T-DNA deletion has occurred subsequent to T-DNA integration, or there has been an independent insertion of a shorter T-DNA segment. Furthermore, the isolation of these 'transformed' shoots is a low frequency event (1/300).

Another class of Ti plasmids (*roi*) which induce root formation have been shown recently to also transform plant cells and allow normal plant differentiation (Barton *et al.*, 1983); and in this case, the entire T-DNA was even transmitted to progeny plants. However, it is not known why this particular T-DNA does not interfere with regeneration and whether or not its non-oncogenic phenotype in infected plant cells will be reproducible in all conditions of plant cell growth.

Based on these findings, we tried systematically to design a modified T-region of the Ti plasmid which should have the following properties: (i) T-DNA border regions; (ii) no genes which prevent normal differentiation of transformed plant cells; and (iii) a genetic marker to monitor the presence of the T-DNA. Using the nopaline Ti plasmid (pTiC58), we have constructed a deletion which removes all the T-DNA except

[4]Present address: Centro de investigaciones sobre fijacion de nitrogeno, Ap. postal 565-A, Cuernavaca, Morelos, Mexico.

*To whom reprint requests should be sent.

for the borders and which also contains the nopaline synthase gene, a T-DNA-specific marker. In addition, this Ti plasmid contains the widely used cloning vehicle pBR322 between the borders of the mutated T-DNA: this makes this Ti plasmid an extremely versatile acceptor for the introduction of any foreign gene contained in a pBR-like plasmid. The construction of this Ti plasmid vector and its properties are described below.

Results

Construction of Ti plasmid vector pGV3850

Figure 1 shows the construction of the T-DNA deletion mutant pGV3850. This was possible due to the existence of a clone pAcgB which only contained the left and right border regions of the nopaline T-DNA. This clone has been previously described (Zambryski *et al.*, 1980,1982); it was obtained by reisolating portions of the T-DNA from transformed tobacco DNA. pAcgB contains the junction of two T-DNA copies which are arranged in tandem so that it contains the left and right borders of the T-DNA. In addition, pAcgB contains the nopaline synthase gene which maps immediately adjacent to the right T-DNA border (Holsters *et al.*, 1980; Depicker *et al.*, 1982). An interesting application of this clone is to use it to create a deletion of the internal oncogenic portion of the T-DNA.

To select for the genetic events which lead to the formation of the T-DNA deletion, we used an acceptor Ti plasmid pGV3839 (Joos *et al.*, 1983) which contains the kanamycin-resistance (Km^R) gene near the centre of its T-DNA (Figure 1A). Thus, pAcgB which carries a ColE1-specific *bom* site in its pBR322 portion is directly mobilized to an *Agrobacterium* strain containing pGV3839 using helper plasmids as described recently (Van Haute *et al.*, 1983). The ampicillin resistance (Ap^R) of pBR322 is used to select for the first single cross-over event with the nopaline Ti plasmid. As pBR322 cannot replicate in *Agrobacterium*, the only way that the ampicillin resistance can be stabilized in *Agrobacterium* is by a cross-over event upon homologous recombination with pGV3839 through one of the homology regions near the T-DNA borders. By a second cross-over event through the other homology region at the other border, the central portion of the T-DNA of pGV3839 including the Km^R gene is replaced by the pBR322 sequences of the clone pAcgB. The DNA molecule containing the central portion of the T-DNA including the Km^R marker is lost as it cannot replicate. Second recombinants are thus ampicillin-resistant and kanamycin-sensitive (Figure 1B). These recombination events were verified by restriction enzyme analysis and Southern blot hybridization (data not shown), and the detailed restriction map of the modified T-DNA of pGV3850 is shown in Figure 1C.

Phenotype of pGV3850 after infection of plant tissue

We tested whether *Agrobacterium* carrying pGV3850 could still transfer its T-DNA to plant cells in the absence of the internal T-DNA functions. All the DNA which encodes RNA transcripts involved in the production of the undifferentiated phenotype as well as additional sequences which encode transcripts of as yet unknown function (Joos *et al.*, 1983; Willmitzer *et al.*, 1983) are deleted in pGV3850. Only the DNA encoding nopaline synthase (*nos*) and a transcript of unknown function which maps in *Hin*dIII fragment 10 remain. While it has been shown that removing one or more transcripts does not interfere with T-DNA transfer, an

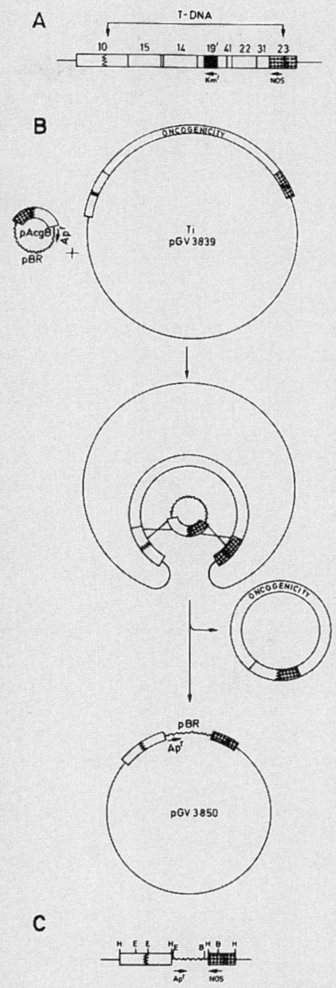

Fig. 1. Construction of Ti plasmid vector pGV3850. (**A**) The T-DNA of nopaline Ti plasmid pGV3839 (Joos *et al.*, 1983). The numbers refer to the *Hin*dIII fragments of the C58 nopaline Ti plasmid (Depicker *et al.*, 1980) and the vertical bars represent the *Hin*dIII restriction sites. One of the *Hin*dIII fragments labelled 19′ (normally fragment 19 in the wild-type map) contains a substitution of the nopaline Ti plasmid *Sma*I fragment 24 (Depicker *et al.*, 1980) for the *Hin*dII fragment of pKC7 (Rao and Rogers, 1979) which contains the Km^R gene: this deletion substitution is indicated by the black area. The two *Hin*dIII fragments 10 and 23 contain the T-DNA borders indicated by the jagged lines. In addition, the map position of the *nos* gene in *Hin*dIII fragment 23 is indicated, sequences homologous to *Hin*dIII-23 are indicated by the stippled regions throughout the figure. (**B**) The double cross-over events between pGV3839 and pAcgB leading to the formation of pGV3850. pAcgB contains only a portion of the left and right *Hin*dIII T-DNA border fragments 10 and 23 joined together at the T-DNA borders as indicated by the jagged line; these DNA sequences were cloned as a *Hin*dIII fragment in pBR322 as described (Zambryski *et al.* 1980,1982). The pBR322 sequences are indicated as a wavy line. The orientation of the DNA sequences in pAcgB and in pGV3839 are from left to right as they normally occur in the Ti plasmid. The T-DNA portion internal to the *Hin*dIII border fragments 10 and 23 is indicated as 'oncogenicity'. The double cross-over event results in the production of pGV3850 and a DNA circle containing the T-DNA of pGV3839 which is lost as it is not a replicon. Again, the T-DNA borders are indicated by jagged lines. The orientation of the two border fragments in pAcgB is: left border fragment 10 – pBR322 – right fragment 23; this results in the deletion of the internal portion of the T-DNA region and its replacement with pBR322. (**C**) A detailed restriction map of the modified T-DNA of pGV3850. The abbrevations used are: Ap^r, ampicillin resistance; B, *Bam*HI; E, *Eco*RI; H, *Hin*dIII; NOS, nopaline synthase.

analysis of the effect of a complete deletion of the internal portion of the nopaline T-DNA had not yet been tested. We assayed for the presence of nopaline in pGV3850-infected plant tissues as evidence that T-DNA transfer had occurred. Potato and carrot discs, as well as tobacco and petunia plantlets inoculated with *Agrobacterium* containing pGV3850 all produced nopaline-positive tissue (data not shown). These results agree with previous ones using large T-DNA deletions of both the nopaline and octopine Ti plasmids which also produce opine-positive tissue (Leemans *et al.*, 1982; Joos *et al.*, 1983). They provide conclusive evidence that a deletion of the entire internal portion of the nopaline T-DNA does not affect the transfer of the remaining border regions of the mutant T-DNA. Furthermore, all these data suggest that it is unlikely that the T-DNA itself encodes products involved in T-DNA transfer, integration and stabilization; these functions are likely encoded elsewhere, either by the Ti plasmid, other *Agrobacterium* DNA, or by the plant genome.

Nopaline-positive carrot and potato tissues were obtained when the discs were incubated on solid Murashige and Skoog (MS) medium (Murashige and Skoog, 1962) in the presence of growth-regulating substances, either auxin alone [naphthalene acetic acid (NAA) (1 mg/l), or auxin and cytokinin (NAA, 1 mg/l; benzylaminopurine (BAP), 0.2 mg/l]. Without added hormones, there is no response of the potato disc to infection with pGV3850. However, small callus growths are produced when the disc is infected with pGV3850 in the presence of hormones. This callus tissue is nopaline-positive. It is interesting that the callus growth in the presence of hormones is in part due to the presence of the Ti plasmid as this response does not occur with an *Agrobacterium* strain cured of its Ti plasmid. This is evidence that there may be regions outside the T-DNA encoding products which affect plant cell growth.

Nopaline-positive tobacco and petunia tissues were also obtained by analyzing the response of decapitated plantlets to infection with *Agrobacterium* harbouring pGV3850. Tumors are not produced as expected; however, tiny calli are observed at the wound site 4–8 weeks after infection. This latter effect is never seen when *Agrobacterium* cured of its Ti plasmid is used to infect plants; thus, this response may be due to some functions outside the T-DNA region of the Ti plasmid as discussed above for the potato response. This slight though significant response to pGV3850 allowed us to obtain transformed tissue and to further analyze the phenotype when the T-DNA of pGV3850 is stably integrated into the plant cell genome.

Regeneration of normal plants containing the non-oncogenic T-DNA of pGV3850

As pGV3850-transformed plant tissues are not tumorous in their growth pattern, we tested whether these transformed cells could regenerate whole plants which still contain the minimal T-DNA of pGV3850. This result would be particularly useful since foreign DNAs of interest could be inserted within the T-DNA of pGV3850 in order to transform plant cells, and subsequently produce normal plants containing these foreign DNAs. Tobacco plantlets infected with pGV3850 were used as a model system.

Four to eight weeks after infection of decapitated tobacco stems with *Agrobacterium* carrying pGV3850, tiny calli develop at the wound surface. This wound surface is removed from the plantlet and incubated further on Linsmaier and Skoog (LS) medium (Linsmaier and Skoog, 1965) containing

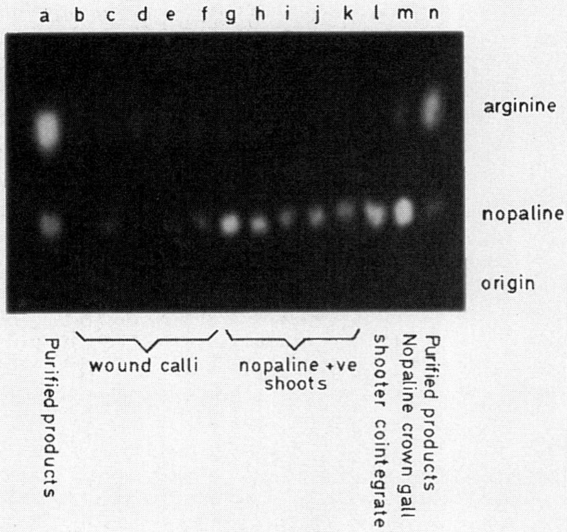

Fig. 2. Nopaline tests of pGV3850-infected tobacco tissues. Results of nopaline tests are shown of plant tissues isolated upon infection with the *Agrobacterium* strain C58C1 (pGV3850). The samples (also 2 μl) in **a** and **n** were derived from a control solution containing arginine and nopaline each at 0.5 mg/ml. Samples **b**–**f** are derived from callus grown on LS medium (Linsmaier and Skoog, 1965) with auxin (NAA, 1 mg/l) and cytokinin (BAP, 0.2 mg/l) which was derived from the wound surface of decapitated tobacco plantlets infected with *Agrobacterium* containing pGV3850. Samples **g**–**k** were derived from differentiating shoots obtained after nopaline-positive pGV3850-infected wound callus was placed on LS regeneration medium containing 1 mg/l BAP. Sample **l** is from callus tissue growing on hormone-free medium obtained from infection with *Agrobacterium* carrying a co-integrate between pGV3850 and the oncogenic genes of the octopine TL-DNA (those contained on *Eco*RI fragment 7 (De Vos *et al.*, 1981) of the octopine B6S3 Ti plasmid) which were cloned in a plasmid containing pBR322 sequences for homologous recombination with pGV3850 (J. Leemans, R. Deblaere, unpublished results). Sample **m** was derived from tumor callus produced by infection with *Agrobacterium* containing wild-type Ti plasmid C58.

auxin (NAA, 1 mg/l) and cytokinin (BAP, 0.2 mg/l). After further growth, a portion of the growing callus was tested for *nos* activity and a clear, but weak response was obtained (Figure 2d). This result is probably due to the presence of both transformed and untransformed cells in the callus tissue since there is no selection for transformed cells. A portion of this *nos*-positive tissue was transferred to shoot regeneration medium (LS medium plus BAP, 1 mg/ml). Subsequently, the shoots (Figure 3A) were transferred to hormone-free medium; all grew in a normal fashion and developed roots. Finally, the shoots were tested for *nos* activity: 11 out of 130 were positive for nopaline (see for example, Figure 2g, h, i, j, k). All 11 showed a strong *nos* signal indicating that the shoots are probably composed entirely of transformed cells. These *nos*-positive shoots have been grown further into whole plants and one is displayed in Figure 3B.

The *nos* activity is a useful marker to assay for the presence of pGV3850-transformed cells. Figure 2 shows the *nos* tests for the tissue shown in Figure 3, and also additional pGV3850-transformed tissues. For example, Figure 2b, c, d, e and f show the amount of nopaline present in various primary calli obtained from the wound surface of different tobacco plantlets and which were subsequently grown on solid agar medium containing auxin and cytokinin. These different calli produce different amounts of nopaline, from zero (Figure 2b) to significant amounts (Figure 2f). Furthermore,

Fig. 3. Regeneration of whole plants containing the shortened T-DNA of pGV3850. (**A**) The appearance of normal shoots which begin to form 4 weeks after nopaline-positive wound callus (obtained following infection of tobacco stems with *Agrobacterium* containing pGV3850) is placed on regeneration medium (LS medium plus 1 mg/l BAP). When these shoots are 1 cm long they are removed from the primary callus and placed on LS medium without hormones to allow the formation of roots. When the shoots have grown sufficiently to allow a portion to be removed, a nopaline test is performed. *Nos*-positive shoots are removed to larger pots to allow further growth of the transformed plants. A *nos*-positive pGV3850-transformed plant is shown in **B**.

the amount of nopaline present in the original callus is directly related to the frequency of occurrence of *nos*-positive shoots when the calli are placed on regeneration medium. For example, the calli in Figure 2d and 2f gave rise to *nos*-positive shoots at a frequency of 9 and 78%, respectively. However, not all inoculated plantlets produce *nos*-positive wound callus. In three separate experiments we have found that only 25% of the plants respond: in the first experiment 1 in 4, in the second experiment 2 in 8, and in the third 2 in 7. Nevertheless, even at this lowered frequency it is very easy to detect transformed tissue by a simple test for the presence of nopaline.

We have regenerated >50 plantlets from several independent infection experiments; in all cases the plantlets appear completely normal as in the sample shown in Figure 3B. In addition, these regenerated plants respond as wild-type tobacco plants to superinfection with *Agrobacterium* strains carrying wild-type octopine or nopaline Ti plasmids. This effect may be important for the genetic engineering of plants which already contain a pGV3850-related T-DNA with a gene of interest within the pBR portion of pGV3850 (see below); these plants could then be superinfected with an *Agrobacterium* containing pGV3850 plus another different gene of interest.

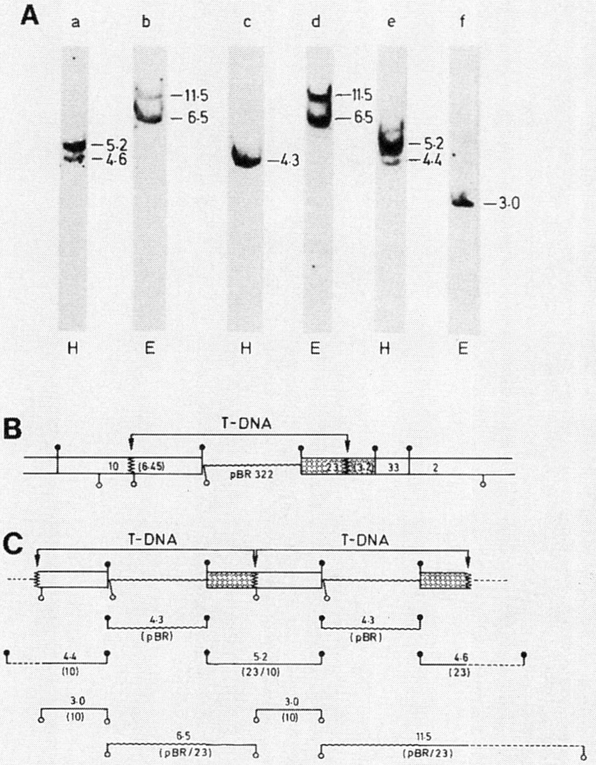

Fig. 4. Analysis of the T-DNA structure in pGV3850-transformed tobacco plants. Total DNA was prepared from the nopaline-positive pGV3850-transformed plant shown in Figure 3B. The purified DNA was digested with either restriction endonuclease *Hin*dIII or *Eco*RI and separated on a 1% agarose gel. The gel was blotted onto nitrocellulose filters and hybridized to radioactive probes homologous to purified restriction fragment *Hin*dIII-23 of pTiC58 (**A**) (**a** and **b**); to pBR322 (**c** and **d**); or to purified restriction fragment *Hin*dIII-10 (**e** and **f**). The numbers refer to the sizes of the radioactive DNA bands in kilobase pairs (kb), and H and E refer to restriction endonucleases *Hin*dIII and *Eco*RI, respectively. The faint upper bands visible in **e** and **f**, which are not labeled in the figure, are not reproducible and likely represent partial digestion products which occurred in the reactions shown in **e** and **f**. (**B**) The T-DNA region of pGV3850. The vertical bars and the numbers refer to *Hin*dIII restriction endonuclease fragments of pTiC58. The sizes of *Hin*dIII fragments 10 and 23 are given in kb in brackets. Fragment 10 is shown as a white box, fragment 23 as a stippled box, and the borders of the T-DNA as jagged lines. Restriction endonuclease sites for *Hin*dIII (●) and *Eco*RI (○) are indicated. (**C**) Results of the Southern blot hybridization analyses shown in **A**. We show two T-DNA copies arranged in a tandem structure since one of the bands (5.2 kb) hybridizes with probes homologous to both *Hin*dIII fragments 10 and 23. The numbers refer to the sizes of the fragments in kb which are also found in **A**. The indications below each fragment in brackets refer to probe(s) which give positive hybridization. All the symbols are as in **B** except that the dashed line indicates plant DNA sequences.

Analysis of the structure of the T-DNA in tobacco cells transformed by pGV3850

As a further check that these nopaline-positive tissues were indeed transformed with the minimal T-DNA of pGV3850, we analyzed total DNA prepared from the pGV3850-transformed plant shown in Figure 3B by restriction endonuclease and Southern blot hybridization analyses. The DNA was digested with either *Hin*dIII or *Eco*RI, and hybridized with probes for either the left border (pTiC58 *Hin*dIII fragment 10 probe), right border (pTiC58 *Hin*dIII fragment 23 probe), or middle (pBR322 probe) of the T-DNA of pGV3850. We were especially concerned that the pBR322 sequences were intact (see below, application of pGV3850), and this was found to be true. In addition, with the left and right probes we obtained border fragments as expected and neither Ti plasmid fragments *Hin*dIII-23 or *Hin*dIII-10 were observed.

Figure 4A shows the Southern blot hybridization patterns; Figure 4B shows the T-DNA of Ti plasmid pGV3850 prior to transfer to plant cells; and Figure 4C summarizes and presents a diagram of the T-DNA structure in transformed plant cells derived from the hybridization results of Figure 4A. The results suggest that the T-DNA is arranged in tandem copies; for simplicity in the drawing we show only two copies, although we estimate there are between two and five copies of pGV3850 T-DNA in the transformed plant cell DNA analyzed here. There are several reasons for proposing a tandem structure; (i) the intensity of hybridization of the internal pBR322 band when transformed plant cell DNA is digested with *Hin*dIII is greater than any of the border fragment hybridizations; (ii) there is a *Hin*dIII border fragment (5.2 kb) which hybridizes with probes homologous to *Hin*dIII fragment 23 and *Hin*dIII fragment 10 of the Ti plasmid; the intensity of hybridization is also greater as would be expected for a fragment present in several copies; (iii) we find only one right border fragment (*Eco*RI fragment of 12 kb or *Hin*dIII fragment of 4.6 kb) and one left border fragment (*Hin*dIII fragment of 4.4 kb); (iv) digestion with *Eco*RI does not reveal a left border fragment as the *Eco*RI site lies within 50 bp of this border (Yadav *et al.* 1980,1982; Zambryski *et al.*, 1982); however, we also obtain the expected internal *Eco*RI fragment of 3 kb. This latter internal fragment also hybridizes with a higher intensity as would be expected for multiple copies of the T-DNA.

These results represent the first analysis of plant cell DNA transformed with a Ti plasmid lacking all the internal portion of the T-DNA and demonstrate that the T-DNA borders are sufficient to allow transfer and stable integration of the DNA contained between these borders.

The use of pGV3850 as a Ti plasmid vector for the introduction of foreign DNA into plant cells

A useful application of pGV3850 is that the pBR322 sequences which are contained between the T-DNA borders can be used directly as an acceptor site for the insertion of foreign DNA. A single cross-over event in *Agrobacterium* will introduce a cloning vector containing any foreign gene of interest which is inserted in pBR322 (or its derivatives) into the modified T-DNA of pGV3850 (Figure 5). It may be of practical advantage that the DNA to be introduced is contained in

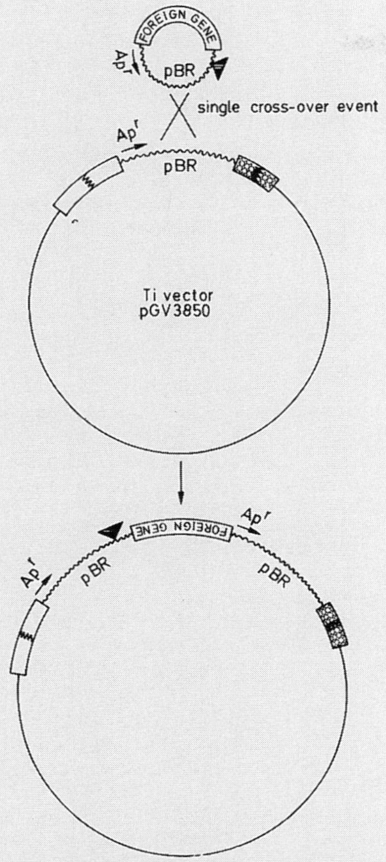

Fig. 5. Usefulness of Ti plasmid vector pGV3850 as a recipient for any foreign gene of interest whose expression is to be monitored in whole plants. A DNA of interest (foreign gene) can be cloned by recombinant DNA techniques into a pBR-type plasmid vehicle in *E. coli*. This recombinant plasmid can be mobilized to *Agrobacterium* using helper plasmids (Van Haute *et al.*, 1983). A single recombination event will result in the co-integration of the recombinant plasmid containing the gene of interest with pGV3850. By using selectable antibiotic resistance marker genes linked to the foreign gene, the co-integrate structure is stably maintained in *Agrobacterium* despite the duplication of the pBR sequences. As shown, the alignment of homologous pBR regions for recombination results in the reversal of the orientation of the foreign gene. The symbols are as in Figures 1 and 4, except the triangle with the diagonal lines indicates a second antibiotic resistance marker gene other than ampicillin resistance gene of pBR322.

a vector with an additional resistance marker gene to the Ap^R already present in pGV3850. Such a marker can be used as a genetic selection for the transfer of the intermediate cloning vector from *Escherichia coli* to *Agrobacterium*. This resistance marker can be contained either within the vector sequences such as chloramphenicol resistance (Cm^R) for pBR325 (Bolivar, 1978) or kanamycin resistance (Km^R) for pKC7 (Rao and Rogers, 1979), or within the foreign DNA which is to be tested in the plant cell. The resulting Ti plasmid contains a duplication of pBR322 sequences. One might expect this to lead to instability of the inserted DNA. However, this is not a problem in *Agrobacterium* because the gene of interest is linked to an antibiotic resistance marker that can be maintained by growth on selective medium. An instability of the sequences in the resulting transformed plant DNA is not expected as repeated DNA sequences are a normal part of the plant genome, and the T-DNA itself is often (also, as shown, in the pGV3850-transformed tissue investigated here) organized as an array of tandemly repeated copies in transformed plant cells (Lemmers *et al.*, 1980; Zambryski *et al.*, 1980, 1982).

An experiment was designed to test the stability of co-integrate derivatives of pGV3850. The *Eco*RI fragment 7 of the T-DNA of the octopine Ti plasmid (De Vos *et al.*, 1981) was cloned in a plasmid vector containing resistance marker genes (ampicillin, tetracycline, streptomycin, spectinomycin and sulfonamide) as well as pBR322 sequences (J. Leemans and R. Deblaere, unpublished results). This *Eco*RI fragment carries several of the oncogenic genes encoding a tumorous phenotype including hormone-independent growth (Garfinkel *et al.*, 1981; Leemans *et al.*, 1982; Willmitzer *et al.*, 1982). The co-integrate between pGV3850 and the intermediate plasmid vector (Leemans *et al.*, 1981) carrying this *Eco*RI fragment in *Agrobacterium* was used to infect tobacco plantlets. The resulting tissue is transformed and has been growing as a tumor on hormone-free medium for 10 months so far without any alteration of its properties; in addition, the tissue remains nopaline-positive (Figure 2l).

Discussion

In the past, the insertion of foreign DNA into the T-DNA region of the Ti plasmid required a further cloning of the DNA of interest into a plasmid vector which also contained a portion of the T-DNA region to serve as a region of homologous recombination with the acceptor Ti plasmid (Leemans *et al.*, 1981; Matzke and Chilton, 1981). The development of techniques for direct mobilization of plasmids containing pBR322 sequences to *Agrobacterium* (Van Haute *et al.*, 1983) makes pGV3850 an extremely versatile acceptor Ti plasmid. Assuming a transmission frequency from *E. coli* to *Agrobacterium* of 4.5×10^{-3} (Van Haute *et al.*, 1983), we have observed a recombination frequency of 2×10^{-2} for the homology region of the pBR322 sequence of pGV3850 (L. Herrera-Estrella, D. Inzé and P. Zambryski, unpublished results). Besides insertion of defined clones containing a pBR322 sequence by a single cross-over event into the acceptor Ti plasmid pGV3850, this Ti plasmid can also be used as a recipient for cloned banks of DNA in pBR322 or its derivatives in a 'shotgun cloning' experiment. The total population of hybrid plasmids in *Agrobacterium* can be used to infect plant cells and can be subsequently screened for expression of the gene(s) of interest.

Recently, foreign genes have been successfully expressed in plant cells using a vector which contains the promoter sequences of the Ti plasmid-specific nopaline synthase gene (Herrera-Estrella *et al.*, 1983a,1983b). Four different coding sequences were inserted behind the *nos* promoter and were shown to be properly expressed following infection of plant cells with wild-type Ti plasmid carrying these chimeric genes. The four different coding sequences used were octopine synthase, and three others for bacterial antibiotic resistance genes: the aminoglycoside phosphotransferase (APH(3')II) conferring kanamycin resistance from Tn*5* (Davies and Smith, 1978), the methotrexate-insensitive dihydrofolate reductase (DHFR Mtx^R) conferring methotrexate resistance from R67 (Fling and Elwell, 1980; O'Hare *et al.*, 1981), and chloramphenicol acetyl transferase conferring chloramphenicol resistance from pBR325 (Bolivar, 1978). Furthermore, the expression of the antibiotic resistance genes could be used as a specific selection for transformed tobacco protoplasts growing as a mixed population in tissue culture (Herrera-Estrella *et al.*, 1983b).

As a further test of the vector properties of pGV3850 as well as of the expressibility of foreign genes in plants, we have recombined these various chimeric constructions, which contain the *nos* promoter linked to antibiotic resistance genes, into pGV3850. The co-integrates have been used to infect tobacco protoplasts and resistant calli have been isolated (M. De Block, L. Herrera-Estrella and P. Zambryski, unpublished results); these calli will be grown further to check for the physical presence of the foreign genes. These tissues can be subsequently regenerated into complete plants to test for the expression of the resistance genes in differentiated tissues. The use of selectable marker genes in combination with non-oncogenic Ti plasmids, such as pGV3850, makes it possible to select for transformed plant cell protoplasts in tissue culture and subsequently regenerate whole plants containing the inserted DNA. For the future, one has to simply link other genes of interest to the selectable marker gene prior to transfer to the acceptor Ti plasmid.

A simple approach to obtain transformed plants by inoculation of decapitated tobacco plantlets with *Agrobacterium* containing pGV3850 is presented. As there is no tumor response, the wound surface is removed and cultivated in tissue culture as a callus. A subsequent transfer of this wound callus to regeneration medium results in the production of transformed plantlets which can be easily distinguished from untransformed plantlets by a nopaline test. Furthermore, the frequency of obtaining transformed shoots is rather high. These results demonstrate that it is not necessary to have a protoplast cell culture system in order to have efficient transformation; it is likely that the method utilized here can be applied to the inoculation of transformed plants from species which are not amenable to protoplast cultivation.

The vector pGV3850 makes use of the natural transfer properties of the Ti plasmid; only those genes which interfere with normal plant differentiation have been removed. Thus, the most important aspect of pGV3850-transformed cells is their capacity to regenerate into complete plants. These plants can be derived from single cells and the regeneration process itself is extremely simple, requiring only minor changes in tissue culture conditions. These results clearly open up several areas for investigation of plant biology. For example, we may now ask questions about tissue-specific regulation of genes. Genes isolated from one cell type can be re-added to plant cells via pGV3850 to investigate whether or not they are expressed again in the appropriate cell; these results should lead to the identification of tissue-specific controlling signals. In addition, we can monitor the effect(s) of completely foreign genes whose properties we may wish to transfer into plants. Basic plant processes, such as chloroplast function or hormone physiology may also be studied. Perhaps the most interesting subject will be the study of genes which are developmentally regulated. Thus, evidence has been presented that the Ti vector has evolved to a point where it is ready to be used to genetically engineer whole plants; it remains for us to turn our efforts toward the isolation of particular genes whose expression we wish to study.

Materials and methods

Conditions of bacterial growth

The *E. coli* and *Agrobacterium* strains to be analyzed were cultured as described by Joos *et al.* (1983).

Bacterial conjugation

Conjugations involving *E. coli* and *Agrobacterium* strains were performed as described by Van Haute *et al.* (1983).

DNA preparation

Total *Agrobacterium* DNA was prepared using techniques described by Dhaese *et al.* (1979).

Total plant DNA was prepared in the manner previously described (Lemmers *et al.*, 1980), except that wet tissue was used directly and the extraction buffer was two times concentrated. The tissue was rapidly frozen in liquid nitrogen and homogenized in a Waring Blender. One volume (equivalent to the wet weight of the plant tissue) of 2 x extraction buffer was added and the DNA was isolated as before (Lemmers *et al.*, 1980).

Restriction digests of DNA

All restriction enzyme digests were incubated in the TA buffer described by O'Farrell *et al.* (1980).

Hybridization conditions

Hybridization and washing of nitrocellulose filters was performed at 68°C in 3 x SSC (1 x SSC: 0.15 M NaCl, 0.015 M Na-citrate) and 0.2% SDS.

Oncogenicity tests

The different *Agrobacterium* strains have been infected on four different test plants in order to test their oncogenic capacities. Tobacco (*Nicotiana tabacum* cv. Wisconsin 38) and *Petunia hybrida* (var. Mitchell) plantlets, and potato (*Solanum tuberosum* var. Bintje) and carrot (*Daucus carota*) slices have been infected as described previously (Leemans *et al.*, 1982; Joos *et al.*, 1983).

In vitro cultivation of plant tissues

Hormone-independent crown gall tumor tissue was grown *in vitro* on plates containing LS medium (Linsmaier and Skoog, 1965) with 1% sucrose and 0.8% agar. Hormone-dependent tissue induced by pGV3850 was grown continuously on the same medium containing 1 mg/l NAA and 0.2 mg/l BAP. Shoots were regenerated from callus tissue by placing it on LS medium containing 1 mg/l BAP. Tissues were made free of *Agrobacterium* by growing primary wound tissue as well as the first few subcultures of this tissue on medium which also contained 500 μg/ml cefotaxime (Claforan, Hoechst).

Nopaline detection in plant tissues

A small amount (50 mg) of either callus tissue or leaf material was crushed with a glass rod in an Eppendorf centrifuge tube. After centrifugation for 5 min, a part of the supernatant (2 μl) was spotted directly onto Whatman paper and the extracts were electrophoresed and stained for the presence of opines as described (Aerts *et al.*, 1979).

Acknowledgements

We would like to thank Luis Herrera-Estrella, Henri De Greve, and Jean-Pierre Hernalsteens for their critical advice during the course of this work. In addition, we thank Ms. Martine De Cock, Karel Spruyt and Albert Verstraete for their help in preparing this manuscript. This research was supported by grants from the 'Kankerfonds van de Algemene Spaar- en Lijfrentekas', from the 'Instituut tot Aanmoediging van het Wetenschappelijk Onderzoek in Nijverheid en Landbouw' (I.W.O.N.L. 3849A), from the Services of the Prime Minister (OOA/12052179), and from the 'Fonds voor Geneeskundig Wetenschappelijk Onderzoek' (F.G.W.O. 3.001.82) to J.S. and M.V.M. H.J. and J.L. were supported by fellowships of the I.W.O.N.L., and P.Z. is indebted to the USDA (grant #5901-0410-9-0374-0) for support and to the EMBO for a long-term fellowship.

References

Aerts,M., Jacobs,M, Hernalsteens,J.P., Van Montagu,M. and Schell,J. (1979) *Plant Sci. Lett.*, **17**, 43-50.

Barton,K.A., Binns,A.N., Matzke,A.J.M. and Chilton,M.-D. (1983) *Cell*, **32**, 1033-1043.

Bevan,M.W. and Chilton,M.-D. (1982) *Annu. Rev. Genet.*, **16**, 357-384.

Bolivar,F. (1978) *Gene*, **4**, 121-136.

Davies,J. and Smith,D.I. (1978) *Annu. Rev. Microbiol.*, **32**, 469-518.

De Beuckeleer,M., Lemmers,M., De Vos,G., Willmitzer,L., Van Montagu, M. and Schell, J. (1981) *Mol. Gen. Genet.*, **183**, 283-288.

De Greve,H., Decraemer,H., Seurinck,J., Van Montagu,M. and Schell,J. (1981) *Plasmid*, **6**, 235-248.

De Greve,H., Leemans,J., Hernalsteens,J.P., Thia-Toong,L., De Beuckeleer,M., Willmitzer,L., Otten,L., Van Montagu,M. and Schell,J. (1982) *Nature*, **300**, 752-755.

Depicker,A., De Wilde,M., De Vos,G., De Vos,R., Van Montagu,M. and Schell,J. (1980) *Plasmid*, **3**, 193-211.

Depicker,A., Stachel,S., Dhaese,P., Zambryski,P. and Goodman,H.M. (1982) *J. Mol. Appl. Genet.*, **1**, 561-574.
De Vos,G., De Beuckeleer,M., Van Montagu,M. and Schell,J. (1981) *Plasmid*, **6**, 249-253.
Dhaese,P., De Greve,H., Decraemer,H., Schell,J. and Van Montagu,M. (1979) *Nucleic Acids Res.*, **7**, 1837-1849.
Fling,M.E. and Elwell,L.P. (1980) *J. Bacteriol.*, **141**, 779-785.
Garfinkel,D.J. and Nester,E.W. (1980) *J. Bacteriol.*, **144**, 732-743.
Garfinkel,D.J., Simpson,R.B., Ream,L.W., White,F.F., Gordon,M.P. and Nester,E.W. (1981) *Cell*, **27**, 143-153.
Hernalsteens,J.P., Van Vliet,F., De Beuckeleer,M., Depicker,A., Engler,G., Lemmers,M., Holsters,M., Van Montagu,M. and Schell,J. (1980) *Nature*, **287**, 654-656.
Herrera-Estrella,L., Depicker,A., Van Montagu,M. and Schell,J. (1983a) *Nature*, **303**, 209-213.
Herrera-Estrella,L., De Block,M., Messens,E., Hernalsteens,J.-P., Van Montagu,M. and Schell,J. (1983b) *EMBO J.*, **2**, 987-995.
Holsters,M., Silva,B., Van Vliet,F., Genetello,C., De Block,M., Dhaese,P., Depicker,A., Inzé,D., Engler,G., Villarroel,R., Van Montagu,M. and Schell,J. (1980) *Plasmid*, **3**, 212-230.
Holsters,M., Villarroel,R., Van Montagu,M. and Schell,J. (1982) *Mol. Gen. Genet.*, **185**, 283-289.
Holsters,M., Villarroel,R., Gielen,J., Seurinck,J., De Greve,H., Van Montagu,M. and Schell,J. (1983) *Mol. Gen. Genet.*, **190**, 35-41.
Joos,H., Inzé,D., Caplan,A., Sormann,M., Van Montagu,M. and Schell,J. (1983) *Cell*, **32**, 1057-1067.
Kahl,G. and Schell,J. (1982) *Molecular Biology of Plant Tumors*, published by Academic Press, NY.
Leemans,J, Shaw,C., Deblaere,R., De Greve,H., Hernalsteens,J.P., Maes, M., Van Montagu,M. and Schell,J. (1981) *J. Mol. Appl. Genet.*, **1**, 149-164.
Leemans,J., Deblaere,R., Willmitzer,L., De Greve,H., Hernalsteens,J.P., Van Montagu,M. and Schell,J. (1982) *EMBO J.*, **1**, 147-152.
Lemmers,M., De Beuckeleer,M., Holsters,M., Zambryski,P., Depicker,A., Hernalsteens,J.P., Van Montagu,M. and Schell,J. (1980) *J. Mol. Biol.*, **144**, 353-376.
Linsmaier,E.M. and Skoog,F. (1965) *Physiol. Plant.*, **18**, 100-127.
Matzke,A.J.M. and Chilton,M.-D. (1981) *J. Mol. Appl. Genet.*, **1**, 39-49.
Murashige,T. and Skoog,F. (1962) *Physiol. Plant.* **15**, 473-497.
Nester,E.W. and Kosuge,T. (1981) *Annu. Rev. Microbiol.*, **35**, 531-565.
O'Farrell,P.H., Kutter,E. and Nakanishi,M. (1980) *Mol. Gen. Genet.*, **179**, 421-435.
O'Hare,K., Benoist,C. and Breathnach,L. (1981) *Proc. Natl. Acad. Sci. USA*, **78**, 1527-1531.
Ooms,G., Klapwijk,P.M., Poulis,J.A. and Schilperoort,R.A. (1980) *J. Bacteriol.*, **144**, 82-91.
Ooms,G., Hooykaas,P.J., Moleman,G. and Schilperoort,R.A. (1981) *Gene*, **14**, 33-50.
Otten,L., De Greve,H., Hernalsteens,J.P., Van Montagu,M. and Schell,J. (1981) *Mol. Gen. Genet.*, **183**, 209-213.
Rao,R.N. and Rogers,S.G. (1979) *Gene*, **7**, 79-82.
Ream,L.W., Gordon,M.P. and Nester,E.W. (1983) *Proc. Natl. Acad. Sci. USA*, **80**, 1660-1664.
Simpson,R.B., O'Hara,P.J., Krook,W., Montoya,A.L., Lichtenstein,C., Gordon,M.P. and Nester,E.W. (1982) *Cell*, **29**, 1005-1014.
Thomashow,M.F., Nutter,R., Montoya,A.L., Gordon,M.P. and Nester, E.W. (1980a) *Cell*, **19**, 729-739.
Thomashow,M.F., Nutter,R., Postle,K., Chilton,M.-D., Blattner,F.R., Powell,A., Gordon,M.P. and Nester,E.W. (1980b) *Proc. Natl. Acad. Sci. USA*, **77**, 6448-6452.
Van Haute,E., Joos,H., Maes,M., Warren,G., Van Montagu,M. and Schell, J. (1983) *EMBO J.*, **2**, 411-418.
Willmitzer,L., Simons,G. and Schell,J. (1982) *EMBO J.*, **1**, 139-146.
Willmitzer,L., Dhaese,P., Schreier,P.H., Schmalenbach,W., Van Montagu, M. and Schell,J. (1983) *Cell*, **32**, 1045-1056.
Yadav,N.S., Postle,K., Saiki,R.K., Thomashow,M.F. and Chilton,M.-D. (1980) *Nature*, **287**, 458-461.
Yadav,N.S., Vanderleyden,J., Bennett,D.R., Barnes,W.M. and Chilton, M.-D. (1982) *Proc. Natl. Acad. Sci. USA*, **79**, 6322-6326.
Zambryski,P., Holsters,M., Kruger,K., Depicker,A., Schell,J., Van Montague,M. and Goodman,H.M. (1980) *Science (Wash.)*, **209**, 1385-1391.
Zambryski,P., Depicker,A., Kruger,K. and Goodman,H. (1982) *J. Mol. Appl. Genet.*, **1**, 361-370.
Zambryski,P., Goodman,H., Van Montagu,M. and Schell,J. (1983) in Shapiro,J. (ed.), *Mobile Genetic Elements*, Academic Press, NY, pp. 505-535.

FOREIGN GENES CAN BE EXPRESSED IN PLANT CELLS

Luis Herrera-Estrella
Departamento de Ingeniería Genética, Unidad de Biotecnología Agrícola
Centro de Investigacion y Estudios Avanzados, Carretera Irapuato-Leon, Guanajuato, Mexico
E-mail: lherrera@ira.cinvestav.mx

Commentary on the papers:

Herrera-Estrella, L., Depicker, A., Van Montagu, M., and Schell, J. 1983. Expression of chimaeric genes transferred into plant cells using a Ti-plasmid-derived vector. Nature 303:209–213.

Fraley, R. T., Rogers, S. G., Horsch, R. B., Sanders, P. R., Flick, J. S., Adams, S. P., Bittner, M. L., Brand, L. A., Fink, C. L., Fry, J. S., Galluppi, G. R., Goldberg, S. B., Hoffmann, N. L., and Woo, S. C. 1983. Expression of bacterial genes in plant cells. Proc. Natl. Acad. Sci. U.S.A. 80:4803–4807.

Bevan, M. W., Flavell, R. B., and Chilton, M.-D. 1983. A chimaeric antibiotic resistance gene as a selectable marker for plant cell transformation. Nature 304:184–187.

My work on *Agrobacterium* and the expression of foreign genes in plant cells started when I joined the group of Jeff Schell and Marc Van Montagu at the end of the summer of 1981 as a Ph.D. student. I had just finished a training period in the laboratory of Francisco Bolivar (the scientist who produced the first useful cloning vectors; pBR322 and derivatives) in which I had received training in gene splicing technology. A common discussion topic in Bolivar's lab was the expression of human genes in bacteria. Therefore, when Marc Van Montagu gave a lecture on the Ti plasmid in Mexico City, I thought at that time that a good and original Ph.D. project would be to use the Ti plasmid to transfer and express foreign genes in plant cells. As a naive student I was far from imagining that several other groups were interested in doing exactly the same work!

When I arrived in Belgium it was already well established that a defined segment of the Ti plasmid, the T-DNA, was transferred to plant cells, that transposons carrying bacterial genes inserted into the T-region were transferred to plant cells along with the T-DNA, and that genes encoded by the T-DNA were functionally expressed in plant cells, including those encoding enzymes responsible for opine synthesis. In fact, the sequence of the nopaline synthase (*nos*) gene had recently been determined and was fortunately available to use in expression vectors for plant cells. With all these tools in hand, we initiated a project to design and construct plant expression vectors using the *nos* gene as a source of 5′ and 3′ transcriptional regulatory sequences. I must remind the reader that at that time PCR technology had not been developed and the construction of expression vectors relied upon the use of exonucleases to produce deletions in the *nos* gene. These types of projects, now routinely carried out daily, were slow and time consuming. After analyzing over 250 deletion plasmids by doing DNA minipreps, restriction enzyme analysis and sequencing a few of them, we were able to produce a DNA fragment containing the promoter and 30 bp of the 5′-untranslated leader region of the *nos* gene, flanked by *Bam*HI and *Hind*III restriction sites. A fragment containing the polyadenylation and transcription termination signals was included in the expression vector, in which target-coding sequences could be inserted in a *Bam*HI restriction site, directly downstream of the *nos* promoter. This tedious work was greatly facilitated by a young student, June Simpson, who joined Marc Van Montagu's laboratory as a summer undergraduate student and who years later became my wife. I later learned that Mike Bevan, then a postdoctoral fellow in Mary-Dell Chilton's group, had used an almost identical strategy to produce a very similar expression vector and that a group at the Monsanto Company was also working on a very similar project in St. Louis, Missouri, U.S.A.

To test whether our plant expression vector was indeed functional, we used the coding sequence of the octopine synthase (*ocs*) gene, the chloramphenicol acetyl transferase (*cat*) gene from pBR325 and the neomycin phosphotransferase II (*npt II*) gene derived from the transposon Tn*5,* to produce chimeric genes potentially functional in plant cells. The *ocs* gene was known to be functionally transcribed and translated in tumor cells and the other two coding sequences, although of bacterial origin, had been previously shown to be functionally transcribed and translated in mammalian cells. A single recombination event, mediated by the homology provided by the *nos* sequences present in the plasmids containing the chimeric genes, was used to insert these genes into the T-DNA of a nopaline type Ti plasmid. Analysis of tobacco tumors induced with *Agrobacterium* strains containing the chimeric *nos-ocs* and *nos-cat* genes showed that both chimeric genes were functional in plant cells.

Agrobacterium strains containing the *nos-npt II* gene were used by Mike Bevan, by ourselves, and by the Monsanto group to identify transformed cells that could grow in media containing either kanamycin or G-418. These experiments were successful and together provided conclusive evidence that foreign genes can be expressed in plant cells when provided with regulatory sequences recognized by the plant transcriptional machinery and that the *nos-npt II,* is a very effective selectable marker to identify transformed cells.

In December 1982, I had a meeting with Marc Van Montagu and Jeff Schell in which they reviewed the results of the experiments testing the activity of chimeric genes in plant cells. Fortunately at that time we already had good results showing that the *nos* expression vector that we had constructed was functional. My Ph.D. advisors were very pleased because there were rumors that Monsanto was going to present spectacular results on the successful genetic modification of plant cells at the Miami Winter Symposium in January 1983. In that meeting our results, and those of Mary-Dell Chilton's and the Monsanto's groups, were presented publicly and it is considered to be the official date when the age of plant genetic engineering was born.

After the Miami Winter symposium the race to complete and publish the results on the expression of foreign genes in plant cells began. The first paper on these experiments, reporting the expression of *ocs* and *cat* under control of the *nos* regulatory sequences was submitted on February 8, and appeared in the May 19 issue of *Nature.* On March 21st we submitted a second manuscript describing the use of the *nos-npt II* gene as a dominant selectable marker gene for plant cells; that paper was published in the June issue of *EMBO Journal* (Herrerra-Estrella et al. 1983). Almost at the same time, the Monsanto group led by Rob Fraley, Steve Rogers and Rob Horsch communicated on April 25th a manuscript describing the use of *nos-npt II* as a selectable marker that was published in the August 1983 issue of the *Proceedings of the National Academy of Science U.S.A.,* and Mike Bevan, Dick Flavell and Mary-Dell Chilton submitted a paper on May 4, 1983 to *Nature* that was published on July 14th of that year. Who did what first is still a matter of debate; 20 years after these discoveries, the ownership of the *Agrobacterium* transformation patent is still an unsettled matter.[1]

The following year, 1984, two other very important papers describing the functional inheritance of foreign genes in plants transformed, using Ti plasmid derivatives devoid of tumor inducing genes, appeared in *Science* and *EMBO Journal* by the Monsanto and the Belgiun groups, respectively (Horsch et al. 1984; De Block et al. 1984).

It is important to note that the technology described in these papers not only served for the development of transgenic plants for commercial and agricultural purposes but also provided the essential tools that paved the way for the wealth of studies, that have been published to date, aimed at analyzing the processes that regulate gene expression, protein targeting into different subcellular compartments, gene function and plant development. It must also be said that although I have only referred to the key players in the race to express foreign genes in plants, many other researchers both in the laboratories mentioned above and in other groups in different parts of the world contributed knowledge, information and tools that were fundamental for the development of plant transformation technology.

References

De Block, M., Herrera-Estrella, L., Van Montagu, M., Schell, J., and Zambryski, P. 1984. Expression of foreign genes in regenerated plants and their progeny. EMBO J. 3:1681–1689.

Herrera-Estrella, L., De Block, M., Messens, E., Hernalsteens, J.-P., Van Montagu, M., and Schell, J. 1983. Chimeric genes as dominant selectable markers in plant cells. EMBO J. 2:987–995.

Horsch, R. B., Fraley, R. T., Rogers, S. G., Sanders, P. R., Lloyd, A., and Hoffmann, N. 1984. Inheritance of functional foreign genes in plants. Science 223: 496–498.

[1] After 20 years of dispute over patent rights, at the end of 2003 the Board of Patent Appeals and Interferences for the U.S. Patent and Trademark Office and the European Patent Office awarded priority of invention to the Belgian team (Patricia Zambryski, Josef Schell, Jean-Pierre Hernalsteens, Marc Van Montagu, Luis Herrera-Estrella and Jan Leemans) for a modified version of the tumor-inducing plasmid of *Agrobacterium* and extended to related claims concerning plant transformation processes and transgenic products which can be obtained using such processes. This resolution implies that this patent will be valid for the next 17 years.

LUIS HERRERA-ESTRELLA: In 1978, I graduated B.Sc. in Biochemical Engineering from the National Polytechnic Institute of Mexico. I obtained my M.Sc. in Genetics and Molecular Biology, from the Centro de Investigación y Estudios Avanzados, Mexico City, in 1980. I studied for a Ph.D. at Rijksuniversiteit, Ghent, Belgium, and graduated in 1984.

I have had several mentors that even today greatly influence my scientific career: Dr. Jose Ruiz Herrera, my M.Sc. thesis advisor, who introduced me to the wonderful world of molecular biology; Dr. Francisco Bolivar, in whose laboratory I learned gene-splicing technology and who has continuously supported my career; Dr. Patricia Zambryski, who was a friend and guide during my Ph.D. studies; Dr. Marc Van Montagu, a true maestro, who was interested not only in our academic preparation but also in our wider preparation as constructive individuals in society; and Dr. Jeff Schell, a brilliant scientist, a source of inspiration and the leader of many stimulating discussions.

Expression of chimaeric genes transferred into plant cells using a Ti-plasmid-derived vector

Luis Herrera-Estrella*, Ann Depicker*, Marc Van Montagu* & Jeff Schell*†

* Laboratorium voor Genetica, Rijksuniversiteit Gent, B-9000 Gent, Belgium
† Max-Planck-Institut für Züchtungsforschung, D-5000 Köln 30, FRG

Foreign genes introduced into plant cells with Ti-plasmid vectors are not expressed. We have constructed an expression vector derived from the promoter sequence of nopaline synthase, and have inserted the coding sequences of the octopine synthase gene and a chloramphenicol acetyltransferase gene into this vector. These chimaeric genes are functionally expressed in plant cells after their transfer via a Ti-plasmid of Agrobacterium tumefaciens.

CROWN gall formation on dicotyledonous plants by *Agrobacterium tumefaciens* is the result of the transfer and covalent integration of a segment (called T-region) of the Ti-plasmid into the chromosomal DNA of plant cells (for reviews see refs 1-4). Insertion of foreign DNA sequences within the T-region of Ti-plasmids leads to their co-transfer and integration into the plant genome[5]. To date, inserts of up to 50 kilobases (kb) have been successfully co-transferred with the T-DNA (unpublished results of this laboratory). The upper limit of the size of the DNA segments that can be transferred to plant genomes with this system has not been determined. The transferred T-region (or T-DNA) is expressed in plant cells[6] and codes for several host polymerase II-dependent polyadenylated transcripts[7-10]. Some of these transcripts were shown to be responsible for the tumorous mode of growth of the transformed plant cells, while others code for enzymes that synthesize novel compounds (termed opines), which are amino acid or sugar derivatives. None of these transcripts was found to be essential for T-DNA transfer[11-13].

To understand the mechanism by which T-DNA-linked genes are expressed in plants, the nucleotide sequence and the precise location of 3′ and 5′ ends of the transcripts of two different opine synthase genes were determined, the octopine synthase gene carried by pTiB6S3 (ref. 14) and the nopaline synthase from pTiT37 (refs 15, 16). Although both genes are encoded by plasmids of bacterial origin, they share more characteristics with eukaryotic genes than with prokaryotic genes. Both octopine and nopaline synthase genes, designated *ocs* and *nos* respectively, have a sequence similar to the so-called 'TATA' or 'Goldberg–Hogness' box[17] in the 5′ region upstream of the start of transcription, and a sequence 'AATAA', similar to the polyadenylation signal of eukaryotic genes, near the 3′ end[18,19].

The opine synthase genes provide useful tools for studying gene expression in plant cells, because they are constitutively expressed in callus cells as well as in all tissues of plants regenerated from T-DNA-transformed cells[20,21]. Of practical importance is the fact that very rapid and sensitive assays for the detection of both nopaline and octopine have been developed[22,23]. Various attempts to express a number of bacterial, animal and some unrelated plant genes (for example, leghaemoglobin) introduced and integrated into the tobacco genome with the use of Ti-plasmid gene vectors[24], have thus far failed (unpublished results of this laboratory), because these foreign genes were not transcribed in the transformed tobacco cells. This lack of expression was presumably due to the fact that the plant transcription machinery could not recognize the transcription signals (promoter sequences) carried by these foreign genes.

0028-0836/83/200209—05$01.00

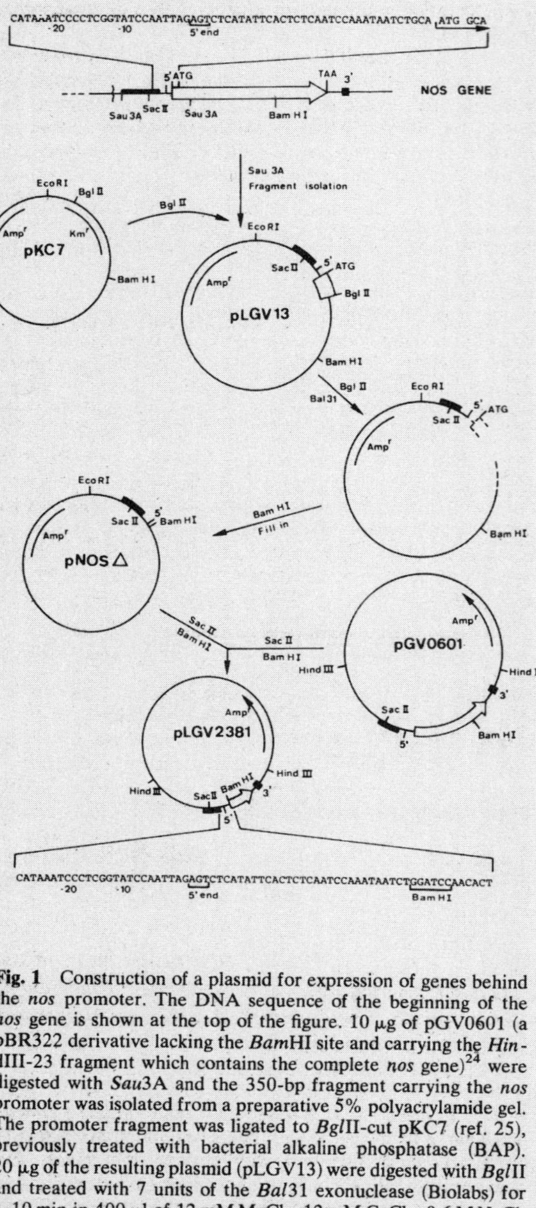

Fig. 1 Construction of a plasmid for expression of genes behind the *nos* promoter. The DNA sequence of the beginning of the *nos* gene is shown at the top of the figure. 10 μg of pGV0601 (a pBR322 derivative lacking the *Bam*HI site and carrying the *Hin*-dIII-23 fragment which contains the complete *nos* gene)[24] were digested with *Sau*3A and the 350-bp fragment carrying the *nos* promoter was isolated from a preparative 5% polyacrylamide gel. The promoter fragment was ligated to *Bgl*II-cut pKC7 (ref. 25), previously treated with bacterial alkaline phosphatase (BAP). 20 μg of the resulting plasmid (pLGV13) were digested with *Bgl*II and treated with 7 units of the *Bal*31 exonuclease (Biolabs) for 4–10 min in 400 μl of 12 mM $MgCl_2$, 12 mM $CaCl_2$, 0.6 M NaCl, 1 mM EDTA and 20 mM Tris-HCl, *p*H 8.0, at 30 °C. During this time ~20–50 bp of DNA were removed. The *Bal*31-treated molecules were digested with *Bam*HI and incubated with DNA polymerase (Klenow fragment) and the 4-deoxynucleotide triphosphates (at 10 mM each) to fill in the ends. Plasmids were screened for a regenerated *Bam*HI site derived from the ligation of a filled-in *Bam*HI end and the end of the *Bal*31 deletion. The sizes of the *Bam*HI–*Sac*II fragment of several candidates were estimated in a 6% urea-polyacrylamide gel, and the nucleotide sequences of the candidates with sizes ranging over 160–190 nucleotides were determined by the method of Maxam and Gilbert[42]. A pNOSΔ clone lacking the *nos*-coding sequence and two nucleotides of the leader sequence was used to substitute the *Sac*II–*Bam*HI fragment in the *nos* gene in pGV0601; the final promoter vector is called pLGV2381. All the recombinant plasmids were selected by transformation of the *E. coli* strain HB101. The bases marked 5′ and 3′ constitute the start and stop of transcription, and the AUG and TAA refer to the codons used to start and stop translation. The heavy black line indicates the *nos* promoter region, and the white area indicates the *nos*-coding region. Amp^r, ampicillin resistance; Km^r, kanamycin resistance.

To overcome the potential barrier to the expression of foreign genes in plants, we have constructed a number of chimaeric genes consisting of promoter sequences derived from the *nos* gene and of coding sequences derived from foreign genes. These chimaeric genes were introduced in the genome of tobacco with Ti-plasmid vectors and the resulting transformed tobacco cells were shown to express these chimaeric genes and to produce functional proteins.

Expression vector

Figure 1 outlines the construction of an expression vector using the *nos* promoter region. Fragment *Hin*dIII-23 from plasmid pTiC58 carries the complete coding sequence and the sequences responsible for faithful expression of the *nos* gene in plant cells (ref. 15 and C. Konsz *et al.*, in preparation). DNA from pGV0601 (ref. 24) (a pBR322 derivative lacking the *Bam*HI site and carrying the fragment *Hin*dIII-23) was digested with *Sau*3A and a fragment, carrying the transcription signals and the sequences coding for the first 15 amino acids of the *nos* gene, was subcloned into the *Bgl*II site of the cloning vector pKC7 (ref. 25), resulting—because of the specific sequences involved—in the regeneration of a *Bgl*II site near the initiation codon of *nos* (pLGV13).

Eukaryotic ribosomes seem to use preferentially the AUG most proximal to the 5′ end of a messenger RNA to initiate translation[26,27]. Therefore, pLGV13 was digested with *Bgl*II and treated with the exonuclease *Bal*31 to delete the *nos*-coding sequence remaining in the plasmid. To place a *Bam*HI site at the end of the *Bal*31 deletions, the treated plasmid was digested with *Bam*HI, and the protruding ends were filled in with DNA polymerase. Self-ligation of these molecules followed by screening for the presence of a *Bam*HI site, allowed us to identify a number of pNOSΔ plasmids with *Bam*HI-ended *nos* promoter sequences. Three of these, containing *Sac*II–*Bam*HI fragments of sizes 150–180 base pairs (bp), were chosen and subjected to nucleotide sequence analysis, and shown to have 3′ ends in different positions of the 5′-untranslated sequence of the *nos* gene. A candidate with a deletion removing the *nos*-coding sequence and two nucleotides of the untranslated 5′ leader was chosen for further work (Fig. 1).

The *Sac*II–*Bam*HI fragment from this pNOSΔ plasmid was subsequently cloned back into the plasmid pGV0601 (ref. 24), resulting in a vector which has a *Bam*HI restriction site directly behind the *nos* promoter. This vector contains a deletion of 800 bp of the *nos*-coding sequence, removing 300 of the 400 amino acids of the *nos* protein while retaining the putative signal for polyadenylation and termination of transcription.

An *ocs* 'cassette' fragment

To determine whether the constructed expression vector (pGV2381) would promote expression of coding sequences of other genes, we chose to combine it first with the coding sequence of the octopine synthase gene. Indeed, previous work had established that this enzyme is stable in cells of different plants, and that its activity can readily be assayed; moreover *ocs* activity has never been found to be present in extracts of any of the untransformed plants tested so far[22]. We therefore constructed a 'cassette' fragment consisting primarily of the coding sequence of the octopine synthase, in such a way that this fragment would be convenient for insertion between this and other potential promoter sequences.

The construction of the octopine cassette is outlined in Fig. 2. A *Bam*HI–*Sma*I fragment from pTiB6S3 carrying the complete coding sequence and part of the promoter of the *ocs* gene[14] was cloned into pBR322 cut with *Bam*HI and *Pvu*II. The resulting plasmid, pAGV828, was digested with *Bam*HI and treated with the exonuclease *Bal*31 to remove the 5′-upstream flanking sequences from the structural part of the *ocs* gene, then digested with *Hin*dIII, filled in and self-ligated. A set of plasmids were obtained carrying the complete coding sequences of the *ocs* gene with different deletion end points in the 5′-nontranslated leader sequence. To bracket the *ocs*-

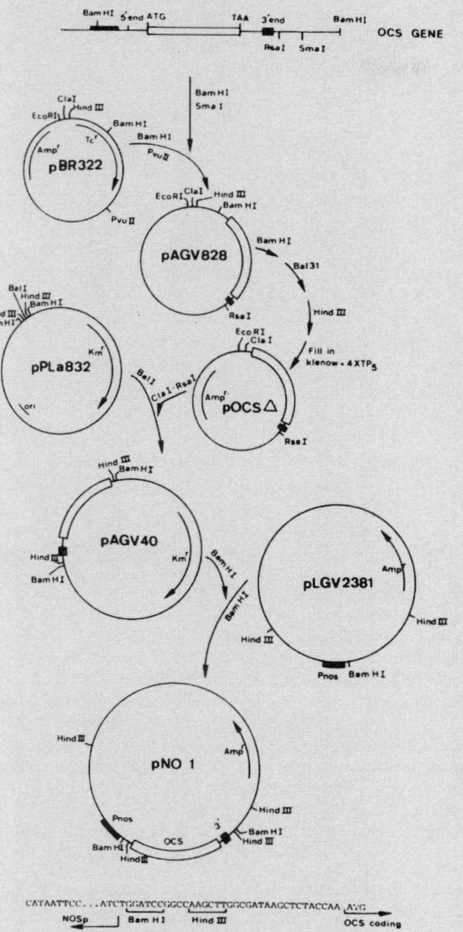

Fig. 2 Construction of a cassette containing the complete *ocs*-coding sequence, and its insertion in plasmid pLGV2381 behind the *nos* promoter. 10 μg of *Bam*HI fragment 17a of the octopine Ti-plasmid B6S3 were digested with *Bam*HI and *Sma*I, the fragment containing the *ocs*-coding sequence was isolated from a 1% agarose gel, and ligated to the large *Bam*HI–*Pvu*II fragment of pBR322; 20 μg of the resulting plasmid, pAGV828, were digested with *Bam*HI, treated with the exonuclease *Bal*31 as described in Fig. 1 legend, subsequently digested with *Hin*dIII, and the ends were filled-in and self-ligated. The sizes of the *Bal*31 deletions were estimated in a 6% polyacrylamide gel. The nucleotide sequences of several candidates were determined, and a candidate having only 7 bp remaining of the 5′-untranslated leader sequence was chosen for further work. To bracket the *ocs* sequence with *Bam*HI sites, the *Cla*I–*Rsa*I fragment was filled in and subcloned into the *Bal*I site of pPLa832. The resulting plasmid pAGV40 was digested with *Bam*HI, the fragment carrying the *ocs* sequence isolated by electroelution from a preparative 1% agarose gel, and ligated to pLGV2381 previously cut with *Bam*HI, and treated with bacterial alkaline phosphatase (BAP). The insertion of the *ocs* sequence in pLGV2381 was obtained in both orientations (pNO1 and pNO2). The junction sequence between the *ocs*-coding sequence and the *nos* promoter in pNO1 is shown. The heavy black line refers to the *ocs* promoter region and the white area to the *ocs*-coding region. Other notations are as for Fig. 1.

coding sequence with *Bam*HI sites, the *Cla*I–*Rsa*I fragment of the resulting pOCSΔ plasmids was isolated, the *Cla*I site was filled in with DNA polymerase and the fragment was cloned into the *Bal*I site of pPLa832 (ref. 28). One of these plasmids (pAGV40) containing only 7 nucleotides of the 5′-untranslated leader sequence and the complete coding sequence, as well as the 3′-untranslated sequence, including the AATAAA signal from the *ocs* gene, was chosen for further work.

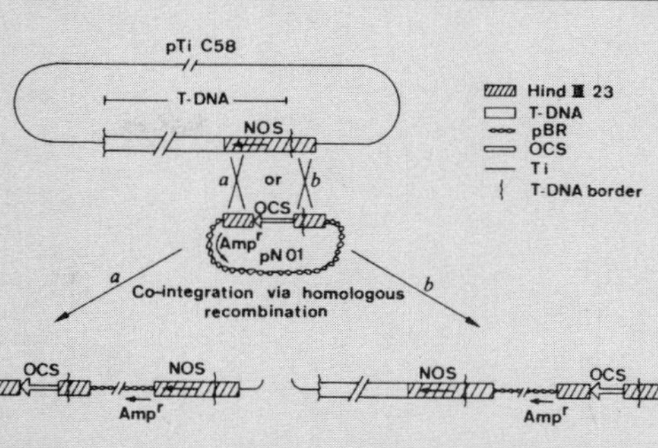

Fig. 3 Possible structure of the co-integrates between pTiC58 and pNO1 in *Agrobacterium*. The helper plasmids that provide the transfer and mobilization functions (R64*drd*11 and pJG28, respectively)[29] were transferred to *E. coli* HB101 (pNO1) by conjugation. The resulting strain carrying all three plasmids were conjugated to *Agrobacterium* (pTiC58) for 16 h on solid Luria broth (LB) medium. Exconjugants carrying the co-integrate pNO1 :: pTiC58 were selected by plating on LB medium containing 100 μg ml^{-1} of ampicillin. The structure of the co-integrates was determined by Southern blotting analysis (data not shown), and the two possibilities for co-integration via homologous recombination between the Ti-plasmid and pNO1 to the 1-kb regions of *Hin*dIII-23 present in pNO1 is shown.

Fig. 4 Detection of octopine in plant tissue transformed with the chimaeric *nos–ocs* gene. Axenic tissue derived from tumours induced with the chimaeric genes were obtained by cutting the tumour into small pieces and subsequently growing *in vitro* on hormone-free Murashige and Skoog (MS) medium in the presence of the antibiotic HR-756 (Hoechst) for several weeks, and then transferring to medium without antibiotic[24]. 10 mg of the bacterium-free tissue was incubated for 12 h in liquid MS medium containing 0.1 M arginine, ground with a glass rod and centrifuged for 3 min at 12,000 r.p.m. 3 μl of the supernatant were spotted onto 3 MM Whatman paper and subjected to electrophoresis[22]. Lane 1 shows the control octopine and nopaline (3 μl from a solution of 1 mg ml^{-1} each), and extracts obtained from tissues infected with *Agrobacterium* carrying pTiB6S3 (lane 2), pTiC58 (lane 3), co-integrates pNO1 :: pTiC58 with structure-type *a* (lanes 4, 5), co-integrates pNO1 :: pTiC58 structure-type *b* (lanes 6, 7), and co-integrates pNO2 :: pTiC58 (lanes 8–10). Lanes 1 and 3–10 show the presence of nopaline, and lanes 2 and 4–7 show the presence of octopine.

A chimaeric *nos–ocs* gene

A chimaeric gene linking the *nos* promoter sequence to the coding sequence of the octopine synthase gene was constructed as shown in Fig. 2. The *Bam*HI fragment of pAGV40 was cloned into the *Bam*HI site of pLGV2381 in both orientations pNO1 and pNO2, respectively (Fig. 2).

Plasmids pNO1 and pNO2 were transferred by conjugation to *Agrobacterium* and recombined with an acceptor pTiC58 plasmid by a method that allows direct mobilization of pBR322 derivatives or any other ColE1-like plasmid to *Agrobacterium*[29]. As the origin of replication of pBR322 is not functional in *Agrobacterium*[30], all the exconjugants carrying the Amp^r marker are derived from homologous recombination between pNO1 and pTiC58. The structure of such a co-integrate is illustrated in Fig. 3. The co-integration of pNO1 and the Ti-plasmid results in a plasmid having two right T-DNA borders[31] in tandem, due to a duplication of the part of *Hin*dIII-23 carried by pNO1. Southern blotting, and DNA hybridization analysis (data not shown) of eight independent co-integrates

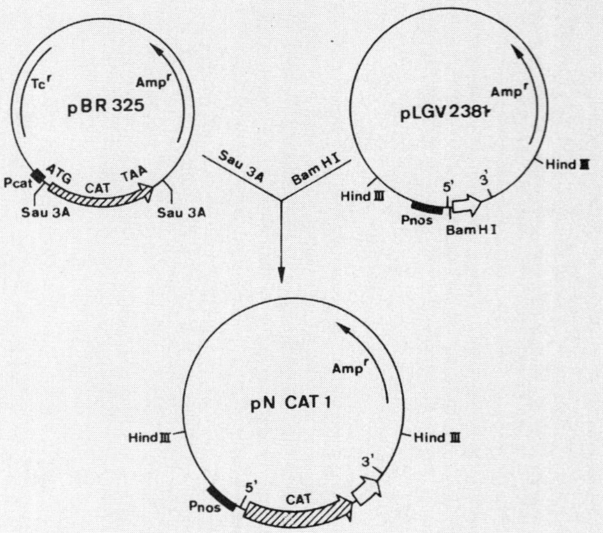

Fig. 5 Construction of a chimaeric gene that directs the expression of the chloramphenicol acetyltransferase in plant cells. 20 μg of pBR325 were digested with *Sau*3A and the fragment containing the complete coding sequence of the *cat* gene was isolated by electroelution from a 2% agarose gel. The isolated fragment was cloned into the *Bam*HI site of pLGV2381, previously treated with BAP. The orientation of the insert was determined by *Eco*RI and *Hin*dIII digestion. pNCAT1 carries the *cat* gene in the proper orientation to be transcribed from the *nos* promoter, and pNCAT2 carries the gene in opposite orientation. The small blackened box represents the *cat* promoter, the long blackened box the *nos* promoter, and the cross-hatched box represents the structural part of the *cat* gene.

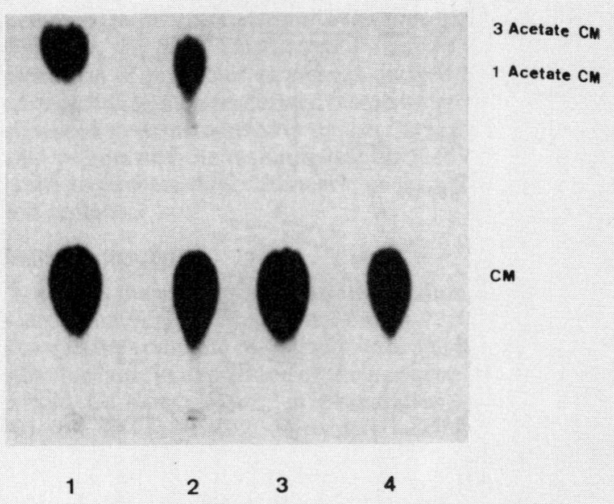

Fig. 6 Assay of *cat* activity in plant cells. 100 mg of tumour tissue was extracted by grinding manually with a glass rod in the presence of an equal volume of buffer containing 0.5 M sucrose, 0.25 M Tris-HCl, 1% ascorbic acid, 0.5 mM leupeptine (Sigma), 10 mM EDTA, 1% cysteine-HCl, *p*H 7.5, 0.5 mM acetyl CoA and 1 μCi ^{14}C-chloramphenicol (50 μCi mmol^{-1}, NEN), and the mix was incubated at 37 °C for 30 min (ref. 34), followed by centrifugation (5 min at 12,000 r.p.m.) and removal of the supernatant for analysis. A sonicated extract of *E. coli* containing pBR325 (ref. 32) was used as a positive control. Plant and bacterial reaction mixes were extracted with ethyl acetate, concentrated by evaporation, and subjected to ascendent chromatography in a silica gel thin-layer plate with chloroform/methanol (95:5) as eluant. The autoradiogram shown was exposed for 48 h at room temperature. Lane 1 shows the bacterial enzyme control, lane 2 shows the *cat* activity found in cells transformed with *Agrobacterium* containing pNCAT1::pTiC58. No *cat* activity was found in tissue transformed with the strain carrying pNCAT2::pTiC58 (lane 3), even after much longer exposure (data not shown). Lane 4 shows the reconstruction experiment where 10^9 agrobacteria carrying pTiC58::pNCAT1 were added to nopaline tumour tissue (induced by agrobacteria carrying wild-type pTiC58) before extraction.

revealed that five had the orientation *ocs*–border–*nos*–border (*a*), and three the orientation *nos*–border–*ocs*–border (*b*).

Tumours were induced by infecting decapitated tobacco seedlings with eight independently isolated *Agrobacterium* strains carrying the co-integrates pNO1::pTiC58, and four carrying co-integrates pNO2::pTiC58. Tumours were subsequently cultivated *in vitro* in hormone-free medium[24] and 10 mg of bacterium-free tissue were used to test for the presence of octopine and nopaline. As shown in Fig. 4, octopine was present in all the tumours induced with strains carrying the co-integrate pNO1::pTiC58 in amounts slightly higher than those found in tumours induced by the octopine wild-type Ti-plasmid pTiB6S3. Furthermore, the presence of both nopaline and octopine in tumours transformed with the pNO1::pTiC58 co-integrates of either type *a* or type *b* (see Fig. 3) suggests that both genes are transferred in spite of their position on either side of the internal right border of the T-DNA. This might also suggest that either the most external borders are preferentially used to transfer the DNA positioned between them, or that the presence of two borders flanking any DNA sequence may create a new 'T-DNA' that is independently transferred to the plant genome. As the tumour tissue (used for these experiments) was not cloned, the results may reflect several independent integration events, and further experiments are needed to clarify the structure of the T-DNA present in tumour tissue transformed with the co-integrates used in this study. On the other hand, tumours induced with pNO2::pTiC58, which contains the octopine synthase in opposite orientation to the transcription direction of the *nos* promoter, while positive for nopaline, did not contain octopine.

These results indicate that the *Hin*dIII–*Bam*HI fragment of pLGV2381 contains a functional *nos* promoter able to promote transcription of the octopine synthase coding sequence, and, therefore, presumably of any other coding sequence properly oriented downstream from this region.

A chimaeric *nos–cat* gene

To test further the applicability of the constructed expression vector, the *nos* promoter was combined with the coding sequence of chloramphenicol acetyltransferase (EC 2.3.1.28) (*cat*) from pBR325 (ref. 32). This enzyme of bacterial origin provides a very sensitive and simple assay system, and no similar endogenous activity is present in plant cells. Furthermore, this enzyme has been shown to be expressed in yeast and mammalian cells[33,34].

A *Sau*3A fragment from pBR325, containing the complete coding sequence of the *cat* gene, and 56 bp of the untranslated leader sequence where no other ATG than the initiation codon is present[35,36], was cloned in two orientations into the *Bam*HI site of pLGV2381 as outlined in Fig. 5. Both pNCAT1 and pNCAT2 (opposite orientation) were mobilized from *Escherichia coli* to *Agrobacterium* as described for pNO1 (Fig. 2), and exconjugants containing pTiC58::pNCAT1 and pTiC58::pNCAT2 co-integrates were selected on ampicillin-containing media. Southern blotting DNA hybridizations (data not shown) demonstrated that strain GVCAT1 contained a pTiC58::pNCAT1 co-integrate, and strain GVCAT2 a pTiC58::pNCAT2 co-integrate, which both resulted from a cross-over similar to situation *a* illustrated in Fig. 3.

Strains GVCAT1 and GVCAT2 were used to induce tumours on tobacco seedlings and the resulting tumour tissues were grown in axenic cultures[24]. Extracts from such axenic tissue cultures were tested for *cat* activity. As shown in Fig. 6, the tissue derived from tumours induced by the strain GVCAT1, whose T-DNA contains the *cat* gene-coding sequence properly oriented relative to the *nos* promoter, did indeed contain *cat* activity, as seen by the presence of the same chloramphenicol derivatives as in the control incubated with an extract of *E. coli* (pBR325). The tumour line induced with GVCAT2, whose *cat* gene-coding sequence is in the wrong orientation, was negative. The fact that tumours induced by GVCAT1 were shown to contain nopaline (data not shown) as well as the observed *cat* activity, suggests that the external border was involved in the integration of the T-region of the co-integrate pTiC58::pNCAT1 in the plant DNA, as discussed above for the *nos–ocs* chimaeric gene.

Agrobacterium strains have been found to contain some endogenous *cat* activity[37]. To ensure that bacterial *cat* activity cannot account for the observed activity in tobacco tumours induced by GVCAT1, we took particular care to demonstrate that the tumour lines were completely devoid of any surviving bacteria before extraction. Furthermore, wild-type C58 tobacco crown gall tissue was mixed with 10^9 cells from a fresh culture of GVCAT1, and the extraction procedure described in Fig. 6 legend was repeated. No *cat* activity was observed in these circumstances, demonstrating that the very gentle mechanical extraction procedure used to open the plant cells could not lead to a contamination of the plant cell extract with bacterial *cat* activity.

Discussion

The successful expression of the octopine synthase and chloramphenicol acetyltransferase coding sequence as chimaeric genes linked to the transcription-promoting sequence of the nopaline synthase gene demonstrates that this approach can be used to transfer and express the coding sequences of foreign genes in plants. The *nos* promoter is particularly useful because of its wide host range: it is functional in calli derived from all dicotyledonous plants so far tested, and it appears to be expressed in most tissues of regenerated *nos*-containing plants[38].

The easy and sensitive assay for the presence of octopine in plant cell extracts makes the *ocs* cassette a very versatile tool for studying the structure–function relationships of various promoter sequences in plants, especially with regard to their role in the regulation of gene expression and specificity.

Another important application is the construction of general selectable marker genes for plants. Using the *nos* promoter expression vector (pLGV2381), we constructed two such chimaeric genes, one containing the coding sequence of the neomycin phosphotransferase of Tn*5* (ref. 39) and the other the methotrexate-resistant dihydrofolate reductase of plasmid R67 (ref. 40). Both these genes were shown to be expressed in tobacco, and able to be used as selectable marker genes[41], thus confirming the general applicability of the constructions reported here. On the basis of these results it can be expected that gene-splicing techniques, so far applied to bacterial and mammalian cells, can be used for plant cells also. Moreover, the availability of simple and rapid techniques for the regeneration of plants from single cells, makes plants an especially interesting system for the study of development and differentiation, as well as for genetic engineering.

We thank Paul Tenning, Ms June Simpson and Jef Seurinck for technical assistance, and Albert Verstraete and Ms Martine De Cock for help in preparing this manuscript. This research was supported by grants from the 'Kankerfonds van de Algemene Spaar- en Lijfrentekas', from the 'Instituut tot Aanmoediging van het Wetenschappelijk Onderzoek in Nijverheid en Landbouw' (IWONL3849A), from the Services of the Prime Minister (OOA/12052179), and from the 'Fonds voor Geneeskundig Wetenschappelijk Onderzoek' (FGWO3.0001.82) to J.S. and M.V.M. L.H.-E. is indebted to CONACYT México for a PhD fellowship.

Received 8 February; accepted 30 March 1983.

1. Bevan, M. W. & Chilton, M.-D. *A. Rev. Genet.* **16,** 357–384 (1982).
2. Ream, L. & Gordon, M. P. *Science* **218,** 854–858 (1982).
3. Van Montagu, M. & Schell, J. *Curr. Topics Microbiol. Immun.* **96,** 237–245 (1981).
4. Schell, J. *et al. Cell Fusion* (Raven, New York, in the press).
5. Hernalsteens, J. P. *et al. Nature* **287,** 654–656 (1980).
6. Drummond, M. H., Gordon, M. P., Nester, E. W. & Chilton, M.-D. *Nature* **269,** 535–536 (1977).
7. Willmitzer, L., Simons, G. & Schell, J. *EMBO J.* **1,** 139–146 (1982).
8. Willmitzer, L. *et al. Cell* **32,** 1045–1056 (1983).
9. Bevan, M. W. & Chilton, M.-D. *J. molec. appl. Genet.* **1,** 539–546 (1982).
10. Gelvin, S. B., Gordon, M. P., Nester, E. W. & Aronson, A. I. *Plasmid* **6,** 17–29 (1981).
11. Garfinkel, D. J. *et al. Cell* **27,** 143–153 (1981).
12. Leemans, J. *et al. EMBO J.* **1,** 147–152 (1982).
13. Joos, H. *et al. Cell* **32,** 1057–1067 (1983).
14. De Greve, H. *et al. J. molec. appl. Genet.* **1,** 499–512 (1982).
15. Depicker, A., Stachel, S., Dhaese, P., Zambryski, P. & Goodman, H. M. *J. molec. appl. Genet.* **1,** 561–574 (1982).
16. Bevan, M., Barnes, W. M. & Chilton, M.-D. *Nucleic Acids Res.* **11,** 369–385 (1983).
17. Breathnach, R. & Chambon, P. *A. Rev. Biochem.* **50,** 349–383 (1981).
18. Fitzgerald, M. & Shenk, T. *Cell* **24,** 251–260 (1981).
19. Dhaese, P. *et al. EMBO J.* **2,** 419–426 (1983).
20. De Greve, H. *et al. Nature* **300,** 752–755 (1982).
21. Otten, L. *et al. Molec. gen. Genet.* **183,** 209–213 (1981).
22. Otten, L. *Pl. Sci. Lett.* **25,** 15–27 (1982).
23. Aerts, M., Jacobs, M., Hernalsteens, J. P., Van Montagu, M. & Schell, J. *Pl. Sci. Lett.* **17,** 43–50 (1979).
24. Leemans, J. *et al. J. molec. appl. Genet.* **1,** 149–164 (1981).
25. Rao, R. N. & Rogers, S. G. *Gene* **7,** 79–82 (1979).
26. Kozak, M. *Nucleic Acids Res.* **9,** 5233–5252 (1981).
27. Lomedico, P. T. & McAndrew, S. J. *Nature* **299,** 221–226 (1982).
28. Remaut, E., Stanssens, P. & Fiers, W. *Gene* **15,** 81–93 (1981).
29. Van Haute, E. *et al. EMBO J.* **2,** 411–418 (1983).
30. Leemans, J. *et al. Gene* **19,** 361–364 (1982).
31. Zambryski, P., Depicker, A., Kruger, K. Goodman, H. M. *J. molec. appl. Genet.* **1,** 361–370 (1982).
32. Bolivar, F. *Gene* **4,** 121–136 (1978).
33. Cohen, J. *et al. Proc. natn. Acad. Sci. U.S.A.* **77,** 1078–1082 (1980).
34. Gorman, C. M., Moffat, L. F. & Howard, B. H. *Molec. cell. Biol.* **2,** 1044–1051 (1982).
35. Marcoli, R., Iida, S. & Bickle, T. A. *FEBS Lett.* **110,** 11–14 (1980).
36. Prentki, P., Karch, F., Iida, S. & Meyer, J. *Gene* **14,** 289–299 (1981).
37. Zaidenzaig, Y., Fitton, J. E., Packman, L. C. & Shaw, W. V. *Eur. J. Biochem.* **100,** 609–618 (1979).
38. Wöstemeyer, A. *et al. Genetic Engineering in Eukaryotes* (ed. Lurquin, P.) (Plenum, New York, in the press).
39. Beck, E., Ludwing, G., Averswald, E. A., Reiss, B. & Schaller, H. *Gene* **19,** 329–336 (1982).
40. Fling, M. E. & Elwell, L. P. *J. Bact.* **141,** 779–785 (1980).
41. Herrera-Estrella, L. *et al. EMBO J.* (in the press).
42. Maxam, A. M. & Gilbert, W. *Meth. Enzym.* **65,** 499–559 (1980).

Proc. Natl. Acad. Sci. USA
Vol. 80, pp. 4803–4807, August 1983
Genetics

Expression of bacterial genes in plant cells

(plant protoplasts/transformation/foreign DNA/antibiotic resistance/selectable markers)

ROBERT T. FRALEY, STEPHEN G. ROGERS, ROBERT B. HORSCH, PATRICIA R. SANDERS, JEFFERY S. FLICK, STEVEN P. ADAMS, MICHAEL L. BITTNER, LESLIE A. BRAND, CYNTHIA L. FINK, JOYCE S. FRY, GERALD R. GALLUPPI, SARAH B. GOLDBERG, NANCY L. HOFFMANN, AND SHERRY C. WOO

Monsanto Company, 800 North Lindbergh Boulevard, St. Louis, Missouri 63167

Communicated by Howard A. Schneiderman, April 25, 1983

ABSTRACT **Chimeric bacterial genes conferring resistance to aminoglycoside antibiotics have been inserted into the *Agrobacterium tumefaciens* tumor-inducing (Ti) plasmid and introduced into plant cells by *in vitro* transformation techniques. The chimeric genes contain the nopaline synthase 5′ and 3′ regulatory regions joined to the genes for neomycin phosphotransferase type I or type II. The chimeric genes were cloned into an intermediate vector, pMON120, and inserted into pTiB6S3 by recombination and then introduced into petunia and tobacco cells by cocultivating *A. tumefaciens* cells with protoplast-derived cells. Southern hybridization was used to confirm the presence of the chimeric genes in the transformed plant tissues. Expression of the chimeric genes was determined by the ability of the transformed cells to proliferate on medium containing normally inhibitory levels of kanamycin (50 μg/ml) or other aminoglycoside antibiotics. Plant cells transformed by wild-type pTiB6S3 or derivatives carrying the bacterial neomycin phosphotransferase genes with their own promoters failed to grow under these conditions. The significance of these results for plant genetic engineering is discussed.**

The transformation of plant cells by virulent strains of *Agrobacterium tumefaciens* has been studied extensively by several laboratories (1–4). A small fragment of the tumor-inducing (Ti) plasmid, called transferred DNA (T-DNA), is known to be transferred to and stably incorporated in the nuclear DNA of transformed plant cells (5–7). The T-DNA is actively transcribed in plant cells (8–10) and specific gene products have been shown to be responsible for the observed phytohormone-independent growth characteristics (11, 12) and novel metabolic capacities (13) exhibited by crown gall tumor cells. The transfer and insertion of T-DNA into plant DNA is thought to involve repeated nucleotide sequences present near the T-DNA "borders" (14, 15) as well as other genes of unknown function located in specific virulence regions outside of T-DNA (16, 17).

In spite of our considerable understanding of the *A. tumefaciens*-Ti plasmid system, several problems remain which limit its use as a vector for genetically modifying higher plants. Because of the high levels of phytohormones produced by crown gall tumor cells (18) they have generally proven recalcitrant to attempts to induce regeneration into whole plants (19, 20). Exceptions to this are cases in which, as a result of aberrant integration or spontaneous deletion events, transformed cells have lost all or part of the Ti plasmid tumor genes and can now be regenerated (21, 22). In addition, transformation of cells by weakly virulent, mutant Ti plasmids (23) and transformation by root-inducing (Ri) plasmids (24, 25) have been shown to produce callus that can be regenerated into whole plants. However, these plants often display morphological aberrations and may retain certain tumorous properties (26). Another obstacle has been the failure to obtain expression from a variety of foreign genes that have been introduced into plants (23, 27). Reasons for this include the fact that, up to now, most studies have utilized either heterologous genes from bacteria, fungi, and mammalian cells whose regulatory regions may not be recognized by the plant RNA polymerases or highly regulated plant genes which are normally expressed in specialized tissues and which may not be transcribed in undifferentiated crown gall tumor tissue.

To bypass the dependence on tumor genes for identifying transformed plant cells and to overcome the barriers to gene expression in plants, chimeric genes that function as dominant selectable markers have been assembled. These contain the neomycin phosphotransferase (NPTase) coding sequences from the bacterial transposons Tn*5* (type II) or Tn*601* (type I) joined to the 5′ and 3′ regulatory regions of the nopaline synthase gene from the Ti plasmid. This paper describes the construction of these chimeric genes and their introduction and expression in plant cells.

MATERIALS AND METHODS

DNA Preparation. Plasmid pBR322 and its derivatives or M13 replicative form DNAs were purified by using either a Triton-X-100/CsCl procedure (28) or a large-scale alkaline lysis procedure (29), followed by purification on hydroxylapatite (30).

DNA fragments were isolated by electroelution into dialysis bags after polyacrylamide gel electrophoresis and band excision or by adsorption onto NA-45 DEAE membrane (Schleicher & Schuell) after agarose gel electrophoresis (31).

The *Bam*HI synthetic DNA linkers (5′ C-C-G-G-A-T-C-C-G-G) were purchased from Collaborative Research (Waltham, MA). Other synthetic DNAs were synthesized by using a modification of the phosphite procedure (32).

Enzymes. All restriction endonucleases and the large Klenow fragment of DNA polymerase I were obtained from New England Biolabs or Bethesda Research Laboratories and were used according to the instructions of the supplier. Phage T4 DNA ligase was prepared as in ref. 33. DNA fragment assembly was carried out as described (31).

Transformation of *Escherichia coli* Cells. Plasmid DNAs were introduced into *E. coli* cells by using $CaCl_2$-treated or $RuCl_2$-treated cells (31). The recipient *E. coli* K-12 strains were SR200 = C600 *thr pro recA56 hsdR*(r^-m^+) (34); LE392 = ED8554 *hsdR* (r^-m^+) (31); SR20 = GM42 = *his dam-3* (35); and the M13

The publication costs of this article were defrayed in part by page charge payment. This article must therefore be hereby marked "*advertisement*" in accordance with 18 U.S.C. §1734 solely to indicate this fact.

Abbreviations: bp, base pair(s); kb, kilobase(s); NPTase I and NPTase II, neomycin phosphotransferase, types I and II, respectively; Ti plasmid, tumor-inducing plasmid; T-DNA, transferred DNA; Ri plasmid, root-inducing plasmid.

phage host, JM101 (36). Cells carrying recombinant plasmids were selected or grown (or both) on Luria medium plates or broth at 37°C containing appropriate antibiotics (ampicillin, 200 μg/ml; spectinomycin, 50 μg/ml; and kanamycin, 40 μg/ml).

Introduction of pMON120 Derivatives into *A. tumefaciens*. Plasmid pMON120 or its derivatives were transferred to a chloramhpenicol-resistant *A. tumefaciens* strain GV3111 = C58C1 Cm^R carrying pTiB6S3tra^c (37) by using a triparental plate mating procedure (ref. 38; R. Riedel, personal communication). Briefly, 0.2 ml of a fresh overnight culture of LE392 carrying pMON120 or its derivative was mixed with 0.2 ml of an overnight culture of HB101 (31) carrying the pRK2013 (38) plasmid and 0.2 ml of an overnight culture of GV3111 cells. The mixture of cells was spread on an LB plate and incubated for 16–24 hr at 30°C to allow plasmid transfer and recombination. The cells were resuspended in 3 ml of 10 mM $MgSO_4$ and a 0.2-ml aliquot was then spread on an LB plate containing 25 μg of chloramphenicol per ml and 100 μg each of spectinomycin and streptomycin per ml to select *A. tumefaciens* carrying the pMON120 derivatives. After incubation for 48 hr at 30°C, ≈10 colonies per plate were obtained. Control matings between HB101/pRK2013 cells and GV3111 cells never gave rise to colonies after this selection. Typically, one colony was chosen and grown at 30°C in LB medium containing chloramphenicol, spectinomycin, and streptomycin at the same concentrations given above.

Protoplast Isolation and Culture. "Mitchell" petunia plants were grown in environmental chambers under fluorescent and incandescent illumination (≈5,000 lux, 12 hr/day) at 21°C in a 50:50 mixture of vermiculite and Pro-mix BX (Premier Brands, Quebec, PQ, Canada). Leaves were surface sterilized, cut into 2-mm strips, and enzymatically digested as described (39). The resulting protoplasts were purified by passage through stainless steel meshes and by density floatation as described (39). The protoplasts were plated in tissue culture flasks (T75, Falcon; 6 ml per flask) at a cell density of 10^5 cells per ml in culture medium [MS salts (GIBCO), B-5 vitamins, 3% (wt/vol) sucrose, 9% (wt/vol) mannitol, 1 μg of 2,4-D per ml, and 0.5 μg of benzyl adenine per ml, pH 5.7].

Cocultivation of *A. tumefaciens* Cells with Plant Protoplasts. On day 2 after protoplast isolation, aliquots (10–50 μl) of an overnight culture of *A. tumefaciens* cells were added to each flask (final bacterial cell density = 10^8 cells per ml) and cocultivation with plant cells was carried out for 24–30 hr essentially as described (40). On day 3, 6 ml of culture medium (lacking phytohormones and mannitol) containing carbenicillin (1.5 mg/ml) was added to each flask (final concentration = 500 μg/ml) to prevent further bacterial growth. On day 4, an additional 6 ml of the above medium (containing carbenicillin at 500 μg/ml) was added. On day 6, 0.5 ml of the cell mixture was transferred to and spread in a thin layer on the surface of double-filter feeder plates (41). These consisted of agar medium (MS salts, B-5 vitamins, 3% sucrose, 3% mannitol, 0.1 μg of indole acetic acid per ml, and 500 μg of carbenicillin per ml at pH 5.7), a layer of *Nicotinia tabacum* suspension cells, a tight fitting 8.5-cm Whatman filter paper disc (guard disc), and a 7.0-cm Whatman filter paper disc (transfer disc). After 7–10 days, microcolonies (≈0.5 mm) were observable on the feeder plates and the transfer disc was removed and placed on selection medium (MS salts, B-5 vitamins, 3% sucrose, 500 μg of carbenicillin per ml at pH 5.7) lacking phytohormones. Within 2 wk, hormone-independent transformants could be readily distinguished as green colonies against a background of dying, brown nontransformed cells. The transformation frequency in these experiments was $\approx 10^{-1}$. The hormone-independent transformants were then transferred to medium (MS salts, B-5 vitamins, 3% sucrose, 500 μg of carbenicillin per ml at pH 5.7) containing kanamycin (50 μg/ml).

Analysis of Transformants. Octopine and nopaline synthase activities were determined as in ref. 42 with the substitution of [^{14}C]arginine (Amersham, 0.5 μCi/2.5-μl assay; 1 Ci = 3.7 × 10^{10} Bq) for the unlabeled arginine in the assay buffer. The conditions for electrophoresis were as described (42) and the resulting electrophoretograms were exposed to x-ray film (Kodak, XAR-5) for 16–24 hr. The positions of octopine, nopaline, and arginine were established by their comigration with authentic standards.

Callus for NPTase assays were frozen in liquid N_2 and extracted by using a mortar and pestle in a minimal volume of buffer (0.2 M Tris·HCl/2 mM EDTA/7.5% polyvinylpolypyrrolidone). The crude extract was clarified by centrifugation (Eppendorf; Brinkmann) and assays were performed as described (43).

RESULTS

NPTase coding sequences were used in the initial chimeric gene constructions described in this study because plant cells were determined to be sensitive to various aminoglycoside antibiotics (unpublished data), and the expression of NPTase in yeast (43) and mammalian cells (44, 45) has been previously shown to confer resistance to the antiobiotic, G418. The nopaline synthase gene promoter and 3′-nontranslated regions were selected because this gene has been well characterized (9, 46) and it is known to be expressed constitutively in most plant tissues transformed with the *A. tumefaciens* Ti plasmid (47).

Construction of Chimeric Genes. The nopaline synthase promoter region, obtained on a 350-base-pair (bp) *Sau*3A fragment from the *Hin*dIII-23 fragment of pTiT37 (Fig. 1; ref. 46), was engineered to remove the entire nopaline synthase coding sequence. The resulting promoter fragment that extends from base −264 to base 35 of the nopaline synthase sequence (46) was positioned next to the *Bgl* II site located just outside the NPTase II coding sequence (49). In addition, a 260-bp *Mbo* I fragment, extending from base 1,297 to base 1,554 of the published nopaline synthase sequence (46), was isolated from the *Hin*dIII-23 fragment. This *Mbo* I fragment contains the nopaline synthase 3′-nontranslated region and polyadenylylation site. This fragment was ligated together with the *Eco*RI–*Bam*HI fragment that contained the nopaline synthase promoter and NPTase II structural gene to yield the intact chimeric gene on a 1.5-kilobase (kb) *Eco*RI fragment (Fig. 1). A second chimeric gene, containing the nopaline synthase promoter and 3′-nontranslated region joined to the NPTase I coding sequence (Fig. 2), was constructed in a similar fashion. As controls, plasmids were constructed that contained an intact NPTase II promoter and structural sequence with the nopaline synthase 3′-nontranslated region (pMON139 and pMON140; Fig. 2).

Introduction of Chimeric Genes into the Ti Plasmid. The vector pMON120 used for the transfer of the chimeric genes into *A. tumefaciens* cells is shown in Fig. 2. Its essential features include (*i*) a segment of pBR322 DNA for replication in *E. coli*, (*ii*) a segment from pTiT37 that contains a functional nopaline synthase gene to facilitate the rapid identification of transformants, (*iii*) a segment of Tn7 carrying the spectinomycin/streptomycin-resistance determinant for selection in *A. tumefaciens*, (*iv*) a DNA segment obtained from the pTiA6 T-DNA fragment *Hin*dIII-18c (see T-DNA map, ref. 11), which is included to provide homology for recombination with a resident octopine-type Ti plasmid in *A. tumefaciens*, and (*v*) unique restriction sites (*Eco*RI and *Hin*dIII) for insertion of the chimeric genes. The pMON120 plasmid and derivatives were in-

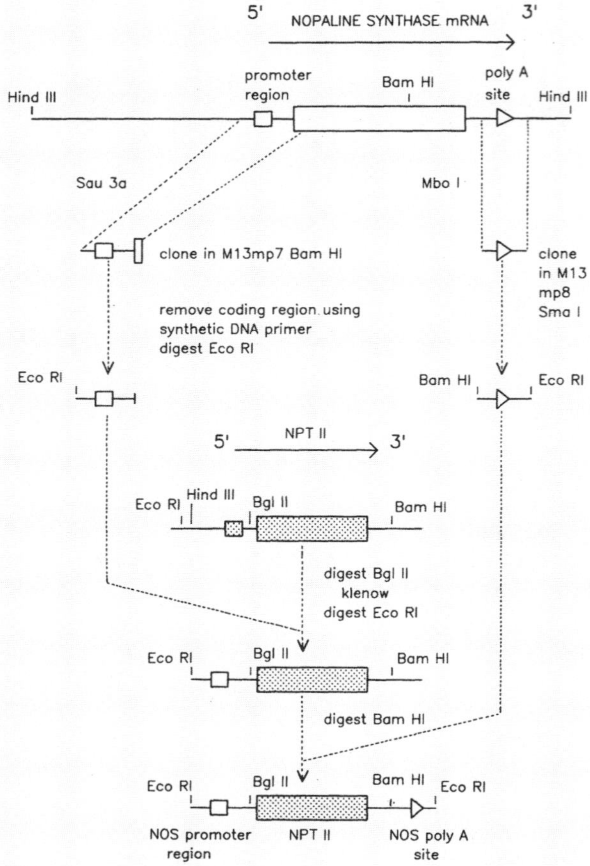

FIG. 1. Isolation and assembly of a chimeric gene containing nopaline synthase (NOS) promoter-NPTase II (NPT II) coding sequence-nopaline synthase 3′-nontranslated region. The nopaline synthase promoter was isolated on a 350-bp *Sau*3A fragment that also contained the first 44 bp of the nopaline synthase coding sequence. The sense strand of this fragment was cloned into the *Bam*HI site in M13 mp7 (36), the 44 bp was removed by using a modification of a published synthetic primer procedure (48) with a primer complementary to bases 22–35 of the published nopaline synthase sequence (46), and a 308-bp promoter fragment was obtained after digestion with *Eco*RI. The flush-end of the promoter fragment was joined to a 1-kb *Bgl* II–*Bam*HI fragment carrying the NPTase II coding sequence (a *Bam*HI linker had been inserted at the *Sma* I site) at the filled-in *Bgl* II site (49). This fusion regenerates the *Bgl* II site. The chimeric gene was completed by the addition of a 260-bp *Mbo* I fragment that contained the nopaline synthase 3′-nontranslated region. This fragment, which contains a polyadenylylation signal (46), was converted to a flush-ended fragment with Klenow polymerase and cloned into the *Sma* I site of a M13 mp8 (50) to introduce *Bam*HI and *Eco*RI sites at the 5′ and 3′ ends, respectively. The resulting 280-bp fragment was joined to the 1,300-bp *Eco*RI–*Bam*HI nopaline synthase promoter-NPTase II coding sequence fragment to generate the complete chimeric gene.

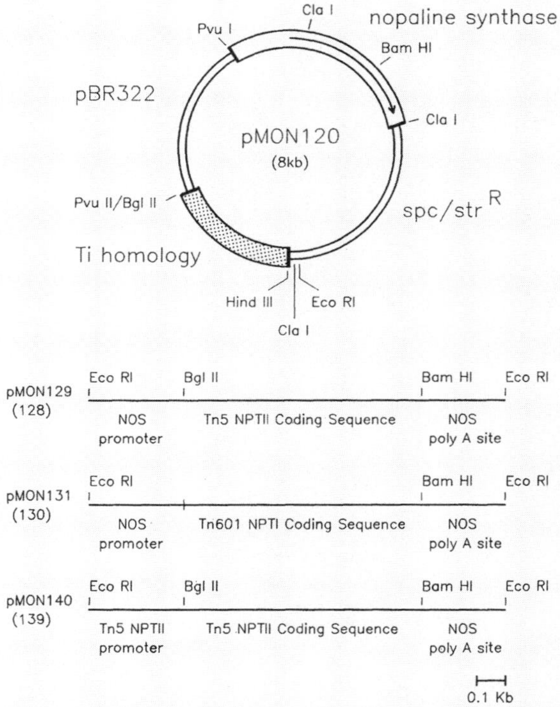

FIG. 2. Structures of the pMON120 intermediate vector and chimeric genes introduced into plant cells. Plasmid pMON120 contains the following segments of DNA: the 1.7-kb pBR322 *Pvu* II to *Pvu* I fragment that carries the origin of replication and *bom* site (51), a 2.2-kb partial *Cla* I to *Pvu* I fragment of pTiT37 DNA that encodes an intact nopaline synthase (NOS) gene, a 2.7-kb *Cla* I-*Eco*RI fragment of Tn7 (37) DNA carrying the determinant for spectinomycin/streptomycin resistance, and the 1.6-kb *Hin*dIII–*Bgl* II fragment from the *Hin*dIII-18c fragment of the pTiA6 plasmid. This T-DNA fragment is known to specify two transcripts that are not essential for tumorous growth (8, 12). At the bottom are three chimeric genes inserted at the unique *Eco*RI site of pMON120. The chimeric nopaline synthase-NPTase II-nopaline synthase gene was inserted to give pMON129 and pMON128. In all of these examples, the first plasmid carries the inserted gene as it is drawn in the figure. The second plasmid carries the insert in the opposite orientation to that drawn. Plasmids pMON131 and pMON130 carry a chimeric nopaline synthase-NPTase I-nopaline synthase gene. The final chimeric gene is carried in plasmids pMON140 and pMON139. The bacterial NPTase II promoter and coding sequence have been joined to the nopaline synthase 3′-nontranslated region.

troduced into *A. tumefaciens* as described in *Materials and Methods.*

Selection of Kanamycin-Resistant Petunia Transformants. Several hundred hormone-independent calli (1–2 mm in diameter) obtained from cocultivation experiments with *A. tumefaciens* strains carrying pTiB6S3::pMON120 (or derivatives) recombinant plasmids were pooled and analyzed by DNA blot hybridization for the presence of the chimeric genes (Fig. 3). The results confirm the presence of the expected 1.6-kb *Eco*RI fragment, which carries the chimeric nopaline synthase-NPTase II-nopaline synthase gene in pMON128 and pMON129 transformants, and the control NPTase II-NPTase II-nopaline synthase construct in pMON139 and pMON140 transformants (Fig. 3*a*).

Similar results were obtained for pMON130 and pMON131 transformants, which contain the chimeric nopaline synthase-NPTase I-nopaline synthase gene on a 1.5-kb *Eco*RI fragment (Fig. 3*b*). No hybridization with either the Tn*5*- or Tn*601*-specific probe was detected in transformants containing only the pMON120 vector. Other minor bands of hybridization are present in the pMON129 and pMON140 transformants; these may be attributable to partial digestion or aberrant integration events and their assignment awaits further analysis of clonal tissue. Blot hybridization analysis of DNA from these transformants using T-DNA-specific probes confirmed the presence of the expected internal T-DNA fragments in the transformed tissues and ruled out any possibility that the plant tissue was contaminated by *A. tumefaciens* cells (data not shown).

Other transformed, hormone-independent calli from these experiments were transferred to agar medium containing kana-

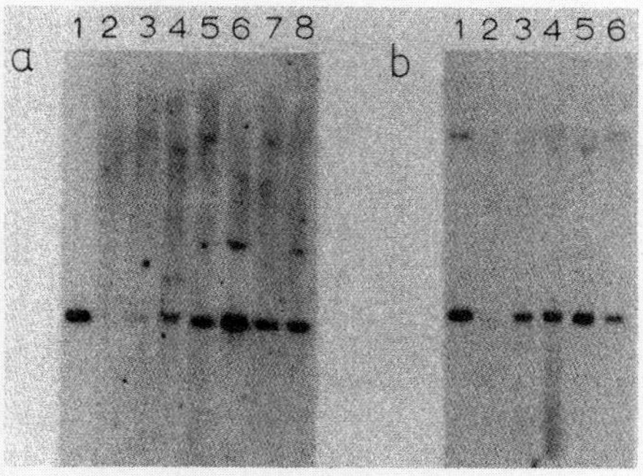

FIG. 3. DNA blot hybridization analysis of *in vitro* transformants. Several hundred hormone-independent *in vitro* transformants from each experiment were pooled and total DNA was extracted (52). The DNAs were digested with *Eco*RI and the fragments were separated by electrophoresis and transferred to nitrocellulose (53). (*a*) Hybridization with NPTase II-specific probe. A gel-purified 3.3-kb *Hin*dIII fragment from Tn*5* (54) was used as probe. Lane 1, pMON128::Ti plasmid marker; lane 2, pMON120 transformants; lane 3, pMON139 transformants; lane 4, pMON140 transformants; lane 5, pMON128 transformants; lane 6, pMON129 transformants; lane 7, pMON128 transformants; and lane 8, pMON129 transformants. Lanes 2–6 represent transformants selected for hormone-independent growth prior to scoring for kanamycin resistance; lanes 7 and 8 represent transformants selected only for kanamycin resistance on medium containing phytohormones. (*b*) Hybridization with NPTase I-specific probe. A gel-purified 1.2-kb *Ava* II fragment from Tn*601* (55) was used as a probe. Lane 1, pMON130::Ti plasmid marker; lane 2, pMON120 transformants; lane 3, pMON130 transformants; lane 4, pMON131 transformants; lane 5, pMON130 transformants; and lane 6, pMON131 transformants. Lanes 2–4 represent transformants selected for hormone-independent growth prior to scoring for kanamycin resistance; lanes 5 and 6 represent transformants selected only for kanamycin resistance on medium containing phytohormones.

mycin (50 μg/ml) and these were scored after 2–3 wk for resistance to the antibiotic. All transformants obtained from experiments utilizing pMON120, pMON139, or pMON140 failed to grow on medium supplemented with kanamycin, whereas all the transformants from experiments utilizing pMON128, pMON129, pMON130, or pMON131 grew on medium containing the antibiotic at rates comparable to growth on normal medium. A quantitative assessment of the level of resistance conferred by the chimeric genes is shown for pMON120,

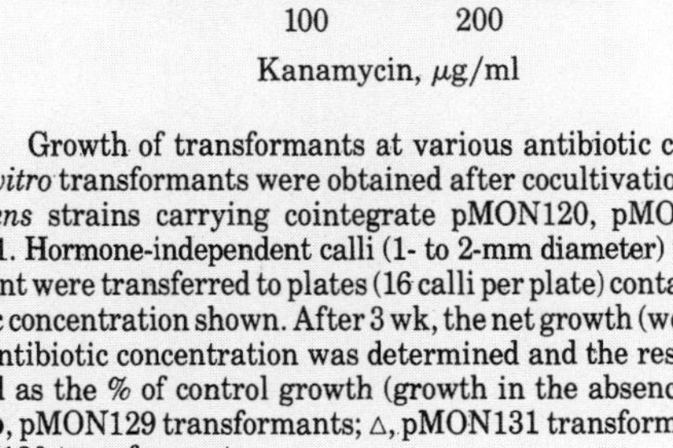

FIG. 4. Growth of transformants at various antibiotic concentrations. *In vitro* transformants were obtained after cocultivation with *A. tumefaciens* strains carrying cointegrate pMON120, pMON129, or pMON131. Hormone-independent calli (1- to 2-mm diameter) from each experiment were transferred to plates (16 calli per plate) containing the antibiotic concentration shown. After 3 wk, the net growth (wet weight) at each antibiotic concentration was determined and the results were expressed as the % of control growth (growth in the absence of antibiotics). ●, pMON129 transformants; △, pMON131 transformants; and ○, pMON120 transformants.

pMON129, and pMON131 (Fig. 4). The results are based on the net growth of independent transformants on medium containing the levels of antibiotic shown in the figure, compared to growth in the absence of antibiotics. It is apparent that transformants containing the chimeric nopaline synthase-NPTase II-nopaline synthase gene (pMON129) require ≈20-fold higher levels of kanamycin to depress net growth by 50% in comparison to transformants lacking the chimeric gene (pMON120). Similar results were obtained for pMON128, which contains the chimeric gene in the opposite orientation in the pMON120 vector (not shown). Transformants containing pMON139 and pMON140 have dose responses identical to pMON120. Transformants containing pMON130 or pMON131 (chimeric nopaline synthase-NPTase I-nopaline synthase gene) are less resistant to kanamycin than those containing pMON128 or pMON129 (results shown for pMON130). However, this level of resistance (≈3-fold greater than control cells) is still quite adequate for selection (see below).

Additional cocultivation experiments were carried out without hormone-independent selection (i.e., medium supplemented with phytohormones which support the growth of nontransformed cells). The resulting microcolonies (≈1 mm) were transferred to phytohormone-supplemented medium containing kanamycin (50 μg/ml) and within 2–3 wk, growing colonies were readily observable on plates containing cells that were transformed with pMON128, pMON129, pMON130, or pMON131. The frequency of transformation obtained by using antibiotic selection was comparable to that obtained by using hormone-independent selection. Opine (data not shown) and Southern hybridization analysis (Fig. 3*a*, lanes 7 and 8; Fig. 3*b*, lanes 5 and 6) of the kanamycin-resistant colonies confirmed that they were indeed transformants. No growing colonies were observable on plates containing cells transformed by pMON120, pMON139, or pMON140 plasmids.

DISCUSSION

The expression of the prokaryotic NPTase I and NPTase II enzymes in plant cells by using the intermediate vector pMON120 probably depends on transcription from the nopaline synthase promoter. Support for this comes from the facts that (*i*) the prokaryotic genes with their own promoters do not confer antibiotic resistance to petunia cells (Figs. 2 and 4) and (*ii*) all of the constructions function identically in either orientation in the pMON120 vector, suggesting that transcription does not initiate elsewhere in the vector. RNA blot hybridization experiments have confirmed the presence of NPTase II-specific mRNA in the transformed tissues and nuclease S1 mapping experiments demonstrate the expected 5′ and 3′ ends for the chimeric NPTase II mRNA (data not shown). In addition, low levels of neomycin-dependent NPTase II activity have been reproducibly observed in crude cell extracts from tissues transformed with pMON128 or pMON129 (no activity has been detected in extracts from control cells or cells transformed with pMON120, pMON139, or pMON140).

The useful range of these chimeric antibiotic resistance genes appears to be quite broad. In addition to the results presented for petunia, successful selection of aminoglycoside-resistant transformants has also been demonstrated for tobacco, sunflower, and carrot (results not shown). It seems likely that most plants within the host range of *A. tumefaciens* could be transformed and identified in this manner. Those plant cells that are not particularly sensitive to kanamycin may be killed by other aminoglycoside antibiotics. In this respect pMON128 (or pMON129) and pMON130 (or pMON131) also function to confer resistance to G418 and neomycin on petunia, carrot, sunflower, and tobacco (unpublished data).

The availability of dominant selectable markers on small plasmids such as pMON120 should facilitate the development of alternate, non-*A. tumefaciens*-mediated methods for transforming plant cells such as spheroplast fusion (56) or the use of liposomes (57) or calcium-phosphate (58) techniques. These chimeric genes should also prove useful as markers in somatic hybridization experiments or as sensitive probes for studying promoter function. Finally, two obvious but significant aspects of the results presented in this paper are (*i*) it should now be possible, by using Ti plasmids that have the tumor genes (i.e., tms and tmr loci, 12) deleted, to obtain kanamycin-resistant transformants that can be readily and reproducibly regenerated into phenotypically normal plants, and (*ii*) there is no reason to believe that NPTase I and NPTase II are unique in their ability to be expressed in plant cells and it is quite likely that other bacterial, fungal, or mammalian genes, including those whose products could be expected to modify plant properties in a useful manner, could also be successfully engineered and expressed.

We gratefully acknowledge the contributions and strains provided by Drs. M.-D. Chilton and J. Schell. We also thank Dr. M. Bevan for letting us compare our sequence data on the nopaline synthase gene with his prior to its publication, Dr. S. Gelvin for T-DNA probes, P. Kelly for helpful comments on the manuscript, and Ms. D. Lam and P. Guenther for preparing the manuscript. Finally, we would like to thank Dr. E. Jaworski for his support and encouragement throughout the course of this project.

1. Chilton, M.-D., Drummond, M. H., Merlo, D. J., Sciaky, D., Montoya, A. L., Gordon, M. & Nester, E. (1977) *Cell* **11**, 263–271.
2. Van Larebeke, N., Engler, G., Holsters, M., Van der Elsacker, S., Zaenen, I., Schilperoort, R. & Schell, J. (1974) *Nature (London)* **252**, 169–170.
3. Kerr, A., Manigault, P. & Tampé, J. (1977) *Nature (London)* **265**, 560–561.
4. Braun, A. (1956) *Cancer Res.* **16**, 53–56.
5. Chilton, M.-D., Saiki, R., Yadav, N., Gordon, M. & Quétier, F. (1980) *Proc. Natl. Acad. Sci. USA* **77**, 4060–4064.
6. Yadav, N., Postle, K., Saiki, R., Thomashow, M. & Chilton, M.-D. (1980) *Nature (London)* **287**, 458–461.
7. Willmitzer, L., DeBeuckeleer, M., Lemmers, M., Van Montagu, M. & Schell, J. (1980) *Nature (London)* **287**, 359–361.
8. Willmitzer, L., Simons, G. & Schell, J. (1982) *EMBO J.* **1**, 139–146.
9. Bevan, M. & Chilton, M.-D. (1982) *J. Mol. Appl. Genet.* **1**, 539–546.
10. Gelvin, S., Gordon, M., Nester, E. & Aronson, A. (1981) *Plasmid* **6**, 17–29.
11. Leemans, J., Deblaere, R., Willmitzer, L., DeGreve, H., Hernalsteens, J., Van Montagu, M. & Schell, J. (1982) *EMBO J.* **1**, 147–152.
12. Garfinkel, D., Simpson, R., Ream, R., White, F., Gordon, M. & Nester, E. (1981) *Cell* **27**, 143–155.
13. Holsters, M., Silva, B., Van Vliet, F., Genetello, C., DeBlock, M., Dhaese, P., Depicker, A., Inze, D., Engler, G., Villarael, R., Van Montagu, M. & Schell, J. (1980) *Plasmid* **3**, 212–230.
14. Zambryski, P., Depicker, A., Kruger, K. & Goodman, H. (1982) *J. Mol. Appl. Genet.* **1**, 361–370.
15. Yadav, N., Vanderleyden, J., Bennet, D., Barnes, W. & Chilton, M.-D. (1982) *Proc. Natl. Acad. Sci. USA* **79**, 6322–6326.
16. Hille, J., Klasen, I. & Schilperoort, R. (1982) *Plasmid* **7**, 107–118.
17. Klee, H., Gordon, M. & Nester, E. (1982) *J. Bacteriol.* **150**, 327–331.
18. Akiyoski, D., Morris, R., Hinz, R., Mischke, B., Kosuge, T., Garfinkel, D., Gordon, M. & Nester, E. (1983) *Proc. Natl. Acad. Sci. USA* **80**, 407–411.
19. Braun, A. & Wood, H. (1976) *Proc. Natl. Acad. Sci. USA* **73**, 496–500.
20. Yang, F., Montoya, A., Merlo, D., Drummond, H., Chilton, M.-D., Nester, E. & Gordon, M. (1980) *Mol. Gen. Genet.* **177**, 707–714.
21. Otten, L., DeGreve, H., Hernalsteens, J., Van Montagu, M., Schieder, O., Straub, J. & Schell, J. (1981) *Mol. Gen. Genet.* **183**, 209–213.
22. Wullems, G., Molendijk, L., Ooms, G. & Schilperoort, R. (1981) *Cell* **24**, 719–727.
23. Barton, K., Binns, A., Matzke, A. & Chilton, M.-D. (1983) *Cell* **32**, 1033–1043.
24. Chilton, M.-D., Tepfer, D., Petit, A., David, C., Casse-Delbart, F. & Tempé, J. (1982) *Nature (London)* **295**, 432–434.
25. White, F., Ghidossi, G., Gordon, M. & Nester, E. (1982) *Proc. Natl. Acad. Sci. USA* **79**, 3193–3198.
26. Spano, L. & Costantino, P. (1982) *Z. Pflanzenphysiol.* **106**, 87–92.
27. Chilton, M.-D., Bevan, M., Yadav, N., Matzke, A., Byrne, M., Grula, M., Barton, K., Vanderleyden, J., DeFramond, A. & Barnes, W. (1981) *Stadler Genet. Symp.* **13**, 39–51.
28. Davies, R. W., Botstein, D. & Roth, J. R. (1980) *Advanced Bacterial Genetics* (Cold Spring Harbor Laboratory, Cold Spring Harbor, NY), p. 116.
29. Ish-Horowicz, D. & Burke, J. F. (1981) *Nucleic Acids Res.* **9**, 2989–2998.
30. Colman, A., Beyers, M. J., Primrose, S. B. & Lyons, A. (1978) *Eur. J. Biochem.* **91**, 303–310.
31. Maniatis, T., Fritsch, E. F. & Sambrook, J. (1982) *Molecular Cloning* (Cold Spring Harbor Laboratory, Cold Spring Harbor, NY), p. 504.
32. Adams, S. P., Holder, S. B., Wykes, E. J., Kavka, K. S. & Galluppi, G. R. (1983) *J. Am. Chem. Soc.* **105**, 661–663.
33. Murray, N. E., Bruce, S. A. & Murray, K. (1979) *J. Mol. Biol.* **132**, 493–505.
34. Rogers, S. G. & Weiss, B. (1980) *Gene* **11**, 187–195.
35. Bale, A., d'Alarcao, M. & Marinus, G. M. (1979) *Mutat. Res.* **59**, 157–165.
36. Messing, J., Crea, R. & Seeburg, P. (1981) *Nucleic Acids Res.* **9**, 309–321.
37. DeGreve, H., Decraemer, H., Seurinck, J., Van Montagu, M. & Schell, J. (1981) *Plasmid* **6**, 235–248.
38. Ditta, G., Stanfield, S., Corbin, D. & Helinski, D. (1980) *Proc. Natl. Acad. Sci. USA* **77**, 7347–7351.
39. Ausubel, F., Bahnsen, K., Hanson, M., Mitchell, A. & Smith, H. (1980) *Plant Mol. Biol. Newsl.* **1**, 26–32.
40. Wullems, G., Molendijk, L., Ooms, G. & Schilperoort, R. (1981) *Proc. Natl. Acad. Sci. USA* **78**, 4344–4348.
41. Horsch, R. & Jones, G. (1980) *In Vitro* **16**, 103–108.
42. Otten, L. & Schilperoort, R. (1978) *Biochim. Biophys. Acta* **527**, 497–500.
43. Jimenez, A. & Davis, J. (1980) *Nature (London)* **287**, 869–871.
44. Colbere-Garapin, F., Horodniceanu, F., Kourilsky, P. & Garapin, A.-C. (1981) *J. Mol. Biol.* **150**, 1–14.
45. Southern, P. & Berg, P. (1982) *J. Mol. Appl. Genet.* **1**, 327–341.
46. Depicker, A., Stachel, S., Dhaese, P., Zambryski, P. & Goodman, H. (1982) *J. Mol. Appl. Genet.* **1**, 561–574.
47. Tempé, J. & Goldmann, A. (1982) in *Molecular Biology of Plant Tumors*, eds. Kahl, G. & Schell, J. (Academic, New York), pp. 427–449.
48. Goeddel, D., Shepard, H., Yelverton, E., Leung, D., Crea, R., Sloma, A. & Pestka, S. (1980) *Nucleic Acids Res.* **8**, 4057–4074.
49. Beck, E., Ludwig, G., Auerswald, E., Reiss, B. & Schaller, H. (1982) *Gene* **19**, 327–336.
50. Messing, J. & Vieira, J. (1982) *Gene* **19**, 269–276.
51. Covarrubias, L., Cervantes, L., Covarrubias, A., Soberon, X., Vichido, I., Blanco, A., Kuperztoch-Portnoy, Y. & Bolivar, F. (1981) *Gene* **13**, 25–35.
52. Nagao, R., Shah, D., Eckenrode, V. & Meagher, R. (1981) *DNA* **1**, 1–9.
53. Southern, E. (1975) *J. Mol. Biol.* **98**, 503–517.
54. Berg, D., Davies, J., Allet, B. & Rochaix, J. (1975) *Proc. Natl. Acad. Sci. USA* **76**, 3628–3632.
55. Oka, A., Sugisaki, H. & Takanami, M. (1981) *J. Mol. Biol.* **147**, 217–226.
56. Hasezawa, S., Nagata, T. & Syono, K. (1981) *Mol. Gen. Genet.* **182**, 206–210.
57. Fraley, R. & Papahadjopoulos, D. (1982) in *Current Topics in Microbiology and Immunology*, eds. Hofschneider, P. & Goebel, W. (Springer, New York), pp. 171–192.
58. Krens, F., Molendijk, L., Wullems, G. & Schilperoort, R. (1982) *Nature (London)* **296**, 72–74.

A chimaeric antibiotic resistance gene as a selectable marker for plant cell transformation

Michael W. Bevan & Richard B. Flavell

Plant Breeding Institute, Maris Lane, Trumpington, Cambridge CB2 2LQ, UK

Mary-Dell Chilton

Department of Biology, Washington University, St Louis, Missouri 63130, USA

The T-DNA region of *Agrobacterium tumefaciens* tumour-inducing plasmids of the nopaline type[1] contains a gene coding for the enzyme nopaline synthase. This gene is expressed constitutively in host plant cells to which it is transferred during tumour induction[2]. We have exploited the regulatory elements of this gene to construct a chimaeric gene that confers antibiotic resistance on transformed plant cells. The chimaeric gene encodes the expected chimaeric transcripts in plant cells, and confers on transformed cells the ability to grow in the presence of normally lethal levels of the antibiotic G418 (ref. 3). Experiments using *in vitro* transformation techniques on single plant cells indicate that this antibiotic resistance can be used as a selectable marker, and can therefore be used in selecting cells transformed by T-DNA vectors that have had the genes for hormone autotrophy deleted[4]. Plant cells transformed by such 'disarmed' T-DNA vectors can be regenerated into entire plants, whose sexual progeny contain unaltered copies of the inciting T-DNA[5]. The availability of this dominant selectable marker should allow a wider range of experiments to be undertaken using different host plants.

The antibiotic resistance gene used in our experiments (*neo*) is derived from Tn*5* and codes for neomycin phosphotransferase II. The *neo* coding region, when spliced to viral gene regulatory sequences, confers on animal cells resistance to the aminoglycoside antibiotic, G418 (ref. 3). We have placed the coding region of *neo* under the transcriptional control of the nopaline synthase promoter, which has been identified by DNA sequencing and S_1 nuclease mapping experiments[6,7]. We included in our construct the 5′-untranslated region of the nopaline synthase gene, but nothing of the 5′ end of the coding region. At a position 3′ to the *neo* coding region our construct contains the last portion of the nopaline synthase-coding region, a 3′-untranslated region, and termination and polyadenylation signals.

The 5′ end of the chimaeric *neo* gene was constructed as outlined in Fig. 1. A 370-base pair (bp) *Sau*3a fragment containing the nopaline synthase promoter region and the first 16 codons of the coding region was treated with *Bal*31 to remove the coding region including the initiator methionine codon. The resulting population of molecules was ligated to *Bam*HI linkers and inserted into the *Bam*HI site of pBR322. A 270-bp *Bam*HI fragment with all 48 nucleotides of the nopaline synthase-coding region deleted was identified by DNA sequencing[8] after subcloning into M13mp7 (ref. 9). The plasmid containing this insert was designated pJFΔ166.

To construct a nopaline synthase gene containing the appropriate deletion and a unique *Bam*HI site for insertion of the *neo* coding region, the 270-bp promoter fragment was digested with *Bam*HI and *Sst*II and cloned into *Bam*HI/*Sst*II-digested pJF28 (see Fig. 1 legend). This produced a deletion of 814 bp in the nopaline synthase gene in the resulting plasmid, pJF28Δ166 (Fig. 1, step *C*). The resected gene in this plasmid has a unique *Bam*HI site adjacent to the promoter, followed by 423 bp of coding region and the polyadenylation site.

In order to be able to transfer the reconstructed gene to an *Agrobacterium* Ti plasmid by double recombination, it is necessary to have extensive regions of homology on both sides of the gene, and a tightly linked prokaryotic selectable marker[10]. We therefore subcloned the deleted nopaline synthase gene, together with the closely linked kanamycin resistance (Km^r) marker from pACYC177 (ref. 11), into pJFXhoX to create pJFXhoXΔ166 (Fig. 1, step *D*).

Construction of the chimaeric *neo* gene started with the plasmid pAM1 (ref. 12), which contains a *Hin*dIII/*Sal*I fragment of Tn*5* (ref. 13). The coding region of the *neo* gene is found on a 1-kilobase (kb) *Bgl*II/*Sma*I fragment[14]. To insert this into the deleted nopaline synthase gene, the *Sma*I site adjacent to the 3′ end of the *neo* gene was first converted to a *Bam*HI site using linkers to create pAM1 (B). The coding region was isolated from pAM1 (B) by digestion with *Bam*HI and *Bgl*II, and ligated into *Bam*HI-digested pJFXhoXΔ166 (Fig. 1, step *E*). The orientation of the *neo* coding region in relation to the nopaline synthase promoter was verified in several plasmids by digestion with *Bam*HI and *Sst*II, *Hin*dIII and *Sst*II, and *Pst*I. Clones with the AUG initiation codon adjacent to the promoter were designated pJFneo$^+$, and those in the reverse orientation pJFneo$^-$.

0028-0836/83/280184—04$01.00

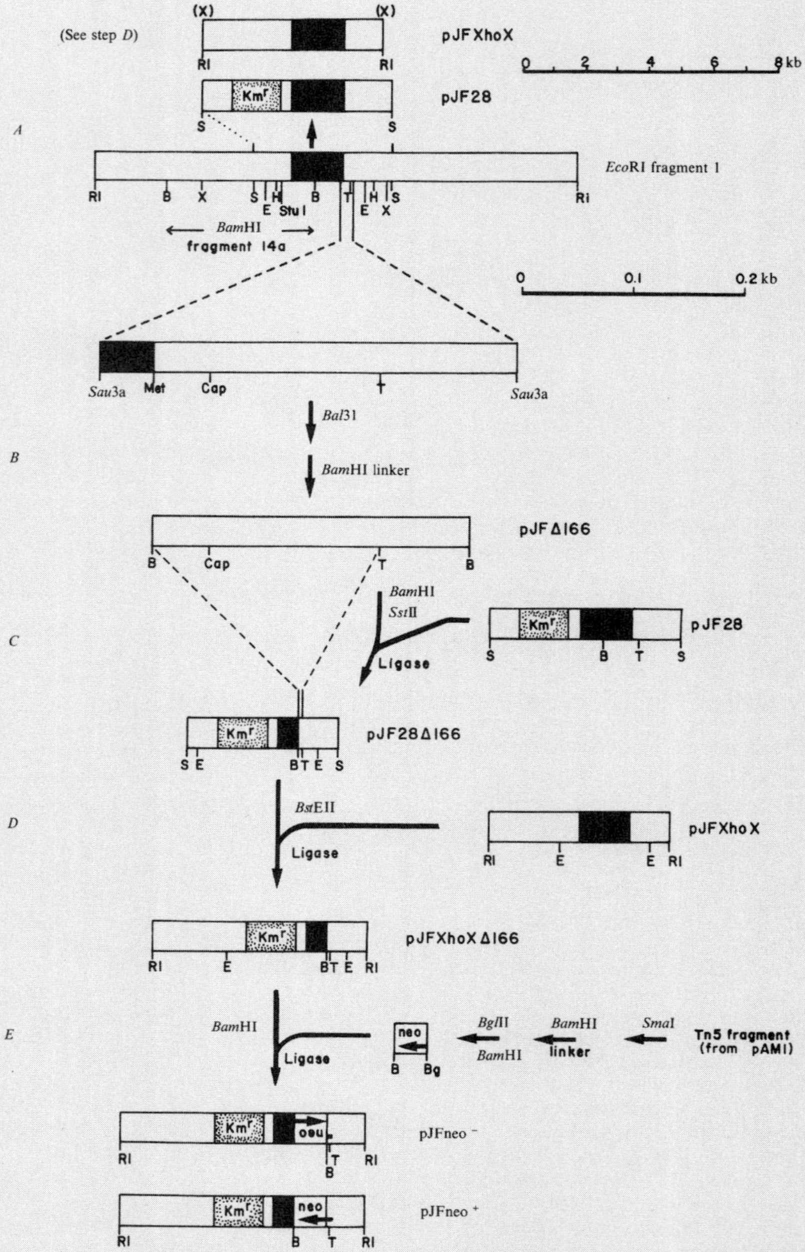

Fig. 1 Construction of the chimaeric *neo* gene. ***A***, Construction of pJF28, a plasmid containing Km^r closely linked to the nopaline synthase gene. A recombinant plasmid containing *Sal*I fragment 10 of pTiT37 was partially digested with *Stu*I, and to this was ligated a 1.5-kb *Hae*II fragment of pACYC177 (ref. 11) (encoding the Km^r gene). Plasmid pJF28 contained the Km^r insert in the *Stu*I site adjacent to the 3′ end of the nopaline synthase gene, as determined by analysis of minipreps. ***B***, Isolation of the nopaline synthase promoter. A 370-bp *Sau*3a fragment that contains the first 16 codons of the nopaline synthase gene and extends nearly to the right border of T-DNA was isolated from *Hin*dIII fragment 23 by preparative gel electrophoresis in 4% polyacrylamide; 5 μg of this fragment was treated with limiting *Bal*31 exonuclease, and the resulting fragments were ligated with 100-fold molar excess of *Bam*HI linkers[26] (25 μl, 18 h, 4 °C, 10 U T_4 DNA ligase). After heating at 70 °C for 10 min the product was digested with *Bam*HI (100 U, 4 h, 37 °C). After preparative electrophoresis (5% polyacrylamide gel), molecules of ~270 bp were excised and recovered by electroelution, DEAE-Sephadex chromatography and ethanol precipitation. This population of molecules was cloned into *Bam*HI-linearized pBR322 (ref. 27). Minipreps of recombinant plasmids (Ap^rTc^s) were digested with *Bam*HI and *Sst*II, and analysed for the presence of the desired 192-bp *Bam*HI–*Sst*II fragment. Twelve candidates were subcloned into M13mp7 (ref. 9) and their inserts sequenced using the dideoxy method[8]. One plasmid, pJFΔ166, was found to have been deleted to the nucleotide just 5′ to the ATG initiation codon of the nopaline synthase gene. ***C***, Deletion of the 5′ end of the coding region of the nopaline synthase gene. Plasmid pJF28 (1 μg) and plasmid pJFΔ166 (5 μg) were digested with *Bam*HI and *Sst*II, and ligated together (25 μl, 10 U T_4 DNA ligase). Km^r transformants were screened for the presence of the 192-bp *Bam*HI–*Sst*II fragment. The desired plasmid was designated pJF28Δ166. ***D***, Substitution of the truncated nopaline synthase gene into a plasmid containing flanking homology with the Ti plasmid. Plasmid pJFXhoX was made by isolating a 6-kb *Xho*I fragment containing the nopaline synthase gene from *Eco*RI fragment 1, and converting the *Xho*I sites to *Eco*RI sites using molecular linkers[26]. The desired *Eco*RI fragment of 6 kb was cloned into pBR325-B (ref. 27; pBR325 that has had the *Bam*HI site destroyed by blunting and recircularization). The resulting plasmid, pJFXhoX (1 μg), was digested with *Bst*EII to completion and ligated with 1 μg of *Bst*EII-digested pJF28Δ166 that had been treated with alkaline phosphatase. Substitution in the correct orientation of the kanamycin resistance-conferring *Bst*EII fragment for the *Bst*EII fragment of pJFXhoX was verified by digestion of minipreps with *Bst*EII, *Sma*I and *Eco*RI plus *Bam*HI. The desired recombinant molecule was designated pJFXhoXΔ166. ***E***, Construction of chimaeric nopaline synthase/*neo* genes. The coding region of the *neo* gene lies within a *Bgl*II/*Sma*I fragment of Tn*5*; this fragment includes 30 bp upstream and 180 bp downstream from the coding region[14]. Plasmid pAM1 contains a *Hin*dIII–*Sal*I fragment of Tn*5*[12] that includes the *neo* gene. The *Sma*I site adjacent to the 3′ end of the *neo* gene was converted to a *Bam*HI site using linkers as described[26], producing plasmid pAM1(B). The *neo* coding region was excised from pAM1(B) by digestion with *Bam*HI and *Bgl*II and purified. Plasmid pJFXhoXΔ166 (1 μg) was digested to completion with *Bam*HI, and treated with alkaline phosphatase as described above. To this was ligated 0.1 μg of the gel-purified *Bam*HI–*Bgl*II fragment containing the *neo* coding region. Recombinant plasmids with the *Bgl*II–*Bam*HI fusion adjacent to the *Sst*II site were designated pJFneo⁺ and those with the reverse orientation were called pJFneo⁻. The *Escherichia coli* host for DNA constructions was LE392. Drug concentrations used for selective media were: for *E. coli*, ampicillin (Ap), 100 μg ml^{-1}; kanamycin (Km), 25 μg ml^{-1}; tetracycline (Tc), 25 μg ml^{-1}; chloramphenicol (Cm), 25 μg ml^{-1}; for *A. tumefaciens* A208, Km, 100 μg ml^{-1}; gentamycin, 50 μg ml^{-1}. The scale bar at the top of the figure refers to all constructions except those in step *B*, which are enlarged (see separate scale bar). Cap is the transcription initiation site, and met the initiation of translation codon. Restriction endonuclease sites used in constructions are designated as follows; B, *Bam*HI; E, *Bst*EII; H, *Hin*dIII; R1, *Eco*RI; S, *Sal*I; T, *Sst*II; X, *Xho*I; (X), destroyed *Xho*I site. The black box represents the coding region of the nopaline synthase gene. The stippled box represents the prokaryotic Km^r gene from pACYC177. The coding region of neomycin phosphotransferase II from Tn*5* is designated *neo*, and the arrow on this DNA fragment indicates the orientation of the gene.

To transfer the relevant segments of pJFneo⁺ and pJFneo⁻ to *A. tumefaciens*, the plasmids were digested with *Eco*RI and ligated into *Eco*RI-digested pRK290 (ref. 28). The resulting DNAs were cloned in *E. coli* and subsequently transformed into *A. tumefaciens* strain A208 (ref. 15) using kanamycin selection. Double recombinants (homogenotes) that incorporated the engineered fragment onto pTiT37 were isolated as described previously[10]. The structure of the homogenotes was confirmed by Southern blot analysis of total bacterial DNA. Clones harbouring the desired recombinant pTiT37 plasmids were designated A208neo⁺ and A208neo⁻. These bacteria were used to incite tumours on sterile stem explants of *Nicotiana tabacum* var. Turkish. The resultant tumorous overgrowths were excised, rendered axenic by growth in MS medium[16] containing 500 μg ml^{-1} carbenicillin, and cloned as described elsewhere[17]. Normal teratomatous tumour growth occurred 1 month after cloning. To look for the presence of T-DNA, and to determine its structure in the tumours, DNA was isolated

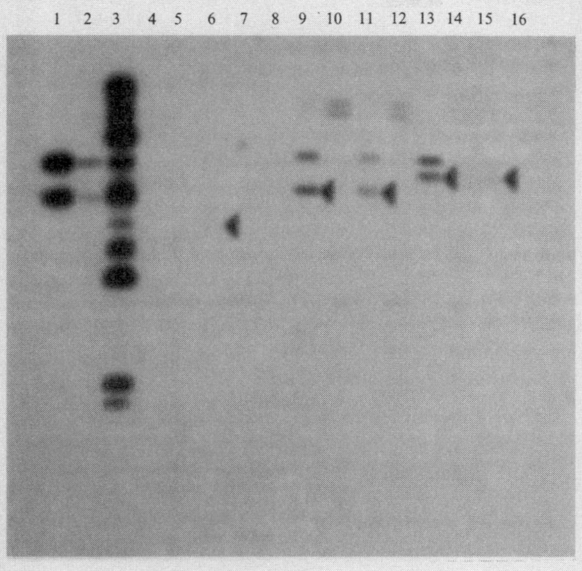

Fig. 2 Southern blot analysis of DNA from cloned tumours. The hybridization probe used is a clone of the entire T-DNA region of pTiT37 cloned in pBR322. Lanes contained the following DNA samples: 1, 10 copy reconstruction: 2, 1 copy reconstruction; 3, molecular weight marker; 4, normal, *Bam*HI; 5, normal, *Eco*RI; 6, HT37, *Bam*HI; 7, HT37, *Eco*RI; 8, no DNA; 9, neo$^+$ clone 14, *Bam*HI; 10, neo$^+$ clone 14, *Eco*RI; 11, neo$^+$ clone 8, *Bam*HI; 12, neo$^+$ clone 8, *Eco*RI; 13, neo$^-$ clone 2, *Bam*HI; 14, neo$^-$ clone 2, *Eco*RI; 15, neo$^-$ clone 1, *Bam*HI; 16, neo$^-$ clone 1, *Eco*RI. The arrows in the autoradiogram indicate the position of the modified *Bam*HI fragment 14a adjacent to the *neo* gene.
Methods: DNA was isolated[18] from representative cloned tobacco tumour lines incited by A208neo$^+$ and A208neo$^-$. Positive and negative control DNAs were isolated from HT37-8-70 tumour tissue which contains an unaltered T-DNA insert (G. Jen, unpublished data), and from *N. plumbaginifolia* leaves. Ten μg of each DNA sample were digested to completion with either *Eco*RI or *Bam*HI and electrophoresed through a 0.7% agarose gel. Southern blotting, hybridization and washing were performed as described elsewhere[29]. Labelled probe DNAs ($1.5-3\times 10^8$ c.p.m. μg^{-1}) were prepared by nick translation[20] using [α-^{32}P]dATP (NEN). The molecular weight marker was λ DNA digested with *Eco*RI and *Hin*dIII, and the reconstruction lanes contained 15 pg and 150 pg of pJFneo$^+$ DNA digested with *Eco*RI. The Southern blots were exposed to X-ray film with two intensification screens for 2 days at −70 °C.

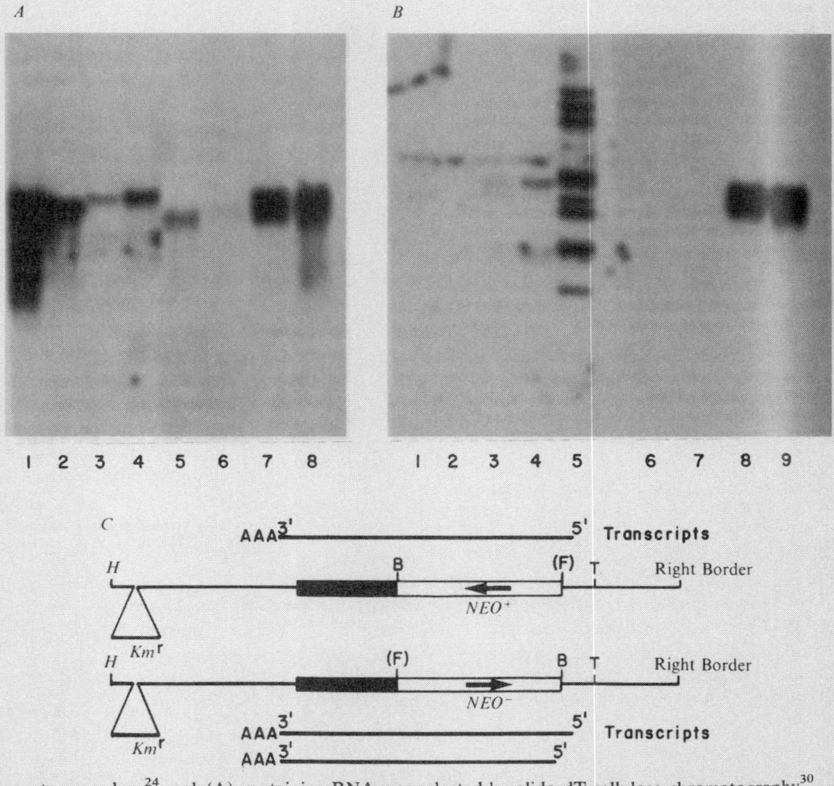

Fig. 3 Northern blot analysis of transcripts. ***A***, Hybridization probe specific for 3′ end of nopaline synthase gene. Lanes contained the following RNA samples: 1, 20 μg HT37; 2, 10 μg HT37; 3, 5 μg neo$^+$ clone 14; 4, 5 μg neo$^+$ clone 8; 5, 5 μg HT37; 6, 1 μg HT37; 7, 20 μg neo$^-$ clone 2; 8, 20 μg neo$^-$ clone 2. ***B***, Hybridization probe specific for *neo* coding region. Lanes contained the following RNA samples: 1, 20 μg HT37; 2, 10 μg HT37; 3, 5 μg neo$^+$ clone 14; 4, 5 μg neo$^+$ clone 8; 5, molecular weight marker; 6, 5 μg HT37; 7, 1 μg HT37; 8, 20 μg neo$^-$ clone 2; 9, 20 μg neo$^-$ clone 1. ***C***, Structure and transcription of chimaeric genes. (F) is the *Bam*HI–*Bgl*II fusion. Lanes 1–4 of *A* and 1–5 of *B* are from the same gel, and lanes 5–8 of *A* and 6–9 of *B* are from another gel. Direct size comparisons between these different gels was not possible. Transcript sizes were calculated from standards run in each gel. The diagram showing a truncated chimaeric transcript in neo$^-$ tumours shows only one possible RNA molecule. Another possible transcript may be truncated at the 3′ end.
Methods: RNA was isolated from the four tumour clones whose DNA was analysed by Southern blotting (Fig. 2). Control RNA was isolated from HT37-8-70 tumour tissue. After purification by a guanidine thiocyanate procedure[24], poly(A)-containing RNA was selected by olido-dT cellulose chromatography[30]. RNA was denatured and electrophoresed in a 1.5% agarose–formaldehyde gel[22], and transferred to nitrocellulose[23]. Hybridization probes were nick-translated *Bam*HI–*Bgl*II fragment containing the *neo* coding region excised from pAM1(B) and a 1.1-kb *Bam*HI–*Hin*dIII fragment complementary to the 3′ end of the nopaline synthase gene (see Fig. 1 for location). Molecular weight markers were *Eco*RI and *Hin*dIII fragments of λ DNA that had been end-labelled with [α-^{32}P]dATP using Klenow polymerase. The blot was hybridized for 16 h at 41 °C, washed as described[29], and exposed to X-ray film at −70 °C for 2 days with two intensification screens.

from representative tobacco clones[18], digested with *Eco*RI or *Bam*HI and Southern-blotted[19]. We used nick-translated[20] MINI-Ti plasmid[12] containing the entire T-DNA of pTiT37 (ref. 21) as a hybridization probe for analysis of T-DNA structure. An autoradiogram showing the pattern of hybridization is presented in Fig. 2. Hybridization with MINI-Ti revealed that our tumour lines contained 1–2 copies of T-DNA identical to that of natural HT37 tumour cells except for the size of *Bam*HI fragment 14a and the diverse border fragments. In HT37 tumours *Bam*HI fragment 14a is 4.5 kb (arrowed in Fig. 2). As expected, in neo$^+$ tumours, the corresponding fragment is nearly 6 kb due to the presence of the 1.5-kb pACYC177

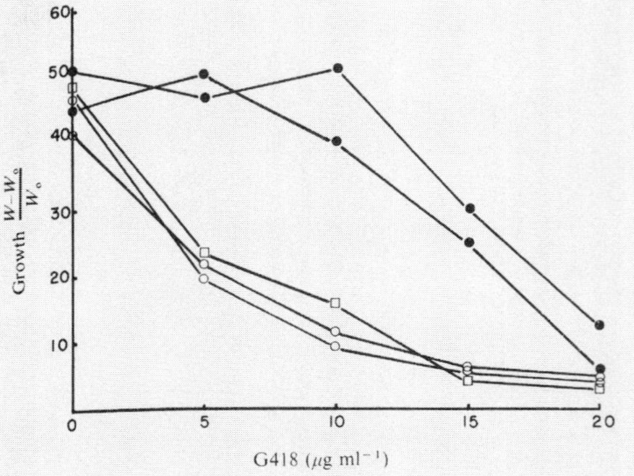

Fig. 4 Growth of tumours on G418-containing medium. Open circles, neo$^-$ clone 1 and neo$^-$ clone 2 tumours; closed circles, neo$^+$ clone 8 and neo$^+$ clone 14 tumours; open squares, HT37-8-70 tumour. W_0, initial weight (30 mg); W, final weight after 25 days.

Methods: Approximately 20 clones of neo$^-$ and neo$^+$ tumours were analysed for growth on MS medium[16] containing 3% sucrose and various concentrations of G418. Ten pieces of tissue from each clone, each weighing ~30 mg, were transferred to the indicated concentrations of G418 and incubated at 25 °C in a 12 h light/dark cycle for 30 days. Tissues were collected and weighed, and the mean weight increase calculated. The data presented here are for the cloned tumour lines whose T-DNA and transcripts are described above (Figs 2 and 3). The remaining 18 clones of each type (neo$^+$ and neo$^-$) gave similar data.

kanamycin resistance gene (arrowed in Fig. 2). In neo$^-$ tumours, the counterpart of *Bam*HI fragment 14a is 7 kb due to the insertion of the *Bgl*II/*Bam*HI coding region of *neo* (arrowed).

The transcriptional activity of the chimaeric gene in the cloned tobacco tumours was studied by northern blot analysis[22,23]. Polyadenylated RNA isolated from HT37, neo$^+$ and neo$^-$ tumours[24] was denatured, subjected to electrophoresis and transferred to nitrocellulose. The RNA blots were hybridized with the *neo* coding region fragment, which hybridized to a 1,800-base transcript in neo$^+$ tumours (Fig. 3*B*, lanes 3 and 4) and to 1,700- and 1,800-base transcripts in neo$^-$ tumours (Fig. 3*B*, lanes 8 and 9). As expected, this probe did not hybridize significantly to HT37 RNA (Fig. 3*B*, lanes 1, 2, 6 and 7). A 1.1-kb *Bam*HI/*Hin*dIII fragment homologous to the 3′ end of the nopaline synthase gene[6,7] also hybridized to a 1,800-base transcript in neo$^+$ tumours (Fig. 3*A*, lanes 3 and 4) and to 1,700- and 1,800-base transcripts in neo$^-$ tumours (Fig. 3*A*, lanes 7 and 8). This probe, as expected, hybridized to an unaltered 1,600-base nopaline synthase transcript in HT37 RNA (Fig. 3*A*, lanes 1, 2, 5 and 6). The 1,800-base transcript observed in neo$^+$ and neo$^-$ tumours that hybridized to both probes is expected of a transcript that originates at the normal nopaline synthase initiation (cap) site, traverses the 1,010-bp *Bgl*II/*Bam*HI *neo* coding region (which replaces 814 bp of the nopaline synthase gene), and terminates at the nopaline synthase polyadenylation site. As the 1,700-base transcript observed in neo$^-$ tumours is polyadenylated, it seems likely that it arises from either aberrant initiation of transcription 100 bp from the normal cap site or polyadenylation at the unused poly(A) signal observed 100 bp upstream of the 3′ end[7].

The abundance of neo$^-$ and neo$^+$ transcripts relative to the natural nopaline synthase transcript can be estimated roughly from these hybridization data. When the different levels of contamination with rRNA are taken into account (visible in ethidium bromide-stained gels), the neo$^+$ transcript and both neo$^-$ transcripts are approximately as abundant as the nopaline synthase transcript. A diagram of the transcription of neo$^+$ and neo$^-$ genes is presented in Fig. 3*C*.

Having established that the chimaeric *neo* gene is transcribed in the tumours, we investigated its effect on the plant cell phenotype. The growth rates of neo$^+$ and neo$^-$ cloned tumours were determined on MS medium without phytohormones, but containing various concentrations of G418. Tumour explants known to contain T-DNA were collected and weighed after 30 days. Figure 4 shows the results obtained for two neo$^+$ and two neo$^-$ clones. Tumours containing the neo$^-$ construction exhibited almost complete cessation of growth on medium containing more than 5 μg ml^{-1} G418, as did natural HT37 tumours. In contrast, tumours harbouring the neo$^+$ gene showed no decrease in growth rate until 15–20 μg ml^{-1} G418. We tested approximately 20 additional clones of neo$^-$ and neo$^+$ tumours, and these displayed almost identical growth curves to their counterparts in Fig. 4. As the growth rate of tumours is strictly dependent on the orientation of the coding region relative to the nopaline synthase promoter, we conclude that the observed differences in growth rate are due to differences in the activity of the *neo* gene in these tumours.

The results presented here demonstrate that a chimaeric gene composed of T-DNA regulatory, bacterial coding and T-DNA termination elements is transcribed in plant cells, and that the expected enzyme activity associated with the translation of the chimaeric transcript is manifest by an enhanced resistance of those plant cells to the antibiotic G418. Preliminary cocultivation experiments[25] using *Nicotiana plumbaginifolia* and A208neo$^+$ and A208neo$^-$ indicate that transformants can be selected using 10 μg ml^{-1} G418 in the growth medium of the regenerating cells (data not shown). Thus, this G418 resistance will be useful in the selection of plant cells transformed by 'disarmed' T-DNA vectors containing no natural selectable marker[4].

This research was supported by DOE contract AC02-81ER 10889. M.W.B. gratefully acknowledges postdoctoral fellowship funds from Monsanto during his stay at Washington University, and NRDC funds during his work at the PBI.

Note added in proof: Recently Herrera-Estrella *et al.*[31] have demonstrated the expression of chloramphenicol transacetylase from nopaline synthase gene constructions nearly identical to those reported here.

Received 4 May; accepted 8 June 1983.

1. Leemans, J. *et al. J. molec. appl. Genet.* **1,** 149–164 (1981).
2. Wullems, G. J., Molendijk, L., Ooms, G. & Schilperoort, R. A. *Cell* **24,** 719–727 (1981).
3. Colbere-Garapin, F., Horodniceanu, F., Kourilsky, P. & Garapin, A.-C. *J. molec. Biol.* **150,** 1–14 (1981).
4. Leemans, J. *et al. J. molec. appl. Genet.* **1,** 149–164 (1981).
5. Barton, K. A., Binns, A. N., Matzke, A. J. M. & Chilton, M.-D. *Cell* **32,** 1033–1043 (1983).
6. Depicker, A., Stachel, S., Dhaese, P., Zambryski, P. & Goodman, H. M. *J. molec. appl. Genet.* **1,** 561–573 (1982).
7. Bevan, M., Barnes, W. M. & Chilton, M.-D. *Nucleic Acids Res.* **11,** 369–385 (1983).
8. Sanger, F., Nicklen, S. & Coulson, A. R. *Proc. natn. Acad. Sci. U.S.A.* **74,** 5463–5467 (1977).
9. Messing, J., Crea, R. & Seeburg, P. H. *Nucleic Acids Res.* **9,** 309–323 (1981).
10. Matzke, A. J. M. & Chilton, M.-D. *J. molec. appl. Genet.* **1,** 39–49 (1981).
11. Oka, A., Sugisaki, H. & Takanami, M. *J. molec. Biol.* **147,** 217–226 (1981).
12. de Framond, A. J., Barton, K. A. & Chilton, M.-D. *Bio/Technology* **1,** 262–269 (1983).
13. Rothstein, S. J., Jorgenson, R. A., Postle, K. & Reznikoff, W. S. *Cell* **19,** 795–805 (1980).
14. Auerswald, E. A., Ludwig, G. & Schaller, H. *Cold Spring Harb. Symp. quant. Biol.* **45,** 107–113 (1980).
15. Holsters, M. *et al. Molec. gen. Genet.* **163,** 181–187 (1978).
16. Murashige, T. & Skoog, F. *Pl. Physiol.* **15,** 473–497 (1962).
17. Meins, F. & Binns, A. *Proc. natn. Acad. Sci. U.S.A.* **74,** 2928–2932 (1977).
18. Chilton, M.-D. *et al. Nature* **295,** 432–434 (1982).
19. Southern, E. M. *J. molec. Biol.* **98,** 503–517 (1975).
20. Maniatis, T., Jeffrey, A. & Kleid, D. G. *Proc. natn. Acad. Sci. U.S.A.* **72,** 1184–1188 (1975).
21. Depicker, A. *et al. Plasmid* **3,** 193–211 (1980).
22. Rave, N., Ckvenjakov, R. & Boedtker, H. *Nucleic Acids Res.* **6,** 3559–3567 (1979).
23. Thomas, P. S. *Proc. natn. Acad. Sci. U.S.A.* **77,** 5201–5205 (1981).
24. Bevan, M. W. & Chilton, M.-D. *J. molec. appl. Genet.* **1,** 539–546 (1982).
25. Wullems, G. J., Molendijk, L., Ooms, G. & Schilperoort, R. A. *Proc. natn. Acad. Sci. U.S.A.* **78,** 4344–4348 (1981).
26. Green, M. R. & Roeder, R. G. *Cell* **22,** 231–242 (1980).
27. Bolivar, F. *Gene* **4,** 121–136 (1978).
28. Ditta, G., Stansfield, S., Corbin, D. & Helinski, D. *Proc. natn. Acad. Sci. U.S.A.* **77,** 7347–7351 (1980).
29. Thomashow, M. F. *et al. Proc. natn. Acad. Sci. U.S.A.* **77,** 6448–6452 (1980).
30. Aviv, H. & Leder, P. *Proc. natn. Acad. Sci. U.S.A.* **69,** 1408–1412 (1972).
31. Herrera-Estrella, L., Depicker, A., Van Montagu, M. & Schell, J. *Nature* **303,** 209–213 (1983).

THERE WERE MULTIPLE TRANSFORMANTS ON EACH PLATE

Kenneth A. Feldmann
Ceres, Inc., Malibu, California, United States
E-mail: feldmann@ceres-inc.com

Commentary on the paper:

Feldmann, K. A., and Marks, M. D. 1987. *Agrobacterium*-mediated transformation of germinating seeds of *Arabidopsis thaliana:* A non-tissue culture approach. Mol Gen. Genet. 208:1–9.

My interest in developing a T-DNA insertion mutagenesis system for *Arabidopsis* was sparked in 1981 when, as one of my graduate requirements, I presented a non-thesis seminar on the T-DNA of *Agrobacterium* to the Genetics Department at Ohio State. As a young plant geneticist very much intrigued by mutants, I could already see the possibility of using T-DNA as an insertion mutagen. Three years later, when I was interviewing for a postdoctoral position with Michael Christianson at Zoecon Corporation, this idea took root as Michael not only encouraged me to pursue this goal but also had many ideas on how to accomplish it.

During my Ph.D. with Randy Scholl, I isolated a large number of mutants in nitrogen metabolism. In addition, I observed hundreds of interesting morphological and biochemical mutants that piqued my interest. However, because of the mutagen that I was using, EMS, there was no way to isolate the mutated genes. It would be more than ten years before it became efficient to clone genes of interest; there had to be a better way to identify and isolate the genes affected. For me, this created the need to develop an insertion mutagen, but how?

From my Masters research and some early tissue culture work with *Arabidopsis* by myself and a fellow postdoc at Zoecon, M. David Marks, it was clear that somaclonal variation in regenerated plants was going to make it difficult to distinguish between a gene knock-out and a phenotype due to the tissue culture process. In fact, today, few genes have been isolated from mutants produced from tissue culture protocols. Still, I had set out to do insertion mutagenesis in *Arabidopsis* with a tissue culture approach. David and I developed a procedure that allowed us to regenerate plants essentially from intact tissues in two weeks, avoiding many of the problems associated with long term exposure to exogenous hormones. However, when we added transformation to it, longer periods on media were required to regenerate plants. Before we could work on optimizing this protocol, we hit on the generation of transformants using an *in planta* procedure (see following paper).

The idea for an *in planta* transformation system came from a number of different experiments that Michael and I were doing in the lab and from several published reports. As there was no transformation system for *Arabidopsis* in 1985, I began my work on gene tagging by going after the *Sulfur* gene in tobacco using T-DNA. The *Sulfur* gene is unusual in that a heterozygote can be distinguished from a homozygote visually by the amount of chlorophyll in the leaves under appropriate conditions. Michael and I tried various unusual experiments in a vain attempt to increase the insert number in tobacco to make it more efficient to disrupt this gene. What I learned from this effort was that wounded cells, in the way that we thought about them then, were not necessary for transformation and that *Agrobacterium* could survive for long periods of time in tissues. In early 1986, I became familiar with the work of L. Ledoux and his colleagues on the integration of naked DNA into imbibing seeds of *Arabidopsis.* The authors had treated imbibing seeds of *Arabidopsis* with naked DNA and reported that the DNA became integrated into the genome. Although I knew there were serious concerns about the interpretation of these experiments, over the next month I wondered whether it would be possible to treat imbibing seeds with *Agrobacterium.* The bacteria might enter through the micropylar end of the seed and transform rapidly expanding cells in a subepidermal layer that would give rise to the sporogenous tissues. I set up an experiment where I imbibed seeds for different lengths of time and then co-cultivated the imbibed seeds with *Agrobacterium* (6 flats of plants). After growing these plants to maturity and testing the progeny on kanamycin (the selectable marker) we were looking at our first positive transformants on 16 May 1986. One of the treatments resulted in multiple transformants on each plate! Over the next several months, David and I showed that not only did all the transformants segregate for the kanamycin-resistance marker, up to four unlinked insertions, but also using Southern analyses we showed that all had unique insertion patterns, suggesting, most unexpectedly, that each transformation event was unique. Michael and I spent hours discussing the mechanisms for how we could have generated unique transformants with this procedure because it was not imagined that *Agrobacterium* could be carried along with the plant, presumably internally, as it developed. In fact, when we presented this research at a Fallen Leaf Lake conference in 1986, Mary-Dell Chilton remarked that these experiments suggested that we might need to look at a new paradigm for how *Agrobacterium* transfers genes in nature. Imagine how surprised we would all have been if we had known that we could just spray *Agrobacterium* onto plants, at least *Arabidopsis,* to get stable transformation (Clough and Bent 1998)!

After proving that the transformants were all unique, we moved on to show that the progeny from our transformed lines were segregating for mutant phenotypes and that the kanamycin marker co-segregated with these mutations (Feldmann et al. 1989). I continued to try to increase the frequency of transformants and to figure out what cells were being transformed. These experiments led me to hypothesize that *Agrobacterium* infected the plant during the co-cultivation regime and that sometime during "sporogenesis, gametogenesis, or fertilization, the *Agrobacterium* transforms a cell that will give rise to a zygote" (Feldmann 1991).

In 1987, David moved to a faculty position at the University of Nebraska to continue to work on cellular differentiation as exemplified by trichomes and I moved to DuPont where I generated another much larger population of transformants. Many of these tagged mutants were distributed to academics. More than 50 colleagues from around the world came to Wilmington, DE to help screen this population for mutants in 1990, but the population was never made publicly available. As such, when I took a faculty position at The University of Arizona in 1990, I generated a new population and these were distributed through newly established biological resource centers. Again, I want to acknowledge the large group of academics who came to the University to help me with a large phenotypic screen of these lines in 1991–92. Many of the T-DNA tagged mutants that were published in the 1990s came from either the DuPont or the Arizona population.

By 1992 I had to turn my attention to traditional academic research problems and I spent the rest of my academic time characterizing brassinosteroid dwarfs, epicuticular waxes and cytochromes P450. Others, who were successful in generating transformants using the seed transformation protocol, but at very low frequency, would develop *in planta* procedures that were more reliable than seed transformation (Bechtold et al. 1993; Chang et al. 1994; Clough and Bent 1998). These protocols have been further modified such that it is now possible to simply spray *Agrobacterium* onto plants to get reasonable transformation frequencies. The publicly available populations of T-DNA tagged mutants in *Arabidopsis* are now so large that nearly every gene in the genome is tagged, many genes more than once. The plant DNA flanking the T-DNA has been identified for more than 100,000 lines and these, as well as the corresponding seeds, are available from the resource centers or from companies. Finally, reverse genetic strategies on DNA from populations of T-DNA mutants allow a researcher to start with a gene sequence and to identify a mutant line.

Two more items are worthy of note here. The first is that these *in planta* techniques have been extended to other plant species most notably *A. lasiocarpa, Brassica campestris* (pakchoi), *Raphanus sativus,* and *Medicago truncatula.* In addition, progress has been made in ascertaining which cells are transformed during *in planta* transformation. Two groups have shown that it is the maternal chromosome set that is the target of transformation and that the ovules are the likely site of transformation. These studies also suggest that *Agrobacterium* must be interior to the gynoecium before the locules fuse. Ideas that evolve from this knowledge may lead to the transformation of additional species.

References

Bechtold, N. J., Ellis, J., and Pelletier, G. 1993. *In planta Agrobacterium*-mediated gene transfer by infiltration of adult *Arabidopsis thaliana* plants. C. R. Acad. Sci. Paris Life Sci. 316:119–199.

Chang, S. S., Park, S. K., Kim, B. C., Kang, B. J., Kim, D. U., and Nam, H. G. 1994. Stable genetic transformation of *Arabidopsis thaliana* by *Agrobacterium* inoculation *in planta.* Plant J. 5:551–558.

Clough, S. J., and Bent, A. F. 1998. Floral dip: A simplified method for *Agrobacterium*-mediated transformation of *Arabidopsis thaliana.* Plant J. 16:735–743.

Feldmann, K. A. 1991. T-DNA insertion mutagenesis in *Arabidopsis:* Mutational spectrum. Plant J. 1:71–83.

Feldmann, K. A., Marks, M. D., Christianson, M. L., and Quatrano, R. S. 1989. A dwarf mutant of *Arabidopsis* generated by T-DNA insertion mutagenesis. Science 243:1351–1354.

KENNETH FELDMANN: I graduated from the University of Northern Iowa with a B.A. in 1977 and an M.A. in 1979. I undertook graduate studies at the Ohio State University and obtained my Ph.D. in Genetics in 1985.

I had many wonderful mentors who inspired me to work on *Arabidopsis* beginning in 1977. Jean Amos taught me *Arabidopsis* tissue culture, Randy Scholl got me excited about mutants and biochemical genetics, Michael Christianson challenged me to try new ideas and gave me the opportunity to pursue genetic engineering approaches and my friend David Marks taught me molecular biology. A required graduate seminar that I presented to the Genetics Department at Iowa State University in 1981 on the T-DNA of *Agrobacterium* had a big effect on my later interests in using this insertion element for generating mutants.

Mol Gen Genet (1987) 208:1–9

MGG

Agrobacterium-mediated transformation of germinating seeds of *Arabidopsis thaliana*: A non-tissue culture approach

Kenneth A. Feldmann and M. David Marks
Zoecon Research Institute, Sandoz Crop Protection Corp., 975 California Ave., Palo Alto, CA 94304, USA

Summary. Germinating seeds of *Arabidopsis thaliana* were cocultivated with an *Agrobacterium tumefaciens* strain (C58Clrif) carrying the pGV3850:pAK1003 Ti plasmid. This Ti plasmid contains the neomycin phosphotransferase II gene (*NPT II*) which confers resistance to kanamycin and G418. Seeds (T1 generation) imbibed for 12 h before a 24 h exposure to *Agrobacterium* gave rise to the highest number of transformed progeny (T2 generation). Over 200 kanamycin-resistant T2 seedlings were isolated. Some of the T2 seedlings and T3 families were characterized for genetic segregation of functional *NPT II* gene(s), NPT II activity, and the presence of T-DNA inserts (Southern analysis). Ninety percent of the T2 individuals transmitted the resistance factor to the T3 families in a Mendelian fashion. Of the T3 families segregating in a Mendelian fashion ($n=111$), 62% segregated for one functional insert, 29% for two unlinked or linked functional inserts, 5% for three unlinked inserts, 1% for four unlinked inserts, whereas 3% appeared to be homozygous for the insert(s). The 13 families that did not exhibit Mendelian segregation ratios fell into 2 classes, both of which had a deficiency of kanamycin-resistant seedlings. In the Group I T3 families ($n=6$) only 0%–2% of the seedlings were resistant to kanamycin (100 mg/l), whereas in the Group II families ($n=7$) 8%–63% of the seedlings were resistant. All of the kanamycin-resistant plants that were tested were found to possess NPT II activity. Southern analysis revealed that all of the resistant plants contained at least one copy of the T-DNA and that the majority of the plants had multiple inserts. Explants from kanamycin-resistant plants survived and formed callus when cultured on callus-inducing medium containing G418.

Key words: *Agrobacterium tumefaciens* – Transformation – *Arabidopsis thaliana* – Mendelian segregation

Introduction

There have been numerous reports describing the transformation of protoplasts or wounded tissue from a variety of plant species employing either *Agrobacterium*-mediated or direct gene transfer techniques (e.g., Otten et al. 1981; De Block et al. 1984; Hernalsteens et al. 1984; Budar et al. 1986; Wallworth et al. 1986). Whole transformed plants are obtained from these transformed tissues after employing regeneration procedures that usually require the transformed tissue to be in culture for a substantial period of time. The resulting cultures frequently exhibit a loss in morphogenetic potential. Further, plants regenerated from cultures are often abnormal in appearance (Larkin and Scowcroft 1981) and can exhibit a loss in fertility (Scholl et al. 1981). In this regard, regenerated transformed plants may not be useful for studying the expression of an introduced gene. Transformation procedures that avoid tissue culture would be highly advantageous.

The development of such a procedure is facilitated by an understanding of classical mutagenesis. The stable Mendelian inheritance of a mutation induced by chemical, or physical means requires that the alteration occurs in a cell of an organism that will give rise to the gametes. Such cells are referred to as genetically effective cells (Li and Redei 1969). As with mutagenesis, the goal of treatments with transforming agents is the stable transformation of the genetically effective cells of an intact embryo or plant. Such a procedure would avoid the problems encountered in the culturing process and might also save considerable time.

Over the past two decades numerous reports of transformation procedures using naked DNA to target genetically effective cells have appeared (*Lycopersicon*, Stroun et al. 1966; *Arabidopsis*, Ledoux et al. 1971; *Petunia*, Hess 1969, 1973; *Hordeum*, Ledoux and Huart 1968). This early work has been reviewed by Hess (1977), and Kleinhofs and Behki (1977). Some of the reports have met with considerable skepticism (Kleinhofs and Behki 1977). A series of recent reports have renewed interest in this concept (Ohta 1986; Graves and Goldman 1986; Zhou et al. 1983).

In this paper we describe a procedure for transforming germinating seeds of *A. thaliana* with a disarmed *Agrobacterium* strain that carries a Ti plasmid containing the neomycin phosphotransferase II (*NPT II*) gene between the T-DNA borders (Velten and Schell 1985). Plants resulting from treated seeds (T1) give rise to progeny (T2) which are resistant to kanamycin. The kanamycin resistance trait segregates in a Mendelian fashion in most of the T3 seed families arising from resistant T2 plants. NPT II activity and Southern analysis confirm the presence of T-DNA inserts in resistant plants. In addition to demonstrating that *Agrobacterium*-mediated transformation has taken place, the analyses also indicate that a simple transformation of the genetically effective cells in the embryo rarely occurred. The ramifications of the results will be discussed.

Offprint requests to: K.A. Feldmann

Materials and methods

Genetic material. Seeds from the geographic race, Wassilewskija (WS), of *A. thaliana* (L.) Heynh. were originally obtained from the collection of Bruce Griffing at The Ohio State University, Columbus, Ohio. Transformation was carried out with an *A. tumefaciens* strain, C58C1rif, containing the 3850:1003 Ti plasmid which was kindly provided by J. Velten (New Mexico State University, Las Cruces, NM). This plasmid contains the bacterial *NPT II* gene under the control of the 1′ promoter from the T-DNA region of pTiAch5 (Velten and Schell 1985).

Media for plant and bacterial growth. Three media were used for plant culture and growth. A liquid basal medium, BM, was used for imbibition and cocultivation of seeds and bacteria. It consisted of Murashige and Skoog (M&S) salts (1962), 4% (w/v) sucrose, and 10.0., 0.5, 0.5, and 100.0 mg/l of thiamine, pyridoxine, nicotinic acid, and inositol, respectively. An agar-solidified medium (100K) was used for selection of resistant plants under aseptic conditions. It consisted of M&S salts, 1% (w/v) sucrose, 100 mg/l kanamycin sulphate (Sigma Chemical Co), 1.0, 0.5, 0.5, and 100.0 mg/l of thiamine, pyridoxine, nicotinic acid, and inositol, respectively. The medium was solidified with 0.8% Difco Bacto agar. The pH of both media was adjusted to 6.0 before autoclaving. A complete nutrient solution was used for growth of plants in vermiculite. It consisted of 1.1 gm $Ca(NO_3)_2 \cdot 4H_2O$, 738 mg $MgSO_4 \cdot 7H_2O$, 505 mg KNO_3, 120 mg NH_4NO_3, 8 mg KH_2PO_4, 15 mg chelated iron (supplied as "Sequestrine 138-Fe" obtained from Ciba-Geigy Corp.), 2.1 mg H_3BO_3, 1.4 mg $MnCl_2 \cdot 4H_2O$, 165 µg $ZnSO_4 \cdot 7H_2O$, 68 µg $H_3MoO_4 \cdot H_2O$, and 60 µg $CuSO_4 \cdot 5H_2O$ per liter. The medium was adjusted to pH 5.3.

A. tumefaciens was cultured in YEB medium (Lichtenstein and Draper 1985; YEB consisted of 5 gm Bacto Beef Extract (Difco), 5 gm Bacto-Peptone (Difco), 1 gm Bacto Yeast Extract (Difco), 240 mg $MgSO_4$, and 5 gm sucrose per liter (pH 7.2 with NaOH)) and grown overnight at 28° C. *Agrobacterium* carrying the appropriate plasmid was selected by supplementing the medium with 25 µg/ml kanamycin (Sigma) and 100 µg/ml rifampicin (Sigma). An overnight incubation produced a bacterial suspension with a density of approximately 10^9 cells/ml.

Transformation and growth of plants. Batches of seeds (5000) were surface sterilized [8 min in 5.25% (w/v) sodium hypochlorite and 0.15% (v/v) Tween 20] and washed 5 times with sterile water. The seeds were transferred aseptically to 125 ml Ehrlenmeyer flasks containing 50 ml of a liquid BM. The seeds were either shaken (120 rpm) at 22° C in constant fluorescent light (750 lx) for 6, 12, or 24 h (one-step imbibition; treatments 1–3), or pretreated at 4° C (250 lx) for 24 h, without shaking, to allow for imbibition. This chilling treatment promotes uniform germination in *A. thaliana* (Langridge 1957). Following the cold treatment seeds were shaken as previously described for 6, 12, or 24 h before infection (two-step imbibition; treatments 4–6). Infection consisted of adding 3 ml of an overnight culture of *A. tumefaciens* (approximately 3×10^9 bacteria in YEB medium) to the seeds in 50 ml of BM. Following a 24 h cocultivation (120 rpm, 250 lx, 28° C) the seeds (designated T1) were washed three times with sterile water.

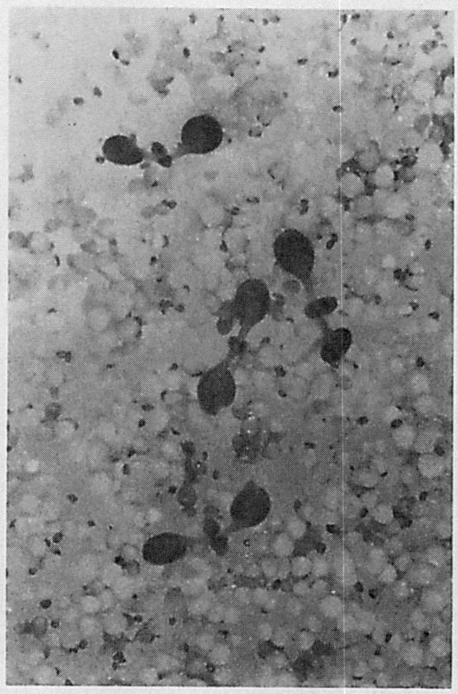

Fig. 1. Isolation of kanamycin-resistant T2 seedlings on a medium containing 100 mg/l kanamycin

Treated seeds were allowed to dry briefly and sprinkled on vermiculite (Peters, medium grade) pre-soaked with a complete nutrient solution. To obtain a uniform number of seeds from each T1 plant, the seeds were sowed to a density that allowed little or no secondary branching (approximately 1 plant/cm^2). Flats containing the seeds were covered with Saran wrap for 6–8 days to prevent seedling desiccation. The plants were grown in a shaded greenhouse at 20°–24° C with the addition of 7000 lx of constant fluorescent light. The flats were subirrigated with the complete nutrient solution as necessary.

Isolation of kanamycin-resistant plants. After the T1 plants had reached maturity (6 weeks) the T2 seeds were collected (180–230 T2 seeds per T1 plant) in bulk from batches of 1000–1500 T1 plants. The T2 seeds were air dried and cleaned. Small batches of T2 seeds ($1–2 \times 10^4$) were surface sterilized as previously described and plated in petri dishes (100 × 15 mm; approximately 1000 seeds/dish) containing 30 ml of 100K medium. The plates were wrapped with parafilm and transferred to 4° C (dark) for 24 h to improve germination. The plates were transferred to constant fluorescent light (7500 lx, 22°–24° C) to allow for germination.

The seeds from non-transformed plants germinated in the presence of kanamycin but 2 days after germination the cotyledons began to bleach. Sensitive seedlings never developed primary leaves and root growth was completely retarded. The resistant seedlings germinated, remained green, and continued to display wild-type growth except that root growth was slightly retarded (Fig. 1). After 12–14 days the resistant T2 seedlings were selected and transferred to either aseptic glass jars (100 × 80 mm) containing 50 ml of 100K medium for continued aseptic growth or to presoaked vermiculite to allow for growth and seed set. The growth conditions for vermiculite-grown plants are the same as those described for T1 plants. A few of the aseptically grown T2 plants were assayed for NPT II activity and for the number of physical inserts (Southern blot analysis). When the plants growing on vermiculite matured, T3 seed families were collected from single T2 plants.

Characterization of progeny. The T3 seeds were dried and sterilized as previously described and sprinkled on 100K medium (30 ml in 10 cm petri dishes; typically 200–400 seeds/dish). After 12 days of growth the ratio of resistant to sensitive progeny was ascertained. NPT II activity and Southern analysis were performed on both aseptic single T3 plants and bulked resistant plants. Resistant progeny were also transferred singly to vermiculite to study the segregation of the inserts and NPT II activity in subsequent generations.

NPT II activity. NPT II activity was detected by the in situ gel assay of Reiss et al. (1984). Briefly, 50 to 100 mg of leaf and stem tissue from 12- to 20-day-old plants was mixed with 100 μl of extraction buffer in a 1.5 ml eppendorf tube. The tissue was ground for 15 s with a Konte pestle and centrifuged (14000 *g*) for 5 min at 4° C. A 30 μl aliqout of the supernatant was resolved on a non-denaturing acrylamide gel. The gel was covered with a second gel containing 1% agarose, 600 μg/ml kanamycin, and 300 μCi γ-[^{32}P]ATP. The agarose gel was covered by a P 81 filter (Whatman Ltd., Maidstone, England) and a stack of blotting paper. After blotting overnight, the P 81 filter was removed, washed several times in water at 80° C, and autoradiographed.

DNA isolation and Southern analysis. DNA was isolated from 200–500 mg of stem and leaf tissue by a slight modification of the procedure described by Weeks et al. (1986). Inclusion of a chloroform:octanol (24:1) extraction step before the addition of CsCl resulted in a cleaner gradient. DNA samples were resolved by electrophoresis on a 0.7% agarose gel in a 1 × TBE buffer (0.089 M Tris, 0.089 M boric acid, 2 mM EDTA). Resolved DNA was transferred to nitrocellulose by the method of Southern (1975) as described by Maniatis et al. (1984). Probes were labeled with ^{32}P by nick translation (using a kit from BRL, Bethesda, Md) to a specific activity of approximately 7×10^7 cpm/μg. Hybridization was carried out at 42° C in 5 × SSC (1 × SSC = 0.15 M NaCl, 0.015 M trisodium citrate), 50% formamide, 0.5% SDS and 0.04% each of bovine serum albumin, Ficoll, and polyvinylpyrrolidone. Filters were washed in 1 × SSC and 0.5% SDS at 60° C and exposed to Kodak XAR-5 Film using DuPont Cronex Quanta IIB intensifier screens at −80° C.

Results

Traditional mutagenesis studies have shown that imbibition of dry seeds followed by chemical mutagenesis provides a high rate of mutations in progeny resulting from such treated seeds. In a similar fashion we explored whether there was a time during imbibition in which the germinating seedlings of *A. thaliana* were maximally susceptible to *Agrobacterium*-mediated transformation. Four of six imbibition regimes yielded observable transformants (Table 1). These

Table 1. Transformation efficiency of *Arabidopsis thaliana* seeds imbibed for various lengths of time before infection with *Agrobacterium*

Imbibition (h)	Proportion of T2 seedlings resistant to kanamycin			
	Treatment # [b]	One-step imbibition	Treatment #	Two-step imbibition [a]
6	1	0/63800 [c,d]	4	0/63800
12	2	204/63800	5	29/63800
24	3	1/63800	6	1/63800

[a] These seeds were imbibed for 24 h at 4° C before transfer to 22° C
[b] Number indicates an arbitrary treatment number
[c] Number of kanamycin-resistant T2 seeds per T2 seeds tested
[d] The total number of T2 seeds in each treatment was between $2.5–3.5 \times 10^5$. A portion of the seeds was tested for resistance to kanamycin

were a 12 or 24 h imbibition at 22° C prior to inoculation with *Agrobacterium* (treatments 2 and 3; one-step imbibition) and a 24 h imbibition at 4° C preceding a 12 or 24 h imbibition at 22° C prior to inoculation (treatments 5 and 6; two-step imbibition). A 12 h imbibition before cocultivation resulted in the highest rate of transformants (Table 1, 0.32%). A 24 h imbibition at 22° C, in the one-step or two-step imbibition regime, resulted in a low, but detectable, number of transformants (1/63,800). The experiment (with modifications) has been repeated four times and each time the highest transformation rate was observed following a 12 h imbibition of seeds (data not shown). An imbibition time less than 9 h has produced no transformants (Table 1 and data not shown).

Some of the aseptically grown T2 plants showing resistance to kanamycin were sacrificed for NPT II activity and Southern blot analyses; most of the plants were transferred to vermiculite and grown to maturity. Progeny were collected from 124 individual T2 plants (103 from treatment 2 and 21 from treatment 5), and assayed for kanamycin resistance on 100K medium (examples of segregation ratios are shown in Table 2). The majority of the T3 families from both treatments (52%–56% for treatments 2 and 5, respectively) segregated 3 resistant to 1 sensitive, which indicated that the respective parents were heterozygous for one functional *NPT II* gene (Table 3). Ratios indicating the presence of two linked and two, three and four unlinked functional *NPT II* genes were also observed. Only 3 of 124 T2 parents appeared to be homozygous for the *NPT II* gene (see note a, Table 3). Genetic and molecular analyses are in progress to ascertain whether these three families are homozygous. T4 seed families from some resistant T3 plants have also been tested for resistance to kanamycin and both segregating families and non-segregating families (100% kanamycin resistant) have been identified (unpublished results). Thus, the introduced kanamycin resistance gene has been transmitted through two or three rounds of meiosis in *A. thaliana* as a stable Mendelian factor.

Ten percent of the T3 families did not segregate for kanamycin resistance in a Mendelian fashion. These T3 families exhibited a deficiency in the number of kanamycin-resistant seedlings (Table 4). The segregation patterns fell into two groups: those in which less than 2% of the seedlings were resistant to kanamycin ($n = 6$) and those in which

Table 2. Segregation analysis (kanamycin resistance) of representative T3 families from treatment 2

T2 plant	Resistant	Sensitive	Chi-square test against ratios			
			3R:1S	15R:1S	63R:1S	255R:1S
1	196	67	ns[a]	**[b]	**	**
60	148	42	ns	**	**	**
59	122	20	** ⟨–⟩[c]	**	**	**
77	193	31	** ⟨–⟩	**	**	**
12	313	22	**	ns	**	**
70	365	25	**	ns	**	**
14	618	16	**	**	ns	**
66	322	7	**	**	ns	**
86	635	2	**	**	*	ns
45	848	0	**	**	**	ns
108	524	0	**	**	*	ns
41	279	290	**	**	**	**
10	173	362	**	**	**	**
13	2	125	**	**	**	**
76	2	172	**	**	**	**

[a] ns indicates a chi-square value in which there is no significance at the 0.05 level
[b] One asterisk indicates a chi-square value that is outside the 5% level and two asterisks a value outside the 1% level
[c] This designation indicates that the ratio fell between 3:1 and 15:1 resistant to sensitive

Table 3. Summary of chi-square analysis of T3 families

Class	Treatment 2	Treatment 5
One insert (3:1)	58	11
Two unlinked inserts (15:1)	20	6
Two linked inserts	5	1
Three unlinked inserts (63:1)	5	1
Four unlinked inserts (255:1)	1	0
100% Resistant	3[a]	0
Other	11[b]	2
Total	103	21

[a] Chi-square analysis shows that these three families could be segregating for four to five linked or unlinked inserts
[b] See Table 4

8%–63% were resistant ($n=7$). Surprisingly, many of the T4 families resulting from these resistant T3 plants segregated in a normal Mendelian fashion for the kanamycin resistance trait (data not shown).

Resistant plants from a number of the lines were assessed for NPT II activity with the in situ gel assay of Reiss et al. (1984). We assayed individual resistant T2 and T3 plants as well as T3 families. In every case, kanamycin resistance was found to be associated with a positive NPT II assay (Fig. 2). Non-transformed control plants contain two endogenous phosphotransferases, found in all of the plants assayed; however, they did not exhibit any phosphotransferase activity in the region of the gel to which NPT II migrates (Fig. 2, lane C). The resistant plants showed considerable variation in the level of NPT II activity which may be due to differences in source of tissue (stem or leaf), age of tissue, or to actual differences at the level of gene expression.

To demonstrate the presence of T-DNA inserts in the genome, Southern hybridization analyses were performed.

Table 4. Segregation of T3 families (treatments 2 and 5) that failed to segregate in a Mendelian ratio

T2 plant	Resistant	Sensitive
Group I		
13	2	125
21	1	50
50	0	80
76	2	172
82	4	380
91[a]	1	120
Group II		
10	173	362
39	160	95
41	279	290
62	13	155
84	86	58
102[a]	107	89
122	43	36

[a] Families that came from treatment 5; all others are from treatment 2

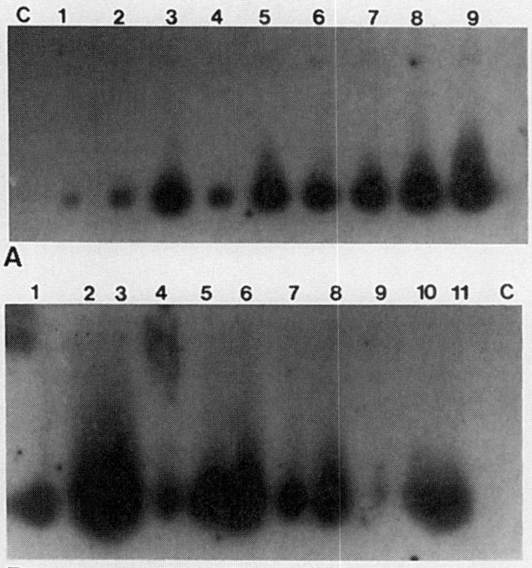

Fig. 2A, B. Neomycin phosphotransferase II (NPT II) analysis of kanamycin-resistant plants. Extracts from 50 mg of stem and leaf tissue were assayed for NPT II activity by the in situ gel assay of Reiss et al. (1984) as described in Materials and methods. **A** C, individual sensitive control plant; 1–4, individual resistant T2 plants; 5–9, individual resistant plants from the same T3 family. **B** 1–11, assay on tissue from pooled resistant plants of individual T3 families; C, control sensitive plant

In the first analysis (Fig. 3) DNA was isolated from resistant seedlings in a number of T3 families. This DNA was left intact or cleaved with *Hin*dIII before being resolved by electrophoresis (Fig. 3). DNA was transferred to a nitrocellulose filter and hybridized to ^{32}P-labeled pBR322 plasmid. The probe should hybridize to the interior of the T-DNA region which contains the pBR322 sequence (Fig. 3C). In the lanes containing uncut DNA, the probe hybridized only to the high molecular weight DNA. The

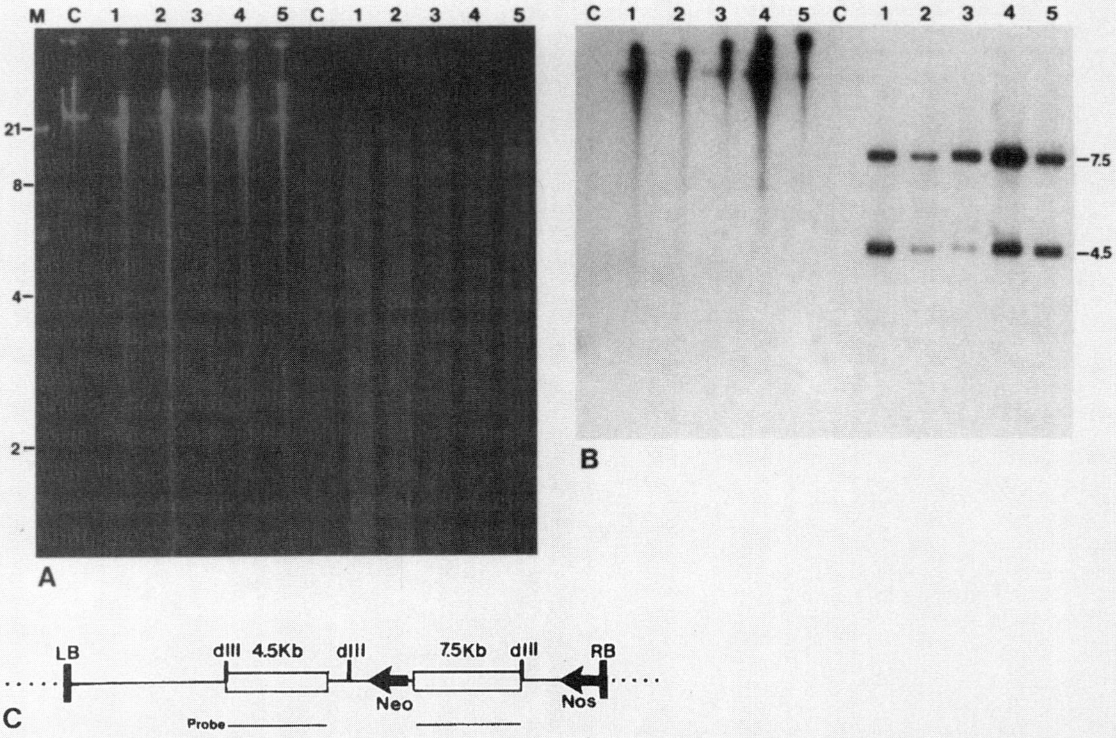

Fig. 3A–C. Southern blot analysis of DNA isolated from kanamycin-resistant progeny of transformed T2 plants. **A** Ethidium bromide stained agarose gel (0.7%) containing: M, 1 kb ladder (BRL, Bethesda, Md) plus *Eco*RI-cleaved lambda DNA; C, 0.5 μg DNA non-transformed *Arabidopsis* plants; 1–5, 0.5 μg DNA from pooled resistant plants of individual T3 families. The first set of lanes (C, 1–5) contains uncut DNA and the second set contains *Hin*dIII-cut DNA. Sizes of selected marker fragments (to the nearest 1 kb) are indicated along the left edge. **B** Autoradiograph (36 h exposure) of Southern blot of **A** that was hybridized to the plasmid pBR322 (labelled with ^{32}P by nick translation using a kit from BRL, Bethesda, Md). Size of the *Hin*dIII-cleaved fragments (in kb) that hybridized to the probe are indicated on the right edge. **C** Diagram of T-DNA region of pGV3850:pAK1003. LB, left border; RB, right border; dIII, *Hin*dIII sites; Nos, nopaline synthase gene; Neo, neomycin phosphotransferase gene; *boxed region*, pBR322 sequences. Region of homology to probe used to hybridize to filter in **B** is shown across the bottom and the sizes of internal *Hin*dIII fragments are shown across the top. Not drawn to scale

probe hybridized to fragments of the expected size in the lanes containing *Hin*dIII-cleaved DNA which is consistent with integration of intact T-DNA sequences. No hybridization was detected in the lanes containing non-transformed control DNA (Fig. 3B, lanes C).

In a second analysis, a probe that contains DNA homologous to the right border as well as to the internal region of the T-DNA was hybridized to a nitrocellulose filter containing *Hin*dIII-cleaved DNA from an individual kanamycin-resistant T2 plants (Fig. 4, lane 1) and from resistant plants of individual T3 families (Fig. 4, lanes 2–9). In addition to the two internal *Hin*dIII fragments, this probe hybridized to a number of T-DNA right border fragments attached to variable lengths of genomic DNA. Most of the lanes contained a DNA fragment with a length (approximately 5 kb) that corresponds to that expected for a direct repeat of the T-DNA. Integration of direct repeats of T-DNA has also been observed in *Lycopersicon* and *Petunia* (Johnathan Jones and Richard Jorgensen, Advanced Genetic Sciences, Oakland, Calif, personal communication). In most of the lanes a fragment with a length slightly less than 4.5 kb also hybridized to the probe. The origin of this fragment is unknown. Interestingly, only one of five lanes containing DNA from families segregating 3:1 (resistant:sensitive) for resistance to kanamycin contained a single border fragment (see Fig. 4 legend).

Although selection for kanamycin resistance was originally performed with whole plants, the *NPT II* gene functioned in cultured tissue as well. Explants from kanamycin-resistant T3 plants had the ability to grow on callus-inducing medium containing 25 μg/ml of the antibiotic G418 (an aminoglycoside related to kanamycin; Fig. 5). The explants remained green and proliferated as callus after 1 week on callus-inducing medium. Explants from wild-type or kanamycin-sensitive plants fail to grow on medium containing G418.

Discussion

Our recent aim has been to develop a reliable and rapid transformation system for a higher plant species so that the expression of introduced genes can rapidly be characterized. In addition, we are attempting to develop a gene tagging system (via insertional mutagenesis) that will be useful in the isolation and characterization of selectable mutations. *A. thaliana* is a species in which both of these goals

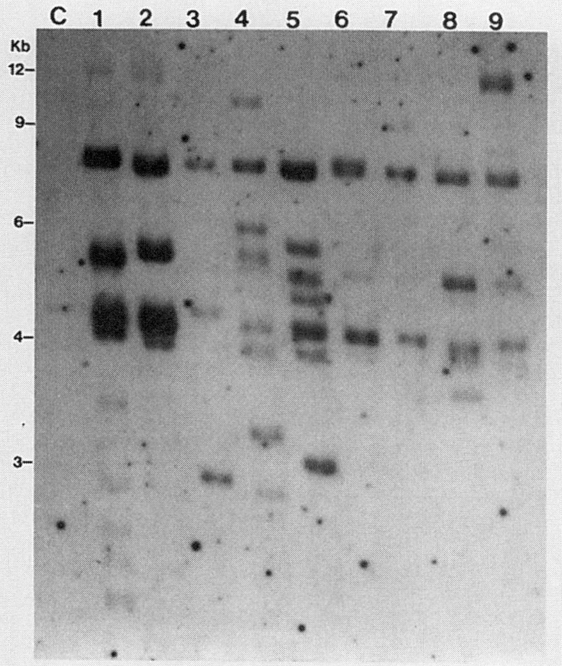

Fig. 4. Southern blot analysis of DNA isolated from kanamycin-resistant plants using a probe containing sequences homologous to the right T-DNA border. Between 0.1 and 0.5 μg *Hin*dIII-cleaved DNA from: C, non-transformed *Arabidopsis*; 1, individual T2 plant; 2–9, pooled resistant plants of individual T3 families that were segregating 3:1 (lanes 2–6), 15:1 (lanes 7–8), and 63:1 (lane 9) were resolved on an agarose gel (0.7%) and transferred to a nitrocellulose filter. Sizes of markers are shown along the left edge of an autoradiograph (4 day exposure) of the filter which was hybridized to ^{32}P-labeled pNOSNEO (a gift from Mike Fromm, Stanford University). This probe is a pBR322 plasmid that contains a chimeric neomycin phosphotransferase II (NPT II) gene. The chimeric gene contains 1.3 and 1.1 kb fragments from the 5′ and 3′ non-coding region of the *Nos* gene, respectively, and the 1.0 kb coding region of the *NPT II* gene from Tn*5*. As can be seen in Fig. 3C this probe should hybridize to the two internal *Hin*dIII fragments as well as to the right border region. *Asterisks* indicate the position of hybridizing *Hin*dIII fragments that have the same size as the internal *Hin*dIII fragments in the T-DNA. *Star* highlights approximate position of an internal fragment that would be generated by the integration of an inverted repeat of the T-DNA

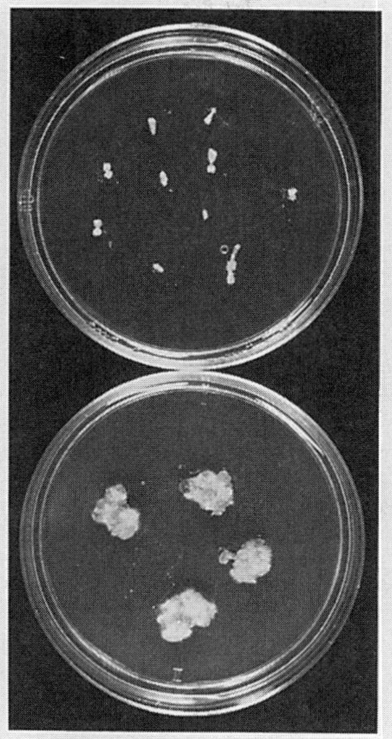

Fig. 5. Growth of 4-week-old stem explant calluses in the presence of G418. Calluses were cultured from a wild-type plant (*upper plate*) and a kanamycin-resistant T3 plant (*lower plate*) on the callus-inducing medium, R3 (Christianson 1977) containing 25 μg/ml G418 sulfate (Geneticin, Gibco Laboratories, Chagrin Falls, Ohio)

can reasonably be accomplished. This plant is well recognized as a model plant system for molecular biologists (Somerville et al. 1985; Meyerowitz and Pruitt 1985). Its many attributes include a very small genome (70000 kb/haploid genome) that contains 86%–90% non-repetitive DNA, a short life cycle (seed to seed in less than 6 weeks), self-fertilization, and a large seed set (> 10000 seeds/plant). Recently, we have developed a rapid and efficient regeneration procedure for this plant (Feldmann and Marks 1986). This protocol allows for the recovery of fertile whole plants from leaf or stem explants in as little as 7–8 weeks. However, when we adapted this procedure to a leaf-disc type *Agrobacterium*-mediated transformation scheme we found that it took 3–5 months to recover seed from transformed plants. Further, a considerable amount of somaclonal variation was observed in the regenerated transformants (unpublished observations). A transformation method that could circumvent the use of tissue culture and at the same time capitalize on the short life cycle of *Arabidopsis* would have distinct advantages.

We have presented definitive evidence that germinating seeds of *A. thaliana* can be transformed with *A. tumefaciens*. This evidence came from an analysis of kanamycin-resistant progeny resulting from seeds cocultivated with an *A. tumefaciens* strain containing an *NPT II* gene within the T-DNA of a Ti plasmid. The analysis revealed that: (1) in most cases the resistance factor was transmitted in a Mendelian fashion; (2) resistant plants expressed the NPT II enzyme activity; (3) resistant plants contained integrated T-DNA; and (4) explants from resistant plants had the ability to form callus on medium containing the antibiotic G418. Although the evidence demonstrates that transformation has taken place, it does raise some questions concerning "how" and "when" the transformational event actually occurred.

Two aspects of the data suggest that the process is not a straight-forward, simple transformation of the genetically effective cells in the embryo followed by Mendelian transmission of the resistance gene to the subsequent generations.

First, we found that very few, if any, of the resistant T2 plants were homozygous for the *NPT II* gene (see Table 4). The genetically effective cell number in *Arabidopsis* has been estimated by Li and Redei (1969) to be two. Each of these meristematic cells gives rise to approximately half of the reproductive portion of the plant and thus half of the resulting seed (Li and Redei 1969). If the transformational event occurs in one of the two genetically effective cells then three-eighths of the progeny from a transformed T1 plant would be resistant to kanamycin; kanamycin resistance is a dominant selectable marker. Furthermore, one-third of the resistant T2 plants should be homozygous for the resistance factor and produce only progeny (T3) that are resistant to kanamycin. However, this is not the case.

Second, all of the single T2 plants and T3 families examined had unique hybridization patterns indicating that there were many transformational events. If we suppose that the transformational event occurred in one of the two genetically effective cells of the embryo, we can estimate how many T1 seeds were transformed. For treatment 2, progeny (325000) were collected from 1420 T1 plants (230 T2 seed/T1 plant). There were approximately 1040 kanamycin-resistant T2 seedlings (3.2/1000) among these progeny. If 1 of the 2 genetically effective cells had a single functional insert, 86 of the 230 progeny (3/8) from a T1 plant would be resistant to kanamycin (kanamycin resistance is a dominant selectable marker). This would mean that only 12 (1040/86) unique families of insertional patterns would be identified. However, Southern hybridization analysis has revealed the presence of many more than 15 different families of insertional patterns (Fig. 4 and data not shown). This fact strongly argues that the genetically effective cells of the embryo are not the main target for *Agrobacterium* transformation.

Outcrossing between a transformed plant and sensitive plants cannot explain these results. The low frequency of outcrossing in this species ($<10^{-4}$; Redei 1975), and the large number of unique insertional patterns make it unlikely that one or a few transformed plants in a particular treatment could have contributed transformed pollen to sensitive plants.

The deficiency of homozygotes and large number of unique insertional patterns could be explained in two ways. There might be selection against the formation of seeds homozygous for the *NPT II* gene (selection against a gamete carrying an insert). It is, however, unlikely that selection against seeds homozygous for the *NPT II* gene occurs; normal numbers of homozygous plants have been identified in most T3 families. Additionally, this mechanism would not explain the large numbers of unique insertional patterns that were identified. A more likely explanation to account for the higher than expected number of unique transformants is that the transformational event occurred later in development.

We offer two hypotheses as to how the transformation of the genetically effective cells might occur later in development. The first hypothesis (for which we can offer no support), is that during cocultivation the agrobacteria enter the extracellular spaces and remain "quiescent" until they are activated by the production of a particular cell type in one gametophytic tissue or the other. The second hypothesis is that during cocultivation or shortly thereafter the transformation occurs in an epidermal cell (Layer 1 of the tunica). Layer I gives rise to the epidermis of the plant via anticlinal divisions (Esau 1965). Later in development, prior to the initiation and development of a microgametophyte or megagametophyte, there is a replacement (Stewart 1978) of a kanamycin-resistant cell from Layer I to Layer II (the layer that will give rise to the sporogenous tissue; Redei 1969) via a periclinal division. There are numerous examples of this event in the literature using a variety of markers (Esau 1965; Stewart and Burk 1970; Stewart and Dermen 1970). Stewart and Burk (1970) estimate that this replacement of cells from Layer I to Layer II occurs at a frequency of 5×10^{-4} in *Nicotiana*. If replacement of a cell from Layer I to Layer II has to occur to allow for the transmission of the *NPT II* gene to resulting progeny and if this rate is the same as in tobacco, then the majority of the treated T1 seedlings from treatment 2 should be chimeric for kanamycin resistance. Both mechanisms (transport of bacteria or cell replacement) would result in progeny that were predominantly heterozygous. Experiments are in progress to distinguish between these two possibilities.

The results of varying lengths of imbibition clearly show that a 12 h imbibition of T1 seeds prior to cocultivation with *Agrobacterium* resulted in the highest percentage (0.32%) of kanamycin-resistant T2 seedlings (Table 1). As the length of imbibition (at 22° C) moves away from 12 h the transformation frequency decreases ($<1/5000$ T2 seeds; Table 1 and data not shown). Under the described conditions no transformants resulted when seeds were imbibed for less than 9 h and few transformants resulted from seeds imbibed for more than 15 h at 22° C prior to cocultivation with *Agrobacterium*. These experiments have been repeated four times and each time maximal numbers of transformed progeny have been obtained from those treatments in which the seeds were imbibed for approximately 12 h before infection. Thus, there appears to be a 6 h window in which the imbibing seeds of *A. thaliana* are most susceptible to *Agrobacterium*. This window may be physical and/or chemical in nature. For example, germination may involve a wounding process in which cells of the embryo become susceptible to infection only after the testa has broken open. However, the cells may be susceptible to transformation by *Agrobacterium* for only a very limited time due to healing of the tissue or other processes (Stachel et al. 1985) occurring during the imbibition process. It is unclear as to why seeds imbibed for less than 9 h before infection yield no transformants since the *Agrobacterium* are present and continue to divide rapidly during imbibition.

The initial experiment and the four repeat experiments were carried out over an 8 month period (Feb–Aug, 1986). The T1 plants were grown in growth chambers and in greenhouses. The light quality and intensity, as well as the temperature, would have fluctuated somewhat over this period. We have not been able to detect any significant difference in the rate of transformation by unintentional variation of these physical parameters.

Budar et al. (1986) have recently published the genetic results of 44 independently transformed and regenerated *N. tabacum* plants. Ninety percent of the characterized plants (40/44) transmitted the antibiotic resistance gene in a Mendelian fashion. Thirty-five of the 40 clones (88%) segregated for 1 functional insert (progeny from a plant heterozygous for the *NPT II* gene segregated 3 kanamycin-resistant to 1 kanamycin-sensitive) and 5 (12%) for 2 inserts (15 resistant to 1 sensitive). We examined 124 kanamycin-resistant lines in the present study and we also found that 90% ($n=111$) of the T3 families displayed normal Mendelian segregation ratios (Table 3). The majority of these (62%) had one functional *NPT II* gene (T3 family segregation ratios of 3 resistant:1 sensitive). Twenty-three percent segregated for 2 unlinked inserts (15 resistant:1sensitive). The 6 families that were classified as having 2 linked inserts (Table 3) yielded T3 family segregation ratios that were approximately 6 resistant:1 sensitive. Chi-square analysis showed that these ratios do not fit either a 3:1 or 15:1 segregation ratio which would be expected for 1 or 2 unlinked functional inserts, respectively. Genetic and molecular analysis of T4 families from resistant plants will confirm or deny the linkage hypothesis (in progress). Six families segregated for 3 unlinked inserts (T3 family segregation ratios of 63 resistant:1 sensitive). A single family segregated for 4 unlinked functional inserts (255 resistant:1 sensitive). Three families have been classified as being pure (homozygous for resistance). However, to distinguish between homozygosity and 4 unlinked functional inserts in a T2 plant, 983 T3 seedlings need to be scored (Table 2). The high number of functional inserts in many plants in the present study may have resulted because of a high initial inoculum concentration or may be due to the nature of the interaction between germinating *Arabidopsis* seeds and *Agrobacterium*.

The non-Mendelian segregation patterns (observed here and in tobacco (Budar et al. 1986)) fall into two groups which are both characterized by an unexpectedly high proportion of sensitive plants in the T3 families (Table 4). The prevalence of sensitive seedlings in both groups appears most likely to be due to an instability of the insert, a loss of expression of the *NPT II* gene, or to the induction of a recessive mutation for gamete or embryo lethality. Genetic analysis of T4 families from resistant plants confirms the lack of chloroplast transformation in all lines examined (data not shown). Sensitive seedlings from these groups are being examined to confirm or deny the presence of physical inserts. In addition, T4 and T5 families from resistant plants are being genetically analyzed for kanamycin resistance.

A question of particular interest is "do the number of functional inserts correlate positively with the number of physical inserts?". The answer appears to be "no". Even when the genetic analysis indicates that the transformed families (lanes 2–6) contain only one functional insert, most of the plants contain multiple T-DNA inserts (Fig. 3). For example, transformant T2–130 segregates for one functional insert (3:1) but Southern analysis indicates seven or more physical inserts (lane 4, Fig. 4B). This anomaly could be due to the integration of many inserts into a small region of the genome or, integration of many inactive inserts at several points in the genome (inactive due to rearrangements or to insertion into unexpressed regions of the genome). We do not know which insert is functional; however, genetic analyses are now in progress which should allow for the identification of the functional insert. If a large number of the inserts are inactive then it is probable that some proportion of the sensitive T2 plants have inactive inserts.

In tissue cultures of *A. thaliana*, G418 acts as a selectable marker for the identification of transformed tissue (transformed with *NPT II*), and as expected, explants cultured from kanamycin-resistant T3 plants grow and form callus in the presence of G418 (25 µg/ml; Fig. 5). Additionally, transformed seedlings can be isolated from T2 seed populations on G418 (25 mg/l) but the contrast between resistant and sensitive seedlings is not as great as that for kanamycin selection. On G418 the sensitive seedlings remain green and short in stature while the resistant seedlings elongate. We have isolated 17 seedlings from T2 populations on G418 and all have tested positive for transformation (genetic segregation of T3 families).

The combination of this plant model system with this particular transformation protocol should be useful in the identification of genes from *Arabidopsis* by insertional mutagenesis. There are a number of reasons for our optimism in this regard. This technique allows for the rapid identification of non-wild-type phenotypes in the T2 (dominant only) and T3 (dominant or recessive) generations; plants are not regenerated from tissue culture so there is no somaclonal variation which could affect the phenotype of the plant; the average number of physical inserts in T2 plants or T3 families is three to four. In contrast to Budar et al. (1986), who maintain that the insertion at "a single locus stresses the advantage of the Ti plasmid as a gene delivery system and insertion mutagen", we would argue that the best gene tagging system for identifying mutations is one that yields multiple inserts. This would eliminate working with such a large "mutant" population. In addition, the insert, functional or silent, and the altered phenotype will rapidly cosegregate. Given an average of 3 inserts per plant we estimate that 50–70000 unique transformants would contain enough inserts to ensure a 95% probability that any single gene in the genome would contain an insert. For *Arabidopsis* this is not an unmanageable number of plants.

We are presently applying this technique to other geographic races of *A. thaliana* and other plant genera, e.g., *Nicotiana*, *Lycopersicon*, *Brassica*, *Capsicum*, and *Zea*. Preliminary data suggest that other geographic races of *A. thaliana* are susceptible to *Agrobacterium*-mediated seed transformation.

Acknowledgements. We thank M.L. Christianson for numerous insightful discussions, in particular, those on plant development. We also thank D.A. Warnick, S.M.E. Smith and M. Motoyama for technical assistance and D.A. Warnick, M.L. Christianson, D.P. Weeks, K.J. Brunke, and R.M. Sinibaldi for critically reading this manuscript. We thank J. Velten and J. Schell for providing us with the *Agrobacterium* strain and M. Fromm for the pNosNeo plasmid contribution. This work was funded by the Zoecon Research Institute of Sandoz Crop Protection Corporation and by an NSF grant (# DCB-8609623) awarded to M.D.M. and D.P. Weeks.

References

Budar F, Thia-Toong L, Van Montagu M, Hernalsteens JP (1986) *Agrobacterium*-mediated gene transfer results mainly in transgenic plants transmitting T-DNA as a single Mendelian factor. Genetics 114:303–313

Christianson, ML (1977) Zinc sensitivity in *Phaseolus*: expression in cell culture. Environ Exp Bot 19:217–221

De Block M, Herrera-Estrella L, Van Montagu M, Schell J, Zambryski P (1984) Expression of foreign genes in regenerated plants and in their progeny. EMBO J 3:1681–1689

Esau K (1965) Plant Anatomy. 2nd edn. John Wiley and Sons, New York, pp 448–450

Feldmann KA, Marks MD (1986) Rapid and efficient regeneration of plants from explants of *Arabidopsis thaliana*. Plant Science 47:63–69

Graves AC, Goldman SL (1986) The transformation of *Zea mays* seedlings with *Agrobacterium tumefaciens*: Detection of T-DNA specific enzyme activities. Plant Mol Biol 7:43–50

Hernalsteeens JP, Thia-Toong L, Schell J, Van Montagu M (1984) An *Agrobacterium*-transformed cell culture from the monocot *Asparagus officinalis*. EMBO J 13:3039–3041

Hess D (1969) Versuche zur Transformation an höheren Pflanzen: Wiederholung der Anthocyan-Induktion bei *Petunia* und erste Charakterisierung des transformierenden Prinzips. Z Pflanzenphysiol 61:286–298

Hess D (1973) Transformationsversuch an höheren Pflanzen: Untersuchungen zur Realisation des Exosomenmodells der Transformation bei *Petunia hybrida*. Z Pflanzenphysiol 68:432–440

Hess D (1977) Cell modification by DNA uptake. In: Reinert J, Bajaj YPS (eds) Applied and fundamental aspects of plant cell, tissue, and organ culture. Springer-Verlag, New York, pp 506–577

Kleinhofs A, Behki R (1977) Prospects for plant genome modification by nonconventional methods. Annu Rev Genet 11:79–101

Langridge J (1957) The asceptic culture of *Arabidopsis thaliana* L. Heynh. Aust J Biol Sci 10:243–252

Larkin PJ, Scowcroft WR (1982) Somaclonal variation – a novel source of variability from cell cultures for plant improvement. Theor Appl Genet 60:197–214

Ledoux L, Huart R (1968) Integration and replication of DNA of *M. lysodeidticus* in DNA of germinating barley. Nature 218:1256–1259

Ledoux L, Huart R, Jacobs M (1971) Fate of exogenous DNA in *Arabidopsis thaliana*. Translocation and integration. Eur J Biochem 23:96–108

Li SL, Redei GP (1969) Estimation of mutation rate in autogamous diploids. Radiat Bot 9:125–131

Lichtenstein CP, Draper J (1985) Genetic engineering of plants. In: Glover DM (ed) DNA cloning: a practical approach, vol 2. IRL Press, Washington, DC p 78

Maniatis T, Fritsch EF, Sambrook J (1982) Molecular cloning: a laboratory manual. Cold Spring Harbor Laboratory, New York

Meyerowitz EM, Pruitt RE (1985) *Arabidopsis thaliana* and plant molecular genetics. Science 229:1214–1218

Murashige T, Skoog F (1962) A revised medium for rapid growth and bioassays with tobacco tissue cultures. Physiol Plant 15:473–497

Ohta Y (1986) High-efficiency genetic transformation of maize by a mixture of pollen and exogenous DNA. Proc Natl Acad Sci USA 83:715–719

Otten L, De Greve H, Hernalsteens JP, Van Montagu M, Schieder O, Straub J, Schell J (1981) Mendelian transmission of genes introduced into plants by the Ti plasmids of *Agrobacterium tumefaciens*. Mol Gen Genet 183:209–213

Redei GP (1969) *Arabidopsis thaliana* (L.) Heynh.: A review of the genetics and biology. Bibliographia Genetica 21:1–151

Redei GP (1975) *Arabidopsis* as a genetic tool. Annu Rev Genet 9:111–151

Reiss B, Sprengel R, Will H, Schaller H (1984) A new sensitive method for qualitative and quantitative assay of neomycin phosphotransferase in crude cell extracts. Gene 30:211–218

Scholl RL, Keathley DE, Baribault TJ (1981) Enhancement of root formation and fertility in shoots regenerated from anther- and seedling-derived callus cultures of *Arabidopsis thaliana*. Z Pflanzenphysiol Bd 104:225–231

Southern E (1975) Detection of specific sequences among DNA fragments separated by gel electrophoresis. J Mol Biol 98:503–517

Somerville CR, McCourt P, Caspar T, Estelle M, Keith K (1985) *Arabidopsis thaliana* as a model system for plant genetics and molecular biology. In: Freeling M (ed) Plant genetics. Alan R. Liss, New York, pp 651–660

Stachel SE, Messens E, Van Montagu M, Zambryski P (1985) Identification of the signal molecules produced by wounded plant cells that activate T-DNA transfer in *Agrobacterium tumefaciens*. Nature 318:624–629

Stewart RN (1978) Ontogeny of the primary body in chimeral forms of higher plants. In: Subtelny S, Sussex IM (eds) The clonal basis of development. Academic Press, New York, pp 131–160

Stewart RN, Burk LG (1970) Independence of tissues derived from apical layers in ontogeny of the tobacco leaf and ovary. Am J Bot 57:1010–1016

Stewart RN, Dermen H (1970) Determination of number and mitotic activity of shoot apical initial cells by analysis of mericlinal chimeras. Am J Bot 57:816–826

Stroun M, Anker P, Charles P, Ledoux L (1966) Fate of bacterial deoxyribonucleic acid in *Lycopersicon esculentum*. Nature 121:397–398

Velten J, Schell J (1985) Selection-expression plasmid vectors for use in genetic transformation of higher plants. Nucleic Acids Res 13:6981–6998

Wallroth M, Gerats AGM, Rogers SG, Fraley RT, Horsch RB (1986) Chromosomal localization of foreign genes in *Petunia hybrida*. Mol Gen Genet 202:6–15

Weeks DP, Beerman N, Griffith OM (1986) A small-scale five-hour procedure for isolating multiple samples of CsCl-purified DNA: Application to isolations from mammalian, insect, higher plant, algal, yeast, and bacterial sources. Anal Biochem 152:376–385

Zhou GY, Weng J, Zeng Y, Huang J, Qian S, Liu G (1983) Introduction of exogenous DNA into cotton embryos. Methods Enzymol 101:433–481

Communicated by G.R. Fink

Received October 16, 1986

Note added in proof

The approximately 4.3 kb fragments that hybridized to the pNeoNos probe in Fig. 4 are of the correct size to have been formed by an inverted repeat about the right border. Additional data, to be reported elsewhere, indicates that inverted repeats about the left border also occurred and that the complex hybridization patterns in Fig. 4 are the result of tandem arrays of T-DNAs inserting into one or a few sites and not from large numbers of independent insertions. Complex arrays of T-DNA have also been observed in transformed Lycopersicon (Jorgensen R, Snyder D, and Jones JDG (1987) T-DNA is organized predominantly in inverted repeat structures in plants transformed with *A. tumefaciens* C58 derivatives. Mol Gen Genet, in press) and Petunia lines (Jones JDG, Gilbert DE, Grady KL, Jorgensen RA (1987) T-DNA structure and gene expression in petunia plants transformed by *A. tumefaciens* C58 derivatives. Mol Gen Genet, in press).

AGROBACTERIUM INFECTS MAIZE, AFTER ALL!

Barbara Hohn
Friedrich Miescher Institute, Basel, Switzerland
E-mail: barbara.hohn@fmi.ch

Commentary on the paper:

Grimsley, N., Hohn, T., Davies, J. W., and Hohn, B. 1987. *Agrobacterium*-mediated delivery of infectious maize streak virus into maize plants. Nature 325:177–179.

"Barbara, we have symptoms!" This was Nigel Grimsley on the phone, calling from the BL3 containment unit of the Friedrich Miescher Institute where our maize plantlets were growing. This call reached me on June 1, 1986, a Sunday, in my home near Basel where Thomas and I were just celebrating our older son's birthday with a lovely fondue. Why this excitement?

Since the earlier days of plant transformation, cereals have always been considered as prime targets. However, as their inoculation with wild-type strains of *Agrobacterium tumefaciens* never yielded real tumors, these plants of special scientific and agricultural importance were termed recalcitrant to transformation. Over the years, several apparent deficiencies were considered responsible for the resistance of cereals to *Agrobacterium*'s tricks, which are so irresistible to dicotyledonous plants: perhaps cereals do not produce chemicals acting as inducers of bacterial virulence; or the pathogen cannot attach to cereal cells; or perhaps the T-DNA cannot integrate into chromosomal DNA of monocots or cannot be expressed there; maybe the cereals cannot respond to increased levels of plant hormones produced under the direction of genes located on the T-DNA. Opine tests of whole tissue from some monocotyledonous plants, including maize inoculated with *Agrobacterium,* indicated that expression of the T-DNA encoded nopaline synthase gene can occur. However, the presence in the recipient plant cells of T-DNA could not be demonstrated. Clearly, a rigorous test was needed.

Nigel, then one of my postdocs, had previously demonstrated that *Agrobacterium*-mediated delivery of viral genomes to plants, which he termed agroinfection, is a very sensitive procedure (Grimsley et al. 1986). For this work he used a virus of known infectivity, *Cauliflower mosaic virus* (CaMV) and the plant *Brassica rapa* which was a known host for both the virus and *Agrobacterium.* Partial or complete dimers of the viral genome were placed in tandem orientation into the T-DNA of a binary vector and turnip plants were inoculated with agrobacterial cells containing such plasmids. With high efficiency and reliability the plants came down with CaMV symptoms. "Why not use an agroinfection device to test if an interesting plant such as maize, which is apparantly resistant to *Agrobacterium,* can at least be a recipient for T-DNA transfer?" said my husband Thomas Hohn, plant virus specialist. This assay would be independent of T-DNA integration (at least in theory), would not require tumor formation as a symptom and the assay for viral disease should be straight forward. "Great idea, but is there a maize virus clone with demonstrated infectivity?" "We have to ask Jeffrey Davies at the John Innes Institute."

In October and November 1984 several letters went back and forth between the continent and Great Britain. It became clear that the John Innes group had a clone of *Maize streak virus* (MSV), which had never been demonstrated to be infectious. In nature the virus, originating from South Africa, is transmitted by a leafhopper with the beautiful name *Cicadulina mbila.* In fact it was not even clear whether it would ever be infectious, because the genomes of gemini viruses known up to then, consisted of two circular components of about 2.7 kb of single-stranded DNA each. However, in MSV infected tissue the isolated geminate particles contained only one unit of circular DNA (Mullineaux et al. 1984).

Jeffrey Davies agreed to send the clone, but success of our experiments was based on a combination of four assumptions: Maize is a host for transfer of T-DNA from *Agrobacterium;* MSV DNA, located on the T-DNA, is able to recombine out of it in the cell nucleus, to produce DNA able to replicate; MSV infection is independent of factors such as insect vector compounds essential for natural delivery of the virus; and the provided clone is a faithful replica of the viral DNA and contains all genetic constituents for viral infectivity. In addition, there were two further, not entirely negligible obstacles to overcome. MSV work at John Innes was financially supported by a company claiming rights to the clone; since viral sequences on a T-DNA in an agrobacterial strain may be hazardous for the environment, the experiments, according to the Swiss Biosafety Officers, had to be conducted under BL3 conditions. The first obstacle was "easily" overcome by year-long discussions by lawyers, the second by installing strong lamps in one of our BL3 containment cabinets at the FMI, which were otherwise devoted to animal experiments.

Finally Nigel Grimsley could begin constructing another one of his famous pEAP vectors: a binary plasmid which can move from *E. coli* to *Agrobacterium* to Plants, in this case containing a complete duplicated sequence of MSV in the T-DNA. With enthusiastic help from our technical assistant Cynthia Ramos, maize seedlings were inoculated on various locations with an octopine-strain carrying pEAP vectors. Nothing happened. Nigel tried again—still no symptoms. I almost gave up after numerous further trials yielded the same negative outcomes; one or more of our assumptions must be wrong, we thought. Our hopes had faded.

However, Nigel had NOT given up. As a last rescue experiment he transformed C58 with his pEAP's and then, on June 1, 1986, "Barbara, we have symptoms."

As none of us in Basel had seen symptoms of MSV before, we had to make sure MSV DNA could be detected in non-inoculated leaves of infected plants. A day or two after the symptoms appeared Nigel extracted DNA, ran a gel and hybridized it with labeled MSV-DNA: the Southern blot was positive. To ensure that *bona fide* T-DNA transfer was the cause of viral infection, two control experiments were conducted. A T-DNA borderless plasmid called pEA2 was used and the theoretically infectious pEAP clone was introduced into an *Agrobacterium* strain which was avirulent due to a mutation in the *virA* gene. Neither of those strains, when inoculated onto maize seedlings, gave rise to symptoms, nor did inoculation with purified plasmid DNA containing a tandem MSV dimer. The publication appeared in early 1987 (following paper).

This was the successful end of one story, but also the beginning of many follow-up research activities. A whole battery of virulence mutants was tested and it was found that transfer of T-DNA with MSV sequences to maize was dependent on the same bacterial activities required for agroinfection of turnips with CaMV-T-DNA (Grimsley et al. 1989). The potential of MSV as a vector for delivery of sequences to maize plants could be evaluated and viral functions could be analyzed. An interesting analysis consisted of testing naturally MSV-resistant maize cultivars.

Once the agroinfection system was established, the use of viral sequences on the T-DNA was no longer necessary for the establishment of a transformation system. Only highly devoted work on the benches of plant tissue culture laboratories led to final success in the stable transformation of maize (Ishida et al. 1996). It is to be hoped that transgenic plants expressing important input—but more importantly output—traits will contribute to the food supply.

I would like to acknowledge colleagues who contributed to the work described: Thomas Hohn, Cynthia Ramos, Zdena Koukolikova-Nicola and our collaborators in the UK: Jeffrey Davies, Margaret Boulton and Phil Mullineaux. Very special thanks go to Nigel Grimsley for his stubbornly optimistic experimentation and the many good discussions we had, not only about science but also life in general, in the BL3 laboratory. However, this work was not done in isolation but rather during a general pioneering atmosphere in plant molecular biology. Special contributions came from plant scientists including Jeff Schell, Marc van Montegu, Gene Nester, Mary-Dell Chilton, Patti Zambryski, Paul Hooykaas, Cal Kado and Stanton Gelvin.

References

Grimsley, N., Hohn, B., Hohn, T., and Walden, R. 1986. Agroinfection, a novel route for plant viral infection using Ti plasmid. Proc. Natl. Acad. Sci. U.S.A. 83: 3282–3286.

Grimsley, N., Hohn, B., Ramos, C., Kado, C., and Rogowsky, P. 1989. DNA transfer from *Agrobacterium* to *Zea mays* or *Brassica* by agroinfection is dependent on bacterial virulence functions. Mol. Gen. Genet. 217: 309–316.

Ishida, Y., Saito, H., Ohta, S., Hiei, Y., Komari, T., and Kumashiro, T. 1996. High efficiency transformation of maize (*Zea mays* L.) mediated by *Agrobacterium tumefaciens*. Nat. Biotechnol. 14:745–750.

Mullineaux, P. M., Donson, J., Morris-Krsinich, B. A., Boulton, M. I., and Davies, J. W. 1984. The nucleotide sequence of maize streak virus DNA. EMBO J. 3:3063–3068.

BARBARA HOHN: I studied chemistry in Vienna and Tübingen and obtained my Ph.D. for work on bacteriophage fd. From 1967–1968 I was a postdoc at Yale, working with Don Marvin, a biophysicist and an important mentor. I did another postdoc with David Korn at Stanford Medical School, where I worked on the *E. coli* episome F. I returned to Europe and from 1970–1977 worked with my husband Thomas Hohn on morphogenesis of bacteriophage λ at the Biocenter, Basel University. My λ work moved into DNA packaging and later to the establishment of the cosmid cloning system. Ken Murray from Edinburgh became a very special mentor and adviser. From 1977 to the present, I have been studying *Agrobacterium* and homologous recombination at the Friedrich Miescher Institute in Basel. Ingo Potrykus, a botanist with a special vision for plant gene technology, introduced me to working with plants. I had great help in establishing my *Agrobacterium* work from Marc Van Montagu and Jeff Schell. I also had a very good contact in Mary-Dell Chilton and did collaborative work with Gene Nester, Cal Kado and Stan Gelvin.

Reprinted from Nature, Vol. 325, No. 7000, pp. 177-179, 8 January 1987

Agrobacterium-mediated delivery of infectious maize streak virus into maize plants

Nigel Grimsley*, Thomas Hohn*, Jeffrey W. Davies† & Barbara Hohn*

* Friedrich Miescher-Institut, PO Box 2543, CH-4002 Basel, Switzerland
† John Innes Institute, Colney Lane, Norwich NR4 7UH, UK

Cells of certain strains of *Agrobacterium* colonize plants by transferring a portion of their DNA (the T-DNA) into a host plant cell, so causing it to proliferate and produce substances (opines) which the bacteria can use as food[1]. Most dicotyledonous plants can act as hosts, but most monocotyledonous species (including the economically important gramineae) are thought not to be susceptible[2]. We have used agroinfection (*Agrobacterium*-mediated virus infection[3]) as a very sensitive assay to test whether DNA transfer at least occurs during *Agrobacterium* inoculations of maize, a graminaceous plant. Naked DNA of the geminivirus maize streak virus (MSV) is not infectious to plants by itself and the intact virus can only infect if transmitted by an insect vector[4]. We report here that whole maize plants develop symptoms of viral infection if inoculated with strains of *Agrobacterium* carrying tandemly repeated copies of MSV genomes in their T-DNA. Mutant *Agrobacterium* strains defective in transfer of T-DNA cannot transmit MSV DNA in this test. We conclude that cloned MSV DNA is biologically active and that *Agrobacterium* can transfer DNA to maize.

During transformation of plants by *Agrobacterium* the Ti-plasmid-encoded virulence genes are induced by compounds excreted from wounded plant tissues[5,6]. As a consequence, T-DNA, a stretch of the Ti-plasmid delimited by border sequences 25 base pairs (bp) long, is transferred to the plant cell where it becomes integrated into nuclear DNA[1,7]. Any DNA between the border sequences in the *Agrobacterium* becomes transferred to plants as well. If the T-DNA contains partially or completely duplicated genomes of viruses or viroids, single copies of virus genomes can escape and, at least for competent host-virus combinations, replicate and spread systemically[8-10]. As this process, termed 'agroinfection', is many orders of magnitude more efficient than infection by inoculation with nucleic acids[3], and because the assay for successful transfer is especially sensitive due to amplification of the transferred viral DNA and resultant symptom formation, we sought to test a host-virus combination which has been refractory to analysis until now. MSV is a leafhopper-transmitted geminivirus believed to contain one circle of single-stranded DNA (refs 11, 12). The viral genome replicates to a high copy number in the nucleus of proliferating cells (M. Boulton, P. Mullineaux and J.W.D., unpublished data), causing stunting and formation of yellow-streaked leaves in infected plants. Geminiviruses are thought to replicate by DNA intermediates only and are thus potentially interesting for use as plant expression vectors. The MSV genome, however, has never been successfully reintroduced into its host plant as native or cloned DNA[4].

A series of plasmids containing a tandemly repeated dimer of MSV DNA was constructed (Fig. 1*a*), and maize plants were infected with *Agrobacterium tumefaciens* containing MSV dimers in the T-DNA. If the strain contained intact T-DNA transfer functions, viral symptoms appeared within 2 weeks of plant inoculation (Table 1; Fig. 2). No symptoms of viral infection were obtained from control inoculations with naked MSV DNA, with *Agrobacterium* carrying MSV DNA in a broad host-range plasmid lacking T-DNA borders, or in *Agrobacterium*

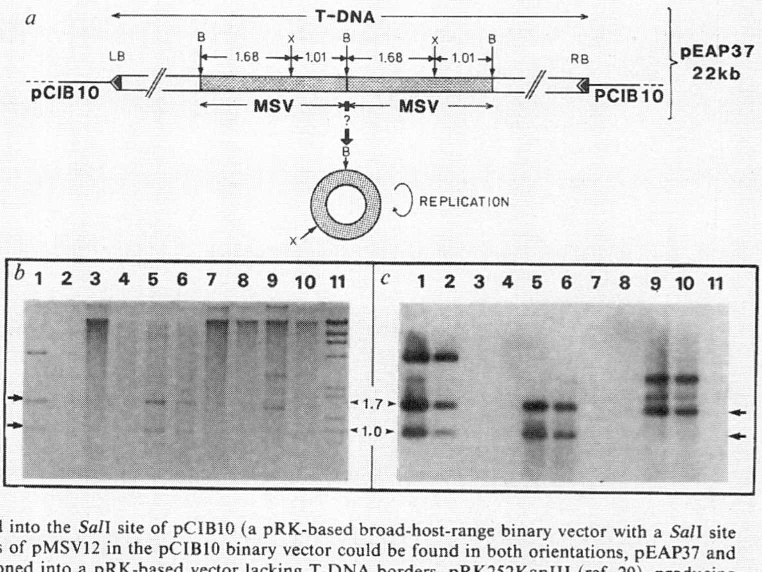

Fig. 1 *a*, Schematic representation of the input and output molecules of MSV agroinfection. LB, left border; RB, right border. Restriction enzyme sites: B, *Bam*HI; X, *Xho*I. Sizes are in kb. *b*, Electrophoretic analysis of DNA of infected and control maize plants. Lanes 1 and 2: 500 and 50 ng of pMSV 12 digested with *Bam*HI + *Xho*I; lanes 3-10: aliquots (1 μl) of DNA preparations (see below) of uninfected B73 (lanes 3 and 7) and GB (lanes 4 and 8) and infected B73 (lanes 5 and 9) and GB (lanes 6 and 10) plants. Samples in lanes 3 to 6 were restricted with *Bam*HI and *Xho*I and those in lanes 7-10 were left undigested. Lane 11, marker DNA (λ*Hin*dIII + ΦX174 RF *Hae*III). *c*, Southern transfer of the gel in *b* and hybridization to nick-translated pMSV12 DNA.

Methods. For plasmid construction (see also Table 1), MSV DNA was cut at its unique *Bam*HI site[11] and inserted into the unique *Bam*HI site of pACYC184 (ref. 26). A tandem dimer clone of MSV, pMSV12, found by screening bacterial mini-plasmid preparations[27], cut at its unique *Sal*I site in the pACYC184 part of the molecule, was introduced into the *Sal*I site of pCIB10 (a pRK-based broad-host-range binary vector with a *Sal*I site between the T-DNA border sequences[28]). Clones of pMSV12 in the pCIB10 binary vector could be found in both orientations, pEAP37 and pEAP38 respectively. Similarly, pMSV12 was cloned into a pRK-based vector lacking T-DNA borders, pRK252KanIII (ref. 29), producing the control binary vector pEA2. Mobilizations into *Agrobacterium* (C58, GV3101, ref. 30) were performed by triparental mating[31]. As a further control, pEAP37 was mobilized to a strain of *Agrobacterium* carrying a *vir*A mutation[13]. Verification of constructs in *Agrobacterium* was done by restriction and Southern blotting analysis[32] (data not shown). All *Agrobacterium* strains containing MSV dimers were produced and handled in a BL3 laboratory. For plant DNA extractions, ~400 mg fresh weight of young leaf tissue was homogenized in a pestle and mortar on ice in 0.5-1 ml of STEN (15% sucrose, 50 mM Tris-HCl, 50 mM Na_3EDTA, 0.25 M NaCl, *p*H 8), using sand (~50 mg) to aid tissue breakage. The homogenate was transferred to a tube (1.5 ml) and centrifuged for 5 min at 4 °C at top speed in a clinical centrifuge. The supernatant was discarded and the crude nuclear pellet resuspended in 0.5 ml ice-cold SET (15% sucrose, 50 mM Na_3EDTA, 50 mM Tris-HCl, *p*H 8) by stirring with a sterile toothpick, then vortexing for 5 s. Then 10 μl of 20% SDS and 100 μl of proteinase K (20 mg ml^{-1}) was added, mixed, and the tube was heated to 68 °C for 10 min. After 3 M sodium acetate was added (1/10th volume), the lysate was extracted twice with phenol/chloroform (3:1). Nucleic acids were precipitated by addition of 2 vols ethanol and leaving overnight at −20 °C. Centrifugation at top speed in a clinical centrifuge (10 min) gave a DNA-containing pellet; the supernatant was discarded and the pellet redissolved in 40 μl TE (10 mM Tris-HCl, 1 mM Na_3EDTA, *p*H 8). Aliquots (1 μl) were used for restriction mapping in a final volume of 20 μl assay buffer.

carrying MSV DNA as pEAP37, but with a mutation in the *vir*A locus[13]. Contamination by an insect vector as a source of infection can be ruled out, because plants were kept under insect-free growth conditions and the specific insect vectors (*Cicadulina mbila*) are not found in Switzerland in nature. DNA was extracted from infected and uninfected plants and analysed on agarose gels (Fig. 1*b*); only DNA from infected plants showed distinct bands. Undigested DNA showed two strong bands, migrating at the expected positions for supercoiled and open circular double-stranded replicative form (RF) MSV DNA. DNA cut with *Xho*I and *Bam*HI gave fragments of 1.7 kb and 1.0 kb, characteristic of MSV RF. The observed bands cannot result from lysed *Agrobacterium* cells because (1) the pEAP37 binary vector is much larger (22 kb) and there are no bands visible in this region of the gel in undigested samples; (2) the probe containing MSV and pACYC184 sequences does not detect pEAP37-specific restriction fragments other than those from MSV and (3) the leaves used for DNA isolation were distant from the site of bacterial inoculation. Southern analysis, using cloned MSV DNA as a probe, confirmed the identity of the virus (Fig. 1*c*). Two independent clones of this DNA (from different plants) are identical to the cloned DNA used for agroinfection as shown by restriction maps made with the frequently cutting enzymes *Sau*IIIA and *Alu*I (data not shown); thus no major changes have occurred. The apparent absence of single-stranded DNA on gels (Fig. 1) most probably arises because of the simple DNA extraction procedure used, which does not specifically purify the virion DNA. However, a cell fraction enriched for virus particles[14] yielded, upon proteinase digestion, an S_1-sensitive DNA species of lower relative molecular mass which we presume to be single-stranded viral DNA (data not shown). Further work is necessary for detailed characterization of the nature of the infection, but our observations suggest that the cloned viral DNA encodes all of the information necessary for production of a systemically spreading virus.

Tumour formation provides a sensitive assay of T-DNA transfer to host cells, because transformants acquire the capacity for persistent proliferation. This may not be valid for monocotyledonous plants, as integration of the transforming DNA, expression of the tumour genes, and response of the host plant to the hormones expressed, are conditions which may not be

Table 1 Maize streak virus symptom appearance after various inoculations of maize plants

Inoculum	T-DNA borders	Virulence	No. of plants with symptoms/number of inoculated plants
C58(pTiC58, pEAP37)	+	+	38/43
C58(pTiC58, pEAP38)	+	+	14/15
C58(pTiC58, pEA2)	−	+	0/24
A2034(pTiA6NC*vir*A, pEAP37)	+	−	0/19
MSV DNA (pMSV12)	−	−	0/12

Plasmids carried by bacterial strains are given in parenthesis. Growth of bacteria and all manipulations with whole plants were performed under BL3 containment. Plasmid DNA prepared from pMSV12 was dissolved in 10 mM Tris, 1 mM EDTA at a concentration of 125 $\mu g\ ml^{-1}$. Strains of *Agrobacterium* were grown in YEB[9] in shaking liquid cultures supplemented with appropriate antibiotics at 28 °C for 48 h, then resubcultured in the same medium using a 1:20 dilution, and grown for a further 20 h. Cells were then harvested by centrifugation and resuspended in 1/20 volume MSSP medium[5]. For inoculation of plants, 10-day-old seedlings (~10 cm high) of *Zea mays*, varieties 'Golden Cross Bantam' (GB) and 'B73', were used; inoculation and subsequent incubation of plants was at 22 ± 2 °C under ~5000 lx continuous white light (Philips 400 W G/92/2). Naked DNA or bacterial cells (40 μl of either) prepared as above were inoculated onto each plant. The inoculum was divided: 20 μl was injected into the region of the plant comprising stem plus leaf sheaths using a 25G hypodermic syringe and 20 μl was rubbed onto a single 2–6 cm-long upper leaf which had previously been dusted with carborundum powder to provide abrasion/wounding. Gentle rubbing was continued until the leaf appeared wet all over. Symptoms appeared 7–18 days following inoculation. Observation continued for a total of 6–7 weeks, but no further plants developed symptoms. Recent experiments indicate that stem inoculations are much more effective than leaf inoculations, and that the kind of medium in which the bacteria are resuspended prior to inoculation is not critical (data not shown).

Fig. 2 Infected and uninfected leaves of maize. Photographs were taken 12 days after inoculation with (left) C58(pTiC58, pEAP37) and (right) C58(pTiC58, pEA2).

fulfilled. Opine tests of whole plant tissue inoculated with *Agrobacterium* indicate that, in the case of two non-graminaceous monocotyledonous plants[15,16] and maize[17] expression of the T-DNA nopaline synthase gene can occur, but the presence of the transforming DNA in the recipient plant cells could not be demonstrated. Agroinfection, on the other hand, is an extremely sensitive test, theoretically requiring the presence of only one infectious DNA molecule, whereby amplification by replication and systemic spread allows recloning and analysis of part of the transferred DNA.

The requirement for T-DNA borders[18] suggests that the transfer proceeds by the same route as it does in dicotyledonous plants. The T-DNA is most probably transferred to the nucleus, for this is the cellular compartment in which MSV replicates. To test the involvement of *vir* gene functions we used a *vir*A[13] strain of *Agrobacterium* carrying pEAP37, which did not agroinfect plants. However, this mutation is not in the C58 chromosomal background; detailed analysis of the effects of *vir* mutants, as performed for dicotyledonous plants using another agroinfection system[8], and of different strains of *Agrobacterium* must await further studies. Whatever the mechanism, because agroinfection of maize is possible when virulence genes have not been induced by an exogenously added compound[5], we must assume that either *Zea mays* produces substance(s) which can induce the virulence genes or that the uninduced level of virulence gene expression is sufficient.

We do not know the mechanism by which MSV DNA escaped from the T-DNA. Two routes[19], not mutually exclusive, can be considered: (1) homologous recombination between the tandemly repeated MSV genomes produces a circular, unit length MSV DNA molecule and (2) a replicative intermediate is produced directly from the T-DNA. It is also not known whether the T-DNA had integrated in the maize genome, since only the amplified event could be observed. In a different system[10], tandemly repeated genomes of tomato golden mosaic virus integrated in a plant genome have been observed to escape and give rise to infection, demonstrating that it is at least possible for a geminivirus to be replicated from an integrated form.

T-DNA has been shown to become rearranged upon induction of bacterial *vir* genes by plant cells[20] and some data suggest that only a single strand of DNA may be transferred to the plant cell[21]. If true, this would mean that agroinfection would lead to delivery of either the (+) strand or the (−) strand of a virus, depending on the orientation of the viral sequences with respect to the T-DNA borders. The infectivity of viral (+) or (−) strands may differ, so the orientation of the virus in the T-DNA might affect the frequency of agroinfection. However, pEAP37 and pEAP38 (a vector identical to pEAP37 except for the orientation of the MSV genomes) were equally agroinfectious (Table 1), indicating either that it does not matter which strand of MSV is transferred, or that transfer of double-stranded DNA occurs.

This is being investigated further (G. Bakkeren, Z. Koukoliková-Nicola, B.H. and N.G., unpublished data).

Although it is not clear why native and cloned MSV DNA is not infectious, the experiments described show that the cloned DNA has the potential to replicate and produce symptoms by itself and is not dependent on other viral or insect components. Agroinfection should allow the study of mutant strains of MSV produced *in vitro* after introduction of the cloned DNA into whole plants, enabling the evaluation of such viruses as vectors for propagation of foreign DNA. Viruses also provide defined nucleic acid sequences which can be used to study host-encoded functions such as recombination and gene expression. The expression of hybrid marker genes in undifferentiated monocotyledonous plant cells has been demonstrated by direct gene transfer[22-25]: *Agrobacterium*-mediated transformation should now allow the study of such constructs in whole maize plants and possibly in other gramineae.

We thank Michael Bevan, Eugene Nester, Georgia Helmer, Bret A. M. Morris-Krsinich, Philip M. Mullineaux, Jonathan Donson, Margaret Boulton and Peter Markham for contributing materials and for valuable discussions, Cynthia Ramos for expert technical help, and Mike Saul and Fred Meins for critical reading of the manuscript. The Agrigenetics Corporation and the Advanced Science Company provided financial support for the cloning and characterization of MSV at the John Innes Institute.

Received 1 September; accepted 14 November 1986.

1. Gheysen, G., Dhaese, P., Van Montagu, M. & Schell, J. in *Genetic Flux in Plants* (eds Hohn B. & Dennis, E. S.) 12-27 (Springer, New York, 1985).
2. DeCleene, M. *Phytopath. Z.* **113,** 81-89 (1985).
3. Grimsley, N. & Bisaro, D. in *Plant DNA Infectious Agents* (eds Hohn, T. & Schell, J.) (Springer, New York, in the press).
4. Davies, J. W., Townsend, R. & Stanley, J. in *Plant DNA Infectious Agents* (eds Hohn, T. & Schell, J.) (Springer, New York, in the press).
5. Stachel, S. E., Messens, E., Van Montagu, M. & Schell, J. *Nature* **318,** 624-629 (1985).
6. Bolton, G. W., Nester, E. W. & Gordon, M. P. *Science* **232,** 983-985 (1986).
7. Chilton, M-D., Saiki, R. K., Yadav, N., Gordon, M. P. & Quetier, F. *Proc. natn. Acad. Sci. U.S.A.* **77,** 4060-4064 (1980).
8. Gardner, R. & Knauf, V., *Science* **231,** 725-727 (1985).
9. Grimsley, N., Hohn, B., Hohn, T. & Walden, R. *Proc. natn. Acad. Sci. U.S.A.* **83,** 3282-3286 (1986).
10. Rogers, S. G. *et al. Cell* **45,** 593-600 (1986).
11. Mullineaux, P. M., Donson, J., Morris-Krsinich, B. A. M., Boulton, M. I. & Davies, J. W. *EMBO J.* **3,** 3063-3068 (1984).
12. Howell, S. *Nucleic Acids Res.* **12,** 7359-7375 (1978).
13. Klee, H. J., White, F. F., Iyer, V. N., Gordon, M. P. & Nester, E. W. *J. Bact.* **153,** 878-883 (1983).
14. Bock, K. R., Guthrie, E. J. & Woods, R. D. *Ann. appl. Biol.* **77,** 289-296 (1974).
15. Hernalsteens, J. P., Thia-Toong, L., Schell, J. & Van Montagu, M. *EMBO J.* **3,** 3039-3042 (1984).
16. Hooykaas-Van Slogteren, G. M. S., Hooykaas, P. J. J. & Schilperoort, R. A. *Nature* **311,** 763-764 (1984).
17. Graves, A. C. F. & Goldman, S. L. *Plant molec. Biol.* **7,** 43-50 (1986).
18. Wang, K., Herrera-Estrella, L., Van Montagu, M. & Zambryski, P. *Cell* **38,** 455-462 (1984).
19. Grimsley, N., Hohn, T. & Hohn, B. *EMBO J.* **5,** 641-646 (1986).
20. Koukolíková-Nicola, Z. *et al. Nature* **313,** 191-196 (1985).
21. Stachel, S. E., Timmerman, B. & Zambryski, P. *Nature* **322,** 706 (1986).
22. Potrykus, I., Saul, M. W., Petruska, J., Paszkowski, J. & Shillito, R. D. *Molec. gen. Genet.* **199,** 183-188 (1985).
23. Lörz, H., Baker, B. & Shell, J. *Molec. gen. Genet.* **199,** 178-182 (1985).
24. Fromm, M. E., Taylor, L. P. & Walbot, V. *Nature* **319,** 791-793 (1986).
25. Uchimiya, H. *et al. Molec. gen. Genet.* **204,** 204-207 (1986).
26. Chang, A. C. Y. & Cohen, S. N. *J. Bact.* **134,** 1141-1146 (1978).
27. Birnboim, H. C. & Doly, J. *Nucleic Acids Res.* **7,** 1513-1523 (1979).
28. Rothstein, S. J. *et al. Gene* (in the press).
29. Bevan, M. *Nucleic Acids Res.* **12,** 8711-8721 (1984).
30. Holsters, M. *et al. Plasmid* **3,** 212-230.
31. Rogers, S. G., Horsch, R. B. & Fraley, R. T. *Meth. Enzym.* **118,** 627-640 (1986).
32. Dhaese, P. *et al. Nucleic Acids Res.* **7,** 1837-1849 (1979).

THE UPS AND DOWNS OF *Bt* GENE EXPRESSION IN PLANTS

David Fischhoff
Monsanto Company, St Louis, Missouri, United States
E-mail: david.a.fischhoff@monsanto.com

Commentary on the papers:

Vaeck, M., Reynaerts, A., Höfte, H., Jansens, S., De Beuckeleer, M., Dean, C., Zabeau, M., Van Montagu, M., and Leemans, J. 1987. Transgenic plants protected from insect attack. Nature 328:33–37.

Perlak, F. J., Fuchs, R. L., Dean, D. A., McPherson, S. L., and Fischhoff, D. A. 1991. Modification of the coding sequence enhances plant expression of insect control protein genes. Proc. Natl. Acad. Sci. U.S.A. 88:3324–3328.

The development of insect resistant transgenic plants is one of the major achievements in the application of *Agrobacterium* to the genetic modification of agronomic traits in plants. Transgenic plants expressing genes from *Bacillus thuringiensis* (*Bt*) to confer insect resistance have become significant and important commercial products for insect control in agriculture. Work in this area occurred in multiple laboratories around the world beginning in the 1980s, with some important precursors to the work beginning even earlier. Transgenic insect resistant crop plants were first commercialized in the mid-1990s. Along the path from the earliest work to commercialization, there were several important milestones. In this commentary I will reflect on two publications, one from my colleagues and myself, and one from another laboratory, both of which help to mark some key results on this research path.

The development of insect resistant plants represents the convergence of two lines of research. One of these was the discovery and characterization of genes that could produce insecticidal proteins. The second was the development of effective means to transform plants with foreign genes that could be expressed in plants especially through the use of *Agrobacterium;* this second topic is extensively covered in this volume.

For background it is important to review a bit of the history of *Bt,* which had long been known as an insecticidal bacterium. By the 1950s it was recognized that the insecticidal activity of *Bt* was typically contained in the remarkable (and visually striking in micrographs) parasporal protein crystals that *Bt* forms upon sporulation. By the 1960s, *Bt* microbial preparations made by fermentation were in use by farmers as insecticides due to their ability to control some important insect pests. The most common isolates of the *Bt* bacterium were active against lepidopteran pests (i.e., the caterpillar larvae of moths and butterflies), which include some of the most damaging insect pests in agriculture such as the bollworms of cotton, corn borers and armyworms. Later work showed that the protein crystals that *Bt* formed were composed of one or a few protein subunits, and it was further shown that isolated individual protein subunits from the crystals were insecticidal. These proteins showed a high degree of specificity. For example, typical lepidopteran-active *Bt* proteins had no activity against other orders of insects including most beneficial insects and no activity against non-insects including invertebrates such as worms or vertebrates such as fish, birds and mammals.

The isolation and characterization of individual *Bt* proteins set the stage for genetic engineering because single isolated insecticidal proteins were found to be encoded by single genes (Schnepf and Whiteley 1981) which made the subsequent work on plant genetic engineering easier. With the advent of recombinant DNA technology, individual *Bt* genes encoding insecticidal proteins were isolated and characterized. Importantly, it was shown that the isolated genes could be expressed in heterologous systems like *E. coli* and that the proteins produced from isolated genes retained their insecticidal activity and specificity.

There was one additional observation from *Bt* gene characterization that would be important in some of the plant genetic engineering. The typical lepidopteran-active *Bt* proteins were quite large, about 130 kDa. In the test tube, these could be digested with proteases to create a protease resistant fragment of about half that size. This fragment was derived from the N-terminal half of the protein, and still retained insecticidal activity. Susceptible insects also seemed to digest the full-length *Bt* proteins to this protease resistant portion in their guts after ingesting the crystals.

With *Bt* genes in hand, it was shown that a truncated *Bt* gene that contained somewhat more than half the gene from the 5′-end could make a truncated *Bt* protein that remained insecticidally active (Schnepf and Whiteley 1985).

Among the laboratories that early on isolated and characterized *Bt* genes were Monsanto (where my colleague Fred Perlak led this work) and the group of Mark Vaeck at Plant Genetic Systems (PGS) in Belgium. At both of these institutions, the techniques for transforming plants with *Agrobacterium* and for expressing foreign genes in transformed plants were also in hand. So, the stage was set to begin the experiments to determine if we could develop insect resistant plants through the expression of *Bt* genes.

A key year in documenting this work was 1987. Mark Vaeck and his coworkers at PGS published the results of their work in *Nature* in July of 1987. My colleagues and I published our work the following month (Fischhoff et al 1987), and Ken Barton and his coworkers published their work later in that year (Barton et al. 1987). There are certain important similarities in the results reported by all of these groups. The paper by Vaeck et al. documents the significant results: transgenic tobacco plants can express truncated *Bt* genes at levels sufficient to kill some lepidopteran insects. Their paper included a striking photograph of a tobacco plant expressing *Bt* that was then challenged with tobacco hornworm (*Manduca sexta*) larvae. This plant showed no apparent damage from the larvae while a companion plant with no *Bt* gene had leaves that were nearly completely consumed by the insects. In addition to the photograph, this paper also provided an analysis of *Bt* gene expression in these tobacco plants, and this analysis made two important points. First, truncated *Bt* genes were critical to the results obtained; full-length *Bt* genes were not expressed in their tobacco plants. Second, while the truncated *Bt* genes were expressed at insecticidal levels in these plants, the absolute level of expression was relatively low, in terms of both protein and mRNA produced. Our paper, published just after Vaeck's, showed similar results, but in tomato. (However, in tomato we were able to detect some expression from a full-length gene as well as a truncated gene.)

The 1987 publications showed that some insect control was possible with certain forms of *Bt* genes in transgenic plants, but they also pointed to new questions: Why were *Bt* genes so poorly expressed in plants? Could plants with these low levels of expression be commercially useful insect control products in agriculture? What might be done to increase the level of expression?

Fred Perlak and I and our coworkers at Monsanto turned our attention to these questions. Different insect species have different levels of sensitivity to a given *Bt* protein. Some insects such as tobacco hornworm, which was the primary insect used in the papers published in 1987, are quite sensitive. Other insects including many of the most important pests in major crops such as the cotton bollworm (*Helicoverpa zea,* also known as the corn earworm or the tomato fruit worm depending on which crop it is infesting) are many times less sensitive to *Bt.* We believed that a commercially successful insect-resistant transgenic plant would have to provide insect protection that was comparable to chemical insecticides. While we had shown, for example in our 1987 publication, that our initial transgenic tomato plants with truncated *Bt* genes could provide some protection against less sensitive insects like tobacco budworm (*Heliothis virescens*) and tomato fruitworm (*H. zea*), our analysis led us to conclude that a key to making commercially successful insect resistant plants was to substantially increase the expression of *Bt* genes in plants.

Our analysis also led us to the view that the problem in expressing *Bt* genes in plants was the *Bt* coding sequence itself. The *Bt* genome has a very high A+T base composition, quite unlike that of plants. A consequence of this is that *Bt* genes contain multiple sequence motifs that are high in A+T. Many of these sequence motifs resemble sequences that are not typically found in coding regions of plant genes. Moreover, some of these sequence motifs can resemble certain types of regulatory sequences of plant genes such as polyadenylation signals, signals involved in mRNA degradation, transcription termination signals, and others. We believed that these types of sequence motifs when present in native *Bt* genes led to very low accumulation of stable *Bt* mRNA which in turn led to very low production of *Bt* protein. An additional consequence of the high A+T content of *Bt* genes is that the codon usage of *Bt* genes is unlike that of plant genes.

So, having identified the *Bt* coding sequence as a putative problem and having identified what we believed were some of the specific problems in the *Bt* genes, Fred Perlak and I set about designing new coding sequences for *Bt.* The new sequences would reduce or eliminate these problematic sequences while still encoding insecticidal proteins, which in many cases remained identical to native *Bt* proteins. Our program of *Bt* gene redesign took two approaches: Selective removal/modification of sequences that we believed were deleterious for plant expression, and complete coding sequence redesign that we often implemented via gene synthesis. We then worked with our colleagues at Monsanto to transform plants with our redesigned genes and to assess the transgenic plants for expression of Bt protein and mRNA and for their insect controlling activity.

The results of all this work were dramatic. As shown in our 1991 publication, *Bt* protein levels in these plants increased up to several-hundred fold compared to plants with native *Bt* genes. Along with this increased protein expression, the plants showed a very large increase in insecticidal activity including a high level of control of agriculturally important pest insects. In parallel with the work reported in the 1991 paper, which described tobacco and tomato plants, we had also transformed cotton plants with a redesigned *Bt* gene. In an initial publication of these results we showed cotton plants that were highly resistant to the cotton bollworm (Perlak et al. 1990).

In further extensions of this work, we applied the same gene redesign approach to a wholly different gene from *Bt* var. *tenebrionis* and obtained excellent control of the Colorado potato beetle in transgenic potato (Perlak et al. 1993), and this approach has since been applied to many other *Bt* genes by our group and by other laboratories. In fact, even full length *Bt* genes, which could not be expressed from native genes in tobacco (for example, in the Vaeck et al. 1987 paper), were expressed from our constructs at very high levels with redesigned synthetic *Bt* genes and have been used in successful commercial products.

The synthetic gene approach was the key for our subsequent development of commercial insect resistant plant products such as Bollgard® cotton and YieldGard® corn. In addition to Fred Perlak and myself, many of our Monsanto colleagues contributed to this work over the years. They all helped to make this an exciting and successful research venture which turned that research into important tools for farmers around the world.

References

Barton, K. A., Whiteley, H. R., and Yang, N.-G. 1987. *Bacillus thuringiensis* δ-endotoxin expressed in transgenic *Nicotiana tabacum* provides resistance to lepidopteran insects. Plant Physiol. 85:1103–1109.

Fischhoff, D. A., Bowdish, K. A., Perlak, F. J., Marrone, P. G., McCormick, S. M., Niedermeyer, J. G., Dean, D. A., Kusano-Kretzmer, K., Mayer, E. J., Rochester, D. E., Rogers, S. G., and Fraley, R. T. 1987. Insect tolerant tomato plants. Bio/Technology 5:807–813.

Perlak, F. J., Deaton, R. W., Armstrong, T. A., Fuchs, R. L., Sims, S. R., Greenplate, J. T., and Fischhoff, D. A. 1990. Insect resistant cotton plants. Bio/Technology 8:939–943.

Perlak, F. J., Stone, T. B., Muskopf, Y. M., Petersen, L. J., Parker, G. B., McPherson, S. A., Wyman, J., Love, S., Reed, G., Biever, D., and Fischhoff, D. A. 1993. Genetically modified potatoes: Protection from damage by Colorado potato beetles. Plant. Mol. Biol. 22:313–321.

Schnepf, H. E., and Whiteley, H. R. 1981. Cloning and expression of the *Bacillus thuringiensis* crystal protein gene in *Escherichia coli.* Proc. Natl. Acad. Sci. U.S.A. 78:2893–2897.

Schnepf, H. E., and Whiteley, H. R. 1985. Delineation of a toxin-encoding segment of a *Bacillus thuringiensis* crystal protein gene. J. Biol. Chem. 260:6273–6280.

DAVID A. FISCHHOFF: In 1974, I graduated with a B.S. in Biology from the Massachusetts Institute of Technology and in 1980, I completed my Ph.D. at Rockefeller University. My education and training were greatly enriched by the opportunity to work with some truly wonderful scientists. I completed my undergraduate research with David Botstein and Nancy Kleckner and graduate training was carried out in the laboratory of Norton Zinder with Peter Model and other tremendous colleagues. My postdoctoral research was accomplished with Maynard Olson. At Monsanto I had the great fortune to collaborate with Fred Perlak on *Bt* genes and their expression in plants, and the opportunity to work with a remarkable community of scientists in biotechnology.

Reprinted from Nature, Vol. 327, No. 6125, pp. 33-37, 2 July 1987

Transgenic plants protected from insect attack

Mark Vaeck, Arlette Reynaerts, Herman Höfte, Stefan Jansens, Marc De Beuckeleer, Caroline Dean*, Marc Zabeau, Marc Van Montagu & Jan Leemans

Plant Genetic Systems NV, Jozef Plateaustraat 22, B-9000 Gent, Belgium

The Gram-positive bacterium Bacillus thuringiensis *produces proteins which are specifically toxic to a variety of insect species. Modified genes have been derived from* bt2, *a toxin gene cloned from one* Bacillus *strain. Transgenic tobacco plants expressing these genes synthesize insecticidal proteins which protect them from feeding damage by larvae of the tobacco hornworm.*

MODERN agriculture uses a wide variety of insecticides to control insect damage. Most of them are chemically synthesized. Notable exceptions are the insect toxins produced by *Bacillus thuringiensis*: spore preparations of this Gram-positive bacterium have been used for more than 20 years as a biological insecticide[1]. The insecticidal activity resides in crystalline inclusion bodies produced during sporulation of the bacteria, which are composed of proteins (termed delta endotoxins) specifically toxic against a variety of insects. Different strains of *B. thuringiensis* differ in their spectra of insecticidal activity. Most are active against Lepidoptera, but some strains specific to Diptera[2,3] and Coleoptera[4,5] have been identified. The crystals dissolve in the alkaline conditions of the insect midgut and release proteins of relative molecular mass 65,000–160,000 (M_r 65K–160K)[2,5,6] which are proteolytically processed by midgut proteases to yield smaller toxic fragments[7]. *B. thuringiensis* insect toxins are highly specific, in that they are not toxic to other organisms. Hence, they are safe insecticides and present an interesting alternative to chemical control agents. Their commercial use however is limited by high production costs and the instability of the crystal proteins when exposed in the field.

We have used *Agrobacterium*-mediated T-DNA transfer[8] to express chimaeric *B. thuringiensis* toxin genes in tobacco plants with the objective of protecting the plants from insect attack. We show here that a defence mechanism against phytophagous insects can be devised by genetic engineering.

* Permanent address: Advanced Genetic Sciences Inc., San Pablo Avenue, 6701, Oakland, California 94608, USA.

Modified *Bacillus* toxins

We have reported[9] the cloning of the *bt2* gene from *B. thuringiensis* strain berliner 1715 and the characterization of the recombinant polypeptide expressed in *Escherichia coli*. This protein, termed Bt2, is 1,155 amino acids long and is a potent toxin to several lepidopteran larvae, such as those of *Manducta sexta*, a pest of tobacco. Bt2 is a protoxin and generates a smaller polypeptide of M_r 60K which retains full toxic activity[9]. The smallest fragment of Bt2 that is still fully toxic has been mapped in the NH_2-terminal half of the protein, between amino-acid positions 29 and 607 (ref. 9) (and see Fig. 1*a*).

In plant transformation experiments, we used chimaeric genes containing the entire coding sequence of *bt2* as well as truncated genes. A diagram of the chimaeric genes is shown in Fig. 1*b*. Some of our T-DNA constructs include a chimaeric neomycin phosphotransferase gene (*neo*) as a marker selectable in plants[10]. Others carry translational fusions between fragments of *bt2* and the *neo* gene. Fusions to the 5′ end of the *neo* gene still confer kanamycin resistance in bacteria[11] and in plants[12]. Plasmid pLB884 (ref. 9) contains the truncated gene *bt884* and encodes an NH_2-terminal fragment of Bt2 up to amino-acid position 610. In *E. coli* it produces a polypeptide of the expected size which is fully toxic towards insect larvae[9].

Plasmids pLBKm860 and pLBKm23 (H.H. *et al.*, in preparation) contain fusion genes *bt* : *neo860* and *bt* : *neo23* encoding NH_2-terminal fragments of Bt2 up to amino-acid position 683 and 724, respectively. *E. coli* cells harbouring these plasmids are resistant to kanamycin and produce fusion proteins Bt : NPT860 and Bt : NPT23 which have the expected M_r of 106K

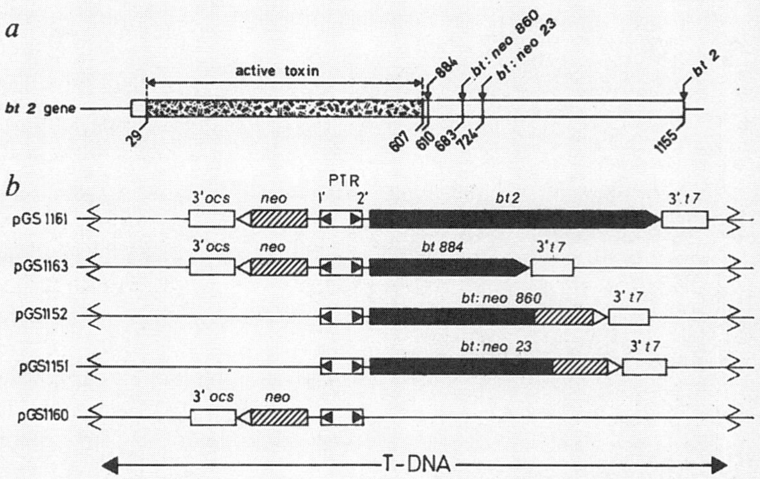

Fig. 1 *a*, Structure of the *bt2* gene. The smallest gene fragment encoding an active insect toxin is indicated, and the 3′ end positions (codon numbers) of the different truncated genes. *b*, Chimaeric genes derived from *bt2* present in plant expression vectors. The 5′ end of the *bt2* coding sequence is fused to the 2′ promoter fragment of the TR DNA. pGS 1161 contains the intact *bt2* gene. The *bt2* segment in pGS1163 ends at nucleotide position 1,830 of the *bt2* coding sequence. In pGS1151 and pGS1152, the *neo* gene has been fused to 5′ fragments of the *bt2* gene at positions 2,173 and 2,050 of the *bt2* coding sequence, respectively. PTR, a 482-base pair (bp) fragment containing the TR DNA 1′ and 2′ promoters, isolated from pOP443 (ref. 15); 3′t7, a 211-bp fragment containing the polyadenylation site of the TL DNA gene 7 (ref. 23); 3′ocs, a 706-bp *Pvu*II fragment containing the polyadenylation site of the octopine synthase gene[24]. Plasmids pGS1160, pGS1161 and pGS1163 were made using the intermediate plasmid pGSH160 which contains a chimaeric *neo* gene[14]. For pGS1151 and pGS1152, containing the *bt2*: *neo* fusion genes pGSH150, a derivative of pGSH160, lacking the *neo* gene and 3′ ocs, was used.

and 110K respectively and which react with anti-Bt2 (Fig. 2A) and anti-NPTII antibodies. At least 50% of the fusion proteins is present in a soluble form in the bacterial cells. The neomycin phosphotransferase activity of the fusion proteins, as determined in an *in situ* assay[13], is comparable to wild-type NPTII activity. Little or no enzymatic activity was exhibited by polypeptides of lower M_r (Fig. 2*b*). We conclude that the fusion proteins are relatively stable and responsible for the kanamycin-resistant phenotype. Insect assays revealed that, on a molar basis, the fusion proteins exhibit the same toxicity towards *M. sexta* larvae as intact Bt2 protein which has an LD_{50} value on first instar larvae of 4 ± 2 ng per larva.

Plant expression vectors

The intact and modified *B. thuringiensis* toxin genes were inserted between the T-DNA borders of plant expression vector pGSH160 (for *bt2* and *bt884*) or pGSH150 (ref. 14) (for *bt*:*neo860* and *bt*:*neo23*). These plasmids contain the promoter of the 2′ gene, a constitutive promoter which directs expression of mannopine synthase in the TR DNA of plasmid pTiA6 (ref. 15). The *Bacillus* genes are followed by a termination signal provided by the 3′ end of gene 7 of pTiA6. The resulting plasmids were mobilized into the *Agrobacterium* recipient C58C1 Rif^R pGV2260. The latter contains an octopine Ti plasmid from which the whole T-DNA region has been deleted and replaced by pBR322 (ref. 14). Recombination between pGV2260 and the expression vector through the homologous pBR322 sequences produced Ti plasmids pGS1161, pGS1163, pGS1151 and pGS1152 containing *bt2*, *bt884*, *bt*:*neo23* and *bt*:*neo860*, respectively (Fig. 1).

Two approaches were used to increase the probability of obtaining high levels of toxin expression in plants. First, the expression levels directed from the 1′ and 2′ promoters from the TR DNA were found to be coordinated[15]. Consequently we expected that expression of the *neo* gene controlled by the 1′ promoter in plants transformed with pGS1161 and pGS1163, would be correlated with transcription of the toxin gene. Second, we anticipated that plant cells transformed with pGS1151 or pGS1152, containing *bt*:*neo* fusions, would produce fusion proteins expressing NPTII activity. Selection for high levels of kanamycin resistance would allow us to select directly for transformed clones producing substantial amounts of *B. thuringiensis* protein.

Transformation of tobacco plants

Transgenic tobacco plants were obtained by leaf disk infection[14,16] of *Nicotiana tabacum* var. Petit Havana SR1 (ref. 17). Shoots resistant to 50, 100 or 200 μg ml^{-1} kanamycin were

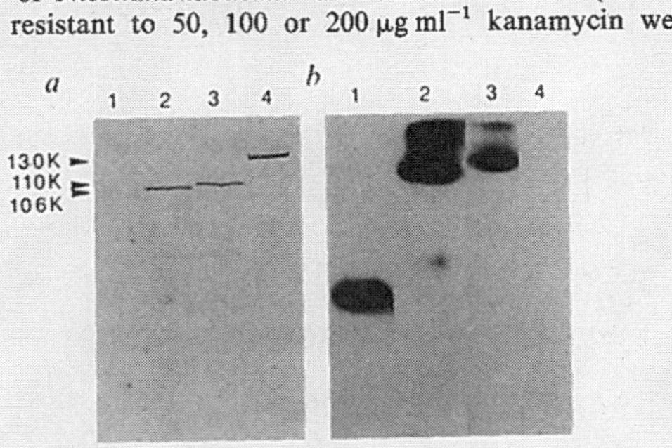

Fig. 2 Antigenic properties and enzymatic activity of Bt:NPTII fusion proteins. *a*, Western blot analysis[25] of crude extracts of *E. coli* clones producing intact NPTII enzyme (lane 1), Bt:NPT860 (lane 2) or Bt:NPT23 fusion protein (lane 3). Lane 4, purified Bt2. Blots were incubated with a diluted anti-Bt2 serum and subsequently with alkaline phosphatase-labelled anti-rabbit immunoglobulin. Substrates were 5-bromo-4-chloro-3-indolyl phosphate-toluidine salt and *p*-nitro-blue tetrazolium chloride (Sigma). *b*, Detection of NPTII activity in samples identical to those in *a*, by *in situ* phosphorylation of kanamycin with ^{32}P-ATP (ref. 13). All samples in *a* and *b* contain 0.25 μg protein per lane.

selected in all transformation experiments, indicating that the fusion genes indeed confer NPTII activity on transformed plant cells. Individual transformed plants were grown up and subsequently assayed for kanamycin resistance by testing their ability to produce callus from leaf disks on increasing concentrations of kanamycin and the more toxic aminoglycoside antibiotic G418. Most of the transgenic plants expressing an intact NPTII protein produced highly resistant calli, growing on 1,000 μg ml^{-1} kanamycin and on 100 μg ml^{-1} G418. In contrast, plants that expressed a Bt:NPTII fusion protein readily fell into different classes of kanamycin resistance (Table 1). Because the specific enzymatic activity of the fusion proteins is comparable to that of intact NPTII, we presume that these fusion proteins are present in lower amounts in the plant cells than intact NPTII protein.

Insecticidal activity in transgenic plants

Leaves of transgenic plants containing the four types of *B. thuringiensis* gene constructs were fed to *M. sexta* larvae in order to evaluate whether the levels of toxin in the plants would be insecticidal. Mortality rates of *M. sexta* larvae were monitored after 6 days of feeding on leaves of transformed plants (Fig. 3). We found that high toxicity to insects, resulting in 75–100% mortality of the larvae, was observed in about one quarter of the plants that expressed the longer fusion protein Bt:NPT23 and in about two-thirds of those with the shorter Bt:NPT860. Thus, the Bt:NPTII fusions allowed us to select transformants that express levels of the toxin sufficiently high to be insecticidal. Second, insect toxicity caused by the fusion proteins is directly correlated with the level of kanamycin resistance of the transformed plant. In addition, the short fusion generates a larger fraction of transformants expressing high kanamycin resistance and insect toxicity than the longer fusion, suggesting that the shorter *bt*:*neo860* gene provides higher levels of biologically active protein than the longer *bt*:*neo23*. Clear insecticidal activity was also detected in most of the 15 plants expressing the truncated *bt884* gene, of which two-thirds induced more than 75% larval death. None of the plants transformed with the full length *bt2* gene produced insect killing activity above levels obtained in NPTII-expressing control plants. These experiments indicate that for the promoter gene constructs we used, only truncated *bt2* genes give rise to expression levels that are strongly insecticidal in transgenic tobacco.

Protection from insect damage

To test whether expression of modified *bt2* genes in plants results in effective protection against insect damage, selected transgenic plants were grown in the greenhouse and were infested with freshly hatched larvae of *M. sexta*. Plants were kept under conditions that were optimal for survival and growth of the insects. Plants N21-11 and N28-16 were highly protected, because the larvae stopped feeding within 18 hours and all were killed within three days. The damage caused by single larvae was limited to areas of only a few square millimetres (Fig. 4). Other plants, such as N21-53 or N28-6, suffered slightly more damage. On these plants however all the larvae were killed after six days. Control plants such as N21-110, transformed with pGS1160, or untransformed SR1, were severely damaged within 4–6 days and were entirely consumed after 12 days.

Expression of chimaeric genes in plants

Quantitative detection of *B. thuringiensis* toxin in the leaves of transformed plants was performed using a sensitive ELISA, with a mixture of monoclonal antibodies specific for the NH_2-terminal region of the protein. A correlation was found between the quantity of *Bacillus* protein and insecticidal activity in the transgenic plants. Plants transformed with the truncated *bt2* gene or the fusion constructs contain approximately ten times more *Bacillus* protein than those transformed with the complete *bt2* sequence (Table 2). Thus, the failure to obtain insect-resistant plants using the intact *bt2* gene is most probably due to

Table 1 Kanamycin and G418 resistance in transgenic tobacco plants

Agrobacterium strain	Expression products of chimaeric genes	No. of plants resistant: Kanamycin ($\mu g\ ml^{-1}$) <50	50	100	200	500	1000	G418 ($\mu g\ ml^{-1}$) <10	10	100	Total
pGS1151	Bt:NPT23	3	1	7	17	14	10	51	1	0	52
pGS1152	Bt:NPT860	0	0	0	0	5	31	35	1	0	36
pGS1161	Bt2, NPTII	0	0	0	0	0	14	0	0	14	14
pGS1163	Bt884, NPTII	0	0	0	0	1	14	1	0	14	15
pGS1160	NPTII	0	0	0	0	2	12	2	0	12	14

Plants were scored according to the highest concentration of antibiotic on which callus could be induced from leaf disks from *in vitro* grown plants[21].

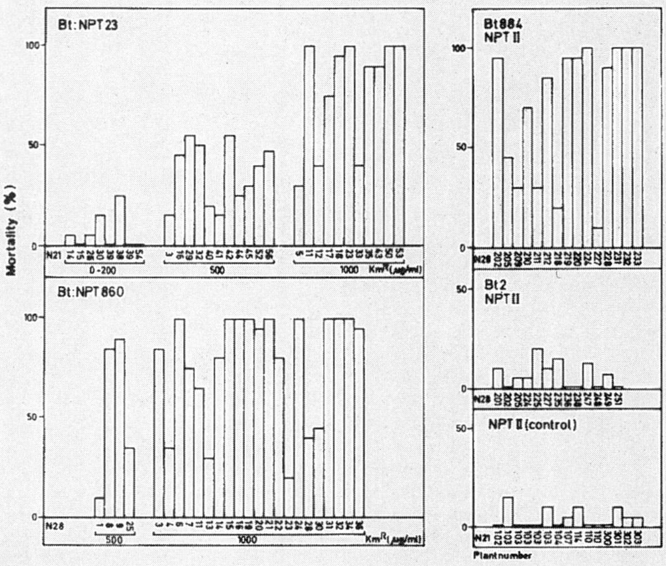

Fig. 3 Insect toxicity of transgenic tobacco plants expressing *B. thuringiensis* protein. Mortality of *Manduca sexta* larvae after 6 days of feeding on tobacco leaves is shown. The serial number of each plant is given below each column, and in the left-hand panels plants are grouped according to the concentration (in $\mu g\ ml^{-1}$) of kanamycin to which they are resistant. Each panel is labelled with the protein expressed (see Fig. 1). Plants were tested 4–6 weeks after transfer to the greenhouse, when they were about 40-cm high and had between six and eight full grown leaves. Two leaf disks of 4 cm in diameter were placed on wet filter paper in Petri dishes and infested with two batches of 10 first instar larvae of *M. sexta*. Leaf disks were replaced daily. Tests were conducted in a growth chamber at 25 °C, 75% relative humidity and under a 16 h light/8 h dark cycle.

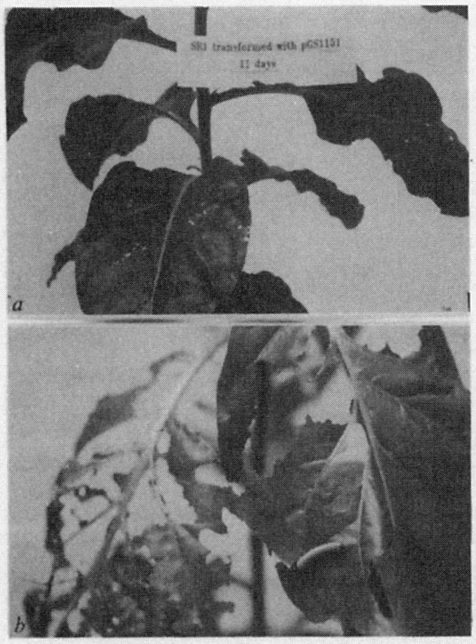

Fig. 4 Protection from insect feeding damage in transgenic plants expressing *B. thuringiensis* protein. Plants 40-50-cm high were infested in the greenhouse with fifteen *M. sexta* larvae per plant. *a*, On N21-11, expressing Bt:NPT23, all larvae died within three days. Leaf damage is very limited. *b*, Damage on a control plant expressing NPTII. Pictures were taken after 11 days.

inefficient protein synthesis in these transformed plant cells.

Transgenic plants that express the shorter Bt:NPT860 protein are on average more effective in killing insects and express higher levels of toxin than those expressing the longer Bt:NPT23 protein (Table 2). The Bt:NPTII fusion proteins detected in leaf extracts of the transformed plants had the expected size, as determined in Western blots. NPTII enzyme activity was exclusively associated with these fusion proteins as determined by *in situ* NPTII assays. Production of incomplete proteins or degradation in the plant cells was not observed. Thus, transformed tobacco plants produced fully functional fusion proteins of the size of the intact gene product.

The steady-state *Bacillus* messenger RNA levels in the transgenic plants were low and could not be reliably detected in Northern blot analysis. Therefore, they were quantified using ribonuclease protection experiments[18]. A probe containing the 5′ fragment of the *bt2* coding sequence up to nucleotide position 186 was synthesized using the SP6 transcription system and annealed to total RNA from the leaves of transgenic plants. After ribonuclease digestion, the protected fragments were run on a denaturing polyacrylamide gel. The results showed that RNA levels in the leaves of the transformed plants were correlated with *B. thuringiensis* protein levels (data not shown). The *B. thuringiensis* mRNA in N28-16, the plant that produces the highest level of protein, corresponds to ~0.0001% of the poly(A)$^+$ mRNA. The *Bacillus* protein detected in this plant represents 0.02% of the total soluble protein, or 3 μg of this protein per gram fresh leaf tissue. A fivefold lower level, as present in plant N28-34, was sufficient to induce 100% killing in a 6-day insect assay (Table 2). For comparison, the expression levels of the intact *neo* gene also driven by the TR 2′ promoter, varied between 10 and 40 μg of NPTII per gram leaf tissue (0.07–0.27% of total protein). This difference in amount of protein is consistent with the lower kanamycin resistance levels of plants transformed with the *bt* : *neo* fusions compared to those transformed with the intact *neo* gene (Table 1).

Inheritance of the protection

Eighty-five transgenic tobacco plants, transformed with the four types of chimaeric *B. thuringiensis* genes and expressing various levels of active insect toxin, were grown in the greenhouse. All grew normally and were indistinguishable from controls in morphology and vigour of growth. We analysed the inheritance of the kanamycin resistance in eleven plants expressing *bt* : *neo* genes. Most of these plants contained one (N21-23, N21-50, N28-31, N28-34, N28-21) or two (N21-35, N28-19, N28-24, N28-32) kanamycin resistance loci, as confirmed by Southern blot analysis. Interestingly, some plants that were recognized as producing a large amount of *Bacillus* protein, such as N21-11 and N28-16, generated exclusively kanamycin-resistant F_1 seedlings. DNA analysis showed the presence of at least five copies of the T-DNA in these plants.

F_1 progeny from some transgenic plants were assayed for the expression of insecticidal activity. Insect toxicity was correlated with the kanamycin resistance marker in the F_1 progeny of plants N21-23, N21-50 and N28-34. Insecticidal activity was similar to that observed in the parental plants. Approximately 15 F_1 progeny of N21-11 and N28-16 were analysed. They all induced 100% killing of *M. sexta* larvae in the standard 6-day assay. Toxin levels in the F_1 progeny of N21-11 varied between 20 and 50 ng per mg of total protein, comparable to the 30 ng per mg in the parental plant.

Discussion

Four chimaeric genes containing modified *Bacillus* toxin genes under the control of the 2′ promoter of the *Agrobacterium* TR DNA, have been transferred into tobacco plants. All contain the toxic core of the Bt2 protein; *bt2* encodes the complete M_r 130K protoxin, *bt884* is a 5′ fragment of *bt2* up to codon 610. *Bt* : *neo23* and *bt* : *neo860* encode fusion proteins which are relatively stable, both in bacteria and plants, and which retain full insect toxicity and NPTII enzyme activity.

Insecticidal levels of toxin were produced when truncated *Bacillus* genes or fusion constructs were expressed in transgenic plants. Mortality rates among *M. sexta* larvae feeding on transgenic leaf material depended on the amount of toxic polypeptide produced. Typically, greenhouse grown plants producing more than 0.004% of their protein as the toxin produced 100% mortality in 6-day feeding assays. Some of the plants we have selected, contain toxin at three to five times this level (N21-11 and N28-16). In greenhouse tests, these plants were well protected from leaf damage caused by insects. Quantification also

Table 2 Insecticidal activity and *B. thuringiensis* protein content of transformed plants

Toxic protein	Plant	*Bacillus* protein detected (ng per mg total protein)	Insecticidal effect (% mortality) Day 3	Day 4	Day 6	Weight reduction in surviving larvae (%)
Bt : NPT23	N21-11	33 (132)	90	100	100	—
	N21-35	6.9 (71)	0	40	90	57
	N21-17	2.6	0	15	75	71
	N21-18	5.7	25	50	90	80
	N21-32	2.5	5	40	50	39
	N21-41	4.3	0	0	15	34
	N21-43	4.7	10	40	80	66
Bt : NPT860	N28-16	42 (190)	100	100	100	—
	N28-34	6.9 (42)	70	85	100	—
	N28-6	13	80	90	100	—
	N28-15	10	75	95	100	—
	N28-19	6.2	35	65	100	—
	N28-21	7.0	45	75	100	—
	N28-24	12	60	90	100	—
	N28-31	6.3	55	80	100	—
	N28-32	14	85	90	100	—
Bt2	N21-105	1.3 (5.5)	0	5	15	7
	N21-225	1.2 (2.1)	5	10	20	17
	N21-201	<1	0	5	10	23
	N21-236	1.3	0	0	0	0
	N21-238	<1	0	0	0	0
	N21-249	1.8	0	10	10	10
Bt884	N28-212	30 (125)	100	100	100	—
	N28-220	11 (40)	60	95	100	—
	N28-219	11	55	65	95	85

Values in parentheses refer to amounts detected in greenhouse grown plants. Extracts were prepared from leaves of propagated *in vitro* plants or from leaves of plants grown in a greenhouse that had between six and eight fully expanded leaves. Leaf tissue was ground up and subsequently sonicated (10 s at 50 W) in extraction buffer (Na_2CO_3, 50 mM at *p*H10; dithiothreitol (DTT), 5 mM; leupeptin, 1 mg ml^{-1}; Triton X-100, 0.05%; EDTA 50 mM; phenylmethylsulphonyl fluoride (PMSF), 0.19 mg ml^{-1}). The extract was cleared by centrifugation and *B. thuringiensis* polypeptides in the supernatant were quantified using an indirect enzyme-linked immunosorbent assay[22] (ELISA). Polyvinyl microtitre plates were coated with a goat antibody against *B. thuringiensis* crystal protein. Plant extract dilutions were incubated at 4 °C for 2 h in the coated wells. After rinsing, bound antigen was reacted with a mixture of four distinct monoclonal antibodies against Bt2 and subsequently with an alkaline phosphatase conjugated goat anti-mouse immunoglobulin antibody. The bound enzyme conjugate was detected by adding *p*-nitrophenyl phosphate as a substrate, and relative quantities were determined by measuring absorbance values at 405 nm. The monoclonal antibodies used specify antigenic epitopes located between amino-acid positions 29 and 222 in the NH_2-terminal region of the Bt2 protein (H. Vanderbruggen *et al.*, in preparation). To quantify the *Bacillus* protein levels in transgenic plants, ELISA binding curves of leaf extracts were compared to standard binding curves, obtained by diluting known quantities of the purified homologous protein in control extracts from non-transformed SR1 tobacco plants. The detection limit of the test for purified, solubilized Bt2, was 0.1–1.0 ng ml^{-1}. Toxin levels in plant leaves are expressed as ng toxin per mg of total soluble protein. Insecticidal activity was determined using *Manduca sexta* larvae feeding on leaf material. Mortality after 3, 4, 6 days was determined and weight reduction in the surviving larvae measured after 6 days.

showed that the toxin expressed in plants has the same specific activity as in a bacterial host.

No significant insecticidal activity could be obtained using the intact *bt2* coding sequence, despite the fact that the same promoter was used to direct its expression. Intact Bt2 protein and RNA amounts in the transgenic plant leaves were 10–50 times lower than those for the truncated *B. thuringiensis* polypeptide or the fusion proteins. Expression levels were not significantly influenced by fusing the *neo* gene to the *bt2* sequence, but rather by the length of the *bt2* fragment. Why the complete *bt2* gene is not expressed at an equally high level in plant cells, is not known. Several parameters, such as differential RNA stability and translation efficiency might be important.

We observed in transgenic plants containing the *bt*:*neo* fusion constructs a correlation between insecticidal activity and resistance to kanamycin. Three-quarters of all plants resistant to 1,000 $\mu g\ ml^{-1}$ kanamycin induced 75–100% insect mortality. Such fusion genes can be used to select efficiently for transformed plants expressing strong insecticidal activity through direct selection for high kanamycin resistance.

Our experiments illustrate the feasibility of engineering plants that defend themselves against lepidopteran insects which are sensitive to the *B. thuringiensis berliner* insect toxin. However some species, such as *Heliothis* and *Spodoptera* which belong to the *Noctuidae*, an important group of pest insects, are less sensitive to common strains of *B. thuringiensis*, including berliner 1715 (ref. 19). To protect plants fully against these insects, higher levels of expression will be required. This might be achieved using chimaeric *Bacillus* genes containing stronger plant-specific promoters. The 35S promoter of cauliflower mosaic virus[20], for example directs a 10–50-fold higher expression than the regular T-DNA promoters in plants. Alternatively, it may be possible to construct chimaeric toxin genes with higher specific activity against important target insects. Transfer of different chimaeric genes into a variety of crops may provide a new and environmentally safer method of controlling destructive insect pests.

We thank Hilde Van De Wiele, Anne De Sonville, Veronique Gosselé, Jan Dockx, Annemie Van Houtven and Carine Matthijs for their contribution to the experiments. We also thank Leonor Fernandez for her contribution to the initial experiments, Danny Degheele for advice on insect tests, Karin Tenning for preparation of the manuscript, and Karel Spruyt and Stefaan Van Gijsegem for the figures.

Received 3 March; accepted 31 April 1987.

1. Dulmage, H. T. *et al.* in *Microbial Control of Pest and Plant Diseases 1970–1980* (ed. Burges, H. D.) 193–222 (Academic, London, 1980).
2. Yamamoto, T. & McLaughlin, R. E. *Biochem. biophys. Res. Comm.* **103,** 414–421 (1981).
3. Goldberg, L. J. & Margalit, J. *Mosquito News* **37,** 355–358 (1977).
4. Krieg, V. A., Huger, A. M., Langenbruch, G. A. & Schnetter, W. Z. *Angew. Ent.* **96,** 500–508 (1983).
5. Herrnstadt, C., Soares, G. G., Wilcox, E. R. & Edwards, D. L. *Biotechnology* **4,** 305–308 (1986).
6. Calabrese, D. M., Nickerson, K. W. & Lane, L. C. *Can. J. Microbiol.* **26,** 1006–1010 (1980).
7. Lilley, M., Ruffell, R. N. & Sommerville, H. *J. gen. Microbiol.* **118,** 1–11 (1980).
8. Gheysen, G., D'haese, P., Van Montagu, M. & Schell, J. in *Genetic Flux in Plants* (eds Hohn, B. & Dennis, E. S.) 11–47 (Springer, New York, 1985).
9. Höfte, H. *et al. Eur. J. Biochem.* **161,** 273–280 (1986).
10. Herrera-Estrella, L. *et al. EMBO J.* **2,** 987–995 (1983).
11. Reiss, B., Sprengel, R. & Schaller, H. *EMBO J.* **3,** 3317–33222 (1984).
12. Schreier, P., Seftor, E. A., Schell, J. & Bohnert, H. J. *EMBO J.* **4,** 25–32 (1985).
13. Reiss, B., Sprengel, R., Will, H. & Schaller, H. *Gene* **30,** 217–223 (1984).
14. Deblaere, R. *et al. Meth. Enzym.* (in the press).
15. Velten, J., Velten, L., Hain, R. & Schell, J. *EMBO J.* **3,** 2723–2730 (1984).
16. Horsch, R. *et al. Science* **227,** 1229–1231 (1985).
17. Maliga, P., Sc. Breznovitis, A. & Marton, L. *Nature new Biol.* **244,** 29–30 (1973).
18. Zinn, K. *et al. Cell* **34,** 865–879 (1983).
19. Kalfon, A. R. & de Barjac, H. *Entomophaga* **30,** 177–186 (1985).
20. Odell, J. T., Nagy, J. & Chua, N. M. *Nature* **313,** 810–812 (1985).
21. Deblaere, R. *et al. Nucleic Acids Res.* **13,** 4777–4788 (1985).
22. Clark, M. F. & Adams, A. N. *J. gen. Virol.* **34,** 475–483 (1977).
23. Velten, J. & Schell, J. *Nucleic Acids Res.* **13,** 6981–6998 (1985).
24. Gielen, J. *et al. EMBO J.* **3,** 835–845 (1984).
25. Towbin, H., Staehelin, T. & Gordon, J. *Proc. natn. Acad. Sci. U.S.A.* **76,** 4350–4354 (1979).

Proc. Natl. Acad. Sci. USA
Vol. 88, pp. 3324–3328, April 1991
Biochemistry

Modification of the coding sequence enhances plant expression of insect control protein genes

(*Bacillus thuringiensis*/insect-resistant plants/synthetic genes)

FREDERICK J. PERLAK*, ROY L. FUCHS, DUFF A. DEAN, SYLVIA L. MCPHERSON†, AND DAVID A. FISCHHOFF

Plant Sciences, Monsanto Company, 700 Chesterfield Village Parkway, Chesterfield, MO 63198

Communicated by Jeff Schell, January 14, 1991

ABSTRACT **Increased expression of the insect control protein genes of *Bacillus thuringiensis* in plants has been critical to the development of genetically improved plants with agronomically acceptable levels of insect resistance. The expression of the *cryIA(b)* gene was compared to partially modified (3% nucleotide difference) and to fully modified (21% nucleotide difference) *cryIA(b)* and *cryIA(c)* genes in tobacco and tomato. The modified genes increased the frequency of plants that produced the proteins at quantities sufficient to control insects and dramatically increased the levels of these proteins. Among the most highly expressing transformed plants for each gene, the plants with the partially modified *cryIA(b)* gene had a 10-fold higher level of insect control protein and plants with the fully modified *cryIA(b)* had a 100-fold higher level of CryIA(b) protein compared with the wild-type gene. Similar results were obtained with the fully modified *cryIA(c)* gene in plants. Specific sequences of the partially modified *cryIA(b)* gene were analyzed for their ability to affect *cryIA(b)* gene expression in tobacco. The DNA sequence of a single region was identified as important to the improvement of plant expression of the *cryIA(b)* gene. The increased levels of *cryIA(b)* mRNA were not directly proportional to the increased levels of CryIA(b) protein in plants transformed with the modified *cryIA(b)* genes, indicating that the nucleotide sequence of these genes had an effect in improving their translational efficiency in plants.**

Insect control proteins from a prokaryotic source, *Bacillus thuringiensis* var. *kurstaki* (*B.t.k.*; ref. 1) are specific for lepidopteran insects and exhibit no activity against humans, other vertebrates, and beneficial insects (2). These properties have made the genes of these insect-specific proteins attractive candidates for genetic modification of crops for protection against lepidopteran pests. Genes encoding lepidopteran-specific insect control proteins have been cloned and sequenced. Truncated genes, which produce insecticidally active protein, have been expressed in tomato (3), tobacco (4), and cotton (5). Field tests of these plants revealed that higher levels of insect control protein in the plant tissue would be required to obtain commercially useful plants (6).

The insect control proteins are highly expressed in their natural host, *B. thuringiensis*. Up to 50% of the total protein in sporulated cultures of *B.t.k.* are the insect control proteins deposited as crystals within the cell. Insect control protein genes are expressed well in *Escherichia coli* (7) or *Pseudomonas* (8). Poor expression in plants is a well-reported characteristic of the *B.t.k.* insect control proteins. Truncating the gene, keeping essentially the N-terminal half of the protein intact, results in improved expression of the gene in plants to barely detectable levels (3, 4). The use of different promoters, fusion proteins, and leader sequences has not significantly increased insect control protein gene expression (4, 9).

We hypothesized that a gene with a sequence adapted for a Gram-positive prokaryote may not be the appropriate coding sequence for efficient plant expression. Examination of the insect control protein gene coding sequence indicated that it differs significantly from plant genes in G+C content. Multiple sequences motifs that are not common in the coding region of plant genes were found to be common in the wild-type (WT) *cryIA(b)* sequence. These included localized regions of A+T richness resembling plant introns (10), potential plant polyadenylylation signal sequences (11), ATTTA sequences, which have been shown to destabilize mRNA in other systems (12), and codons rarely used in plants (13).‡

Two approaches were initiated to increase the level of CryIA(b) and CryIA(c) insect control proteins in genetically modified plants. First, DNA sequences predicted to inhibit efficient plant gene expression at both the translational and mRNA level were selectively removed throughout the coding sequence by site-directed mutagenesis to partially modify the gene (PM gene) without changing the amino acid sequence. Wholesale changes in the DNA characterized the second approach, which required the use of a fully modified (FM) synthetic gene. The FM genes encode proteins nearly identical in amino acid sequence to the WT gene. These constructs were a more rigorous application of the principles used to generate the PM gene, taking into account additional factors such as codon usage in plants, potential secondary structure of mRNA, and potential regulatory sequences. Insect control protein levels in plants increased as a consequence of modification of the DNA sequence, up to 100-fold over levels seen with the wild-type gene in plants. The increased gene expression appears to be the result of increased translational efficiency. The higher level of insect control protein in these plants was directly correlated to increased insecticidal activity. The demonstration of the utility of these genes to provide protection from insects has far-reaching implications for the future of insect-resistant plants and for the application of these gene modification principles to the design of heterologous genes for high-level expression in plants.

MATERIALS AND METHODS

Modifications of the Coding Sequence of Insect Control Genes. The coding sequence of the WT gene in pMON9921

Abbreviations: WT, wild type; PM, partially modified; FM, fully modified; THW, tobacco hornworm; *B.t.k.*, *Bacillus thuringiensis* var. *kurstaki*.

*To whom reprint requests should be addressed.
†Present address: Center for AIDS Research, University of Alabama–Birmingham, Birmingham, AL 35294.

‡The sequence reported in this paper has been deposited in the GenBank data base (accession no. M60856).

was identical in DNA sequence and similar to the truncated version of *cryIA(b)* (3). Modifications in the DNA did not alter the amino acid sequence of the CryIA(b) protein. Partial modification of the WT sequence was by site-directed mutagenesis (14) with 9 oligonucleotides, labeled A–I as they appear in the sequence. The nucleotides that differ from the WT sequence are shown in Fig. 1. Regions of the coding sequence that were targeted for modification were identified by scanning the DNA sequence for 4 or more consecutive adenine (A) or thymine (T) nucleotides. These regions were analyzed for the presence of potential polyadenylylation signal sequences (11). If there was more than 1 signal sequence within 10 nucleotides, the sequence was altered. If any region of 15–30 nucleotides was >80% A+T, it was altered if there were 2 or more ATTTA sequences or if there was a major potential plant polyadenylylation signal sequence (AATAAA or AATAAT). Nucleotides used to modify the DNA sequence increased the overall G+C content, increased plant preferred codons (13) without changing the amino acid sequence, and did not generate consecutive A+T or G+C strings (>6).

The changes of the FM gene [FM *cryIA(b)*] required complete synthesis (British Biotechnology, Oxford, U.K.). The DNA sequence was designed to maintain the amino acid sequence and avoid ATTTA sequences and potential polyadenylylation signal sequences. Plant preferred codons were used. Consecutive strings of A+T or G+C nucleotides of 5 or more were avoided. The FM *cryIA(b)* gene is compared to the WT in Fig. 1. The FM *cryIA(c)* gene (5) is a synthetic truncated version of the *cryIA(c)* gene and encodes amino acids 1–615. A second FM *cryIA(c)* gene, described in Table 2, encodes amino acids 29–615.

Plant Expression Vectors. The genes for insect control were delivered as cointegrates of a disarmed nopaline plasmid (15). These intermediate vectors all contain the insect control protein genes [WT, PM, FM, or FM *cryIA(c)*] driven by the cauliflower mosaic virus 35S promoter with a duplicated enhancer region (16).

Plant Transformation and Regeneration. *Agrobacterium tumefaciens* was used to obtain transgenic tobacco (17) and tomato (18) plants.

Immunological Analysis. Western blot analysis was done as described (5). Antibodies prepared in rabbits to truncated, denatured insect control protein [trypsin-treated CryIA(c) protein] isolated from a SDS/polyacrylamide gel were used. The blots were sensitive to 1–2 ng per 50 μg of total plant protein. ELISA analysis of transgenic plant material for CryIA(b) protein utilized a direct double antibody sandwich with rabbit polyclonal antibody raised to the *cryIA(b)* gene product as expressed in *E. coli*, purified as described (19). The ELISA was sensitive to 0.5 ng per 50 μg of total protein.

mRNA Analysis. S1 nuclease protection assays (20) were used to quantitate the levels of *cryIA(b)*-specific mRNA. A ^{32}P kinase-labeled 45-mer synthetic oligonucleotide with 30 nucleotides of homology to the 5′ untranslated sequence common to all of the insect control protein gene plant mRNAs was hybridized with 40 μg of total RNA. An internal control of a labeled 55-mer with 40 nucleotides of homology to the coding sequence of the endogenous tomato 3-*enol*-pyruvoylshikimate-5-phosphate synthase (EPSPS) mRNA (21) was included in each reaction mixture. Each lane was scanned with a laser densitometer for quantitation and was normalized with respect to the levels of endogenous EPSPS mRNA. Several exposures were scanned to account for film saturation.

Insect Bioassay. Five neonate larvae of tobacco hornworm (THW) were placed on a leaf of a plant that had been selected in tissue culture for kanamycin resistance. Insect control was defined as plant tissue that exhibited 100% mortality to THW and sustained no visible damage after 3 days.

RESULTS

Partial modification of the WT coding sequence focused on potential plant polyadenylylation signal sequences and ATTTA sequences in regions of high local A+T content. The resulting gene, referred to as the PM gene, contained a subset of the changes made in the FM gene (Table 1 and Fig. 1). The design of the FM gene eliminated ATTTA sequences, almost all potential plant polyadenylylation sequences, greatly increased the G+C nucleotide content of the gene, and minimized the use of rare plant codons. The PM gene encodes amino acids 29–607 of the CryIA(b) protein with an additional two amino acids, methionine and alanine, at the N terminus. The FM gene encodes amino acids 1–615. Both genes have been expressed in *E. coli* and the proteins show equivalent activity against lepidopteran insects (data not shown).

Insect Control Protein Levels in Genetically Modified Plants. The plants expressing the WT, PM, or FM genes were identified from the transformed kanamycin-selected plants by a primary screen bioassay against THW (Table 2). No observable damage to the leaf by THW indicated that ingestion of small amounts of leaf tissue by THW was sufficient to kill the insects. The frequency of transgenic plants that exhibited insect control was higher in the PM and FM transformed plants than in the plants with the WT *cryIA(b)* gene (Table 2). The modified genes (PM and FM) were routinely expressed above the biological threshold set for insect control in tomato and tobacco. This occurrence was less frequently observed with the WT gene, especially in tobacco.

Immunological analysis of plants expressing either the WT, the PM, or the FM *B.t.k.* insect control protein genes demonstrated that modification of the *B.t.k.* gene resulted in increased insect control protein levels in both tobacco and tomato. The distribution of these plants by the levels of insect control protein found in leaf tissue revealed that the PM and FM genes had a greater percentage of plants with higher levels of detectable protein compared to transgenic plants with the WT gene (Fig. 2). Over 200 transgenic tomato plants with the WT *cryIA(b)* gene were screened; 53 showed insect control and in only 3 plants were the levels of protein detectable by Western blot analysis. Insect control protein was barely detectable in 2 plants, near the limit of detection in this assay (≈1 ng per 50 μg of total protein).

Tomato plants expressing the PM gene showed an increase in the percentage of insecticidal plants (Table 2) and in the levels of insect control protein detected (Fig. 2*A*). The majority of the plants transformed with the PM gene fall into the 1- to 10-ng level and a small percentage (2 of 25 plants) of the plants are in the lower end of the 10- to 30-ng range (0.02%).

Tomato plants that expressed the FM genes contained a similar percentage of insecticidal plants compared to plants with the PM gene (Table 2). A substantial number of plants (10 of 29; 34%) were found to contain high insect control protein levels (Fig. 2*A*). Slightly higher overall levels of insect control protein were seen with tomato plants transgenic with

Table 1. Summary of changes made in the PM and FM *cryIA(b)* genes compared to the WT *cryIA(b)* gene

	WT	PM	FM
No. of bases different from WT	—	62 of 1743	390 of 1845
No. of codons different from WT	—	55 of 579 (9.5%)	356 of 615 (60%)
G+C content	37%	41%	49%
No. of potential PPSS	18	7	1
No. of ATTTA sequences	13	7	0

PPSS, plant polyadenylylation signal sequences.

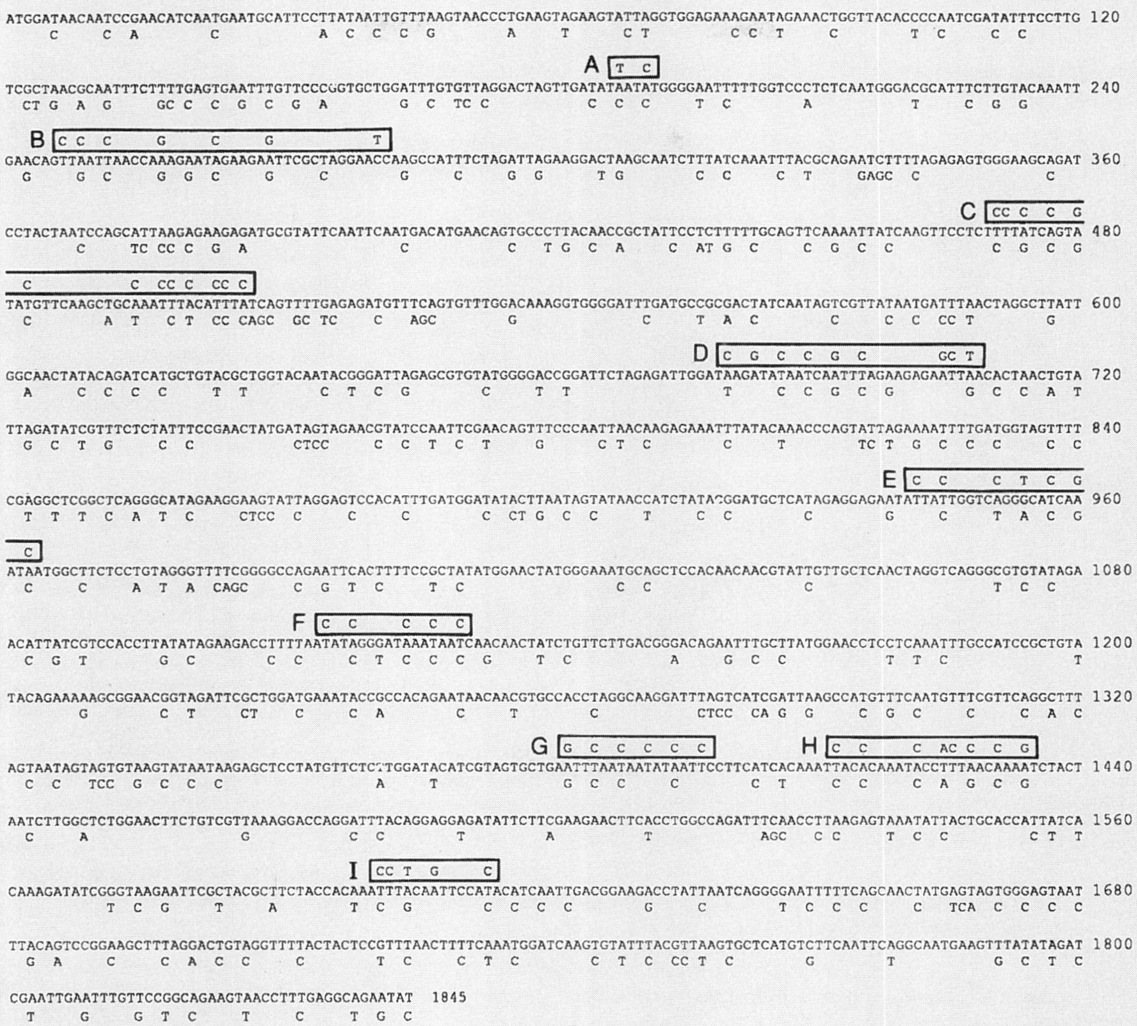

FIG. 1. DNA sequence of the *cryIA(b)* gene on the numbered line with the modifications found in the PM gene on the line above and the FM gene on the line below. The differences between the WT and PM genes are within the labeled boxed areas (A–I).

pMON5383, a truncated *cryIA(c)* FM gene with substantial homology to the *cryIA(b)* FM gene depicted in Fig. 1. Among the 52 FM *cryIA(c)* transgenic tomato plants that were immunologically analyzed, 10 (20%) were at or above the 30- to 100-ng range with several plants approaching 100 ng per 50 μg of total protein. Among the most highly expressing transgenic tomato plants for each insect control protein gene, the FM genes appeared to be expressed 5- to 10-fold higher than the PM gene and up to 100-fold higher than the WT gene.

Table 2. Comparison of the fraction of plants exhibiting insect control with the WT, PM, and FM genes

Gene	Plants exhibiting insect control/ total kanamycin-selected plants	
	Tobacco	Tomato
WT *cryIA(b)*	3/54 (6%)	53/204 (26%)
PM *cryIA(b)*	25/42 (60%)	40/63 (63%)
FM *cryIA(b)*	12/22 (55%)	32/55 (58%)
FM *cryIA(c)*	—	52/97 (54%)
FM *cryIA(c)**	48/95 (51%)	—

*The coding region of the N-terminal amino acids of this FM *cryIA(c)* gene has been deleted as in the WT and PM *cryIA(b)* genes.

In tobacco plants, the WT gene was more poorly expressed than in tomato, below the detection limit of the Western blot analysis in the 3 plants showing insect control against THW. Analysis of transgenic tobacco plants (Fig. 2*B*) with the PM and FM genes revealed results similar to those found in tomato. There was an increase in the frequency of plants with insect control (Table 2) and the levels of insect control protein in plants with the PM gene compared to the WT. As in tomatoes, several tobacco plants (2 of 25; 8%) with the PM gene contained 10–30 ng of insect control protein per 50 μg of total plant protein.

Transgenic tobacco plants with the FM gene demonstrated improved expression of the *B.t.k.* protein over the PM gene plants. Several tobacco plants with the FM genes contained detectable levels of insect control protein >50 ng per 50 μg of total protein (>0.1%). Fig. 2*B* shows a high percentage of tobacco plants with the FM *cryIA(b)* gene in the 30- to 100-ng range but only 13 plants with this construct were analyzed. Extensive analysis was done with pMON5390, a FM trun-

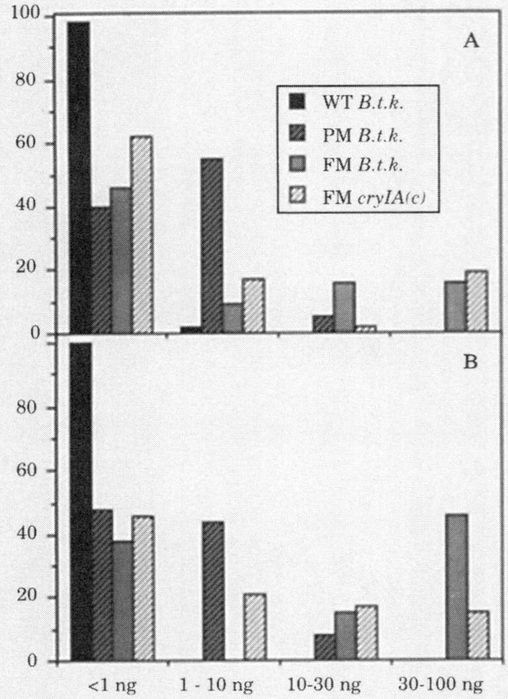

FIG. 2. Distribution of tomato (*A*) and tobacco (*B*) plants that exhibited insect control. The numbers of tomato plants that exhibited insect control used to calculate the percentage of *B.t.k.* protein levels are as follows: WT gene, 53; PM gene, 63; FM gene, 55; FM *cryIA(c)*, 97. The numbers of tobacco plants are as follows: WT gene, 3; PM gene, 42; FM gene, 22; FM *cryIA(c)*, 95. Levels of insect control protein are per 50 μg of total plant protein.

cated *cryIA(c)* gene similar to the *cryIA(c)* gene found in pMON5383 except that the coding sequence for the first 28 amino acids had been deleted as in the WT and PM genes. A high percentage of the plants (50%) that exhibited insect control contained protein at detectable levels. As with the FM gene in tomato, one-third (31%) of the plants were in the 10- to 100-ng range of insect control protein per 50 μg of total protein.

The higher frequency of transgenic plants with the PM and FM genes exhibiting insect control compared to the plants with the WT gene was a good indication that the modifications to the *B.t.k.* coding sequence had improved gene expression. Screening of a large number of plants with the WT gene failed to yield any that produced >1 ng of insect control protein per 50 μg of total protein, while >10% of the plants with the FM genes produced 30–100 ng of insect control protein in both tomato and tobacco. Because of the large numbers of plants analyzed, the overall similarity of the genetic constructs and the large variance in both average and peak expression obtained with the WT, PM, and FM genes, we believe that the data reflected inherent gene expression differences. Gene copy number did not appear to differ for the different coding sequences. The most highly expressing tomato plants containing either the WT, PM, or FM gene were analyzed genetically and the insect control protein gene segregated as a single dominant locus. Highly expressing tobacco and tomato plants with the FM *cryIA(b)* and FM *cryIA(c)* genes were identified that contained insect control protein at levels of 100–150 ng (0.2–0.3%).

mRNA Analysis. The levels of *cryIA(b)*-specific mRNA in plants did increase in plants with the PM gene and the FM gene compared to plants with the WT gene (data not shown). Quantitative S1 nuclease mRNA analysis revealed that selected tomato plants with the PM gene had a 10-fold increase in insect control protein levels and an average 2.5-fold increase in *cryIA(b)* mRNA over tomato plants with the WT gene. Tomato plants with the FM *cryIA(b)*, which exhibited a 50-fold increase in insect control protein levels over tomato plants with the WT gene, had a 5-fold increase of *cryIA(b)* mRNA levels. For two selected plants, levels of mRNA from the plant with the PM gene differed by <20% from that in a plant with the FM gene, yet the plant with the FM gene contained a 4-fold higher level of insect control protein. Northern blot analysis of the best plants with the WT gene and plants with the PM and FM *cryIA(b)* genes indicated the *cryIA(b)* mRNA was of the expected size.

Identification of Important Regions for *B.t.k.* Gene Expression in the PM Gene. The FM gene contains extensive differences compared to the WT gene (Table 1). The PM gene differs from the WT gene by 3% of the nucleotides (Fig. 1), yet these changes account for at least a 10-fold increase in insect control protein levels in tobacco. Plasmids containing a fraction of the total changes (Table 3) were used to obtain transformed, kanamycin-selected tobacco plants that were screened by insect bioassay. *B.t.k.* protein levels comparable to the PM gene (pMON5370) were obtained when the changes in the 5′ half of the coding sequence were present (pMON10707). The gene with the changes in the 3′ half alone (pMON10706) did not differ from the WT gene; there was no immunologically detectable insect control protein. Changes with oligonucleotide B (pMON10537) appeared to be the most significant section to elicit the increased insect control protein levels.

The changes made by oligonucleotide B were not the sole effector of increased levels of insect control protein in plants. A construct that contained a fully modified 5′ one-third of the gene, including oligonucleotide B (pMON5378), resulted in plants with levels of insect control protein comparable to plants with the PM gene. By comparison, replacing the first one-third of the FM gene with the homologous WT section (pMON10538) resulted in lower levels of insect control protein than previously observed for the FM gene. The expression of the hybrid gene of pMON10538 indicated that it is possible to achieve increased *B.t.k.* expression compared to the WT gene without altering the region of oligonucleotide B. These results indicate that a single region is probably not solely involved and that changes throughout the sequence as found in the FM gene were required for the highest levels of expression of the insect control protein gene in plants.

Table 3. Analysis of the contributions of individual regions to increased expression of the *B.t.k.* PM gene in tobacco

Construct	Oligonucleotides	Plants with insect control/plants screened	Protein level, % of total protein
pMON5370	A–I*	23/48 (48%)	0.02–0.03†
pMON10707	A,B,C,D	19/48 (40%)	0.028
pMON10706	E,F,G,H,I	1/53 (2%)	ND‡
pMON10539	A	2/55 (4%)	ND‡
pMON10537	B	17/57 (30%)	0.016
pMON10540	A,B	23/88 (26%)	0.015
pMON10705	C	1/47 (2%)	ND‡
pMON10538	1/3 WT, 2/3 FM	19/62 (30%)	0.016
pMON5378	1/3 FM, 2/3 WT	9/20 (45%)	0.02–0.03†

*The locations of the individual regions (A–I) are depicted in the sequence shown in Fig. 1.
†The levels of CryIA(b) were determined by Western blot analysis.
‡The level of CryIA(b) in these plants was too low to be detected by ELISA or Western blot.

DISCUSSION

The PM and FM genes have dramatically increased insect control protein levels in plants. This appears to be one of the largest increases in gene expression in any system obtained solely through modification of the coding sequence. The rationale for these designs was based on the elimination of sequences such as potential polyadneylylation signal sequences, ATTTA sequences, and A+T-rich regions. The increased expression was generic across several plant species—tobacco, tomato, cotton (5)—and was obtained with two distinct genes, *cryIA(b)* and *cryIA(c)*.

The PM gene contained a subset of the changes found in the fully modified gene, yet produced large increases in insect control protein levels in tobacco and in tomato (at least 10-fold). The lack of extensive changes in the PM gene (3% of the total nucleotides) and the localization of a region critical to increased expression (oligonucleotide B) argues against the hypothesis that the presence of rare plant codons is primarily responsible for the previously reported low level expression of WT insect control protein genes in plants. Analysis of the region of oligonucleotide B revealed no inverted repeats, palindromic sequences, or other apparent secondary structures that distinguished this region from other regions of the structural gene. This region does contain three potential polyadenylylation signal sequences (two AACCAA and one AATTAA). Northern blot analysis of total RNA did not detect truncated *cryIA(b)* mRNA in plants expressing the WT gene. It is possible that the incomplete functioning of a polyadenylylation signal led to processing without polyadenylylation causing instability. Another possibility is that the region of oligonucleotide B contains an as yet undefined regulatory sequence dependent on the context of its surrounding sequence.

The region of oligonucleotide B is not the sole suppressor of insect control protein gene expression in plants. The fully modified C-terminal two-thirds of the truncated insect control protein gene plus an N-terminal WT one-third containing a sequence unchanged in oligonucleotide B (pMON10538) resulted in levels of insect control protein in plants equivalent to the PM gene or ≈10-fold lower than for the FM *cryIA(b)* gene. One interpretation of these results is that the presence of the WT sequence in the 5′ third of the gene, including the region of oligonucleotide B, depresses the expression of the FM gene ≈10-fold. Alternatively, the aggregate of the changes contained in the 3′ two-thirds of the FM gene could partially compensate for the presence of the WT sequences in the 5′ one-third of the gene. These observations suggest that several factors and regions of the gene are involved in the low levels of insect control protein observed with the WT gene in plants. The highest levels of insect control protein that we have observed in plants were obtained with the FM genes, both *cryIA(b)* and *cryIA(c)*.

The major block to efficient WT *cryIA(b)* gene expression in plants appeared to be translational and not at the level of stable accumulation of *cryIA(b)* mRNA. mRNA levels were increased in plants with the PM and FM genes, but the level of mRNA increase was not directly proportional to the level of protein increase in the best of the PM and FM plants. It is not clear that the increased mRNA levels must be directly proportional to the increased protein expression. Analysis of the mRNA levels of some of the PM and FM plants indicated that comparable amounts of FM mRNA led to the production of more insect control protein than PM mRNA, suggesting that the presence of predominantly plant preferred codons in the FM gene has improved its overall translational efficiency.

The impact of the exceptional levels of expression that the PM and FM genes provide is most apparent in the efficacy these plants exhibit to withstand insect attack. The use of these plants in an integrated pest management system will expand the options of farmers to protect their harvest from insect destruction.

We thank N. Hoffman, J. Layton, M. Reedy, and S. Berberich for technical assistance and E. Wong, J. Schell, and R. Fraley for critical comments.

1. Hofte, H. & Whiteley, H. R. (1989) *Microbiol. Rev.* **53,** 242–255.
2. Aronson, A. I., Beckman, W. & Dunn, P. (1986) *Microbiol. Rev.* **50,** 1–24.
3. Fischhoff, D. A., Bowdish, K. S., Perlak, F. J., Marrone, P. G., McCormick, S. H., Niedermeyer, J. G., Dean, D. A., Kusano-Kretzmer, K., Mayer, E. J., Rochester, D. E., Rogers, S. G. & Fraley, R. T. (1987) *Bio/Technology* **5,** 807–813.
4. Vaeck, M., Reynaerts, A., Hofte, H., Jansens, S., De Beuckeleer, M., Dean, C., Zabeau, M., Van Montagu, M. & Leemans, J. (1987) *Nature (London)* **328,** 33–37.
5. Perlak, F. J., Deaton, R. W., Armstrong, T. A., Fuchs, R. L., Sims, S. R., Greenplate, J. T. & Fischhoff, D. A. (1990) *Bio/Technology* **8,** 939–943.
6. Delannay, X., La Valle, B. J., Proksch, R. K., Fuchs, R. L., Sims, S. R., Greenplate, J. T., Marrone, P. G., Dodson, R. B., Augustine, J. J., Layton, J. G. & Fischhoff, D. A. (1989) *Bio/Technology* **7,** 1265–1269.
7. Schnepf, H. E., Wong, H. C. & Whiteley, H. R. (1987) *J. Bacteriol.* **169,** 4110–4118.
8. Obukowicz, M. G., Perlak, F. J., Kusano-Kretzmer, K., Mayer, E. J. & Watrud, L. S. (1986) *Gene* **45,** 327–331.
9. Barton, K. A., Whiteley, H. R. & Yang, K. S. (1987) *Plant Physiol.* **85,** 1103–1109.
10. Goodall, G. J. & Filipowicz, W. (1989) *Cell* **58,** 473–483.
11. Dean, C., Tamaki, S., Dunsmuir, P., Favreau, M., Katayama, C., Dooner, H. & Bedbrook, J. (1986) *Nucleic Acids Res.* **14,** 2229–2240.
12. Shaw, G. & Kamen, R. (1986) *Cell* **46,** 659–667.
13. Murray, E. E., Lotzer, J. & Eberle, M. (1989) *Nucleic Acids Res.* **17,** 477–498.
14. Kunkle, T. A. (1985) *Proc. Natl. Acad. Sci. USA* **82,** 488–492.
15. Hinchee, M. A. W., Connor-Ward, D. V., Newell, C. A., McDonnell, R. E., Sato, S. J., Gasser, C. S., Fischhoff, D. A., Re, D. B., Fraley, R. T. & Horsch, R. B. (1988) *Bio/Technology* **6,** 915–922.
16. Kay, R., Chan, A., Daly, M. & McPherson, J. (1987) *Science* **236,** 1299–1302.
17. Horsch, R. B., Fry, J. E., Hoffmann, N. L., Eichholtz, D., Rogers, S. G. & Fraley, R. T. (1985) *Science* **227,** 1229–1231.
18. McCormick, S. M., Niedermeyer, J., Fry, J., Barnason, A., Horsch, R. & Fraley, R. (1986) *Plant Cell Rep.* **5,** 81–84.
19. MacIntosh, S. C., Stone, T. B., Sims, S. R., Hunst, P. L., Greenplate, J. T., Marrone, P. G., Perlak, F. J., Fischhoff, D. A. & Fuchs, R. L. (1990) *J. Invert. Pathol.* **56,** 258–266.
20. Hanley-Bowdoin, L., Elmer, J. S. & Rogers, S. G. (1988) *Nucleic Acids Res.* **16,** 10511–10528.
21. Gasser, C. S., Winter, J. A., Hironaka, C. M. & Shah, D. M. (1988) *J. Biol. Chem.* **263,** 4280–4289.

TRANSFORMATION OF CEREALS BY *AGROBACTERIUM*

Toshihiko Komari
JT Plant Innovation Centre, Iwata, Shizuoka, Japan
E-mail: toshihiko.komari@ims.jti.co.jp

Commentary on the paper:

Hiei, Y., Ohta, S., Komari, T., and Kumashiro, T. 1994. Efficient transformation of rice (*Oryza sativa*) mediated by *Agrobacterium* and sequence analysis of the boundaries of the T-DNA. Plant J. 6:271-282.

Efficient transformation of cereals by *Agrobacterium tumefaciens* had been a fantasy until the mid 1990s, but a limited number of people still believed it would be possible some day. "We don't see any essential difference between monocotyledons and dicotyledons." "It may be just a matter of difference in response to phytohormones." "With an appropriate selection marker, it can be done." Things like this were being said by a few colleagues of the crown gall research team of Eugene W. Nester and Milton P. Gordon at University of Washington, where I spent two years from 1983 to 1985 under a fellowship program of my employer, Japan Tobacco Inc. (JT). We turned out to be right about monocot transformation, but it took a long time for us to prove it.

I studied strain A281 of *Agrobacterium* in the crown gall team. This is a super-virulent strain, which can transform plants more efficiently than others (Jin at al. 1987). After returning to a JT laboratory at Iwata, Japan, I continued the work to develop efficient transformation vectors exploiting the capability of A281. Our focus was on the *virB* and *virG* genes of A281, and we designed new vectors that carried these genes which resulted in super binary vectors (Komari et al. 1996) These vectors gave a higher frequency of transformation than conventional vectors in dicotyledonous species. Although the transformation of monocots was not my immediate objective, I was hoping that some day these vectors would be useful for cereals.

In 1988, JT initiated biotechnology programs in rice and maize. At that time protoplast electroporation was the method of choice for transformation in cereals. Our tissue culture scientists tried very hard to establish methods for isolating and culturing protoplasts in rice and maize. It was very difficult technology, and procedures were unbelievably tedious. You must first obtain good calli and then rapidly growing cells in suspension cultures. It takes a long time to generate a culture of friable cells suitable for protoplast isolation, and during extended culture, cells lose regenerating capability and mutations accumulate. Many, many factors are involved in successful preparation of viable protoplasts. The subsequent culture and regeneration of protoplasts are not straightforward either. Sometimes the techniques work well, and other times they do not. Worst of all, many of the plants regenerated from protoplasts were sterile. Yukoh Hiei, a rice tissue culture scientist, was so frustrated and sick of working with protoplasts that he was ready to try anything other than protoplasts.

The early 1990s happened to be the time when the resources and elements necessary for cereal transformation by *Agrobacterium* were being brought together in our laboratory. Basic tissue culture techniques in rice and maize had been worked out during the study of protoplasts. Dedifferentiated cells like calli and suspension cell cultures were routinely prepared and plants were regenerated. Numerous varieties of rice and maize were evaluated by breeders. Greenhouses were well maintained for the preparation of fresh tissues from rice and maize. The hygromycin resistance gene was shown to be an excellent selection marker in rice. Transformation of Solanaceae species by *Agrobacterium* was a routine process. Acetosyringone, a potent inducer of virulence genes, was commonly used and highly efficient super binary vectors were available.

Another important factor was the development at Nagoya University in Japan of the "famous" *intron-GUS* gene, which has a catalase intron in the coding sequence. The value of the *intron-GUS* was two-fold. It is not expressed in *Agrobacterium* at all and it is useful for monitoring transgene expression in plants. Furthermore, the intron dramatically enhanced expression in monocotyledons. Shozo Ohta, a molecular biologist and the inventor of the *intron-GUS,* joined JT and introduced us to a very efficient vector containing the gene.

It was a natural choice for Yukoh Hiei to test *Agrobacterium.* Active cell division is a requirement for successful transformation with DNA in any organism. His first attempt was to use cells in suspension cultures, but that did not yield positive results. He occasionally saw some blurred expression of GUS in cell suspensions, which could be considered background expression. This was not yet an official project of the Institute and he was supposed to fully dedicate his time to improving protoplast technology. People around him tried to discourage him from such futile efforts with *Agrobacterium.*

He then co-cultivated rice calli with *Agrobacterium* on solid media. This changed everything. I will never forget the moment he showed me the GUS assay of the rice calli that were co-cultivated with *Agrobacterium;* there were tiny blue spots all over. Expression was clearly dependent on the addition of acetosyringone to the co-cultivation medium. It did not take long from that moment before the first rice plants transformed by *Agrobacterium* were regenerated and grown in our greenhouse. The transgenic rice was mostly normal in morphology and fully fertile. Molecular evidence for transformation was added by Shozo Ohta and me, and the results were published in 1994.

Maize was the next target. Earlier, we had started a maize research project in which transformation was a prerequisite. Two maize tissue culture scientists, Yuji Ishida and Hideaki Saito, were having a hard time. Extensive trials with protoplasts, electroporation, and particle bombardment were taking them nowhere. After seeing the experiments in rice, they jumped at *Agrobacterium.* They examined calli, suspension cultures, immature embryos, mature embryos, shoot apices, germinating seeds, and other maize tissues. At first, it looked straightforward. Strong transient expression was observed in some of the tissue pieces co-cultivated with *Agrobacterium.* Massive blue staining was often seen.

However, there was a trap. It turned out that rice and maize were different in one very important aspect. In rice, conditions for high transient expression and high frequency integrative transformation were well correlated, and it was possible to optimize parameters based on transient expression data. This was not the case in maize Conditions very good for transient expression were often not good at all for integrative transformation. Nobody knew this and they were adjusting conditions in the wrong direction. In addition, the preferred selection marker genes were different for rice and maize. The herbicide resistance gene, *bar,* is very good for transformation in maize and other cereals, but was not available in our laboratory in the early stages of our work. After all of these challenges, it took two years for them to obtain transgenic maize.

It was necessary to use immature embryos at a particular stage of development from a particular germplasm of maize. Immature embryos must be obtained from very healthy plants grown in a greenhouse The marker genes, vectors, strains of *Agrobacterium,* the composition of the medium, and the optimal concentration of bacteria, must all be ideal. The multiplicity of factors involved was tremendous, and my colleagues ultimately found the right combinations through painstaking work. Finally, maize transformation became an efficient process (Ishida et al. 1996). The transgenic maize plants were normal in morphology and fully fertile. Shozo Ohta then created a convincing molecular data package.

It should be noted that the active interaction of tissue culture scientists and molecular biologists supervised by the research team leader Takashi Kumashiro and diligent work by a number of tissue culture technicians underpinned these studies. I was pleased to see that the value of super binary vectors was evident; maize transformation would not have been possible without them.

Let me discuss why for a long time monocotyledonous plants remained resistant to infection by *Agrobacterium.* First, too much attention and effort were devoted to the methods routinely used for the transformation of dicotyledons (such as leaf disc transformation) with the aim of using similar approaches in monocotyledons However, tissue culture of dicotyledons depends heavily on wound-induced cell division, and these are absent in many monocotyledons. Transformation requires actively dividing, dedifferentiated or dedifferentiating cells capable of regenerating plants. Scutella of mature or immature embryos are the best sources of such cells in monocotyledons. Second, many factors are of critical importance and this multiplicity of factors probably also explains why it was initially so difficult to apply this technology to cereals. The importance of tissue culture techniques stands out here.

Agrobacterium-mediated cereal transformation spread quickly and widely; reports soon followed in barley (Tingay et al. 1997), wheat (Cheng et al. 1997), sorghum, sugar cane, and turf grass. Rice is now among the top three species, along with tobacco and *Arabidopsis,* in terms of the number of recent reports in which specific species were transformed with *Agrobacterium.* Numerous genes have been introduced into cereals using this technology, and this is likely to continue, contributing to various studies in genomics, molecular biology, and biotechnology of cereals in the coming years.

References

Cheng, M., Fry, J. E., Pang, S., Zhou, H., Hironaka, C. M., Duncan, D. R., Conner, T. W., and Wan, Y. 1997. Genetic transformation of wheat mediated by *Agrobacterium tumefaciens.* Plant Physiol. 115:971–980.

Ishida, Y., Saito, H., Ohta, S., Hiei, Y., Komari, T., and Kumashiro, T. 1996. High efficiency transformation of maize (*Zea mays* L.) mediated by *Agrobacterium tumefaciens*. Nat. Biotechnol. 14:745–750.

Jin, S., Komari, T., Gordon, M. P., and Nester, E. W. 1987. Genes responsible for the supervirulence phenotype of *Agrobacterium tumefaciens* A281. J. Bacteriol. 169:4417–4425.

Komari, T., Hiei, Y., Saito, Y., Murai, N., and Kumashiro, T. 1996. Vectors carrying two separate T-DNAs for co-transformation of higher plants mediated by *Agrobacterium tumefaciens* and segregation of transformants free from selection markers. Plant J. 10:165–174.

Tingay, S., McElroy, D., Kalla, R., Fieg, S., Wang, M., Thornton, S., and Brettell, R. 1997. *Agrobacterium tumefaciens*-mediated barley transformation. Plant J. 11:1369–1376.

TOSHIHIKO KOMARI: I graduated with a B.S. in Agricultural Biology from the University of Tokyo in 1979 and an M.S. in Botany from the University of Washington, Seattle, in 1985. I received my Ph.D. from the University of Tokyo in 1992.

My mentors included Kouu Yamazaki, Professor Emeritus, University of Tokyo. He was the adviser during my undergraduate studies and taught me the basics of scientific research Eugene W. Nester, Professor, University of Washington, Seattle, introduced me to crown gall and gave me a fundamental understanding of *Agrobacterium tumefaciens*. Gynheung An, Professor, Pohang University of Science and Technology in Korea, was a leading scientist in the Nester–Gordon team and trained me as a molecular biologist.

The Plant Journal (1994) **6**(2), 271–282

TECHNICAL ADVANCE

Efficient transformation of rice (*Oryza sativa* L.) mediated by *Agrobacterium* and sequence analysis of the boundaries of the T-DNA

Yukoh Hiei*, Shozo Ohta, Toshihiko Komari and Takashi Kumashiro
Plant Breeding and Genetics Research Laboratory, Japan Tobacco Inc., 700 Higashibara, Iwata, Shizuoka 438, Japan

Summary

A large number of morphologically normal, fertile, transgenic rice plants were obtained by co-cultivation of rice tissues with *Agrobacterium tumefaciens*. The efficiency of transformation was similar to that obtained by the methods used routinely for transformation of dicotyledons with the bacterium. Stable integration, expression and inheritance of transgenes were demonstrated by molecular and genetic analysis of transformants in the R_0, R_1 and R_2 generations. Sequence analysis revealed that the boundaries of the T-DNA in transgenic rice plants were essentially identical to those in transgenic dicotyledons. Calli induced from scutella were very good starting materials. A strain of *A. tumefaciens* that carried a so-called 'super-binary' vector gave especially high frequencies of transformation of various cultivars of japonica rice that included Koshihikari, which normally shows poor responses in tissue culture.

Introduction

Methods for transformation of higher plants employing *Agrobacterium* have been well established for dicotyledonous species but not for monocotyledonous species, except in a few cases (Bytebier *et al.*, 1987). Transformation of plants in Gramineae with *Agrobacterium*, including *Agrobacterium*-mediated infection of plants with viral genomes, has been attempted in several laboratories (Chan *et al.*, 1992, 1993; Gould *et al.*, 1991; Grimsley *et al.*, 1988; Mooney *et al.*, 1991; Raineri *et al.*, 1990; Schläppi and Hohn 1992; Shen *et al.*, 1993). Raineri *et al.* (1990) obtained transformed rice cells that expressed neomycin-phosphotransferase (NPT) and β-glucuronidase (GUS) activities and they suggested that T-DNA had been transferred to, integrated in, and expressed in, rice cells. Gould *et al.* (1991) described the transfer of genes for NPT and GUS into shoot apices of corn, subsequent regeneration of plants, and detection of the transferred genes in the F_1 progeny by Southern hybridization.

These early studies of *Agrobacterium*-mediated transformation of monocotyledons have, however, been controversial. For example, Potrykus (1990) presented a critical review of the cited reports and suggested that the various authors might have overlooked the possibility of gene expression by *Agrobacterium* attached to inoculated tissues and to the plantlets regenerated from them, as well as transformation of microorganisms that were silently infecting the host plant tissues. Potrykus concluded that there was no unequivocal evidence for stable transformation of monocotyledons with *Agrobacterium*.

Appropriate evidence for such transformation would be a demonstration of random integration of transgenes into chromosomes in a number of independent transformants, with Mendelian segregation of transgenes in the progeny. Recently, Chan *et al.* (1993) obtained a few transgenic rice plants by inoculating immature embryos with a strain of *A. tumefaciens*. They proved the inheritance of the transferred DNA to the progeny by Southern hybridization, although the progeny of only one plant was analysed.

We now present further evidence in an attempt to resolve the controversy. Our evidence is based on molecular and genetic studies of a large number of transgenic rice plants and the analysis of T-DNA junctions in rice. We also show that co-cultivation of calli, derived from scutella, with *A. tumefaciens* can produce rice transformants with an efficiency similar to that of transformation in dicotyledons.

Results

Production of hygromycin-resistant, GUS-expressing plants from rice tissues inoculated with A. tumefaciens

Various tissues from rice, namely, shoot apices and segments of roots from young seedlings, scutella, immature embryos, calli induced from young roots and scutella, and cells in suspension cultures induced from scutella, were

Received 4 January 1994; revised 25 April 1994; accepted 29 April 1994.
*For correspondence (fax +81 538 32 8700).

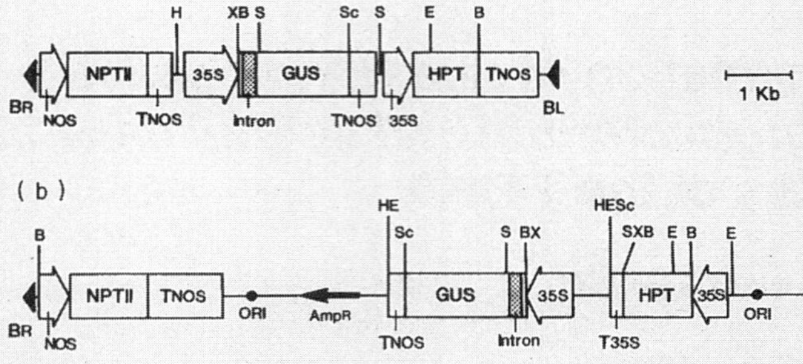

Figure 1. T-DNA regions of pIG121Hm (a) and pTOK233 (b).
Abbreviations: BR, right border; BL, left border; NPTII, neomycin phosphotransferase; GUS, β-glucuronidase; HPT, hygromycin phosphotransferase; NOS, nopaline synthase promoter; 35S, 35S promoter; TNOS, 3' signal of nopaline synthase; T35S, 3' signal of 35S RNA; ORI, origin of replication of ColE1; AmpR, ampicillin-resistance gene active in *E. coli*; B, *Bam*HI; E, *Eco*RI; H, *Hin*dIII; S, *Sal*I; Sc, *Sac*I; X, *Xba*I.

Table 1. Expression of GUS in rice tissues (cv. Tsukinohikari) immediately after co-cultivation with *A. tumefaciens* EHA101(pIG121Hm) and the recovery of GUS-expressing (GUS+), hygromycin-resistant (HygR) cells

		Number of pieces of tissue					
		Immediately after co-cultivation			After 3 weeks on selective medium		
Tissue	Duration of co-cultivation (days)	GUS+	Co-cultivated	(%)	Produced GUS+, HygR cells	Co-cultivated	(%)
Shoot apex	2	40	80	(50)			
Shoot apex	5	69	77	(90)	0	77	(0)
Root segment	3	0	100	(0)			
Scutellum	3	8	89	(9)	0	128	(0)
Immature embryo	3	66	198	(33)	3	51	(6)
Root callus	3	24	115	(21)			
Scutellum callus	3	497	533	(93)	169	743	(23)
Suspension cells[a]	3	61	247	(25)	22	254	(9)

[a]Cluster of 1–2 mm in diameter was counted as a piece of tissue.

co-cultivated with *A. tumefaciens* EHA101(pIG121Hm) (Figure 1) and GUS expression was examined immediately (Figure 2a, Table 1). Expression of GUS was detected in all of the tissues examined apart from the root segments. The calli derived from scutella gave the highest ratio of GUS-expressing tissue to inoculated tissue, and GUS expression in immature embryos was observed mainly in scutella.

Rice tissues that had been co-cultivated with *A. tumefaciens* EHA101(pIG121Hm) were cultured on hygromycin-containing medium. Proliferation of cells on the selective medium was observed in the case of the calli induced from scutella (Figure 2b), immature embryos and cells in suspension culture (Table 1). The calli gave the highest frequency (23%; 169/743) of tissue pieces that produced hygromycin-resistant cells. The proliferating regions showed evidence of the uniform expression of GUS (Figure 2c).

Calli derived from scutella were co-cultivated with four strains of *A. tumefaciens* and then cultured on selective medium. LBA4404(pTOK233) (Figure 1) gave the highest frequency of callus pieces that produced hygromycin-resistant cells (Table 2). The superiority of LBA4404(pTOK233) was especially evident in the case of cv. Koshihikari.

Most of the colonies of cells that recovered from the first round of selection proliferated on the second selective medium. Plants were readily regenerated when the resistant cells were transferred to a regeneration medium that contained hygromycin (Figure 2d). The frequency of regeneration ranged from 50 to 80% of selected colonies. Hundreds of hygromycin-resistant plants that stably expressed GUS activity were obtained. The most efficient strain, LBA4404(pTOK233), produced such plants from all of the cultivars tested, at frequencies between 12.8

Figure 2. Scutellum-derived calli of rice cv. Tsukinohikari that had been co-cultivated with *A. tumefaciens* EHA101(pIG121Hm), and transgenic cells and plants derived from the calli.
(a) Expression of GUS after infection. The calli were stained with 5-bromo-4-chloro-3-indolyl glucuronide (X-Gluc) after 3 days of co-cultivation.
(b) Hygromycin-resistant colonies of cells. The calli were plated on a selective medium after co-cultivation. The photograph was taken after 3 weeks of selection.
(c) Expression of GUS in hygromycin-resistant colonies of cells. Colonies of cells proliferated from the calli were stained with X-Gluc after 3 weeks of selection.
(d) Plant regeneration. The photograph was taken 4 weeks after selected cells had been transferred to regeneration medium.
(e) Expression of GUS in the leaf of a transformant. Segments of leaves of a transgenic plant (upper) and a non-transgenic plant (lower) were stained with X-Gluc.
(f) Expression of GUS in the root of a transformant. A segment of the root of a transgenic plant was stained with X-Gluc.
(g) A transformant at maturity.
(h) Test of the progeny for resistance to hygromycin. Seeds from a non-transgenic (left) and a transgenic (right) plant were plated on selective medium and the photograph was taken 3 days later.
Bars in the photographs represents 1 mm.

Table 2. Production of GUS-expressing (GUS+), hygromycin-resistant (HygR) cells from scutellum-derived calli inoculated with various strains of *A. tumefaciens*

		Calli that produced GUS+, HygR cells / calli inoculated (%)			
Rice cultivar	Experiment	LBA4404 (pIG121Hm)	EHA101 (pIG121Hm)	LBA4404 (pTOK233)	EHA101 (pTOK233)
Tsukinohikari	1		10/521 (2)	141/540 (26)	22/605 (4)
	2		56/187 (30)	114/324 (35)	44/254 (17)
	3		20/325 (6)	100/349 (29)	3/314 (1)
	4	91/338 (27)	139/301 (46)	159/305 (52)	
	5	59/421 (14)	66/425 (16)	110/360 (31)	
	6	64/295 (22)	66/251 (26)	108/263 (41)	30/260 (12)
	7	63/436 (14)	172/394 (44)	229/397 (58)	45/420 (11)
	8	61/236 (26)	65/241 (27)	65/143 (45)	27/219 (12)
Koshihikari	1	6/269 (2)	7/271 (3)	65/283 (23)	
	2	3/262 (1)	18/310 (6)	60/273 (22)	3/251 (1)
	3	5/197 (3)	11/176 (6)	34/138 (25)	
	4	2/177 (1)	16/140 (11)	49/215 (23)	3/162 (2)

Table 3. Efficiency of rice transformation by *A. tumefaciens* LBA4404(pTOK233)

		Number of scutellum-derived calli				
Rice cultivar	Experiment	Inoculated (*A*)	Produced HygR cells	Produced HygR plants	Produced HygR and GUS+ plants (*B*)	Frequency (*B*/*A*, %)
Tsukinohikari	1	42	20	15	12	28.6
	2	112	35	19	15	13.4
	3	64	48	24	18	28.1
	4	65	29	23	13	20.0
	5	45	18	10	7	15.6
	6	88	24	21	16	18.2
Asanohikari	1	86	19	11	11	12.8
	2	77	23	17	15	19.5
Koshihikari	1	74	16	13	13	17.6
	2	138	34	26	26	18.8
	3	215	49	40	40	18.6

HygR, hygromycin-resistant.
GUS+, GUS-positive.

and 28.6% relative to the number of pieces of scutellum-derived calli that had been co-cultivated with bacterial cells (Table 3).

Many of the hygromycin-resistant cells showed uniform blue staining in the histochemical assay for GUS. However, some of the colonies of cells included unstained areas which suggested that some of the hygromycin-resistant colonies were chimeric. GUS assays of segments of leaves or roots from regenerated plants resulted in blue staining of entire samples in most cases (Figure 2e and f), but chimerism (periclinal chimeras and dot-like chimeras) was occasionally observed.

Characterization of the plants in the R_0 generation

Hygromycin-resistant, GUS-expressing plants were grown in a greenhouse, and evaluated for ploidy, morphology and fertility. A total of 103 plants (cv. Tsukinohikari) from co-cultivation of scutellum-derived calli with EHA101(pIG121Hm) and 196 plants (93 of cv. Tsukinohikari and 103 of cv. Koshihikari) from that with LBA4404(pTOK233) were examined. All of the plants were diploid, as revealed by flow-cytometric analysis (data not shown), and none of them exhibited morphological aberrations. In terms of fertility, by contrast, variations from complete sterility to full fertility were observed. Nevertheless, the majority (about 70%) of transformants produced as many seeds as seed-derived control plants (Figure 2g). The progeny of semi-fertile transformants are now under evaluation for fertility, and preliminary data indicate that they are full fertile.

Hygromycin-resistant, GUS-expressing plants were analyzed by Southern hybridization. For example, DNA from 29 plants derived from 20 independent colonies of resistant cells of cv. Tsukinohikari, which were produced by co-cultivation of scutellum-derived calli with EHA101(pIG121Hm), was digested with *Hin*dIII and allowed to hybridize with the *hpt* and *gus* probes (some of the resultant data are shown in Figure 3a and b). Since the T-DNA of pIG121Hm has a single *Hin*dIII site (Figure 1), the number of hybridizing bands reflected the number of copies of integrated genes in the plants unless repeats of multiple copies of the T-DNA had been integrated. Most of the detected bands represented fragments of more than 6.0 kb, which is the minimum size of hybridizing fragments expected from the map of pIG121Hm (Figure 1). Plants regenerated from a given colony of cells gave an identical pattern, indicating that these plants were clonal. Otherwise, the mobilities of the bands differed from plant to plant. The copy number of integrated genes varied from one to six (Table 4). Several

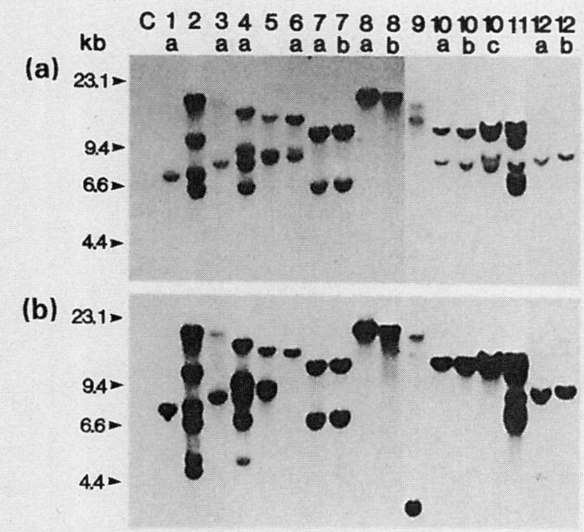

Figure 3. Southern analysis of transformants (R_0 generation).
DNA from a non-transgenic plant (C) and transformants (1a 12b) was digested with *Hin*dIII, fractionated by electrophoresis, transferred to a nylon membrane, and allowed to hybridize to the *hpt* (a) or the *gus* (b) probe.

plants gave signals only with the *hpt* probe, a few others gave signals only with the *gus* probe, and some of the fragments that hybridized with the *gus* probe were smaller than 6.0 kb (data not shown). These unexpected DNA fragments were probably produced by rearrangement of DNA upon transformation.

Inheritance of the marker genes

The selfed progeny were evaluated for resistance to hygromycin and GUS expression. The patterns of segregation of the offspring of 39 plants from the 20 independent colonies of cells described above are shown in Table 4. Resistant and sensitive seedlings were clearly distinguishable on agar plates that contained hygromycin (Figure 2h), and a segregation ratio of 3:1 was observed for 12 out of the 20 independent clones. The majority of the progeny showed clear segregation of GUS-positive and GUS-negative plants, and the estimated numbers of loci were consistent with those estimated on the basis of resistance to hygromycin.

The estimated numbers of loci were smaller than the numbers of copies of the genes that were measured by Southern analysis in some plants. It is likely that more than two copies of genes were integrated close to one another on a chromosome of such a plant. A few plants showed strange segregation patterns (plants 11, 18a, and 19e). For example, the number of hygromycin-resistant plants was smaller than that of sensitive plants in the progeny of plant 19e. Plants that gave unusual segregation ratios for hygromycin resistance gave similar results for GUS activity. Furthermore, plants that were chimeric for GUS activity were observed among such plants (progeny of plants 11, 18a, and 19e).

The R_1 progeny of plants 1b, 3a, 8a, 9, 10c and 12a, which gave segregation patterns of 3:1 or 15:1 for both hygromycin resistance and GUS expression, were analysed by Southern hybridization (parts of the data are shown in Figure 4). All of the hygromycin-resistant, GUS-positive progeny generated bands that hybridized with the *hpt* and *gus* probes whereas all of the sensitive, negative progeny did not.

Clear Mendelian segregation for hygromycin resistance and for GUS expression was also observed in the R_2 progeny of plants 1a, 3a, 8a, 9, 10c and 12a. For example, the R_2 progeny of two-thirds of the hygromycin-resistant, GUS-positive R_1 plants from plant 10c were hygromycin-resistant, GUS-positive plants and hygromycin-sensitive, GUS-negative plants at a ratio of 3:1. The R_2 progeny of the remaining one-third were exclusively hygromycin-resistant and GUS-positive. The R_2 progeny of the hygromycin-sensitive, GUS-negative R_1 plants from plant 10c were all sensitive to the drug and negative for GUS activity.

Table 4. Estimation of copy numbers of transgenes in the transformants produced by EHA101(pIG121Hm), and segregation patterns for hygromycin resistance and expression of GUS in the progeny

	Copy number in R_0 generation		Hygromycin resistance			GUS expression			
			Number of plants in R_1 generation		Number of loci	Number of plants in R_1 generation			Number of loci
Transformant[a]	HPT	GUS	R	S		+	C	–	
1a	1	1	58	22	1	15		5	1
1b	NT	NT	42	18	1	14		6	1
2	4	6	48	22	1	13		7	1
3a	2	2	28	2	2	18		2	1–2
3b	NT	NT	29	1	2–3	19		1	2
4a	4	5	27	3	1–2	18		2	1–2
4b	NT	NT	39	1	2–3	20		0	2–3
5	2	2	30	0	2–3	19		1	2–3
6a	2	1	46	4	2	14		6	1
6b	NT	NT	37	3	2	15		5	1
6c	NT	NT	38	2	2–3	12		8	1
7a	2	2	42	18	1	17		3	1–2
7b	2	2	32	8	1	13		7	1
8a	1	1	76	24	1	15		5	1
8b	1	1	36	14	1	15		5	1
9	2	2	23	7	1	15		5	1
10a	2	1	36	14	1	17		3	1–2
10b	2	1	32	8	1	16		4	1
10c	2	1	72	28	1	14		6	1
11	6	6	89	41	1	2	14	4	?
12a	1	1	40	10	1	15		5	1
12b	1	1	70	20	1	13		7	1
13	2	2	47	13	1	16		4	1
14	4	4	30	10	1	15		5	1
15	1	1	54	26	1	15		5	1
16a	2	2	42	18	1	15		5	1
16b	2	2	31	9	1	15		5	1
16c	NT	NT	21	9	1	14		6	1
16d	NT	NT	31	9	1	15		5	1
17a	2	1	45	5	2	17		3	1–2
17b	NT	NT	40	0	2–3	15		5	1
18a	3	3	29	21	?	7	6	7	?
18b	3	3	48	12	1	16		4	1
19a	3	2	26	14	1	17		3	1–2
19b	3	2	36	14	1	17		3	1–2
19c	3	2	31	9	1	17		3	1–2
19d	NT	NT	27	13	1	15		5	1
19e	NT	NT	8	32	?	0	13	7	?
20	2	2	47	13	1	16		4	1

[a]Transformants with designations that differ only in the letters of the alphabet were regenerated from a single colony of cells.
R, resistant; S, susceptible; C, chimera; NT, not tested.

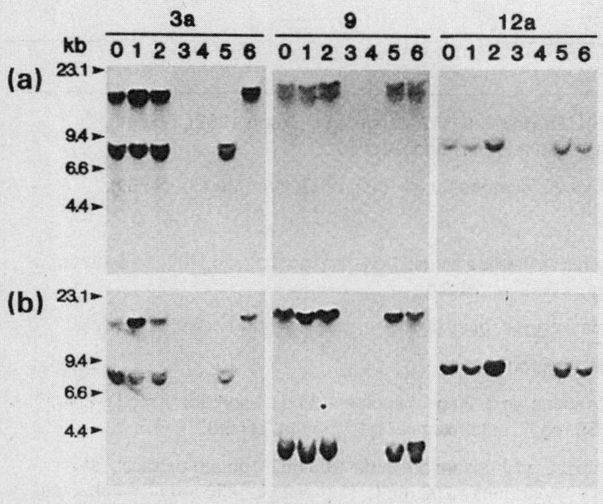

Figure 4. Southern analysis of the R_1 progeny of transformants 3a, 9 and 12a.
DNA from R_0 transformants (0), hygromycin-resistant and GUS-positive R_1 progeny (1, 2, 5, and 6), and hygromycin-sensitive and GUS-negative R_1 progeny (3 and 4) was digested with *Hin*dIII, fractionated by electrophoresis, transferred to a nylon membrane, and allowed to hybridize to the *hpt* (a) or the *gus* (b) probe.

Analysis of the boundaries of the T-DNA

The junction regions between rice genomic DNA and T-DNA were amplified from selective transformants and nucleotide sequences were determined. The sequences of three right boundaries and five left boundaries from rice that had been transformed with strain EHA101(pIG121Hm) are shown in Figure 5 together with the sequence of the T-DNA borders of pTiT37 (Yadav *et al.*, 1982; Zambryski *et al.*, 1982), which is the source of the T-DNA borders in pIG121Hm. One of the right boundaries corresponded precisely to the site found in tobacco transformants (Yadav *et al.*, 1982; Zambryski *et al.*, 1982). The two other right junctions had two more bases that might have originated from T-DNA. The five left boundaries were in the 25 bp repeat region like the left boundaries found in tobacco transformants.

Discussion

The advantages of *Agrobacterium*-mediated gene transfer over other methods that can be used for the transformation of higher plants include the high efficiency of transformation, the transfer of pieces of DNA with defined ends, the transfer of relatively large segments of DNA, and the absence of a requirement for protoplast-culture techniques. Therefore, this type of gene transfer is normally the method of choice when more than one method is available. The ability of *Agrobacterium* to transform monocotyledons has been the subject of a serious debate for some time, since these plants are not natural hosts of this bacterium. The controversy now appears to be resolved, at least for rice, since the present study and that of Chan *et al.* (1993) have clearly demonstrated the stable integration of foreign DNA in rice chromosomes in a process mediated by *A. tumefaciens*, as well as the Mendelian transmission of the DNA to the progeny. We obtained hundreds of independent transformants of three cultivars of japonica rice, and more than 20 plants were characterized in detail.

Various factors affected the efficiency of transformation. Optimization of the conditions of tissue culture and of the condition of co-cultivation was of a critical importance. The composition of the medium was especially important for the koshihikari cultivar, which generally

```
Right boundary   T-DNA        right 25-bp border repeat      outside T-DNA

pTiT37       --TAAACTATCAGTGTTTGACAGGATATATTGGCGGGTAAACCTAAGAGAAAAGAGCGTTTA--
tobacco      --TAAACTATCAGTGTTTnnnnnnnnnnnnnnnnnnnn--
rice 11      --TAAACTATCAGTGTTTGAgctcctcctcctctgtcgcc--
rice 12a     --TAAACTATCAGTGTTTGAtctggtctttttcctttgtt--
rice 30      --TAAACTATCAGTGTTTaaagaggaataaaaaaatta--

Left boundary      T-DNA       left 25-bp border repeat      outside T-DNA

pTiT37         --TAAGCGTCAATTTGTTTACACCACAATATATCCTGCCACCAGCCAGCCAACAGCTCCCC--
rice 2         --TAAGCGTCAATTTGTTTACACCACAATATAgcgaaatattaagtgcgccg--
rice 10a       --TAAGCGTCAATTTGTTTACACCAgtgagacggcaacagctgat--
rice 11*       --TAAGCGTCAATTTGTTTACccccatccacgttccgattt--
rice 11*       --TAAGCGTCAATTTGTTTACACCACAATATAgcagagctaattatggagta--
rice 12a       --TAAGCGTCAATTTGTTTACACCACAATATATCCaacgtcaaagggcgatttat--
```

Figure 5. Sequence analysis of T-DNA/plant DNA junctions.
Sequences of the junctions found in selective rice transformants are shown below the sequences of the T-DNA borders of pTiT37. Sequences presumably originated from rice genomic DNA are shown in lower case letters, and bases that could originate from either T-DNA or rice genomic DNA are shown in italic. A consensus sequence for right junctions in tobacco transformants is shown between the pTiT37 and rice sequences. 'n' stands for any nucleotide.
*Rice plant 11 had multiple copies of T-DNA and the sequences of two left boundaries were determined.

Table 5. Media for culture of rice tissues

Medium	Composition
AAM	AA salts and amino acids (Toriyama and Hinata, 1985), MS vitamins (Murashige and Skoog 1962), 500 mg l^{-1} casamino acids, 68.5 g l^{-1} sucrose, 36 g l^{-1} glucose, 100 μM acetosyringone, pH 5.2.
N6F	N6 major salts, N6 minor salts, and N6 vitamins (Chu 1978), sucrose 20 g l^{-1}, 2 g l^{-1} Gelrite (Merck), pH 5.8.
N6F-H	N6F medium plus 70 mg l^{-1} hygromycin.
N6S3	Half-strength N6 major salts, N6 minor salts, N6 vitamins, AA amino acids, 1 g l^{-1} casamino acid, 20 g l^{-1} sucrose, 0.2 mg l^{-1} naphthaleneacetic acid, 1 mg l^{-1} kinetin, 3 g l^{-1} Gelrite, pH 5.8.
N6S3-AS	N6S3 medium plus 10 g l^{-1} glucose and 100 μM acetosyringone, pH 5.2.
N6S3-CH	N6S3 medium plus 250 mg l^{-1} cefotaxime and 50 mg l^{-1} hygromycin.
N6-7-CH	N6 major salts, N6 minor salts, N6 vitamins, 2 g l^{-1} casamino acid, 20 g l^{-1} sucrose, 30 g l^{-1} sorbitol, 1 mg l^{-1} 2,4-D, 0.5 mg l^{-1} 6-benzyladenine, 100 mg l^{-1} hygromycin, 250 mg l^{-1} cefotaxime, 2 g l^{-1} Gelrite, pH 5.8.
N6-7K-CH	Half-strength N6 major salts, N6 minor salts, N6 vitamins, 2 g l^{-1} casamino acids, 200 ml l^{-1} potato extract[a], 20 g l^{-1} sucrose, 30 g l^{-1} sorbitol, 1 mg l^{-1} 2,4-D, 0.5 mg l^{-1} 6-benzyladenine, 100 mg l^{-1} hygromycin, 250 mg l^{-1} cefotaxime, 2 g l^{-1} Gelrite, pH 5.8.
2N6	N6 major salts, N6 minor salts, N6 vitamins, 1 g l^{-1} casamino acids, 30 g l^{-1} sucrose, 2 mg l^{-1} 2,4-D, 2 g l^{-1} Gelrite, pH 5.8.
2N6L	2N6 medium without Gelrite.
2N6-AS	2N6 medium plus 10 g l^{-1} glucose and 100 μM acetosyringone, pH 5.2.
2N6-CH	2N6 medium plus 250 mg l^{-1} cefotaxime and 50 mg l^{-1} hygromycin.
2N6K-CH	Half-strength N6 major salts, N6 minor salts, N6 vitamins, 1 g l^{-1} casamino acids, 200 ml l^{-1} potato extract, 30 g l^{-1} sucrose, 2 mg l^{-1} 2,4-D, 250 mg l^{-1} cefotaxime, 50 mg l^{-1} hygromycin, 2 g l^{-1} Gelrite, pH 5.8.

[a]Potato extract was prepared as follows: 200 g of potato tubers were peeled, cut into pieces of 1–2 cm^3, mixed with 1 l of water and autoclaved at 120°C for 10 min. Then the supernatant was added to media as indicated.

shows poor responses in tissue culture, during the preliminary stage of this study. Since many high-quality cultivars of japonica rice are related to Koshihikari, the methods described here should be very useful in rice breeding. Important variables related to the conditions of co-cultivation include addition of acetosyringone to the media and a temperature between 22°C and 28°C; transformation was not successful when acetosyringone was omitted or co-cultivation was carried out at a temperature higher or lower than that range (data not shown). Although Chan *et al.* (1993) indicated that the addition of medium from a suspension culture of potato cells to the co-cultivation medium was essential, our methods did not require such additions.

The choice of tissues as starting material was one of the most important factors. Various tissues were screened for their responses to co-cultivation with *A. tumefaciens.* GUS expression in the tissues immediately after infection offered a good indication of preferable tissues. Pretreatment of tissues, for example, by wounding or enzymatic digestion of cell walls, was not necessary. Such pretreatments were found to be essential in other studies (Chan *et al.*, 1993; Moony *et al.*, 1991; Raineri *et al.*, 1990). The *intron-gus* (Ohta *et al.*, 1990) was a convenient marker gene since this gene is strongly expressed in rice cells but not in *A. tumefaciens* cells that attach to the tissues. Early expression of GUS was observed in tissues that included shoot apices and immature embryos, tissues that were successfully transformed in previous studies (Chan *et al.*, 1993; Gould *et al.*, 1991; Raineri *et al.*, 1990). However, only a few transformants were obtained from immature embryos and none were obtained from shoot apices. The results of this study clearly indicate that callus cultures initiated from scutella are excellent materials for transformation of rice by *Agrobacterium*.

The patterns generated by Southern hybridization differed among our transformants, strongly indicating that T-DNAs were randomly integrated in the rice genome. The DNA fragments that hybridized to the *hpt* and *gus* probes clearly did not originate from contaminating *A. tumefaciens* used in the transformation because the vectors in strains employed would have given bands of fragments of defined sizes, for example, a 15 kb band from pIG121Hm, in the analysis. Rearrangement of DNA, for example, deletion of part of the T-DNA, was occasionally observed, as is also the case in *Agrobacterium*-mediated gene transfer in dicotyledons (Deroles and Gardner, 1988; Komari, 1989, 1990a).

Genetic analysis of the R_1 and R_2 progeny also provided conclusive evidence of the incorporation of T-DNA into rice chromosomes. Hygromycin resistance and GUS expression were inherited by the R_1 and R_2 offspring in Mendelian fashions, and the data from Southern analysis of the R_1 generation supported the genetic data. The progeny of a limited number of transformants showed unusual patterns of segregation. Since these lines included plants that were chimeric for GUS expression, it appeared that the chimerism in the R_0 generation affected the segregation ratios. Chimerism in transformants produced by *A. tumefaciens* has also been reported in tobacco (Schmülling and Schell, 1993) but has not been a limiting factor in the application of the technique.

Sequence analysis of the junctions between T-DNA and plant DNA in the rice transformants revealed that the T-DNA boundaries in rice were essentially the same as those in dicotyledons (Yadav *et al.*, 1982; Zambryski *et al.*, 1982). This evidence provides further support for the hypothesis that T-DNAs are transferred from *Agrobacterium* to dicotyledonous and monocotyledonous plants by an identical molecular mechanism.

The performance of so-called 'super-virulent' strains of *A. tumefaciens* has been emphasized in previous reports (Chan *et al.*, 1993; Gould *et al.*, 1991; Raineri *et al.*, 1990), as well as in this study. Since strains that carried pTiBo542, a 'super-virulent' Ti plasmid, were reported to operate very efficiently in transformation (Jin *et al.*, 1987; Komari, 1989), two new types of system based on pTiBo542 have been developed. The first involves strain EHA101 (Hood *et al.*, 1986), which carries a 'dis-armed' version of pTiBo542, and this strain has been popular in trials aimed at transformation of monocotyledons. The second involves what is known as a 'super-binary' vector, in which a DNA fragment from the virulence region of pTiBo542 is introduced into a small, T-DNA-carrying plasmid (Komari, 1990b) used in a binary vector system (An *et al.*, 1988; Hoekema *et al.*, 1983). In this study, we tested combinations of two strains and two vectors. The strains were an 'ordinary' strain LBA4404 (Hoekema *et al.*, 1983), and a 'super-virulent' strain EHA101. The vectors were pIG121Hm, a derivative of the 'normal' binary vector pBIN19 (Bevan, 1984), and pTOK233, a derivative of a 'super-binary' vector pTOK162 (Komari, 1990b).

In transformation experiments, LBA4404(pTOK233) was slightly more effective than both LBA4404(pIG121Hm) and EHA101(pIG121Hm) with cv. Tshukinohikari and it was definitely the most effective with cv. Koshihikari. Contrary to our expectations, EHA101(pTOK233) was, for unknown reasons, not very effective for transformation of rice. These data suggest that an 'ordinary' vector/strain combination is similar to improved combinations in the transformation of cultivars that are 'easy' to grow in tissue culture, while the choice of vectors and strains is important for transformation of 'difficult' cultivars. Since LBA4404(pTOK233) transformed the two types of cultivar equally efficiently, strains such as LBA4404(pTOK233) will be very useful in genetic engineering of rice. We even have preliminary data to suggest that LBA4404(pTOK233) might also be able to transform cultivars of indica rice.

We have developed a method for consistent production of stable transformants of rice of three cultivars at frequencies between 10 and 30% using scutellum-derived calli inoculated with *A. tumefaciens*. It was easy to produce more than 20 independent transformants in a single series of experiments without complicated manipulations in tissue culture. The efficiency of the method seems similar to that of the transformation of dicotyledons with the bacterium. Furthermore, the majority of plants produced by this method were free of morphological aberrations, probably as a result of the fact that the cells were maintained in culture *in vitro* for a short time, about 4 months from the initiation of callus to the transfer of regenerants to soil, as compared with the protoplast-mediated methods, which require at least 6 months. Therefore, *Agrobacterium*-mediated gene transfer is now available as a straightforward and routine method for the genetic modification of rice.

Experimental procedures

Rice cultivars and culture media

Cultivars of japonica rice, namely, *Oryza sativa* L. Tsukinohikari, Asanohikari and Koshihikari, were used in transformation. Media used for tissue culture are listed in Table 5.

Shoot apices, roots and calli derived from roots

Mature seeds were dehusked, sterilized with 70% ethanol for 1 min and then with 1.5% (w/v) NaClO for 30 min. They were rinsed thoroughly with sterile water and allowed to germinate on N6F medium at 25°C in darkness for 3 days. Segments of 0.5 mm × 1 mm, consisting of shoot apices plus two or three leaves, and segments (5–10 mm in length) of roots were excised from the germinating seeds and used for transformation experiments. Calli were induced by culturing some of the root segments on 2N6 medium at 30°C in darkness for 3 weeks. Actively growing pieces of calli (about 1 mm in diameter) were used for transformation experiments.

Scutella, calli derived from scutella, and suspension cultures

Mature seeds, dehusked and sterilized as described above, were cultured on 2N6 medium at 30°C in darkness. After 5 days, scutella were excised from some of the seeds and used for transformation experiments. After 3 weeks, proliferated calli derived from scutella of the remaining seeds were subcultured on fresh 2N6 medium for 4 days. Actively growing pieces of calli

(1–2 mm in diameter) were used for transformation experiments. Pieces of calli derived from scutella were suspended in 2N6L medium and cultured on a rotary shaker (125 r.p.m.) at 25°C in darkness. The medium was changed every 7 days. Cells in the logarithmic phase of growth (4 days after the third or fourth subculture) were used for transformation experiments.

Immature embryos

Developing seeds were collected 10 days after anthesis, hulled, and sterilized with 70% ethanol and 1.5% NaClO. Then immature embryos (about 1 mm in length) were excised aseptically and used for transformation experiments.

Bacterial strains and plasmids

Agrobacterium tumefaciens strains LBA4404 (Hoekema *et al.*, 1983) and EHA101 (Hood *et al.*, 1986) have been described previously. Plasmids were introduced into these strains by electroporation (Sambrook *et al.*, 1989). *A. tumefaciens* was grown on AB medium (Chilton *et al.*, 1974) at 28°C.

A gene for GUS, which has an intron in the N-terminal region of the coding sequence and is connected to the 35S promoter of cauliflower mosaic virus, was constructed by Ohta *et al.*, (1990). This *intron-gus* reporter gene expresses GUS activity in plant cells but not in the cells of *A. tumefaciens* (Ohta *et al.*, 1990). pIG121Hm (Figure 1) is a binary vector that contains a kanamycin-resistance gene (*npt*), a hygromycin-resistance gene (*hpt*), and the *intron-gus* in the T-DNA region. pIG121Hm was introduced into LBA4404 and EHA101.

pIG221 (Ohta *et al.*, 1990) was digested with *Eco*RI, treated with T4 DNA polymerase, and circularized with a *Hin*dIII linker (5′-CAAGCTTG-3′). The resultant plasmid was digested with *Hin*dIII and the 3.1 kb fragment, which contained the *intron-gus*, was inserted into the *Hin*dIII site of pGL2, a derivative of pUC18 (Yanisch-Perron, *et al.*, 1985) that contained *hpt*, to generate pGL2-IG. pTOK162 (Komari, 1990b), a binary vector that contained *virB* and *virG* from Ti-plasmid pTiBo542, was introduced into LBA4404 and EHA101. pGL2-IG was subsequently introduced into LBA4404(pTOK162) and EHA101(pTOK162). The resultant strains contained a co-integrated form of pTOK162 and pGL2-IG generated by homologous recombination. The co-integrated form is referred to as pTOK233 (Figure 1). Thus, pTOK233 contained *npt*, *hpt* and *intron-gus* in the T-DNA region as did pIG121Hm.

Transformation

A. tumefaciens strains LBA4404(pIG121Hm), EHA101(pIG121Hm), LBA4404(pTOK233) and EHA101(pTOK233) were grown for 3 days on AB medium supplemented with 50 mg l^{-1} hygromycin and 50 mg l^{-1} kanamycin. The bacteria were collected with a small spoon, and suspended at a density of 3–5 × 10^9 cells ml^{-1} in AAM medium.

The rice tissues described above were immersed in the bacterial suspension for several minutes and then transferred without rinsing on to 2N6-AS medium (all tissues except for shoot apices) or N6S3-AS medium (shoot apices), and incubated at 25°C in darkness for 3 days. After the co-cultivation, the materials were rinsed thoroughly with 250 mg l^{-1} cefotaxime in sterile water and placed on 2N6-CH medium (all tissues from Asanohikari and Tsukinohikari apart from shoot apices), N6S3-CH (shoot apices), or 2N6K-CH (all tissues from Koshihikari), and cultured for 3 weeks. Colonies of cells that had proliferated on the first selection medium were excised and cultured on N6-7-CH medium (tissues from Asanohikari and Tsukinohikari) or N6-7K-CH (tissues from Koshihikari) for 10 days. Colonies of cells that had proliferated were plated on a regeneration medium, N6S3-CH, and incubated at 25°C under continuous illumination (about 2000 lux). Regenerated plants (R_0 generation) were eventually transferred to soil in pots and grown to maturity in a greenhouse.

Test of the progeny for resistance to hygromycin

Selfed seeds (R_1 and R_2 generations) of transformants were sown on N6F medium and, 2 or 3 days later, the seeds were transferred to N6F-H medium and cultured under continuous illumination at 25°C. Drug resistance was scored 7 days after the transfer; the shoots and roots of resistant seedlings grew normally on the selection medium.

Assay of GUS activity

Expression of GUS in rice cells was assayed essentially as described by Rueb *et al.* (1989) with 5-bromo-4-chloro-3-indolyl glucuronide (X-Gluc) as substrate. Segments of rice tissues were incubated in phosphate buffer (50 mM $NaPO_4$, pH 6.8) that contained 1% Triton X-100 at 37°C for 1 h. The buffer was removed and fresh phosphate buffer containing 1.0 mM X-Gluc and 20% methanol was added to the segments. The reaction mixture was placed under a mild vacuum for 5 min, incubated overnight at 37°C, and then tissues were examined visually. Cells that expressed GUS were dark blue. Leaf segments were washed twice in 99% methanol for 2 h before the visual examination.

Isolation of DNA and Southern hybridization

DNA was extracted from leaf tissues by the procedure described by Komari *et al.* (1989). DNA (5–20 μg) was digested with *Hin*dIII and fractionated on a 0.65% agarose gel by electrophoresis at 1.5 V cm^{-1} for 15 h. Southern hybridization was carried out as described by Sambrook *et al.* (1989). The probe for *hpt* was the 1.1 kb *Bam*HI fragment from pGL2-IG and the *gus* probe was the 1.9 kb *Sal*I–*Sac*I fragment from pGL2-IG.

Sequencing of border regions of the inserted T-DNA

Junction regions of the introduced T-DNA and rice genomic DNA were analysed using cassette-ligation-mediated polymerase chain reactions (PCR), as described by Isegawa *et al.* (1992). The cassette and cassette primers were purchased from Takara (Kyoto, Japan).

DNA isolated from the transformants was digested with *Mbo*I and ligated to a *Sau*3AI cassette that consisted of the sequences 5′-GTACATATTGTCGTTAGAACGCGTAATACGACTCACTATAGGGA-3′ and 3′-CATGTATAACAGCAATCTTGCGCATTATGCTGAGTGATATCCCTCTAG-5′.

The reaction mixture (50 μl) for the first PCR consisted of an aliquot of the ligase reaction mixture (50 ng of template DNA), 50 mM KCl, 10 mM Tris–HCl (pH 8.3), 1.5 mM $MgCl_2$, 0.01% gelatin, 0.2 mM each of dGTP, dATP, dTTP and dCTP, 1 unit of

Taq DNA polymerase and 10 pmol each of cassette primer C1 (5'-GTACATATTGTCGTTAGAACGCG-3') and primer LS1 (5'-TCAGTACATTAAAAACGTCCGCA-3', for the left border) or primer RS1 (5'-GATCAGATTGTCGTTTCCCG-3', for the right border). Thermal cycling for 1 min at 94°C, 1 min at 60°C and 1 min at 72°C was performed for 30 cycles.

The second PCR was carried out using 1 µl of the reaction mixture from the first PCR as template, cassette primer C2 (5'-TAATACGACTCACTATAGGGAGA-3'), and primer LS2 (5'-ACCATGGATCCGCAATGTGTTATTAAGTT-3', for the left border) or primer RS2 (5'-TAGTCAGATCTGTCGTTTCCCGCCT-3', for the right border). The composition of the reaction mixture and the details of thermal cycling were the same as those for the first PCR apart from the template and primers. The products of PCR were cloned into a cloning vector, pCRII (Invitrogen, San Diego, CA), and sequenced by a standard procedure (Sambrook *et al.* 1989)

Acknowledgements

The authors thank Dr K. Nakamura for the gift of pIG121Hm, Dr E. Hood for the gift of EHA101, and Dr J. Paszkowski for the gift of pGL2. The technical assistance of Ms C. Ikegaya is gratefully acknowledged.

References

An, G., Evert, P.R., Mitra, A. and Ha S.B. (1988) Binary vectors. In *Plant Molecular Biology Manual* A3 (Gelvin, S.B. and Schilperoort R.A., eds). Dordrecht: Kluwer Academic Press, pp. 1–19.

Bevan, M. (1984) Binary *Agrobacterium* vectors for plant transformation. *Nucl. Acids Res.* **12**, 8711–8721.

Bytebier, B., Deboeck, F., DeGrave, H., van Montagu, M. and Hernalsteens, J.-P. (1987) T-DNA organization in tumor cultures and transgenic plants of monocotyledon *Asparagus officiallis. Proc. Natl Acad. Sci. USA*, **84**, 5345–5349.

Chan, M.-T., Lee, T.-M. and Chang, H.-H. (1992) Transformation of indica rice (*Oryza sativa* L.) mediated by *Agrobacterium tumefaciens. Plant Cell Physiol.* **33**, 577–583.

Chan, M.-T., Chang, H.-H., Ho, S.-L., Tong, W.-F. and Yu, S.-M. (1993) *Agrobacterium*-mediated production of transgenic rice plants expressing a chimeric α-amylase promoter/β-glucuronidase gene. *Plant Mol. Biol.* **22**, 491–506.

Chilton, M.-D., Currier, T.C., Farrand, S.K., Bendich, A.J., Gordon, M.P. and Nester, E.W. (1974) *Agrobacterium tumefaciens* DNA and PS8 bacteriophage DNA not detected in crown gall tumors. *Proc. Natl Acad. Sci. USA*, **71**, 3672–3676.

Chu, C.-C. (1978) The N6 medium and its applications to anther culture of cereal crops. In *Proc. Symp. Plant Tissue Culture.* Peking: Science Press, pp. 43–50.

Deroles, S.C. and Gardner, R.C. (1988) Analysis of the T-DNA structure in a large number of transgenic petunias generated by *Agrobacterium*-mediated transformation. *Plant Mol. Biol.* **11**, 365–377.

Gould, J., Devery, M., Hasegawa, O., Ulian, E.C., Peterson, G. and Smith, R.H. (1991) Transformation of *Zea mays* L. using *Agrobacterium tumefaciens* and the shoot apex. *Plant Physiol.* **95**, 426–434.

Grimsley, N.H., Ramos, C., Hein, T. and Hohn, B. (1988) Meristematic tissues of maize plants are most susceptible to Agroinfection with maize streak virus. *Bio/Techniques*, **6**, 185–189.

Hoekema, A., Hirsch, P.R., Hooykaas, P.J.J. and Schilperoort, R.A. (1983) A binary plant vector strategy based on separation of *vir*- and T-region of the *Agrobacterium tumefaciens* Ti-plasmid. *Nature*, **303**, 179–180.

Hood, E.E., Helmer, G.L., Fraley, R.T. and Chilton, M.-D. (1986) The hypervirulence of *Agrobacterium tumefaciens* A281 is encoded in a region of pTiBo542 outside of T-DNA. *J. Bacteriol.* **168**, 1291–1301.

Isegawa, Y., Sheng, J., Sokawa, Y., Yamanishi, K., Nakagomi, O. and Ueda, S. (1992) Selective amplification of cDNA sequence from total RNA by cassette-ligation-mediated polymerase chain reaction (PCR): application to sequencing 6.5-kb genome segment of hantavirus strain B-1. *Mol. Cell. Probes*, **6**, 467–475.

Jin, S., Komari, T., Gordon, M.P. and Nester, E.W. (1987) Genes responsible for the supervirulence phenotype of *Agrobacterium tumefaciens* A281. *J. Bacteriol.* **169**, 4417–4425.

Komari, T. (1989) Transformation of callus cultures of nine plant species mediated by *Agrobacterium. Plant Sci.* **60**, 223–229.

Komari, T., Saito, Y., Nakakido, F. and Kumashiro, T. (1989) Efficient selection of somatic hybrids in *Nicotiana tabacum* L. using a combination of drug-resistance markers introduced by transformation. *Theor. Appl. Genet.* **77**, 547–552.

Komari, T. (1990a) Genetic characterization of a double-flowered tobacco plant obtained in a transformation experiment. *Theor. Appl. Genet.* **80**, 167–171.

Komari, T. (1990b) Transformation of cultured cells of *Chenopodium quinoa* by binary vectors that carry a fragment of DNA from the virulence region of pTiBo542. *Plant Cell Rep.* **9**, 303–306.

Li, X.-Q., Liu, C.-N., Ritchie, S.W., Peng, J.-Y., Gelvin, S.B. and Hodges, T.K. (1992) Factors influencing *Agrobacterium*-mediated transient expression of *gusA* in rice. *Plant Mol. Biol.* **20**, 1037–1048.

Mooney, P.A., Goodwin, P.B., Dennis, E.S. and Llewellyn, D.J. (1991) *Agrobacterium tumefaciens*-gene transfer into wheat tissues. *Plant Cell Tissue Organ Cult.* **25**, 209–218.

Murashige, T. and Skoog, F. (1962) A revised medium for rapid growth and bioassays with tobacco tissue cultures. *Physiol. Plant.* **15**, 473–497.

Ohta, S., Mita, S., Hattori, T. and Nakamura, K. (1990) Construction and expression in tobacco of a β-glucuronidase (GUS) reporter gene containing an intron within the coding sequence. *Plant Cell Physiol.* **31**, 805–813.

Potrykus, I. (1990) Gene transfer to cereals: an assessment. *Bio/Technology*, **8**, 535–542.

Raineri, D. M., Bottino, P., Gordon, M. P. and Nester, E. W. (1990) *Agrobacterium*-mediated transformation of rice (*Oryza sativa* L.) *Bio/Technology*, **8**, 33–38.

Rueb, S. and Hensgens, L.A.M. (1989) Improved histochemical staining for β-D-glucuronidase activity in monocotyledonous plants. *Rice Genet. Newsl.* **6**, 168–169.

Sambrook, J., Fritsch, E. F. and Maniatis, T. (1989) *Molecular Cloning: A Laboratory Manual*, 2nd Edn. Cold Spring Harbor, NY: Cold Spring Harbor Laboratory Press.

Schläppi, M. and Hohn, B. (1992) Competence of immature maize embryos for *Agrobacterium*-mediated gene transfer. *Plant Cell*, **4**, 7–16.

Schmülling, T. and Schell, J. (1993) Transgenic tobacco plants regenerated from leaf disks can be periclinal chimeras. *Plant Mol. Biol.* **21**, 705–708.

Shen, W.-H., Escudero, J., Schläppi, M., Ramos, C., Hohn, B. and Koukolikova-Nicola, Z. (1993) T-DNA transfer to maize cells: histochemical investigation of β-glucuronidase activity in maize tissues. *Proc. Natl Acad. Sci. USA*, **90**, 1488–1492.

Toriyama, K. and Hinata, K. (1985) Cell suspension and protoplast culture in rice. *Plant Sci.* **41**, 179–183.

Yadav, N.S., Vanderleyden, J., Bennett, D.R., Barnes, W.M. and Chilton, M.D. (1982) Short direct repeats flank the T-DNA on a nopaline ti plasmid. *Proc. Natl Acad. Sci. USA*, **79**, 6322–6326.

Yanisch-Perron, C., Vieira, J. and Messing, J. (1985) Improved M13 phage cloning vectors and host strains: nucleotide sequences of the M13mp18 and pUC19 vectors. *Gene*, **33**, 103–119.

Zambryski, P., Depicker, A., Kruger, K. and Goodman, H.M. (1982) Tumor induction by *Agrobacterium tumefaciens*: analysis of the boundaries of T-DNA. *J. Mol. Appl. Genet.* **1**, 361–370.

TraR AND AAI: AUTOINDUCTION GOES MAINSTREAM

Stephen K. Farrand[1] and Lian-Hui Zhang[2]

**[1]Departments of Crop Sciences and Microbiology,
University of Illinois at Urbana-Champaign, Urbana, Illinois, United States
E-mail: stephen@life.uiuc.edu**

**[2]Laboratory of Biosignals and Bioengineering, Institute of Molecular Agrobiology,
National University of Singapore, Singapore
E-mail: lianhui@imcb.a-star.edu.sg**

Commentary on the papers:

Zhang, L., Murphy, P. J., Kerr, A., and Tate, M. E. 1993. *Agrobacterium* conjugation and gene regulation by *N*-acyl-L-homoserine lactones. Nature 362:446–447.

Piper, K. R., Beck von Bodman, S., and Farrand, S. K. 1993. Conjugation factor of *Agrobacterium tumefaciens* regulates Ti plasmid transfer by autoinduction. Nature 362:448–450.

Our discovery of TraR and AAI, its acyl-homoserine lactone quormone[1], represents one of those golden moments in science when several groups come together to form a symbiotic collaboration in which each contributes unique strengths and expertise towards solving a problem. The collaboration began in earnest following a chance conversation between one of us (SKF) and Allen Kerr during a IS-MPMI[2] meeting in Interlaken, Switzerland, in 1990. Allen described to me the work of one of his students (LHZ) in which they had discovered that induction of conjugal transfer of octopine-type Ti plasmids required a diffusible factor produced by the donor bacteria as well as a signal from the crown gall tumors induced by the bacteria.

The diffusible conjugation factor was discovered purely by chance. I (LHZ) joined Allen Kerr's group at the University of Adelaide as a postgraduate student in March 1989 and decided to work on the molecular biology of how conjugation of *Agrobacterium tumefaciens* Ti plasmids is controlled. It had been known for many years that induction of Ti plasmid conjugation requires a specific opine, called the conjugal opine, that is produced by crown gall tumors induced by the pathogenic bacteria (Farrand 1998). However, how the opine-induced transfer occurred was not understood. In my first experiments, I screened a series of isolates harboring octopine-type Ti plasmids for a suitable donor strain with which to begin my studies on conjugation. I set out to select as a standard donor, a strain that displayed a clear transfer phenotype following growth with the inducer opine at a minimal concentration. Four different octopine-type donor strains were separately patched as several replicates in the four quadrants of a plate of agar medium containing a low concentration of octopine. Following incubation to induce the transfer system, the donor patches were replica-plated onto the surface of a lawn of recipient agrobacteria spread on an agar plate containing media that allowed growth only of recipients that had inherited a Ti plasmid from the donors. After 4–5 days of incubation, transconjugant colonies appeared but in an unexpected and fascinating pattern. All of the replicate patches of three donor strains, K608 (a derivative of strain B6, partially constitutive for conjugation), NCPPB1001, and A6 tranferred their Ti plasmid, yielding corresponding transconjugants on the selective plate. However, strain B6S3 produced just a few patches of transconjugant growth and only from donor patches located in the region of its quadrant adjacent to that of strain K608. It appeared to me that something interesting had surfaced! Allen declared that it must be a "diffusion phenomenon" and predicted that something being produced by K608 was influencing B6S3. His prediction was verified; using similar plates of the donors in which a small strip of agar between the quadrants was removed prior to growth, none of the patches of strain B6S3 yielded any transconjugants when replicated onto a lawn of recipients. However, such "ditches" had no effect on

[1]Quormone: any signal molecule that is used by an organism to sense its population size.
[2]IS-MPMI = International Society for Molecular Plant-Microbe Interactions.

transfer from the other three donors (Zhang and Kerr 1991).

My (SKF) laboratory had been interested for many years in the conjugal transfer system of the Ti plasmids and had ongoing projects to identify the genes involved in the process and the mechanism by which opines regulated induction of the conjugal transfer process. At the time of the conversation with Allen, we had identified the master transcriptional regulator, a repressor called AccR, of the nopaline/agrocinopine-type Ti plasmid, pTiC58 that responds to the conjugal opine agrocinopine (Farrand 1998). However, our earlier genetic studies suggested the existence of a second level of control probably mediated by an activator. In her hunt for mutants affected in transfer, Susan Beck von Bodman, a postdoctoral in the laboratory, had isolated a *trans*-dominant Tn*5* insertion mutation, Tn*5* 10-23, in wild-type pTiC58 that caused constitutive transfer even in the absence of the conjugal opine (Beck von Bodman et al. 1989). The mutation did not affect expression of the *acc* operon and did not map to *accR*. At about this time, a Microbiology rotation student, Scott Allen, had constructed a *lacZ*-based reporter fusion in one of the *tra* genes. We showed that this fusion, carried on pJM749, was strongly expressed in a strain containing a transfer-constitutive mutant of pTiC58, but not in a strain harboring the wild-type Ti plasmid. The reporter fusion also was expressed at high levels in the 10-23 mutant that is wild-type for *accR*. Clearly this second level of regulation directly controlled expression of the *tra* genes. But what was the nature of the activator defined by the 10-23 mutation?

I was strongly influenced in my thinking about this problem by another event. Shortly after my return from the Swiss meeting, I read a review by Paul Dunlap and Pete Greenberg describing the autoinduction system of *Vibrio fischeri* (Dunlap and Greenberg 1991). In this system a small diffusible substance called VAI (*Vibrio* auto-inducer) serves as a signal to the bacterium to activate expression of the *lux* operon, the products of which cause the production of light. Expression of the *lux* operon is controlled by an activator, LuxR, and the VAI signal, an acyl homoserine lactone, that is produced by the bacteria themselves. The conversation with Allen, coupled with my new appreciation of the *Vibrio fischeri* system and its diffusible autoinducer, VAI, spurred me to ask Kevin Piper, a graduate student in the laboratory, to have a closer look at the nature of the *trans*-dominant 10-23 mutation. Kevin had already sub-cloned this region and he immediately began sequencing the active fragment from the mutant and from the wild-type Ti plasmids using our manual Sanger sequencing system.

I will not soon forget the day that Kevin showed me the results of his sequence analysis. The fragment contained a single open reading frame, we called *traR,* which could encode a protein strongly related to LuxR of the *V. fischeri lux* system. The Tn*5* insertion in the 10-23 mutant was located just upstream of the *traR* gene, consistent with our original suggestion that the transposon was promoting expression of a positively acting regulator (Beck von Bodman et al. 1989). Further genetic and molecular experiments confirmed that TraR, in conjunction with conjunction factor (CF) is responsible for activating expression of the *tra* genes. Kevin also showed that TraR responded to the *V. fischeri* autoinducer, VAI. There could be no doubt; TraR is a LuxR-like activator, and like LuxR, requires a small, diffusible signal molecule for activity. But what was the nature of CF?

For the solution to this problem, the story shifts back to Adelaide. Purifying and identifying CF was an unforgettable experience, testing both patience and determination. For this phase of the work, I (LHZ) enlisted the cooperation of Max Tate, a chemist at the Waite Institute with whom Allen had maintained a long scientific collaboration and a close friendship. We established early on that CF was a highly active, low molecular weight compound exhibiting both hydrophobic and hydrophilic properties. However, as Max would say, there was *un petit problème* with respect to large-scale purification required for structure determination; the plate conjugation method used to monitor CF activity was too slow, especially for Max, who was used to almost instantaneous chemical assays. Moreover, the amount of CF in culture supernatants of the donor strains was very low.

The breakthrough came with the news from Stephen's laboratory that TraR is a homolog of LuxR and that the activator responded to CF and to VAI, suggesting that the signal was an acyl-homoserine lactone. Moreover, the pJM749 reporter provided a quantitative assay for CF, which to Max's delight could be read within 24 hours. Having this reporter strain greatly simplified development of a large-scale purification protocol. Analytical HPLC in which fractions were monitored using the reporter, yielded the active component in amounts suitable for structure analysis. Mass spectroscopy, coupled with a positive ninhydrin reaction with the products of acid hydrolysis, convinced Max that CF was an acylated homoserine lactone. Tandem MS analysis along with IR and proton and carbon NMR analyses, were consistent with this conviction and established that the active compound was *N*-(3-oxo-octanoyl)-L-homoserine lactone. The candidate compound was quickly synthesized and its identity with the chemical structure and biological activity of CF was established. Thus CF is similar in structure to VAI, differing only in the length of its acyl side chain. Based on its structural similarity to VAI, we renamed CF AAI standing for *Agrobacterium* autoinducer.

The gene for production of AAI by the donors subsequently was identified in another collaboration between the three laboratories. Ingyu Hwang and Dave Cook from the Farrand laboratory located, cloned and sequenced the *traI* gene from pTiC58. The product of this gene is closely related to LuxI, the acyl-homoserine lactone synthase of *V. fischeri* (Hwang et al. 1994). Moreover, the cloned *traI* gene specified production of an acyl-HSL, identified by LHZ in work done with Max Tate and Allen Kerr (who had since retired) as AAI. *TraI* is the first gene of the *trb* operon, one of three expression units coding for the conjugal transfer system of this Ti plasmid (Li et al. 1998). As such, expression of *traI* is itself controlled by TraR (Hwang et al. 1994), just as expression of *luxI,* which is the first gene of the *lux* operon, is controlled by LuxR (Dunlap and Greenberg 1991).

The work presented in these two papers has had considerable impact on studies concerning *Agrobacterium* and other members of the family Rhizobiaceae and their plasmids. The phenomenon of gene regulation by acyl-homoserine lactone-mediated autoinduction of the *lux* system of *V. fischeri* first described by Ulitzer and Hastings (1979) has been extended to a number of other Gram-negative proteobacteria (Whitehead et al. 2001). Many of these studies were made possible by the availability of the *Agrobacterium* acyl-homoserine lactone reporter strain. Remarkably, although the strategy of gene regulation utilizing a transcription factor and its cognate acyl-homoserine lactone signal is conserved among these systems, the genes regulated by autoinduction vary widely, depending upon the bacterium being studied. Among bacteria that interact with plants, such systems include not only transfer of the Ti plasmids but also plasmids of *Rhizobium etli* and *Rhizobium leguminosarum,* phenazine production by soil pseudomonads, capsular polysaccharide synthesis by *Pantoeae stewartii* and production of antibiotics and pathogenicity determinants by *Erwinia carotovora.*

References

Beck von Bodman, S., McCutchan, J. E., and Farrand, S. K. 1989. Characterization of conjugal transfer functions of *Agrobacterium tumefaciens* Ti plasmid pTiC58. J. Bacteriol. 171:5281–5289.

Dunlap, P. V., and Greenberg, E. P. 1991. Role of intercellular chemical communication in the *Vibrio fischeri*–monocentrid fish symbiosis. Pages 219–253 in: Microbial Cell-Cell Interactions. M. Dworkin, ed. American Society for Microbiology, Washington, D.C.

Farrand, S. K. 1998. Conjugal plasmids and their transfer. Pages 199–233 in: The Rhizobiaceae: Molecular Biology of Model Plant-Associated Bacteria. H. P. Spaink, A. Kondorosi, and P. J. J. Hooykaas, eds. Kluwer Academic Publishers, Dordrecht, The Netherlands.

Hwang, I., Li, P.-L., Zhang, L.-H., Piper, K. R., Cook, D. M., Tate, M. E., and Farrand, S. K. 1994. TraI, a LuxI homologue, is responsible for production of conjugation factor, the Ti plasmid *N*-acylhomoserine lactone autoinducer. Proc. Natl. Acad. Sci. U.S.A. 91: 4639–4643.

Li, P.-L., Everhart, D. M., and Farrand, S. K. 1998. Genetic and sequence analysis of the pTiC58 *trb* locus, encoding a mating-pair formation system related to members of the type IV secretion family. J. Bacteriol. 180:6164–6172.

Ulitzer, S., and Hastings, J. W. 1979. Autoinduction in a luminous bacterium: A confirmation of the hypothesis. Curr. Microbiol. 2:345–348.

Whitehead, N. A., Barnard, A. M. L., Slater, H., Simpson, J. J. L., and Salmond, G. P. C. 2001. Quorum-sensing in Gram-negative bacteria. FEMS Microbiol. Rev. 25:365–404.

Zhang, L.-H., and Kerr, A. 1991. A diffusible compound can enhance conjugal transfer of the Ti plasmid in *Agrobacterium tumefaciens.* J. Bacteriol. 173:1867–1872.

LIAN-HUI ZHANG: In 1982, I obtained my B.S. in Biology from South China Agricultural University, Guanzhou, China, and an M.S. in Biology from the same university in 1985. For my Ph.D., I studied at the Waite Institute, University of Adelaide, Australia, and graduated in 1993. I spent four years as a postdoc at the University of Queensland, Brisbane, Australia, where I studied bacterial leaf scald disease of sugarcane caused by *Xanthomonas albilineans.* One year was spent as a Principal Investigator, heading the Laboratory of Molecular Plant Protection in the Plant

Biotechnology Centre, Agriculture Victoria, Australia. I joined the National University of Singapore in 1998.

Six scientists have been key forces in fostering and shaping my development as a scientist. They are Kung-Hsun Lin, my M.S. adviser, Allen Kerr who was my mentor during my Ph.D. studies, Peter Murphy who co-supervised my Ph.D. project in Adelaide, Robert Birch who was a great stimulus during my postdoc period, and Max Tate and Stephen Farrand, who continue to influence my appreciation of the wonders and complexity of chemistry and genetics.

STEPHEN K. FARRAND: I completed an A.B. in Biology from Whitman College, Walla Walla, Washington, in 1967. My graduate studies were at the University of Rochester School of Medicine, Rochester, N.Y., where I studied the role of quinones in *Bacillus subtilis* and obtained my Ph.D. in 1973. A three year period as a postdoc in the Department of Microbiology at the University of Washington, Seattle, introduced me to *Agrobacterium* research.

Seven scientists have played substantial roles in shaping my development and progression as a scientist. They are Harry W. Taber, my Ph.D. adviser, Mary-Dell Chilton and Eugene W. Nester who were my mentors during my postdoctoral years, Jacques Tempé and Allen Kerr who introduced me to the wonders of the biology of the *Agrobacterium* system, and Max Tate and Scott Chilton who constantly remind me that the results from genetic experiments have to be interpreted in terms of chemistry.

Agrobacterium conjugation and gene regulation by *N*-acyl-L-homoserine lactones

Lianhui Zhang*, Peter J. Murphy*, Allen Kerr* & Max E. Tate†‡

Departments of * Crop Protection and † Plant Science,
Waite Agricultural Research Institute, The University of Adelaide,
Glen Osmond, South Australia, Australia 5064

CONJUGAL opines secreted by crown gall tumours induce strains of *Agrobacterium tumefaciens* that are donors of Ti plasmids to produce a diffusible conjugation factor[1]. This enhances the conjugal transfer efficiency of the Ti plasmid in other strains of *A. tumefaciens*. This factor behaves as a secondary messenger, transmitting the environmental information to *tra* genes. Here we report the use of spectrometry to show that this factor is identical to synthetic *N*-(β-oxo-octan-1-oyl)-L-homoserine lactone and confirm that the synthetic compound is biologically active. *N*-(Hexan-1-oyl)-L-homoserine lactone has also been detected. A closely related molecule, *N*-(β-oxo-hexan-1-oyl)-L-homoserine lactone, autoinduces bioluminescence in the distantly related bacterium, *Vibrio fischeri*[2]. *N*-Acyl-homoserine lactones thus seem to be conserved molecules in which the length and nature of the lipophilic acyl chain determines the biological function to be regulated. Mutants that do not produce the factor fail to conjugate unless supplied with it in the induction medium (our unpublished data). These data indicate that the conjugation factor is an autoinducer and a key signal molecule in the conjugation system of *A. tumefaciens*. It is, to our knowledge, the first example of a second messenger molecule in a bacterial conjugation system.

‡ To whom correspondence should be addressed.

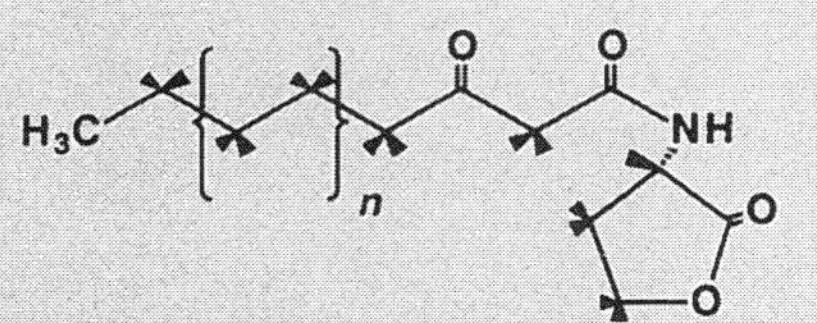

FIG. 1 Structural formulae for *N*-β-oxo-acyl-amides of L-homoserine lactone, $n=0$, 1 and 2 correspond to the *N*-β-oxo-hexanoyl, -octanoyl and -decanoyl amides. For isolation of the conjugation factor CF1 ($n=1$), strain K794, a transfer constitutive octopine strain of *A tumefaciens*, was grown in Petit's liquid minimal medium[7] for 2 days at 28 °C, using proline and mannitol as the nitrogen and carbon sources. Presence of CF activity was determined by induction of a *Tra::lacZ749* reporter gene fusion in strain NT-1 (supplied by, S. Farrand[3]). CF was extracted from K794 bacterial culture filtrates by ethyl acetate. Partial purification of the concentrated extracts was achieved by flash chromatography on silica gel using ethyl acetate as eluant[8], followed by C18 reversed-phase open column flash chromatography eluted isocratically with 30% (v/v) methanol/water. Evaporation of the active fractions gave a yellowish white solid (3 mg L^{-1} culture). Further purification by C18 reversed-phase HPLC (analytical 25 cm, 4 mm inner diameter) eluted (1 mL min^{-1}) with a linear gradient of 30–45% methanol, gave two separate biologically active components, labelled CF1 (retention time 13.38 min) and CF2 (retention time 11.4 min). A quantitative β-galactosidase assay[9], was used to measure CF activity. Synthetic CF1 ($n=1$) was obtained in 49% yield in a one-step procedure using L-homoserine lactone HCl (1 mmol) in pyridine (10 ml) at 130 °C for 45 min with the acyl Meldrum's acid intermediate of Oikawa[10] (1.5 mmol). Cognate preparations proceeded smoothly with the appropriate intermediate[10]. Saturated acyl amides were prepared from the L-homoserine lactone HCl and corresponding acyl chloride in pyridine under the same conditions.

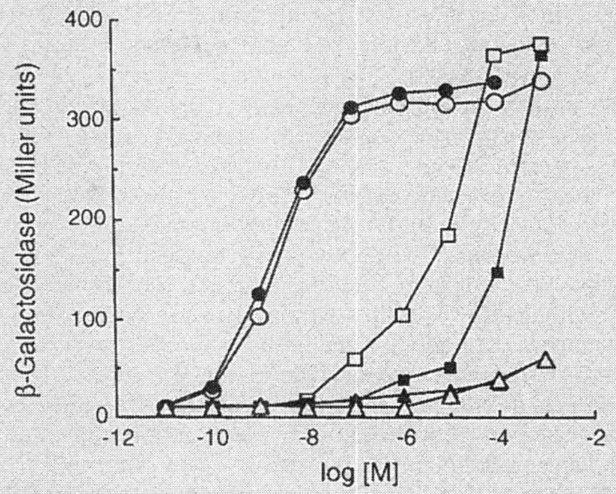

FIG. 2 Selective induction of *tra* gene activity by *N*-acyl-amides of L-homoserine lactone. Natural: CF1 (●); synthetic: *N*-β-oxo-octanoyl- (○), *N*-β-oxo-hexanoyl- (□), *n*-β-oxo-decanoyl- (■), *N*-octanoyl- (△) and *N*-hexanoyl-amides (▲) of L-homoserine lactone. β-Galactosidase activity was expressed as Miller units[9]. *N*-acyl-amides of L-homoserine lactone were added to the minimal medium[7] plus $(NH_4)_2SO_4$ (0.2, w/v) and mannitol (0.2%, w/v) and then inoculated with a 1:40 dilution of a late log phase culture (absorbance at 620 nm ≅ 0.6) of NT-1 (*traR, tra::lacZ749*)[3]. After 22 h of incubation at 25 °C, cells were collected by centrifugation and assayed for β-galactosidase activity[11].

Conjugation factor (CF) is present in the culture supernatant of transfer constitutive strains of *A. tumefaciens*[1]. Its purification using high-performance liquid chromatography (HPLC) yielded two active components, CF1 and CF2. Examination of these fractions by chemical ionization quadrupole mass spectrometry showed strong quasimolecular (M + H) ions, with an m/z of 242 for CF1, and for the less homogeneous CF2 fraction, ions at m/z 274, 214 and 200. All of these even M + H ions were consistent with the presence of nitrogenous components. For CF1 this was confirmed by infrared spectrometry, which showed amide I and II carbonyl absorptions at 1,646 and 1,540 cm^{-1} as well as characteristic absorptions near 1,780 cm^{-1} and 1,714 cm^{-1} assignable to the presence of γ lactone and ketonic carbonyls[2]. Acid hydrolysis of CF1 yielded an amino acid which comigrated with homoserine when separated by high-voltage paper electrophoresis. Daughter ion tandem mass spectrometry of the M + H ions at 242 m/z in CF1, and M + H ions at 274, 214 and 200 in CF2 fraction showed related fragmentations after helium collisional activation. Except for the 214 parent ion, which showed a daugher ion at 113 instead of 99 as well as ions at 102 and 141, each daughter ion set from the parent ions at 242, 274 and 200 m/z, displayed prominent 99, 102 and 141 m/z ions. Comparison of the corresponding electron impact data with those reported[2] suggested a closely related set of homoserine lactone derivatives. These data taken together with the reported[3] homology of TraR in *A. tumefaciens* with LuxR in *Vibrio fischeri*, which requires *N*-(β-oxo-hexan-1-oyl)-L-homoserine lactone for gene induction, indicate that a common homoserine lactone moiety is associated with gene regulation in these distantly related species.

The known luciferase autoinducer β-oxo-hexanol homoserine of *V. fischeri*[2] (Fig. 1, structure $n = 0$) was synthesized to compare its mass spectrum with that of natural CF. The comparison confirmed its relatedness, but nonidentity with the tandem ion mass spectrometry daughter ion spectra from the 214 m/z parent ion in the CF2 fraction. But data from the homogeneous CF1 fraction were consistent with a -$(CH_2)_2$- extension of the lipophilic side chain of the luciferase autoinducer, i.e. β-oxo-octanoyl homoserine lactone (Fig. 1, $n = 1$). This was also synthesized. These molecules have very similar infrared spectra with the four characteristic carbonyl signals[2]. In the case of CF1, sufficient homogeneous material was obtained to show that all the measured properties (ultraviolet, infrared, ^{1}H nuclear magnetic resonance (NMR), ^{13}C NMR; chromatographic parameters data not shown) and biological activity (Fig. 2) were virtually indistinguishable from those of the synthetic sample. The tandem mass spectrometric fragments of the M+H ion of m/z 200 present in the CF2 fraction suggested it was the fully reduced hexanoyl amide of homoserine lactone and the daughter ion fragmentation patterns were indistinguishable from those of a synthetic sample of the hexanoyl amide.

Figure 2 shows *tra* gene activity induced by the natural CF1 from *A. tumefaciens* and by synthetic *N*-β-oxo-octanoyl-, *N*-β-oxo-hexanoyl-, *N*-β-oxo-decanoyl-, *N*-octanoyl- and *N*-hexanoyl amides of L-homoserine lactone. Only the synthetic *N*-β-oxo-octanoyl-L-homoserine lactone shows identity in gene induction activity with the natural product and accounts fully for the observed activity of fraction CF1. Although masses corresponding to the *N*-hexanoyl amide were detected by mass spectrometry in the biologically active CF2 fraction from the HPLC fractionation, the synthetic product had insufficient activity on a molar basis to account for the (autoinducing) activity of this fraction. The close relationship of the daughter ions from the 274 parent ion to those of synthetic conjugation factor m/z 242 and in particular the presence of a strong fragment ion at 134 instead of the ion corresponding to a protonated homoserine lactone at 102, suggested that this fraction contained the methyl ester of the conjugation factor derived as a methanolic-solvent transesterification artefact. This was confirmed by the chromatographic behaviour, CI daughter ion mass spectra, infrared spectra and biological activity of the synthetic acyclic methyl ester of CF1.

The remarkable feature of the data in Fig. 2 is that extension (Fig. 1, $n=2$) or diminution (Fig. 1, $n=0$) of the chain length or conversion of the beta keto groups (Fig. 1, $n=0$ and $n=1$) to methylene groups is sufficient to reduce the inducing activity by four orders of magnitude or more compared to that of the β-oxo-octanoyl derivative ($n=1$). A similar specificity was reported for analogues of the autoinducer of bioluminescence in *V. fischeri*[4].

The facts that *N*-β-oxo-hexanoyl-homoserine lactone is the autoinducer for luminescence activity in *V. fischeri*[2] and for antibiotic biosynthesis in *Erwinia*[5], that *N*-β-hydroxy-butanoyl-L-homoserine lactone is the autoinducer for luminescence in *V. harveyi*[6] and that *N*-β-oxo-octanoyl-L-homoserine lactone is the autoinducer for conjugation in *A. tumefaciens*, suggest that *N*-acyl homoserine lactones may be a widely conserved signal for gene regulation. □

Received 24 September 1992; accepted 21 January 1993.

1. Zhang, L & Kerr, A. *J. Bact.* **173,** 1867–1872 (1991).
2. Eberhard, A. *et al. Biochemistry* **20,** 2444–2449 (1981).
3. Piper, K. R., Beck von Bodman, S. & Farrand, S. K. *Nature* **362,** 448–450 (1993).
4. Eberhard, A., Widrig, C. A., McBath, P. & Schineller, J. B. *Arch. Microbiol.* **146,** 35–40 (1986).
5. Bainton, N. J. *et al. Gene* **116,** 87–91 (1992).
6. Cao, J.-G. & Meighen, E. A. *J. biol. Chem.* **264,** 21670–21676 (1989).
7. Petit, A. & Tempé, J. *Molec. gen. Genet.* **167,** 147–155 (1978).
8. Still, W. C., Kahn, M. & Mitra, A. *J. org. Chem.* **43,** 2923–2925 (1978).
9. Miller, J. H. *Experiments in Molecular Genetics* 352–359 (Cold Spring Harbor Laboratory Press, New York, 1972).
10. Oikawa, Y., Sugano, K. & Yonemitsu, O. *J. org. Chem.* **43,** 2087–2088 (1978).
11. Stachel, S. E., An, G., Flores, C. & Nester, E. W. *EMBO J.* **4,** 891–898 (1985).

Conjugation factor of *Agrobacterium tumefaciens* regulates Ti plasmid transfer by autoinduction

Kevin R. Piper*, Susanne Beck von Bodman† & Stephen K. Farrand*†‡

Departments of † Plant Pathology and * Microbiology, University of Illinois at Urbana/Champaign, Urbana, Illinois 61801, USA

CONJUGAL transfer of Ti plasmids from *Agrobacterium* donors to bacterial recipients is controlled by two types of diffusible signal molecules. Induction is mediated by novel compounds, called opines, that are secreted by crown gall tumours. These neoplasias result from infection of susceptible plants by virulent agrobacteria. The second diffusible signal, called conjugation factor, is synthesized by the donor bacteria themselves. Production of this factor is induced by the opine. Here we show that conjugation is regulated directly by a transcriptional activator, TraR, which requires conjugation factor as a coinducer to activate *tra* gene expression. TraR is a homologue of LuxR, the *lux* gene activator from *Vibrio fischeri* which also requires an endogenously synthesized diffusible coinducer. The two regulatory systems are related; the two activator proteins show amino-acid sequence similarities and the *lux* system cofactor, autoinducer, will substitute for conjugation factor in the TraR-dependent activation of Ti plasmid *tra* genes.

Agrobacterium tumefaciens strains harbouring pTiC58 induce plant tumours that synthesize the opines agrocinopines A and B[1,2]. Either opine induces a Ti plasmid-encoded system required for their catabolism by the bacterium, and also a conjugation system that transfers the Ti plasmid to bacterial recipients[3,4]. Coordinate regulation of these systems operates through a single transcriptional repressor, AccR, encoded by the *accR* gene[5]. A Ti plasmid with a deletion in *accR* constitutively expresses both phenotypes. But two lines of evidence indicate that conjugation is regulated at a second level. First, a Tn*5* insertion mutant of pTiC58, 10-23, is constitutive for conjugal transfer (Tra^c), but remains inducible for agrocinopine catabolism[6]. The 10-23 insertion is not located in *accR* but maps to *Eco*RI fragment 33 in a region encoding conjugal transfer functions (Fig. 1). Furthermore, the mutation is dominant suggesting that it results in a gain of function[6]. Second, conjugation requires a diffusible substance called conjugation factor (CF)[7]. This signal, produced by the donor population itself, accumulates in culture supernatants after induction by the opines. *Agrobacterium* donors that do not produce CF on induction with opines fail to conjugate unless exposed to the factor produced by a CF$^+$ strain[7].

To examine this second level of regulation, we tested *Agrobacterium* strains with or without Ti plasmids for their ability to express the *tra*::*lacZ749* reporter fusion carried on pJM749 (Fig. 1). This clone lacks *accR* and also the region defined by the 10-23 mutation. Strain NT1(pJM749), which lacks a Ti plasmid and therefore the *accR* gene, did not produce β-galactosidase (Table 1). But strains harbouring the Δ*accR* or the 10-23 Tra^c Ti plasmids expressed the *lacZ* fusion at high levels. Failure of the Ti plasmidless strain to express the reporter fusion indicates that the *tra* genes are positively regulated. Expression in the 10-23 mutant suggested that the Tn*5* insertion promotes expression of the regulatory gene, we call ***traR***, responsible for *tra* gene activation. To test this, we cloned the Tn*5*-containing *Eco*RI fragment 33 from mutant 10-23 to form pSVB33-23 (Fig. 1). In two experiments, a donor strain harbouring this plasmid *in trans* to pTiC58 transferred the Ti plasmid by conjugation to the *A. tumefaciens* recipient C58C1RS[6] at frequencies of 4.8×10^{-2} and 6.7×10^{-2} per input donor without induction by agrocinopines. These values are similar to transfer frequencies observed for the Δ*accR* and the 10-23 Tra^c mutant Ti plasmids[6]. In the absence of pSVB33-23, transfer of pTiC58 was undetectable ($<10^{-8}$ per input donor). We conclude that *traR*, expressed from pSVB33-23, activates conjugal transfer of pTiC58 in the absence of the inducer opine.

Because *traR* on *Eco*RI fragment 33 *trans*-activates conjugation, we determined if this fragment is sufficient to induce expression of the *tra*::*lacZ749* fusion on pJM749 in the absence of a Ti plasmid. Plasmids pSVB33-23, and pSVB33 which contains *Eco*RI fragment 33 from wild-type pTiC58 (Fig. 1), were introduced into strain NT1(pJM749). Neither strain produced significant levels of β-galactosidase (Table 1). But unlike the pJM749 merodiploids harbouring Tra^c Ti plasmids, these strains do not produce conjugation factor. When the two reporter strains were incubated with a crude preparation of CF, β-galactosidase levels were elevated 400-fold (Table 1). This response required pSVB33 or pSVB33-23. The *tra*::*lacZ* reporter fusion was not expressed when the cultures were supplemented with a preparation from a strain that does not produce CF (data not shown). Zhang *et al.*[8] show in the accompanying letter that purified CF stimulates TraR-dependent expression of the reporter fusion in a manner indistinguishable from that observed with our crude preparation. These results indicate that the *trans*-acting

‡ To whom correspondence should be addressed.

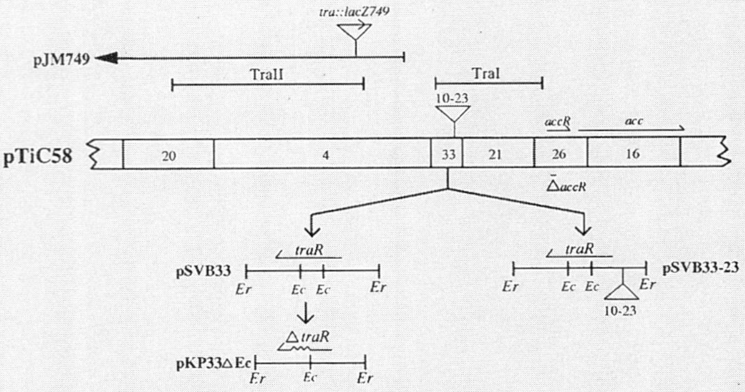

FIG. 1 Physical map of the conjugal transfer-opine catabolism region of pTiC58. The map shows the locations of *Eco*RI sites between plasmid kilobase coordinates 111 and 134, the limits of Tra regions I and II (ref. 6), and the recombinant clones used in this work. The locations of genes encoding catabolism of agrocinopines A and B (*acc*)[4] and the master conjugal transfer-opine catabolism repressor (*accR*)[5] are shown above the map. The small line labelled Δ*accR* shows the location of the 5 bp deletion in *accR* responsible for the Tra^c/Acc^c phenotype of the spontaneous mutant plasmid, pTiC58ΔΔ*accR*[5]. The triangle labelled 10-23 depicts the site of the Tn5 insertion in wild-type pTiC58 that results in a Tra^c/Acc-inducible phenotype[6]. pJM749, containing the *tra*::*lacZ749* reporter fusion, was derived by Tn*3*HoHo1 mutagenesis[22] of pTHB58, a cosmid clone containing an insert from pTiC58[4]. When marker-exchanged into pTiC58Δ*accR* the *lacZ749* insertion results in a Tra$^-$ phenotype[5]. The direction of transcription of the *lacZ* fusion is indicated by the arrowhead. pSVB33 and pSVB33-23 were produced by cloning *Eco*RI (*Er*) fragment 33 from pTiC58 and pTiC58Tra^{c}10-23, respectively, into the IncW vector pSa152 (ref. 23). The location and transcriptional direction of the *traR* ORF is indicated for the two plasmids. pKP33ΔEc was produced by deleting a 121 bp *Ecl*XI (*Ec*) fragment internal to *traR* in pSVB33. The allele is labelled Δ*traR* with the wavy line indicating that portion of the TraR amino-acid sequence altered by the internal deletion.

function encoded by *traR* requires CF to activate expression of *tra* genes.

Addition of the CF preparation to strain C58 harbouring pJM749 did not stimulate expression of the *tra*::*lacZ* fusion (Table 1). This is consistent with the observation that CF will not replace opines for induction of transfer.[7] That the reporter fusion is expressed in strains harbouring the Δ*accR* Tra^c mutant Ti plasmid but not in wild-type strain C58, even when exposed to exogenous CF, indicates that expression of *traR* is regulated by the opines through AccR.

*Eco*RI fragment 33 was sequenced by the dideoxynucleotide chain termination method[9] (EMBL accession no. Z15003). The 1,805 base pair (bp) fragment contains a 702 bp open reading frame (ORF) between an ATG at bp 1,244 and TGA at bp 542. This corresponds to an anticlockwise transcriptional orientation on the standard map of pTiC58 (ref. 10). The 10-23 Tn5 insertion is located between bp 1,482 and 1,483, 238 bp upstream of the ATG initiation codon. This location is consistent with our hypothesis that the Tra^c phenotype results from promoted expression of the ORF by the transposon. Dependence on the intact ORF was determined by deleting an internal 120 bp *Ecl*XI fragment (Fig. 1). Assuming that the ORF represents *traR* and encodes a protein, the deletion removes amino acids 107–146 and alters the remaining sequence of the protein. This clone, pKP33ΔEc, does not activate β-galactosidase expression from the *tra*::*lacZ749* reporter fusion even in the presence of CF (Table 1).

The *traR* ORF can encode a protein of 234 amino acids with an M_r of 26,506. The protein, we name TraR, exhibits 20% amino-acid identity and 62% amino-acid similarity over a 234 residue overlap with the bacterial protein LuxR (Fig. 2). LuxR is a transcriptional activator that regulates bioluminescence gene expression in the marine bacterium, *Vibrio fischeri*[11]. Activation of *lux* expression also requires a diffusible cofactor called autoinducer (AI)[12]. This factor is produced by *V. fischeri* cells themselves and, along with LuxR, regulates bioluminescence by a mechanism called autoinduction[13]. Given the similarity between TraR and LuxR, we determined whether AI can substitute for CF in the TraR-dependent activation of the *tra*::*lacZ749* fusion on pJM749. Cells incubated with a preparation containing AI produced levels of β-galactosidase activity 40-fold higher than those in cells receiving no additions (Table 1). But these levels were 10-fold lower than those obtained with preparations containing CF.

These results show that conjugal transfer of the *Agrobacterium* plasmid pTiC58 is controlled by autoinduction. CF clearly functions as a second message to transmit the opine signal to the conjugation regulatory apparatus. Furthermore, CF is produced by *Agrobacterium* strains harbouring any one of several Ti plasmid types[7] indicating that this mode of regulation is common to the *tra* systems of these plasmids. CF may, as with AI, serve to stimulate donors in a population-dependent manner. Alternatively, because conjugation is induced by opines, CF may function to amplify the response produced by suboptimal concentrations of these signal molecules. Of more general significance, the system is remarkably similar to that which controls bioluminescence in *V. fischeri*. The activators, LuxR and

TABLE 1 Conjugation factor is required for TraR-mediated induction of Tra gene expression

Test plasmid	β-Galactosidase activity from *tra*::*lacZ749** No addition	+CF†	+AI‡
None	<1.0	<1.0	NT
pTiC58	<1.0	<1.0	NT
pTiC58Δ*accR*	200.0	200.3	NT
pTiC5810-23	128.5	133.1	NT
pSa152	<1.0	<1.0	<1.0
pSVB33	<1.0	419.2	43.5
pSVB33-23	<1.0	460.7	34.5
pKP33ΔEc	<1.0	<1.0	<1.0

Agrobacterium tumefaciens strain NT1 (ref. 16) harbouring the *tra*::*lacZ* reporter plasmid pJM749 was used for all constructions. Test plasmids were introduced into this strain by transformation[17] or by electroporation[18].

* Expressed as Miller units[19]. Values less than 1.0 are insignificant. Strains were grown in AB minimal mannitol medium[20] at 28 °C to mid-exponential phase (~5×10^8 cells per ml). Cultures were adjusted to an optical density at 600 nm of 0.4 (~2×10^8 cells per ml) and split in half. The inducer (CF or AI) to be tested was added (25 μl per ml) to one subculture, and the pairs of subcultures were incubated for an additional 6 h. Cells were collected by centrifugation and assayed for β-galactosidase activity as previously described[4].

† A crude preparation of CF was prepared from *A. tumefaciens* strain NT1(pTiC58Δ*accR*). The strain was grown to saturation in AB mannitol medium at 28 °C. Cells were removed by centrifugation, and the culture supernatant was sterilized by filtration. 25 μl of this preparation elicited a maximum response when added to 1.0 ml of a culture of a reporter strain harbouring pJM749 and pSVB33.

‡ A crude preparation of AI was prepared from *E. coli* strain JM109 harbouring pJE202 (Gift of E. P. Greenberg). This recombinant plasmid encodes the entire *lux* regulon of *V. fischeri*[21]. The strain was grown to saturation in LB medium at 28 °C and the culture was processed as described above for CF. NT indicates not tested.

```
                    10v       20v       30v       40v
TraR               MQHWLDKLTDLAAIQGDECILKDGLADLAEHFGFTGYAYLHI
                     :   .  :   :. : ::C L.D    :   :: :  :  Y H
LuxR      MKNINADDTYRIINKIKACRSNNDINQC-LSDMTKMVHCEYYLLAIIYPHS
                   10^       20^       30^       40^       50^

             50v       60v       70v       80v       90v
TraR      QHKHTIAV-TNYHRDWRSAYFENNFDKLDPVVKRAKSRKHVFAWSGEQERS
            K .I:: .NY.:.WR  Y :.N: K DP:V. ::S..  :.W:  ::.:
LuxR      MVKSDISILDNYPKKWRQYYDDANLIKYDPIVDYSNSNHSPINWNIFENNA
                 60^       70^       80^       90^       100^

            100v      110v      120v      130v      140v
TraR      RLSKEERAFYAHAADFGIRSGITIPIKTANGSMSMFTLASERPAIDLDREI
           ::K....    :A . G: :G:::PI.TAN.:.:M:::A .   .  :D. :
LuxR      -VNKKSPNVIKEAKTSGLITGFSFPIHTANNGFGMLSFAHSEKDNYIDSLF
                110^      120^      130^      140^      150^

            150v      160v      170v      180v      190v
TraR      DAAAAAGAVGQLHARISFLQTTPTVEDAAW-LDPKEATYLRWIAVGMTMEE
            A  . ::         :: :.. : :.:.  L. :E ..L W   G.:  :
LuxR      LHACMNIPLIVPSLVDNYRKINIANNKSNNDLTKREKECLAWACEGKSSWD
               160^      170^      180^      190^      200^

             200v      210v      220v      230v
TraR      VADVEGVKYNSVRVKLREAMKRFDVRSKAHLTALAIRRKLI
          ::.: G . .:V  .L :A .:::. :: : .: AI    I
LuxR      ISKILGCSERTVTFHLTNAQMKLNTTNRCQSISKAILTGAIDCPYFKN
              210^      220^      230^      240^      250^
```

FIG. 2 Amino-acid sequence similarity between TraR and LuxR. The entire sequences of the two proteins are presented using the single-letter amino-acid code. Identities between the two sequences are shown by letter. (Colon indicates a conserved change at a given position, dot indicates a neutral change.)

METHODS. The derived amino-acid sequence of the *traR* ORF was determined from the primary DNA sequence using DNA Strider and compared with all peptide and protein sequences in the GenBank, EMBL, SWISS-PROT, and GenPept data banks using the FASTA program[24]. Similarities were detected between TraR and the LuxR protein, and the two amino-acid sequences were aligned optimally using the DNA Star program.

TraR are related, and the cofactors, AI and CF are chemically similar[8,14] and physiologically crossfunctional. Recently elastase production by *Pseudomonas aeruginosa* also was shown to be regulated by a LuxR-like activator[15]. Although the structure of the soluble coinducer is not known, it is probably similar to AI and CF. This degree of relatedness between the systems of three dissimilar bacteria suggests that autoinduction mediated by a gene activator protein and a cognate substituted homoserine lactone coinducer constitutes an environmentally responsive regulatory mechanism that may be common among microorganisms. □

Received 24 September 1992; accepted 21 January 1993.

1. Ellis, J. G. & Murphy, P. J. *Molec. gen. Genet.* **181,** 36–43 (1981).
2. Ryder, M. H., Tate, M. E. & Jones, G. P. *J. biol. Chem.* **259,** 9704–9710 (1984).
3. Ellis, J. G., Kerr, A., Petit, A. & Tempé, J. *Molec. gen. Genet.* **186,** 269–274 (1982).
4. Hayman, G. T. & Farrand, S. K. *J. Bact.* **170,** 1759–1767 (1988).
5. Beck von Bodman, S., Hayman, G. T. & Farrand, S. K. *Proc. natn. Acad. Sci. U.S.A.* **89,** 643–647 (1992).
6. Beck von Bodman, S., McCutchan, J. E. & Farrand, S. K. *J. Bact.* **171,** 5281–5289 (1989).
7. Zhang, L. & Kerr, A. *J. Bact.* **173,** 1867–1872 (1991).
8. Zhang, L., Murphy, P. J., Kerr, A. & Tate, M. E. *Nature* **362,** 446–448 (1993).
9. Sanger, G., Nicklen, S. & Coulson, A. R. *Proc. natn. Acad. Sci. U.S.A.* **74,** 5463–5467 (1977).
10. Depicker, A. *et al. Plasmid* **3,** 193–211 (1980).
11. Engebrecht, J. & Silverman, M. *Nucleic Acids Res.* **15,** 10455–10467 (1987).
12. Nealson, K. H., Platt, T. & Hastings, J. W. *J. Bact.* **104,** 313–322 (1970).
13. Dunlap, P. V. & Greenberg, E. P. in *Microbial Cell-Cell Interactions* (ed. Dworkin, M.) 219–253 (Am. Soc. Microbiol., Washington DC, 1991).
14. Eberhard, A. *et al. Biochemistry* **20,** 2444–2449 (1981).
15. Gambello, M. J. & Iglewski, B. H. *J. Bact.* **173,** 3000–3009 (1991).
16. Watson, B., Currier, T. C., Gordon, M. P., Chilton, M.-D. & Nester, E. W. *J. Bact.* **123,** 255–264 (1975).
17. Holsters, M. *et al. Molec. gen. Genet.* **163,** 181–187 (1978).
18. Cangelosi, G. A., Best, E. A., Martinetti, G. & Nester, E. W. *Meth. Enzymol.* **204,** 384–397 (Academic, San Diego, 1991).
19. Miller, J. H. *Experiments in Molecular Genetics* 352–359 (Cold Spring Harbor Laboratory Press, New York, 1972).
20. Chilton, M.-D. *et al. Proc. natn. Acad. Sci. U.S.A.* **71,** 3672–3676 (1974).
21. Engebrecht, J., Nealson, K. & Silverman, M. *Cell* **32,** 773–781 (1983).
22. Stachel, S. E., An, G., Flores, C. & Nester, E. W. *EMBO J.* **4,** 891–898 (1985).
23. Tate, R. C. *et al. Bio/Technology* **1,** 269–275 (1983).
24. Pearson, W. R. & Lipman, D. J. *Proc. natn. Acad. Sci. U.S.A.* **85,** 2444–2448 (1988).

ACKNOWLEDGEMENTS. We thank A. Salyers for her review of the manuscript. Portions of this work were funded by a grant from the United States Department of Agriculture to S.K.F.

GOING AFTER THE WHOLE GENOME FROM TWO DIFFERENT ANGLES

Brad Goodner[1] and Derek Wood[2]

[1]**Department of Biology, Hiram College, Hiram, Ohio, United States**
E-mail: Goodnerbw@hiram.edu

[2]**Department of Microbiology, University of Washington**
Seattle, Washington, United States
E-mail: derekw@u.washington.edu

Commentary on the papers:

Wood, D. W., Setubal, J. C., Kaul, R., Monks, D. E., Kitajima, J. P., Okura, V. K., Zhou, Y., Chen, L., Wood, G. E., Almeida, N. F., Jr., Woo, L., Chen, Y., Paulsen, I. T., Eisen, J. A., Karp, P. D., Bovee, D., Sr., Chapman, P., Clendenning, J., Deatherage, G., Gillet, W., Grant, C., Kutyavin, T., Levy, R., Li, M.-J., McClelland, E., Palmieri, A., Raymond, C., Rouse, G., Saenphimmachak, C., Wu, Z., Romero, P., Gordon, D., Zhang, S., Yoo, H., Tao, Y., Biddle, P., Jung, M., Krespan, W., Perry, M., Gordon-Kamm, B., Liao, L., Kim, S., Hendrick, C., Zhao, Z.-Y., Dolan, M., Chumley, F., Tingey, S. V., Tomb, J.-F., Gordon, M. P., Olson, M. V., and Nester, E. W. 2001. The genome of the natural genetic engineer *Agrobacterium tumefaciens* C58. Science 294:2317–2323.

Goodner, B., Hinkle, G., Gattung, S., Miller, N., Blanchard, M., Qurollo, B., Goldman, B. S., Cao, Y., Askenazi, M., Halling, C., Mullin, L., Houmiel, K., Gordon, J., Vaudin, M., Iartchouk, O., Epp, A., Liu, F., Wollam, C., Allinger, M., Doughty, D., Scott, C., Lappas, C., Markelz, B., Flanagan, C., Crowell, C., Gurson, J., Lomo, C., Sear, C., Strub, G., Cielo, C., and Slater, S. 2001. Genome sequence of the plant pathogen and biotechnology agent *Agrobacterium tumefaciens* C58. Science 294:2323–2328.

The sequence and subsequent analyses of the first *Agrobacterium* genome represents a major milestone in the history of plant transformation and the study of crown gall. The papers that follow are the results of two independent analyses of the genome sequence of *A. tumefaciens* C58. As interesting as the conclusions and insights gained by these analyses are, the story below relates how such similar projects unfolded led by two independent groups.

Serendipity is an important aspect of scientific discovery and it has certainly played a role in my (BG) career. For example, as an undergraduate at Texas A&M University in 1982, I heard about crown gall, *A. tumefaciens,* and the "transforming principle" in an introductory plant pathology course. I remember talking with Jim Wild, my advisor and research mentor, about this fascinating pathogen and this unknown means of genetic transformation. "They know the bacterium transfers some of its own DNA into the plant cell," Jim told me and fished out the critical 1977 paper by Chilton and coworkers for me to read (see earlier reprint in this anthology). I was hooked. I went on to work with Stan Gelvin at Purdue University for my doctorate where I mainly focused on getting novel genes into plant cells, using *A. tumefaciens* as a delivery device, in order to influence the plant's physiology. That was a heady time in crown gall research—major discoveries on *vir* gene identification and regulation, T-DNA transfer, and generation of transgenic plants seemed to occur monthly in several labs around the world. After Purdue, I moved into plant developmental biology during my postdoc at the University of North Carolina and initial years as a faculty member at the predominantly undergraduate institution, the University of Richmond.

Serendipity struck again in early 1996 as I was teaching a special topics course on plant-microbe interactions. While reviewing the crown gall literature, I ran across a paper by Allardet-Servent et al. (1993) that showed strong physical evidence for two different mega-size DNA molecules in *A. tumefaciens* strain C58, one a circle and the other a linear; also that each contained rRNA genes. It immediately struck me as odd because the genetic maps available at that time supported a single circular chromosome genome (See Goodner et al. 1999 for references). The paradox between the genetic and physical data struck two related chords in me. One, it reminded me of how little we knew about *A. tumefaciens* outside of its unique DNA transfer ability. Two, here was a great project much more practical for undergraduate students than my current research. I proposed to the students in the course that we figure out how to generate a combined genetic and physical map of the *A. tumefaciens* strain C58 genome and by the end of the course I was convinced this project was worth pursuing. Over the next two years, this project became the dominant focus of my lab and we generated the combined genetic and physical maps we were after, using a combination of transposon mutagenesis, biochemical complementation, rare-cutting restriction enzymes, and pulsed-field gel electrophoresis. The resulting paper, with eight undergraduates as coauthors, proved conclusively that the two mega-sized DNAs are truly chromosomes (Goodner et al. 1999). Moreover, our results showed that the two chromosomes are asymmetric in their distribution of essential genes, suggesting that the two molecules did not result from a simple breakage of a larger ancestral chromosome. It was this very hypothesis that led us into a genome sequencing project.

We originally set out to sequence just a portion of the linear chromosome, but decided to construct both molecule-specific and whole genome sequencing libraries in the summer of 1999. With a small seed grant from the University of Richmond, we sequenced a hundred or so clones to test the quality of our libraries and presented our results at the 20th Crown Gall Conference in Houston in November of that year. In a final stroke of luck, it was at that meeting that I first met Steve Slater from Cereon Genomics, a subsidiary of Monsanto Company. Steve proposed that we join efforts to sequence the entire genome of *A. tumefaciens* strain C58. Cereon/Monsanto would provide the vast majority of sequencing and my group would help with the assembly and annotation phases of the project. Over the next two years, with funding by Monsanto and through a NIH AREA grant from the National Institute of General Medical Sciences, we had the fortune to work with a great group of corporate scientists to see the project through to publication. Our major role was to augment the computer-based assembly efforts going on at Cereon Genomics using two approaches. One, we made use of the same transposon-induced mutants from our earlier genetic/physical mapping. Total genomic DNA was isolated from many of these mutants and sent to Fidelity Systems in Rockville, Maryland, where scientists had perfected sequencing reactions directly from genomic DNA templates. The 100–200 bp of good sequence we could read out from one end of the transposon was usually enough to clearly identify the site of transposon insertion in one of the contiguous segments (contigs) of the initial genome assembly and allowed us to place that contig on our combined genetic/physical maps. In some cases, more than one transposon insertion identified the same contig allowing for both genome map position and orientation to be determined. Two, we used BLAST analysis of the contig ends to search for putative gaps that fell within a gene or between genes in an operon. Once putative adjoining contigs were identified by either method, we designed and used appropriate PCR primers to clone the gap region for subsequent sequencing. Our other role came in the annotation phase of the project where we focused on the annotation of the two plasmids, pTiC58 and pAtC58, and analyzed the distribution of essential genes between the two chromosomes.

As you will see in the reprinted paper, our genome sequence analysis reflects both the basic and applied interests of our academic/corporate collaboration. For me, the genome sequence highlights four areas for future experimentation: [1] the evolutionary origin and current maintenance of a complex genome, [2] the evolutionary relationships within the genus *Agrobacterium* and its relationships with its closest sister genera, [3] the full set of interactions, pathogenic and nonpathogenic, that occur between the bacterium and plants, and [4] the interactions between the bacterium and other non-plant organisms.

My (DW) story began in 1998 when Eugene Nester and Brad Goodner first discussed the possibility of sequencing the genome of *Agrobacterium.* Following a number of inquiries by Gene Nester into potential public funding sources, it appeared that this project might never escape the idea phase. While all agencies agreed that the project was worthwhile, none had the funding to support the project at that time. Unknown to either group, this dilemma ultimately led to the formation of two independent projects. Fortunately, neither researcher lost sight of the importance of sequencing the *Agrobacterium* genome. On the University of Washington (UW) side, Gene Nester continued to pursue potential private funding sources and contacted researchers at both Monsanto and DuPont in an effort to generate scientific and financial interest in this project. After more than a year, Forrest Chumley at DuPont offered to provide 6x coverage of the genome with the understanding that the UW would pursue federal funding for the completion of the genome sequence. It was at the beginning of this stage that I became deeply involved in the genome project.

Following a number of years of working on the molecular genetics of biological control with Sandy Pierson at the University of Arizona, I relocated to the UW to pursue my postdoctoral training. As my formal training was in plant pathology and microbiology it was only natural that I sought out Gene and began to work in his group in 1997. My goals at the time were to develop skills in plant transformation that could be applied to my broader interest of plant-microbe interactions. As my work proceeded, however, Gene and I would often discuss his interests in sequencing the *Agrobacterium* genome. These discussions led to my initial interest in genomics and to my eventual role as project leader. This opportunity was immensely rewarding as it both focused my interest in genomics and taught me new skills relating to the management of large collaborative projects in the sciences.

One of the advantages of leading a large multinational collaboration involving both academic and industry groups is that one has the opportunity to meet so many outstanding scientists. While I cannot mention everyone involved in the project, I would like to point out a few key contributors whose stories relate to how the project eventually came together.

Shortly after DuPont began sequencing, our group began to identify key gene products that might relate to virulence or transformation. Two persons played a key role in this phase. The first was Lishan Chen whose extraordinary grasp of the literature made him an outstanding gene 'miner'. Early on, he identified a new type IV secretion system which we subsequently found controlled the conjugal transfer of pAtC58 (Chen et al. 2002). The second person was Shiping Zhang from DuPont. Shiping organized the setup and operation of the initial *Agrobacterium* genome database at DuPont and was a skilled bioinformaticist. He was a tireless worker who would respond to our requests for bioinformatics support or alterations to the database with uncanny speed and intuition. In the course of this project I was fortunate to work with many such people who are both creative scientists and display a phenomenal dedication to their work.

While the initial shotgun-sequencing phase was proceeding at DuPont, we contacted Maynard Olsen at the UW Genome Center (UWGC) to see if he would be willing to finish the genome sequence. Previously, Maynard and his group had sequenced the *Pseudomonas aeruginosa* genome (Stover et al. 2000) in addition to their considerable involvement in the human genome project. Maynard was quite enthusiastic about this project and signed on as the Principal Investigator for the finishing phase that was eventually funded by the National Science Foundation. Once funding was acquired, and with the sequence data acquired from DuPont, the team at UWGC finished the genome in the record time of just over four months.

The final major player in the project became involved via a string of lucky coincidences. As the project proceeded, I realized that we would need a dedicated annotation database and full-time bioinformatics support as the DuPont phase of the project came to a close. At the time, the UWGC was not positioned to provide these services as their focus was on the acquisition and finishing of sequence data. Fortuitously, a few weeks prior to this point, I had attended a seminar by Joao Setubal in which he described his role as the bioinformatics director for the Brazilian project that sequenced the genome of the first plant pathogen, *Xylella fastidiosa* (Simpson et al. 2000). Joao was at the UW on sabbatical leave working with Phil Green, the developer of the phred/phrap sequence analysis packages. When I approached Joao with our dilemma he graciously offered the assistance of himself and his team back in Brazil. Joao and his team leapt to the task of assembling an annotation database for C58, responded with remarkable efficiency to all requests for genome analyses, and provided key computational insights that led to most of our major conclusions. Like Shiping, Joao and his team composed of Joao Kitajima, Vagner Okura and Nalvo Alameida were phenomenally responsive and intuitive bioinformaticists who made the success of the annotation and analysis phases of this project possible.

The final piece in this story of course is how we came to realize that Brad Goodner was working with Cereon on an independent analysis of the genome. We first learned of this duplicated effort when Brad and his collaborators presented their early results at a Crown Gall meeting in 2000. Our subsequent presentation at the International Society of Molecular Plant-Microbe Interactions made both groups aware of the competing projects. While one would expect that the outcome of such a competition would result in publication by the group that finished first, we instead got together and published the papers back to back in the same issue of *Science.* This was due in large part to the efforts of Steve Slater at Monsanto who contacted our group and proposed this idea in mid 2001. Again, luck played a role here, since, as you will see in the subsequent papers, we have learned much from the independent efforts of both groups. Of note is the divergence of the two C58 strains that were sequenced. The original C58 strain was isolated by Richard Dickey from a cherry gall in New York and acquired by Gene Nester in the early 1970s. This strain and the one sequenced at Cereon were separated at this early stage and passed among several labs. Our joint analyses indicate that the two strains show substantial agreement as expected but have acquired small deletions and insertions as well as a few point mutations.

In the end, the efforts of both groups came together and provided us with a look inside the genome of this key biotechnological agent and plant pathogen. The work presented here reflects the efforts of many talented people who contributed their expertise and time not only to tease out the biological insights presented, but also to make this resource available to the larger scientific community. We hope you will see the extent of these efforts reflected in the following papers.

Addendum: For those interested in further information on these two projects, the authors refer you to an interesting sociological examination of the impact of public-private scientific collaborations that was done using this project as the focus (Harvey and McMeekin 2003).

References

Allerdet-Servent, A., Michaux-Charachon, S., Jumas-Bilak, E., Karayan, L., and Ramuz, M. 1993. Presence of one linear and one circular chromosome in the *Agrobacterium tumefaciens* C58 genome. J. Bacteriol. 175:7869–7874.

Chen, L., Chen, Y., Wood, D. W., and Nester, E. W. 2002. A new type IV secretion system promotes conjugal transfer in *Agrobacterium tumefaciens.* J. Bacteriol. 184:4838–4845.

Goodner, B. W., Markelz, B. P., Flanagan, M. C., Crowell, C. B., Racette, J. L., Schilling, B. A., Halfon, L. M., Mellors, J. S., and Grabowski, G. 1999. Combined genetic and physical map of the complex genome of *Agrobacterium tumefaciens.* J. Bacteriol. 181:5160–5166.

Harvey, M., and McMeekin, A. 2003. Symbiosis or pathogenicity: Interactions between academia and industry in the *Agrobacterium tumefaciens* genome race. ESRC Centre for Research in Innovation and Competition. Discussion Paper 63. University of Manchester. Online publication.

Simpson, A. J., Reinach, F. C., Arruda, P., Abreu, F. A., Acencio, M., Alvarenga, R., Alves, L. M., Araya, J. E., Baia, G. S., Baptista, C. S., Barros, M. H., Bonaccorsi, E. D., Bordin, S., Bove, J. M., Briones, M. R., Bueno, M. R., Camargo, A. A., Camargo, L. E., Carraro, D. M., Carrer, H., Colauto, N. B., Colombo, C., Costa, F. F., Costa, M. C., Costa-Neto, C. M., Coutinho, L. L., Cristofani, M., Dias-Neto, E., Docena, C., El-Dorry, H., Facincani, A. P., Ferreira, A. J., Ferreira, V. C., Ferro, J. A., Fraga, J. S., Franca, S. C., Franco, M. C., Frohme, M., Furlan, L. R., Garnier, M., Goldman, G. H., Goldman, M. H., Gomes, S. L., Gruber, A., Ho, P. L., Hoheisel, J. D., Junqueira, M. L., Kemper, E. L., Kitajima, J. P., Krieger, J. E., Kuramae, E. E., Laigret, F., Lambais, M. R., Leite, L. C., Lemos, E. G., Lemos, M. V., Lopes, S. A., Lopes, C. R., Machado, J. A., Machado, M. A., Madeira, A. M., Madeira, H. M., Marino, C. L., Marques, M. V., Martins, E. A., Martins, E. M., Matsukuma, A. Y., Menck, C. F., Miracca, E. C., Miyaki, C. Y., Monteriro-Vitorello, C. B., Moon, D. H., Nagai, M. A., Nascimento, A. L., Netto, L. E., Nhani, L. E., Jr., Nobrega, F. G., Nunes, L. R., Oliveira, M. A., de Oliveira, M. C., de Oliveira, R. C., Palmieri, D. A., Paris, A., Peixoto, B. R., Pereira, G. A., Pereira, H. A., Jr., Pesquero, J. B., Quaggio, R. B., Roberto, P. G., Rodrigues, V., de M Rosa, A. J., de Rosa, V. E., Jr., de Sa, R. G., Santelli, R. V., Sawasaki, H. E., da Silva, A. C., da Silva, A. M., da Silva, F. R., da Silva, W. A., Jr., da Silveira, J. F., Silvestri, M. L., Siqueira, W. J., de Souza, A. A., de Souza, A. P., Terenzi, M. F., Truffi, D., Tsai, S. M., Tsuhako, M. H., Vallada, H., Van Sluys, M. A., Verjovski-Almeida, S., Vettore, A. L., Zago, M. A., Zatz, M., Meidanis, J., and Setubal, J. C. 2000. The genome sequence of the plant pathogen *Xylella fastidiosa.* Nature 406:151–157.

Stover, C. K., Pham, X. Q., Erwin, A. L., Mizoguchi, S. D., Warrener, P., Hickey, M. J., Brinkman, F. S., Hufnagle, W. O., Kowalik, D. J., Lagrou, M., Garber, R. L., Goltry, L., Tolentino, E., Westbrock-Wadman, S., Yuan, Y., Brody, L. L., Coulter, S. N., Folger, K. R., Kas, A., Larbig, K., Lim, R., Smith, K., Spencer, D., Wong, G. K., Wu, Z., and Paulsen, I. T. 2000. Complete genome sequence of *Pseudomonas aeruginosa* PA01, an opportunistic pathogen. Nature 406:959–964.

BRAD GOODNER: I grew up in Texas and earned a B.S. in biochemisty in 1983 at Texas A&M University. Jim Wild, my faculty advisor and research mentor, always prodded me to explore and test my limits. I moved on to Purdue University where I worked with Stan Gelvin and a great group of grad students and postdocs on a variety of projects involving *Agrobacterium.* My biology dissertation in 1991 focused on the use of antisense RNA to down-regulate the expression of the T-DNA oncogenes in plant cells. It was at Purdue that I realized how much I enjoyed teaching as well as doing research. Ralph Quatrano gave me a tremendous opportunity to pursue my research and teaching interests as a postdoc and visiting instructor at the University of North Carolina. From 1994 to 2001, I was an assistant professor of biology at University of Richmond, and since then I have been at Hiram College in northeastern Ohio where I am currently an associate professor of biology.

DEREK WOOD: After spending four years in the U.S. Army as an intelligence analyst, I returned to my studies at Washington State University in 1988 and began to work with Mike Kahn investigating the role of TCA cycle enzymes in the *Rhizobium*-legume interaction. My time with Mike gave me confidence in my approach to science and taught me how to critically plan and interpret experiments. Mike is an outstanding researcher and fantastic mentor who enjoys interacting with students and shows them the value of their contribution. This experience set the standards for my future work and I will always be grateful for the running start Mike gave me. After moving to the University of Arizona for graduate studies in 1991 I was fortunate to connect with Sandy Pierson and worked on a project that became the first to show that quorum sensing systems function in a natural setting. Sandy helped hone my critical analysis skills and allowed me access to all the various roles we need to fill as scientists, from student instruction to grant preparation. It was in this setting that I came to realize the integral role that undergraduate researchers can play. Following my graduate work I moved to the University of Washington in 1998 where I began to work with Eugene Nester. Gene gives all his postdocs free rein to pursue their research interests and I was no exception. As I became interested in genomics, Gene fostered this interest by placing me in a leadership role during the sequencing of the C58 genome. I am certainly grateful to each of these persons, both for the insights they have given me, and the continuing support they provide.

The Genome of the Natural Genetic Engineer *Agrobacterium tumefaciens* C58

Derek W. Wood,[1] Joao C. Setubal,[2,4] Rajinder Kaul,[5] Dave E. Monks,[1] Joao P. Kitajima,[2,3] Vagner K. Okura,[2] Yang Zhou,[5] Lishan Chen,[1]* Gwendolyn E. Wood,[1] Nalvo F. Almeida Jr.,[6] Lisa Woo,[1] Yuching Chen,[1]† Ian T. Paulsen,[7] Jonathan A. Eisen,[7] Peter D. Karp,[8] Donald Bovee Sr.,[5] Peter Chapman,[5] James Clendenning,[5] Glenda Deatherage,[5] Will Gillet,[5] Charles Grant,[5] Tatyana Kutyavin,[5] Ruth Levy,[5] Meng-Jin Li,[5] Erin McClelland,[5] Anthony Palmieri,[5] Christopher Raymond,[5] Gregory Rouse,[5] Channakhone Saenphimmachak,[5] Zaining Wu,[5] Pedro Romero,[8] David Gordon,[9] Shiping Zhang,[10] Heayun Yoo,[10] Yumin Tao,[11] Phyllis Biddle,[10] Mark Jung,[10] William Krespan,[10] Michael Perry,[10] Bill Gordon-Kamm,[11] Li Liao,[10] Sun Kim,[10] Carol Hendrick,[11] Zuo-Yu Zhao,[11] Maureen Dolan,[10] Forrest Chumley,[10]‡ Scott V. Tingey,[10] Jean-Francois Tomb,[10] Milton P. Gordon,[12] Maynard V. Olson,[5] Eugene W. Nester[1,13]§

The 5.67-megabase genome of the plant pathogen *Agrobacterium tumefaciens* C58 consists of a circular chromosome, a linear chromosome, and two plasmids. Extensive orthology and nucleotide colinearity between the genomes of *A. tumefaciens* and the plant symbiont *Sinorhizobium meliloti* suggest a recent evolutionary divergence. Their similarities include metabolic, transport, and regulatory systems that promote survival in the highly competitive rhizosphere; differences are apparent in their genome structure and virulence gene complement. Availability of the *A. tumefaciens* sequence will facilitate investigations into the molecular basis of pathogenesis and the evolutionary divergence of pathogenic and symbiotic lifestyles.

Agrobacterium tumefaciens is an α-proteobacterium of the family Rhizobiaceae and a member of the diverse *Agrobacterium* genus. A ubiquitous soil organism and etiological agent of the plant disease crown gall (*1*), *A. tumefaciens* infects more than 90 families of dicotyledonous plants, resulting in major agronomic losses (*2, 3*). The gall results from the transfer, integration, and expression of a discrete set of genes (T-DNA) located on the tumor-inducing (Ti) plasmid. Expression of these genes leads to biosynthesis of plant growth hormones as well as a bacterial nutrient source called opines (*4*). The processing and transfer of the T-DNA is mediated by the Ti plasmid virulence (*vir*) genes, and several virulence determinants initially characterized in *A. tumefaciens* have been found in plant symbionts and animal pathogens (*5–7*).

The genes within the T-DNA can be replaced by any DNA sequence, making *A. tumefaciens* an ideal vehicle for gene transfer and an essential tool for plant research and transgenic crop production. The research and commercial potential of *A. tumefaciens* has been broadened under laboratory conditions to include the transfer of T-DNA to recalcitrant plants, fungi (*8*), and human cells (*9*).

A. tumefaciens shares a similar habitat and close evolutionary relationship with the nitrogen-fixing symbionts of the Rhizobiaceae (*10*). Indeed, the introduction of a symbiotic plasmid from *Rhizobium phaseoli* into *A. tumefaciens* results in the weak but measurable formation of nitrogen-fixing root nodules (*11*), suggesting a shared genetic background. The recent publication of the genome sequences of two Rhizobiaceae, *Sinorhizobium meliloti* (*12*) and *Mesorhizobium loti* (*13*), allowed a genome-wide comparison with *A. tumefaciens*. We present the results of this comparison as well as a detailed analysis of the genome of *A. tumefaciens* strain C58 (*14, 15*).

General features of the genome. The 5.67-Mb genome of *A. tumefaciens* C58 (*16*) comprises four replicons (*17*): a circular chromosome, a linear chromosome, and the AtC58 and TiC58 plasmids (Table 1 and Fig. 1). The genome contains 5419 predicted protein-coding genes (*14*), of which we have assigned a putative function to 3475 (64.1%). The remaining 1944 genes (35.9%) include 1236 conserved hypothetical genes (22.8%) whose predicted products are similar to proteins of unknown function in other genomes, and 708 hypothetical genes (13.1%) with no significant matches in the sequence databases (Table 1). Our analysis assigns the *A. tumefaciens* genes to 501 paralogous families containing from 2 to 206 members (*14*). The two largest families are composed of genes belonging to the adenosine triphosphatase (ATPase) and membrane-spanning components of the ATP binding cassette (ABC) transport family.

The overall GC content of the *A. tumefaciens* genome is 58%. The TiC58 plasmid has two regions of distinctive GC content: the T-DNA (46%) and the *vir* region (54%) (Fig. 1). Low GC content was noted previously in the T-DNA of a related Ti plasmid (*18*). Reduced GC content (53%) is also seen within a 24-kb segment of pAtC58 (AT island, Fig. 1). This region includes 17 conserved hypothetical or hypothetical genes, an ATP-dependent DNA helicase, and an insertion sequence (IS) element. These genes are flanked by a phage integrase and a second IS element. The genes in these three regions have a distinct codon usage as compared to the rest of the genome, consistent with their recent evolutionary acquisition (*14*).

The genome contains 53 transfer RNAs (tRNAs) that represent all 20 amino acids (Table 1). These tRNAs are distributed unevenly between the circular and linear chromosomes. Transfer RNA species corresponding to the most frequently represented ala-

[1]Department of Microbiology, University of Washington, 1959 NE Pacific Street, Box 357242, Seattle, WA 98195, USA. [2]Bioinformatics Laboratory, Institute of Computing, [3]Center for Molecular Biology and Genetic Engineering (CBMEG), University of Campinas, CP 6176, Campinas SP 13083-970, Brazil. [4]Department of Genome Sciences, University of Washington, Box 357730, Seattle, WA 98195, USA. [5]Genome Center, University of Washington, Fluke Hall on Mason Road, Box 352145, Seattle, WA 98195, USA. [6]Department of Computing and Statistics, Federal University of Mato Grosso do Sul, CP 549, Campo Grande MS 79070-900, Brazil. [7]The Institute for Genomic Research, 9712 Medical Center Drive, Rockville, MD 20850, USA. [8]Bioinformatics Research Group, SRI International, 333 Ravenswood Avenue, Menlo Park, CA 94025, USA. [9]Howard Hughes Medical Institute, University of Washington, Box 357730, Seattle, WA 98195, USA. [10]E. I. du Pont de Nemours Company, 1 Innovation Way, Newark, DE 19714, USA. [11]Pioneer Hi-Bred International Inc., 7300 NW 62nd Avenue, Post Office Box 1004, Johnston, IA 50131, USA. [12]Department of Biochemistry, University of Washington, 1959 NE Pacific Street, Seattle, WA 98195, USA. [13]Department of Botany, University of Washington, 1959 NE Pacific Street, Box 355325, Seattle, WA 98195, USA.

*Present address: Department of Pathology, University of Washington, Box 357470, Seattle, WA 98195, USA.

†Present address: Gene Function & Target Validation, Celltech R&D Inc., Bothell, WA 98021, USA.

‡Present address: Department of Plant Pathology, Kansas State University, 113 Waters Hall, Manhattan, KS 66506, USA.

§To whom correspondence should be addressed. E-mail: gnester@u.washington.edu

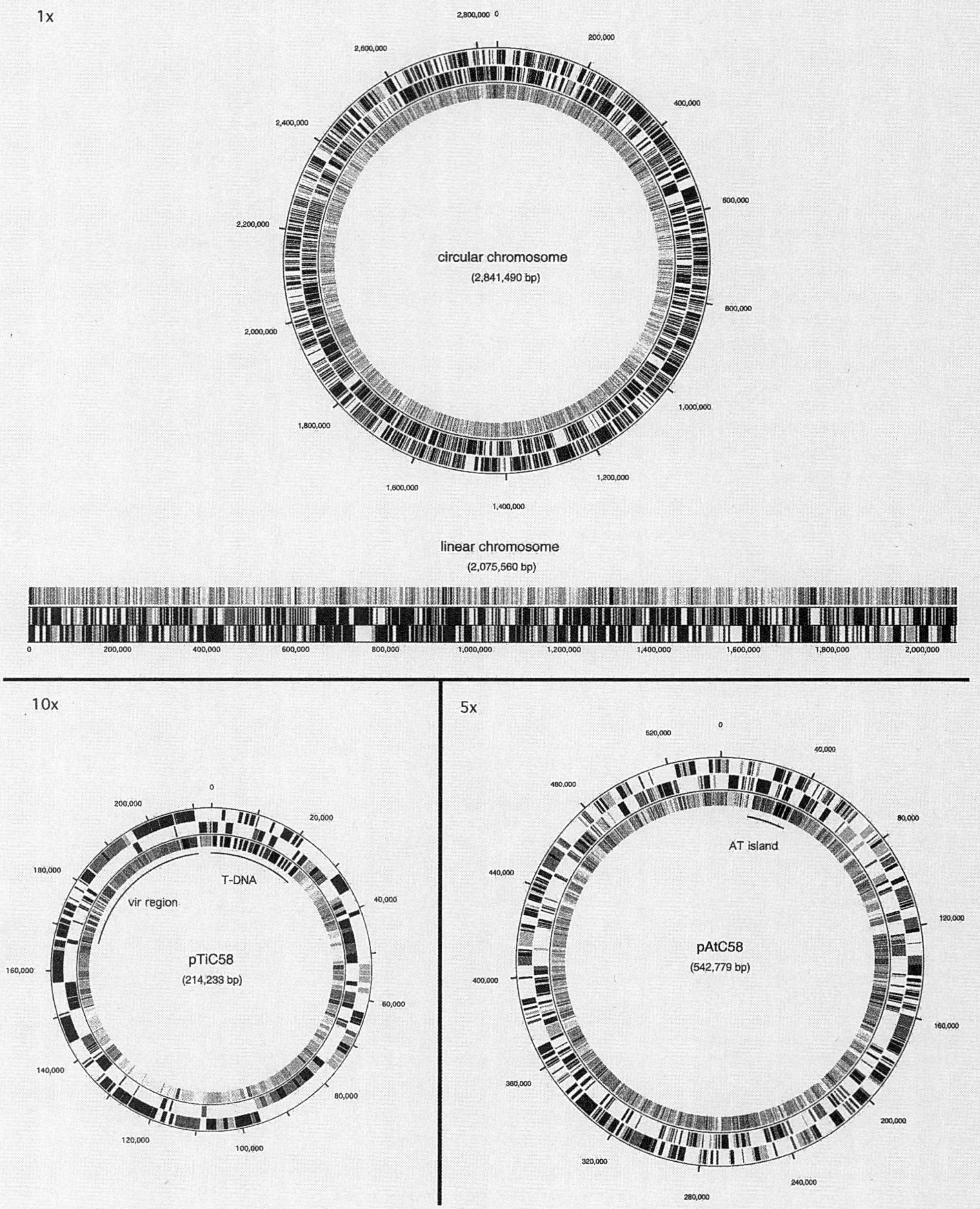
1x
2,800,000
0
200,000
400,000
600,000
800,000
1,000,000
1,200,000
1,400,000
1,600,000
1,800,000
2,000,000
2,200,000
2,400,000
2,600,000
circular chromosome
(2,841,490 bp)
linear chromosome
(2,075,560 bp)
0
200,000
400,000
600,000
800,000
1,000,000
1,200,000
1,400,000
1,600,000
1,800,000
2,000,000
10x
0
20,000
40,000
60,000
80,000
100,000
120,000
140,000
160,000
180,000
200,000
T-DNA
vir region
pTiC58
(214,233 bp)
5x
0
40,000
80,000
120,000
160,000
200,000
240,000
280,000
320,000
360,000
400,000
440,000
480,000
520,000
AT island
pAtC58
(542,779 bp)

nine, glutamine, and valine codons are found only on the linear replicon. The genome contains 25 predicted IS elements representing eight different families (*14*). The largest is the IS3 family comprising 10 IS elements. The IS elements are not equally distributed among the replicons but are located preferentially on the linear chromosome and pAtC58 (Table 1). The adjacent *virH1* and *virH2* genes of the Ti plasmid, encoding p450 mono-oxygenases (*19*), are flanked by IS elements, which suggests that they arrived in *A. tumefaciens* as part of a compound transposon. Twelve genes of probable phage origin were identified, most of which are on the circular chromosome (Table 1). Many of these genes cluster in two discrete regions and thus may represent prophage remnants. None of these clustered phage-related genes are shared with *S. meliloti*, which implies that they were lost from *S. meliloti* or entered the *A. tumefaciens* genome after these organisms evolutionarily diverged.

Phylogeny and whole-genome comparison. A comparison with all sequenced organisms reveals that the *A. tumefaciens* proteome is most similar to that of two rhizobial species, *S. meliloti* and *M. loti* (*14*). This result was obtained by cataloging top BLAST hits of predicted *A. tumefaciens* proteins and by classifying predicted proteins into clusters of orthologous groups (Fig. 2A) (*20*). Of the two rhizobial species, the *A. tumefaciens* proteome is most similar to that of *S. meliloti*. Phylogenetic analyses of broadly conserved proteins indicate that this similarity results from *A. tumefaciens* and *S. meliloti* sharing a recent common ancestor, and not from gene loss or branch rate variation (Fig. 2, B and C).

Sinorhizobium meliloti has a circular chromosome (3.65 Mb) and two plasmids (1.68 Mb and 1.35 Mb), with a total genome size 1.1 Mb larger than that of *A. tumefaciens* (*12*). The circular chromosomes of these organisms show extensive nucleotide colinearity and gene order conservation (Fig. 3) (*14*). Previously, such extensive colinearity has only been seen between members of the same genus. Chromosome-wide conservation of gene order is less pronounced between *S. meliloti* and *M. loti* (*14*). The comparison of the circular chromosomes of *A. tumefaciens* and *S. meliloti* also reveals major rearrangements near the putative replication origin and termini (Fig. 3, regions A and B). Similarly located rearrangements are commonly seen between closely related bacteria (*21*).

A comparison of the other replicons of *A. tumefaciens* with all replicons of *S. meliloti* reveals a mosaic pattern of ortholog distribution (Table 2 and Fig. 1). These orthologs are distributed across the *A. tumefaciens* elements as individual genes and small regions of gene order conservation. Two regions of the linear replicon exhibit extensive conservation of gene order with a segment of the *S. meliloti* chromosome (Fig. 3, region C). The first comprises 46 genes (44 kb) and the second contains 65 genes (89 kb). These regions are partially conserved in the *M. loti* chromosome. The large number of orthologs and the lack of extensive gene order conservation suggest that the smaller *A. tumefaciens* replicons underwent substantial rearrangement since the organisms diverged. This finding is consistent with differential evolutionary pressures acting on these elements. The nonorthologous genes, many of which are seen on the Ti plasmid, reflect lineage-specific gene loss or acquisition from other species. Taken together, these data support the recent evolutionary divergence of *A. tumefaciens* and *S. meliloti*.

Genus-specific genes. Comparison of the genomes of *A. tumefaciens*, *S. meliloti*, and *M. loti* identified genes in each organism that likely contribute to genus-specific biology (*14*). Of the 5419 predicted *A. tumefaciens* proteins, 853 (16%) are not found in these other organisms. Of these, 97 have an assigned function, whereas 756 are hypothetical or conserved hypothetical. The predicted products of these genes are diverse and include proteins involved in cellulose production, plasmid maintenance, cell growth, transcriptional regulation, and cell wall synthesis. Several additional proteins are predicted to catabolize plant cell wall materials, sugars, and exudates. These include polygalacturonases, a glycosidase, an endoglucanase, a myo-inositol catabolism protein, and a cell wall lysis–associated protein. Additional genes, predictably found on the Ti plasmid, include those encoding virulence, T-DNA, and conjugal transfer–associated proteins. With 756 open reading frames (ORFs) yet to characterize, much remains to be elucidated regarding the genetic distinction between *A. tumefaciens* and its Rhizobiaceae relatives.

Linear chromosome. Linear replicons, the predominant genetic element in eukaryotes, have been identified in only a few prokaryotes. These include members of the genera *Borrelia* and *Streptomyces* (*22, 23*). Although sequence analysis did not reveal distinct features associated with terminal secondary structures, Goodner *et al.* found that the termini of this replicon are covalently linked (*15*). This covalent linkage did not prevent nearly complete sequencing of the replicon termini as confirmed by Southern analysis (*14*). Proteins associated with the maintenance of linear ends in other systems, such as telomerases or the *Streptomyces tpg* proteins (*24*), are absent in *A. tumefaciens*. One notable feature of the replicon termini is the presence of IS elements near each end. The evolutionary origin of this replicon awaits investigation, as does the mechanism that *A. tumefaciens* uses to maintain it in a linear form.

There are 1882 protein-coding genes on the linear replicon, including those encoding ribosomal and DNA replication proteins, as

Fig. 1. (facing page) Schematic representation of the *A. tumefaciens* genome. Chromosomes are drawn to scale with plasmids represented at 5× or 10× magnification, as indicated. The outer two bands indicate opposing transcriptional orientations of predicted genes. Colors indicate orthology to proteins in the *S. meliloti* replicons: blue, chromosome; green, pSymA; gold, pSymB; red, nonorthologous. The inner circle depicts GC content for each coding region, with lower GC content indicated by darker shading. The *vir* and T-DNA regions of pTiC58 and the AT island of pAtC58 are indicated. Orthologs were identified by comparison of predicted proteins for each *A. tumefaciens* replicon with the genome of *S. meliloti*. Two proteins were considered orthologous if their BLASTP alignment covered at least 60% of each protein at an expect value of less than or equal to 10^{-5}. Proteins that did not match these criteria were considered nonorthologous (*14*).

Table 1. General features of the *A. tumefaciens* C58 genome.

Feature	Circular	Linear	pAtC58	pTiC58	Total
Size (bp)	2,841,490	2,075,560	542,779	214,233	5,674,062
G+C content (%)	59.4	59.3	57.3	56.7	58.13
Protein-coding genes					
Assigned function	1715	1286	333	141	3475 (64.1%)
Conserved hypothetical	710	353	128	45	1236 (22.8%)
Hypothetical	364	243	89	12	708 (13.1%)
Total	2789	1882	550	198	5419
Average ORF size (bp)	892	988	843	925	922
Coding (%)	87.9	89.9	85.4	85.5	88.3
Regulators (%)	7.7	10.4	11.8	5.1	9.0
ABC transport	47	80	20	6	153
RNA					
rRNA	2	2	0	0	4
tRNA	40	13	0	0	53
tmRNA	1	0	0	0	1
IS elements	2	10	10	2	24
Phage-related	10	1	1	0	12

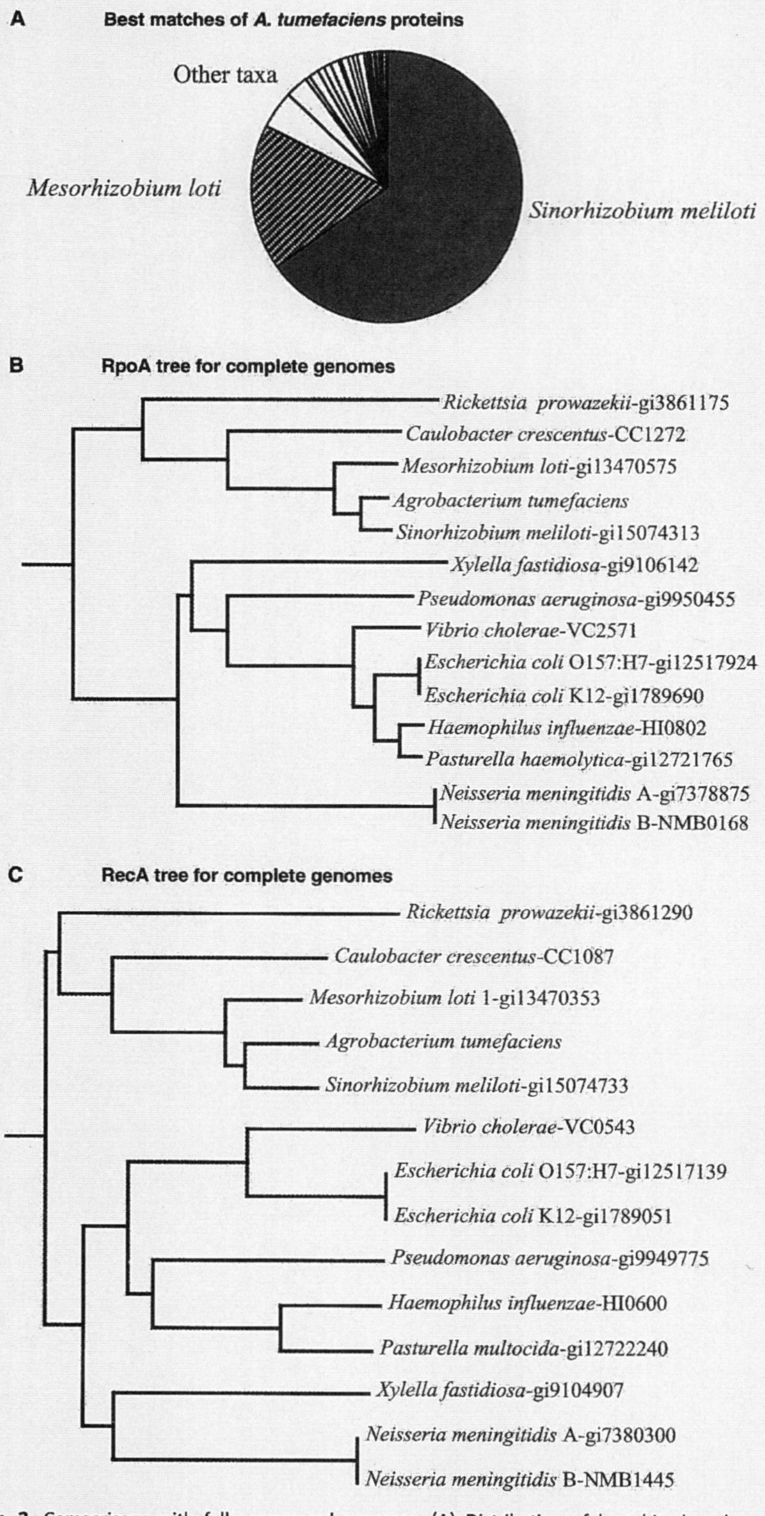

Fig. 2. Comparisons with fully sequenced genomes. (**A**) Distribution of best hits based on a comparison of predicted proteins of *A. tumefaciens* with proteins from all published genomes. (**B** and **C**) Phylogenetic trees generated using two broadly conserved proteins. The trees were generated using PAUP distance methods and a distance calculation based on PAM matrices (*14*).

well as 21 complete metabolic pathways. The presence of these genes confirms the chromosomal identity of this replicon. Additional features, however, resemble those traditionally associated with plasmids. For example, genes whose products are similar to the conjugative proteins TraA, MobC, and TraG are present, although an *oriT* is not apparent. Further, an intact and highly conserved *repABC* operon, the definitive element of the RepABC-type replicator family of circular plasmids, is located near the center of the linear chromosome. The presence of this operon, coupled with a colocalized GC-skew inversion, indicates a bidirectional plasmid-like mode of replication. If experimentally verified, this replication mechanism would prove unique among known linear replicons.

Plasmid replication and transfer. Replication of both pTiC58 and pAtC58 is mediated by RepABC-type systems commonly found in plasmids of the Rhizobiaceae. It is likely that the origin of replication for these plasmids is adjacent to the *repC* gene (*25*). In contrast to the pSymB plasmid of *S. meliloti* (*12*), both *A. tumefaciens* plasmids contain all necessary machinery for conjugation and do not contain essential genes. A new conjugal transfer system belonging to the Type IV secretion family (AvhB) (*26*) was identified on pAtC58. In contrast to the tight control of Ti plasmid conjugal transfer mediated by specific opines that activate quorum sensing (*27*), the conjugal transfer of pAtC58 appears to be constitutive.

Transport. Transporters constitute 15% of the *A. tumefaciens* genome, 87% of which are found on the chromosomes (*14*). These systems are predicted to confer broad capabilities for the transport of common nutrients found in the rhizosphere, including sugars, amino acids, and peptides. In addition, there are 11 LysE/RhtB amino acid efflux proteins, almost double the number seen in any bacterium outside of the Rhizobiaceae (*12, 28*). These transporters may function in the export of homoserine lactones or other signal molecules. There are also a large number of high-affinity tripartite ATP-independent periplasmic (TRAP) dicarboxylate transporters (*29*). Our analyses indicate that *A. tumefaciens* and the other sequenced members of the Rhizobiaceae have similar transport capabilities.

Like both *S. meliloti* and *M. loti*, *A. tumefaciens* has an abundance of ABC transporters, constituting 60% of its total transporter complement. There are 153 complete systems plus additional "orphan" subunits. The number of ABC transporters found in these organisms is greater than that found in any sequenced eukaryote and more than double the number found in any sequenced bacterium (*28, 30*). Predicted substrates of these ABC transport systems include sugars (53 systems), amino acids (29 systems), and peptides (25 systems). Other organisms with large ABC transporter complements include photosynthetic bacteria such as *Synechocystis* PCC6803 and organisms that lack a tricarboxylic acid (TCA) cycle and an electron transfer chain, like the mycoplasmas and *Thermotoga maritima* (*28*). The generation of large ATP pools in these organisms, via photosynthesis or F-type ATPases, may explain their preference for ATP-driven transport. In contrast, the preference for ABC transporters in *A. tumefaciens* may reflect a need for high-affinity uptake systems for the acquisition of nutrients in the highly competitive soil and rhizosphere environments.

Regulation. Bacteria that inhabit diverse environments tend to have large complements of regulatory genes (*31*). Consistent with this, regulatory genes constitute a substantial proportion (9%) of the *A. tumefaciens* genome (Table 1) (*14*). This regulatory capacity likely facilitates survival of *A. tumefaciens* within the dynamic soil and rhizosphere environments. The genome encodes 11 extracellular sigma factors, proteins implicated in stress responses in other organisms (*32*). In addition, although several LuxR family motifs are evident, only one previously identified acyl-homoserine lactone synthase (*traI*) known to be involved in quorum sensing was detected (*4*). Several proteins are similar to eukaryotic signal transduction proteins rarely found in bacteria, including four regucalcin-like calcium-binding regulators and a serine-threonine kinase. As is true of other α-proteobacteria, no *rpoS* gene was identified. However, *A. tumefaciens* does have a homolog of the HF-1 protein known to regulate stationary phase and oxidative stress responses in *Escherichia coli* and *Brucella abortus* (*33*).

Our analysis identified numerous nucleotide cyclases in the plant symbionts *S. meliloti* and *M. loti* (25 and 12, respectively) and in the evolutionarily distinct human pathogen *Mycobacterium tuberculosis* (12). These cyclases are rarely found in other bacterial genomes. The nucleotide cyclases in *S. meliloti* have been noted previously and were postulated to function in signal transduction (*12*). Contrary to our expectation, there are only three nucleotide cyclases in *A. tumefaciens*. It is unclear why the nitrogen-fixing plant symbionts share similarly large numbers of nucleotide cyclases with a human pathogen, whereas few such genes are found in the evolutionarily related *A. tumefaciens*.

Attachment, cell surface, and secretion. The initial interaction of *Agrobacterium* with its plant hosts is mediated by several attachment-related genes (*34*). These include the *chvA*, *chvB*, *exoC*, and cellulose synthesis genes as well as the pAtC58-localized *att* region. Several additional genes encode proteins similar to adhesins in mammalian pathogens, including BfrA of *E. coli* and PsaA of *Streptococcus pneumoniae*.

Pili are extracellular appendages often required for bacterial association with their hosts. Although only the pilus encoded by the *virB* operon has been experimentally confirmed (*35*), the *trb* operon required for Ti plasmid conjugation likely produces a pilus (*36*). The *avhB* and *ctp* clusters, identified by our analyses, may also produce pili. Additional surface components include exopolysaccharides, lipopolysaccharides, and capsular polysaccharides, whose biosynthetic genes are primarily located on the linear chromosome. Such surface polysaccharides are commonly involved in invasion, growth, and survival of plant-associated bacteria.

Five protein secretion systems are found among Gram-negative bacteria (*37*), at least three of which are represented in *A. tumefaciens*. These include four potential type I secretion systems. Although components of the main terminal branch of type II secretion appear to be absent, the Sec system for protein secretion across the inner membrane is intact. There is, however, a type IV pilus biogenesis system with components similar to those of type II secretion systems. Similar to *S. meliloti* (*12*), no type III secretion system was identified. *Agrobacterium tumefaciens* encodes three type IV secretion systems: VirB, Trb (*27*), and AvhB. The genome also contains the twin arginine targeting system, Tat/Mtt (*38*).

Virulence. To date, most virulence determinants of *A. tumefaciens* have been found on the Ti plasmid. Other than the *virB* operon, these genes are not found in *S. meliloti*. The TiC58 plasmid contains a single T-DNA region, in contrast to the two found in a number of other strains (*18*), and the 25–base pair (bp) border regions that delineate the T-DNA are not present elsewhere in the genome.

The availability of the genome sequence has enabled the identification of genes whose products are similar to plant pathogen virulence proteins required for host cell wall degradation. These include pectinase (*kdgF*), ligninase (*ligE*), and xylanase as well as regulators of pectinase and cellulase production (*pecS/M*); *A. tumefaciens* may use such enzymes to breach the cell wall of its host before T-DNA transfer.

In addition, we have identified numerous orthologs of animal virulence genes. Examples include those involved in host survival, such as the *bacA* locus of *Brucella* (*39*) and two members of the widely conserved HtrA family of serine proteases implicated in response to oxidative stress in *Salmonella* and *Yersinia* (*40*). Interestingly, a *bacA* homolog is involved in *S. meliloti* symbiosis (*41*). Invasion-related homologs include the *ialA* and *ialB* genes of *Bartonella henselae* (*42*) as well as five hemolysin-like proteins with associated type I secretion systems. The highly conserved *mviN* gene, implicated in *Salmonella* virulence (*43*), is also present.

Metabolism. *Agrobacterium tumefaciens* grows on minimal medium and therefore possesses all pathways required for pro-

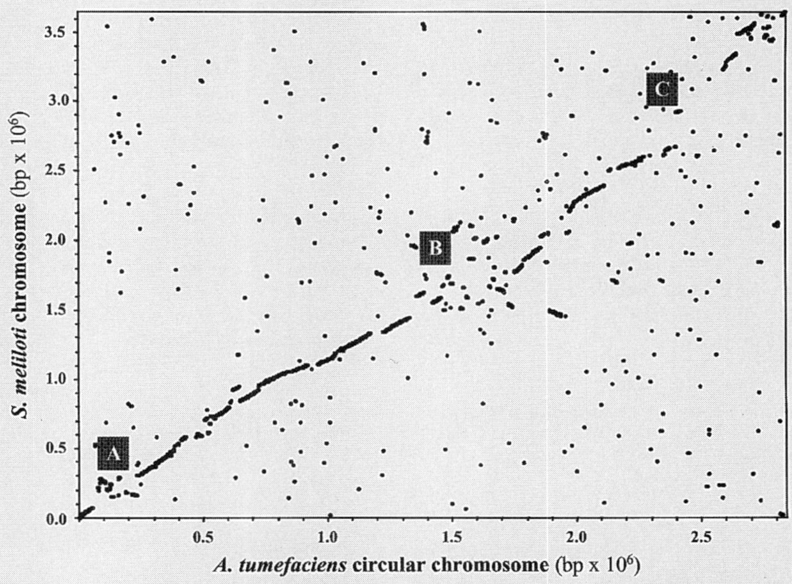

Fig. 3. Alignment of the proteomes of the *S. meliloti* chromosome and the *A. tumefaciens* circular chromosome. Each point in the figure is a bidirectional best hit. These hits were obtained by pairwise BLASTP searches of predicted *A. tumefaciens* proteins against those of *S. meliloti* with a maximum expect value of 10^{-4} (*14*). Putative origins (region A) and termini (region B) of replication are indicated, as well as a sizable region lacking colinearity (region C).

totrophic growth, an observation confirmed by our computational pathway analysis (*14*). These metabolic pathways are dispersed among the four replicons. Unlike their organization in *E. coli*, most genes of these pathways are not tightly clustered, which suggests that they are not present in operons. We identified pathways for the synthesis of all 20 amino acids as well as numerous enzyme cofactors. At least one nonribosomal protein synthesis system for the production of polyketides was identified. Encoded energy metabolism pathways include glycolysis, TCA cycle, and Entner-Doudoroff. *Agrobacterium tumefaciens* can catabolize 17 amino acids, including *S*-adenosylhomocysteine and 4-hydroxyproline. Pathways for use or degradation of plant metabolites typically found in the rhizosphere were also detected. These include sugars such as glucose, fructose, sucrose, ribose, xylose, xylulose, and lactose as well as compounds such as myo-inositol, hydantoin, urea, and glycerol. The capacity to metabolize glucuronate, galactonate, galactarate, gluconate, ribitol, glycogen, quinate, L-idonate, creatinine, stachydrine, ribosylnicotinamide, and 4-hydroxymandelate was also detected. Chemotaxis systems responding to many of these compounds are present in *A. tumefaciens* (*44*).

Agrobacterium tumefaciens encodes a variety of proteins that may protect against toxic compounds in the environment. Examples include four cytochrome p450s, two of which have been previously identified. One of these has been shown to modify ferrulic acid, an inducer of the *vir* genes (*45*). These highly oxidative enzymes may also detoxify or modify plant-derived compounds, including phytoalexins (*46*) and protocatechuate, and xenobiotics such as 1,2-dichloroethane, cyanate, 1,4-dichlorobenzene, and octane. In addition, antibiotic resistance genes targeted against tetracycline (*47*) and chloramphenicol are present.

Many components of nitrogen metabolism are conserved between *A. tumefaciens* and the nitrogen-fixing symbionts *S. meliloti* and *M. loti*. Examples include components of the nitrogen regulation (Ntr) system such as *ntrBC*, *ntrXY*, *ntrA*, *glnE*, *glnD*, *glnB*, and *glnK*. *Agrobacterium tumefaciens* harbors seven glutamine synthetase (GS) genes, which encode GS types I, II, and III. The presence of multiple GS genes may relate to the observation that *A. tumefaciens* requires high concentrations of glutamate for optimal growth. Other members of the Rhizobiaceae also contain multiple GS genes. In addition, *A. tumefaciens* has a gene predicted to encode the large hexameric adenosine monophosphate (AMP)–dependent glutamate dehydrogenase (*48*), but a gene encoding the AMP-independent glutamate dehydrogenase was not identified. In contrast, *S. meliloti* and *M. loti* contain both genes. The *A. tumefaciens* linear chromosome carries denitrification genes, including a periplasmic dissimilatory nitrate reductase (*nap*), nitrite reductase (*nir*), and nitric oxide reductase (*nor*), but lacks nitrous oxide reductase (*nos*). In contrast, all of these genes are present in *S. meliloti*. Nitrate transport genes are also located on the linear chromosome. Although *A. tumefaciens* is considered an aerobe, the existence of these genes implies that it could use nitrate as an electron acceptor under anaerobic conditions. As expected, *A. tumefaciens* lacks the subunits of nitrogenase and its cofactors. Most of the *nod* genes are also absent, except for three genes similar to those involved in nod factor production, *nodL*, *nodX*, and *nodN*.

Conclusions. The combination of a linear and a circular chromosome is found in only a few members of the genus *Agrobacterium* (*49*). This observation represents a key evolutionary distinction between *A. tumefaciens* and *S. meliloti*. On the basis of 16*S* ribosomal DNA phylogenetic analyses, it has been proposed that the genus *Agrobacterium* be reclassified into the genus *Rhizobium* (*10*). Combining what has been elucidated regarding genome structure with the complete genome sequence should allow a more accurate definition of the taxonomic position of *A. tumefaciens* in the Rhizobiaceae.

One striking finding from our analysis is the extensive similarity of the circular chromosomes of *A. tumefaciens* and the plant symbiont *S. meliloti*, which supports the view that these bacteria originated from a recent common ancestor. Galibert *et al.* speculate that the *S. meliloti* chromosome was present in a progenitor that later acquired pSymA and pSymB (*12*). The mosaic structure of the *A. tumefaciens* linear chromosome and plasmids, predominantly composed of orthologs found on each of the *S. meliloti* replicons, suggests that these organisms diverged after acquisition of the pSymA and pSymB ancestral molecules by this progenitor.

Recent models of bacterial evolution suggest that the differential acquisition and loss of genes in organisms that inhabit the same environment allows divergence into symbiotic and pathogenic lifestyles (*50*). The acquisition of such elements is apparent in both *A. tumefaciens* and *S. meliloti*. The *nod* genes of *S. meliloti* (*12, 51*), as well as the *vir* genes and T-DNA of *A. tumefaciens*, display GC content and codon usage distinct from the rest of the genome, which suggests recent evolutionary acquisition. In the case of the T-DNA, reduced GC content may facilitate expression in the plant host, where lower GC content is common. Moreover, none of the T-DNA and few of the *vir* genes of *A. tumefaciens* have orthologs in *S. meliloti*, and most *nod* genes are not found in *A. tumefaciens*. Differential selection and maintenance of such horizontally acquired genes likely led to the divergence into pathogenic and symbiotic states. Thus, these organisms provide a rich model system for further investigations into the evolutionary divergence of pathogens and symbionts.

As the central biological tool in the generation of transgenic plants for research and agriculture, *A. tumefaciens*, and the availability of its genome sequence, will continue to have an impact on plant biotechnology. Detailed studies, supplemented by this sequence, should lead to a directed refinement of plant transformation that increases both the host range and transformation efficiency of this versatile genetic tool. Genes likely to be targeted by such work include potential virulence factors that are shared between plant and animal pathogens. Examination of these genes in the genetically tractable *Agrobacterium* system may also serve to elucidate the molecular role they play in animal pathogens. It is our hope that this work will broaden the scientific foundation from which to address the worldwide debate over the production, use, and safety of genetically modified organisms.

Table 2. Number of orthologous genes of *A. tumefaciens* with respect to *S. meliloti*. The number of orthologous genes is shown in bold, with the percentage of each *A. tumefaciens* replicon they represent shown in square brackets. The remainder of the genes, which are not orthologs, are shown in the last row. Numbers of putative protein coding genes for each replicon are shown in parentheses.

		A. tumefaciens			
		Circular (2789)	Linear (1882)	pAtC58 (550)	pTiC58 (198)
S. meliloti	Chromosome (3341)	**1867** [67%]	**673** [36%]	**114** [21%]	**30** [15%]
	pSymA (1293)	**104** [4%]	**218** [12%]	**118** [21%]	**34** [17%]
	pSymB (1570)	**221** [8%]	**478** [25%]	**108** [20%]	**23** [12%]
	Nonorthologous	**597** [21%]	**513** [27%]	**210** [38%]	**111** [56%]

References and Notes

1. E. F. Smith, C. O. Townsend, *Science* **25**, 671 (1907).
2. M. DeCleene, J. DeLey, *Bot. Gaz.* **42**, 389 (1976).
3. L. Moore, in *Biology of the Rhizobiaceae*, K. L. Giles, A. G. Atherly, Eds. (Academic Press, New York, 1981), pp. 15–46.
4. J. Zupan, T. R. Muth, O. Draper, P. Zambryski, *Plant J.* **23**, 11 (2000).
5. P. J. Christie, *Mol. Microbiol.* **40**, 294 (2001).
6. N. Inon de Iannino, G. Briones, M. Tolmasky, R. A. Ugalde, *J. Bacteriol.* **180**, 4392 (1998).
7. A. Sola-Landa *et al.*, *Mol. Microbiol.* **29**, 125 (1998).
8. M. J. de Groot, P. Bundock, P. J. Hooykaas, A. G. Beijersbergen, *Nature Biotechnol.* **16**, 839 (1998).

9. T. Kunik *et al.*, *Proc. Natl. Acad. Sci. U.S.A.* **98**, 1871 (2001).
10. J. M. Young, L. D. Kuykendall, E. Martinez-Romero, A. Kerr, H. Sawada, *Int. J. Syst. Evol. Microbiol.* **51**, 89 (2001).
11. E. Martinez, R. Palacios, F. Sanchez, *J. Bacteriol.* **169**, 2828 (1987).
12. F. Galibert *et al.*, *Science* **293**, 668 (2001).
13. T. Kaneko *et al.*, *DNA Res.* **7**, 331 (2000).
14. The complete annotated sequence, detailed methods, and all supplementary data, which will be periodically updated, are available at our Web site (www.agrobacterium.org). Supplemental data are also available at *Science* Online (www.sciencemag.org/cgi/content/full/294/5550/2317/DC1). The sequence has been deposited at GenBank (accession numbers: circular chromosome, AE008688; linear chromosome, AE008689; pAtC58, AE008687; pTiC58, AE008690).
15. At a late phase of this project, we became aware of an independent effort by Goodner *et al.* (*52*) to sequence the C58 strain of *A. tumefaciens*. Comparison of the two independently finished sequences indicates excellent overall agreement. There are two small insertions in the other sequence, one in the circular chromosome and one in pAtC58, relative to our sequence. Single nucleotide discrepancies are far below current accuracy standards for finished genomic sequence (99.99%). Further work will determine which of the observed discrepancies reflect strain differences and which reflect sequencing errors.
16. R. H. Hamilton, M. Z. Fall, *Experientia* **27**, 229 (1971).
17. A. Allardet-Servent, S. Michaux-Charachon, E. Jumas-Bilak, L. Karayan, M. Ramuz, *J. Bacteriol.* **175**, 7869 (1993).
18. J. Zhu *et al.*, *J. Bacteriol.* **182**, 3885 (2000).
19. V. S. Kalogeraki, S. C. Winans, *J. Bacteriol.* **180**, 5660 (1998).
20. R. L. Tatusov *et al.*, *Nucleic Acids Res.* **29**, 22 (2001).
21. J. A. Eisen, J. F. Heidelberg, O. White, S. L. Salzberg, *Genome Biol.* **1**, Research 0011.1 (2000).
22. C. M. Fraser *et al.*, *Nature* **390**, 580 (1997).
23. J. N. Volff, J. Altenbuchner, *FEMS Microbiol. Lett.* **186**, 143 (2000).
24. K. Bao, S. N. Cohen, *Genes Dev.* **15**, 1518 (2001).
25. M. A. Ramirez-Romero, N. Soberon, A. Perez-Oseguera, J. Tellez-Sosa, M. A. Cevallos, *J. Bacteriol.* **182**, 3117 (2000).
26. L. Chen, unpublished data.
27. S. K. Farrand, in *The Rhizobiaceae*, H. P. Spaink, A. Kondorosi, P. J. J. Hooykaas, Eds. (Kluwer Academic, Dordrecht, Netherlands, 1998), pp. 199–233.
28. I. T. Paulsen, L. Nguyen, M. K. Sliwinski, R. Rabus, M. H. Saier Jr., *J. Mol. Biol.* **301**, 75 (2000).
29. J. A. Forward, M. C. Behrendt, N. R. Wyborn, R. Cross, D. J. Kelly, *J. Bacteriol.* **179**, 5482 (1997).
30. R. Sanchez-Fernandez, T. G. Davies, J. O. Coleman, P. A. Rea, *J. Biol. Chem.* **276**, 30231 (2001).
31. C. K. Stover *et al.*, *Nature* **406**, 959 (2000).
32. D. Missiakas, S. Raina, *Mol. Microbiol.* **28**, 1059 (1998).
33. G. T. Robertson, R. M. Roop Jr., *Mol. Microbiol.* **34**, 690 (1999).
34. A. G. Matthysse, G. W. Kijne, in *The Rhizobiaceae*, H. P. Spaink, A. Kondorosi, P. J. J. Hooykaas, Eds. (Kluwer Academic, Dordrecht, Netherlands, 1998), pp. 235–249.
35. K. J. Fullner, J. C. Lara, E. W. Nester, *Science* **273**, 1107 (1996).
36. P. L. Li, I. Hwang, H. Miyagi, H. True, S. K. Farrand, *J. Bacteriol.* **181**, 5033 (1999).
37. J. R. Harper, T. J. Silhavy, in *Principles of Bacterial Pathogenesis*, E. A. Groisman, Ed. (Academic Press, San Diego, CA, 2001), pp. 43–74.
38. J. H. Weiner *et al.*, *Cell* **93**, 93 (1998).
39. K. LeVier, R. W. Phillips, V. K. Grippe, R. M. Roop II, G. C. Walker, *Science* **287**, 2492 (2000).
40. M. J. Pallen, B. W. Wren, *Mol. Microbiol.* **26**, 209 (1997).
41. A. Ichige, G. C. Walker, *J. Bacteriol.* **179**, 209 (1997).
42. S. A. Coleman, M. F. Minnick, *Infect. Immun.* **69**, 4373 (2001).
43. M. Carsiotis, B. A. Stocker, D. L. Weinstein, A. D. O'Brien, *Infect. Immun.* **57**, 3276 (1989).
44. Y. Dessaux, A. Petit, S. K. Farrand, P. J. Murphy, in *The Rhizobiaceae*, H. P. Spaink, A. Kondorosi, P. J. J. Hooykaas, Eds. (Kluwer Academic, Dordrecht, Netherlands, 1998), pp. 173–197.
45. V. S. Kalogeraki, J. Zhu, A. Eberhard, E. L. Madsen, S. C. Winans, *Mol. Microbiol.* **34**, 512 (1999).
46. K. J. Miller, J. M. Wood, *Annu. Rev. Microbiol.* **50**, 101 (1996).
47. Z. Q. Luo, S. K. Farrand, *J. Bacteriol.* **181**, 618 (1999).
48. B. Minambres, E. R. Olivera, R. A. Jensen, J. M. Luengo, *J. Biol. Chem.* **275**, 39529 (2000).
49. E. Jumas-Bilak, S. Michaux-Charachon, G. Bourg, M. Ramuz, A. Allardet-Servent, *J. Bacteriol.* **180**, 2749 (1998).
50. H. Ochman, N. A. Moran, *Science* **292**, 1096 (2001).
51. J. A. Downie, J. P. Young, *Nature* **412**, 597 (2001).
52. B. Goodner *et al.*, *Science* **294**, 2323 (2001).
53. We thank P. Green for useful discussions, R. Gibson for providing the genome_plot program, and J. Staley for critical review of the manuscript. Supported by NSF grant 0092815 (M.V.O.), NIH Public Health Service grant GM32618 (E.W.N.), NIH doctoral fellowship GM19642 (D.W.W.), Sao Paulo State Research Foundation (FAPESP) sabbatical fellowship 1999/11876-5 (J.C.S.), a FUNDECT-MS grant (N.F.A.), and the E. I. du Pont de Nemours Company.

3 October 2001; accepted 9 November 2001

Genome Sequence of the Plant Pathogen and Biotechnology Agent *Agrobacterium tumefaciens* C58

Brad Goodner,[1,2] Gregory Hinkle,[3] Stacie Gattung,[4] Nancy Miller,[4] Mary Blanchard,[4] Barbara Qurollo,[3] Barry S. Goldman,[3,4] Yongwei Cao,[3] Manor Askenazi,[3] Conrad Halling,[3] Lori Mullin,[3] Kathryn Houmiel,[4] Jeffrey Gordon,[3] Mark Vaudin,[4] Oleg Iartchouk,[3] Andrew Epp,[3] Fang Liu,[3] Clifford Wollam,[4] Mike Allinger,[4] Dahlia Doughty,[2] Charlaine Scott,[2] Courtney Lappas,[2] Brian Markelz,[2] Casey Flanagan,[2] Chris Crowell,[2] Jordan Gurson,[2] Caroline Lomo,[2] Carolyn Sear,[2] Graham Strub,[2] Chris Cielo,[2] Steven Slater[3]*

Agrobacterium tumefaciens is a plant pathogen capable of transferring a defined segment of DNA to a host plant, generating a gall tumor. Replacing the transferred tumor-inducing genes with exogenous DNA allows the introduction of any desired gene into the plant. Thus, *A. tumefaciens* has been critical for the development of modern plant genetics and agricultural biotechnology. Here we describe the genome of *A. tumefaciens* strain C58, which has an unusual structure consisting of one circular and one linear chromosome. We discuss genome architecture and evolution and additional genes potentially involved in virulence and metabolic parasitism of host plants.

Agrobacterium tumefaciens is a plant pathogen with the unique ability to transfer a defined segment of DNA to eukaryotes, where it integrates into the eukaryotic genome. This ability to transfer and integrate DNA is used for random mutagenesis and has been adapted into a powerful tool for production of transgenic plants, including soybean, maize, and cotton (*1*, *2*). *A. tumefaciens* was identified early in the 20th century as the causal agent of crown gall disease in plants (*3*). Pathogenesis is initiated when *Agrobacterium* detects small molecules released by actively growing cells in a plant wound. These molecules induce a series of virulence (*vir*) genes whose encoded products export the single-stranded transferred DNA (T-DNA) to the plant cell, where it integrates into the genome at an essentially random location. Once integrated, T-DNA gene expression alters plant hormone levels, leading to cell proliferation typical of a gall tumor. The T-DNA also encodes enzymes for the synthesis of opines, a class of nutrient molecules used almost exclusively by *A. tumefaciens* (*4*–*7*).

A. tumefaciens strains fall into three biovars, which differ in their host range, metabolic characteristics, relationships with other genera in the family Rhizobiaceae, and potentially their chromosome structure (*4*–*13*). The taxonomy of the Rhizobiaceae family is not without controversy, but we expect that

[1]Department of Biology, Hiram College, Hiram, OH 44234, USA. [2]Department of Biology, University of Richmond, Richmond, VA 23173, USA. [3]Cereon Genomics, LLC, 45 Sidney Street, Cambridge, MA 02139, USA. [4]Monsanto Company, 800 North Lindbergh Boulevard, St. Louis, MO 63167, USA.

*To whom correspondence should be addressed. E-mail: steven.c.slater@cereon.com

phylogenetic analysis that includes genome sequences should help resolve this issue (*12*, *13*). Until that time, we believe that the biovar concept is the most valuable for understanding the diversity within the genus *Agrobacterium*. We analyzed *A. tumefaciens* strain C58, a biovar 1 strain with an unusual genome structure known only in this genus: one circular chromosome and one linear chromosome. All pathogenic strains also harbor a tumor-inducing (Ti) plasmid that encodes the T-DNA and virulence genes, and they are classified by the types of opines induced and metabolized (*4–7*). Some strains also harbor additional plasmids, but previous studies focused heavily on Ti plasmids, including the recently completed sequences of a nopaline-agrocinopine–type plasmid, pTi-SAKURA, and a consensus plasmid based on several closely related octopine-type plasmids (*4–6*, *14–16*). *A. tumefaciens* C58, isolated from a cherry tree (*Prunus*) tumor, carries two plasmids: pAtC58 and the nopaline-agrocinopine–type pTiC58 (*17*). This strain has been intensively studied and is the parent of many strains used for the genetic transformation of plants (*2*, *4–6*).

Agrobacterium is an α-proteobacterium within the Rhizobiaceae family, an agriculturally important group that includes the nitrogen-fixing symbiotic partners of legumes (*18*). Six complete genomes are now available within the α-proteobacteria: *Sinorhizobium meliloti* and *Mesorhizobium loti* (both nitrogen-fixing symbionts), *Caulobacter crescentus*, two *Rickettsia* species, and *A. tumefaciens* (*19–24*). The mammalian pathogens *Brucella* and *Bartonella* are also related, so comparative genomics may reveal commonalities and differences between pathogens and symbionts (*25*, *26*).

Overall genome structure. The *A. tumefaciens* C58 genome consists of a circular chromosome, a linear chromosome, and two plasmids: the tumor-inducing plasmid pTiC58 and a second plasmid, pAtC58 (*8*, *10*). The genome was sequenced and assembled with standard methods (*27*). The sequence of all four DNA molecules is available on GenBank, along with chromosomal diagrams and analysis tools (*28*). The essential features of the four molecules are summarized in Table 1 and in the accompanying Web materials (*27*). The two chromosomes contain all of the genes for stable RNAs and housekeeping proteins involved in essential cellular functions and prototrophic growth.

The circular chromosome contains a putative origin of replication (*Cori*) similar to the known *Cori* of *Caulobacter crescentus* (*29*). The linear chromosome, on the other hand, has a plasmid-type replication system of the same type found on pTiC58 and pAtC58. This system, encoded by the *repABC* genes, expresses a pair of segregation proteins (RepA and RepB) and an origin-binding replication initiation protein (RepC) (*30*). Thus, we hypothesize that the linear chromosome is evolutionarily derived from a plasmid. The plasmid origin of an "extra" chromosome had been predicted for multichromosome genomes of the α-proteobacteria and has been found in more distantly related organisms such as *Vibrio cholerae* (*19*, *31–33*).

Gene density is very similar between the two chromosomes. However, genes involved in most essential processes are significantly overrepresented on the circular chromosome (*34*) (Fig. 1). This asymmetry is consistent with direct descent of the circular chromosome from the primordial α-proteobacterial genome, with a minority of essential genes moving to the linear chromosome (*10*). Consistent with lateral transfer between chromosomes over a long period, the overall dinucleotide signatures of the two chromosomes are essentially identical but are significantly different from those of the two plasmids (*27*, *35*). The dinucleotide signatures of the two plasmids are quite similar to each other and to related plasmids from other members of the Rhizobiaceae family (*27*).

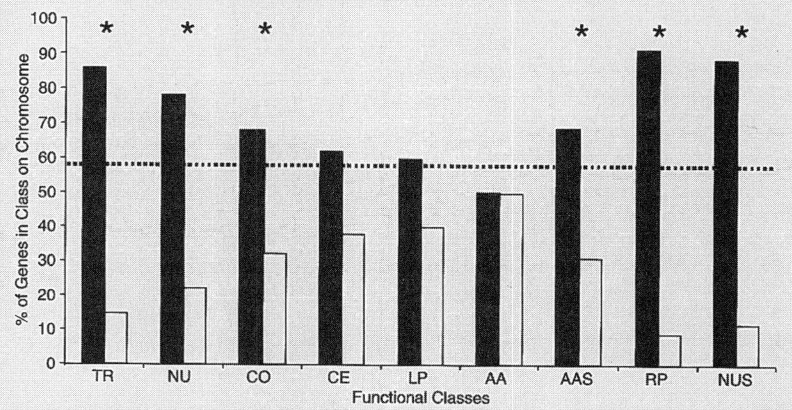

Fig. 1. Distribution of genes from different functional classes on the circular chromosome (solid bars) versus the linear chromosome (open bars). Clusters of orthologous groups (COG) (*34*) classes are as follows: TR, translation; NU, nucleotide metabolism; CO, coenzyme metabolism; CE, cell envelope; LP, lipid metabolism; and AA, amino acid metabolism. Other classes are AAS, amino acid synthesis; RP, ribosomal proteins; and NUS, nucleotide synthesis. Asterisks indicate significant differences in gene distribution (goodness of fit test). The dotted line represents the null hypothesis of gene distribution based on chromosome size for the circular chromosome.

Table 1. General features of the *A. tumefaciens* genome.

	Circular chromosome	Linear chromosome	pAtC58	pTiC58	Total
Size (bp)	2,841,581	~2,075,600	542,869	214,233	~5,674,300
G+C content	59.4%	59.3%	57.3%	56.7%	
Stable RNAs					
rRNAs	2 *rrn* operons	2 *rrn* operons	–	–	
tRNAs	40 species	13 species	–	–	
Coding regions*	2722 (89.1%)	1833 (91.2%)	547 (85.6%)	197 (86.2%)	5299 (89.8%)
Known function†‡	60 (2.2%)	29 (1.5%)	51 (9.3%)	88 (44.7%)	228 (4.3%)
Probable function†§	1708 (62.7%)	1268 (69.2%)	272 (49.7%)	79 (40.1%)	3327 (62.8%)
Possible function†‖	124 (4.6%)	78 (4.3%)	25 (4.6%)	7 (3.6%)	234 (4.4%)
Conserved unknown†¶	152 (5.6%)	79 (4.3%)	29 (5.3%)	3 (1.5%)	263 (5.0%)
Novel†#	678 (24.9%)	379 (20.7%)	170 (31.1%)	20 (10.1%)	1247 (23.5%)

*Number of predicted open reading frames for that DNA molecule (percentage of DNA involved in coding). †Number of predicted open reading frames for that DNA molecule (percentage of total). ‡Known *A. tumefaciens* genes with demonstrated function. §Overall similarity ($P < 1 \times 10^{-15}$) to genes with demonstrated or probable function from other organisms. ‖Genes where limited similarity or short motifs suggest a function. ¶Strong similarity ($P < 1 \times 10^{-15}$) to genes with unknown function from other organisms. #Novel genes.

More than 6000 base pairs of near-perfect sequence identity extend across the two ribosomal RNA (rRNA) gene clusters on each of the two chromosomes. The chromosomes also share some shorter regions of greater than 90% sequence identity with pAtC58 (*27*). Transcription of all rRNA gene clusters is oriented away from the DNA replication origins, with those on the linear chromosome in the same orientation. A number of housekeeping genes are located between the linear chromosome's rRNA operons, and one might expect frequent recombination resulting in lethal events.

The telomeres of the linear chromosome are covalently closed. The linear chromosome is a covalently closed linear molecule, apparently possessing hairpin loops at the telomeres (Fig. 2). Our sequence comes within several hundred bases of the telomeres, and additional sequence is presented by Wood *et al.* (*24*). However, the sequence of the putative hairpin loop is not yet available. The proximal regions of both telomeres are similar in overall architecture but very different in sequence. These regions contain portions of several insertion sequence (IS) elements with intervening DNA of additional repeated and unique sequence. They are rich in potential secondary structure and contain numerous short sequence repeats.

Precedence for covalently closed telomeres is established in *Escherichia coli* phage N15 and spirochaetes such as *Borrelia burgdorferi* (*36*, *37*). DNA replication proceeds through the hairpin, and telomere duplicates are separated to permit chromosome segregation. Resolution of duplicated telomeres is best characterized in the phage N15 system, where protelomerase separates the telomere copies, then rejoins the ends of individual molecules to create hairpin loops (*36*). No ortholog of N15 protelomerase was identified in the *A. tumefaciens* C58 genome. However, several putative transposases are encoded near the telomeres, and these could separate daughter telomeres in a reaction analogous to transposon excision.

DNA replication and the cell cycle. DNA replication is synchronized with the *A. tumefaciens* cell cycle (*11*, *38*, *39*), a feat requiring coordination of four DNA molecules and two different classes of replication origin: *Cori* and *repABC*. The precise signal that initiates DNA replication is not yet clear, although in the related α-proteobacterium *Caulobacter*, many proteins are subject to cell cycle–synchronized transcriptional control and proteolysis (*40*). Because the initiation signal must be interpretable by both types of replication origin, it is unlikely to be transduced by a single origin-specific binding protein.

Processive DNA replication is performed by DNA polymerase III (Pol III), and *A. tumefaciens* carries four paralogs of the *dnaE* gene encoding the Pol III α (polymerase) subunit (*41*). These *dnaE* genes fall into two distinct sequence families, designated as categories A and B (*27*). The category A gene of the circular chromosome is conserved in all sequenced α-proteobacteria and probably encodes the primary replication enzyme (*19–23*, *27*, *41*). Each of the *A. tumefaciens repABC* replicons (linear chromosome, pTiC58, and pAtC58) encodes a category B *dnaE* gene within an operon containing two conserved genes of unknown function. The operon is present in all fully sequenced α-proteobacteria except the *Rickettsia* species and may encode a novel DNA polymerase complex (*19–23*).

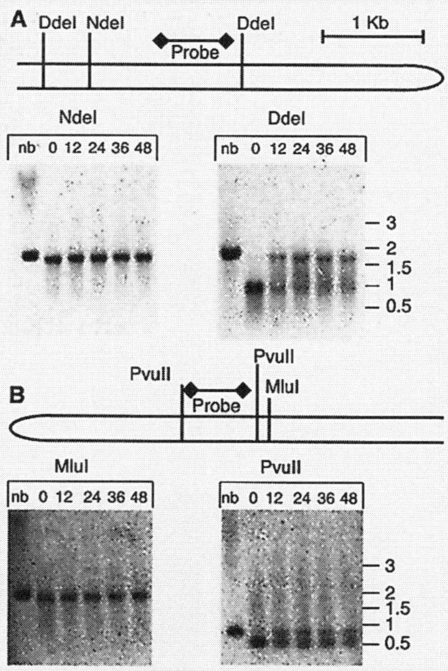

Fig. 2. The linear chromosome telomeres are covalently closed, as shown by Southern blot hybridization using probes near the right (**A**) and left (**B**) telomeres. Chromosomal DNA was digested with specified restriction endonucleases in 100 mM NaCl, 50 mM tris-Cl, 10 mM $MgCl_2$, and 1 mM dithiothreitol, and a portion of each reaction mix was boiled for 10 min, then allowed to cool slowly. After electrophoresis and blot transfer, specific probes identified DNA fragments containing intact telomeres (Nde I and Mlu I digests) or lacking telomeres (Dde I and Pvu II digests). The mobility of fragments possessing telomeres was essentially unchanged by denaturation, suggesting that telomeres contain a covalent hairpin loop. DNA lacking telomeres was denatured by boiling, creating two single-stranded molecules. Slow cooling allowed renaturation of a portion of the denatured molecules. nb, not boiled; 0, 12, 24, 36, and 48 indicate cooling time (in minutes) before freezing in a dry ice and ethanol bath. After 48 min of cooling, the temperature of the DNA samples was approximately 50°C. Numbers on the right of each figure indicate sizes (in kilobases) of double-stranded DNA molecular weight standards.

Synteny analysis of the *A. tumefaciens*, *S. meliloti*, and *M. loti* genomes. Although *Agrobacterium* and *Sinorhizobium* are in different phylogenetic clades (*13*), there is an excellent syntenic relationship between the *A. tumefaciens* circular chromosome and the chromosome of *S. meliloti*. This further supports the idea that the circular chromosome is the original chromosome involved during the evolutionary radiation of the α-proteobacteria, with the subsequent evolution of the linear chromosome taking place from a plasmid (*19*, *42*) (Fig. 3A). Many of the *S. meliloti* orthologs missing from the circular chromosome are present on the linear chromosome, as are a number of gene duplications. The region of *S. meliloti* retaining the most synteny with the *A. tumefaciens* linear chromosome contains about 300 genes, with orthologs arranged in short syntenic groups across the linear chromosome (*19*, *42*) (Fig. 3A). These genes retain the same gene order relative to *S. meliloti*, despite their broad dispersal. If portions of the linear chromosome arose via an excision event from the primordial chromosome, the excision may have originated in this region, with subsequent insertions moving particular sections apart. Many other *S. meliloti* orthologs are also present on the linear chromosome, but their seemingly random location suggests many independent transfers. IS elements are relatively rare on the linear chromosome (with the exception of the telomeres) and cannot explain the highly distributed nature of orthologous genes. The synteny between the circular chromosome of *A. tumefaciens* and the chromosome of *M. loti* is much less pronounced and deserves further analysis in terms of the evolutionary history of the family Rhizobiaceae (*20*) (Fig. 3D).

All four *A. tumefaciens* replicons contain some genes similar to those on the *S. meliloti* pSymA and pSymB plasmids (*19*, *27*, *31*, *43*) (Fig. 3, B and C). Most of these shared genes encode members of large orthologous groups, such as ABC transporters. However, there are exceptions, including the *fixNOQP* operon found on the circular chromosome in *A. tumefaciens* (*44*). In *S. meliloti*, these genes lie on pSymA and encode a special cytochrome oxidase required for the microaerophilic growth found in symbiotic nodules. The role of this cytochrome cbb_3-type oxidase in the biology of *A. tumefaciens* awaits further study. As with comparisons of the *A. tumefaciens* linear chromosome versus the *S. meliloti* chromosome, orthologous genes from the pSym plasmids are spread across both *A. tumefaciens* chromosomes in many short regions of similarity.

Plant transformation and tumorigenesis. Genes involved in plant transformation and tumorigenesis are located on all four genetic elements. The circular chromosome harbors the well-studied *chvAB* genes required for synthesis of the extracellular β-1,2-glucan involved in binding to plant cells; the *chvGI*, *chvE*, and *ros* genes involved in regulation of Ti plasmid *vir* genes; and the *chvD*, *chvH*, and *acvB* genes (*4–6*). The linear chromosome harbors the *exoC* (*pgm*) gene required for synthesis of the extracellular β-1,2-glucan and succinoglucan polysaccharides, and the cellulose synthesis (*cel*) genes involved in binding to plant cells (*4–6*). Our *cel* region sequence differed from the published sequence, and we have reannotated this locus (*27*). Plasmid pAtC58 contains the attachment (*att*) genes involved in initial specific attachment of the bacterium to plant cells, as well as a second, partial *att* locus (*4–6*). pAtC58 is reportedly dispensable for virulence, raising the question of whether there is a virulence requirement for *att* (*14*).

Ti plasmids fall into several opine groups, and the three plasmid sequences now available permit detailed analyses of their relationships (*4–6, 15, 16*) (Fig. 4). The order of genes on the nopaline-agrocinopine–type plasmids pTiC58 and pTi-SAKURA are virtually identical. Major exceptions include one large insertion on pTiC58 and four smaller insertions on pTi-SAKURA. In contrast, the consensus octopine-type plasmid shares only five major gene clusters with the nopaline-type plasmids (Fig. 4). Many regional differences among these plasmids can be circumstantially linked to mobile DNA elements (*45*). Most of the pTiC58-specific genes are involved in metabolism and transport and probably allow the bacterium to scavenge additional nutrients. However, for a few genes unique to a given Ti plasmid, there is direct or circumstantial evidence that they play a role in tumorigenesis on particular hosts (for example, *ligE* of pTi-SAKURA, and *virH1*, *virH2*, and *virJ* of octopine-type plasmids) (*15, 16*). Both pTiC58 and pTi-SAKURA encode a probable NUDIX hydrolase, which may degrade altered nucleotides or other toxic compounds present in the plant wound environment (*46*).

Secretion systems and other pathogenicity factors. *A. tumefaciens* lacks a dedicated type III secretion system for exported pathogenicity factors, but it does have the flagellar type III system, and there is evidence of chemotaxis in response to plant-released compounds (*47, 48*). Flagella are known to secrete pathogenicity factors in *Yersinia enterocolitica*, and flagellar mutants of *A. tumefaciens* exhibit moderately reduced virulence (*49, 50*). However, the *A. tumefaciens* studies do not discriminate between the potential effects of motility and protein secretion (*50*).

A. tumefaciens C58 contains three type IV secretion systems and is the primary model for understanding secretion of DNA and proteins through these transporters. Two systems

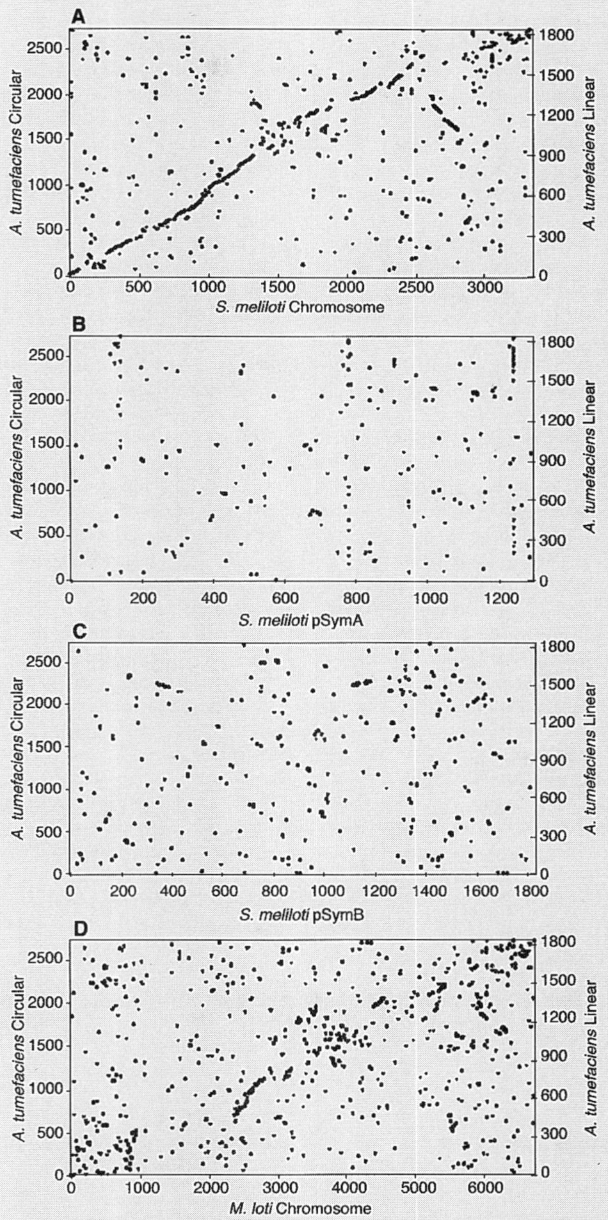

Fig. 3. Synteny among *S. meliloti*, *M. loti*, and the two *A. tumefaciens* chromosomes (*19, 20*). Protein pairs were generated by BLASTP analysis of predicted proteins from each genome, retaining only the best *S. meliloti* or *M. loti* match for each *A. tumefaciens* protein. Each protein pair is graphed according to the location of the corresponding gene on respective DNA molecules. Blue dots, alignment between *A. tumefaciens* circular chromosome and a designated *S. meliloti* or *M. loti* replicon. Red dots, alignment between *A. tumefaciens* linear chromosome and a designated *S. meliloti* or *M. loti* replicon. (**A**) Comparison of *A. tumefaciens* with *S. meliloti* chromosome. (**B**) Comparison of *A. tumefaciens* with *S. meliloti* plasmid pSymA. (**C**) Comparison of *A. tumefaciens* with *S. meliloti* plasmid pSymB. (**D**) Comparison of *A. tumefaciens* with *M. loti* chromosome. Synteny between *A. tumefaciens* and the *M. loti* pML plasmids was extremely low (*27*). To reduce background from members of orthologous groups, only pairs with a BLASTP expectation value ≤ 1×10^{-80} are shown, but full BLASTP data are available (*27*).

encoded by pTiC58 have been heavily studied: the VirB system for T-DNA transfer to plant cells and the Trb/Tra system for conjugal transfer of pTiC58 (*4–6*). A third system is encoded by pAtC58. The utility of this system is unknown, but pAtC58 may be conjugative (*51*). The type IV complexes most similar to the pAtC58 system are from the related animal pathogens *Brucella* and *Bartonella* (*25, 26*).

Several additional loci may play a role in establishing or maintaining a pathogenic relationship with the plant. An autotransporting virulence factor family member is encoded by pAtC58. Such proteins cross the plasma membrane via the signal peptide–dependent pathway and self-insert into the outer membrane (*52*). Typically, a large extracellular domain is exposed, where it modifies cell adhesion or host cell functions. Other genes similar to known virulence factors include orthologs of *bacA* (macrophage survival in *Brucella abortus* and legume symbiosis in *Sinorhizobium meliloti*), putative adhesins, *icmF* (macrophage killing in *Rickettsia*), and as many as six different iron uptake systems (*53, 54*). Iron acquisition is always a priority for pathogens, and *A. tumefaciens* strains vary in their production of siderophores (*55*). Although strain C58 does not produce detectable siderophore activity in low-iron medium, the diversity of iron uptake systems may allow it to co-opt some catechol, hydroxymate, or citrate siderophores produced by other microbes or by plants (*55*).

Other aspects of metabolism and signaling. *A. tumefaciens* establishes a proprietary carbon and nitrogen source by genetically engineering its host to produce opines (*4–7*). The genes for nopaline and agrocinopine utilization lie on pTiC58, but strain C58 derivatives lacking pTiC58 can take up octopine and nopaline without subsequent hydrolysis, and spontaneous mutants of these derivatives have been found with an octopine or mannopine/mannopinic acid utilization phenotype (*56, 57*). Genome analysis provides some clues about the basis of these activities. There are many putative amino acid, dipeptide, and oligopeptide ABC transport systems that may permit scavenging a variety of nitrogenous compounds. Also, the linear chromosome and pAtC58 both encode strong homologs of AgaE, the Ti plasmid–encoded version of which catalyzes the breakdown of mannopine to mannose and glutamate (*58*).

A. tumefaciens also exploits many native plant metabolites, such as sucrose, tannins, and cell wall polymers. Sucrose is the major form by which organic carbon is transported in most plants. Like its relative *S. meliloti*, *A. tumefaciens* has an α-glucoside utilization (*agl*) operon and an additional orphan α-glucosidase, as well as sucrose hydrolase (invertase) (*59*). A third route unique to *A. tumefaciens* involves the periplasmic oxidation of sucrose to 3-ketosucrose, transport of 3-ketosucrose, and cytoplasmic cleavage into fructose and 3-ketoglucose. With no homologs for comparison, we cannot positively identify the genes encoding the enzymes of this pathway, but mutants affecting the pathway are available (*60*). Plasmid pAtC58 encodes a protein with strong similarity to fungal tannases. This enzyme might allow the bacterium to use tannins as nutrients or to defend itself against the antimicrobial activities of many tannins (*61*). Finally, there are numerous genes that may allow degradation of cellulose, hemicellulose, pectin, and/or lignin. These include a β-endoglucanase encoded within the cellulose synthesis gene cluster, putative xylan esterases and xylanases, a previously identified inducible polygalacturonase (pectinase), and several enzymes for degrading monomeric and dimeric components of plant lignin (LigE, VirH2, and the β-ketoadipate pathway) (*4–6, 15, 16, 62, 63*).

The *A. tumefaciens* genome encodes at least 25 different two-component regulatory systems and a wide array of other regulatory proteins, including two new bacterial phytochromes (*64*). Typically, bacterial phytochromes contain the sensory portion of the protein, including the tetrapyrrole chromophore-binding site, attached to a histidine kinase domain. One of the *A. tumefaciens* phytochromes has this structure, and its gene is in a putative operon with a partner response regulator. The other phytochrome is itself a response regulator. There are no data in the literature as to the effect of light on *A. tumefaciens*.

Concluding remarks. The analysis presented here and in the accompanying paper by Wood *et al.* illuminate many new avenues to further explore the biology and biotechnological utility of *A. tumefaciens* (*24*). These include the maintenance of a complex genome, new potential virulence mechanisms, and many additional ways in which the bacterium may parasitize a plant host. At a larger level, the availability of six α-proteobacterial genomes, with more on the way, provides a wealth of comparative data for further understanding the evolutionary history of the group, the evolution and maintenance of multichromosome genomes, and the evolution of mechanisms that interact with and exploit animal and plant hosts (*19–26*).

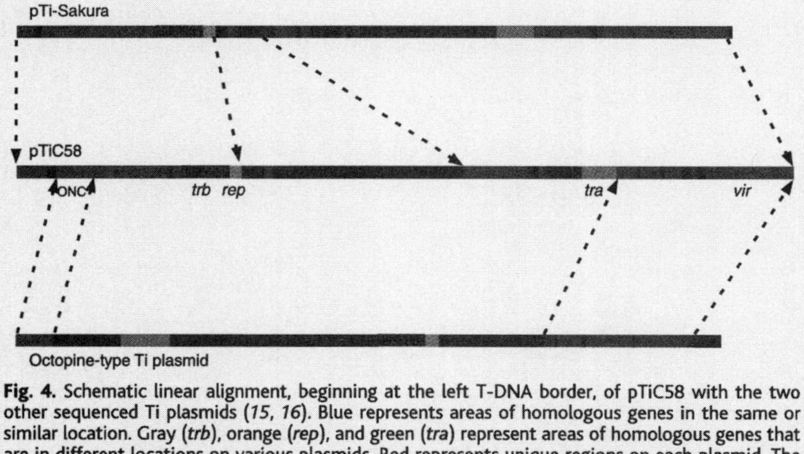

Fig. 4. Schematic linear alignment, beginning at the left T-DNA border, of pTiC58 with the two other sequenced Ti plasmids (*15, 16*). Blue represents areas of homologous genes in the same or similar location. Gray (*trb*), orange (*rep*), and green (*tra*) represent areas of homologous genes that are in different locations on various plasmids. Red represents unique regions on each plasmid. The positions of the T-DNA oncogenes (ONC), *trb* operon, *rep* operon, *tra* operons, and *vir* operons are shown for pTiC58. Arrows delineate regions of identical gene order among the plasmids.

References and Notes

1. J. F. Topping, W. Wei, M. C. Clarke, P. Muskett, K. Lindsey, *Methods Mol. Biol.* **49**, 63 (1995).
2. C. A. Newell, *Mol. Biotechnol.* **16**, 53 (2000).
3. E. F. Smith, C. O. Townsend, *Science* **25**, 671 (1907).
4. A. N. Binns, M. F. Thomashow, *Annu. Rev. Microbiol.* **42**, 575 (1988).
5. Y. Dessaux, A. Petit, S. K. Farrand, P. J. Murphy, in *The Rhizobiaceae: Molecular Biology of Model Plant-Associated Bacteria*, H. P. Spaink, A. Kondorosi, P. J. J. Hooykaas, Eds. (Kluwer Academic, Dordrecht, Netherlands, 1998), pp. 173–197.
6. J. Zupan, T. R. Muth, O. Draper, P. Zambryski, *Plant J.* **23**, 11 (2000).
7. M.-D. Chilton *et al.*, *Cell* **11**, 263 (1977).
8. A. Allardet-Servent, S. Michaux-Charachon, E. Jumas-Bilak, L. Karayan, M. Ramuz, *J. Bacteriol.* **175**, 7869 (1993).
9. E. Jumas-Bilak, S. Michaux-Charachon, G. Bourg, M. Ramuz, A. Allerdet-Servent, *J. Bacteriol.* **180**, 2749 (1998).
10. B. W. Goodner *et al.*, *J. Bacteriol.* **181**, 5160 (1999).
11. K. Suzuki, K. Iwata, K. Yoshida, *DNA Res.* **31**, 141 (2001).
12. A. Willems, M. D. Collins, *Int. J. Syst. Bacteriol.* **43**, 305 (1993).
13. J. M. Young, L. D. Kuykendall, E. Martinez-Romero, A. Kerr, H. Sawada, *Int. J. Syst. Evol. Microbiol.* **51**, 89 (2001).
14. C. Rosenberg, T. Huguet, *Mol. Gen. Genet.* **196**, 533 (1984).
15. K. Suzuki *et al.*, *Gene.* **242**, 331 (2000).
16. J. Zhu *et al.*, *J Bacteriol.* **182**, 3885 (2000).
17. R. H. Hamilton, M. Z. Fall, *Experientia* **27**, 229 (1971).
18. V. Oke, S. R. Long, *Curr. Opin. Microbiol.* **2**, 641 (1999).
19. F. Galibert *et al.*, *Science* **293**, 668 (2001).
20. T. Kaneko *et al.*, *DNA Res.* **7**, 381 (2000).

21. W. C. Nierman *et al.*, *Proc. Natl. Acad. Sci. U.S.A.* **98**, 4136 (2001).
22. S. G. Andersson *et al.*, *Nature* **396**, 133 (1998).
23. H. Ogata *et al.*, *Science* **293**, 2093 (2001).
24. At a late phase of this project, we became aware of an independent effort to sequence the *A. tumefaciens* C58 genome. The work of these authors is published jointly in this issue [D. Wood *et al.*, *Science* **294**, 2317 (2001)]. The two independently derived sequences are in excellent overall agreement, with single nucleotide discrepancies far below current accuracy standards (99.99%) for finished sequence. Our sequence contains two small insertions relative to that of Wood *et al.*; one each in the circular chromosome and pAtC58. Further work will determine which of the observed discrepancies reflect strain differences and which reflect errors.
25. M. L. Boschiroli, V. Foulongne, D. O'Callaghan, *Curr. Opin. Microbiol.* **4**, 58 (2001).
26. K. L. Karem, C. D. Paddock, R. L. Regnery, *Microbes Infect.* **2**, 1193 (2000).
27. Supplementary Web material is available on *Science* Online at www.sciencemag.org/cgi/content/full/294/5550/2323/DC1.
28. DNA sequence and protein predictions are available at the National Center for Biotechnology Information's Entréz Genomes Server (www.ncbi.nlm.nih.gov/PMGifs/Genomes/micr.html). The GenBank accession numbers are NC_003062 (circular chromosome), NC_003063 (linear chromosome), NC_003064 (pAtC58) and NC_003065 (pTiC58).
29. A. K. Brassinga, R. Siam, G. T. Marczynski, *J. Bacteriol.* **183**, 1824 (2001).
30. P. L. Li, S. K. Farrand, *J. Bacteriol.* **182**, 179 (2000).
31. T. M. Finan *et al.*, *Proc. Natl. Acad. Sci. U.S.A.* **98**, 9889 (2001).
32. E. Moreno, *FEMS Microbiol. Rev.* **22**, 255 (1998).
33. J. F. Heidelberg *et al.*, *Nature* **406**, 477 (2000).
34. R. L. Tatusov, *Nucleic Acids Res.* **29**, 22 (2001).
35. S. Karlin, *Trends Microbiol.* **9**, 335 (2001).
36. J. Deneke, G. Ziegelin, R. Lurz, E. Lanka, *Proc. Natl. Acad. Sci. U.S.A.* **97**, 7721 (2000).
37. S. Casjens *et al.*, *Mol. Microbiol.* **26**, 581 (1997).
38. L. S. Kahng, L. Shapiro, *J. Bacteriol.* **183**, 3065 (2001). Kahng and Shapiro analyzed the appearance of hemimethylated DNA, upon passage of a replication fork, at the *att* locus in *A. tumefaciens*. At the time of their study, *att* was believed to reside on the circular chromosome, but we correctly locate it on pAtC58. Thus, at least one *repABC* origin, and probably all three, is coordinated with the cell cycle.
39. K. Skarstad (personal communication) has shown by flow cytometry that all DNA replication is restricted to a relatively short time span within the cell cycle of *A. tumefaciens*.
40. I. J. Domian, A. Reisenauer, L. Shapiro, *Proc. Natl. Acad. Sci. U.S.A.* **96**, 6648 (1999).
41. Z. Kelman, M. O'Donnell, *Annu. Rev. Biochem.* **64**, 171 (1995).
42. D. Capela *et al.*, *Proc. Natl. Acad. Sci. U.S.A.* **98**, 9877 (2001).
43. M. J. Barnett *et al.*, *Proc. Natl. Acad. Sci. U.S.A.* **98**, 9883 (2001).
44. A. Schluter, S. Ruberg, M. Kramer, S. Weidner, U. B. Priefer, *Mol. Gen. Genet.* **247**, 206 (1995).
45. L. Otten *et al.*, *Mol. Plant Microbe Interact.* **5**, 279 (1992).
46. M. J. Bessman, D. N. Frick, S. F. O'Handley, *J. Biol. Chem.* **271**, 25059 (1996).
47. G. V. Plano, J. B. Day, F. Ferracci, *Mol. Microbiol.* **40**, 284 (2001).
48. C. H. Shaw, *Bioessays* **13**, 25 (1991).
49. G. M. Young, D. H. Schmiel, V. L. Miller, *Proc. Natl. Acad. Sci. U.S.A.* **96**, 6456 (1999).
50. E. M. Lai, O. Chesnokova, L. M. Banta, C. I. Kado, *J. Bacteriol.* **182**, 3705 (2000).
51. M. Van Montagu, J. Schell, in *Plasmids of Medical, Environmental & Commercial Importance*, K. M. Timmis, A. Puhler, Eds. (Elsevier, Amsterdam, 1979), pp. 71–95.
52. I. R. Henderson, F. Navarro-Garcia, J. P. Nataro, *Trends Microbiol.* **6**, 370 (1998).
53. K. LeVier, R. W. Phillips, V. K. Grippe, R. M. Roop 2nd, G. C. Walker, *Science* **287**, 2492 (2000).
54. M. Purcell, H. A. Shuman, *Infect. Immun.* **66**, 2245 (1998).
55. R. Penyalver, P. Oger, M. M. Lopez, S. K. Farrand, *Appl. Environ. Microbiol.* **67**, 654 (2001).
56. M. Krishnan, J. W. Burgner, W. S. Chilton, S. B. Gelvin, *J. Bacteriol.* **173**, 903 (1991).
57. G. LaPointe, C. S. Nautiyal, W. S. Chilton, S. K. Farrand, P. Dion, *J. Bacteriol.* **174**, 2631 (1992).
58. S. M. Lyi, S. Jafri, S. C. Winans, *Mol. Microbiol.* **31**, 339 (1999).
59. L. B. Willis, G. C. Walker, *J. Bacteriol.* **181**, 4176 (1999).
60. P. L. Schuerman, J. S. Liu, H. Mou, A. M. Dandekar, *Appl. Microbiol. Biotechnol.* **47**, 560 (1997).
61. T. K. Bhat, B. Singh, O. P. Sharma, *Biodegradation* **9**, 43 (1998).
62. S. Subramaniyan, P. Prema, *FEMS Microbiol. Lett.* **183**, 1 (2000).
63. D. Parke, *J. Bacteriol.* **177**, 3808 (1995).
64. R. D. Vierstra, S. J. Davis, *Semin. Cell Dev. Biol.* **11**, 511 (2000).
65. Supported by funds from Cereon Genomics; Monsanto Company; the University of Richmond; and the National Institutes of Health, National Institute of General Medicine (grant R15 GM61690-01 to B.W.G.). We thank the University of Richmond students of the 1999 and 2000 genetics course and the 2000 and 2001 microbiology course for assistance; S. Salzberg, C. Frasier, and The Institute for Genome Research for providing GLIMMER; and L. S. Kahng, J. Elhai, R. Last, J. Hirschberg, A. Matthysse, S. Winans, S. Farrand, and the manuscript's reviewers for helpful discussions, well-reasoned criticisms, and corrections to our original annotations.

3 October 2001; accepted 19 November 2001

WE HAVE A STRANGE NEW BACTERIOCIN THAT CONTROLS CROWN GALL

Allen Kerr[1] and Max Tate[2]

[1]Emeritus Professor of Plant Pathology,
University of Adelaide, Adelaide, Australia
E-mail: kerra@senet.com.au

[2]Visiting Research Fellow, School of Agriculture and Wine,
University of Adelaide, Adelaide, Australia
E-mail: max.tate@waite.edu.au

Commentary on papers:

Htay, K., and Kerr, A. 1974. Biological control of crown gall: Seed and root inoculation. J. Appl. Bacteriol. 37:525–530.

Roberts, W. P., Tate, M. E., and Kerr, A. 1977. Agrocin 84 is a 6-*N*-phosphoramidate of an adenine nucleotide analogue. Nature 265:379–381.

Biological control of crown gall caused by *Agrobacterium tumefaciens* is probably the best known and most efficient example of biological control of any plant disease. It is so efficient, in fact, that in many areas, the disease has disappeared as a commercial problem. As with many other scientific advances, it was not planned.

My (AK) work on crown gall started as an ecological study of the causal pathogen, *A. tumefaciens.* An early discovery was that there were two distinct biovars of *Agrobacterium* that we named biovars 1 and 2, now considered to be separate species. Although a highly selective medium was available for biovar 1 strains, none was available for biovar 2 strains. My graduate student, Peter New, proceeded to develop one (New and Kerr 1971), the main selective component being erythritol, which can be utilized by biovar 2 strains but not by biovar 1 strains, nor by most soil-inhabiting bacteria. Biovar 2 agrobacteria were readily isolated from soil diluted with water and added to the medium. However, both pathogenic and non-pathogenic strains grew. Previous work by others (Lippincott and Lippincott 1969) prompted us to see what happened when pathogenic and non-pathogenic strains were mixed. We were expecting a reduced size of tumor, when the non-pathogen was present. Much to our surprise, the first non-pathogen we tested (strain K84) completely prevented gall induction when mixed in a ratio of 1:1 with a pathogen. The phenomenon was quite different from the competition for infection sites described by the Lippincotts. Originally, we demonstrated biological control on tomato seedlings and later, on seedlings of peach and almond growing in soil artificially infested with pathogenic bacteria.

Before we could recommend the treatment to commercial growers, we had to be satisfied that control worked in naturally infested soil. A young female graduate student, Daw Khin Htay, had just arrived from Burma (now Myanmar). We agreed that she would work on this problem. She was very quiet but also very determined. I knew of an orchard where many crown gall–affected almond trees had just been removed because they were either dead or no longer productive. I suggested she should go there and bring back several tons of infested soil. She organised it all herself. She contacted the grower, arranged with the University farm staff (a rough lot) to provide a truck and driver to take her to the orchard and she brought back several tons of infested soil. The soil was added to large pots in the open; peach seeds and seedlings were treated with strain K84 and planted in the soil. Results of the experiment are presented in the following paper. They were so clear cut, we recommended that growers should test the treatment under commercial conditions. We supplied the inoculant. Success was dramatic; nurserymen who had been plagued by crown gall for years could hardly find a diseased plant following treatment. Word spread quickly and by 1975, most nurserymen wanted to use it.

A culture of strain K84 was supplied free of charge to anyone who requested it, including two commercial laboratories in Australia and two in America. Commercial use of strain K84 spread to many other countries, including China, most of Europe, Israel, Japan, Korea, Nepal, Russia and South Africa.

Although biological control of crown gall by strain K84 was very effective, we had little understanding of the process at a fundamental level. By 1974, we had established that strain K84 produced a highly specific antibiotic or bacteriocin, later called agrocin 84 by the Ghent group. Synthesis of agrocin 84 was encoded by a plasmid (Ellis et al. 1979), not the Ti plasmid, but another plasmid called pAgK84.

In 1978, it was reported from Greece that in a field experiment involving a treatment in which strain K84 and a pathogen were mixed, new strains appeared which combined both pathogenicity and agrocin 84 synthesis. This finding threatened the effective biological control of crown gall, because such strains are immune to agrocin 84 and cannot be controlled by strain K84. We established that these strains arose because pAgK84 was being transferred from strain K84 to the pathogen. At that time, we were not confident that we could do anything about it but as techniques in molecular biology became simpler, we, along with Stephen Farrand the in U.S. decided to study the molecular biology of pAgK84. The physical characteristics of the plasmid were determined (Slota and Farrand 1981). Farrand spent about six months in my laboratory in 1983 and brought with him a large number of Tn*5* mutants of pAgK84. This enabled us to locate the genes involved in the transfer of pAgK84. All mapped to a small part of the plasmid (Farrand et al. 1985). With the aid of a post-doctoral fellow, David Jones, we succeeded in deleting a large part of the transfer region (Jones et al. 1988). The process was very simple in principle but the necessary manipulations were quite complicated. The final step involved a double recombination event between pAgK84 and a small part of the same plasmid, containing a deletion in the transfer region. We screened 7000 separate colonies and found only one where the desired recombination had occurred. If we had not found any, I think we would have abandoned the project. The new strain was called strain K1026; it shows normal agrocin 84 production, is an effective biocontrol agent but is incapable of plasmid transfer at a detectable frequency. It is now widely used for the biological control of crown gall and was the first genetically engineered organism to be released in 1988 for commercial use. Unfortunately, many countries still preclude its use, even though it contains no foreign DNA.

Along with these studies on the genetics of agrocin 84 biosynthesis, another post-graduate student, Bill Roberts, decided to work on its chemical structure. Bill walked into my (MT) laboratory one day with a Petri dish showing a small colony about 2 mm in diameter surrounded by a large 50 mm inhibition zone and said: "We have a strange new bacteriocin that controls crown gall, which is a plant cancer". My interest was immediately aroused. My first question as a chemist was "What is a bacteriocin?" to which Bill replied, "They are generally considered to be macromolecules for example proteins or defective phage paricles. Unlike broad spectrum antibiotics like penicillin, they selectively inhibit the growth of closely related strains of bacteria." My next question was: "How long has that overlay been growing?" To which he replied, "Over-night". I replied, "I can tell you straight away, there is no way a macromolecule could diffuse that quickly." I suggested he might grow a liquid culture so we could apply a sample for paper electrophoresis at various pH's to examine the charge and size properties.

It was immediately apparent from an overnight bioassay of 1 cm segments of an electrophoretogram run at pH 1.7, that we were dealing with a low molecular weight anion of a strong acid. In natural products, phosphorus and sulfur are the only two atoms that repeatedly give rise to strongly acidic species (pKa's < 2). Labelling with ^{32}P quickly established that biological activity was always associated with both ^{32}P and a UV absorbing component. Acid hydrolysis led to the identification of adenine, an unusual deoxypentose, D-glucose, an unusual diol and two phosphorus atoms. These clues immediately suggested we were dealing with an unusually substituted nucleotide. At the time, Allen was attending a meeting in France and was distressed to hear that it had already been proposed (McCardell and Pootjes 1976) that agrocin 84 was a protein. I faxed back to him saying I was confident we were dealing with a nucleotide not a protein. This was because macromolecules like proteins adhere to the origin of paper electrophoretograms and by contrast, Bill Roberts's biologically active compound moved rapidly at about the same rate as glucose phosphate. We were therefore convinced that it could not possibly be a protein, but must instead, be some unusual type of nucleotide. This became evident once sufficient material was purified by paper electrophoresis so that the unusual spectrum of the modified adenine nucleotide chromophore could be observed. The lability of agrocin 84 to even mild changes in pH meant that purification was a major task. An old supply of charcoal, very efficient for adsorption of the nucleotide from the crude media of strain K84 cultures, was soon exhausted. Recovery of the bacteriocin from new supplies of commercial charcoal was uniformly awful and despair abounded. To counter the poor recovery, the cultures were scaled up to 3 × 150 L aerated batches and passed through two meter columns of charcoal. Anion exchange chromatography, followed by a final clean-up by paper electrophoresis, ultimately yielded sufficient material to demonstrate by phosphorus analysis, that we were dealing with a nucleotide with two phosphorus atoms and one adenine base, but not a dinucleotide. The unusual chromophore was due to a phosphoramidate linkage to N^6 of the adenine ring and the pentosyl was a 3-deoxy arabino-furanose (see paper after next). To our surprise it became apparent that the 5′phosphorus was not linked as the usual phosphodiester, but as a second *O*-phosphoramidate linkage to a 5′-carbon branched chain amide. The basis of the remarkable specificity of agrocin 84 was tracked down to the N^6-glucose-phosphoramidate substituent by another graduate student, Peter Murphy (Tate et al. 1979). Agrocin 84 was the first non-protein bacteriocin to be described.

References

Ellis, J. G., Kerr, A., Van Montagu, M., and Schell, J. 1979. *Agrobacterium*: Genetic studies on agrocin 84 production and the biological control of crown gall. Physiol. Plant Pathol. 15:311–319.

Farrand, S. K., Slota, J. E., Shim, J.-S., and Kerr, A. 1985. Tn*5* insertions in the agrocin 84 plasmid: The conjugal nature of pAgK84 and the locations of determinants for transfer and agrocin 84 production. Plasmid 13:106–117.

Jones, D. A., Ryder, M. H., Clare, B. G., Farrand, S. K., and Kerr, A. 1988. Construction of a Tra^- deletion mutant of pAgK84 to safeguard the biological control of crown gall. Mol. Gen. Genet. 212:207–214.

Lippincott, J. A., and Lippincott, B. B. 1969. Bacterial attachment to a specific wound site as an essential stage in tumor initiation by *Agrobacterium tumefaciens*. J. Bacteriol. 97:620–628.

McCardell, B., and Pootjes, C. F. 1976. Chemical nature of agrocin 84 and its effect on a virulent strain of *Agrobacterium tumefaciens*. Antimicrob. Agents Chemother. 10:498–502.

New, P. B., and Kerr, A. 1971. A selective medium for *Agrobacterium radiobacter* biotype 2. J. Appl. Bacteriol. 34:233–236.

Slota, J. E., and Farrand, S. K. 1981. Genetic isolation and physical characterization of pAgK84, the plasmid responsible for agrocin 84 production. Plasmid 8:175–186.

Tate, M. E., Murphy, P. J., Roberts, W. P., and Kerr, A. 1979. Adenine N^6-substituent of agrocin 84 determines its bacteriocin-like specificity. Nature 280: 697–699.

ALLEN KERR: See the commentary on page 30.

MAX E. TATE: In 1942, at the age of nine, there were no fireworks to celebrate Queen Victoria's birthday, so I decided to make my own. I ploughed through the chemistry section of the local library and succeeded in my endeavor. In the process, I picked up enough chemistry, to make me fascinated by its universality ever since. I obtained a B.Sc. Honors. from the University of Sydney in 1954 and a M.Sc. from the same university in 1955. After a short stint in industry, I commenced a Ph.D. at the University of New South Wales and graduated in 1961, my thesis research concerned the synthesis of inositol phosphates as well as ^{14}C labeled inositol. I spent two postdoctoral years in Ottawa with the Canadian National Research Council where I demonstrated what a powerful tool Thin Layer Chromatography was going to be for organic chemists.

In 1964, I moved to the Waite Institute, University of Adelaide, where I have been involved in elucidating the structures of biologically active molecules.

My most influential mentors have been, in chronological order, John Cymerman-Craig, who impressed on me the importance of appropriate methodology. Then Professor Stephen Angyal who showed me the importance of data interpretation and stereochemical reasoning. Finally, Allen Kerr, who convinced me of the power of genetic reasoning and the importance of interdisciplinary collaboration.

HTAY, KHIN & KERR, A. (1974). *J. appl. Bact.* **37**, 525–530.

Biological Control of Crown Gall: Seed and Root Inoculation

KHIN HTAY AND A. KERR

Waite Agricultural Research Institute, University of Adelaide, Glen Osmond, South Australia

(*Received* 18 *December* 1973 *and accepted* 23 *April* 1974)

SUMMARY. Experiments on the control of crown gall by inoculating susceptible plants with a non-pathogenic strain of *Agrobacterium radiobacter* have continued. In all experiments, highly significant disease control was achieved. In one experiment, 42% of untreated plants growing in soil heavily infested with *A. radiobacter* var. *tumefaciens* died; inoculation of seed with the non-pathogenic strain reduced this to nil. Combined seed and root inoculation was more efficient than seed inoculation alone. In naturally infested soil, combined seed and root inoculation at transplanting gave 99% control of gall formation (as dry weight). A significant increase in plant growth resulted from combined seed and root inoculation. At transplanting, roots should probably be inoculated within 2 h of lifting. This method of biological control is now widely practised by commercial growers in South Australia.

SUCCESSFUL BIOLOGICAL CONTROL of crown gall caused by *Agrobacterium radiobacter* var. *tumefaciens* (Keane, Kerr & New, 1970) has previously been reported (Kerr, 1972; New & Kerr, 1972). The process involves inoculating susceptible plants with a non-pathogenic strain of *A. radiobacter* var. *radiobacter* before growing them in soil containing pathogenic strains. The mechanism of control is through the production of a bacteriocin by the controlling organism (Kerr & Htay, 1973).

This paper describes 2 experiments in which the effectiveness of inoculating peach seed and roots is measured. Preliminary results of one of these experiments have been published (Kerr, 1972). A third experiment, using tomato seedlings, investigated the timing of application of the controlling organism.

Materials and Methods

Isolates

Two South Australian isolates of *A. radiobacter* biotype 2 were used: strain 27 (*A. radiobacter* var. *tumefaciens*), a virulent tumour-inducing pathogen isolated from a peach gall; and strain 84 (*A. radiobacter* var. *radiobacter*), a non-pathogen isolated from soil round a peach gall.

Soils

The soil from Nildottie used in Expt 1 has been described (Kerr, 1972). The soil used in Expt 2 was a sandy loam collected from an almond orchard at Angle Vale, South Australia. Naturally infested soil was dug from a circle 0·3 m radius and 0·3 m deep round each of 15 galled trees. Soil not infested with pathogenic agrobacteria was taken from the same orchard at 12 randomly chosen positions between the rows of trees which were 6·1 m apart. This soil was later inoculated artificially with strain 27. The tomato seedlings used in Expt 3 were grown in John Innes potting compost.

Isolation of agrobacteria

The selective medium of Schroth, Thompson & Hildebrand (1965) was used for isolating biotype 1 agrobacteria and that of New & Kerr (1971) for biotype 2 agrobacteria.

Peach seeds

Wight variety peach stones were used in Expts 1 and 2. For Expt 1, the stones were stratified by burying in soil in the open for 3 months; for Expt 2, the seeds were removed from the stones and artificially stratified by storing them for 3 months at 2°.

Inoculation of peach with strain 84

Strain 84 was grown on yeast-mannitol agar slopes for 48 h and the cells suspended in unsterilized rain water (*c.* 10^8 cells/ml). Seeds or roots were dipped in the bacterial suspension immediately before sowing or transplanting.

Inoculation of tomato seedlings

Stems of 4 weeks-old tomato plants (cv. S.A. Early Dwarf Red) were wounded to a depth of 3 mm with a blunt cylindrical rod 2 mm diameter. A suspension of strain 27 (0·004 ml) was deposited immediately in each wound and this was followed by 0·004 ml of a suspension of strain 84 at various times after wounding. Both suspensions contained *c.* 10^8 cells/ml.

Results

Seed inoculation

The design of Expt 1 has been described previously (Kerr, 1972). Briefly, 125 metal drums (22·7 l capacity) were placed in the open and filled with field soil infested artificially with strain 27. Fifty drums were sown with untreated seed, 50 with seed inoculated with strain 84 and 25 with seed treated with Thiram. Five seeds were sown in each drum in August 1971 and the seedlings thinned to one/drum in January 1972. These were budded with a commercial variety and cut back to the bud in August 1972. Final results were assessed in May 1973. Treatment with Thiram had no significant effect and the results are included with those of the uninoculated treatment to give a simple comparison between inoculated and uninoculated plants (Table 1). Various methods of assessment were made and, where statistical analysis was possible, all showed that inoculation with strain 84 significantly ($P<0·001$) reduced disease incidence. The most marked effect was on the percentage of dead plants; 42% of the plants grown from uninoculated seed died compared with no deaths in the plants grown from inoculated seed.

Death of nursery trees is not often associated with crown gall infection in the field and it seems probable that too high an infestation of *A. radiobacter* var. *tumefaciens* was established in the soil.

TABLE 1

Effect of seed inoculation with strain 84 on incidence of crown gall on peach trees 21 months after inoculation

Treatment	No. of plants treated*	% of plants: With galls	% of plants: With galls on crowns	% of plants: Dead	Mean dry wt (g) of gall tissue on living plants
Inoculated with strain 84	44	61	18	0	12·0
Not inoculated	66	96	92	42	23·3
Significant difference (*P*)	—	<0·001	<0·001	†	<0·001

* 15 plants died for a reason unconnected with the experiment.
† Statistical analysis not possible because of zero value.

Seed and root inoculation

Experiment 2 was also run over 2 growing seasons. In August 1971, peach seed was either inoculated with strain 84 or uninoculated and sown in soil in which no biotype 2 agrobacteria could be detected (<3·3 cells/g of soil). After 12 months, the plants which developed were removed from the soil and the root inoculation treatment applied. This gave 4 treatments: (1) no inoculation; (2) seed inoculation; (3) root inoculation;

TABLE 2

Effect of inoculation with strain 84 on the incidence of crown gall on peach trees growing in naturally and artificially infested soil

Inoculation treatment	Naturally infested soil. % plants* with galls on				Artificially infested soil. % plants* with galls on			
	Crown	Primary root	Lateral roots	Any part	Crown	Primary root	Lateral roots	Any part
None	31	77	100	100	62	54	96	100
Seed	0	4	96	96	8	8	85	88
Root	4	4	31	35	12	12	73	77
Seed and root	0	8	23	31	0	0	54	54

* There were 26 plants in each of the 8 treatment–soil combinations. Because nil and 100% infection were frequently recorded, valid statistical analyses were not possible.

(4) seed and root inoculation. The plants were then transplanted into naturally, or artificially, infested Angle Vale soil in metal drums (18·2 l capacity) to give a total of 8 treatments with 26 replicates of each. The naturally infested soil contained $9{\cdot}0 \times 10^3$ cells of pathogenic biotype 1 agrobacteria/g of soil and $4{\cdot}4 \times 10^3$ pathogenic biotype 2/g of soil; the number of pathogens in the artificially infested soil was not measured. In the rhizosphere of plants grown from inoculated seed, there were $6{\cdot}4 \times 10^6$ cells of

non-pathogenic biotype 2 agrobacteria/g of soil (presumably strain 84); none was detected in the rhizosphere of plants grown from uninoculated seed.

The incidence of crown gall was determined in June 1973 by several methods (Tables 2 & 3). The effect of crown gall infection on growth of young trees was also

TABLE 3

Effect of inoculation with strain 84 on mean number of galls and mean dry weight of gall tissue/peach tree

Inoculation treatment	Mean number of galls/plant			Mean dry weight of gall tissue/plant (g)		
	Naturally infested soil	Artificially infested soil	Overall treatment means	Naturally infested soil	Artificially infested soil	Overall treatment means
None	18·08	16·00	17·04	11·64	17·09	14·37
Seed	3·04	8·25	5·64	2·50	9·50	5·87
Root	0·46	2·58	1·52	0·59	5·69	3·14
Seed and root	0·35	2·31	1·33	0·14	2·33	1·24
L.S.D.* (P=0·05)	2·81					2·58

* In number of galls/plant, there was a significant interaction between treatments and soils. Therefore, the L.S.D. of 2·81 applies to any two treatment—soil means. In mean dry weight of gall tissue/plant, there was no significant interaction between treatments and soils. Therefore the L.S.D. of 2·58 applies to the overall treatment means.

TABLE 4

Effect of inoculation with strain 84 on mean stem diameter and mean root dry weight of peach trees

Inoculation treatment	Mean dry weight of root (g)			Mean stem diameter (mm)		
	Naturally infested soil	Artificially infested soil	Overall treatment mean	Naturally infested soil	Artificially infested soil	Overall treatment mean
None	24·68	24·55	24·62	1·37	1·27	1·32
Seed	29·95	28·91	29·43	1·45	1·35	1·40
Root	51·36	34·55	42·96	1·53	1·40	1·47
Seed and root	42·72	31·18	37·00	1·47	1·48	1·47
L.S.D. (P=0·05)*	8·12					0·07

* In dry weight of roots there was a significant interaction between treatments and soils. Therefore the L.S.D. of 8·12 applies to any two treatment–soil means. In stem diameter, there was no significant interaction between treatments and soils. Therefore the L.S.D. of 0·07 applies to the overall treatment means.

assessed by measuring the diameter of the main stem 2 cm above the junction of stem and root and by measuring dry weight of roots after removal of galls (Table 4). All inoculation treatments markedly reduced disease incidence and root inoculation increased plant growth. Significantly more and larger galls were formed on trees growing in artificially infested soil compared with those in naturally infested soil

($P < 0{\cdot}05$ for number of galls/plant; $P < 0{\cdot}001$ for dry weight of galls/plant). Presumably artificial inoculation resulted in higher numbers of pathogenic bacteria in the soil than occurred under natural conditions. Unfortunately the level of infestation in the artificially inoculated soil was not measured.

The nearest approach to field conditions was in the naturally infested soil and probably the most satisfactory method of assessing disease incidence is by measuring dry weight of gall tissue/plant because this integrates percentage infection and number and size of galls formed. The data on dry weight of gall tissue/plant in naturally infested soil show that combined seed and root inoculation achieved 99% control

$$\left(\frac{11{\cdot}64 - 0{\cdot}14}{11{\cdot}64} \times 100\right).$$

Timing of root inoculation

Until recently, it was accepted that wounding of plant cells is a prerequisite for crown gall induction (Braun, 1954, 1962). Recent evidence (Kerr, 1972), however, indicates that galls on peach trees can develop from unwounded lenticels and it was suggested that the basic prerequisite for gall induction is pathogenic bacteria in contact with dividing cells. Wounding of plant cells is only one way in which this can be achieved.

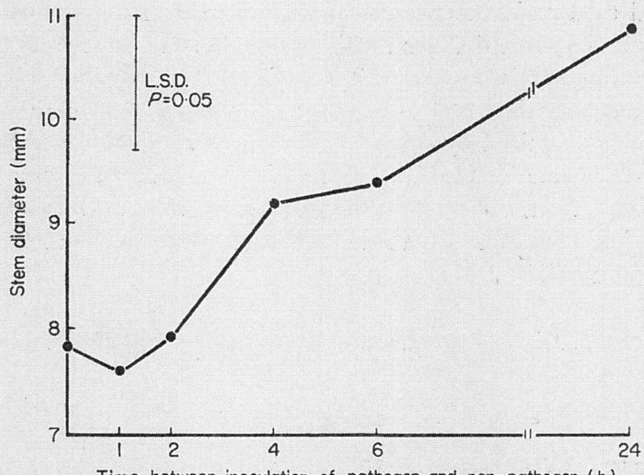

Fig. 1. Stem diameter of wounded tomato seedlings inoculated with pathogenic strain 27 immediately after wounding and then with non-pathogenic strain 84 at various times after wounding.

Nevertheless, it is a very important way. When young stone fruit trees are dug up prior to transplanting, wounding of roots inevitably occurs and if pathogenic bacteria are present on the root surface, the probability of gall induction will be increased. Thus, it is important to know the protective effect of strain 84 applied at various times after wounding. This was investigated using tomato seedlings as indicators of gall induction.

Forty-eight tomato seedlings were wounded and all were inoculated with the pathogenic strain 27. Strain 84 was then added to the wounds immediately or 1, 2, 4, 6 or 24 h after inoculation with the pathogen. There were 8 replicates of each treatment and results are given in Fig. 1. No galls developed on plants treated with strain 84 immediately or 1 or 2 h after inoculation with the pathogen but galls developed on all other plants, the size of gall increasing as application of strain 84 was delayed.

If results of this experiment with tomato seedlings can be extrapolated to stone fruit trees, strain 84 should be applied to transplants within 2 h of lifting.

Discussion

In South Australia, the most serious damage from crown gall is caused to peach and almond trees. In the propagation of these crops, there are only 2 occasions when inoculation is feasible; when sowing the seed of the rootstock and when young trees are transplanted one year, or sometimes 2 years, later.

If seed is sown in soil infested with *A. radiobacter* var. *tumefaciens*, prior seed inoculation with strain 84 provides marked protection. If inoculated seed is sown in uninfested soil and the plants which develop are transferred to infested soil 12 months later, protection still operates but at a lower level than if the roots are inoculated at transplanting (Table 3). Although root inoculation alone was not significantly different from combined seed and root inoculation, the former treatment cannot be recommended unless it is known that seed will be sown in soil free from *A. radiobacter* var. *tumefaciens*. In commercial nurseries, it would be practically impossible to determine this. So it seems clear that both seed and root inoculation should be used. In fact, this is now widely practised by commercial growers in South Australia and there has been a spectacular reduction in the incidence of crown gall. In several nurseries which have had a serious crown gall problem for 10 or more years, the incidence of crown gall was negligible following inoculation. However, results of adequately replicated field trials are not yet available.

We wish to thank Miss Yvonne Hoppenbrouwers for technical assistance.

References

Braun, A. C. (1954). The physiology of plant tumors. *Ann. Rev. pl. Physiol.* **5**, 133.

Braun, A. C. (1962). Tumor inception and development in the crown gall disease. *Ann. Rev. pl. Physiol.* **13**, 533.

Keane, P. J., Kerr, A. & New, P. B. (1970). Crown gall of stone fruit. II. Identification and nomenclature of *Agrobacterium* isolates. *Aust. J. biol. Sci.* **23**, 585.

Kerr, A. (1972). Biological control of crown gall: seed inoculation. *J. appl. Bact.* **35**, 493.

Kerr, A. & Htay, Khin (1974). Biological control of crown gall through bacteriocin production. *Physiol. pl. Pathol.* **4**, 37.

New, P. B. & Kerr, A. (1971). A selective medium for *Agrobacterium radiobacter* biotype 2. *J. appl. Bact.* **34**, 233.

New, P. B. & Kerr, A. (1972). Biological control of crown gall: field measurements and glasshouse experiments. *J. appl. Bact.* **35**, 279.

Schroth, M. N., Thompson, J. P. & Hildebrand, D. C. (1965). Isolation of *Agrobacterium tumefaciens—A. radiobacter* group from soil. *Phytopathology* **55**, 645.

(Reprinted from Nature, Vol. 265, No. 5592, pp. 379–381, January 27, 1977)

Agrocin 84 is a 6-*N*-phosphoramidate of an adenine nucleotide analogue

WE wish to draw attention to the natural occurrence of an extraordinarily selective antibiotic, agrocin 84, which is a 'fraudulent' member (structure I) of the hitherto unreported 6-*N*-phosphoramidate series of adenine nucleotides (structure II), of which 6-*N*-AMP (II, R=H) is the first member.

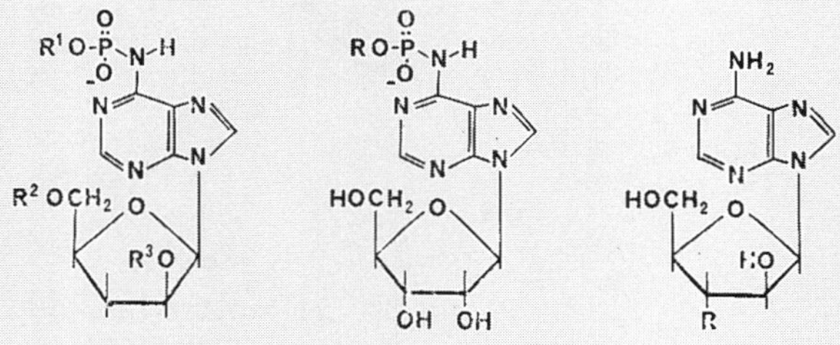

I II III

Kerr and Htay[1] have shown that the biological control of crown gall, achieved by the inoculation of susceptible plants with the non-pathogenic strain 84 of *Agrobacterium radiobacter*, is due to the production of a bacteriocin by the controlling organism. This bacteriocin, now known as agrocin 84, selectively inhibits the growth of converted pathogenic strains of *A. radiobacter*, while the growth of most non-pathogenic strains is unaffected[2]; this property is utilised in the bioassay, which was originally developed by Stonier[3] for another bacteriocin. Heip *et al.*[4] have outlined a partial purification procedure for agrocin 84 from solid media, but the degree of purification was not established and no structural studies were reported.

We have found that isolation is greatly facilitated by culturing a rough colony variant of strain 84 in a defined liquid medium. After a 10,000-fold purification of the agrocin 84 had been achieved by gradient elution techniques involving charcoal adsorption and anion exchange chromatography on DEAE Sephadex, a correlation between biological activity and the presence of a nucleotide with an absorption spectrum maximum at 264 nm was noted and found to hold for all the chromatographic and electrophoretic systems which have been investigated.

After further purification by electrophoresis on glass-fibre filter paper agrocin 84 has been characterised as an adenine nucleotide with the characteristic spectra shown in Fig. 1*a*. The purified agrocin contained adenine, phosphorus, 3′-deoxy-D-arabinose and D-glucose in the molar ratio of 1:2:1:1. It also seems to contain an as yet unidentified reducing sugar acid, and the presence of other constituents cannot be excluded at the present stage of the purification. An upper limit for the molecular weight of the triethylammonium salt is set at 1,100 ± 100 by an $A_{1\,cm}^{1\%} = 180$ and an ε value of 19,860 at 264 nm and neutral *p*H for the purest solid samples obtained to date.

By defining an arbitrary agrocin 84 unit as that amount of agrocin necessary to inhibit totally a zone of radius 1.0 cm within a lawn of pathogenic indicator strain 24, using the Stonier[3] bioassay, we have found at *p*H 7, that one 264 nm absorbance unit of the purified agrocin 84 contains $8 \pm 1 \times 10^3$ agrocin units. On a molar basis this corresponds to 1.6×10^{11} units mol^{-1}.

The infrared spectrum of agrocin 84 showed the presence of a P–O stretching frequency at 1,225 cm^{-1} and the absence of simple carbonyl groups and the pyrophosphate grouping. The electrophoretic mobility/*p*H profile indicated the absence of any secondary dissociation corresponding to a phosphate monoester.

Glucose and 3′-deoxy-D-arabinose were identified as the main carbohydrate components of an acid hydrolysis of agrocin 84 by chromatographic and electrophoretic comparison with authentic samples; the pentose reference sample was obtained from a crystalline sample of methyl 2,5-di-*O*-acetyl-3-deoxy-β-D-*threo*-pentofuranoside by the procedure of Casini and Goodman[5]. Alkaline hydrolysis (1 M NH_4OH, 100 °C 30 min) of the agrocin 84 produced a crystalline adenine nucleoside containing this unusual pentose, with the typical ultraviolet spectral behaviour ($\varepsilon_{max}^{260.5}$ 14,300, *p*H 12.55; $\varepsilon_{max}^{260.0}$ 14,400, *p*H 6.92; $\varepsilon_{max}^{258.0}$ 14,200, *p*H 1.06) for a 9-substituted adenine nucleoside.

It is now a decade since the corresponding 'fraudulent' nucleoside 9-(3′-deoxy-β-D-*threo*-pentofuranosyl) adenine (structure III, R = H) was first synthesised by Martinez, Lee and Goodman[6] in the course of a cancer chemotherapy programme. Although the original sample is no longer available, the required nucleoside has been synthesised by desulphurising the penultimate intermediate 9-(3-S-ethyl-3-thio-β-D-arabinopentofuranosyl) adenine (structure III, R=CH_3CH_2S–) with Raney Nickel.

The synthetic nucleoside (structure III, R = H) and the nucleoside from the alkaline degradation of agrocin 84 were found to be indistinguishable in all the properties examined, including melting point, mixed melting point, optical rotation, paper and thin-layer chromatographic systems, electrophoretic systems, gas chromatography of the trimethylsilyl derivatives, mass spectra of the nucleoside and its silyl derivative, ultraviolet spectral behaviour, and infrared spectra.

With the establishment of the structure of the nucleoside fragment of agrocin 84 as structure III (R = H), attention was next directed to the nature of the substitution of the nucleoside, which would result in the ultraviolet spectra of the intact agrocin as shown in Fig. 1*a*. Four properties of the agrocin which are particularly relevant, are first, the initial bathochromic shift in the absorption maximum from 264 nm to 270 nm on acidification (Fig. 1*a*); second the relatively rapid hydrolysis of the agrocin in 0.1 M HCl at room temperature to produce a nucleotide with an absorption maximum at 260 nm at neutral *p*H values and whose ultraviolet spectral shape and behaviour are similar to other 'normal' 9-substituted adenine nucleotides; third the extreme thermal lability (110 °C, *p*H 7, 15 min) of agrocin 84 at neutral *p*H to produce a second nucleotide with an absorption maximum at 260 nm; and finally the room temperature hydrolysis of agrocin 84 in alkali (1 M NH_4OH, 25 °C, 3 h) to produce a third nucleotide, which is thermolabile, contains D-glucose, has an absorption maximum at 264 nm, and is similar in ultraviolet spectral behaviour to the intact agrocin. These three nucleotides are readily resolved from agrocin 84 and one another, by electrophoresis in 0.05 M sodium tetraborate buffer at *p*H 9.4.

The reported[7,8] bathochromic shifts for 6-*N*-acylated 9-substituted adenine nucleoside derivative and the known thermolability and acid lability of phosphoramidates led us to postulate the presence of a 6-*N*-phosphoramidate grouping. Model 6-*N*-phosphoramidates of adenosine were therefore synthesised from 2′,3′,5′-tri-*O*-acetyl adenosine[9] and the appropriate phosphate, using dicyclohexylcarbodiimide in pyridine as the condensing agent, under conditions analogous to those described by Tener[10] for the condensation of cyanoethyl phosphate with 2′-deoxyadenosine. Figure 1*b* shows the spectral characteristic of 6-*N*-(cyanoethylphosphoryl) adenosine (structure II, R = $CNCH_2CH_2$–). At *p*H 5.0 the spectral behaviour of the monoprotonated 6-*N*-(hydrogen phosphoryl) adenosine (6-*N*-AMP) (structure II, R = H) with an absorption maximum at 264 nm, ε, 19,500, resembles its cyanoethyl precursor shown in Fig. 1*b*. The non-protonated species of 6-*N*-AMP in 0.1 M sodium hydroxide, shows a shift to 267.5 nm, ε, 19,800.

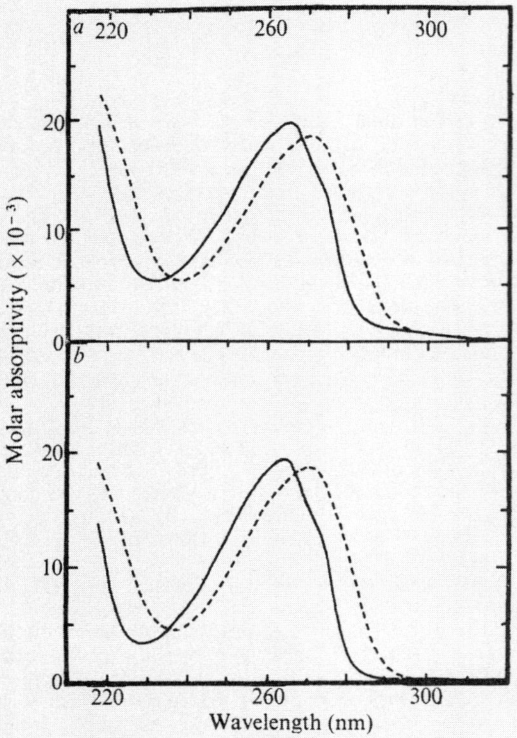

Fig. 1 *a*, Absorption spectrum for agrocin 84 at *p*H 1 (- - -) and 7 (———). *b*, Absorption spectrum for 6-*N*-(cyanoethylphosphoryl) adenosine (structure II, R = $CNCH_2CH_2$–) at *p*H 1(- - -) and 7 (———).

The striking resemblance between the spectra of agrocin 84 in Fig. 1*a* and the 6-*N*-(cyanoethylphosphoryl) adenosine in Fig. 1*b* leaves little doubt that agrocin 84 contains the hitherto unrecognised 6-*N*-phosphoramidate linkage to adenine as shown in structure I, where R^1 is at present considered to contain glucose. R^2 and R^3 represent hydrogen and a second phosphorylated carbohydrate moiety or vice versa.

The current investigation has firmly established the structure of the 'fraudulent' nucleoside fragment and the unique aspect of the phosphoramidate linkage is clearly indicated. If we accept the definition of a bacteriocin as simply 'an antibiotic produced by strains of bacteria and in general active on some other strains of the same species or related species'[11] and do not further restrict it to being 'colicin-like' as mentioned by Reeves[11], then this agrocin 84 should probably be classed as a new type of bacteriocin.

In conclusion, we point out that current attempts to explain all biochemical phenomena involving adenine nucleotides solely in terms of the 2′,3′, and 5′-phosphate esters may be less than adequate and a thorough search for other labile phosphoramidates such as the naturally occurring agrocin 84 and the synthetic 6-*N*-AMP reported here, with their predictably high phosphoryl transfer potentials, should be rewarding.

We thank the Australian Research Grants Committee for support. The award of a Charles John Everard Scholarship to W.P.R. is gratefully acknowledged. The ethylthionucleoside was kindly supplied by Drs Wood and Lee of the Stanford Research Institute.

W. P. Roberts
M. E. Tate

Department of Agricultural Biochemistry,

A. Kerr

Department of Plant Pathology,
Waite Agricultural Research Institute,
University of Adelaide,
Glen Osmond, South Australia 5064, Australia

Received October 7; accepted November 24, 1976

1 Kerr, A., and Htay, Khin, *Physiol. Pl. Path.*, **4**, 37–44 (1974).
2 Roberts, W. P., and Kerr, A., *Physiol. Pl. Path.*, **4**, 81–92 (1974).
3 Stonier, T., *J. Bact.*, **79**, 889–898 (1960).
4 Heip, J., Chatterjee, G. C., Vandekerckhove, J., Van Montague, M., and Schell, J., *Arch. Int. Physiol. Biochim.*, **83**, 974–976 (1975).
5 Casini, G., and Goodman, L., *J. Am. chem. Soc.*, **86**, 1427–1431 (1964).
6 Martinez, A. P., Lee, W. W., and Goodman, L., *J. org. Chem.*, **31**, 3263–3267 (1966).
7 Ralph R. K., and Khorana, H. G., *J. Am. chem. Soc.*, **83**, 2926–2934 (1961).
8 Schweizer, M. P., McGrath, K., and Baczynski, L., *Biochem. biophys. Res. Commun.*, **40**, 1046–1052 (1970).
9 Bredereck, H., and Martini, A., *Chem. Ber.*, **80**, 401–405 (1947).
10 Tener, G. M., *J. Am. chem. Soc.*, **83**, 159–168 (1961).
11 Reeves, P., *The Bacteriocins, Molecular Biology and Biophysics*, **11**, 113 (Chapman and Hall, 1972).

Printed in Great Britain by Henry Ling Ltd., at the Dorset Press, Dorchester, Dorset